The Lost World of the Moa

LIFE OF THE PAST

James O. Farlow, Editor

Trevor H. Worthy & Richard N. Holdaway

The Lost World of the Moa

Prehistoric Life of New Zealand

Principal Photography by Rod Morris

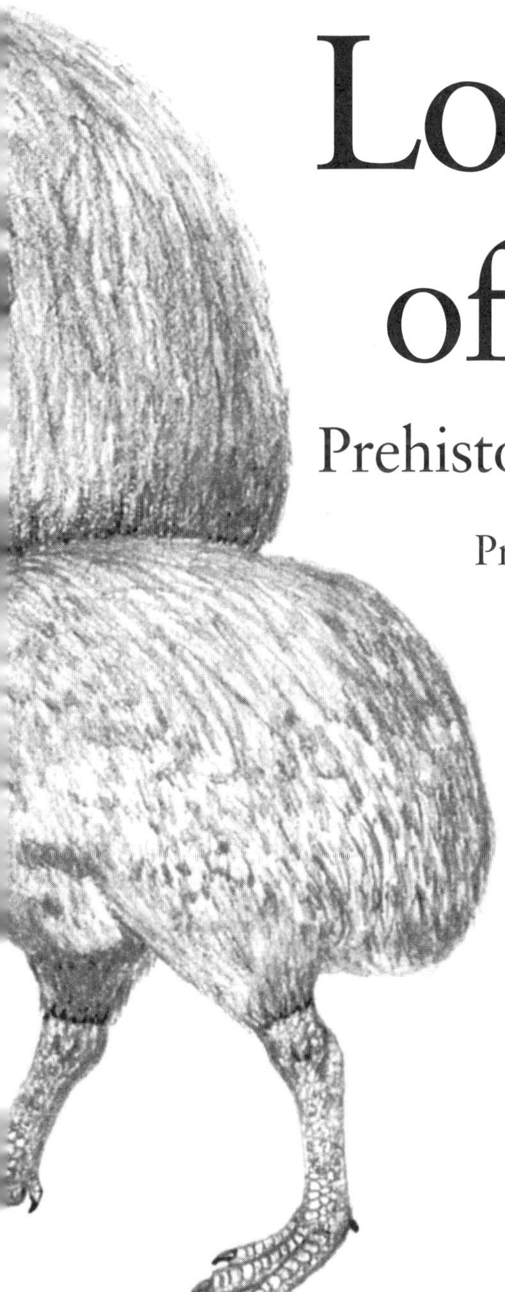

INDIANA University Press
Bloomington and Indianapolis

This book is a publication of

Indiana University Press
601 North Morton Street
Bloomington, Indiana
47404-3797 USA

http://iupress.indiana.edu

Telephone orders 800-842-6796
Fax orders 812-855-7931
Orders by email iuporder@indiana.edu

The paper used in this publication meets the minimum requirements of American National Standard for Information Sciences—Permanence of Paper for Printed Library Materials, ANSI Z39.48-1984.

Manufactured in the United States of America

Library of Congress Cataloging-in-Publication Data

Worthy, T. H.
The lost world of the moa : prehistoric life of New Zealand / Trevor H. Worthy and Richard N. Holdaway ; principal photography by Rod Morris.
p. cm. — (Life of the past)
Includes bibliographical references (p.).
ISBN 0-253-34034-9
1. Moas—New Zealand. 2. Extinct birds—New Zealand. 3. Extinct animals—New Zealand. I. Holdaway, Richard N. II. Title. III. Series.
QE872.D5W67 2002
568'.5—dc21
2001001753

1 2 3 4 5 07 06 05 04 03 02

[Frontispiece] Dinornis *family. The three species (*D. giganteus, D. novaezealandiae, *and* D. struthoides*) differ mainly in size. The takahe (middle) is about the size of a large chicken and stands about 40 cm high. Drawing by J. Winn.*

For

Ron Scarlett,

osteologist at the Canterbury Museum, whose studies over 30 years laid the foundation for our work

CONTENTS

Preface

This book could not have been written a decade ago. Two summaries of the Quaternary avifauna and the effects people had wrought on it appeared at the start of the 1990s. Atholl Anderson began the decade by summarizing knowledge of moa and the people who hunted them. A year later, Ian Atkinson and Phil Millener firmly placed the avifauna in an ecologic setting. Both studies were hampered by the limited number of dated faunas and the continued perception that the extinct assemblages were simply more diverse versions of the surviving forest faunas.

Although significant progress toward describing Quaternary faunas had been made from the mid-1970s, primarily by these authors, since 1991, the amount of information on the Holocene and Pleistocene vertebrate faunas has increased by at least 10-fold. That dramatic increase in data on distribution and in particular in distribution through time, gained through a strong program of radiocarbon dating, has meant that the views of even a decade ago are now outdated. This book is a first attempt at synthesizing the information. It is also a baseline for research over the next decade. Application of new techniques and further and deeper surveys promise to lead to more insights and the resolution of the many problems that have necessarily arisen through resolution of previous ones. Many of the ideas formed during the early 1990s have already had to be rethought and revised in light of more recent findings. The process—and progress—continues.

The new information base has allowed the refinement of former ideas on the composition and organization of the vertebrate fauna, and of the relationships of many of its constituent species. It has also allowed the formulation and testing of new hypotheses on larger issues, such as the causes of the major extinction event that overtook the New Zealand vertebrate fauna in the late Holocene. A key point in the recent progress has been the application of new techniques to problems of the systematics and biology of extinct species. Obviously, much research has gone into elucidating the systematics of the moa. The shakedown of moa taxonomy is nearly complete, as far as present morphologic-based methods can take us, but where will the application of DNA techniques take it? We have taken the opportunity to provide a full summary of previous work on this major group of ratite birds, whose relationships and evolutionary history have ramifications beyond their pivotal role in the New Zealand avifauna. Other groups, such as waterfowl, have not been neglected, but because most of the systematic research has been done as an adjunct to the faunistic and biologic studies, much more work will be necessary before we can be sure that the full extent of the diversity of even the birds themselves is known.

New technology has been able to answer old questions. New insights on the diet of the strange and fascinating adzebills and the systematics of the moa are now available. Analysis of other data suggests that much of the New Zealand ecosystem derived its nutrients from the ocean. Refining old technologies, especially morphologic analyses, has led not only to improved systematics but to an appreciation of evolutionary changes as well. As in other areas, progress has been underpinned by an ability to date faunas and the realization that different faunas occur in the same site. In the future, we may be able to look forward to a high-resolution chronology for the extinctions and much better resolution of dating for older faunas—resolutions based on new techniques that include luminescence dating of sediments laid down over 40,000 years ago.

Last, this volume is the result of a success in the funding of science. Although the changes in science funding in New Zealand that began in the late 1980s have not been universally welcomed and have not been without their problems, it is probably fair to say that this book and the information it is based on would not exist if a system that could adequately fund private, individual research efforts had not been established. Although there has been a trend to fund large research programs, the current New Zealand initiatives have supported small, inexpensive research programs such as ours. Most of the research undertaken, although fundamental, has also depended to some extent on serendipity. The freedom and opportunity to follow the most promising—and unforeseen—veins in the lode have been crucial. Several of the major advances made over the past decade have depended, as paleontological studies must, on the fortunate discovery of sites with deposits that can provide answers to questions or challenge existing understandings.

The necessary chronologic framework within which to set our data has been made possible by the simultaneous development of radiocarbon technologies over the last decade, particularly as it relates to dating of bones with accelerator mass spectrometry. The confident dating of small bones, undoubtedly aided by New Zealand's cool climate, which has favored organic preservation, has been fundamental in some of the new insights gained. It has, for example, resulted in a challenge to the dogma that assumed people's first influence on the New Zealand environment was when they came to settle 800–1000 years ago: it now seems probable that the Pacific rat and its attendant faunal impacts were introduced possibly 2000 years ago by humans who did not settle. Second, we have developed the new view that New Zealand had distinctive terrestrial faunas related to vegetation patterns, rather than one uniform whole.

Our research has been made possible by a recognition that basic information about our environment and the historical development of our faunas is significant, as has been shown by the popular demand for data on extinct animals. More specifically, the recognition that it is significant is important for the current crisis management of the remnants of the fauna. The freedom to follow evolving research directions and to integrate different taxonomic and ecologic objectives has been important in this process. And being private researchers, we have had academic freedom without the overheads and teaching load that come with a university career.

We have been truly fortunate to have been able to grasp the opportunities presented to us and follow the paths veiled in time. We hope that this book provides some reward for the New Zealanders who have funded our work. We are grateful for their support, and we are grateful to a publisher who has allowed us to present the history of the New Zealand experience to the world so that others may wonder at it—and begin to learn from the mistakes.

Acknowledgments

In the development of our experience and advancement of our knowledge we are indebted to many individuals. For their help on this project and in the above we thank the following:

Illustrations: Jo Ogier for penning several figures (Figs. 4.2, 4.14). Jeanette Winn, for the drawings of the moa reconstruction, as well as for penning the frontispiece and a number of figures (Figs. 5.14–5.18, 7.9, 7.11, 10.1, 10.6, 10.11, 10.13).

Photographs: Ian Millar for owlet-nightjar (Fig. 9.13); Jim Palmer for Figs. 4.2, 7.8, 8.14–8.17, 9.1–9.3, 9.5, 10.10; Rod Morris for principal photography.

Ewan Fordyce, for help with the early record chapter. To the collection managers and the following institutions: Brian Gill, Auckland Institute and Museum, Auckland; Jack Grant-Mackie, Palaeontology Collection, Geology Department, Auckland University, Auckland; Peter Dimond, Museum of Caves, Waitomo Caves; Pamela Lovis, Manawatu Science Centre and Museum, Palmerston North; Michelle Houghton, Whanganui Regional Museum, Wanganui; John Yaldwyn, Alan Tennyson, Phil Millener, Raymond Coory, and J. A. (Sandy) Bartle, Museum of New Zealand Te Papa Tongarewa, Wellington; Geoff Tunnicliffe, Beverley McCulloch, and Chris Jacomb, Canterbury Museum, Christchurch; Erena Barker and John Darby, Otago Museum, Dunedin; Michael Forrest, Southland Museum and Art Gallery, Invercargill. Sandra Chapman, Cyril Walker, and Joanne Cooper at the Natural History Museum, London, helped obtain data at one time or another. Storrs Olson and Helen James of the Smithsonian Institution, Washington, D.C., have freely given of advice, for which we are grateful. B. McGowan and A. Kitchener of the Royal Museum of Scotland, Edinburgh, provided data on collections in their care. Manuela Angelo and other staff members at Hector Library, Museum of New Zealand Te Papa Tongarewa, were ever helpful in sourcing references.

Others to whom we are indebted for professional advice include Ewan Fordyce, Michael Winterbourn, Janet Wilmshurst, Matt McGlone, Jack Grant-Mackie, Paul Sagar, and Tony Whitaker.

Private individuals who also helped include Jacqui Fisher, Anne de Bode, Michael Denny, Brian Reeve, Ken Tomlinson, Jim Eyles, Gavin Udy, Mike Endres, Philip Burnaby, Jack Cheyne, Les Cheyne, Ross Tobin, Ted Parr, and Elliot Dawson.

We are indebted to the Department of Conservation and the many private landowners throughout New Zealand for permission to study sites on land in their control. We owe much to our special friends Ngaire and John Dalmer and Deborah and Peter Turner.

Peter Johns, Rowan Emberson, and Willy Kuschel have provided data on the identity of fossil insect remains.

We thank the many assistants who have helped us in our fieldwork through the years, notably Mike Huaki, Anne Melhuish, Jonathan Ravens, Jeanette Winn, Sonia Read, Amanda Freeman, Wendy Gurden, Reg Cotter, the late Max Falconer, and Jenny Jones.

We thank the cavers of New Zealand who noticed fossil bones and collected them or who drew them to our attention, notably Phil Wood, Chris Templar, John and Daphne Hobson, and Dave Smith, among many others.

We thank the New Zealand Lottery Science Board and the Foundation for Research Science and Technology, which have funded much of the work upon which this project is based.

RNH gratefully thanks Maureen and Scott, and THW his family.

Introduction: The Land That Time Forgot

New Zealand, Aotearoa, the Māori "land of the long white cloud," is at the southern apex of the Polynesian triangle in the South Pacific Ocean. It is, in the popular phrase, a land that time forgot. It is now an archipelago, but it was once part of the great southern continent of Gondwana. Before the end of the age of dinosaurs, a fragment of Gondwana broke free, carrying with it a ready-made fauna. Isolated on the new land that came to be called New Zealand, this fauna was protected from overland influxes of faunal innovations elsewhere and took its own unique evolutionary course. Some 80 million years later, primitive forms still dominate the flora and fauna. The forests are still filled with podocarp trees whose ancestors sheltered dinosaurs. Some trees of similar ancient lineage are truly magnificent—for example, the kauri, whose giant boles often reach 30 m before branching. Ferns grow to tree size, and liverworts and mosses, including the world's tallest, abound.

The fauna too is unique. Giant flightless orthopterans (weta) crawl in the forests and share it with peripatus or walking worms and giant land molluscs. Their relatives are found on the other Gondwanan landmasses, such as Australia and South America. Other archaic forms include frogs unchanged since the Jurassic period, and a reptile, the tuatara, which is the last known living sphenodontid. The tuatara lineage, extinct everywhere else since the Cretaceous, lived on in New Zealand, unaware of evolutionary novelties such as mammals. In the southern location, many reptiles could not gain a place in New Zealand. Crocodiles lost out to lowering global temperatures. Whole reptilian groups (terrestrial turtles, iguanids, and snakes) never gained a hold. Molded by the ages, the constituent parts of this fauna were supremely adapted to their land. But now, not uniquely, this fauna is disappearing. Why? How did the minicontinent known as New Zealand come to be? What was it that created a place where evolution and environment could combine to generate "the land of birds"?

Brief Geological History of New Zealand

The Earth is not as stable as people once thought. Earthquakes and volcanoes have long been known to have violent effects on the Earth's surface, and they have always been feared by people. The common origin of earthquakes, volcanism, faults, and mountain-building was only suspected in the mid-20th century. Plate tectonics has become a unifying theory of earth sciences. Now, supported by most earth scientists, plate tectonics views the Earth's crust as being divided into a number of fragments, or plates, that are carried around the surface by convection currents in the Earth's underlying mantle. The Earth's crust mainly comprises denser oceanic rocks, primarily basalts, which are rich in silicon and magnesium, and which, as the term "oceanic rocks" implies, underlie most of the oceans.

The continents are formed of relatively light, granitic, silicon and aluminum–rich rocks that float high on the oceanic rocks. As the different plates are pushed around by convection currents deep in the mantle, the continents ride along with them. At plate margins, one plate usually sinks (subducts) beneath another, or (as in New Zealand and southern California) the edges grind sideways past each other. There is a zone of friction between the plates, which is revealed by numerous earthquakes generated as the stress at the junction is released suddenly. Friction or other factors melt some of the sinking rock, forming magma, whose rise to the surface produces volcanoes. The process is slow—the movement is measured in millimeters per year—but because the Earth is 4.5 billion years old, considerable distances may be traveled over time. New Zealand straddles the Pacific and Australian plates, which are converging at about 42 mm per year, or about the speed our fingernails grow. This is 42 km in a million years.

It is generally thought that until late in the Triassic period, some 245–208 million years ago (Ma), all the continents were united in a single supercontinent called Pangaea. A split into two smaller continents, northern Laurasia and southern Gondwana, then created a warm ocean, the Tethys. About 190 Ma, the Atlantic Ocean began to form, and North America separated from Europe and Africa. Beginning about 120 Ma, Gondwana began to rupture, with Africa separating from South America. By about 90 Ma, Antarctica was moving into its polar position, and India had started north on a solo journey. New Zealand as such did not exist, but its parts were being formed elsewhere. New Zealand was prefabricated and assembled on site—the ultimate Liberty ship. Figure I.1 describes the early formation of New Zealand.

Figure I.1. (opposite page) Four stages in the formation of New Zealand. One hundred twenty million years ago (Ma), New Zealand was part of Gondwana, and Africa and South America had just begun to rift apart. By 90 Ma, India and Africa were separated from Gondwana. By 56 Ma, New Zealand was out on its own and Australia was just starting to break free from Antarctica. By 29 Ma, New Zealand was still drifting alone and Australia was well separated from Antarctica, but the rift between Antarctica and South America was just beginning. After Owen (1984).

Sediments being eroded and washed off the northern shore of Gondwana (from the region that split to become Australia and Antarctica) had been accumulating in a deep nearshore trough. About 100 Ma, a rupture (or plate boundary) formed along the margin of the Australian part of Gondwana, and with the onset of seafloor spreading in the area, a tiny sliver of Gondwana and the huge pile of sediments accumulated in the trough moved east. Together, these sediments formed the New Zealand minicontinent, which extends from New Caledonia

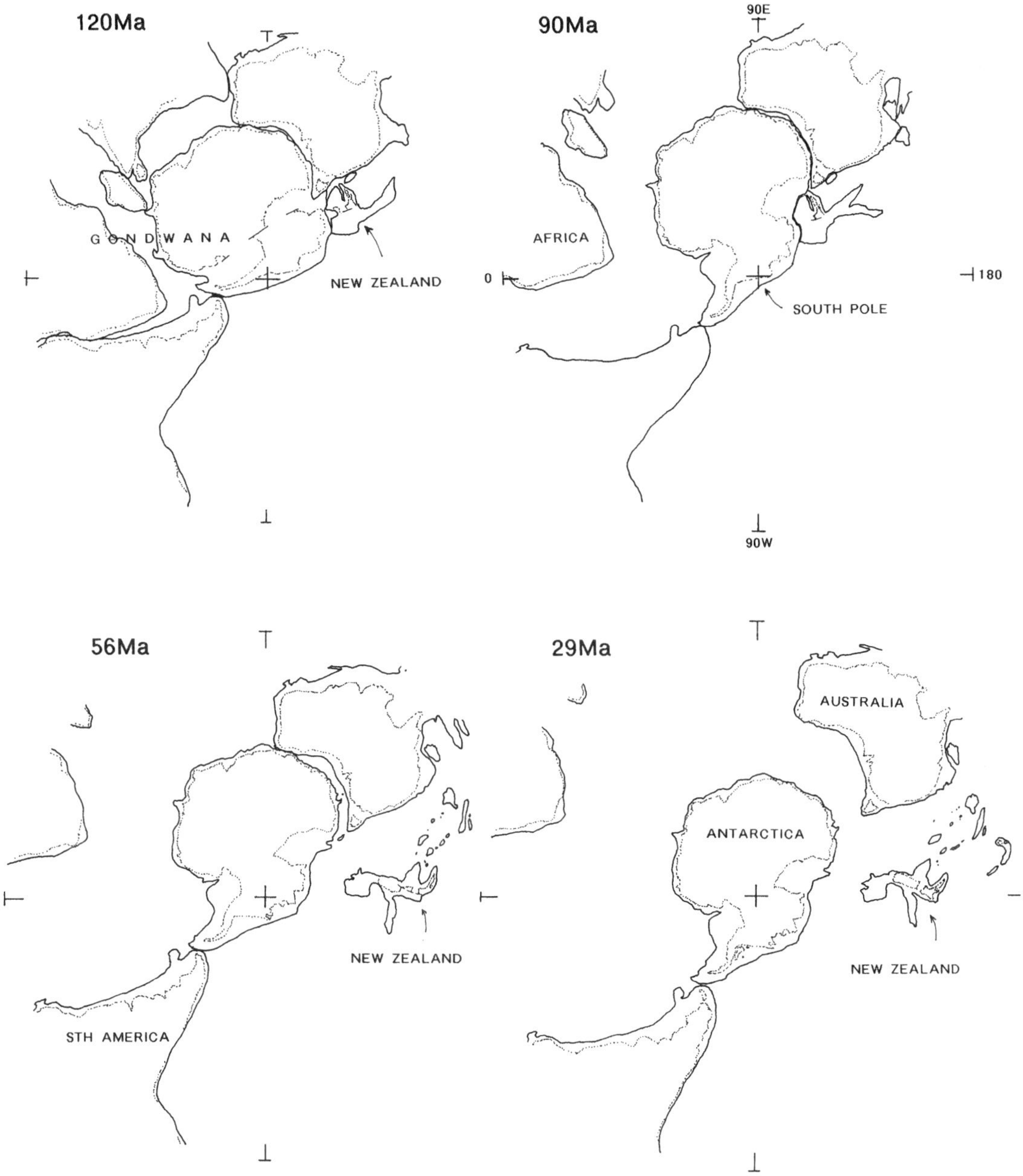

in the north to the south of the present New Zealand archipelago. It includes the Lord Howe Rise, the Norfolk Ridge, the Chatham Rise, and the Campbell Plateau. Disruption was complete and New Zealand was on its own by 82 Ma. Seafloor spreading continued to 60 Ma, by which time New Zealand was 1200 km from Australia (at New Caledonia) and 1500 km away at the South Island. Australia, however, had remained attached to Antarctica all this time and only broke free about 55 Ma; it is still heading north.

Between 38 and 21 Ma, the Tasman plate boundary became inactive and a new plate boundary was initiated near Antarctica. This rupture spread northward along what is now the Macquarie Ridge, and New Zealand found itself straddling the new plate boundary (Stevens 1980a: 458–459). The new edge of the Indian–Australian Plate abutted the Pacific Plate south of New Zealand and formed the Macquarie Ridge. In the northern part of New Zealand, the Pacific Plate sank under New Zealand off Hawke's Bay, and the deep Kermadec Trench formed along the subduction zone to the north. In between, New Zealand was rent asunder, and the spectacular Alpine Fault was born about 20 Ma. A twin to the San Andreas Fault, the Alpine Fault traverses the length of the South Island, and rocks displaced on either side indicate 450 km of lateral movement.

Before the development of the new plate boundary, New Zealand was carried along in a relatively stable tectonic setting. There was no uplift; the landmass of New Zealand was slowly eroded throughout the Tertiary to form a flattened terrain of low relief (a peneplain). By 27 Ma, in the late Oligocene (local Duntroonian Stage), perhaps as little as 18% of the present land area remained above water (Cooper and Cooper 1995). Activity at the new plate boundary made New Zealand the tectonically active place it is today. By the Miocene, the peneplanation of New Zealand ground to a halt and uplift outstripped erosion, so that more and more land was exposed. In the late Miocene (12–5 Ma), continued movement associated with the Alpine Fault resulted in more and more land rising from the sea, and volcanism formed the bulk of what is now Coromandel and Banks and Otago Peninsulas. Only very recently, with the intensified earth movements of the Pliocene (5–3 Ma), did the Southern Alps and other main axial mountain ranges begin to form. Tectonic basins in central New Zealand produced a series of straits and bays in which shallow-water limestone was being deposited as little as 1 Ma. Since then, the seabed has risen nearly 1 km above sea level. The Southern Alps are now rising at 17 mm per year in the south, almost matched by their erosion, but in the north of the South Island, the uplift proceeds at the more sedate rate of 1–1.5 mm per year. At the northern end of the fault, some 20 km of the Australian Plate has been wedged beneath the Pacific Plate.

From even this brief history, it is evident that New Zealand has been isolated from overland immigration of fauna and flora from other continents for at least 80 Ma. Although the land area has fluctuated greatly, there has always been some, however low and swampy, and so a terrestrial fauna could always find a home somewhere in the New Zealand archipelago.

Present Geography

The present outline of New Zealand has attained something of its present form only in the last million years (Fig. I.2). Change has been substantial and rapid during that time. For example, the ranges of the lower North Island have risen from the sea. Cook Strait was bridged by land until the mid-Pleistocene. A seaway probably first formed in an interglacial only about 0.45 Ma (Lewis et al. 1994). Every glacial pe-

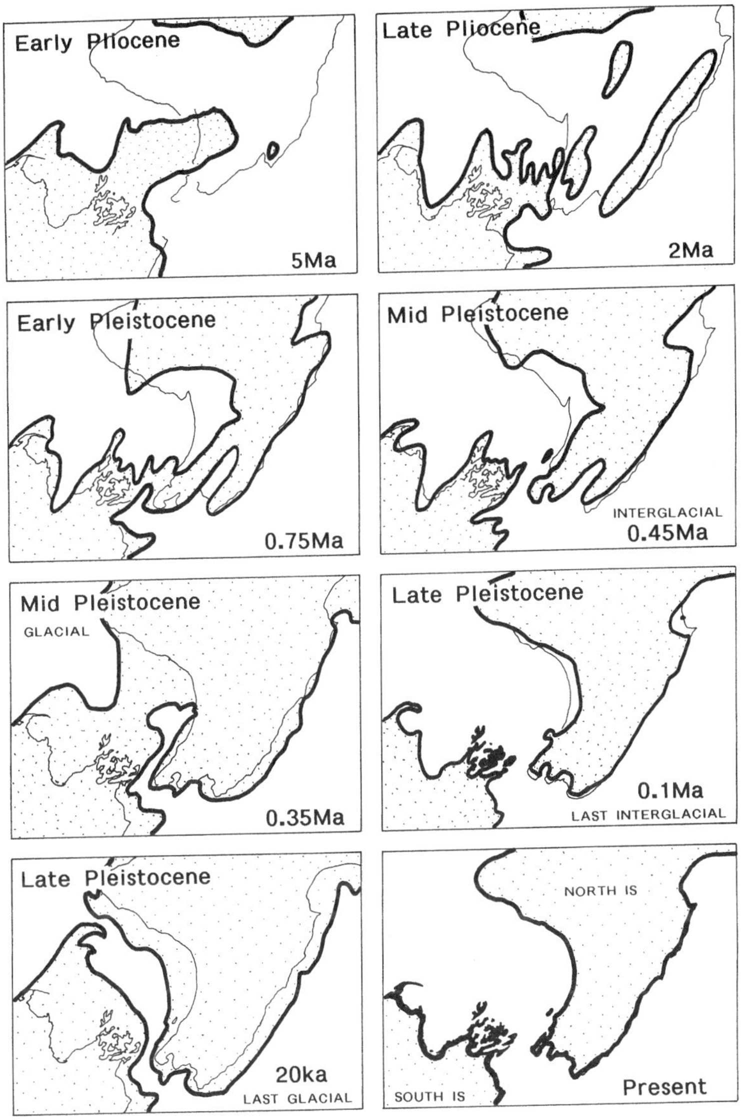

Figure I.2. Maps showing the changing distribution of land (stippled areas) in central New Zealand over the last 5 million years. The area has been subjected to rapid basin lowering and mountain building. The modern coastline is indicated by thin lines. The southeastern corner of the North Island has moved southward along the alpine fault. Something approaching present topography was only achieved during interglacial periods in the mid-Pleistocene, though lowered sea levels then resulted in one landmass. Unlike Lewis et al. (1994), we do not accept that continuous land extended across Cook Strait at any time during the last Glacial period, because fossil faunas from about 35 ky to the present have remained distinct, with endemic moa and other bird species restricted to either the North or South Island. Modified from Lewis et al. (1994).

riod since then may have seen dry land connecting the North and South Islands, but there is substantial faunal evidence that no bridge was formed in the last glaciation. Endemic species of birds and lizards on each side of Cook Strait have maintained discrete distributions and genetic integrity throughout the period from 35 ky to the present.

New Zealand (in the political sense, excluding New Caledonia and Lord Howe Island) is an archipelago of hundreds of islands stretching from Raoul in the Kermadec group (29°15'S) to Campbell Island (52°30'S), a total of just 270,000 km^2 (Fig. I.3). The main islands are the North (114,740 km^2), South (151,120 km^2), and Stewart (966 km^2) Islands, which span 13° of latitude (34°S to 47°S) and account for 98.8% of land area. There are 254 small islands (more than 8 ha) within 50 km of the coast of the North and South islands, and more

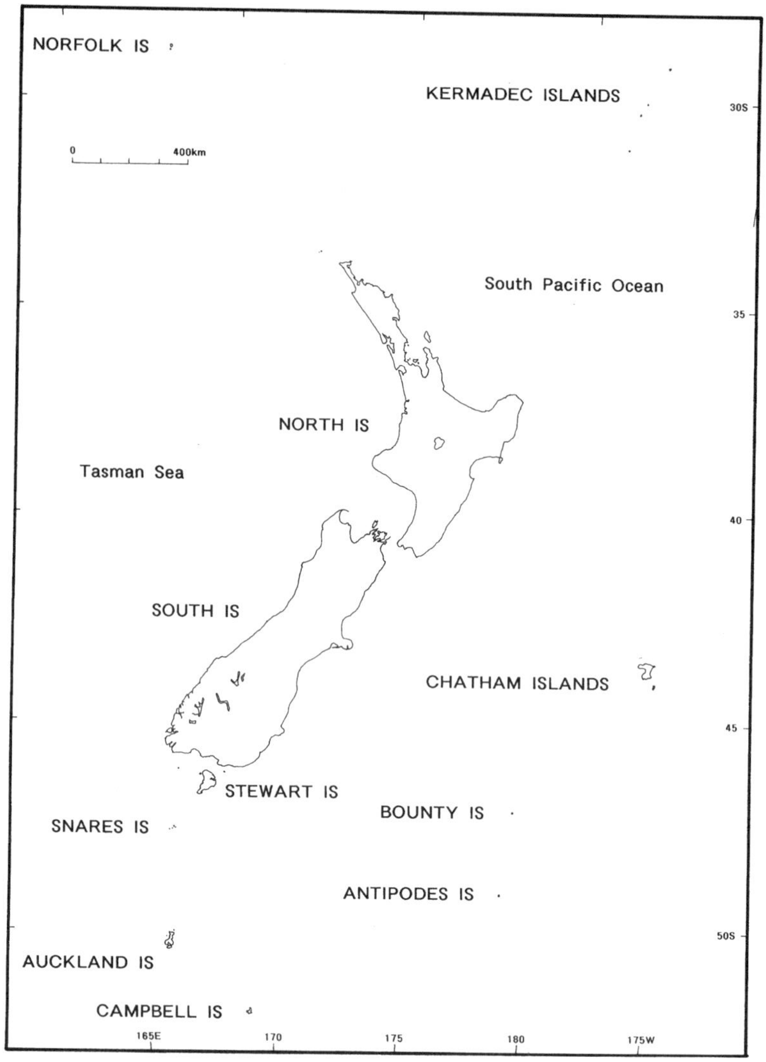

Figure I.3. Map of New Zealand and its outlying island groups.

than 400 islands of area less than 8 ha. The total land area is one thirtieth that of Australia, one third that of Madagascar, and comparable to the area of the British Isles.

About 500 km east of Christchurch, on the Chatham Rise, lie the Chatham Islands. Main Chatham Island (44°S) is about 45 km north to south by 30 km east to west at its widest, but its hourglass shape is dominated by a large central, brackish lagoon, which is periodically open to the sea. The main smaller islands of Pitt, Mangere, and Rangatira lie to the southeast across the 23-km Pitt Strait. The Chatham group in some form has existed separately from New Zealand since about 60 Ma. Although land was present in the area for most of that time, it is likely that there was complete submergence in the Oligocene. Nevertheless, the islands have still had a considerable period of time in which to develop their own unique flora (12% endemism) and fauna (e.g., 20% of insects are endemic), but they lack ancient endemic fauna

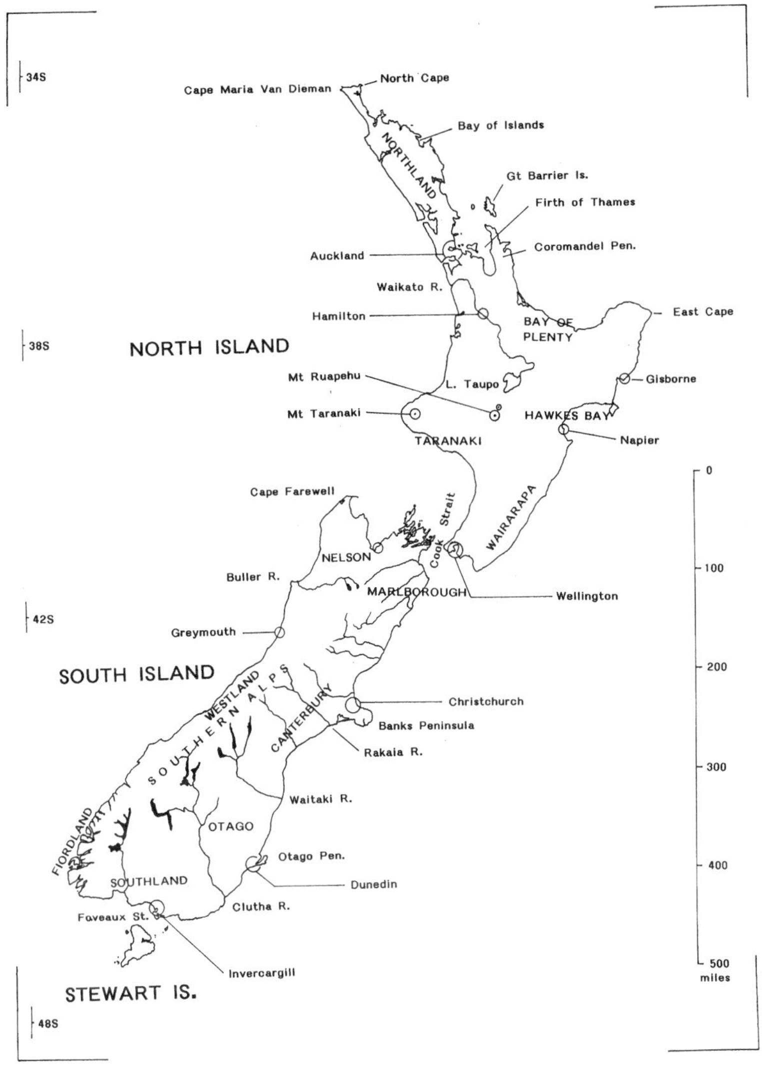

characteristic of the mainland, such as peripatus, giant crickets (weta), though two species of cave weta are present, and birds such as moa (Dugdale and Emberson 1996; Given 1996).

Figure I.4. Map of New Zealand showing major geographic features, regions (capitalized), and urban centers (circled). Latitude (degrees south) is indicated at left.

South and southeast of New Zealand lie a series of small island groups in subantarctic waters in the west wind belt known as the Roaring Forties (Fig. I.3). The Bounty Islands (47°45'S), Snares Islands (48°S), Antipodes Islands (49°40'S), Auckland Islands (50°45'S), and Campbell Island (52°30'S) are all small, volcanic islands that are much younger than the North and South Islands. All are extremely important breeding places for many species of seabirds, mainly petrels and penguins.

New Zealand's main islands present a varied relief (Fig. I.4). The North Island has low rolling hill country north of Auckland and down the western side. Elsewhere, although the relief is not high, the rapidly eroding Tertiary rocks form steep, rugged landscapes. The axial ranges in the southeast from Wellington to East Cape rise to 1300 m, and the steep slopes were heavily forested. The Central Plateau is a volcanic pile, built in the back arc basin of the southern end of the Kermadec Trench. Three active mountain volcanoes (Ruapehu, Tongariro, and Ngauruhoe) lie just south of Lake Taupo, a large, flooded caldera, which conceals the vents of Taupo Volcano. Other volcanoes are major features of the landscape. To the west, the impressive symmetrical peak of Mount Taranaki (Egmont) dominates the western bulge of the North Island. Auckland, New Zealand's largest city, has been built amid 26 scoria cones formed by small-scale basaltic eruptions, the last of which was just 200 years ago.

Volcanism has been a very significant force in shaping the North Island landscape during the Quaternary. For the last million years at least, ignimbrite flows have spewed forth from the eruption center at Taupo, traveled across the landscape at up to 400 km/h, at times reaching the west coast over 100 km from the vents. To the east, flows spilled over the axial ranges. The most recent Taupo eruption was only 1850 years ago. The world's largest recent eruption, it is not surprising that its effects may have been recorded in China and Egypt. The final phase deposited ignimbrites (debris from pyroclastic flows) 100 km from the source on the eastern side of the axial ranges. The debris also changed the course of the North Island's largest river—the Waikato—from a route leading into the Firth of Thames in the north to its present outlet on the west coast.

The South Island has escaped recent volcanism, but instead, tectonism has created the spectacular Southern Alps. Stretching along the length of the western margin of the South Island, the Alps rise to over 3000 m within 25 km of the narrow coastal plain. They are extremely steep and rugged as a result of rapid uplift and the high rainfall and have some of the highest rates of erosion in the world (Griffiths 1979). Their size and aspect, right across the path of the prevailing westerly winds, has major effects on the island's climate, and hence erosion processes and landform development. The orographic uplift induced by the flow over the mountains causes the moisture from the saturated oceanic air to precipitate on the western slopes of the Alps. At lower altitudes in the west, rain often falls at an inch (25 mm) an hour for 24 hours at a time, with peak sustained rates of over 100 mm/h. But because up to 2000 m of the Alps are above the tree line, much of the moisture falls as snow, which forms the basis for valley glaciers and two ice fields. During the Pleistocene glacial periods, these were of much greater extent, and an ice cap covered much of the Alps. Glaciers flowing from the ice fields gouged deep, steep-sided valleys. In the southwest, most are now partly drowned by rising sea levels, forming fjords. To the east of the ranges, rivers have transported huge volumes of greywacke gravels to form the Canterbury Plains, a 100-km-wide region of coalescing fans in central eastern South Island.

One of the by-products of the extreme geologic instability of New Zealand in the past few million years is the lack of potential for fossil sites. Tectonism itself accounts for much of the comparative lack of a fossil record. Rising land erodes more quickly. The processes of uplift and erosion are usually more or less in balance; otherwise, the thousands of meters of uplift would have resulted in some very tall mountains. Small rates can have large consequences over long periods of time. For example, where the land is subject to moderate rates of uplift, say 1.5 mm per year as in the northern South Island, it will rise 1500 m in a million years. So a cave—and potential fossil site—formed a million years ago on the top of one of the marble mountains in northwest Nelson will be lost in well under a million years. In contrast, across the Tasman Sea in Australia, faunas have been accumulating for at least 5 million years in caves in the same 20-m-high hill at Wellington in New South Wales. There are few places in New Zealand with such an old landscape: the major region, in Otago, lacks limestone. There are few terrestrial swamp or lacustrine deposits of pre-Pleistocene age. Only one of Miocene age has so far produced a vertebrate fauna. Volcanism and glaciation compound the problem.

Volcanism has shrouded much of the central North Island in a blanket of tephras from various sources, mostly of recent age. The South Island was heavily glaciated during the Pleistocene. Glaciers gouged away much of the previous landscape and buried much of the rest under many tens of meters of outwash gravels transported by many rivers. As sea levels rose and fell in concert with oscillating Pleistocene climates, and as climate changed from wet to dry and back again, the base level of streams and rivers changed as the coastline receded and advanced and as flow rates fluctuated. The rivers went through corresponding cycles of erosion and deposition. Many caves in New Zealand have evidence of extensive deposition in the last part of the last Glacial period followed by often near complete exhumation of such deposits in the past few thousand years.

Climate

The insular nature and the isolation from large landmasses give New Zealand a decidedly maritime climate. New Zealand lies in the path of a circumpolar air and water circulation, which began to form when Australia broke free of Antarctica and which was completed with the opening of Drake Passage about 26 Ma (Fordyce and Barnes 1994). The climate is characterized by alternating anticyclones and depressions moving eastward across the country. The weather pattern is also strongly controlled by topography and aspect. Mean annual temperatures vary between 15°C in the north and 9°C at the southern end of the South Island. Summers (December–February) are warm, and snow is normal on several days in southern regions in winter (June–August). Most rainfall originates in low-pressure systems from the west. Because of orographic effects, rain falls mainly in the west of both islands. Rainfall is spread throughout the year, with a minimum in summer. Higher summer temperatures cause short-term water deficits or drought, particularly in the rain-shadowed areas east of the Southern Alps and in central Otago, where annual rainfall is as low as 400 mm. Similarly, the Hawke's Bay region of the eastern North Island is in a rain shadow, and some areas there receive less than 1000 mm of rainfall annually. Much of the country receives over 2000 sunshine hours per year. The other, less regular, source of precipitation is tropical cyclones, which move from low latitudes of the tropics across the North Island. Cyclones can bring heavy rainfall to the northern and eastern North Island, but the effects are often patchy and can be extreme locally.

The tree line is presently at about 1200 m in the North Island and lowers southward, reaching about 900 m in Southland. Above this level, illustrated in Figure I.5, snow lies on the ground for a few weeks each year, and vegetation is limited to shrubs, herbs, and grasses.

Vegetation

The flora (taxonomic composition of plant communities) and vegetation (physical structure of the plant communities, e.g., grassland or forest) of New Zealand have changed through time. Through most of the Tertiary, when New Zealand had a low relief and a warm climate, lowland forest was the main vegetation. However, with the onset of mountain-building in the Miocene, new environments appeared that were too cool to support forests. In the new mountains, a shrubland flora proliferated, with much speciation in families such as the Compositae, Pittosporaceae, Papilionaceae, Rubiaceae, and Scrophulariaceae. Similarly, there was rapid evolution of herbs in the Ranunculaceae and Compositae.

The Pleistocene and its series of glaciations eliminated many tropical taxa such as eucalypts and casuarinas, which are still found in Australia. The latest glacial event, locally called the Otiran Glacial, involved a general cooling of about 5°C. A tree line depressed by 800 m pushed all vegetation communities downslope. Many plants probably survived only on land that is now below sea level when sea levels were 100–120 m below present. Lowland forest probably survived in these now-drowned habitats and in the most sheltered sites in southern areas in these coldest of times. In the central South Island, glaciers spreading down from the Alps on the West Coast pushed beech forests aside and eliminated them from 300 km along the mountains. In the 10,000 years since the glaciers retreated, the beech have not reestablished there; they spread slowly because their heavy fruits are not dispersed far from the parent trees.

Vegetation Types

The forests of New Zealand are dominated by uniquely Gondwanan elements. Foremost among these are evergreen gymnosperm trees, especially the ancient podocarps (family Podocarpaceae). Podocarps are conifers named for their fleshy "foot-fruit," which are associated with the female ovule that birds disperse and which have tiny male cones. Podocarps have changed little in the last 190 million years, and lineages of most extant species can be traced back to the Cretaceous. The various species are found only on North, South, and Stewart Islands; they are absent from the Chatham Islands and other subantarctic islands. Some species are creeping shrubs, but several are spectacular trees. Rimu *Dacrydium cupressinum* can attain 60 m on a trunk up to 1.5 m in diameter and ages of 700 years up to 1000 years. It has drooping, soft, yellow-brown foliage and is—or was—the most prized timber tree of wet lowland forests. In virgin forests, crowns may abut to form a closed canopy, but more often, they are emergent over a closed subcanopy. Usually there is a multitiered understory of hardwood species and a dense groundcover of shrubs, lianas, and ferns. In wet gullies, a rich fern flora includes several tree ferns, some of which reach more than 10 m in height. In wetlands, such as behind the levee banks of rivers, the kahikatea *Dacrycarpus dacrydioides* sends straight trunks soaring for 30–40 m to the first branch and are often 60 m tall. In drier eastern areas, matai *Prumnopitys taxifolia* is the dominant podocarp in lowland forests—or at least it was. Only a few precious patches of these forests remain. The blue-gray foliage of short flattened needles of matai is borne by a trunk with a characteristic "hammered" pattern, and the tree commonly reaches 25 m. The other main species of podocarp in New Zealand forests are totara *Podocarpus totara* (30 m), which likes well-drained soil, and miro *Prumnopitys ferruginea* (25 m). Sometimes these trees form a mixed podocarp forest (Fig. I.6). All provide rich but seasonal food supplies for fruit eaters; in New Zealand, that means birds. As with the other main forest type in New Zealand, the southern beeches (*Nothofagus*), podocarps are mast-seeding, and heavy crops of fruit can be several years apart.

Above 38°S in the North Island of New Zealand, the softwood or gymnosperm forests were once dominated by truly giant trees in the ancient family of Araucariaceae. Their ancestry can be traced back to the Jurassic period at 190 Ma. The kauri *Agathis australis* was certainly not the tallest tree in the world, but it is renowned for its straight, parallel-sided trunk. Trees often reached 2000 years of age, and some may have reached 4000 years. The New Zealand kauri holds the record for the greatest timber volume of any tree in the world. Some specimens reached 60 m high, 30 m to the first branch, and 7 m in diameter. The largest tree (in volume) measured was only 21.8 m to the first branch but had the staggering girth of 23.43 m. The foliage is extremely turpene rich and its sap forms kauri gum, which was the subject of an industry in the late 19th and early 20th centuries. Timber merchants felled most kauri forests last century, but a few remnants of these majestic forests are preserved, including some very large trees. In an odd twist, timber merchants now use "fossil" kauri wood up to 40 ky of age recovered from swamps for furniture manufacture. Less than 10% of all lowland forests remain.

Podocarp forests of one kind or other dominated lowland landscapes, but on higher slopes and in montane areas, there was a different type of forest, which now provides the greatest remaining area of native forest. It is an almost monotypic forest of southern beech (Fig. I.7). Southern beeches are of Gondwanan origin, and in New Zealand, the genus *Nothofagus* contains just four species—red, silver, hard, and black beech. They usually form a continuous monospecific canopy above a single lower understory, and the forest floor is open. Red beech (*N. fusca*) may grow over 30 m, but the other species are smaller, and near the tree line, they are reduced to the height of shrubs. They all have small, hard leaves that do not attract folivores. Although they produce beechnuts each year, this favored food varies greatly in abundance. In some years, these fruiting bodies are produced in extraordinary abundance, and such "beechmast" years are instrumental in the breeding success of some birds such as kaka (*Nestor meridionalis*). An unusual feature of beech forests is the presence of a small sap-sucking bug that lives in the bark of the trees in great abundance. As they feed, these bugs extrude sugar water from a narrow tube that projects from the tree's bark about 2–3 cm. The sugar water is an important food for many beech forest animals.

Where beech or podocarps do not dominate, forests are composed of a mixture of hardwood trees. Plant ecologists have long debated whether mature podocarp forests were replaced by hardwood forests or the reverse. The relevance of such debate is questionable because it was probably rare for a New Zealand forest to be undisturbed in succession. Cyclones, fire, and—especially in the North Island—volcanic events have regularly destroyed forests. After such disturbances, hardwood forests are first to colonize the ground; the podocarps, which may take 300–500 years for trees to mature, follow after. Some hardwood species have a distribution controlled by climate. Several, such as puriri (*Vitex lucens*) and taraire (*Beilschmiedia taraire*), are a subtropical element found, like kauri, only north of 38°S. Their rich potential harvest of fruit is thus available only in relatively small areas.

The remaining significant type of vegetation in New Zealand is grassland of clump grasses or tussocks, mainly in the genus *Chionochloa*. There are limited areas of lowland tussock communities, but they dominate above the tree line in the subalpine zone. In areas buried in snow for several weeks every year, the many species of tussock shelter many spectacular herbs, such as the New Zealand buttercups *Ranunculus*, which have beautiful flowers and large, palatable leaves. Daisies are common and vary in form from rosettes in the genus *Celmisia* to others that appear to be furry rocks. In the so-called vegetable sheep (e.g., *Haastia* sp. and *Raoulia* sp.), the bracts are clumped tightly together, and the plants form low mounds, often more than 1 m in diameter, rising from scree slopes.

In the ecotonal zone, between the beech or upland podocarp forests and the subalpine grasslands, are dense, low shrublands. These are also dominated by daisies, but here, they form several genera and many species of woody shrubs up to 3 m high. In the subalpine zone there is a marked seasonal growth in spring when plants grow and flower and often die away again. Food for browsers is plentiful in this vegetation zone, for at least part of the year. A similar shrubland vegetation is found at lower elevations, particularly in the drier eastern areas along forest margins, such as along riparian strips, or where forests are recovering from catastrophic damage.

From this brief summary of vegetation types, it is clear that there were many varied habitats for animals, and for birds in particular. Evergreen, lowland, gymnosperm forests dominated by podocarps provided rich seasonal fruit supplies but little edible foliage. Folivores needed to rely on the leaves of hardwoods. Not all podocarp forests offered the same habitats to birds. Those in wetter western areas dominated by rimu were quite different from those dominated by matai in the seasonally dry eastern districts. The montane beech forests were different. Although floristically poorer, they still supported large numbers of birds. The subalpine zone of shrublands and grasslands is another distinct habitat that is at least seasonally available, although at least two birds are known to overwinter here. One is the tiny rock wren (*Xenicus gilviventris*) and the other a large rail, the takahe (*Porphyrio hochstetteri*). The wren can even survive among the dense vegetation under a snow pack, much as rodents do in the northern continents. Although there are no modern analogs, pollen studies show that dry eastern areas at low altitude presented rich mosaics of vegetation with many edges between forest, shrubland, swamps, and grasslands. In the latest Holocene, matai forest was interspersed with shrublands and grasslands. Such mosaics were doubtless formed at least in part by the migrating beds of braided rivers, with vegetation successions on abandoned channels, and as a result of a variable soil fertility. The mosaics appear to have been characterized by an abundance of forest margins or open areas. Some species that preferred these areas also used the subalpine zone, which, although cooler, had the same vegetation structure.

In a State of Change

New Zealand has had a varied and at times violent history, but there has been a terrestrial fauna since its split from Gondwana. It is otherwise hard to explain the presence of saltwater-intolerant organisms, such as earthworms, peripatus, and many insect groups that are also found in Australia, South Africa, or South America, all of which are fragments of Gondwana. Despite the near-total drowning of the landmass in the Oligocene, these and such vertebrates as leiopelmatid frogs (*Leiopelma* spp.) and the tuatara (*Sphenodon punctatus*) survived. The animals lived in the Gondwana podocarp and beech forests, forming communities characterized by change. Tectonic forces wrenched the land, islands formed and vanished, and mountains were built. Volcanism was locally catastrophic, yet also improved soil fertility and rejuvenated vegetation over wide areas. Pleistocene glaciations waxed and waned rapidly in time with abrupt changes in the climate. Since the Miocene, ecosystems in New Zealand have been in a state of continual flux, continuously adapting to this heady mix of tectonic events, volcanism, and climate change.

The effects on the biota varied. The rise of the mountains stimulated rapid evolutionary development of a host of new forms to occupy new habitats. Changing patterns of islands promoted allopatric speciation in plants and birds. The Pleistocene glaciations succeeded each other sufficiently slowly such that as climate changed, plant communities could follow the moving belts of suitable temperature and humidity, upslope or downslope, and across the landscape. Faunas that depended on particular vegetation communities shifted their ranges to follow their preferred vegetation as the distribution of the vegetation types themselves responded to climate. The catastrophic effects of volcanism destroyed entire forests over wide areas in the central North Island. These always regenerated quickly, though, undoubtedly aided by the transport of fruit by birds. Some taxa, though, particularly some invertebrates such as the giant carnivorous snails of the genus *Powelliphanta* or the *Leiopelma* frogs, do not seem to have been able to spread fast enough to recolonize ground lost to the ignimbrite flows of only 1850 years ago.

We know that the flora changed greatly in the deeper past. Warmth-loving casuarinas and eucalypts became extinct sometime after the onset of glaciations in the late Pliocene. The fauna changed too as the climate cooled. It is now known that dinosaurs sailed off with New Zealand and crocodilians were present in the Miocene. It has long been assumed that because many New Zealand birds are similar to Australian species, they were relatively recent immigrants. In fact, only moa, the archaic wrens in the family Acanthisittidae, and wattlebirds in the Callaeatidae have usually been considered as potentially being part of the original Gondwanan fauna. But exactly how much change there has been is still largely unknown: its violent history has left New Zealand bereft of a fossil record for most of the terrestrial faunas of pre-Quaternary time. What else may have been on the biota inherited from Gondwana can only be guessed at.

Fauna: Invertebrates

A major part of the Gondwanan legacy is in the composition of the invertebrate fauna. As a consequence, New Zealand has a distinctive invertebrate fauna. Groups with great diversity in other countries are poorly represented here—and the reverse is also true. For example, in New Zealand, there are fewer than a dozen species of ants but more than 1000 species of land snails. Some of the latter are several centimeters across and eat earthworms well over 30 cm long. The invertebrate fauna has long evolved in a world without small mammal predators, and so giant flightless forms developed in many groups. Members of the Orthoptera (crickets and grasshoppers) have a long history in New Zealand, and present forms differ little from Triassic fossils 190 million years old found in Australia. Most spectacular are numerous species of giant weta, the biggest of which can attain a body mass of over 70 g—nearly as heavy as an American robin—and its body is about 100 mm long. Another weta, about 85 mm long, has huge mandibles that resemble the tusks of a mammoth. Many large, flightless grasshoppers have evolved in the subalpine herb fields. As in all faunas, beetles are the most diverse insect group in New Zealand, accounting for more than half of all species identified so far. As with the orthopterans, several groups of beetles, particularly stag beetles (*Dorcus* sp.) and weevils, have produced large, flightless forms. There are even giant flightless click beetles (*Amychus* sp.). Most of these large, flightless forms are now either extinct or endangered and are restricted to relict populations, often on small offshore islands.

Other distinctive New Zealand endemic fauna include the peripatus, whose fossil ancestors indicate a remarkable lineage of 500 million years. The several species have relatives in the other Southern Hemisphere Gondwanan remnants. Earthworms, some over 1 m long, also have Gondwanan origins.

Present Terrestrial Vertebrate Fauna

On the warmer tropical islands of Lord Howe and New Caledonia that straddle the northern ridges of the Gondwanan fragment that carries New Zealand, there are distinct vertebrate communities that show some affinity with New Zealand. Lord Howe had a giant horned terrestrial turtle (*Meiolania platyceps*), indicating a considerable antiquity to the fauna, but had, so far as is known, only one bird species with relatives in New Zealand. This was the flightless Lord Howe gallinule (*Porphyrio alba*), a takahe (*Porphyrio hochstetteri*) equivalent. Lord Howe had two bats: a small vespertilionid (*Eptesicus pumilus*) and the larger fossil *Nyctophilus howensis*. New Caledonia has a diverse skink and gecko fauna, many of the latter in genera whose closest relatives are the New Zealand *Hoplodactylus* and *Naultinus*. New Caledonia also had a species of terrestrial turtle *Meiolania mackayi,* a terrestrial crocodilian *Mekosuchus inexpectatus,* and a large varanid lizard *Varanus* cf. *indicus*. Of eight bat species (Flannery 1995), only the vespertilionid *Chalinolobus neocaledonicus* has relatives in the New Zealand region. Many of its birds are dominated by tropical forms, but it had representatives of the New Zealand snipe (*Coenocorypha*) and parakeet (*Cyanoramphus*). Its distinctive kagu (*Rhynochetos jubatus*) has been considered by some to be closely related to the *Aptornis,* a highly specialized endemic gruid of New Zealand (e.g., Cracraft 1982). However, overall, the tropical location and dispersal abilities of birds have removed avian links between New Zealand and New Caledonia, and we will not discuss it further.

The native vertebrate fauna of New Zealand is dominated by birds. In greater New Zealand (including Macquarie, Norfolk, Kermadec, North Island, South Island, Stewart Island, Chatham Island, Bounty Island, Antipodes Island, Campbell Island, Snares Island, and Auckland Island), there were 245 breeding species of birds in the latest Holocene, immediately before humans arrived (Appendix 1). Endemism is high: 174 of 245 species (71.0%) and 35 of 110 genera (31.8%) are endemic. In the core New Zealand (North, South, and Stewart Islands, and their inshore islands—the "mainland"), there were 132 breeding species. The remaining 113 were tropical species in the northern groups, and the many specialist cold-water marine species on one or other of the subantarctic island groups. In addition, many "mainland" species have an equivalent on one or more of the outer island groups. For example, each subantarctic group has its species of pink-footed shag (*Leucocarbo* sp.), or a snipe of the southern genus *Coenocorypha.*

In addition, the core islands, but none of the outlying groups of New Zealand, had frogs (*Leiopelma,* 6 species), tuatara (*Sphenodon*), geckos (*Hoplodactylus* and *Naultinus,* 22 and 7 species, respectively), and bats (*Mystacina,* 2 species) and *Chalinolobus* (1 species). There were at least 30 skinks in two genera (*Cyclodina,* 8 species, and *Oligosoma* = *Leiolopisma,* 22 species).

Outside the core (but not including New Caledonia and Lord Howe; see above), there were few vertebrates other than birds. Chatham Islands had only one other vertebrate: a single species of skink. Norfolk Island had a vespertilionid bat (*Chalinolobus gouldi*), a skink (*Leiolopisma lichenigera*), and a gecko (*Phyllodactylus guentheri*). But among land birds, Norfolk had clear affinities to New Zealand with a *Coenocorypha* snipe, *Hemiphaga* pigeon, and *Nestor* and *Cyanoramphus* parrots. New Zealand lacks salamanders; other lizard groups such as agamids, iguanids, or varanids; snakes (except vagrant sea snakes); land turtles; crocodilians (except a Miocene fossil record); or other mammals, apart from the marine seals and whales.

The birds known historically from New Zealand are only a vestige of the fauna that greeted the first humans to reach its shores. Yet they provide a window on that lost fauna. Incongruously, in view of its dearth of older fossils, New Zealand has one of the best late Quaternary (less than 50,000 years) fossil records in the world. There are many and widespread caves, dunes, and swamps in which material has been preserved (Fig. I.8). Most important, there were no mammalian scavengers to move and destroy carcasses. Once dead in a cave, an animal had a very good chance of remaining there undisturbed. The record, unveiled in only the last 150 years, has revealed a wealth of species. Their loss is one of the most dramatic extinctions of recent times. The North and South Islands are on a par with Hawaii for the sheer number of vertebrate extinctions in proportion to the original fauna. It is now known that on those islands, more than 50% of bird species became extinct. Some 40 species are extinct in the core area of New Zealand alone. If it were not for the fortuitous fact that New Zealand consists of some 800 islands, many of which are near enough the main islands to carry some of their fauna, many more species would be globally extinct by now. The nearshore islands provide what Hawaii did not have: an opportunity for relict populations to survive and hence postpone or prevent global extinction. The catastrophe was precipitated by the arrival of humans and the subsequent establishment of permanent settlement, which occurred about 700 years ago.

In the following pages, we examine this fauna. Particular attention is paid to the more interesting or unique forms known only or largely from the fossil record. We describe the process of their discovery and follow the complex paths that have led, and continue to lead, to an understanding of their biology, ecology, and relationships. Foremost among the evolutionary novelties of New Zealand were the moa (Aves: Dinornithiformes). We will summarize all that is presently known about these birds, describe their skeletons, and reconstruct their life and ecology. Then we discuss the only threat to the survival of an adult moa, the world's largest eagle. Then we place, in turn, the primary herbivores, the top predator, and other species in the total fauna, where ducks, gruids, and even bats all followed an evolutionary path to flightlessness. But the most specialized species were also the most naïve to new players in the environment. After 80 million years of isolated evolution and adaptation to all manner of climatic and geologic changes, the fauna was ill prepared to face mammalian predators. First rats, then humans, and then a growing suite of other rapidly breeding, highly adaptable predators wrought havoc on the fauna. In the past 2000 years, the fauna has been reduced to a mere shadow of its former glory.

Figure I.5. (opposite page top) Upper slopes of Mount Arthur, northwest Nelson, South Island, showing the tree line at 1200 m. Beech forest gives way to tussocks and herbs, which extend to about 2000 m. In the marble of these upper slopes, there are many caves, and moa bones are common. Photograph by T.H.W.

Figure I.6. (opposite page bottom) Example of podocarp forest showing tall rimu trees emergent above the canopy. Photograph by R. Morris.

Figure I.7. (opposite page) Inside a New Zealand Nothofagus *(beech) forest. They are usually found in upper montane habitats, such as that on Mount Arthur in northwest Nelson (Fig. I.5). Photograph by T.H.W.*

Figure I.8. (above) Sand dunes at Tokerau Beach, Northland, North Island, a renowned fossil locality for moa and other birds. Photograph by T.H.W.

The Lost World of the Moa

1 • The Early Record

Cretaceous

The pre-Quaternary record of terrestrial vertebrates in New Zealand is exceedingly poor. In contrast, that of marine vertebrates, including penguins, whales, and dolphins, is rich. This record has been extensively reviewed by Fordyce (1991) and Long (1998), from which the following summary is largely derived. It is presented to put the history of paleontological investigation of the terrestrial ecosystem and birds, the subject of this book, into context. Because they are outside the main scope of this book, we treat the marine reptiles and mammals only briefly, mainly to indicate their diversity. For similar reasons, the rich New Zealand penguin fauna of pre-Quaternary age is also treated briefly. Here we have listed the species present and indicated, where necessary, the existence of so far undescribed further species. Principal references are cited; notably, Fordyce, in Turbott (1990a), lists known taxa, and Fordyce (1991) and Fordyce and Jones (1990) provide good reviews of all pertinent literature on penguins.

Higher vertebrates make their first appearance in New Zealand in marine rocks. Ichthyosaurs have been identified in Late Triassic rocks near Mount Potts, Canterbury, and the Hokonui Hills in Southland (Campbell 1965; Fleming et al. 1971; Long 1998). The material from both sites is not determinable to genus or species. No Jurassic fossils are known, and it is not until the Upper Cretaceous that a rich record is preserved. Figure 1.1 shows a time scale from the Triassic to the present comparing local and international names.

Abundant remains of marine reptiles have been found in the Conway Formation (Maastrichtian) of the northeastern South Island, particularly at the Waipara River in North Canterbury and Haumuri Bluff in Marlborough. Sir Richard Owen, better known in the field of

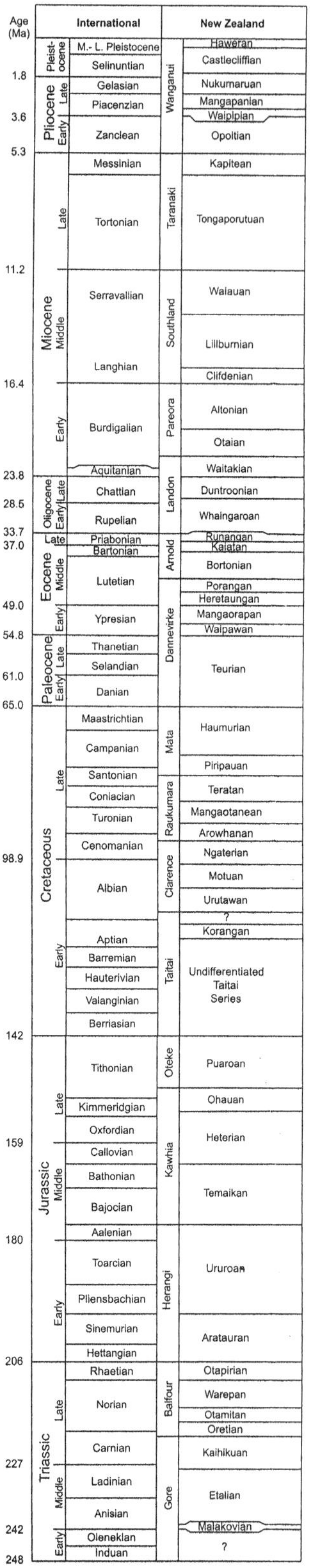

New Zealand paleontology for his work with moa, was instrumental in the identification of the first specimens: in 1861, he announced the first discovery of plesiosaurs from the Waipara River. Throughout the 1860s and 1870s, many discoveries were made, but most of these specimens are now lost. Some went down with the ship *Matoaka* in 1869 en route to England. Two large consignments (one in 1877 and another in 1889 of some 39 cases of material) to the eminent American dinosaur expert Edward Drinker Cope are missing, although it is known that they were received (Welles and Gregg 1971).

The fauna from all these early collections contained several plesiosaurs and mosasaurs. The fauna was extensively reviewed by Welles and Gregg (1971). They accepted two plesiosaurs, an elasmosaur, *Mauisaurus haasti,* and an indeterminate species of pliosaur or polycotylid. The first bones Owen saw from New Zealand were of this pliosaur. Four species of mosasaurs were identified, two as new: *Taniwhasaurus oweni, Tylosaurus haumuriensis, Mosasaurus mokoroa,* and *Prognathodon waiparaensis.* Recently, Bell et al. (1999) reported the discovery of a fossil skull that indicates *Tylosaurus haumuriensis* is not distinct from *Taniwhasaurus oweni* (these taxa were previously known from different and not comparable parts), and so *Tylosaurus* should be deleted from the New Zealand fauna.

The remarkable fossil deposits of the Mangahouanga Stream in Hawke's Bay have been steadily yielding specimens from a late Cretaceous fauna, thanks to the efforts of Mr. M. and Mrs. Joan Wiffen. The late Cretaceous rocks (Campanian–Maastrichtian) in this site have many marine reptile bones. Turtle plastron and carapace fragments of the family Protostegidae and other undetermined marine turtles are common. More significantly, plesiosaur and mosasaur bones are common and include good cranial material. The mosasaurs *Mosasaurus flemingi, Moanasaurus mangahouangae, Prognathodon* cf. *P. overtoni,* and *Rikisaurus tehoensis* are known from Mangahouanga Stream (Long 1998; Wiffen 1980, 1990). Recent reassessment of the Mangahouanga fauna by Bell et al. (1999) has revealed that *Moanasaurus* contains two species and that specimens previously identified as *Rikisaurus, Prognathodon* cf. *P. overtoni,* and *Mosasaurus flemingi* are *Moanasaurus.* The generic diversity of New Zealand mosasaurs is therefore very much less than has been thought. In addition, the fauna includes an indeterminate pliosaur and the plesiosaur *Tuarangisaurus keyesi* (Wiffen and Moisley 1986). However, it is not the marine vertebrates for which this locality is now well known in New Zealand, but dinosaurs.

A few tantalizing fragments and relatively nondescript bones have been discovered (Long 1998; Molnar and Wiffen 1994). As with the Australian dinosaurs, the New Zealand fauna has to be reconstructed from meager remains. Theropods, the large bipedal meat eaters, are represented by a single tail vertebra and a pedal phalanx from an animal the size of a 12-m *Allosaurus.* Sauropods, large herbivores, are known from a rib fragment from an animal about 10–12 m long. The armored herbivorous ankylosaurs are represented by a rib and a caudal vertebra from a 2.5–3-m animal. Ornithopods are represented by the hip bone (right ilium) from an animal about 3 m long, similar to

Dryosaurus. In addition to these, at least two pterosaur bones have been found (Wiffen and Molnar 1988). These scraps, and others currently under study (J. Wiffen, personal communication) show that New Zealand had a diverse dinosaur fauna in the Late Cretaceous, despite its position at high latitudes at this time. Other high-latitude dinosaur faunas of similar age are also known from Victoria in Australia, the Antarctic Peninsula, and the north slope of Alaska, and analysis indicates that dinosaurs as a group had a broad ecologic tolerance. At this time (80–70 million years ago [Ma]), New Zealand was between 60° and 55° south (Lawver et al. 1992), so dinosaurs were probably able to survive periods of darkness and the presumed associated lack of plant growth.

A number of other sites of Late Cretaceous age listed by Fordyce (1991) contain marine reptiles. The known southern Cretaceous avifauna includes two records of undescribed fossils from New Zealand, a tarsometatarsus from the *Lingula* beds in the Waimakariri Gorge (formerly of the New Zealand Geological Survey Collection, but now lost), and fragments of a distal left femur from shallow marine sands of the ?Conway Formation near Cheviot, both in North Canterbury, South Island.

Cenozoic

There are few Paleocene vertebrate fossil bones from New Zealand. Fordyce (1991) listed two, which were of possible turtle and undescribed bird bones. The latter are from the Moeraki Formation and include an indeterminate distal right tibiotarsus more than 30 mm wide. A second, more significant specimen is a primitive penguin from the Waipara Greensands of North Canterbury. The specimen includes the coracoid, part of a humerus, and the scapula, among several elements of a single skeleton (Fordyce 1991; Fordyce and Jones 1990). Jones and Mannering (1997) reported more material of this bird, which is as yet undescribed. Last, Fordyce (1991) mentioned a proximal radius of a large volant bird from the Tahatika Grit (Teurian) of the Chatham Islands.

At least two penguins (*Palaeeudyptes marplesi* and *Pachydyptes ponderosus*) are known from the New Zealand Eocene (Marples 1952; Simpson 1971, 1975). Fordyce (1991) also lists turtle and cetaceans. Recently, Eocene strata have produced several archaeocete whales that represent three or more species: a very large species and probably two dolphin-size animals, including one close to the Northern Hemisphere genus *Zygorhiza* (Fordyce 1985; Köhler and Fordyce 1997). Recently an archaic leatherback turtle has been described from the Eocene of New Zealand (Köhler 1995).

Early Oligocene rocks have produced a few turtles, mysticete cetaceans, and penguins (Fordyce 1991). The first described penguin from New Zealand *Palaeeudyptes antarcticus* is from Kakanui in north Otago. Its source horizon is uncertain, but it is probably from the Ototara Limestone, which is basal Oligocene, 34–32 Ma (Fordyce, personal communication). Hector (1872b) referred large bones of a fossil pen-

Figure 1.1. (opposite page) Time scale for New Zealand from the Triassic to the present comparing local and international names (modified from Rattenbury et al. 1998).

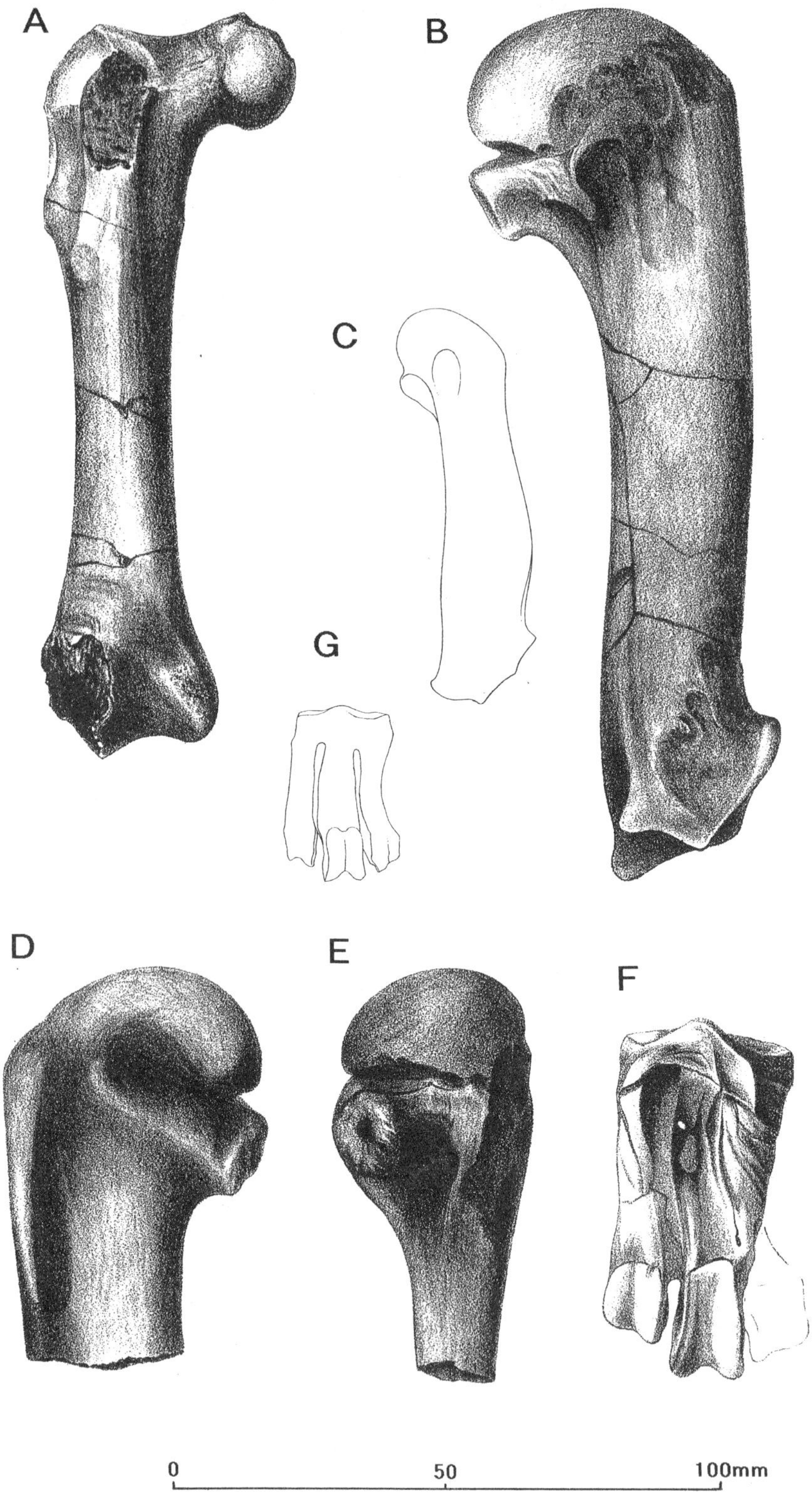
A
B
C
G
D
E
F
0
50
100mm

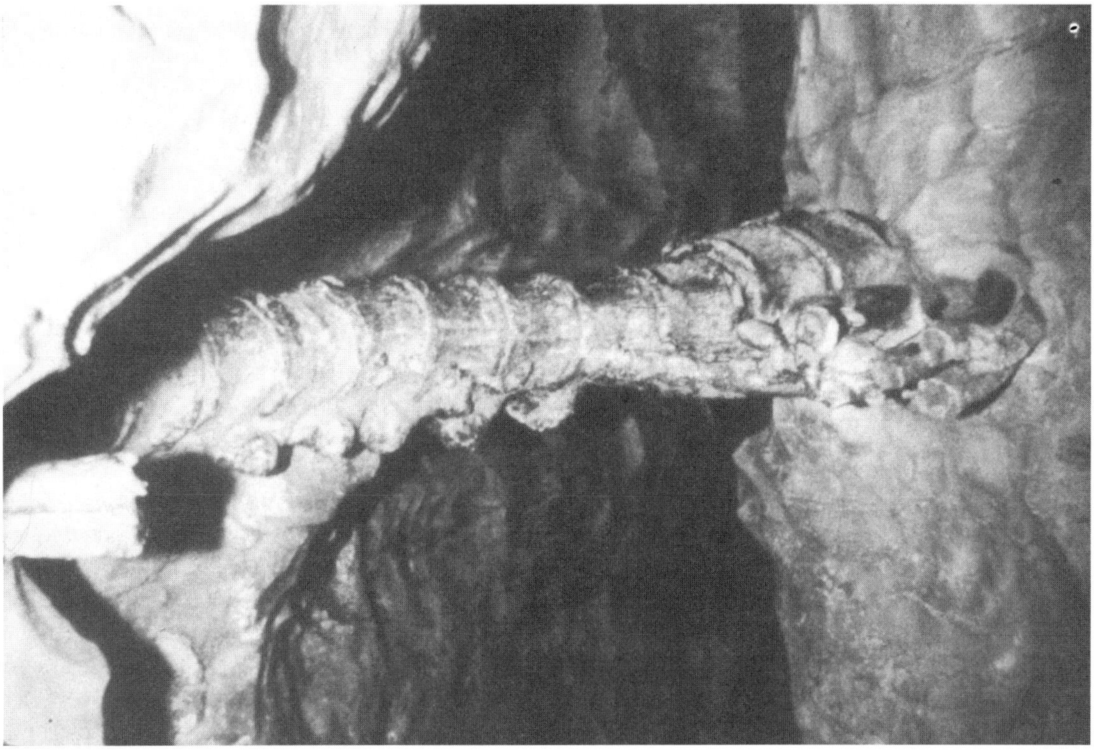

guin from Seal Rock, West Coast, of the South Island to this form (Fig. 1.2), but this association is uncertain because the skeleton from Seal Rock lacked a tarsometatarsus, which is all that is known for *P. antarcticus*.

The late Oligocene to early Miocene is represented by the local upper Whaingaroan, Duntroonian, and Waitakian Stages. Cetacean and penguin fossils are known from the well-cemented, somewhat diagenetically altered limestone of the Te Kuiti Group in the North Island and others in northwest Nelson and on the West Coast (Fig. 1.3). However, most fossil vertebrates of this age come from the soft and less well cemented bioclastic limestones and greensands on the eastern side of the South Island—the Ototara Limestone/Earthquakes Marl, the Kokoamu Greensand, and the Otekaike Limestone. They contain abundant cetacean fossils. Beyond penguins and rare other birds, associated Oligocene vertebrates include chondrichthyans (sharklike fishes) and teleosts (bony fishes), but no sea turtles, pinnipeds (seals), sirenians (manatees), or desmostylians (early marine mammals). Faunas and sediment character indicate that these rocks were formed in a sheltered, shallow, and broad seaway, which was only rarely inhabited by more oceanic or deep-diving species of delphinoids and sperm whales.

Other than penguins, only one bird has been described from Oligocene rocks. *Manu antiquus* was described from a furcula by Marples (1946) and was regarded as similar to albatrosses. Marples also figured two fragmentary femora, which he suggested were referable to *Manu*

Figure 1.2. (opposite page) Giant penguins from the New Zealand Oligocene. (A) Right femur and (B, D, E) right humerus from Seal Rock, West Coast, South Island. (F) Right tarsometatarsus that is the holotype of Palaeeudyptes antarcticus *from Kakanui, north Otago, South Island, to which Hector (1872b) referred the Seal Rock bones. (C) Humerus and (G) tarsometatarsus of the 4-kg* Eudyptes pachyrhynchus *at the same scale. After Hector (1872b: pls. 17, 18).*

Figure 1.3. (above) Fossil whale bones exposed in a cave on the West Coast of the South Island. Photograph by T.H.W.

because they differed from penguins and had some similarity to albatrosses. However, Olson (1985) considered that the furcula differed considerably from living albatrosses and that its affinities should be considered as indeterminate. Fordyce has conducted extensive fieldwork in these rocks and has found only three bones of birds other than penguins—a tibiotarsus and a femur, their affinities both so far unidentified (Fordyce 1991), and a procellariiform (bird of petrel family) ulna in the Otekaike Limestone (Fordyce, personal communication). The penguins *Platydyptes novaezealandiae, P. amiesi, P. marplesi, Archaeospheniscus lowei, A. lopdelli, Duntroonornis parvus,* and *Korora oliveri* are all known from late Oligocene rocks, upper Miocene rocks, or both (Marples 1952, 1962a; Simpson 1971, 1972, 1975). Recent collecting by Fordyce and Jones et al. in the Kokoamu Greensand and equivalents of Duntroonian age has produced at least three new species, two represented by relatively complete skeletons.

Before 1982, there were only 14 named species within Archaeoceti (primitive toothed whales), Odontoceti (dolphinlike whales), and Mysticeti (baleen whales) from the Eocene and Oligocene. Recent work has greatly expanded this fauna. Fordyce (1991) listed 1 archaeocete (*Kekenodon onamata*), at least 20 mysticetes—13 species of cetothere mysticete, or baleen whales, from the Kokoamu Greensands and its lateral equivalents—in north Otago and Canterbury alone, and at least 4 squalodontids (shark-toothed dolphins) from this period. Since then, further new taxa have been found and previous family-level determinations significantly revised. For example, archaic toothed mysticetes of at least two families (Kekenodontidae and Mammalodontidae) were present. The rare *Kekenodon onamata* is now considered to be one such archaic toothed mysticete: archaeocetes occur in Eocene but not in younger strata.

The various archaic fully toothless true baleen whales of this time are presently included in the Cetotheriidae, but this family is in need of revision (Fordyce, personal communication). These larger baleen whales dominate assemblages in these rocks.

Dolphins (odontocetes) are conspicuous in the fauna and mostly belong in the Platanistoidea, a group that was diverse in the Oligocene but that is now represented by only two extant freshwater species. Notable platanistoids include species of shark-toothed dolphins (Squalodontidae), Waipatiidae, Squalodelphidae, and Dalpiazinidae. Archaic true dolphins (Kentriodontidae), sperm whales (Physeteridae), and long-beaked extinct dolphins (Eurhinodelphidae) are rare. Notable absences include archaic odontocetes, such as Agorophiidae and the *Xenorophus* group, and beaked whales (Ziphiidae) (Fordyce 1991, 1994).

Less effort has been put into Early Miocene cetaceans from strata younger than the Otekaike Limestone, but the fauna is known to include baleen whales, kentriodontids, and probable eurhinodelphids (Fordyce, personal communication). In Middle to late Miocene deposits from the Chatham Rise, at least four species of delphinoids (true dolphins) are known (Fordyce 1991).

The Miocene Greta Siltstone from Motunau Beach, North Canterbury, contains penguins, pelagornithids, and cetaceans (Fordyce 1991). However, some of the material from this locality is found as loose rocks on the beach and may be of Pliocene or even younger age. The pelagornithid *Pseudodontornis stirtoni* (a bony-toothed pelicaniform) was based on a crushed skull and femur (Howard and Warter 1969). Harrison and Walker (1976) transferred it to the new genus *Neodontornis,* but the New Zealand checklist does not follow this. Scarlett (1972a) described a proximal right humerus of a pelagornithid from Late Miocene sediments near the mouth of the Waipara River. Harrison and Walker (1976) tentatively identified this as *Pelagornis miocaenus,* which is otherwise known from France. The penguin *Korora oliveri* is known from rocks of early Miocene or possibly latest Oligocene age. A second form, *Marplesornis novaezealandiae,* is known from Motunau and may be either later Miocene or Pliocene in age (Simpson 1972).

However, the most significant record from the Miocene for the terrestrial story is that of waterfowl from the Manuherikia Group near St. Bathans, Central Otago, South Island. At least two species of duck are represented in so far undescribed material from lacustrine beds, which have recently yielded the first crocodilian known from New Zealand. The bone is an isolated angular of an indeterminate taxon (Molnar and Pole 1997). Fordyce (1991) also listed a fragmentary unidentified bird bone from limestone of equivalent age in the Waitaki Valley. Last, two bird footprints of latest Miocene age are known: one, attributed to a "kiwi" (but which could be any similar-size tridactyl bird), is from the Longford Formation in Murchison (Mildenhall 1974), and the other, a "kiwi-like bird," is from Manaroa, Pelorus Sound, northern South Island (Hutton 1899).

Pliocene rocks have produced few fossil vertebrates other than the ubiquitous cetaceans and penguins. Cetaceans occur in several localities, notably the Waipipian beds of South Taranaki, which have yielded delphinoids (at least three species), seals, penguins, and pelagornithids. Pliocene penguins are not common. Apart from the possibility that *M. novaezealandiae* is actually of Pliocene age, two other species of Pliocene age (*Pygoscelis tyreei* and *Aptenodytes ridgeni*) were named in extant genera by Simpson (1972). A single species of penguin *Tereingaornis moisleyi* is so far known from North Island Pliocene deposits from Hawke's Bay and Hawera (McKee 1987; Scarlett 1983). McKee (1985) described a humerus and a radius of a pelagornithid from the Tangahoe Formation near Hawera, but these are not referable to any species with certainty. McKee has continued to collect from the Pliocene formations at Hawera and in Hawke's Bay and has found further pelagornithid bones representing at least four individuals, material of two different shearwaters, and a second, larger Pliocene penguin from the North Island (*Society of Avian Paleontology Newsletter* 2, 4, 5, 6, 10).

A femur similar to that of bony-toothed pelecaniforms has been collected from the Nukumaruan (latest Pliocene–early Pleistocene) of Motunau, North Canterbury (Fordyce 1991), and may be the latest record of this group.

Pleistocene

The most significant feature of the Pliocene–Pleistocene period is that widespread glaciation commenced in oxygen isotope stage 100, about 2.5 Ma, which marks the start of the Nukumaruan Stage in New Zealand (Naish et al. 1998). Since then, there have been about 50 significant glacial–interglacial cycles; the glaciations are the even-numbered isotope stages. The Pliocene–Pleistocene boundary is at about 1.8 Ma (isotope stage 66) but is not marked in New Zealand by any major faunal changes. There were thus at least 32 glacial–interglacial cycles in the Pleistocene (Naish et al. 1998), but their severity was markedly greater in the dozen or so cycles of the last million years. In most of New Zealand, there is physical evidence for glaciations from only the last 350,000 years, because the later, more severe, cycles eroded all traces of the earlier ones. However, they can be traced in the sedimentary record in cores taken from the deep ocean near New Zealand (Suggate 1990). The Pleistocene, or the last 1.8 million years, has a comparatively better fossil record for birds than do the preceding periods, but it is still poor. Very little information is available on the bird faunas that inhabited New Zealand for most of this time, but it is rich for the last glacial–interglacial cycle.

Marton Fauna

The most significant mid-Pleistocene fauna is a very small collection from Marton in southwestern North Island. Fossil bird bones have been recovered at irregular intervals from an old beach deposit (Kaimatira Pumice Sand) dating to about 1 Ma (isotope stage 25–27) by its associated pumice (Naish et al. 1998; Worthy 1997a). Moa bones from two species (*Anomalopteryx didiformis* and possibly a species of *Dinornis*) are present in this deposit. Other birds are represented by a femur and a tibiotarsus of a large rail (*Pleistorallus flemingi*), a small rail tibiotarsus (smaller than *Gallirallus australis*, but bigger than *G. philippensis*), and an indeterminate duck (Worthy 1997a). Most recently, a tarsometatarsus of a new form of kiwi has been recovered (A. Tennyson, personal communication). Sparse as it is, this fauna indicates the probability of substantial faunal change between 1 Ma and the last 100,000 years, as three of the five nonmoa species do not occur in the recent deposits.

Fossil Moa

The Marton fauna illustrates one factor about moa evolution. One million years ago, a moa indistinguishable from the Late Holocene *Anomalopteryx didiformis* was present. In fact, although the fossil record of moa before the last glaciation is sparse, it does support the view that moa taxa as we know them from Late Quaternary deposits are several million years old. When Worthy et al. (1991) reviewed the fossil record of moa, only 33 records of known locality older than the last (Otiran) glaciation could be found. Most came from the Whanganui–Hawke's Bay areas of the North Island and from marine rocks.

Eleven of these were of Haweran age, or no more than 400,000 years old. A further eight records came from the Castlecliffian (0.4–1.0 Ma). Species to which the material could be referred included *Euryap-*

teryx curtus, Euryapteryx geranoides, and *Dinornis novaezealandiae,* plus *Anomalopteryx didiformis* from the Marton site. Since 1996, Nixon has recovered a series of fossil bones and fragments from the beach at Kai-iwi, near Whanganui. Their appearance suggests that they are derived from the Okehu Shell Grit, which is about 1.0 million years old (Naish et al. 1998). *Euryapteryx curtus* and at least two other indeterminate taxa are present.

Worthy et al. (1991) listed five surviving, well-provenanced records from the Tewkesbury Formation of the Nukumaruan, dated to 1.8–2.1 Ma (Naish et al. 1998). Species recorded from this period include *Dinornis novaezealandiae, Euryapteryx geranoides,* and *Euryapteryx curtus.* In 1992, a femur from Nukumaruan limestone on Puketautahi Hill, Hawke's Bay (Fossil Record Number V21/f109) was located in a private collection (A. Beu and I. Keyes, personal communication). It is very worn but has features of *Dinornis* and is the size of the late Quaternary *D. struthoides.*

From Gleniti in the South Island is a single contentious pre-Otiran fossil record of a moa. Hutton (1893e) named *Anomalopteryx antiquus* from fragmentary bones recovered from beneath a basalt flow, which has now been dated by K/Ar at 2.47 Ma (Worthy et al. 1991: Fig. 1.4). Unfortunately, the possibility that the bones were in a postbasalt fis-

Figure 1.4. South Mole Quarry at Gleniti. The bones, referred to Anomalopteryx antiquus, *came from beneath the basalt. Photograph courtesy of the Canterbury Museum.*

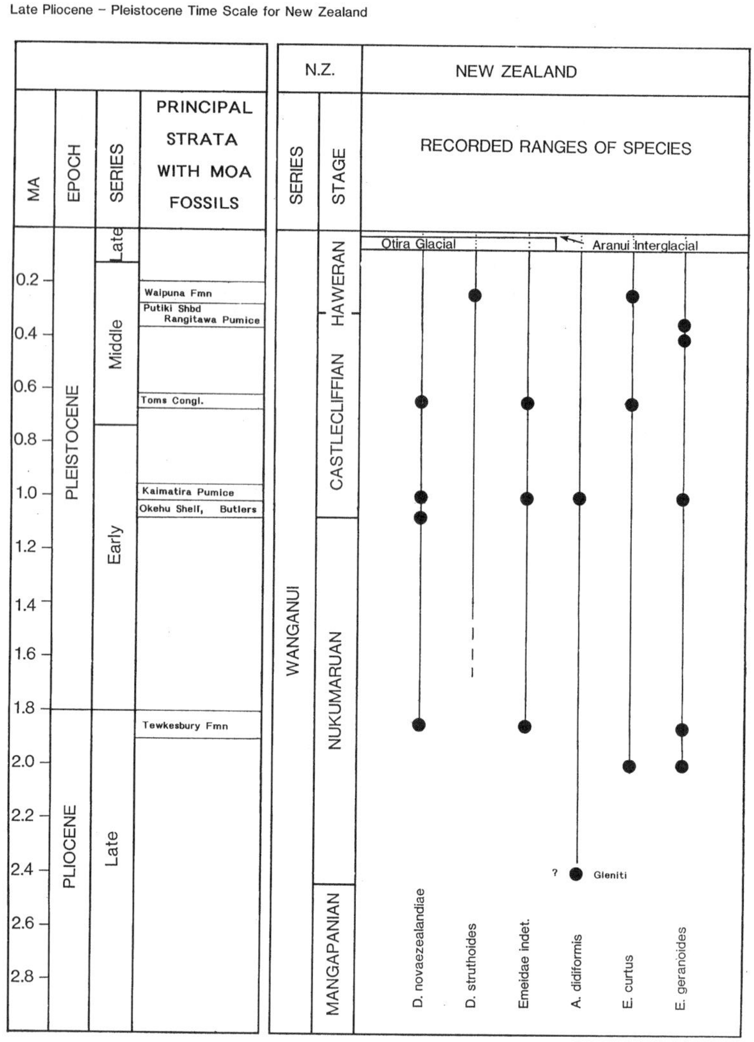

Figure 1.5. The geologic range of some moa taxa based on records from well-dated strata that are mostly in the Wanganui Series. The figure is based on Worthy et al. (1991), but strata are given the revised ages advocated by Naish et al. (1998). The principal strata in which moa fossils have been found are indicated: Waipuna Formation, Toms Conglomerate, Kaimatira Pumice Sand, Okehu Shell Grit, Butlers Conglomerate, and the Tewkesbury Formation. It is doubtful that the record for Anomalopteryx didiformis *at Gleniti is so old.*

sure cannot be ruled out. The bones are referable to *Anomalopteryx didiformis.*

As noted above, the interesting point about all these fossils is that all the identifiable bones can be readily referred to one of the late Holocene species (Fig. 1.5). Moreover, the common species are both in *Euryapteryx,* which is one of the more derived genera of moa (Archey 1941; data herein). The fossil record thus suggests that most species that were present in the late Holocene had evolved by 2 Ma. But this conclusion should be treated with caution because none of the fossils is represented by more than isolated leg bones. If articulated or associated remains are ever found, or if cranial materials are ever found, they may show differences between these early moa and more recent taxa.

2 • The Unveiling

There is a widespread misconception among scientists and the general public that the fossil record of birds is poor. The attitude seems to prevail in New Zealand, if Molloy (1994: 36) is to be believed: "Because the fossil record is so poor for birds in New Zealand, and because so many rapidly became extinct after the arrival of humans, it is difficult to be too definitive about the composition of the bird fauna during the post-glacial period."

In New Zealand, this is patently untrue. The fossil record of birds for the last 30,000 years, through the last glaciation and the postglacial period before the arrival of humans, is possibly the best in the world. In the words of Olson (1985: 217), "Although we can expect to keep hearing that fossils of birds are rare and that the avian fossil record is uninformative, henceforth it should be taken as a mark of ignorance in anyone who continues to parrot this canard." Over the last 150 years, many people have contributed work that reveals that New Zealand has one of the richest late Quaternary avifaunas in the world.

Some Principal Contributors to the Study of New Zealand's Fossil Avifauna

Richard Owen (1804–1892)

Owen was probably the foremost anatomist and vertebrate paleontologist in England in the 19th century. He began his remarkable career as a trainee surgeon, but he quickly abandoned this and dedicated the rest of his life to science and museums. He became assistant curator at the Hunterian Museum of the Royal College of Surgeons at the age of 22 years, where he was employed to prepare catalogs of the extensive anatomic collections. By age 32, he became the first permanent professor of the Hunterian Museum and was its curator from

1842 to 1856, leaving to take charge of the natural history collections in the British Museum at Bloomsbury. It was Owen's efforts, more so than those of anyone else, that resulted in the establishment of England's national museum of natural history, the British Museum (Natural History) in Kensington, London, which was completed in 1881. Owen supervised the removal of the collections from Bloomsbury to Kensington, which was completed in 1883; at age 79, he retired, his lifelong dream of a dedicated natural history museum attained.

Owen was widely acknowledged to be the greatest comparative anatomist of his time. Initially, greatly influenced by the functionalist approach of Cuvier (who argued that all forms were as they were because of functional requirements), Owen's work in comparative anatomy led him to promote the concept that vertebrates had a common underlying structural plan that was modified to enable different functions. This led to the concept that all vertebrates had been derived from a single primitive form—an "archetype." Fundamental to this thesis was the development of the concept of homology and homologous structures. For example, the paddle of a dolphin, foreleg of a horse, arm of a human, and wing of a bat are all homologous structures that developed from equivalent morphologic structures. Owen's extensive work in this field paved the way for Charles Darwin to use Owen's data and make the next great intuitive leap: that the archetype was the "ancestor" and that homologous structures were evidence of ancestry, thus forming the necessary background to the concept of evolution.

Owen's output was prodigious and included descriptions of bird species of New Zealand, including moa, as well as giant marsupials of the Tertiary from Australia and South American fossil mammals such as giant sloths. Also notable was his work on the anatomy of nautilus, belemnites, and the gorilla and his description of the oldest bird *Archaeopteryx*. In addition, he coined the term "dinosaur." But in the development of comparative anatomy, Owen's works on vertebrates in general were also immensely important, specifically *On the Archetype and Homologies of the Vertebrate Skeleton,* published in 1848, and *On the Nature of Limbs,* published in 1849. These texts had a huge impact. He was immensely popular during his lifetime. Although he was not born of gentry or educated in the "right" places, he was elected to the membership of many pivotal society clubs. He published over 600 papers, including more than a dozen book-length works, and he accumulated a total of 100 distinctions, ranging from honorary doctorates from Oxford, Edinburgh, Cambridge, and Dublin to a knighthood (Rupke 1994).

Owen examined the first moa bone in 1839 and published his now-famous announcement of the existence of moa in 1840 (Fig. 2.1). Over the next 46 years, until 1886, he published more than 50 contributions on the osteology of New Zealand birds. A majority of moa species bear his name. Yet he still looked on his work as a mere beginning: "I look back on my own work as the scantest initiative groundwork for a truly philosophical knowledge of the Dinornithidae" (letter, R. Owen to Captain Hutton, August 18, 1876, Canterbury Museum Archives).

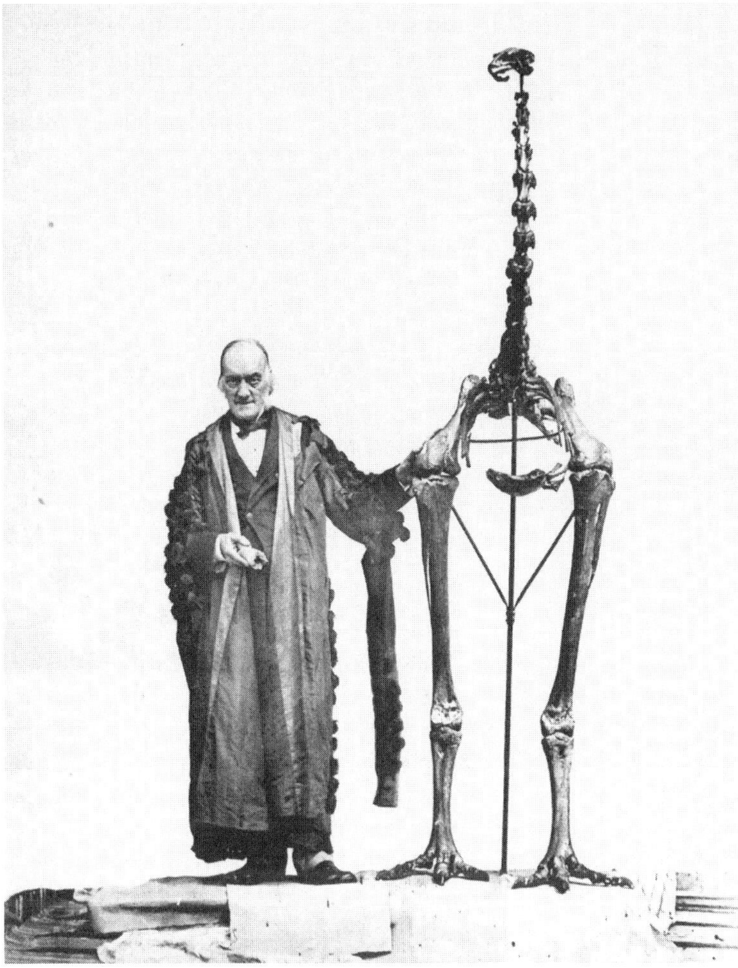

Figure 2.1. *Richard Owen holding the first fragment that revealed to him the presence of moa in New Zealand. He is standing beside the remains of the largest of all moa,* Dinornis giganteus. *From Owen (1879: pl. 47).*

Walter Baldock Durrant Mantell (1820–1895)

The son of Dr. Gideon Mantell and Mary Ann Mantell, who are credited with the discovery of the dinosaur *Iguanodon* in southeastern England, Walter B. D. Mantell arrived in New Zealand in 1840 with the first organized European settlement in the North Island. He embarked on a career in the civil service and politics, with an appointment to the postal service and as clerk to the bench of magistrates. He learned the Māori language, and by 1846–1847, he was acting commissioner for the settlement of native claims in the South Island. His main contribution to Quaternary studies arose from collections made during travels associated with this post.

His first major collection was taken from the Waikouaiti site, discovered by Dr. MacKellar in 1846, which, along with those from other sites in the South Island, totaled more than 500 bones. A few months later, in May 1847, Mantell visited Waingongoro in the North Island, where moa bones had previously been discovered in 1843 by the Reverend R. Taylor. Mantell collected about 400 bones, including the type specimen of the North Island takahe, which Owen named *Notornis mantelli* (now in the genus *Porphyrio*), and the type of *Euryapteryx geranoides*. These two major collections were sent to his father in 1847, who made them available to Owen. They formed the basis for Owen's third memoir on *Dinornis* (Owen 1848a). In the early 1850s, Mantell was also responsible for sending the first two skins of the South Island takahe, which still survives in southern South Island, to his father in England, which caused some excitement as a species that had seemed to have come back from the dead.

Mantell was appointed the commissioner for Extinguishing Native [land] Titles in the South Island in 1848, which means he was responsible for the sale of Māori land. He traveled widely in the South Island. His next most significant discovery of moa remains was at Te Awakokomaka (which he renamed Awamoa) in 1852. These were sent to England in 1853, but because his father Gideon had died in 1852, the cases remained unopened until Walter Mantell himself arrived in 1856 and took them to the British Museum. Material from Awamoa included the tarsometatarsus on which Owen named the moa *Dinornis* [*Pachyornis*] *elephantopus*. On his return to New Zealand, he entered politics in 1860 and was elected to the House of Representatives in 1861 and was a minister of native affairs in three governments. When the Colonial Museum was founded in 1865, he was on its board, and he acted for the director, James Hector, when necessary. Mantell brought some of his father's collections with him back to New Zealand, and these are now in part in the present Museum of New Zealand Te Papa Tongarewa, Wellington, and in part in the paleontology collection of the Institute of Geological and Nuclear Sciences, Lower Hutt (Yaldwyn et al. 1997).

Johann Franz Julius Haast (1822–1887)

Haast studied geology at university in Bonn and in Austria, and then traveled widely. He arrived in New Zealand in 1858 and by chance met Dr. Ferdinand Hochstetter; they immediately struck up a lifelong friendship. They traveled widely in the North and South Islands in the next few months, discovering moa bones in caves at Aorere (Fleming 1959), which were sent to Hochstetter in Vienna, where they remain to this day. After Hochstetter departed with the *Novara,* Haast continued doing exploratory geologic work in the South Island and became Canterbury's provincial geologist in 1861. Settling in Christchurch, he began pressing for the establishment of a provincial museum. This was facilitated by the discovery of a large deposit of moa bones on Glenmark Station 80 km north of Christchurch in 1866. In 1867, the beginnings of the Canterbury Museum were opened in two rooms of the government buildings. In 1868, the building of a museum was authorized, and in 1870, it opened its doors to the public. In 1869, Haast was appointed director and remained so until his death. By trading moa bones widely around the world, Haast obtained many exhibits for the fledgling museum.

Haast made significant contributions in the early exploration of the fledgling New Zealand colony. Several prominent landmarks bear

his name, including the significant Haast Pass. He was not idle in the fields of scientific discovery in natural history and archaeology: he published some 23 contributions on extinct birds during the period 1863–1886. Many had an archaeologic context, but six dealt with the systematics of moa. During this time, his contributions to science included the discovery and naming of the New Zealand eagle *Harpagornis moorei* and discovery of the biggest known kiwi *Apteryx haastii*. His interest in moa continued, and from specimens obtained at Glenmark, he named *Dinornis maximus*. During his career, he named several other moa taxa (e.g., *Megalapteryx hectori* and *Anomalopteryx oweni*) and, most important, he generously provided information and specimens to specialists elsewhere, including Walter Buller and Richard Owen.

Figure 2.2. Julius Haast in Canterbury Museum beside its central moa display in 1870. Photograph by Dr. A. C. Barker; reproduced with the permission of Canterbury Museum.

During this time, he was instrumental in founding the Philosophical Institute of Canterbury and the Canterbury Collegiate Union, which later became Canterbury University. In 1875, he received an Austrian knighthood, becoming "von Haast," and in 1886, he was granted a British knighthood. The Canterbury Museum remains a tribute to his foresight and energy (Fig. 2.2).

Henry Ogg Forbes (1851–1932)

Forbes succeeded Haast as the director of Canterbury Museum in 1887. His main work with New Zealand fossil birds came from excavations at Monck's Cave at Sumner and Enfield Swamp in north Otago and a visit to the Chatham Islands in February 1892. Shortly after that trip, he left for England, taking many specimens from the Canterbury Museum (including the Chatham material paid for by the Canterbury Museum) and from Augustus Hamilton's private collections. In England, he became director of the Liverpool Public Museums from 1894 to 1911. It is known that the collections he took from New Zealand arrived in the zoology department of the British Museum (Natural History) by 1894 because the registers for 1894 and 1895 contain many entries for Forbes's material. However, they remained largely unexamined until Elliot Dawson located them (Dawson 1958b).

Forbes's contributions to the paleoornithology of New Zealand were often little more than abstracts, often repeated in journals such as *Nature, Ibis,* and *Transactions and Proceedings of the New Zealand Institute*. For example, his "Preliminary notice of additions to the extinct avifauna of New Zealand [with twelve new names proposed]," was published "in abstract as it has been found impossible to prepare all the necessary drawings for its proper illustration in time" (Forbes 1892d: 185), despite two years' preparation time—and it never was enlarged. However, unfortunately, in several, there are valid descriptions of species, though none of these are at all adequate, and if it were not for the fact that the bones with "type" written on them were found in the British Museum (Dawson 1958b), most names would be *nomina nuda*.

Frederick Wollaston Hutton (1836–1905)

Hutton worked as an economic geologist early in his career, but in 1877, he was appointed a professor of natural science at Otago University. Three years later, in 1880, he moved to Christchurch as a professor of biology at Canterbury University College. After the departure of Forbes, Hutton became curator at Canterbury Museum in 1892, where he worked until his death in 1905 (Fig. 2.3).

Although his early work was in the field of geology, Hutton was soon to make major contributions in the field of zoology. He wrote numerous papers on birds, insects, and molluscs, both fossil and recent. His contribution to the taxonomy of these groups was enormous. Nevertheless, between the years 1872 and 1906, he also published some 30 substantial works on moa, covering various aspects, such as structure of eggshell, feathers, and skeleton. His works were particularly significant in the development of moa systematics.

Buick (1931: 9) described him as "Professor Hutton, one of the most cautious and cultured of our scientists," and E. G. Turbott, in Gill and Heather (1990: 180), did not understate Hutton's impact when he said, "Hutton was the leading figure in the development of biology in New Zealand in the late 19th century."

The following three letters were part of a series published in the Lyttelton *Times,* May 31, 1892, when, it seems, business was conducted much more in the public eye.

Canterbury Museum, May 28 1892

Sir,—Knowing that there were in the Canterbury Museum, a few months ago, some very rare moa bones belonging to Mr A. Hamilton of the Otago University, and not finding them when I took over the museum, I wrote to Mr Hamilton asking if he had got them back again. I enclose his reply, in which he says that these bones have not been returned to him, but that Mr Forbes had sent him some aphanapteryx bones (from the Chatham Islands) instead. There is no record of these aphanapteryx bones having left the museum, though they must have been part of the collection made at the expense of the museum. The box of moa bones from Te Aute, mentioned in Mr Hamilton's letter, is entered as an exchange for moa bones sent to the Napier Museum from here. It has nothing to do with the moa bones belonging to Mr Hamilton, which I now write about.—I am, your obedient servant,

F. W. Hutton, Acting Curator

Canterbury Museum, May 28, 1892.

Sir,—On looking over the contents of the drawer handed over to me by Mr Forbes as containing the bones of harpagornis obtained at Enfield, I find that there are only two harpagornis bones in it; all the others belong to cnemiornis or to the moa. Mr Sparkes informs me that many more harpagornis bones were brought to the museum, and there is no record of any of them having been sent away.

F. W. Hutton, Acting Curator

University of Otago, May 26 1892

Dear Captain Hutton,—Many thanks for your note. F. had a great number of small bird bones of mine, and I said in a letter that any he wished to figure I would let him take to England if necessary. I have had no letter from him, but a box containing aphanapteryx bones, and inside the cover written "in exchange for moa bones." Now I do not know if this is meant as an exchange from the museum or from him personally, neither am I sure what he means by "Moa bones." Between ourselves I am not very anxious for the aphanapteryx bones, as I am likely to get some from another source. However I should like to know if they are in exchange for a box of Te Aute leg bones which were sent by me to Christchurch Museum some time ago or if they are private exchange. In the latter case I do not wish to part with any of the small moa bones (scap. Cor. or sterna) which were lent to H. O. F. I am, of course, ignorant as to whether the aphanapteryx bones are museum or private property. I was at Warrington on Tuesday, and picked up a very perfect metatarson of the smaller form of harpagornis on the sandhills. I suppose you will resume "the moa" now. Let me know if I can do anything for you. —I remain, &c.,

A. Hamilton

PS: Will you let me have a list of the Harpagornis bones from Enfield. I saw them when I was up, but did not take a note of them.

Figure 2.3. Captain Hutton at work in the Canterbury Museum. Photograph by E. Wheeler and Son; reproduced with the permission of Canterbury Museum.

Augustus H. Hamilton (1854–1913)

Hamilton arrived in New Zealand and took up a career as a teacher and educator. It was when he moved to Petane in Hawke's Bay that he entered the scientific community, as secretary of the local philosophical society. He established the museum in Napier, which was based on his collections of Māori ethnographic material. It is as an ethnologist and collector of Māori and Pacific artifacts that he is most remembered. However, he made important contributions to knowledge of the fossil birds, in a major respect by putting a greater emphasis on the smaller species.

Unfortunately, most of the catalogs from his earlier years are now lost, but his published papers contain important information and analyses of material collected at Castle Rocks, in western Southland in the South Island, and on other sites in both main islands. His first major collections of bird bones were made during the draining of the swamp at Te Aute in Hawke's Bay in 1888–1889, and the last site he published on was at Ngapara in north Otago, in 1903. Fossils collected by Hamilton form part of the collections in museums in Otago and Hawke's Bay, as well as the Museum of New Zealand and the Natural History Museum (formerly called the British Museum [Natural History]), London. Some of the bones are recognizable as being part of his collection only by the characteristic penciled number, for which a key is yet to be found.

At Castle Rocks, Hamilton and his helpers collected the remains of two Haast's eagles, a male and a female. The less complete female specimen remains the largest Holocene specimen ever found; the male was mounted (with some bones of the female included to complete the mount) and was the best skeleton available until the discovery of the Mount Owen bird in 1990. Oliver (1930, 1955) published a photograph of the mounted bird, and the specimen remains, in a somewhat less lifelike condition than seen in the photograph, the only mounted skeleton of the species.

Hamilton was a collector, but he went further than most of his contemporaries in listing the material he found or acquired. That industry and scholarship led to his successive appointments as registrar at the University of Otago and as director of the Colonial Museum (subsequently the Dominion Museum, National Museum of New Zealand, and now Museum of New Zealand Te Papa Tongarewa).

Sir James Hector (1834–1907)

Hector had one of the most colorful careers of any of the Europeans who arrived in New Zealand in the second half of the 19th century and established the scientific community. He was also to oversee the formal establishment of government science and was a major figure in the community of what was then a colony. Fleming (1966) noted that over the course of 38 years, Hector was responsible for founding several institutions that survived well into the second half of the 20th century, including the Meteorological Service, the national museum (under its various guises), and the Royal Society of New Zealand.

Although Hector was trained in medicine, he also studied natural sciences and was selected by Sir Roderick Murchison to be geologist and surgeon for an expedition to western Canada. Kicking Horse Pass on the Canadian Pacific Railroad is named for one of his experiences while exploring and surveying passes across the Rocky Mountains. He accepted the appointment of geologist to the Otago provincial government and arrived in Dunedin in 1862. The first director of the New Zealand Geological Survey, which was set up in 1867, Hector then added director of the Colonial Museum and manager of the fledgling New Zealand Institute to his responsibilities. In these positions, and as a key figure in the life of the capital, Wellington, Hector was the most influential of the first generation of resident New Zealand scientists.

Although he made an early mark as an explorer and geologist in the southwest of the South Island, he published 15 papers on moa between 1865 and 1902. But perhaps his major contributions to the study of New Zealand fossil birds was as the mentor of others and as editor of the *Transactions and Proceedings of the New Zealand Institute*, which was for many years the main outlet for work by local researchers. Under his direction, the Colonial Museum issued catalogs and manuals of New Zealand fauna and flora.

Professor T. Jeffery Parker (1850–1897)

An exception to the rule that scientists with careers in New Zealand had to have an initial "H" was T. Jeffery Parker. But during his education, Parker was influenced by another "H," Thomas Henry Huxley. At the age of 22 years, he worked for and with Huxley as demonstrator in the older man's classes in zoology. It was as an expatriate at the University of Otago that he made his major contributions, including the publication (with W. A. Haswell, in Sydney) of a textbook on zoology that was still in use in the 1960s. The textbook demonstrated Parker's organizational abilities: the authors were in separate countries, and the publisher was in a third.

Parker brought a strong evolutionary and anatomic approach to the study of New Zealand's fauna. He contributed to the study of moa by emphasizing the structure of the cranium and provided comparative information on kiwi and other birds. He published nine significant papers on moa between 1889 and 1895. Parker's brief report on the discovery of a skeleton of Haast's eagle at the Golden Point mine at Macrae's Flat was the only indication of the existence of this important material until it was relocated recently in the collections of the Royal Museum of Scotland in Edinburgh (A. C. Kitchener, personal communication). Parker died young; his contribution to the studies of New Zealand's extinct birds was probably more in evidence in the results of his writing and teaching. He was replaced at the University of Otago by W. B. Benham.

Walter Reginald Brook Oliver (1883–1957)

Oliver was born in Tasmania and reached New Zealand with his parents in 1896. After leaving school, he pursued a career in the New Zealand customs service. An interest in natural history was obvious by 1906, when he joined a group of like-minded individuals in Christchurch. His passion for collecting was first demonstrated by his beginning a lifelong collection of all kinds of information on natural history, often to the detriment of the source publication involved. In 1908, he had been part of an expedition that spent 10 months in the subtropical Kermadec Islands 900 km to the northeast of the North Island. Collections he made there and in Tahiti in 1919 formed the basis of papers that established his credentials to a scientific career. However, he lacked formal scientific education until after he gained the position of senior scientific assistant at what was then the Dominion Museum in Wellington in 1920. He gained a master's degree in zoology and later was awarded a doctorate in science for his *New Zealand Birds* of 1930, which went into an enlarged second edition in 1955 and remains a significant resource on New Zealand birds. Oliver was director of the Dominion Museum from 1928 to 1947.

His interests were wide, and his first major contribution was a revision of Cheeseman's pioneering *Manual of New Zealand Flora*. Other publications on plant groups followed, and Oliver was a member of many scientific societies, but his main contribution to Quaternary research was his monograph on moa, published in 1949. His idiosyncratic views on evolution and taxonomy were presented in *New Zealand Birds* and *The Moas of New Zealand and Australia,* and he published a summary of his views on the evolution of New Zealand birds (Oliver 1945a, 1945b).

It is as a compiler of information and author of *Pachyornis australis* that Oliver made his contribution to New Zealand fossil birds. His philosophy of applying species names to arbitrary divisions within continuous series further cluttered an already confused nomenclature for the group.

William Blaxland Benham (1860–1950)

Benham played a minor direct role in the study of New Zealand's extinct birds. He trained under Sir Ray Lankester at University College, London, and was an active researcher in England before succeeding T. Jeffery Parker as chair of zoology at the University of Otago in 1898. Although his major achievements lay in invertebrate biology, he had a passing interest in other fields. He published material about a moa egg and moa trackways, and he recorded moa material from Stewart Island. His name was given to a supposed large species of *Megalapteryx,* but this has since been shown to have been an Otiran Glacial period cold-climate morph of *M. didinus.*

Gilbert Edward Archey (1890–1974)

Archey was the last of the first generation of immigrant scientists to make his mark in studies of the fossil birds of New Zealand. Unlike most of the others, except his near contemporary Oliver, he was educated in his adopted country, gaining an M.A. and D.Sc. in zoology at Canterbury University College in Christchurch, alma mater of the Nobel prize–winning physicist Lord Rutherford. From 1914 to 1923, he was assistant curator at Canterbury Museum, where he published on topics as diverse as *Daphnia* and *Hydra,* although he also published two papers on living birds. In 1924, he was appointed director of the Auckland Institute and Museum and remained in this post until 1964. There he published a few papers on moa, but his major achievement was *The Moa: A Study of the Dinornithiformes* in 1941. This monograph was probably the most influential work on moa in the 20th century. He was knighted in 1973 for his services to science and the museum movement in New Zealand.

Ron Scarlett (1911–)

Scarlett was born near Nelson but spent much of his early life in the small, rural community at Karamea on the South Island West Coast, and from the earliest time, he was a keen naturalist. Ron moved to Christchurch in 1937 to take up university studies and soon spent much time in the Canterbury Museum. Under the tutelage of Dr. Robert Falla, Scarlett developed his lifelong interest in birds and gained his first part-time job through him at the museum in 1940. So began an association with the Canterbury Museum that has not ended yet, although he retired in 1980 and officially vacated his office in 1997.

While Scarlett was studying at the university, he became involved in the excavations at Pyramid Valley initiated—and funded—by Robert Cushman Murphy in January 1948. When the new museum director, Roger Duff, started systematic excavations at Pyramid Valley on January 31, 1949, Scarlett joined Jim Eyles (the discoverer of the Wairau Bar moa hunter site) as principal excavators. The mass of material entering the museum from the site and Scarlett's burgeoning interest led to a museum appointment. From 1950 to 1981, he was a fixture at Canterbury Museum, first as recorder of collections, then as the osteologist. He single-handedly amassed one of New Zealand's most important collections of bird skeletons, both recent specimens as reference skeletons (his comparative material) and fossil birds. Foremost in this collection was the internationally significant collection of fossils from Pyramid Valley, the results of excavations between 1939 and 1965.

Scarlett encouraged the early members of the New Zealand Speleological Society, who began exploring and mapping caves in earnest after 1949, to collect such fossil remains as they found. As a result, numerous important discoveries were made and substantial collections recovered from the caves of the Waitomo Caves area and the South Island West Coast. As the science of archaeology developed in New Zealand over this period, Scarlett provided an important service in identification of the midden bones, and he stressed the need for archaeologists to collect faunal remains. He was instrumental in demonstrating the degree to which Māori interacted with extinct species.

Scarlett has published over 30 papers on fossil birds and has named or contributed to the naming of several fossil species (*Circus eylesi, Gallinula hodgenorum, Aegotheles novaezealandiae, Pelecanus conspicillatus novaezealandiae,* and *Tereingaornis moisleyi*) and was instrumental in the description of others (e.g., *Capellirallus karamu*).

Robert Cushman Murphy (1888–1973)

Murphy's contribution to the American Museum of Natural History in paleontological research in New Zealand was brief and peripheral, but important. He left a legacy in the research undertaken by one of his part-time employees in the years after his second visit. Although he was best known for research and publications on petrels and other seabirds, he was a museum man as well, and he contributed much to the development of research and display. At the time of his first contact with New Zealand in 1947, he was chair of the bird department at the American Museum of Natural History and became Lamont Curator of Birds in 1948 (Falla 1973).

When Murphy arrived in New Zealand in 1947, he was seeking materials and specimens with which to build dioramas of New Zealand and the subantarctic Snares Islands for the Whitney South Seas Hall at the museum (Falla 1973). He was aware of the deposit at Pyramid Valley, which had been discovered and first excavated in 1939, and he persuaded the University of Canterbury committee, which had control of the site, to permit excavations and the removal of moa and other skeletons. After visiting The Snares and being treated for an injury to his thigh sustained on the island, he was at last able to visit the swamp just before Christmas 1947.

Early the next year, he organized and paid for a week's excavation. Among the other birds were what he believed to be "the first known complete skeletons and perfect skulls of the giant rail *Aptornis*," which he was lucky enough to find in "a pit I opened" (R. Cushman Murphy, in litt. 1948). In that week, the remains of 22 moa were excavated, representing most of the species recorded from the site. He reported that the excavation, and all the specimens, cost only US$152.01, plus half the cost of freight, which, for a collection including at least six moa skeletons—and one of *Aptornis* as well as some smaller birds—might be considered a bargain. On his second visit, in early 1949, Murphy participated in a larger, more organized excavation undertaken by Canterbury Museum, with Scarlett and Eyles providing most of the manpower. He presented some results of the excavations at the International Ornithological Congress in Uppsala in 1950 and published a little on the material, but he took no further part in the field in New Zealand.

History of Discovery

The first discoveries of moa bones in New Zealand were in fluvial deposits at Waiapu, above Gisborne and at Poverty Bay on North Island's East Coast (see Chapter 3). These bones were apparently eroded from either gravel beds or swamp deposits in the catchments of the streams concerned and so were all single bones or mixed bones from more than one individual, and generally they were damaged. These first specimens were forwarded to England, and Owen announced that there were large birds called moa in New Zealand and that he would like more specimens. A widespread search commenced. Unassociated bones of a similar nature to the East Coast specimens continued to be revealed and were found throughout the country. It seems that bones were much more commonly found on the ground surface during the last century, a result of recent erosion, than they are now. For example, Chapman (1885) recorded moa skeletons eroding from a sand flat in the upper reaches of the Waitaki River in 1884, and later, in March 1884, nine moa skeletons and their associated gizzard stones were found on an eroding dune field near Lake Tekapo (unfortunately, these were not collected). Perhaps this was due to enhanced erosion when the land was cleared for pasture, and to the fact that areas that were then unstable, such as dunes, are now stabilized under pasture.

The next major discovery after the East Coast finds was in the middens on the Taranaki coast at Waingongoro by the Reverend R. Taylor in 1843, which was given notice by Owen (1844a) at the end of his first memoir on *Dinornis*. The discovery was brought to Owen's attention by J. R. Gowan of the New Zealand Company, in a letter from Colonel William Wakefield dated September 19, 1843. But it was not until Walter Mantell visited the site in June 1847 that important collections were made that were sent to England and placed at the disposal of Owen and described (Owen 1848a)—although the conditions were somewhat less than ideal, as Walter Mantell describes in a letter reported by G. A. Mantell (1848b: 240–241):

> You will readily imagine how exasperating it was to me to see specimen after specimen destroyed before my eyes, with no possibility of preventing it. From your ignorance of the excessive obstinacy and mulishness of the natives, I fear your indignation will be directed against me; but if so, let me assure you, you are indeed in error. All that man could do I did to dissuade them from turning oryctologists or palaeornithists; but to no purpose. Men, women and children resolutely dashed at every bone that appeared when the sand was removed; and if they listened for a moment to my entreaties and remonstrances, it was but to return with renewed vigour to the work of destruction. Although I am of a forgiving disposition, yet I cannot but hope Mr Hawkins will place these Māoris in the same category with the Vandals who destroyed the Alexandrian library, and the Somersetshire "varmint" who mistook a Cheiroligostinus for a "viery zarp'nt."

As a result of Taylor's tardiness at forwarding specimens to the reputed scientist of the day, the first major deposit to be described was at Waikouaiti, just north of Dunedin on the Otago coast. There, swamp sediments were exposed on the beach when storms removed cover beds of sand, and a substantial collection of bones was made by Dr. MacKellar and Percy Earl. These included the first described associated bones of an individual leg—the foot of a *Dinornis* (Owen 1846a, 1851a)—and were the basis for several new species.

Walter Mantell discovered the middens at Awamoa in December 1852, which contained the well-preserved remains of eggs that Mantell reconstructed, three of which survive in the British Museum (Natural History), and two in the Museum of New Zealand. The stoutest of all moa, *Pachyornis elephantopus,* was revealed first in this site (Owen 1858a, 1858b).

At about this time, caves were beginning to be explored for bones and notable discoveries made. A. S. Thomson visited a cave that fits the description of Masons Dry Cave at Waitomo, North Island, in October 1852 (Thomson 1854) and obtained many bones. Thomson (1854) recorded that he had visited the area three years before and purchased moa bones from the Māori at Parianiwaniwa Village; the Māori found the bones in this same cave. The bones found in 1849 are therefore likely those referred to by Sir George Grey in a letter to Owen dated November 29, 1849 (see Owen 1852: 59), and which included the well-preserved *Dinornis* skull described by Owen (1852).

Buller (1888: xxi) recorded that it was in 1850 that Grey sent his collection to Owen that had been collected "in the district lying under Tongariro Mountain," but the area is probably the karst district now renowned for the Waitomo Caves, some 100 km northwest of that mountain; few fossil sites lie close to Tongariro. A complete skeleton was discovered in the Aorere Caves near Collingwood, South Island, by Julius Haast in 1859. The caves were described by Haast in Ferdinand Hochstetter's *Geologie von Neuseeland,* published in 1864 and translated by Charles Fleming (1959).

However, most earlier finds were overshadowed by the magnitude of the Glenmark deposit, discovered in 1866 and excavated during the

Figure 2.4. Excavations in progress at the Kapua Swamp site in October 1894. Photograph courtesy of the Canterbury Museum.

next few years (Haast 1869). Thousands of bones were obtained, and many skeletons were assembled and distributed around the world. It was at the Glenmark deposit that the eagle was discovered and the largest of moa, *Dinornis maximus,* found. Tibiotarsi of this individual were nearly 1 m long, but the form it represents is best interpreted as the biggest of a large species *Dinornis giganteus.* Discovered within a couple of years was a similar large deposit at Hamilton in Otago (Booth 1875). Discoveries of other large deposits in swamps were foremost in importance through the rest of the 19th century: Te Aute (Hamilton 1889), Enfield (Forbes 1892b, 1892c, 1892d), and Kapua (Hutton 1896a: Fig. 2.4) each revealed thousands of specimens. These great swamp digs produced many bones—but the specimens also created taxonomic confusion because they mostly lacked associated bones of individual specimens.

Polynesian middens also revealed numerous remains, such as at Moa Bone Point Cave at Sumner, Christchurch (Haast 1875b), and Shag Mouth (Haast 1875a, 1877). Investigation of these middens resulted in an at times bitter feud between Haast and Frederick Hutton, James Hector, and Alexander McKay regarding Haast's interpretation that the moa hunters were of a race of people distant in time and separate from the Māori of the present day (Anderson 1989c).

Figure 2.5. Earnscleugh Cave in central Otago. Photograph by T.H.W.

Figure 2.6. (opposite page top) Hamilton's 1905 investigations at Makirikiri near Whanganui. Photograph by Augustus Hamilton, Museum of New Zealand Te Papa Tongarewa, C736.

Figure 2.7. (opposite page bottom) General view of the excavations at Makirikiri in 1936–1937. Photograph courtesy of the Wanganui Regional Museum.

Exploration of the arid region of Otago (Fig. 2.5) for gold led to the discovery of several spectacular individual finds of moa with preserved soft tissue (Chapter 4). These include the *Dinornis* skeleton from Tiger Hill in the Manuherikia, which had tissue over its pelvis and preserved pads on the toes (Hector 1872a; Owen 1869a); the remarkable *Emeus* remains from Earnscleugh Cave (Hutton and Coughtrey 1875b); and the even more remarkable three sets of *Megalapteryx* remains (Hamilton 1895b; Owen 1883b; Vickers-Rich et al. 1995). However, cave findings also revealed substantial faunas of birds other than moa before the end of the 19th century.

Although Owen had described some small birds from Waingongoro and others, such as *Cnemiornis* and *Aptornis* remains from a cave near Timaru (Owen 1866b), the first substantial "small bird" fauna was described by Hamilton, who first recorded an interest in small birds when he described those present in Te Aute (Hamilton 1889). However, the fauna he excavated from Castle Rocks in Southland was much larger and had thousands of specimens (Hamilton 1892b, 1893, 1894b; Worthy 1998d). A few years later, Hamilton (1904a) described another major deposit of s mall birds from Ngapara in Otago, including many rare species. After Hamilton's move to Wellington in 1904–1905, he continued his interest in birds other than moa with his excavations of Te Rauparaha's Pa, at Paremata, near Wellington (Davidson 1978).

There followed a hiatus in the exploration of New Zealand's fossil avifauna. Occasional deposits were excavated at Martinborough and

TABLE 2.1.
A chronicle of discovery of the principal sites pertaining to New Zealand Quaternary paleontology.

Site	Island	Year	Discoverer or investigator	Reference	Various
Poverty Bay	NI	1837	Dr. J. Rule	Owen 1840	Fluvial
Poverty Bay	NI	1841–1843	William Williams; Colenso	Owen 1843a, 1844a, 1844b	Fluvial
Waingongoro	NI	1843; June 1847	Rev R. Taylor; Walter Mantell	Owen 1846b, 1848a	Midden
Waikouaiti River	SI	1846	Dr. MacKellar; Percy Earl; Walter Mantell	Owen 1846a; Buick 1931: 105–108	Natural swamp
Awamoa	SI	Dec 1852	Walter Mantell	Mantell 1873; Buick 1931: 115–123	Midden
Aorere Caves	SI	1858–1859	Hochstetter and Haast	Buick 1931: 125–128	Cave
Glenmark Swamp	SI	1866-	Haast, CM	Haast 1869; Buick 1931: 130–131	Swamp
Hamilton Swamp	SI	1870	B. S. Booth, CM/OM	Booth 1875, 1877; Buick 1931: 131–143	Swamp
Earnscleugh Cave	SI	1871-	Dr. Thomson, Hutton et al. OM	Fraser 1873; Hutton 1875	Cave
Shag Mouth	SI	1872	Haast/Booth, OM	Haast 1875a	Midden
Moa Bone Point Cave	SI	1872?-	Haast, CM	Haast 1875b	Midden
Te Aute	NI	1888	A. Hamilton, MNZ	Hamilton 1889; Buick 1931: 143–148	Swamp
Enfield	SI	1891	Forbes, OM/CM	Forbes 1892b; Hutton 1896b; Buick 1931: 149–156	Swamp
Castle Rocks	SI	1891–1893	A. Hamilton, OM	Hamilton 1893, 1894b; Buick 1931: 168–171	Cave
Kapua	SI	1895	F. Hutton, CM	Hutton 1896a; Buick 1931: 157–159	Swamp
Martinborough	NI	1901; 1920	Hector, MNZ; Thomson, MNZ	McLeod 1902; Buick 1931: 161; Yaldwyn 1956	Caves
Ngapara	SI	1903	Hamilton, OM	Hamilton 1904a	Cave
Clevedon	NI	1912	Cheeseman, AIM	Buick 1931: 159	Swamp
Makirikiri (Upokongaro)	NI	1892–1906; 1936	Field, Hector, A. Hamilton, MNZ; WM	Field 1892; Buick 1931: 160–161; Worthy 1989a	Swamp
Coonoor	NI	1914	MNZ	Oliver 1949	Cave
Martinborough (Aorangi Station)	NI	1916	W. Harrison, MNZ	Buick 1931:162	Caves
Waikaremoana	NI	1930s	Archey, AIM		Caves
Doubtless Bay dunes	NI	1900–present	AIM		Dunes

Site	Island	Year	Discoverer or investigator	Reference	Various
Riverlands, Whanganui	NI	?1940s	WM	Worthy 1989a	Swamp
Pyramid Valley	SI	1937–1950	Scarlett, Duff, CM		Swamp
Herbert	SI	January 1941	MNZ, Oliver	Oliver 1949	Swamp
Southland dunes	SI	1940s	Sorenson, Teviotdale; S M	Worthy 1998d	Dunes
Marfells Beach (Lake Grassmere)	SI	1944–1979	CM	Worthy 1998a	Dune/ midden
Takapau Road Swamp	NI	1961–1980s	MNZ	Worthy 1989a	Swamp
Albury Park Spring	SI	1963, 1972	CM	Worthy 1997b	Swamp
Scaifes Lagoon, Wanaka	SI	1965	CM		Swamp
Lake Poukawa	NI	1966–1975	T. R. Price	Horn 1983	Lake/ swamp
Hamiltons, Winton	SI	1974	SM	Worthy 1998b	Swamp
Paerau Swamp	SI	1987	OM	Worthy 1998b	Swamp
Honeycomb Hill Cave System	SI	1982–1992	MNZ	Worthy and Mildenhall 1989; Worthy 1993b	Cave
Glencrieff	SI	1992–1994	MNZ	Worthy and Holdaway 1996b	Swamp
Kauana Swamp	SI	1994	MNZ, SM	Worthy 1998d	Swamp
Hermits Cave	SI	1992	MNZ, CM	Worthy and Holdaway 1993	Cave
Predator Cave	SI	1993–1994	MNZ	Worthy and Holdaway 1996a; Holdaway and Worthy 1996	Cave

Abbreviations: AIM = Auckland Institute and Museum; CM = Canterbury Museum; NI = North Island; MNZ = (variously) Colonial Museum, National Museum of New Zealand, or Museum of New Zealand; OM = Otago Museum; SI = South Island; SM = Southland Museum; WM = Wanganui Museum.

small sites found elsewhere (Table 2.1), but generally, these were not described. One such site was at Makirikiri near Whanganui, in lower North Island, which Hamilton visited in 1905 (Fig. 2.6). This site was reinvestigated in a large excavation in 1936–1937 (Figs 2.7, 2.8) and many bones recovered, but these were not described until recently (Worthy 1989a).

Archey, after his move to Auckland Museum, reinvigorated paleo-ornithology in New Zealand with his entry into the field in the 1930s. Benham (1934: 88) set the scene:

> To begin overhauling the nomenclature and trying to find out what each author supposed to be the limits of the genera, as has

Figure 2.8. View of the excavations at Makirikiri in 1936–1937. Photograph courtesy of the Wanganui Regional Museum.

> been attempted by such competent men as Lydekker and Hutton, will be futile until we have complete skeletons of the various genera so that the skulls may be correlated correctly with the leg bones. . . . It is a waste of time until we have such skeletons to commence the task of unravelling the tangled skein already much involved by Owen as well as by Hutton.

Archey took on this challenge—to tidy up the taxonomy of moa—and set about looking for associated skeletons. Archey searched caves for new materials of moa and was rewarded by good partial or complete skeletons of the previously not well-known North Island taxa. He made several expeditions to the Lake Waikaremoana area in the Urewera country of northeastern North Island. There, hydroelectric development preparations had resulted in the discovery of numerous small caves between the large blocks of a giant landslip that had blocked the river and so created the lake. At the same time, Archey explored caves in the King Country karst in central western North Island. The combined result was a large series of individual skeletons of *Anomalopteryx didiformis*, which allowed variation to be assessed in this species.

More importantly, the material included several skeletons of the species Archey named *Pachyornis mappini*, thus clearing up confusion present since Owen's time when that anatomist referred leg bones of *P. mappini* to *Euryapteryx geranoides*. Archey also visited northwest Nel-

Figure 2.9. Overview of Pyramid Valley as seen in 1988. Photograph by R.N.H.

son in search of skeletons of *Megalapteryx didinus* and was successful in obtaining several, which allowed him to combine two of the taxa that were then accepted. Archey's monograph was just nearing completion when the first bones were being excavated from Pyramid Valley Swamp.

Oliver, who had previously attempted a new listing of moa (Oliver 1930), was probably stimulated by Archey's work to a major review of his own, which was completed in 1949 (Oliver 1949). Oliver did few field investigations in caves or in swamps and was satisfied with revising the taxonomy of moa on the basis of the collections in the Museum of New Zealand.

The discovery of moa remains in Pyramid Valley Swamp in 1938 was the beginning of a major period of development of paleontology in New Zealand (Figs. 2.9–2.11). Essentially, it marked the introduction of Scarlett to the field, and the continued work of the next few years firmly established Scarlett as the most important paleoornithologist in New Zealand in the 1950s. When the landowner, Joseph Hodgen, and his son Rob buried a dead horse in the swamp, they found three large bones of *Dinornis giganteus*. David Hope (a friend of the Canterbury Museum) saw the bones in the wool shed in late 1938 and took them to the Canterbury Museum, thus informing the museum of a potentially rich site for excavation.

Armed with steel probes, a group of researchers returned to the swamp and quickly established the presence of several skeletons. The excavations began. People involved, apart from D. Hope, included such notable individuals as R. S. Allan and E. Percival (professors at Canter-

Figure 2.10. Start of the January 1965 Pyramid Valley Swamp excavations in squares 54, 119, 120, and 121. Photograph courtesy of Canterbury Museum.

bury University College), E. Stead (a noted ornithologist), C. A. Fleming (who joined the New Zealand Geological Survey in 1940 and became one of New Zealand's most renowned naturalists), E. Sorenson (an amateur archaeologist), and B. Wisely and R. J. Scarlett (both students at the time) (Duff 1951). Excavations continued through summer weekends from 1939 to 1941, when shortages of gasoline and other goods curtailed fieldwork.

The moa remains were preserved in good condition as individual skeletons, which was especially important because it allowed the current generation of scientists to check the attribution of crania to leg bones for a number of species, most of which had been erected on the basis of individual leg bones. The excavations did not resume until late 1947; these later excavations were instigated by Cushman Murphy of the American Museum of Natural History. At first, the site was controlled by Canterbury University, and the university committee allowed him to excavate. Only in 1949 did responsibility for excavations pass to the Canterbury Museum under the new director, Roger Duff. Cushman Murphy excavated at Pyramid Valley in 1947 and 1948, and he helped for two weeks in 1949.

Scarlett and Eyles of the Canterbury Museum continued to excavate the site in the summers of 1952, 1954, 1955, 1956, 1957, 1963, 1965, and 1973. In total, some sixty-one 3.66- by 3.66-m (12- by 12-foot) squares were excavated to depths of 1.22–1.83 m. An additional 330–490 m^3 had been excavated before the establishment of the grid in

Figure 2.11. Excavations in progress in 1965 at Pyramid Valley Swamp in squares 119, 120, and 121. Photograph courtesy of Canterbury Museum.

1949; therefore, somewhere between 1300 and 1960 m^3 were excavated. A total of 183 moa individuals from 5 species was found in the excavations, among 46 bird species. However, bones of small birds were rare—one bone for every 0.55–0.82 m^3 of sediment. The work stimulated a series of systematic studies by Falla and Scarlett, resulting in the addition of the pink-eared duck *Malacorhynchus,* a gallinule *Gallinula hodgenorum,* and a harrier *Circus eylesi* to the fauna of New Zealand (Falla 1941; Scarlett 1953, 1955b). The moa remains were associated with preserved gizzard contents that formed the bases of the analyses in the pivotal studies of Burrows (1980) and Burrows et al. (1981), which revolutionized interpretations of moa ecology.

The 1950s through the 1970s marked a real advance in knowledge due to the enthusiasm of Scarlett, who worked from the Canterbury Museum as he actively solicited specimens from people throughout the country. His working career spanned the development of archaeology as a science and the rise of two strong research schools in New Zealand. So naturally he was called on for all bone identifications, and he often helped in excavations or carried out his own. It also coincided with the emergence of speleology as a recreational pastime in New Zealand. The members of the recently formed New Zealand Speleological Society were dedicated cave explorers, and many new caves were discovered, as were many extensions to known caves, and numerous fossil deposits were found.

From the late 1950s onward, Scarlett was the recipient of much of this material. When major finds were evident, he visited and aided the discoverers in their excavations. Scarlett assiduously cataloged all specimens sent to him and never failed to get a letter in the mail within a short time to inform the finders of what had been in their parcel. When the occasion called for it, he went to the caves himself. For example, when John Hobson sent down bones that included takahe and the rare North Island goose from Haggas Hole at Waitomo, Scarlett was there not too long afterward. In the late 1950s and early 1960s, W. H. Hartree excavated small rockshelters in the inland parts of Hawke's Bay and found moa nests and rich deposits of small bones. Scarlett helped in several of these excavations.

Scarlett's interests were for all types of deposit: swamp, dune, cave, or archaeologic, and most importantly, no specimen was deemed to be of too little importance to catalog. Thus the first complete, rather than filtered, collections of material came to be assembled. Too often in the past, a "stamp collector mentality" had prevailed, and "good" specimens were kept while broken or common ones were discarded. The result of Scarlett's endeavors was an impressive collection that has and will form the raw material for many distributional and morphometric studies. Scarlett discovered and named several species new to science. However, his investigations or interpretations were hindered by his inability to date the deposits. Radiocarbon dating was in its infancy during much of this time, and the New Zealand fauna appeared homogenous: there were no "early" extinct faunas that bespoke earlier times, such as those that characterize, for example, the Irvingtonian in Europe.

Fleming's View of the Age of New Zealand Fauna

In the early 1960s, Charles Fleming, that rare naturalist with an overview of both geology and biology and particularly birds, sought to explain the relationships of the New Zealand fauna. Fleming (1962b) believed that the Holocene in New Zealand had begun with an impoverished biota, which he thought had resulted from extinctions during the Pleistocene ice ages. He noted that among marine organisms, most extinction events took place during the first two glaciations, and he presumed the same to be true for birds. Starting from this premise, Fleming (1962b: 113) believed that the early New Zealand Holocene was characterized by "faunal diversification and active colonisation." He assumed that deposits formed during the Late Holocene (0–4000 years) were the main source of moa and other bird bones. Still without the benefit of an absolute dating technique for small bones, he wrote (Fleming 1962b: 114), "Though it would be rash to be dogmatic, it is doubtful if any *major* source of extinct bird bones so far known in New Zealand is as old as early or even mid Holocene in age, let alone Pleistocene."

In this compression of the fossil record into the later Holocene, we now know from radiocarbon dating that Fleming was wrong; many deposits do indeed extend back more than 20–30 ky. None of the colonizations Fleming referred to happened in the Holocene, and all the birds present and breeding in the fauna when people first reached New Zealand were present 30,000 years ago. It is probable that potential colonizers arrive constantly but that most are unsuccessful. Of several recent examples, attempts at colonization by the Australian avocet *Recurvirostra novaehollandiae* and white-eyed duck *Aythya australis* in the 19th century failed after a few decades. The crucial factor that has allowed recent colonizations to succeed was the formation of new habitats that were not occupied by any resident species. Colonization was also helped by the loss or extinction of half of the land and freshwater birds and the clearing of indigenous vegetation.

In the early 1970s, P. R. Millener became interested in fossil bird bones and began a study of North Island fossil deposits that culminated in 1981 with a Ph.D. thesis at Auckland University entitled "The Quaternary Avifauna of the North Island of New Zealand." He listed faunas from all museum collections and concentrated on comparing the faunas from the sand dunes of the northern tip of New Zealand (the Far North) with those from Waitomo.

Like his predecessors, he collected mostly surface specimens; no stratigraphically controlled excavations were made. But for the first time, an attempt was made to date faunas. By use of the newly available radiocarbon technology, moa bones were dated from numerous dune and cave sites. The series of radiocarbon dates obtained indicated for the first time that cave deposits extended into the glacial age, although it was confirmed that the dunes were of Late Holocene age. Millener was impressed by similarities, and he concluded that a single ubiquitous fauna was common to all of the North Island forests. His mainly taxonomic interest was revealed by publications on, for example, a refutation of the presence of *Tyto alba* in New Zealand (Millener 1983), of differences in the New Zealand coots (Millener 1980, 1981b), and the synonymy of *Anomalopteryx oweni* with *Anomalopteryx didiformis* (Millener 1982).

The Last Decade

Millener moved to the National Museum of New Zealand, where he briefly took the post of curator of subfossil birds. His main interest was taxonomy, and the investigations he made at Honeycomb Hill Cave, or the large collection from the Chatham Islands, were aimed at

*Figure 2.12. Photograph of New Zealand wren (*Xenicus *sp.) bones as found in floor of Honeycomb Hill Cave. Photograph by T.H.W.*

obtaining specimens for this purpose. His main contributions before his departure in 1993 were papers describing the new wrens (Acanthisittidae) *Pachyplichas yaldwyni, P. jagmi* (Millener 1988), and *Dendroscansor decurvirostris* (Millener and Worthy 1991).

The investigations into the fossil deposits in Honeycomb Hill Cave System marked a turning point in the development of Quaternary paleontology in New Zealand. Honeycomb Hill Cave was discovered in 1976 by cavers prospecting the limestone made newly accessible by road construction to facilitate timber milling. In 1982, P. R. Millener and J. A. Bartle (Museum of New Zealand) made a preliminary visit to the cave, and the fossil deposits were recognized to be of high significance. In 1983–1984, Millener carried out large-scale collecting of surface material in the cave, with an emphasis on microvertebrates. This work generated the specimens to describe a new flightless wren (*Pachyplichas*); specimens had been known previously, but they were mainly unassociated leg bones whose relationships were unperceived. Honeycomb Hill Cave provided numerous whole skeletons of this and other wrens—and in fact of many small, terrestrial species (Fig. 2.12).

In 1984, T. H. Worthy mapped the cave in detail, showing the presence of over 14.5 km of passages and 70 entrances. This exercise made Worthy intimately familiar with the cave and its fossil deposits, and in 1986–1987, he returned to the cave to examine two specific issues. He had noted that the spectacularly rich Graveyard deposits

(Figs. 2.13–2.18) were stratified into two layers at least, so an excavation to sample and date the different faunas from each layer was planned. A second issue was motivated by the taxonomy of *Megalapteryx,* which at that time allowed two species, one much larger than the other but rare. It was his contention that this species represented the large specimens of glacial times, so a collection-dating program was planned to examine this hypothesis.

The Graveyard excavation provided the first well-dated fauna from the Otiran (last) glacial period and also showed that there had been changes in faunal composition over time. It also marked the beginning of a new perception of the Quaternary fossil record. Glacial deposits were found to be common, time-averaged deposits (where materials of different ages lie one beside the other) were all too common, faunal composition had changed with time, and the importance of dating became paramount.

Worthy became interested in paleontology after a university career that permitted a sporting interest in speleology. Through extensive explorations of cave systems in New Zealand, he became aware of the still large resource of fossil bones and decided to study them. After a year or two of caving based at the Waitomo Caves Museum, during which time it was his good fortune to be paid to locate fossil bones in caves, he did a short stint at Victoria University. There, the taxonomy and distribution of fossil *Leiopelma* frogs in New Zealand were studied and three species of extinct *Leiopelma* described (Worthy 1987a, 1987b). The National Museum of New Zealand (now the Museum of New Zealand) enabled Worthy to work further on fossil faunas and facilitated the Honeycomb Hill Cave studies and early forays into moa taxonomy.

In 1991, there was a virtual revolution in the funding of science in New Zealand. The Foundation for Research Science and Technology and its associated ministry were born—and private individuals could seek a share of the funding, formerly shared among large crown research organizations. Funding became accountable and outputs—usu-

Figure 2.13. The Graveyard, Honeycomb Hill Cave, northwest Nelson, before excavation in 1984. Photograph by T.H.W.

Figure 2.14. The Graveyard, Honeycomb Hill Cave, during excavation in 1987. Boardwalks protected the fossil deposits. Photograph by T.H.W.

ally published work—became the watchword of the day. Worthy succeeded in getting an initial contract in 1991–1992, and so began a program of research with Holdaway that continues to the present. International issues of the day included the effects of greenhouse gases and global warming on the environment, and the program of research they entered into was to provide a historical framework for the New Zealand fauna from within the last glacial period to the present.

Our approach was to select relatively small areas and document all faunas in museums from them and augment these with an intensive collection program and radiocarbon dating. We specifically sought Holocene versus Pleistocene faunas and sites with various taphonomies.

Figure 2.15. Detail of the fossil moa bones exposed in the wall in The Graveyard, Honeycomb Hill Cave. Photograph by T.H.W.

Figure 2.16. R.N.H. and Ian Gibson in The Graveyard, Honeycomb Hill Cave, after the excavation had ceased, in 1993. Excavated pits can be seen toward the rear of the photograph and also were to the left of where the photograph was taken. Photograph by T.H.W.

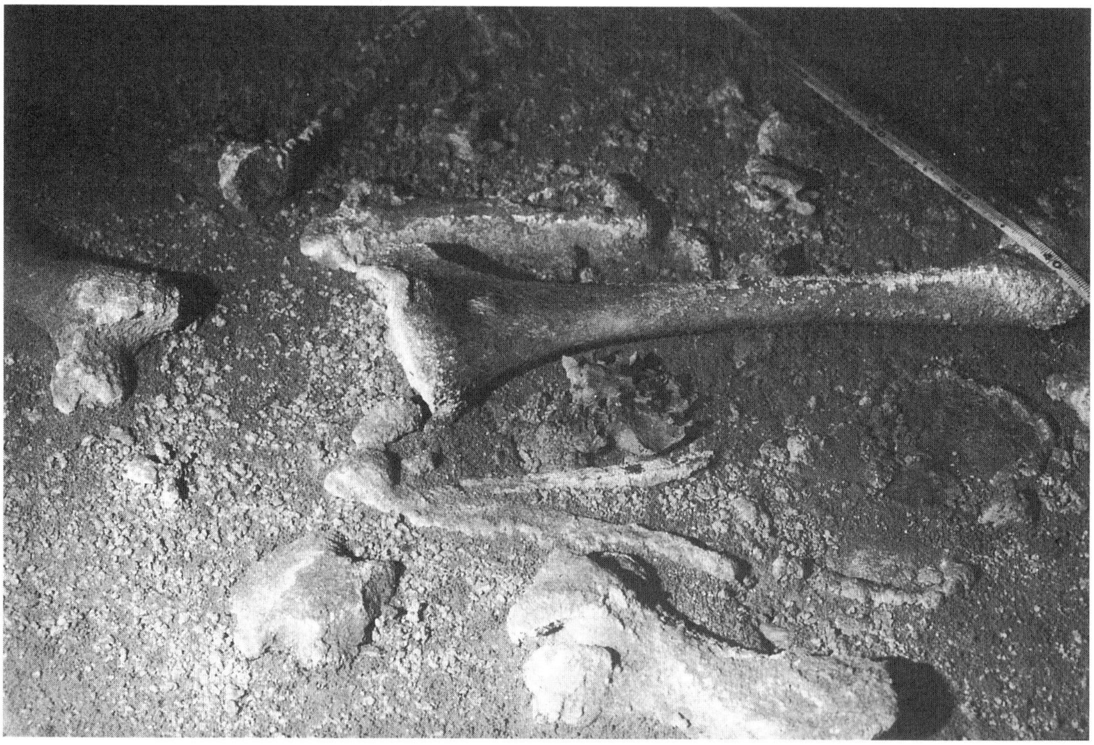

We began in the west in the Punakaiki region of the South Island, and, working clockwise around the South Island, we worked next in the Takaka area of northern South Island, then two areas of North Canterbury, a part of South Canterbury, three areas of Otago, then Southland, and most lately Stewart Island (see Reference List; Fig. 2.19). As of this writing, we are investigating the Hawke's Bay and Waitomo regions of the North Island

Along the way, we found that western areas had experienced marked climatic changes and hence changes in vegetation and faunas between the Pleistocene and the present, whereas the drier eastern, rainshadowed areas have not. Perhaps the single most important discovery of this research program has been the presence of avian predator deposits, the existence of which had not previously been suspected. More than 50 faunas accumulated by the laughing owl (*Sceloglaux albifacies*) and many of the falcon (*Falco novaeseelandiae*) are now documented, providing important insight into the composition of the smaller vertebrates in the faunas. Most recently, deposits attributed to the large extinct harrier (*Circus eylesi*) have been found in the North Island. Although new species are not numerous and most faunas are usually less than 25,000 years old, the data on the former distribution of species have been hugely improved. Now we can confidently state that although there were often regional changes in the distribution of species at the end of the Pleistocene, there were no extinctions. These all came after the arrival of humans in the last 700 years.

Figure 2.17. (opposite page top) Moa Cave, Honeycomb Hill Cave System. View of a late Pleistocene deposit containing the remains of 57 moa. Photograph by M. Strange, Museum of New Zealand Te Papa Tongarewa, B18628.

Figure 2.18. (opposite page bottom) Moa Cave, Honeycomb Hill Cave System. Detail of floor with Pachyornis australis *leg bones and cranium (lower left) and a* Dinornis struthoides *cranium (center). Photograph by T.H.W., Museum of New Zealand Te Papa Tongarewa, B18630.*

Figure 2.19. Map showing important fossil sites or groups of sites (in lowercase letters and indicated with solid circles) and other geographic features, localities, or cities mentioned in the text. At localities such as Punakaiki karst, Takaka Hill, Waitomo Caves, Hawke's Bay, and Central Otago, there are numerous fossil localities.

3 • Discovery of Moa

Origin of the Name "Moa"

Throughout much of Polynesia, the domestic fowl is called "moa." So it is strange that the New Zealand birds, of diverse forms and of such different stature to fowl, should be likened to a chicken. How did this association of the word "moa" with New Zealand's giant extinct birds come about?

Although Joel Polack was the first to describe several large fossil "ossifications," which he concluded belonged to a bird similar to a species of emu or a bird of the genus *Struthio,* he made no mention of the word "moa" (Polack 1838). The first recorded name for the animal to which these fossil ossifications belonged is apparently "movie." The trader John W. Harris forwarded what was to become a significant bone to Dr. J. Rule, accompanied by a letter dated February 28, 1837, in which he indicated that the Māori had told him the enclosed bone fragment came from a movie, which was a giant bird of the eagle kind (Anderson 1987; Andrews 1986: 124; Buick 1936; Owen 1840, 1842).

Anderson (1987) examined the origin of this name and concluded that Rule managed to associate the Māori name for the North Island with the name of the animal to which the bones belonged. Rule believed that the term "movie" came from the French[1] word for seagull (*mauve*) and was the name given to large birds met by the first explorers, which was then adopted for the island they had discovered—hence Ikana movie. The North Island, known as "Te Ika a Maui" in classic Māori, sounds strikingly similar; it means "the fish belonged to Maui." "Movie" is thus most probably a wrongly recorded version of "Maui" and not a name relating to ancient birds.

1. The French are often overlooked as early explorers of New Zealand. First was Jean de Surville on the *St. Jean Baptiste,* who arrived in New Zealand barely two months after Captain Cook and landed first at Doubtless Bay on December 18, 1769. Two important early scientific expeditions were mounted later: that led by Louis-Isidore Duperrey in the *Coquille,* which spent April 3–17, 1824, in New Zealand; and that led by Dumont d'Urville in *L'Astrobale,* which was in New Zealand January 10–March 19, 1827 (Andrews 1986).

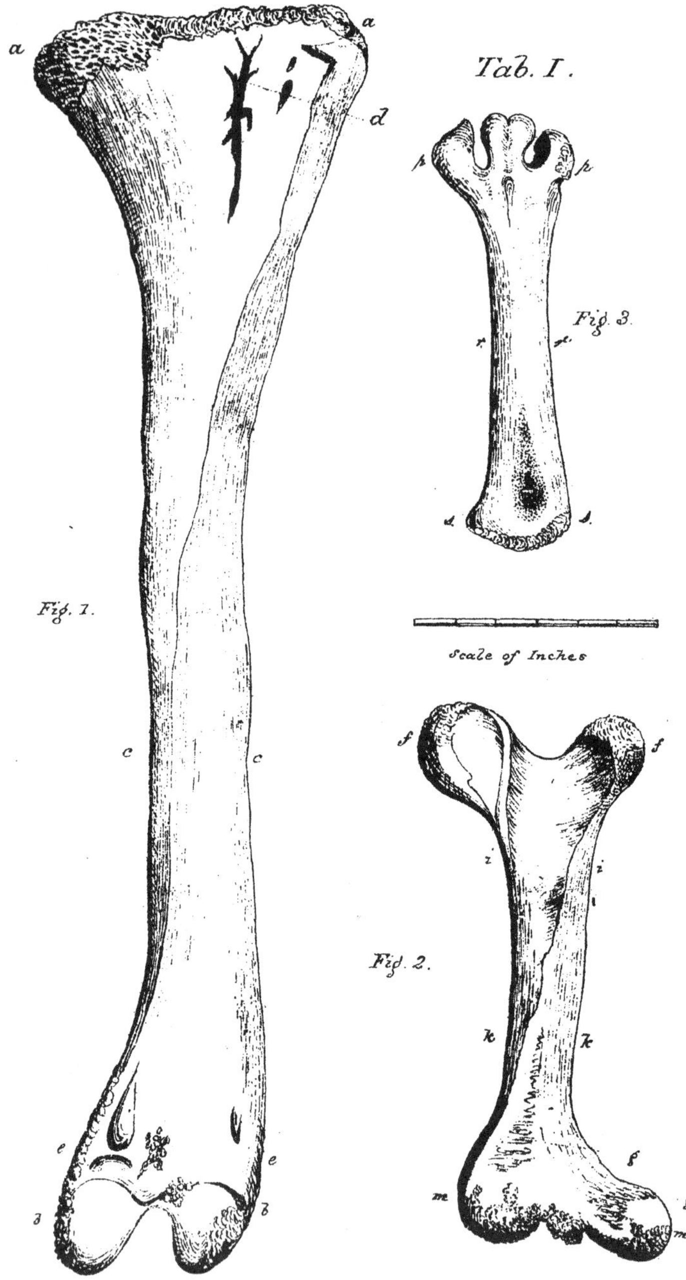

Figure 3.1. First drawings of moa bones. Drawn in 1842; published in Colenso (1843).

Hill (1914) noted that from the time the first missionaries arrived in 1814 until December 1837, there was no mention of the word "moa." In December 1837 and January 1838, the Reverend William Williams and William Colenso vacationed from the Bay of Islands down to the East Coast after completing the publication of the New Testament in Māori. It was during this journey in 1838 that the word "moa" was first heard. Colenso wrote about this occasion in early 1842 and sent the manuscript to the *Tasmanian Journal of Science,* where its receipt was acknowledged in August 1842 (Hill 1914: 336) and where it was published on October 6, 1843 (Plomley 1969), along with two plates illustrating the bones of the moa (Colenso 1843: 81; Fig. 3.1):

> Whilst at Waiapu, a thickly inhabited locality about twenty miles southwest from the East Cape, I heard from the Natives of a certain monstrous animal. Whilst some said it was a bird and others "a person," all agreed that it was called "moa"; that in general appearance it somewhat resembled an immense domestic cock, with the difference, however, of its having a face like a man; that it dwelt in a cavern in a precipitious side of a mountain; that it lived on air; and that it was guarded or attended by two immense tuataras, who, Argus-like, kept incessant watch while the moa slept; also that if any one ventured to approach the dwelling of this wonderful creature he would be invariably trampled on and killed by it.

Furthermore, in this article (reprinted with amendments in the *Annals and Magazine of Natural History* 1844; 14: 81–96 and in Colenso 1880), Colenso noted that many of the natives had occasionally seen bones larger than those of an ox and that these were used to fashion fishhooks.

A year later, Williams returned again to the East Coast with the Reverend R. Taylor, and while there, he sought evidence of the "moa." Eventually Williams obtained a worn fragment (Owen 1879a: 75–76). Taylor also obtained a moa claw on this journey (Colenso 1880) that he told Owen about in a letter in 1844 (Owen 1846a: 327) and that Dieffenbach saw in the Bay of Islands (Hill 1914: 334). However, many years later, when Taylor recalled and recorded these events, a somewhat mixed version appeared (Taylor 1873). He said they noticed a bone in the roof of a native house in the pa at Waiapu, which they thought might be a bird bone:

> They [natives] said it was the bone of a Tarepo, a very large bird which lived on the top of Hikurangi, the highest mountain on the East Coast, and that they made their largest fish-hooks from its bones. I then enquired whether the bird was still to be met with, and was told that there was one of an immense size which lived in a cave, and was guarded by a large lizard, and that the bird was always standing on one leg. (Taylor 1873: 97)

Taylor claimed to have purchased this bone and so claimed to be the discoverer of the first moa bone. This was strongly denied by Colenso (1880: 105–106), who cited Taylor's February 14, 1844, letter to Owen (Owen 1846a: 327), in which Taylor recorded finding moa bones near Whanganui, and in which he told Owen that he had procured a toe bone near Poverty Bay in 1839. Colenso concluded that Taylor's (1873) statements were of many incidents jumbled together in date and sequence, so Taylor's rather belated claim to be the discoverer of the first moa bone is not supported by contemporary authors and can be dismissed.

Taylor's use of the word "tarepo" is likewise to be treated with caution; he first used it in his 1844 letter to Owen: "The Kakapo or Tarepo is about the size of a turkey, and from its habits, nature, and other circumstances, seems to closely resemble the dodo, as to lead me to suppose it is the same" (Taylor, in Owen 1846a: 327–328). "Tarepo" would not seem to be a name of any moa.

After his return to the Bay of Islands, Colenso received some fragmentary fossil bones sent by Māori teachers who had been sent to the East Coast in early 1838. But it was not until Colenso visited the East Coast again in the summer of 1841–1842 that he was able to procure seven fossil moa bones, including five femora and a tibiotarsus (Colenso 1843, 1880: 66) from natives on the Waiapu River. They had been going to cut them up to fashion fishhooks. However, there and during the rest of his journey down the East Coast and back through the middle of the North Island to the Bay of Plenty, he was unable to obtain any information, either legendary or other, about the moa. He concluded that moa no longer lived in New Zealand. But perhaps Colenso searched using the wrong name. What if the ancient birds were not called moa?

Williams, then living in Poverty Bay, had by the time of Colenso's visit obtained an imperfect tibiotarsus, which nevertheless was still some 45.7 cm (18 inches) long. Colenso left a pair of femora with Williams, who sent them to Dr. W. Buckland at Oxford. However, before he could do so, a native arrived with another moa bone, which Williams bought. Hearing of the sudden worth of these relics, a great number of natives searched for moa bones. Bones of more than 30 birds were shortly obtained (Colenso 1880: 68). Williams referred these bones to the bird known as "moa" to the Māori in a letter to Dr. W. Buckland dated February, 28 1842, that accompanied some bones (reprinted in Owen 1844a). It therefore seems probable that after Colenso and Williams's 1838 excursion down the East Coast, the name "moa" was promulgated into general usage for the large bones by the Reverend William Williams and Colenso. As these men scoured the country for "moa" bones, especially in 1842, the name was spread. Thus in 1843, after Taylor had moved to Whanganui and saw bones at Waingongoro Stream and asked the natives what they were, he was told that they were moa bones—what else?

Colenso (1843, 1880: 76) discussed the origin of the word "moa." Colenso thought it strange that the word appeared to have no meaning in Māori—that is, it does not describe some feature of the subject. In addition, such a short name was a rare occurrence indeed in the Māori language. He noted that "moa" was the name for fowl in the Friendly Islands (Tahiti) and other Pacific groups. Furthermore, he speculated that the ancestral Māori may have been familiar with cassowaries, to which he thought the moa was related, and that they gave the moa this name upon their arrival in New Zealand.

Colenso (1880) reviewed all the myths and legends of the Māori and found only one instance where the word "moa" was featured. This legend concerns the fires of Tamatea, which burned all the land and birds and moa therein. Two moa were said to have survived the fires, one in a cave on Whakapunake and one at Te Whaiiti. In no other legend do moa feature, yet other animals, both mythic and real, do. Similarly, moa do not feature in proverbs, charms, or spells. Colenso also noted that moa was the name for a small drill, a raised garden, a type of grass, and a stone or layer of rock. However, the wide use of the word within place names but not elsewhere in the language was taken by Colenso to mean that the bird "moa" was known only to the distant ancestors of Māori.

That the ancient birds of New Zealand may have been formerly called by a name other than moa is attested to by the following account, which Henry Hill (1914: 340) recorded and which was related to him on July 16, 1912, by Urupeni Puhara, aged 88 years, an old Māori chief then living at Pakipaki:

> The "moa" was not the name by which the great bird that lived in this country was known to my ancestors. The name was "Te Kura," or the red bird; and it was only known as "moa" after the pakehas [Europeans] said so. Te Kura was known to all ancient people, and was handed down from father to son, who spoke about the big bird that was as high as the top of the door (pointing to the door of the room in which we were sitting). Its legs were thick as those of a bullock. Neither his father nor grandfather had seen Te Kura, but it had been told to them, and tales were told of what it did. He did not know how it was caught or snared, nor had he seen or heard of moa's eggs, and knew nothing of its footmarks. The moa lived, he had heard, all over the North Island, but they disappeared after the coming of Tamatea, who set fire to the land.

This story has a ring of truth to it. And while Best (1977: 182) stated that "the Māori has preserved little knowledge of [the moa]," he did attribute other names in Māori lore to moa. Our subsequent history of moa investigation has determined that they were indeed widespread and that they were extinct hundreds of years before Urupeni Puhara told his story. We now know there were at least seven species in the North Island. Perhaps all these names had some basis in reality—Te Kura or moa—and these and other names probably all related to various of the different species of these large birds. However, by 1843, the word "moa" was widespread in New Zealand and was used by Owen in 1844 (Owen 1844a) and thereafter.

Moa versus Moas

In this work, we use the word "moa" to mean both singular and plural, because in Māori there is no "s" at the end of the word to denote the plural. Sentence structure makes clear when we are referring to more than one species of moa or to individual moa. (Māori names for other birds are also affected by this grammar, so, for example, the word "kiwi" is used for both singular and plural.)

Taxonomic History of the Moa Species

> So far as my skill in interpreting an osseous fragment may be credited, I am willing to risk the reputation for it on the statement that there has existed, if there does not now exist, in New Zealand, a Struthious bird nearly, if not quite, equal in size to the Ostrich.
>
> Richard Owen, 1840: 170–171

The basis of this famous statement was the shaft of a femur that was acquired some time between 1831 and 1836 by the trader John W. Harris from the East Coast of the North Island of New Zealand. Dr. John Rule, a highly qualified surgeon residing near Sydney, Australia (Buick 1931: 59), was given the bone in 1837 by his friend Harris. Harris left the bone and other curios with a note dated February 28, 1837, which read as follows: "I have left here with Mrs. Pike, for you—one stone club, one small carved wooden box, two wooden war-clubs, and a bone. The natives have a tradition that the bone belongs to a bird of the eagle kind, but which is now become extinct. They call it *A Movie*" (Buick 1931: 61).

In 1839, Rule took the bone to England and compared it with various animal bones. He convinced himself it was from a bird. Rule then tried to sell it to the British Museum and was declined. Next, under the advice of G. R. Gray, he approached the Royal College of Surgeons, who forwarded the bone and a letter from Rule outlining his belief that it was from a bird, to Professor Richard Owen for appraisal (Buick 1931). Owen compared it at length to bones of human, ox, horse, pig, kangaroo, and various other mammals before concluding that it was indeed from a bird. The bone was then purchased by Mr. M. P. Bright from Bristol, and only many years later was it acquired by the British Museum (Buller 1888: xviii). Figure 3.2 shows the first moa bone described by Richard Owen.

The earliest evidence of moa was, however, found a few years earlier. In 1834, another trader, Joel Polack, put into Tolaga Bay on the North Island's East Coast to repair his storm-damaged cutter. While there, he was shown "several large fossil ossifications" by the Māori. He concluded that "a species of emu, or a bird of the genus Struthio had once lived in New Zealand" (Polack 1838: 303). However, his conclusion was published in a book that escaped the attention of scientists of the day. Polack did not keep the bones, and New Zealand's giant birds remained unknown for a few more years.

Owen's deduction that a giant bird had existed on the small islands of New Zealand seemed preposterous to his contemporaries; they were fated to be shocked even further. When Owen described the femur shaft in detail in the *Transactions of the Zoological Society of London*, he ended with the comment that "The present notice, it is hoped . . . may stimulate . . . the collection of the remaining parts of the skeleton" (Owen 1842: 31). He had not long to wait, and so began a lifelong study of the moa that resulted in more than 30 papers. In 1843, he erected the genus *Dinornis* to accommodate the bones from New Zea-

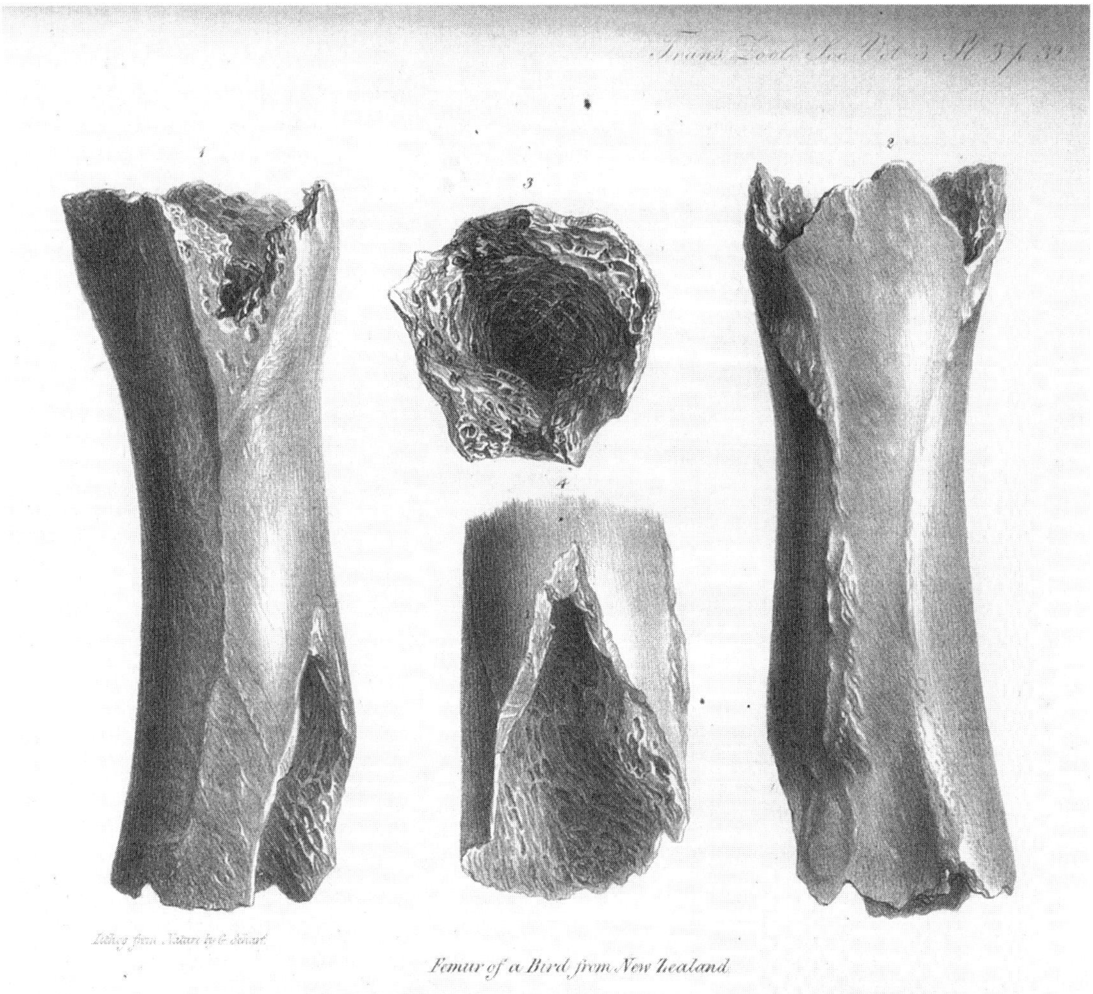

Figure 3.2. First moa bone described by Richard Owen. From Owen (1842: pl. 3).

land (Owen 1843a). *Dinornis* (Greek) can be variously interpreted as "surprising," "prodigious," or "terrible" birds; Owen (1844a) gave the interpretation as "surprising bird." For an apt discussion of Owen's use of *deinos* in the context of the erection of the term "dinosaur," we can do no better than quote Gould (1998: 24): Owen "wished to emphasize the awesome and fearful majesty of such astonishingly large, yet so intricate and well adapted, creatures living so long ago. He therefore chose a word that would evoke maximal awe and respect." "Terrible" at that time implied awe and respect, not the nasty, ugly connotations of the term's present use.

The first substantial collection of moa bones to reach England was acquired by the missionary Reverend W. Williams from streams near Poverty Bay, Wairoa, between 1839 and 1842. He sent them to the eminent geologist Dr. W. Buckland at Oxford. In an accompanying letter dated February 28, 1842 (transcript in Owen 1844a), Williams stated that the bones came from birds known to the Māori as "moa" and concluded that some of the birds were at least 4.2 to 4.8 m (14 to 16 feet) high.

The first notice of this collection came in the *Proceedings of the Zoological Society of London* in July 1843 (Owen 1843a), in which was reported Owen's reading of part 1 of his first memoir. The report was not written by Owen, but nonetheless it provided a valid description of the genus and species *Dinornis novaezealandiae,* with reference to a femur, a tibiotarsus, and a tarsometatarsus, although only the tibiotarsus was adequately described. A further report in the *Proceedings,* in March 1844 (Owen 1844c), summarized the rest of Owen's Memoir as read. Inadvertently, all of Owen's proposed names were used, but, with the exception of *Dinornis giganteus,* for which there was the bare description "the largest tibia, belonging to the *Din. giganteus,* presented the extraordinary dimensions of two feet eleven inches" (0.88 m; Owen 1844c: 145), they are *nomina nuda.*

The procedure of reporting the papers read before the Society nullified Owen's presumed intent to present and describe the species in subsequent memoirs. Throughout the early years of moa taxonomic endeavor, the concept of a holotype (the specimen on which a species is first based and other specimens subsequently are likened to and then referred to) did not exist. Thus in the first substantial paper on the moa bones, Owen (1844a—his Memoir I) described some 47 bones, including 18 femora, 11 tibiotarsi, and 6 tarsometatarsi, within five species of *Dinornis,* naming 4 as new. Interestingly, *D. novaezealandiae* was not discussed in this or subsequent memoirs and was presumably abandoned while Owen was writing the "definitive" memoir.

One bone (tibiotarsus 2 [t2]) of what can be considered the type series of *D. novaezealandiae* was made the type of *D. ingens.* Another of the series, the tarsometatarsus (m3) was made the type of *D. struthoides,* thus creating the first of many nomenclatural problems for the group. This issue was not to be finally resolved until 1954 (Hemming 1954).[2] Owen considered the tarsometatarsus to be the most instructive bone, and intending that his species be based on these, he launched into a description of those elements first. Unfortunately for him, as noted above, there were already two types, those for *giganteus* and *novaezealandiae,* published in the reports. *D. didiformis* was based on the tarsometatarsi m4 and m5. The larger tarsometatarsi (m1 and m2) were called *D. giganteus. Dinornis didiformis* was named for its similarity with the dodo (then known as *Didus ineptus*). *D. struthoides* was so named because it was thought to attain the height of an ostrich. Of

2. Taxonomists have devised a set of rules to guide the naming of organisms. The foremost rule is that of priority—the first published name with an adequate description of the species in question has priority over other names. To "adequately" describe the species means with respect to known members of the group, so many early descriptions are very brief, as are those for these first moa. A species is also based on a holotype, which is a single individual or specimen; all other individuals or specimens are referred to that species. Thus the name refers to the holotype, so if more than one species are found in the suite of specimens referred to a species, only those the same as the holotype retain that name, and the rest revert to the next available name for their type.

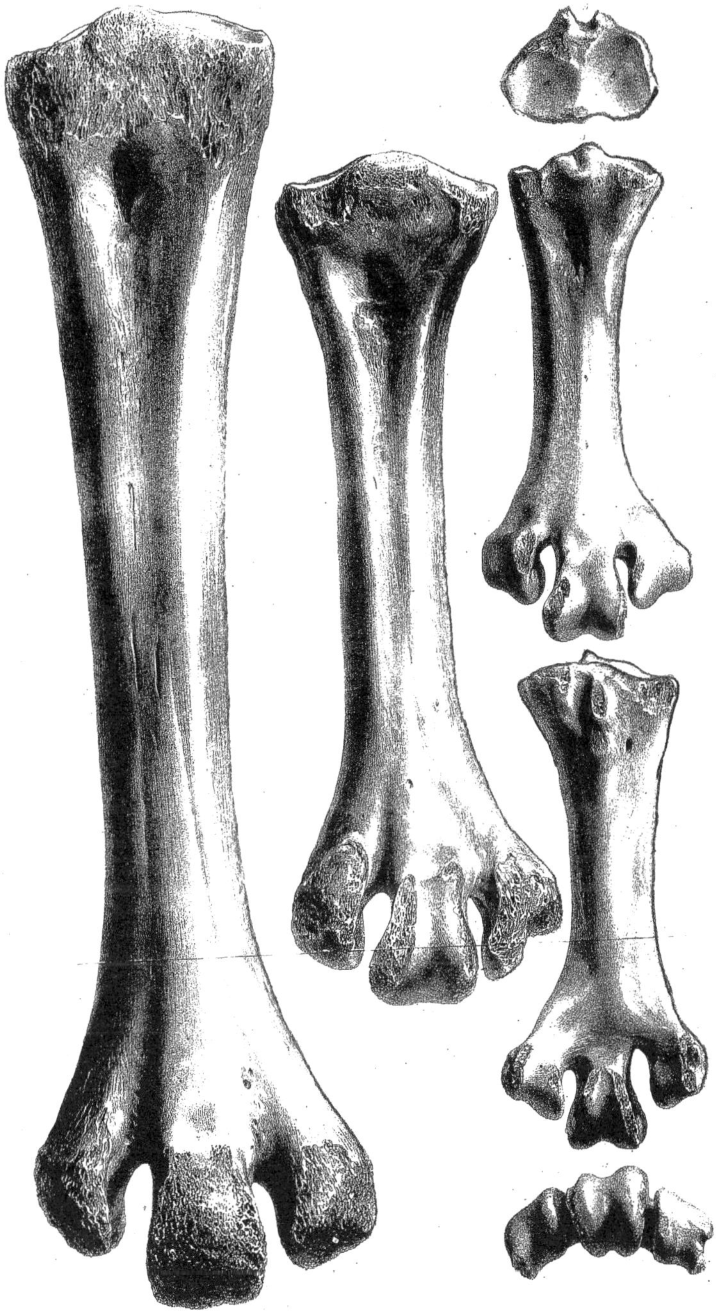

Figure 3.3. Some of the first consignment of moa bones from Poverty Bay. These tarsometatarsi include (from left to right) one referred to Dinornis giganteus, *the type of* Dinornis struthoides, *and the type of* Anomalopteryx didiformis. *From Owen (1879: pl. 27).*

the tibiotarsi, the smallest (t11) was made the basis of the name *D. otidiformis,* being similar in size to that of a bustard; four were referred to *D. didiformis,* and the largest (t1), at 88.9 cm (35 inches) in length, was that on which the name *D. giganteus* had been based. A smaller tibiotarsus, 73.7 cm (29 inches) long, was given the name *D. ingens.*

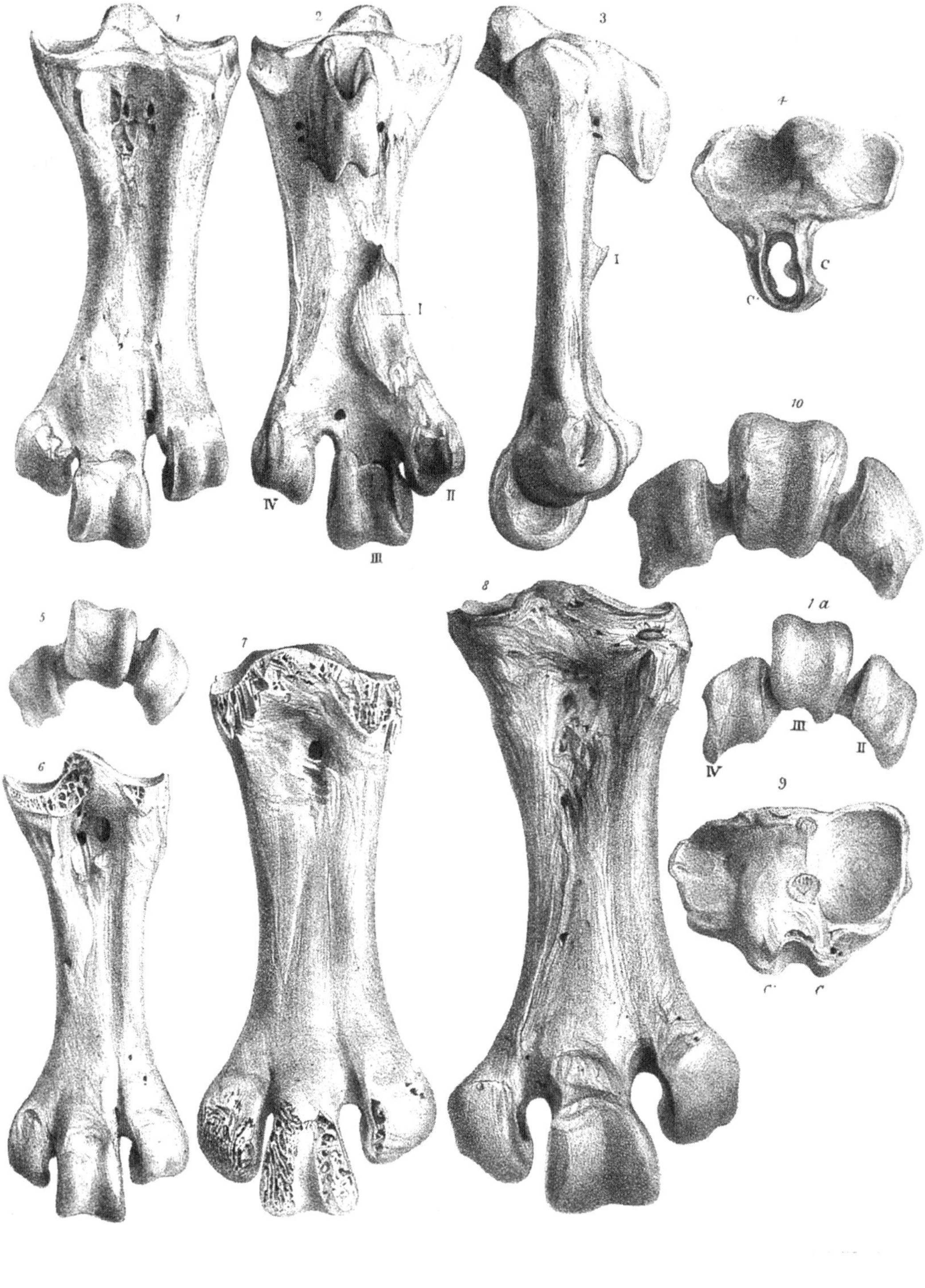
1
2
3
4
I
I
C
C
IV
II
III
10
8
1 a
5
7
6
III
II
IV
9
C
C

Figure 3.4. (opposite page) Tarsometatarsi of Aptornis species (1–6) and of Dinornis curtus (7–10). From Owen (1879: pl. 87).

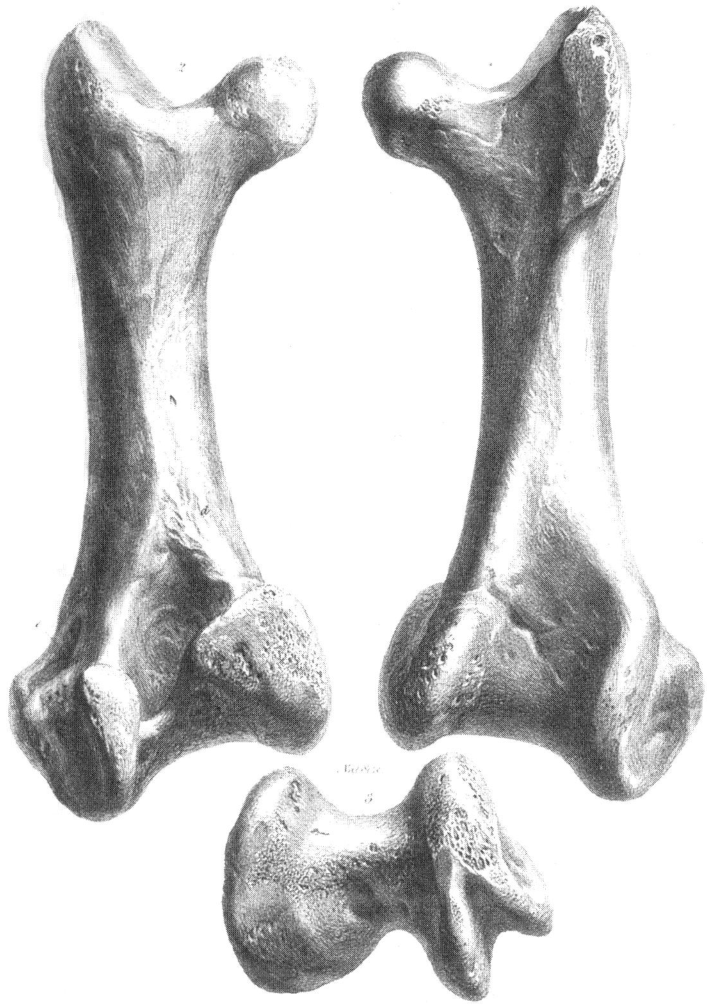

Figure 3.5. Femur Owen referred to Dinornis didiformis, which is actually Euryapteryx curtus. From Owen (1879: pl. 24).

The femora were referred to species partly on the basis of their size; the largest (f1) was referred to *D. giganteus* and the next largest (f2) to *D. ingens*—indeed, the presence of this bone helped Owen decide on the distinctiveness of the type tibiotarsus. Similarly, femora f12 and f13 were referred to *D. struthoides*. Femora f7, f8, and f17 were referred to *D. didiformis* because their size was as expected if one assumed that this small species had similar proportions to the larger taxa. There remained two slightly larger femora not referred to any of the species represented by tarsometatarsi or tibiotarsi. Because they were similar in size to those of emus, Owen gave them the name *D. dromaeoides*.

And so, in the first paper, the confusion of correct associations of elements to species had begun. The primary assumption that smaller species had leg bones in similar proportions as the larger species was false: the femora referred to *D. didiformis* were actually from the as

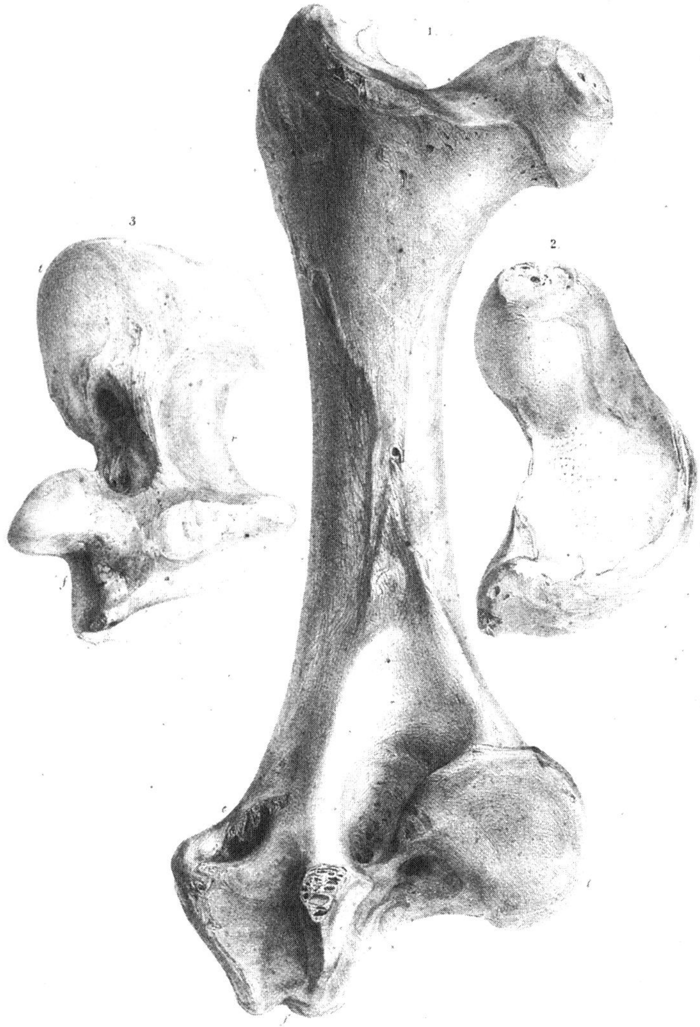

Figure 3.6. Type femur of Dinornis casuarinus, *which is now known as* Emeus crassus. *From Owen (1879: pl. 38).*

yet unnamed *Dinornis curtus,* as figures in Owen (1844a: pl. 24) clearly show. Similarly, those named *D. dromaeoides* belonged to *D. didiformis.* The practice of attaching names to single leg bones and not to those associated with other parts of the skeleton was a source of confusion in moa taxonomy for the next century. Figures 3.3 and 3.4 show lower leg bones (tarsometatarsi) from several moa species.

At the end of the first major paper he published, Owen drew several conclusions. He did not doubt that moa were flightless, but he predicted that they had wings somewhat bigger than those of kiwi. And he noted that *D. giganteus* would have been 1.67 m (5 feet, 6 inches) to the top of the femur when the normal stance of struthious birds was used. By basing the stance and proportions of the neck on that of an ostrich, he estimated the total height of this largest of moa at 3.2 m (10 feet, 6 inches).

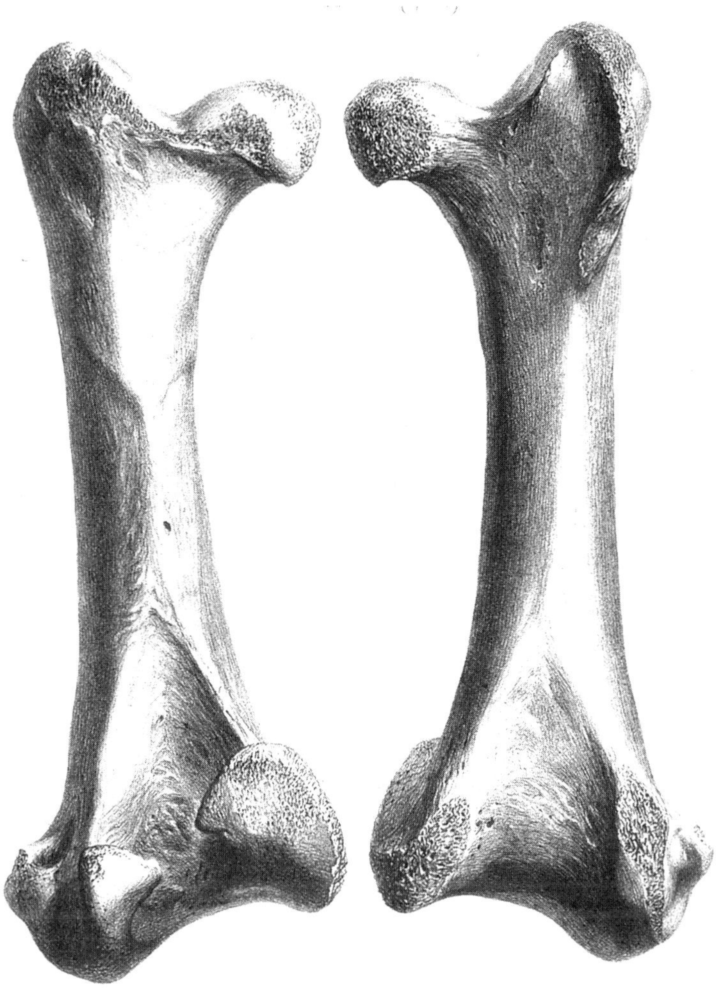

Figure 3.7. Type femur of Dinornis dromioides, *which is a subadult* Anomalopteryx didiformis. *From Owen (1879: pl. 22).*

To his credit, he ventured that this pose was probably wrong and that the cassowary offered the best analog, giving a maximum height of *D. giganteus* of 3.0 m (10 feet). In turn, *D. struthoides* did not exceed 2.1 m (7 feet) and so was equal to a moderate-size ostrich in height, but it was of a somewhat stouter build. Owen noted that the fragment of femur first described belonged to *D. struthoides. D. didiformis,* the smallest moa then known, he thought had been about 1.2 m (4 feet) high, and *D. ingens,* 2.7 m (9 feet). Figures 3.5, 3.6, and 3.7 illustrate moa femora.

Owen also sought an explanation for why New Zealand alone contained more than one genus of struthious birds (although we now know this to be wrong), and why there should be this unusual diversity in *Dinornis*: "It seems at least natural to suppose that some peculiarity in the vegetation of New Zealand adapted that island to be the seat of apterous tridactyle birds, so unusually numerous in species and some of them of so stupendous a size" (Owen 1844a: 268). He pointed to the abundance of ferns in the New Zealand flora, mentioned by Charles Darwin, who had stated that the whole country abounded with fern. Owen suggested that moa were adapted to feed on them.

Unfortunately, Owen was not to know that the fern lands that so impressed Darwin had been induced by humans and that the bracken (*Pteridium esculentum*) had spread only at the time moa became extinct. The cause of the presumed extinction of all moa was not a mystery to Owen either—he blamed humans: "It is not altogether improbable that the species of Dinornis were in existence when the Polynesian colony first set foot on the island; and, if so, such bulky and probably stupid birds, at first without the instinct and always without adequate means of escape and defence, would soon fall prey to the progenitors of the present Maoris" (Owen 1844a: 270).

More evidence of the diversity of the moa family was not long in coming. By 1846, Owen had received bones from other localities in the North Island and also from several in the South Island (or the Middle Island, as it was known then). The collections included the first crania and sterna and also provided evidence for a further new species *Dinornis crassus* (Owen 1846a, 1846b). The memoir "On *Dinornis,* Part II" correctly ascribed a cranium each to *D. struthoides* and *D. dromioides* (= *A. didiformis*). The name *D. gigas* was inadvertently added to the growing synonymy as a misprint for *D. giganteus.*

A new genus, *Palapteryx,* was erected to accommodate *Dinornis* tarsometatarsi for which a metatarsal scar indicated the presence of a hind toe, and *D. ingens* and *D. dromioides* were transferred to it. Bones of *D. giganteus* from the Middle Island were noted as being larger and more robust than those from the North Island.

In addition, two other species were named from the new collections. *Dinornis casuarinus* was erected for a series of 10 femora, 11 tibiotarsi, and 6 tarsometatarsi from "Waikawaite" (= Waikouaiti) in the southeastern South Island. Owen referred some North Island specimens to *D. casuarinus,* although he noted that most examples of this species were from the South Island. Also from Waikouaiti were other larger, more robust bones Owen named as *D. crassus*; he thought that these were restricted to the South Island. Some bones found only in North Island collections represented a third new species he named *D. curtus.*

Owen's third memoir was read January 11, 1848 (Owen 1848b), and published that year (Owen 1848a). It was based on a collection of 700–800 bones made by Walter Mantell from specimens found in sand dunes at Waingongoro in the North Island and made available by his father Gideon Mantell, the noted geologist of *Iguanodon* fame.

A skull from Waingongoro was described as the new species *Palapteryx geranoides.* Also in this lot was a tarsometatarsus for which Owen erected the genus *Aptornis* (Owen 1848a). He transferred *Dinornis otidiformis* to the new genus. As with much of the earlier moa history, *Aptornis* itself has an interesting nomenclatural history; the name *Apterornis* was used by Owen (1848b) on April 13, 1848, before it was published as *Aptornis* in the *Transactions of the Zoological Society of London* of April 22, 1848 (Owen 1848a). Owen thereafter used only *Aptornis.* Recently, a few authors tried to reinstate *Apterornis,* but *Aptornis* was conserved and that of *Apterornis* placed on the list of invalid names (Opinion 1874, *Bulletin of Zoological Nomenclature* 1997; 54: 142–143). But Gideon Mantell had already referred to these bones on February 2, 1848, in the *Quarterly Journal of the Geological Society of London* (vol. 4, p. 233), making the whole recent discussion unnecessary. In fact, because Mantell's words clearly indicate that *D. otidiformis* should be accommodated in the genus *Aptornis,* perhaps Mantell should be the author, as given by Reichenbach in 1852 and overlooked since—such are the problems generated by publication delays:

> 3. *Aptornis*—Among the bones of small size, those for example that are comparable in magnitude to the skeletons of the Bustard and Apteryx, there are several tarso-metatarsals, femora, tibiae, pelves, &c., which indicate a new tridactyle genus, very closely

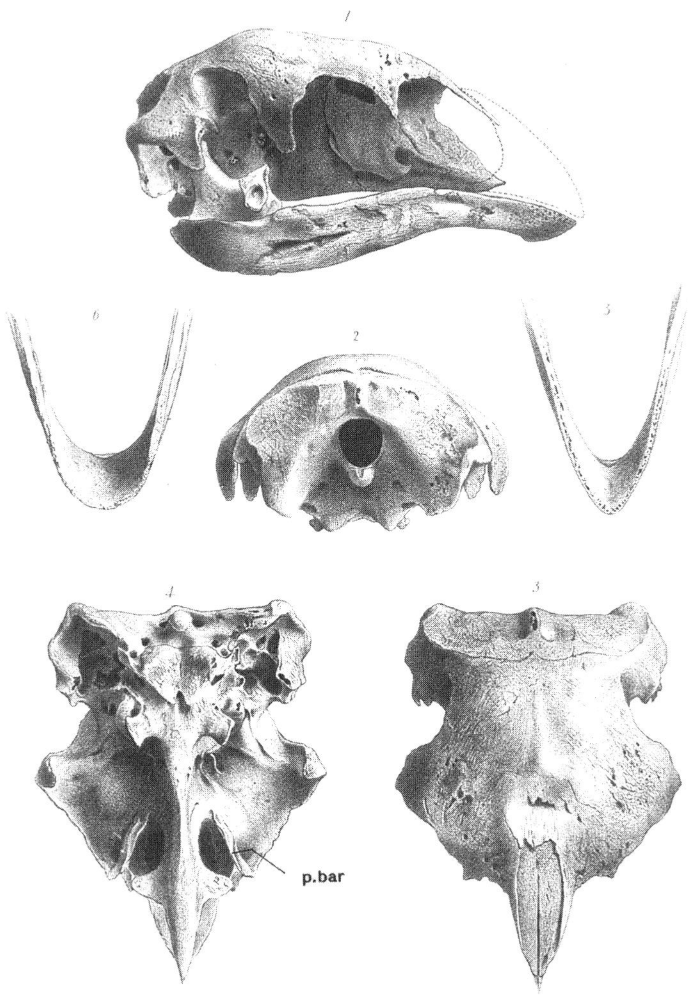

Figure 3.8. Type cranium (1–4) of Dinornis gravis, *which is a small individual of* Euryapteryx geranoides *(from Owen 1879: pl. 81). Note the incorrect placement of the left quadrate facing the rear in the right side. Also, here Owen incorrectly referred the narrower mandible (5), which we now know to be* Emeus *(or* D. crassus, *as Owen knew it), to* D. gravis. *As a result, he also incorrectly associated the broad mandible (6), which is* D. gravis, *with* Dinornis crassus. *The antorbital or postchoanal bar (p.bar) is clearly visible here because the palate bones are missing.*

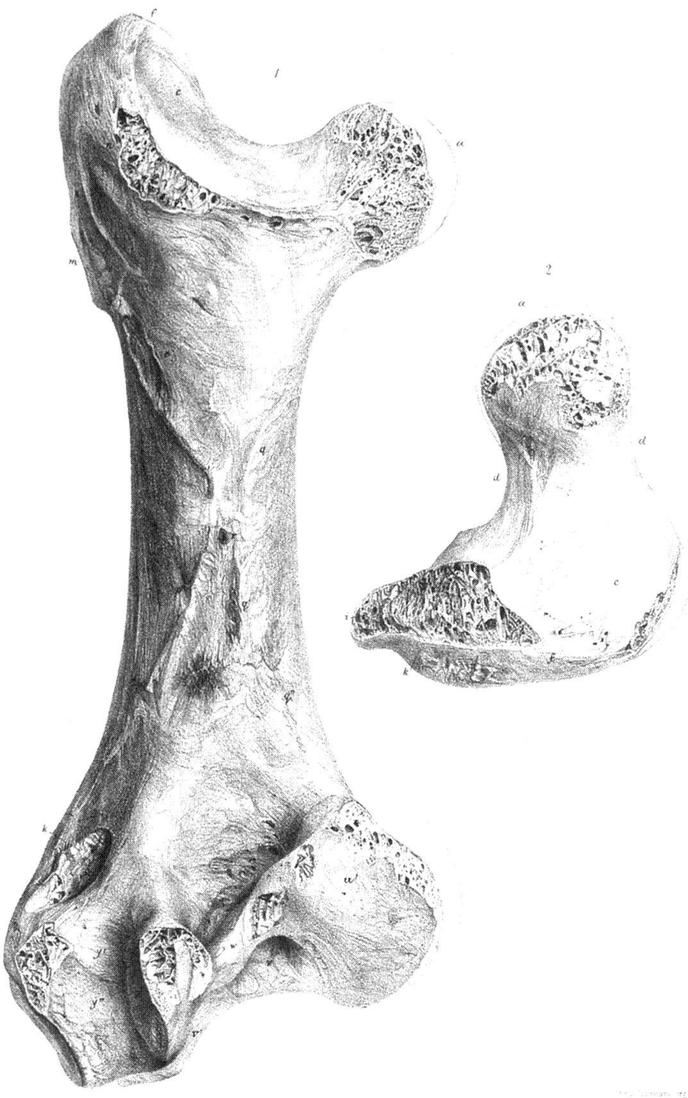

Figure 3.9. (above) Femur and (opposite) tibiotarsus of Dinornis gravis, *now referred to* Euryapteryx geranoides. *From Owen (1879: pls. 41, 42).*

> allied to the living struthious bird of New Zealand, the Apteryx. Some of these bones are referable to the *D. otidiformis,* or Bustard-like Moa, of Professor Owen's second memoir: I believe the name of *Aptornis* will be assigned to this genus. (Mantell 1848a: 233)

To compound the confusion, the skull of *Aptornis* was then attributed to the moa *Dinornis casuarinus* because it was thought to be far too large to be associated with the smaller *Aptornis* material.

There followed a further 22 memoirs and their associated notices in the *Proceedings of the Zoological Society of London* published between 1851 and 1886 on the fossil birds of New Zealand. Memoirs 1–22 were collected into the book *Memoirs on the Extinct Wingless Birds*

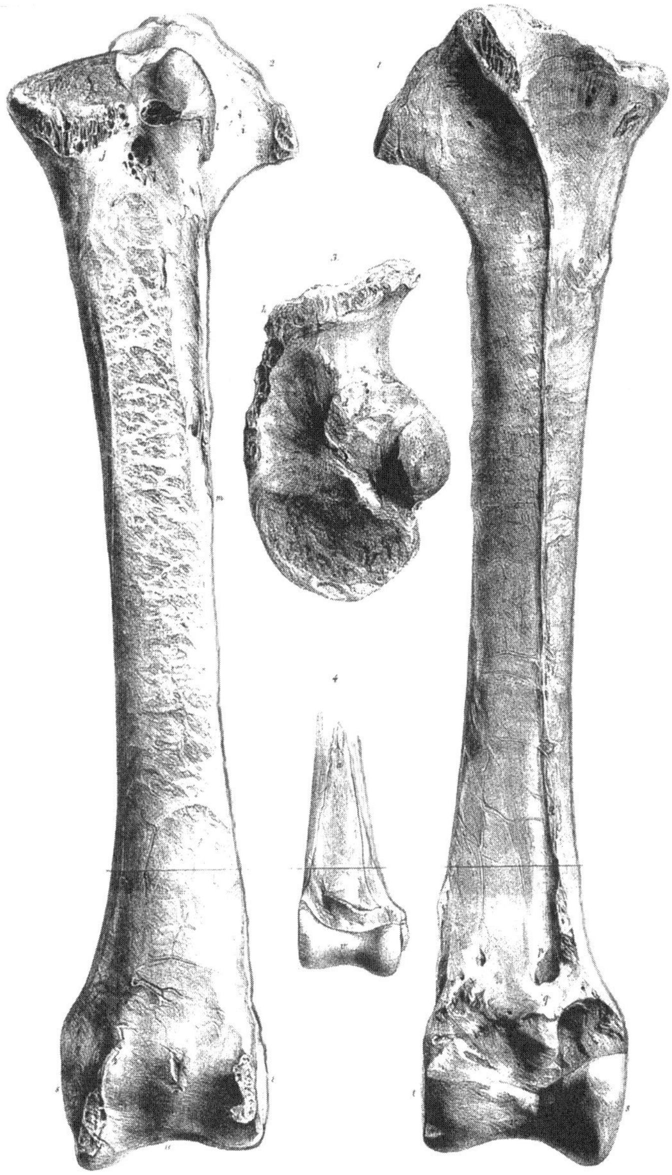

of New Zealand (Owen 1879a), which reprinted all the superb lithographs. Six memoirs (X, XV, XVI, XVII, XX, and XXIII) dealt mainly with birds other than moa—the *Apteryx, Cnemiornis, Aptornis,* and *Harpagornis.* The rest provided detailed descriptions of moa morphology, including their feet, sterna, axial skeleton, crania, and other elements (see Reference List for a complete listing). Periodically, new species were introduced to accommodate new material, e.g., *Dinornis rheides* (Owen 1851a). After earlier establishing *Palapteryx,* Owen then struggled to identify defining cranial characters (Owen 1852). He

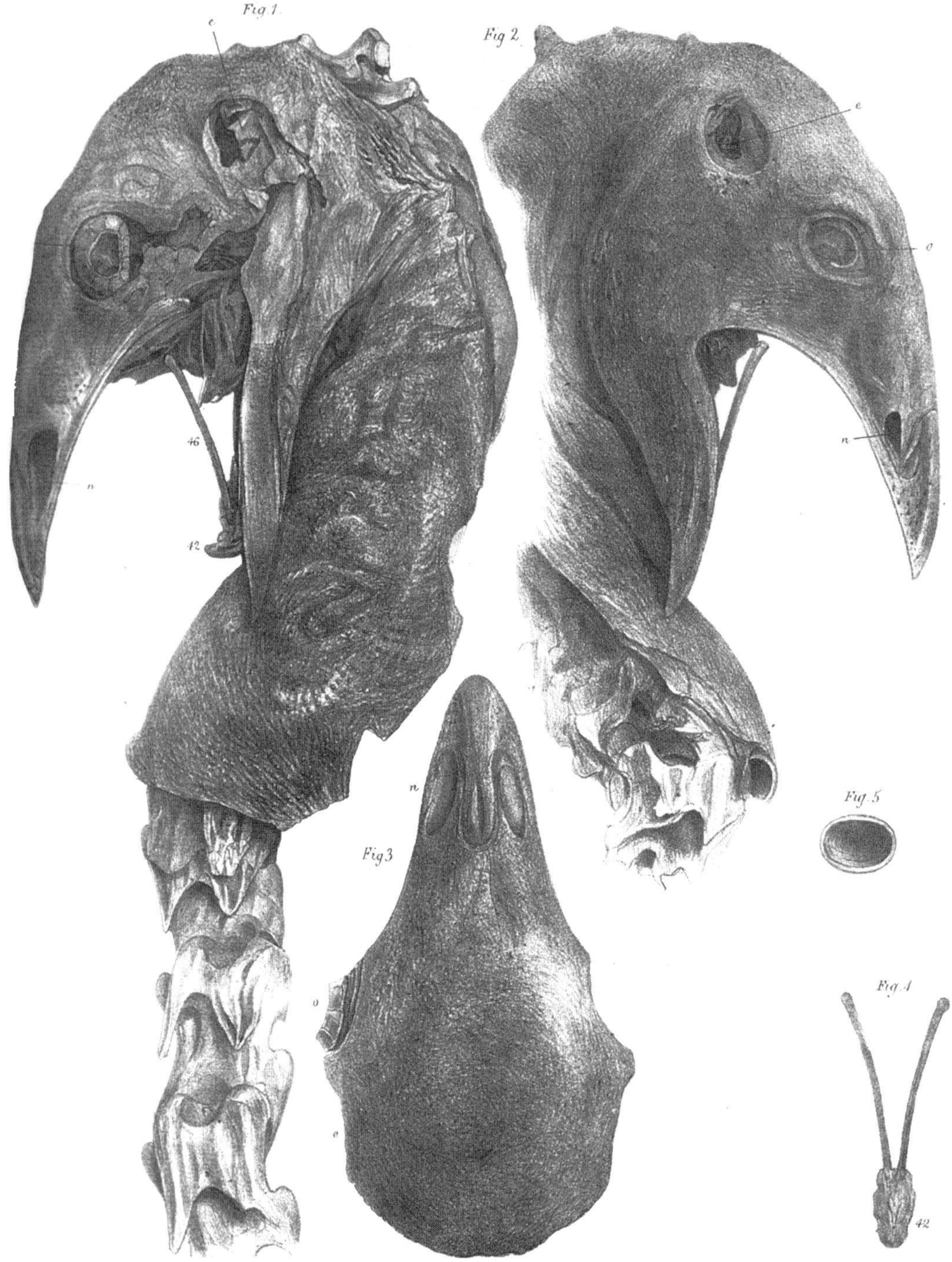
Fig.1.
e
n
46
42
Fig 2.
e
o
n
Fig.3
n
o
o
Fig.5
Fig.4
42

distinguished *Dinornis* by its greater extent and curvature of the beak from the skull of *Palapteryx* (for which the examples described were emeids as now known), which indicates that by this time, he had recognized the major cranial differences between emeids and dinornithids. But *Palapteryx* had been erected initially on the basis of *Dinornis* tarsometatarsi that had a hind toe, and soon it became apparent that all species had one. This was a novel feature for large struthious birds because the extant ones (emu, ostrich, rhea, and cassowary) do not have a hind toe. The distinction was insupportable, and the name was relegated to synonymy in *Dinornis* (Owen 1855, 1856a).

Figure 3.10. (opposite page) Remarkable dried head of Megalapteryx didinus *found near Queenstown in Otago and sent to the British Museum. From Owen (1883b: pl. 59).*

Thereafter, Owen retained all species in the single genus *Dinornis* and did not concern himself with the interrelationships of the species. Owen noted disapprovingly that Reichenbach had divided the species of moa into a variety of nominal genera with no "facts or characters being assigned for that multiplication of generic names" (Owen 1858a: 152). To the end, Owen refused to accept these or any other generic names.

Memoir XIV was notable because in it, Owen described the cranial morphology of *D. elephantopus, D. crassus, D. rheides, D. giganteus, D. casuarinus,* and *D. gravis* (Owen 1870b). *Dinornis gravis* was described as a new species, the first where the skull and almost the entire postcranial skeleton were associated (Figs. 3.8 and 3.9). In terms of modern nomenclature, the skull Owen described as *D. elephantopus* was correctly referred, the one referred to *D. crassus* belonged to *Euryapteryx geranoides,* and those referred to *D. rheides* and *D. casuarinus* were *Emeus crassus.* Later, when he described the postcranial skeleton of *Dinornis gravis,* he seems to have forgotten that he introduced the species *gravis* in 1870. Owen (1873a: 361) began this contribution with, "It is with feelings akin to compunction that I come again before the Society with claims to place the record of a fifteenth species of *Dinornis* in a volume of its Transactions," and then he went on to describe *Dinornis gravis.* Perhaps his huge output meant that details were sometimes missed, and when he added the section on the skull of *D. gravis* to the previously written memoir on crania to complete the set of descriptions, he forgot to amend the introduction of the next memoir.

In Memoirs XXIII and XXIV, Owen described the species *Dinornis parvus* and *Dinornis didinus,* the first based on a whole skeleton and the second on a remarkable dried head, neck, and legs, with skin and claws (Owen 1883a, 1883b: Fig. 3.10).

Beyond Owen

Owen had the field of moa taxonomy to himself for the first 10 years, and then Reichenbach, in his *Avium Systema Naturale,* published in (1849–)1850(–1852), erected several genera (pp. 33–35) to accommodate Owen's species. Despite his giving no justification to any of the genera, a type species was clearly assigned to each, and so when the distinctiveness of these taxa became apparent, Reichenbach's names had priority.

Julius von Haast entered the field in 1869 with the description of *D. maximus,* which unwittingly was published a month before Owen's description of part of the same skeleton under the same name (Haast 1869). In view of the transport time for specimens from New Zealand to Great Britain in the mid-19th century, it seems that Haast had probably delayed his publication in the expectation that Owen would officially name it, but he miscalculated slightly. *D. maximus* was significant because it was only the second species named for which several parts of the same skeleton were known. Haast (1874b), after his researches into the many bones found at Glenmark Swamp in Canterbury, made the first attempt to separate the various species into genera on the basis of shared characters. He used features of the premaxilla, tarsometatarsus, sternum, pelvis, and scapulocoracoid, and he advocated the existence of 4 genera and 11 species.

One of the most productive scientists from New Zealand in the 19th century was Captain F. W. Hutton. To his wide-ranging interests he added the field of moa taxonomy in 1875, when he relied mainly on measurements to distinguish the species in the large bone assemblage recovered from a swamp at Hamilton in Otago (Hutton 1875). He noted the presence of two forms somewhat larger than those of Owen's species they most resembled. For the next 15 years, he spent little time on moa, but in the 1890s, he made the greatest contribution of any New Zealand scientist until that time. He began with a provisional attempt at classifying the species of moa into genera, mainly by use of cranial characters and the presence or absence of the scapulocoracoid. Within the genera, species were defined solely by size (Hutton 1891a).

Then, in 1892, Hutton thoroughly reviewed all the available data on moa, including their classification, origin, and extinction (Hutton 1892c). He noted that "there is considerable confusion in the characters given to the different species" (Hutton 1892c: 93) and that several genera were distinguishable—

> undoubtedly the differences in the crania of the different moas are quite sufficient to indicate the existence of several genera; the difficulty hitherto in defining these genera has been that the cranial differences did not appear to go with other characters. (Hutton 1892c: 100)

The cranial–postcranial problem was exacerbated by crania having been correctly referred to only a few taxa, notably "*D.*" *robustus,* "*D.*" *struthoides,* "*D.*" *parvus,* "*D.*" *didinus,* and "*D.*" *gravis.* He first attempted to clearly identify the types (taking the tarsometatarsi as such) and used those to identify specimens with associated leg bones and other major parts of an individual skeleton. He was able to find partial associated skeletons for 16 of the 20 species then known. Previously, only four species had been described on partial skeletons ("*D.*" *maximus,* "*D.*" *didinus,* "*D.*" *parvus,* and "*D.*" *oweni*). He overlooked "*D.*" *gravis.* By use of the skeletons, he was able to define the "correct" legs and skulls for the nominal taxa. He then "allocated" the maximum and minimum size and robustness allowed in each taxon and so recognized 26 species.

Therefore, although Hutton's genera were based on sound cranial, sternal, and pelvic characters, the species were "not distinguished by any anatomical differences in the leg bones, but merely by size" (Hutton 1892c: 101). The size boundaries were remarkably tight by modern standards. The tibiotarsus of *D. giganteus* was "allowed" a range of 86.9–91.4 cm (34.2–36 inches), and that of *D. robustus* 76.2–83.1 cm (30–32.7 inches), ranges which are about 5% and 8% of length, respectively. Modern observations of ranges for large animals show that long bones can vary by more than 20% between the smallest and largest individuals.

Nevertheless, Hutton's paper was a major step forward and set moa taxonomy on a firm base from which to work out the many problems remaining. For example, Hutton had incorrectly identified the types of some taxa, such as "*D.*" *geranoides,* and his "allowable size" ranges did not reflect normal variation in living populations. Just a year later, in 1893, Hutton refined his classification. Significantly, he noted that specimens of a given species varied slightly and consistently between sites. Further, he said,

> The difficulties involved in making a correct classification of the moas are due partly to the gradation of characters during the long interval between the earliest and latest forms known: partly to their extraordinary number, through which bones belonging to some three or four genera and ten or twelve species are usually found mixed together: and partly to collectors who either do not preserve intact individual skeletons or, what is much worse, add a bone or two to make a skeleton more complete. (Hutton 1893b: 8)

Having said this, he maintained his view of narrow size ranges for species and erected four more taxa on extreme variants (for example, *Anomalopteryx fortis* and *Euryapteryx compacta*), or differences in mean size in combination with different geographic ranges (for example, *D. strenuus*).

Meanwhile, on the other side of the world, Lydekker was working on the same problem in the British Museum (Natural History) in London. His major contribution when cataloging the moa bones in the museum (Lydekker 1891a) was to clarify the application of the rule of nomenclatural priority with respect to all moa and thus identify the types of nominal taxa. He synonymized some taxa and published in the catalog a classification of moa that differed considerably from those of Hutton (1891b, 1892c). Although Lydekker had most of the types and so could solve nomenclatural problems, the relatively small and mixed collections of bones in the British Museum did not allow an adequate understanding of specific variation. Perhaps as a result of the limited material, Lydekker founded a few species on slight mensural differences and also emended several names without justification.

In contrast, Hutton had at hand individual skeletons and large series of bones for many species. As he continued his research into moa systematics into the late 1890s, he was therefore able to build on Lydekker's nomenclature. In 1895, Hutton described the axial skeletons of the various moa and gave keys to the vertebrae, pelves, and

LEG-BONES FROM KAPUA.

Dinornis.

—	Length.	Mid-width of Shaft in Inches.														
		2·9	2·8	2·7	2·6	2·5	2·4	2·3	2·2	2·1	2·0	1·9	1·8	1·7	1·6	1·5
Tibia.	35	1*a*	..	..	..	1*a*	..	1*a*	..	..	..	..	..	..	..	..
	34	1*a*	..	2*a*	..	3*a*	..	..	..	..	..	..	..	..	..	..
	33	..	..	..	4*a*	5*a*	2*a*	..	..	..	..	..	..	..	..	..
	32	..	..	..	2*a*	3*a*	..	..	..	..	..	..	..	..	..	..
	31	..	..	..	..	5*a*	1*a*	1*b*	1*b*	1*b*	..	..	..	..	..	..
	30	..	..	..	..	..	2*b*	6*b*	..	1*b*	1*b*	..	..	..	..	..
	29	..	..	..	..	2*b*	1*b*	2*b*	..	1*b*	..	..	..	..	..	..
	28	..	..	..	..	2*b*	1*b*	..	1*b*	1*b*	1*b*	..	..	..	..	..
	27	..	..	..	..	..	..	..	1*b*	..	4*b*	1*b*	..	..	..	..
	26	..	..	..	..	..	..	..	..	..	1*b*	2*b*	..	..	..	..
	25	..	..	..	..	..	..	..	..	..	..	1*c*	1*c*	..	..	..
	24	..	..	..	..	..	..	..	..	..	1*c*	1*c*	1*c*	5*c*	..	..
	23	..	..	..	..	..	..	..	..	..	..	1*c*	7*c*	5*c*	..	..
	22	..	..	..	..	..	..	..	..	..	..	..	3*c*	2*c*	..	..
	21	..	..	..	..	..	..	..	..	..	..	..	3*c*	..	1*c*	..
Femur.	16	1*a*	1*a*	2*a*	6*a*	6*a*	6*a*	2*a*	1*a*	..	..	..	..	..	..	..
	15	..	..	..	2*b*	4*b*	6*b*	1*b*	..	..	..	..	..	..	..	..
	14	..	..	..	1*b*	4*b*	6*b*	4*b*	2*b*	1*b*	1*b*	..	..	..	..	..
	13	..	..	..	..	..	..	3*b*	..	3*b*	1*b*	1*c*	..	..	..	..
	12	..	..	..	..	..	..	..	..	..	1*c*	6*c*	5*c*	7*c*	1*c*	..
	11	..	..	..	..	..	..	..	..	..	1*c*	2*c*	6*c*	5*c*	1*c*	..
Metatarsus.	19	..	..	..	..	..	2*a*	2*a*	..	..	2*a*	..	..	..	..	..
	18	..	..	..	1*a*	..	4*a*	7*a*	5*a*	1*a*	..	..	..	..	..	..
	17	..	..	..	..	..	..	2*a*	1*a*	3*a*	..	..	..	..	..	..
	16	..	..	..	..	..	..	..	9*b*	4*b*	8*b*	..	..	..	..	..
	15	..	..	..	..	..	..	..	..	1*b*	4*b*	4*b*	2*b*	..	..	..
	14	..	..	..	..	..	..	..	..	..	1*b*	2*b*	1*b*	3*b*	..	..
	13	..	..	..	..	..	..	..	..	..	..	2*c*	1*c*	1*c*	1*c*	2*c*
	12	..	..	..	..	..	..	..	..	..	..	1*c*	3*c*	8*c*	19*c*	9*c*
	11	..	..	..	..	..	..	..	..	..	..	..	..	..	1*c*	7*c*

a. Dinornis maximus.
b. „ robustus.
c. „ torosus.

sterna. In 1896, he made a major advance in his technique for defining the species of moa. Huge collections of bones from two major sites, Kapua and Enfield, had come to hand (Hutton 1896a, 1896b), and in his words, "the present large collections—all from one place—enables me to pursue another system, and to ascertain which bones belong to different species by the method of averages" (Hutton 1896a: 630; Fig. 3.11). He plotted lengths and widths onto cluster diagrams; he had previously sorted the bones to genera by shape features. For *Dinornis,* he obtained three clusters. The method caused him to reject *D. validus, D. potens, D. strenuus, Pachyornis plena, Euryapteryx compacta,* and *P. valgus* as based on bones from within the range of other species. For the first time, he presented generic characters for the leg bones, many of which are accepted in the present taxonomy.

Hutton continued his review of moa species with an examination of the North Island taxa (Hutton 1897c). Although he discussed the application of the rules of priority to the type material, he was still unable to correctly distinguish bones of the small species of *Euryapteryx* and *Pachyornis.* The difficulty was largely a result of a lack of associated skeletons to which the relevant types, all single bones, could be compared, and partly because the small *Pachyornis* was still undescribed. Continuing his search for the identification of the real biologic limits to variation within species, which he had begun with the assessment of the Kapua and Enfield assemblages, Hutton then examined the Glenmark assemblage (Hutton 1897a). By use of the same method of generating dimensional clusters, he obtained the same species array as at Kapua but noted that mean size was slightly different. He commented that "the differences thus shown to exist between the birds from Glenmark and those from Kapua and Enfield are probably due to differences in the age of the deposits" (Hutton 1897a: 559). This major insight was Hutton's final word on the subject and completed the greatest contribution by any New Zealand worker until then. He had come to grips with the rules of nomenclatural priority, and with the large collections of material available to him, he made significant advances in associating the types with other elements of the skeletons. He was able to distinguish and define genera rather better than anyone else. Above all, he realized that large samples from single sites enabled definition of the actual variation within species. Moreover, he realized that moa species varied both geographically and over time.

Professor T. Jeffery Parker of Otago University in the southern city of Dunedin worked in conjunction with Hutton, who was by then curator at Canterbury Museum in Christchurch. He did not name any species, but he was interested in the higher systematics of the group. His approach was to associate crania with named taxa and then determine the relationships of the taxa by comparison of details of the skulls (Fig. 3.12). In his first paper (Parker 1893a), or rather in its postscript, he presented a classification very nearly the same as the presently accepted one. This marked the rectification of a long-standing confusion as to whether *Pachyornis* or the *Euryapteryx/Emeus* pair had a narrow bill (as originally defined by Owen). Owen had been right. Two years later, he published the detailed version of his study of the cranial morphology

Figure 3.11. (opposite page) An example of Hutton's analysis of measurements for Dinornis *bones from Kapua, by which he referred them to species. After Hutton (1896a).*

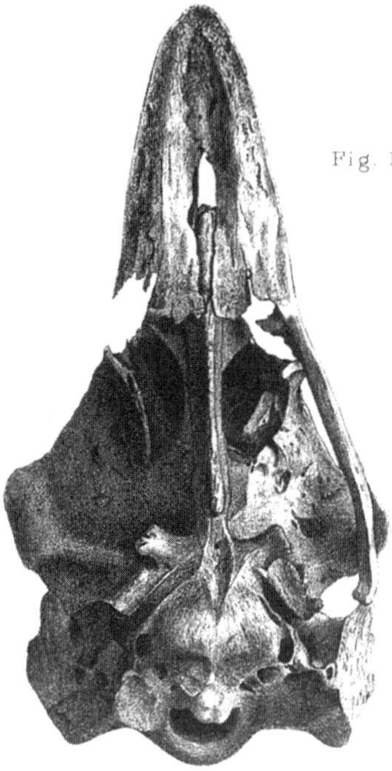

Fig. 19.

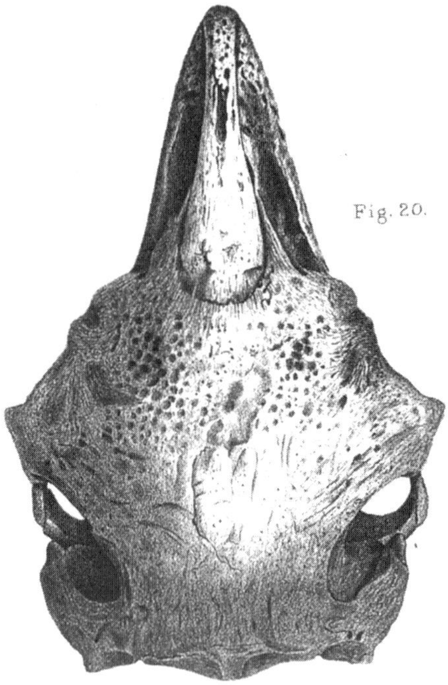

Fig. 20.

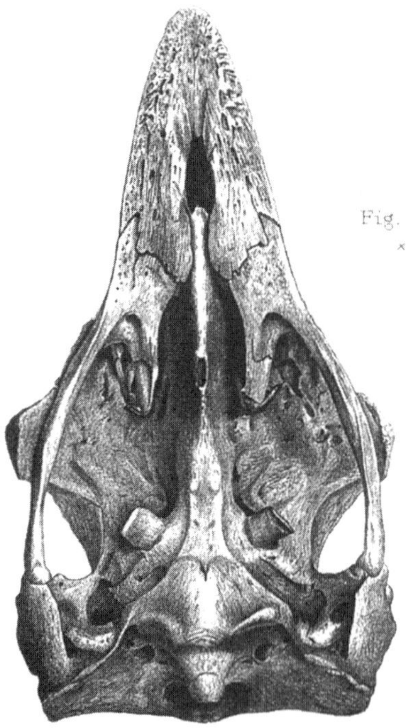

Fig. 22. × 5/6

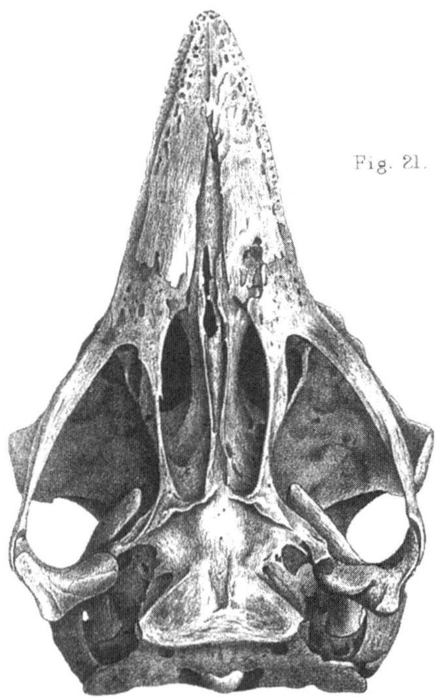

Fig. 21.

of moa (Parker 1895), compared them with other ratites, and presented a classification of the Dinornithidae.

Parker lacked securely associated skeletons for several taxa, as had all his predecessors. For example, he accepted Hutton's assertion that *D. oweni* was a synonym of *D. curtus*. On examining its skull, he found it to be similar to that of *Anomalopteryx*, so he removed *curtus* to *Anomalopteryx*. The type of *D. oweni* is a chimera of at least two species, and only the skull is of *Anomalopteryx* (Millener 1982), as correctly noted by Parker but whose observation was overlooked by Millener (1982). Another difficulty for Parker was that of the long-standing nomenclatural problems. Perhaps the most far-reaching problem was that the name *crassus* was thought to refer to bones now included in *Euryapteryx,* and so the name *Emeus* of Reichenbach was applied to broad-billed forms. All the narrow-billed forms with small temporal fossae were therefore included under the next available generic name, *Mesopteryx*. These complexities were resolved by the next person to examine the issue, but meanwhile Parker had established that moa could be divided clearly into two major groups, which he called subfamilies, and five genera. He had not examined *Megalapteryx,* but he probably would have recognized its distinctiveness if he had.

Rothschild (1907) did not introduce any new criteria for the determination of species, and he added to the nomenclatural problems by naming the forms Parker had called *Emeus* species A, B, and G as *Emeus boothi, E. haasti,* and *E. parkeri,* respectively. Rothschild also resurrected the genus *Palaeocasuarius* and placed three species in it (*P. haasti, P. elegans,* and *P. velox*), as first proposed in a wretched paper by Forbes (1892d). The genus is that of Rothschild because Forbes's (1892d) publication of the name, without designation of a type species or specimens, did not constitute a valid erection of the name. Rothschild's descriptions are based on single bones from the Maniopoto region of Otago in the South Island, said to be in the Liverpool Museum in Liverpool, England, but not found by Oliver in 1938 (Oliver 1949: 155). Dawson (1958b: 235) reported, "The types of the genus of Moa, '*Palaeocasuarius*' Forbes, have been found, and indicate that they belong to the genus *Megalapteryx* Haast 1886," but he gave no indication of why he considered them *Megalapteryx,* nor of where the specimens were. Therefore, now, rather unsatisfactorily, they are classed as *M. didinus*. Rothschild also named *Megalapteryx hamiltoni* for a femur ostensibly from Waingongoro in the North Island. Both Oliver and Archey concurred with the diagnosis that it was a *Megalapteryx,* so this seems to be so, but because the genus is unknown from the North Island, it seems likely that its stated location is wrong. *M. hamiltoni* is now included in the synonymy of *Megalapteryx didinus*.

Oliver (1930: 34) briefly reviewed the classifications of Owen, Reichenbach, Haast, Lydekker, Rothschild, and Hutton but was "unable to reconcile any of these classifications with the specimens at my disposal." He "accordingly proposed a new arrangement of the genera and species. The genera are arranged on the proportions of the leg bones." In doing this, he retreated from the advances made by Hutton and Parker but asserted that "the skull and sternum in every case where

Figure 3.12. (opposite page) Plate LX from Parker (1895) showing some of the superb skulls he worked with. Mesopteryx casuarina *(Parker's fig. 19) is now* Emeus crassus, Mesopteryx *sp. B (Parker's figs. 20, 21) is* Pachyornis australis, *and* Pachyornis elephantopus *(Parker's fig. 22) was correctly named.*

they have definitely been assigned to a species, confirm my limitations of the groups."

The 22 species that Oliver considered to be adequately supported by specimens were arranged into five "groups" in two families. Oliver pointed to the difficulties arising from Owen's treatment of cranial and postcranial material without the support of associated skeletons. He also suggested that Hutton had perpetuated the problem "until some erroneous ideas such as the supposed broad skull of *Emeus* and the supposed broad sternum of *Euryapteryx* (= *Pachyornis*) became well established" (Oliver 1930: 35). Oliver's dominant philosophy on taxonomy is dealt with below, but at this early stage in his interest in the group, he thought that the only course to follow when species were founded on bones of more than one species was to take the first described bone as the type specimen and to ignore the others.

His own ("artificial"; Oliver 1930: 35) key was based on the proportions of the leg bones. He accepted five genera: *Dinornis, Megalapteryx, Anomalopteryx, Emeus,* and *Euryapteryx. Pachyornis* was suppressed. This early effort was altered drastically 20 years later (Oliver 1949, 1955) in response to the contribution by Archey (1941).

Within three years of Oliver's first publication, Lambrecht published his monumental *Handbuch der Palaeornithologie,* and in it he summarized the current taxonomy of moa, listing the 22 species accepted by Oliver (Lambrecht 1933). He followed the taxonomy advocated by Oliver (1930) and included a comprehensive reference list as well as summary data on the biology, distribution, and geologic history of the group.

As has been made plain above, the classification of moa was hindered to a large degree by the absence of associated skeletons, especially of North Island species. So it was fortunate that in the 1930s, caves in the Lake Waikaremoana area yielded good skeletons for the Auckland Museum. Perhaps so stimulated, Archey, with others from the museum, investigated the caves in the large area of karst southwest of Hamilton city, near the famous Waitomo Glowworm Caves, and obtained many fine specimens. At about the same time, almost complete skeletons arrived from the dune sites in the far north of New Zealand, and good skeletons were found in caves in northwest Nelson. Then the excavations in the famous Pyramid Valley moa swamp began revealing its treasure trove of associated skeletons of most of the South Island species. At last, undoubted associations between skulls, sterna, pelves, and legs were possible. It remained for Archey to revisit the types, and he did.

Archey realized that in all moa species, there was a wide range in the length and proportionate width of the leg bones. He concluded that leg bone relative length and breadth were not that useful for species definition, but he observed that the relative length of the leg bones was significant. He found features of basic structure and form—skulls, sterna, and numbers of phalanges—all provided useful generic features but still separated species by size. He divided the species among the six genera now accepted, and for each, he provided an unambiguous definition. Then, starting with *Anomalopteryx didiformis,* he provided a complete synonymy and described all parts of the skeleton. All the other species were then compared with *Anomalopteryx*. He described 20 species in total, two of them new. Notably, the small North Island *Pachyornis mappini* was finally recognized and distinguished from *Euryapteryx curtus* and *Euryapteryx geranoides,* with which it had long been confused. *Emeus* and *Euryapteryx* were separated clearly and unambiguously, although he retained pairs of species where now only one is accepted. What was essentially the modern classification had been achieved. The taxa defined on a few bones (*M. benhami, P. oweni,* and *P. pygmaeus*) would not stand reexamination.

Then Oliver reentered the scene espousing a philosophy of naming all variations: "in dealing with fossil species we should not hesitate to give specific names to forms that differ only slightly from one another because the probability is that had we knowledge of the external characters we would have to admit a greater number of species" (Oliver 1949: 164). Like Archey, he defined genera on characters of the bill, cranium, vertebrae, sternum, pelvis, and phalangeal formulae, but he reverted to ranges of measurements as the primary means of defining species. "Normal" variation was not discussed, nor what this might mean for moa. He erected new species for forms that were larger or smaller than what he considered was within the normal-size range for the well-established species. This was a reversal to the practice advocated by Hutton early in his dealings with moa (e.g., 1893) but which Hutton abandoned later.

Oliver erected the new genus *Zelornis* for forms of *Euryapteryx,* which had a steeper culmen—but he gave no other distinguishing characters. In fact, he stated that the rest of the bones seemed to be indistinguishable from those of similar-size *Euryapteryx* species. In all, Oliver accepted 29 species in 6 genera. Nowhere is there evidence that he considered age differences in the fossil assemblages as possibly contributing to variation. Neither did he recognize geographic variation, other than that between the North Island and South Island forms. Hutton's insights about specific, temporal, and geographic variation of more than 50 years before had been forgotten or dismissed. Table 3.1 outlines the convoluted nomenclatural history of moa.

An Australian Moa

The possibility of an Australian moa was first raised by De Vis (1884) when he described a proximal fragment of a left femur in *Dinornis* as *D. queenslandiae*. However, Hutton (1894) quickly rejected this, finding a cast of the bone to be more similar to emus and cassowaries, suggesting that it was the ancestor of these genera. Oliver (1949), apparently unaware of Hutton's assessment, examined and figured the bone and placed it within *Pachyornis* as *P. queenslandiae*. Miller (1963) did not accept Oliver's referral and placed the fossil with emus as *Dromiceius* (= *Dromaius*) *queenslandiae*. Most recently, Scarlett (1969c) studied the specimen and recognized that the bone exhibited "strikingly different" taphonomic features from the other Kings Creek material, finding it similar to those from moa-hunter middens in

TABLE 3.1.
Nomenclatural history of moa.[a]

Date	Proposal	Reference	Present attribution[b]
1840	Giant struthious bird in New Zealand	Owen 1840	
July 1843	*Dinornis* Owen (*D. novaezealandiae* Ow.)	Owen 1843a	***Dinornis***
	Dinornis novaezealandiae Owen	Owen 1843a	***D. novaezealandiae* Owen**
	Megalornis Owen (name withdrawn)	Owen 1843b	*Dinornis*
March 1844	*Dinornis giganteus* Owen	Owen 1844b	***D. giganteus* Owen**
	Dinornis didiformis Owen (*nomen nudum*)	Owen 1844b	
	Dinornis struthoides Owen (*nomen nudum*)	Owen 1844b	
	Dinornis dromaeoides (*nomen nudum*)	Owen 1844b	
	Dinornis otidiformis Owen (*nomen nudum*)	Owen 1844b	
June 5, 1844	*Dinornis didiformis* Owen	Owen 1844a	***Anomalopteryx didiformis* (Owen)**
	Dinornis struthoides Owen	Owen 1844a	***D. struthoides* Owen**
	Dinornis ingens Owen	Owen 1844a	*D. novaezealandiae* Owen
	Dinornis dromaeoides Owen	Owen 1844a	*A. didiformis* (Owen)
	Dinornis otidiformis Owen	Owen 1844a	***Aptornis otidiformis* Owen**
July 1846	*Palapteryx* Owen (*D. ingens* Ow.)	Owen 1846b	*Dinornis*
	Dinornis crassus Owen	Owen 1846b	***Emeus crassus* (Owen)**
	Dinornis casuarinus Owen	Owen 1846b	*Emeus crassus* (Owen)
	Dinornis curtus Owen	Owen 1846b	***Euryapteryx curtus* (Owen)**
	Dinornis ingens var. *robustus*	Owen 1846b	*D. novaezealandiae* Owen
	Dinornis dromioides	Owen 1846b	*A. didiformis*
1846	*Dinornis gigas* Owen (spelling lapse)	Owen 1846a	*D. giganteus*
April 22, 1848	*Palapteryx geranoides* Owen	Owen 1848b	***Euryapteryx geranoides* (Owen)**
January 1, 1851	*Dinornis rheides* Owen	Owen 1851a	Indeterminate (bones described in Owen 1870a are *Emeus crassus*)
1852	*Emeus* Reichenbach (*D. crassus* Ow.)	Reichenbach [1850], p. xxx	***Emeus***
	Syornis Reichenbach (*D. casuarinus* Ow.)	Reichenbach [1850], p. xxx	*Emeus*
	Cela Reichenbach (*D. curtus* Ow.)	Reichenbach [1850], p. xxx	*Euryapteryx*
	Anomalopteryx Reichenbach (*D. didiformis* Ow.)	Reichenbach [1850], p. xxx	***Anomalopteryx***
	Movia Reichenbach (*D. ingens* Ow.)	Reichenbach [1850], p. xxx	*Dinornis*
	Moa Reichenbach (*D. giganteus* Ow.)	Reichenbach [1850], p. xxx	*Dinornis*

Date	Proposal	Reference	Present attribution[b]
April 11, 1855	*Dinornis gracilis* Owen	Owen 1855	*Dinornis struthoides* Owen
1855	*Owenia* Gray (*D. struthoides* Owen)	Gray 1855	*Dinornis*
July 30, 1856	*Dinornis elephantopus* Owen	Owen 1856b	***Pachyornis elephantopus* (Owen)**
1856	*Graya* Bonaparte (*D. dromaeoides* Owen)	Bonaparte 1856	*Anomalopteryx*
	Celeus Bonaparte (new name for *Cela*)	Bonaparte 1856	*Euryapteryx*
May 1869	*Dinornis maximus* Haast	Haast 1869	*Dinornis giganteus* Owen
June 1, 1869	*Dinornis maximus* Owen	Owen 1869b	*Dinornis giganteus* Owen
January 1870	*Dinornis gravis* Owen	Owen 1870b	*Euryapteryx geranoides* (Owen)
1874	*Meionornis* Haast (*D. casuarinus* Ow.)	Haast 1874b	*Emeus*
	Euryapteryx Haast (*D. gravis* Ow.)	Haast 1874b	***Euryapteryx***
July 1875	*D. crassus* var. *major* Hutton	Hutton 1875	*E. crassus* (Owen)
	D. elephantopus var. *major* Hutton	Hutton 1875	*P. elephantopus* (Owen)
1879	*Dinornis altus* Owen	Owen 1879a	*D. giganteus* Owen
	Dinornis huttonii Owen	Owen 1879a	*Emeus crassus* (Owen)
January 1883	*Dinornis parvus* Owen	Owen 1883a	*A. didiformis* (Owen)
	Dinornis didinus	Owen 1883b	***Megalapteryx didinus* (Owen)**
May 1884	*Megalapteryx* Haast (*M. hectori* Haast)	Haast 1884	***Megalapteryx***
	Megalapteryx hectori Haast	Haast 1884	*M. didinus* (Owen)
1884	*Dinornis queenslandiae* De Vis	De Vis 1884	*Pachyornis elephantopus* (see Scarlett 1969c)
December 1886	*Dinornis oweni* Haast	Haast 1886a	*A. didiformis* (Owen)
April 25, 1891	*Megalapteryx tenuipes* Lydekker	Lydekker 1891: 251	*M. didinus* (Owen)
	Dinornis struthioides Lyd. (emendation of *D. struthoides*)	Lydekker 1891: 242	*D. struthoides*
	Anomalopteryx didina Lyd. (emendation of *D. didinus*)	Lydekker 1891: 277	*M. didinus* (Owen)
	Anomalopteryx parva Lyd. (emendation of *D. parvus*)	Lydekker 1891: 278	*A. didiformis* (Owen)
	Anomalopteryx geranoides Lyd. (reassignation of species)	Lydekker 1891: 288	*Euryapteryx geranoides* (Owen)
	Emeus gravipes Lyd. (new name for legs of *D. gravis*)	Lydekker 1891: 297	*Euryapteryx geranoides* (Owen)
	Pachyornis Lyd. (*D. elephantopus* Ow.)	Lydekker 1891: 316	***Pachyornis***
	Pachyornis immanus Lyd.	Lydekker 1891: 343	*P. elephantopus* (Owen)
November 1891	*Dinornis excelsus* Hutton	Hutton 1891a	*Dinornis giganteus* Owen
	Dinornis validus Hutton	Hutton 1891a	*Dinornis giganteus* Owen
	Dinornis firmus Hutton	Hutton 1891a	*Dinornis giganteus* Owen

Date	Proposal	Reference	Present attribution[b]
	Dinornis potens Hutton	Hutton 1891a	*D. novaezealandiae* Owen
	Tylopteryx Hutton (*D. struthoides* Ow.)	Hutton 1891a	*Dinornis*
	Dinornis torosus Huttton	Hutton 1891a	*D. novaezealandiae* Owen
	Palapteryx plenus Hutton	Hutton 1891a	*Dinornis struthoides* Owen
	Mesopteryx Hutton (*D. huttonii* Ow.)	Hutton 1891a	*Emeus*
	Euryapteryx ponderosus Hutton	Hutton 1891a	*P. elephantopus* (Owen)
	Euryapteryx pygmaeus Hutton	Hutton 1891a	*E. geranoides* (Owen)
April 1892	*Pachyornis rothschildi* Lydekker	Lydekker 1892a	?*P. elephantopus* (Owen) or *Euryapteryx geranoides*
May 1892	*Anomalopteryx antiquus* Hutton	Hutton 1892c	*A. didiformis* (Owen)
1892	*Palaeocasuarius* Forbes (*nomen nudum*)	Forbes 1892d	*Megalapteryx*
May 1893	*Dinornis strenuus* Hutton	Hutton 1893b	*D. struthoides* Owen
	Anomalopteryx fortis Hutton	Hutton 1893b	*A. didiformis* (Owen)
	Euryapteryx compacta Hutton	Hutton 1893b	*Euryapteryx geranoides* (Owen)
	Pachyornis inhabilis Hutton	Hutton 1893b	*P. elephantopus* (Owen)
	Pachyornis valgus Hutton	Hutton 1893b	*P. elephantopus* (Owen)
	Anomalopteryx antiqua Hutton (spelling)	Hutton 1893b	*A. didiformis* (Owen)
June 1897	*Anomalornis* Hutton (vice *Anomalopteryx*)	Hutton 1897c	*Anomalopteryx*
	Euryapteryx exilis Hutton	Hutton 1897c	*Euryapteryx curtus* (Owen)
November 12, 1907	*Megalapteryx hamiltoni* Rothschild	Rothschild 1907: 197	*M. didinus* (Owen)
	Emeus boothi Rothschild	Rothschild 1907: 210	*E. geranoides* (Owen)
	Emeus haasti Rothschild	Rothschild 1907: 210	*E. geranoides* (Owen)
	Emeus parkeri Rothschild	Rothschild 1907: 211	*E. geranoides* (Owen)
	Palaeocasuarius Rothschild (*P. haasti* Rothschild)	Rothschild 1907: 219	*Megalapteryx*
	Palaeocasuarius haasti Rothschild	Rothschild 1907: 220	*M. didinus* (Owen)
	Palaeocasuarius velox Rothschild	Rothschild 1907: 220	*M. didinus* (Owen)
	Palaeocasuarius elegans Rothschild	Rothschild 1907: 220	*M. didinus* (Owen)
August 15, 1927	*Dinornis expunctus* Archey	Archey 1927	*E. geranoides* (Owen)
1930	*Euryapteryx kuranui* Oliver	Oliver 1930	*E. geranoides* (Owen)
1941	*Megalapteryx benhami* Archey	Archey 1941	*M. didinus* (Owen)
	Pachyornis mappini Archey	Archey 1941	***Pachyornis mappini* Archey**
August 1949	*Zelornis* Oliver (*Euryapteryx exilis* Hutton)	Oliver 1949	*Euryapteryx*

Date	Proposal	Reference	Present attribution[b]
	Pachyornis (*Mauiornis*) Oliver (*P. septentrionalis* Oliver)	Oliver 1949: 59	*Pachyornis*
	Pachyornis (*Pounamua*) Oliver (*P. murihiku* Oliver)	Oliver 1949: 59	*Pachyornis*
	Pachyornis (*Mauiornis*) *septentrionalis* Oliver	Oliver 1949: 61	*Pachyornis mappini* Archey
	Pachyornis (*Pounamua*) *murihiku* Oliver	Oliver 1949: 67	*Pachyornis elephantopus* (Owen)
	Pachyornis australis Oliver	Oliver 1949: 70	***Pachyornis australis* Oliver**
	Euryapteryx tane Oliver	Oliver 1949: 105	*Euryapteryx curtus* (Owen)
	Dinornis gazella Oliver	Oliver 1949: 166	*Dinornis struthoides* Owen
	Dinornis hercules Oliver	Oliver 1949: 174	*Dinornis novaezealandiae* Owen
1963	*Dromiceius queenslandiae*	Miller 1963	*Pachyornis elephantopus*

[a] Data from Archey (1941), Brodkorb (1963), Oliver (1949), and Scarlett (1969c). Present attribution follows Turbott (1990a) and is in accordance with the recommendations of Cracraft (1976c), Millener (1982), and Worthy (1987a, 1988c, 1988d, 1989a, 1989b, 1992b, 1994a).

[b] Entries in boldface type indicate the first valid use of that name.

New Zealand. It even had greenstick fractures typical of midden bones. Scarlett made a detailed comparison of the bone with various taxa and found it indistinguishable from those of *Pachyornis elephantopus*. How the bone came to be in the Queensland Museum is unknown, but midden bones were distributed widely out of New Zealand by Haast.

Modern Biologic Approaches to the Classification

In 1976, Cracraft attempted a classification of moa that reflected the biologic structure of the species rather than the typologic thinking of much of the previous century. For the first time in the study of moa, multivariate analytic techniques were employed, specifically principal component and discriminant function canonical analyses. The analyses were based on the length and proximal, shaft, and distal widths of the leg bones (Cracraft 1976a). Cracraft acknowledged several sources of variation that could contribute to overall variability. He thought intrapopulational variation was probably significant, but he did not measure sufficiently large samples from single sites to determine its extent. Because other ratites are significantly sexually dimorphic, he considered that at least some moa were likely to be as well. Geographic variation, particularly that between the North Island and South Island, was likely to be significant, but he believed samples did not exist that would allow him to examine intraisland variation. Temporal variation was excluded because he thought all deposits were no older than 7000–8000 years old, and most were much younger. Of necessity, Cracraft assumed that bones in the museum collections he examined had been correctly identified to species and genera.

Cracraft accepted just 13 species. He found that the species of moa were more variable in size than most other birds and attributed much of this to geographic variation between islands. His small sample sizes may have contributed to this perception. He synonymized species for a variety of reasons. Morphologically more or less identical species previously separated by size were combined if their pooled coefficients of variation[3] were within what he considered an acceptable range (i.e., <10 to 12). By use of this criterion, *Anomalopteryx didiformis* assimilated *A. parvus*; *Megalapteryx didinus* and *M. hectori* were taken to be one species; and *Pachyornis australis* was submerged in *P. elephantopus* because he could find no consistent differences between them. Species previously erected for size outliers—for example, *Pachyornis murihiku* and *Dinornis hercules*—were included in the species nearest in size to them.

Four pairs of taxa were each considered to represent two "populations" that differed only in size (these were interpreted as representing sexual dimorphism). The species pairs that were lumped together were *Pachyornis mappini* and *P. septentrionalis, Euryapteryx geranoides* and *E. gravis, Euryapteryx curtus* and *E. exilis,* and *Emeus crassus* and *E. huttoni.* Cracraft did not recognize *Zelornis* Oliver because he thought that features of the bill alone were insufficient for generic distinction. Moreover, the forms labeled *Zelornis* seemed to be one extreme of a range continuous with that seen in *Euryapteryx.*

Cracraft's study marked a radical departure from nearly all that had appeared before. Although not mentioned by Cracraft (1976a), Hutton in the 1890s had also sought the "biological structure," or what could be termed more appropriately as the sum of the inter- and intrapopulation variation of a species. Moreover, Hutton had large series of specimens from several sites, and most of these collections still exist.

Cracraft was at a disadvantage because many of the specimens he examined were "identified," according to their length and width measurements, by whomever had cataloged them. This was the way most of the 19th-century workers identified moa leg bones, and Oliver also relied heavily on measurements. Only Hutton and Archey relied more on shape characters before measurements. When T. H. Worthy began studying moa in the 1980s, he quickly realized what Hutton and Archey had noted: that different genera overlapped broadly in length and width measurements for the three main leg bones but that there were many characters by which species could be identified.

Examination of bones in collections also showed that even in the early 1980s, many specimens either had identifications attributed to them in the 19th century or newer names that were invoked by subsequent synonymy of the first name. There had been few revisions of

3. Coefficients of variation are statistics that are used here to measure size variation in populations independently of size. Thus, although the standard deviation increases as numeric values increase, making objective comparison difficult, the coefficient of variation enables comparison of the variation between different-size specimens to be made much more easily. It is obtained by the following formula: standard deviation/mean × 100.

identifications of specimens. True, Archey had worked on the Auckland Museum collection, but his influence had not extended to the other major collections. Similarly, Oliver had worked on the Dominion Museum (now Museum of New Zealand) collection, but he had identified few, if any, bones held in the Auckland, Canterbury, Otago, or Southland Museums, or in any of the other collections elsewhere. Curators at most museums, notably Scarlett at Canterbury Museum, followed Archey (1941) and Oliver (1949), but generally they worked only within a single institution. As a result, there were many errors in identification and inconsistent usage of nomenclature. In particular, Haast's "skeletons" from Glenmark, which were all composites and which had been distributed worldwide, often consisted of several species and up to three genera each. Species as obviously different as dinornithids and emeids were mixed in all collections throughout New Zealand. Any studies on moa, such as their ecology, distribution, or variability in form, had to wait for the species to be characterized, because although Archey and Cracraft had more or less established a robust classification, many specimens in collections were still misidentified.

Beginning in the 1980s, the work started by Hutton and Archey was continued, and a set of characters that separated first the genera and then the species was developed. Of necessity, the typologic approach was adopted because the characters of the types had to be described, then representative part skeletons referred to them by use of those characters, so that the characters of the whole skeleton could be determined.

Applying the philosophy of first identifying taxa by characters and then analyzing size variation within them paid immediate dividends. The site at Tangatupura in the North Island contained mainly two taxa, referable to the genera *Pachyornis* and *Euryapteryx*. The large samples present (n = 32 to 88 for various elements), allowed the clear distinction of two discrete populations of animals in each taxon. Multivariate analyses confirmed that these differed only in size (Worthy 1987a). Here was clear support for the hypothesis of sexual dimorphism advocated by Cracraft. But the analysis also clearly showed, as stated by Archey (1941) and Cracraft (1976b), that the measurements on their own did not distinguish the species; form did. Rather, the measurements showed the variation within the species.

It is unlikely that a series of measurements on moa leg bones would ever be able to separate the species that can certainly and unambiguously be determined by a set of a priori characters, against the expectations of Caughley (1977). This is because of the great convergence in gross shape exhibited by these birds, which results in size and relative robustness being the overriding measurable characters, as determined by Cracraft (1976b). The information summarized in Figure 3.13 is particularly instructive. Here, tibiotarsi of a species of moa, an anatid, a gruid, and an accipitrid are compared at the same scale. Traditional measurements of length, and proximal, shaft, and distal widths do not distinguish them. We would not expect them to, so why should we expect the same measurements to distinguish moa taxa?

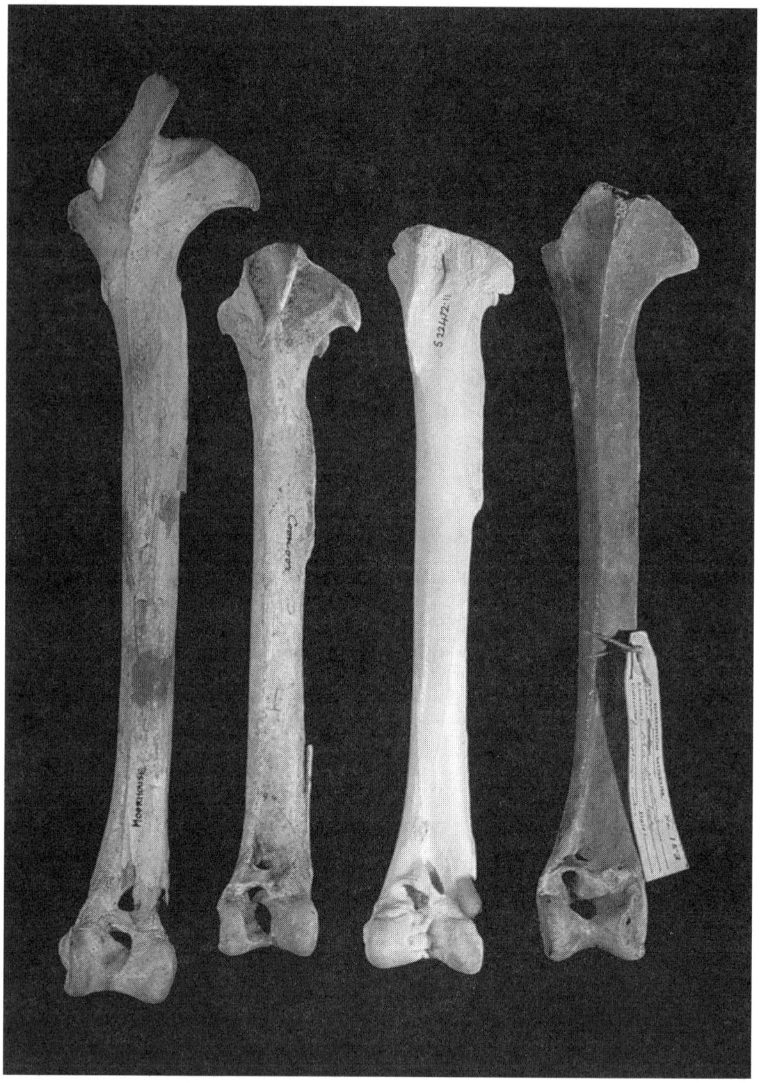

Figure 3.13. Tibiotarsi of four extinct and unrelated birds showing superficial similarity. Left to right: Cnemiornis calcitrans *Museum of New Zealand specimen S38941,* Aptornis otidiformis *MNZ S8037,* Harpagornis moorei *MNZ S22472, and the small moa* Euryapteryx curtus *MNZ S153. Lengths and widths do not discriminate these taxa. Photograph by R. Morris.*

However, the application of radiocarbon dating techniques was beginning to have a profound influence on the interpretation of New Zealand's fossil faunas. Not all were Late Holocene, as previously inferred by Fleming (1962a) and adopted by Cracraft (1976a). Dates on odd bones from caves showed that several were 10,000–20,000 years old, which placed them firmly in the last ice age. These results meant that it was possible that there could be significant temporal variation in size within species, in the same manner as has been demonstrated in large animals elsewhere. The brown bear *Ursus arctos* of Europe was at least 16% bigger in the Pleistocene than in the Holocene, with the reduction considered to have taken place over the course of about 8000 years, and similar variation was found in the cave bear *Ursus spelaeus* (Kurtén 1955). *Macropus titan* is the eastern grey kangaroo (*M. giganteus*) of the Ice Age—a temporal form only (Murray 1991).

So the Tangatupuran moa fauna was doubly significant because it dated from the coldest period of the last ice age, about 18,000 years ago. Comparisons with the late Holocene populations of *Euryapteryx curtus* showed that this species had become somewhat smaller during the intervening millennia. *Pachyornis mappini* was more remarkable. In either the Ice Age or in Holocene collections, two discrete size classes that did not differ at all in other characters were present. But there had been a marked decrease in size over time, such that three size groupings could be discerned. However, the smallest was only present in the Holocene, and the largest was present only in the Ice Age. By the Holocene, the largest form of the Ice Age had shrunk to the size of the small Ice Age form, and the small Ice Age form became a midget by the Late Holocene. The midget was the material formerly known as *P. septentrionalis,* and the middle-size form (which could be either sex, depending on its geologic age) was the form called *Pachyornis mappini.* The largest form, always rare, had been referred to *Pachyornis elephantopus,* which is a South Island species.

In the next few years, first a key was produced to identify leg bones (Worthy 1988c); then various genera were examined in detail, starting with *Megalapteryx* (Worthy 1988d), then *Pachyornis* and the emeiins (Worthy 1989b, 1992b). These analyses, and those of large samples that allowed the limits of specific variation within single sites to be better characterized, allowed some refinement of moa taxonomy (Worthy 1989a, 1989b). Most recently, the genus *Dinornis,* which encompasses a huge range in size but is morphologically the most conservative (invariate, one could say) was studied; only three taxa were distinguished (Worthy 1994a).

By the end of the 1980s, several major deposits had been shown by radiocarbon dating to be of Pleistocene age. Dates for others showed that they had formed at various times in the Holocene. Temporal variation in many genera had been shown to be real. Geographic variation had too, but it was not confined to interisland variation; there was also variation within the main islands, just as Hutton had said. The existence of substantial variation within islands ought not to have been too surprising because each covers several degrees of latitude, and large mountain ranges create markedly different climatic zones in different areas of each island.

Now that distinguishing morphologic characters have been identified for all taxa, analyses of shape by use of measurements can be made without circularity of logic. If measurements of the groups of specimens separated into species are found to be significantly different, we have an independent verification of the ability of the first approach to discern pattern in the data or of the ability to discern the presence of useful phylogenetic data. The large samples available (often more than 30 per site, and hundreds overall) can provide data adequate for the development of allometric relationships, as well as on variation in size and robustness in time and space for each species. Recent collecting has increased the numbers of individual skeletons for most species to levels that will allow more detailed multivariate analyses. Geologic dating is now available for many sites, so temporal variation can be factored into the analyses, a luxury not available to earlier workers.

Present Moa Systematics: Consensus after 150 Years

The first 150 years of taxonomic and systematic studies resulted in a proliferation of names (at least 64 species and 20 genera) being applied to moa. Rothschild, who really did little more than list taxa, accepted 37 species. The two more recent monographs of Archey (1941) and Oliver (1949) accepted 20 and 29 species, respectively. The following arrangement of species is that advocated in the New Zealand bird checklist (Turbott 1990a). Good synonymies and taxonomic notes are present in Archey (1941) and Brodkorb (1963). The 11 species accepted here follow the more recent systematic revisions advocated by Cracraft (1976a), Millener (1982), and Worthy (1987a, 1988c, 1989a, 1989b, 1992b, 1994a).

Order: Dinornithiformes Gadow
Family: Emeidae (Bonaparte)
Subfamily: Anomalopteryginae (Oliver)
Anomalopteryx didiformis (Owen)
Megalapteryx didinus (Owen)
Pachyornis elephantopus (Owen)
Pachyornis mappini Archey
Pachyornis australis Oliver
Subfamily: Emeinae Bonaparte
Euryapteryx geranoides (Owen)
Euryapteryx curtus (Owen)
Emeus crassus (Owen)
Family: Dinornithidae Bonaparte
Dinornis struthoides Owen
Dinornis novaezealandiae Owen
Dinornis giganteus Owen

4 • Moa Anatomy and Evolution

Structure of Moa (Aves: Dinornithiformes)

The material in this chapter is technical, and nonspecialist readers may wish to proceed to Chapter 5.

The use of names to describe parts of the body and skeleton is inescapable; each is explained or identified on a figure. However, some general terms having to do with orientation are used frequently—for example, dorsal indicates the upper surface; ventral, the lower surface; proximal, the end closest to body; distal, the end farthest from body; anterior (= cranial), the side that faces forward; posterior (= caudal), the side that faces backward; medial or mesial, the side that faces the center line of body; and lateral, the side that faces out from body. The positional term for all these orientations ends in the suffix "-ad"; for example, the foramen is mesad of the ridge. Abbreviations used to denote orientation and specific parts are used in the figures and defined in the figure captions.

Moa Skull

The skull (Figs. 4.1, 4.2) of a moa, like that of all ratites, is relatively small in comparison to the size of the bird. It is much shorter than the femur, whereas in all other birds, the skull is longer than the femur. Even the gigantic dromornithids had a skull much longer than the femur (Murray and Megirian 1998). The cranium or braincase is generally wider than high, a feature most accentuated in *Dinornis,* and flattened or only slightly domed on top. Bony crests, such as in cassowaries, are absent. The bill is relatively small compared with that of most birds: in all species, it is shorter than the cranium. The shape of the bill is variable. In some species, it is pointed in dorsal, or bird's-eye, view; in others, it is rounded. Some have a wide bill with a low-angled rostrum or culmen in lateral view; others have a short bill with a steep culmen.

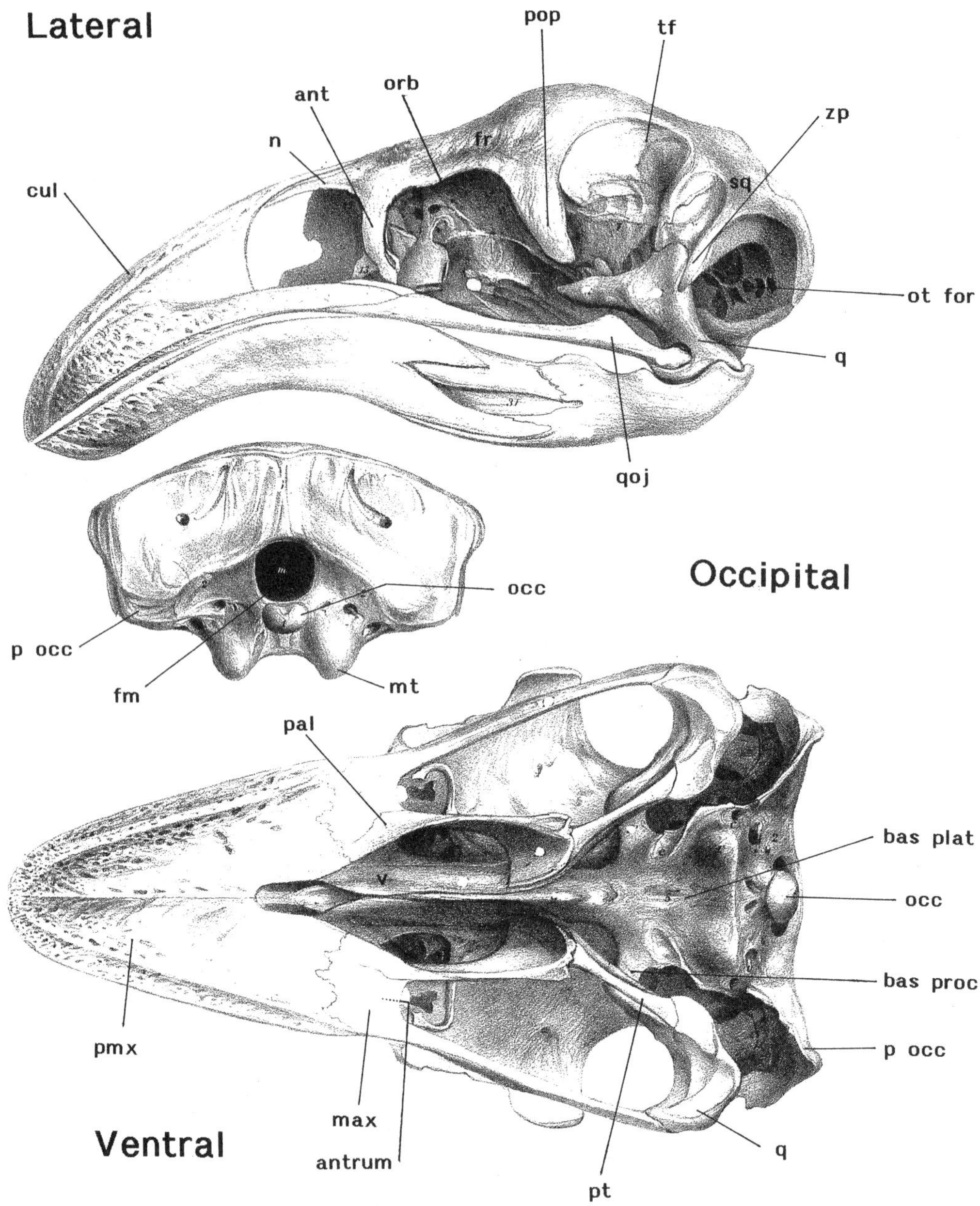

Figure 4.1. Dinornis *skull, after Owen (1879: pl. 82), in left lateral (caudal or occipital) and ventral views. ant = antorbital plate; bas proc = basipterygoid processes; bas plat = basitemporal platform; cul = culmen; fm = foramen magnum; fr = frontal; max = maxilla; mt = mamillar tuberosities; n = nasal; occ = occipital condyle; orb = orbit; pmx = premaxilla; ot for = otic foramina; pal = palatine; p. occ = paroccipital process; pt = pterygoid; q = quadrate; qoj = quadratojugal; sq = squamosal; tf = temporal fossa; v = vomer; zp = zygomatic process. Skull length = ~20 cm.*

All moa have relatively small orbits, which are smaller in proportion to the cranium than the orbits of any other ratite, that allow good lateral but more limited frontal vision. The orbits are bound anteriorly by a descending preorbital plate or antorbital. This is a complex structure formed by a separate lachrymal bone that fuses to the nasal medially and abuts the posterior margin of the frontal. To the lachrymals are fused the modified anterolateral walls of the olfactory chamber. Lying on the front of the lachrymal, a thin descending process of the nasal in most species terminates in a marked expansion that articulates with the maxilla. In one moa, *Megalapteryx,* the maxillary process of the nasal is much reduced and is a sliver that is barely connected dorsally to the nasal, and it lies medial to the lachrymal. In all other ratites, apart from tinamous, the maxilla is separated from the nasal by a wide gap.

Behind the orbit in adult skulls, a well-developed postorbital process curves down toward the quadratojugal. Analysis of juvenile skulls has shown that this is not a separate bone but simply an outgrowth of the frontal that enlarges during the chick's growth. The postorbital process bounds the front side of the deep temporal fossa housing the mandibular retractor musculature. The size and extent of this fossa varies between species in parallel with the strength of the mandible—and hence the bite. The temporal fossa is bounded posteriorly by the squamosal (with a descending zygomatic process), which houses the otic foramina and the articulation of the quadrate.

The caudal end of the skull of all moa is unusual because it is here that the occipital condyle and foramen magnum are found; they are not ventrally directed, as in other birds. The vertebral column articulates with the occipital condyle, which is a prominent, either rounded or dorsoventrally compressed peduncle below the foramen magnum, or the entry of the spinal cord to the brain. The plane of the foramen magnum is usually at 90–100° to the axis of the skull, and it may be either rounded or slightly higher than wide. The position of the foramen magnum and the occipital condyle means that in moa, the vertebral column attaches to the rear of the skull in a snakelike or lizardlike fashion. It is not directed ventrally or ventrocaudally as in most other birds.

Ventrally, the base of the skull is formed by the basicranium (literally, the base of the cranium) (posteriorly), and the palate (anteriorly), which is a complex structure involving the premaxilla, vomer, palatines, maxillae, pterygoids, and quadrates. The mandible articulates with the quadrates, as in all other birds.

The adult skull is a complex structure formed from many discrete elements. But the underlying complexity is even greater. As in other birds, in moa, many parts of the adult skull are themselves actually composed of several separate bones that fuse together during ontogeny, the process of development of an individual. The difference with moa is that the individual elements remain separate for a longer time and ossify over a longer time than in volant birds, in which the fusion and ossification occur so quickly that the separate elements are only discernible in the embryo and the hatchling. The biggest part of the skull, the cranium, which houses the brain, eyes, olfactory apparatus, and

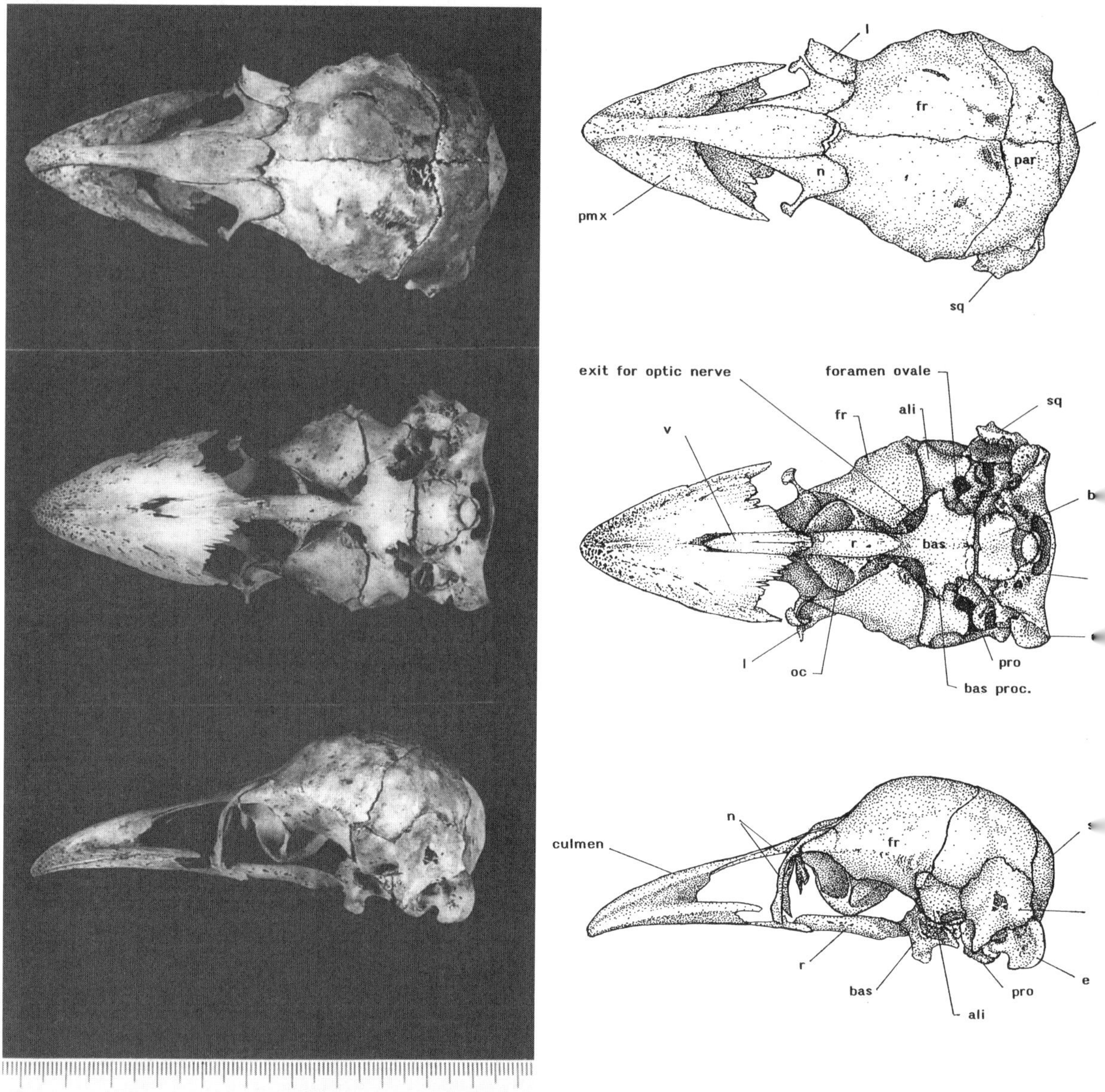

l
fr
par
n
pmx
sq
exit for optic nerve
foramen ovale
ali
fr
sq
v
r
bas
l
oc
pro
bas proc.
culmen
n
fr
r
bas
ali
pro
e

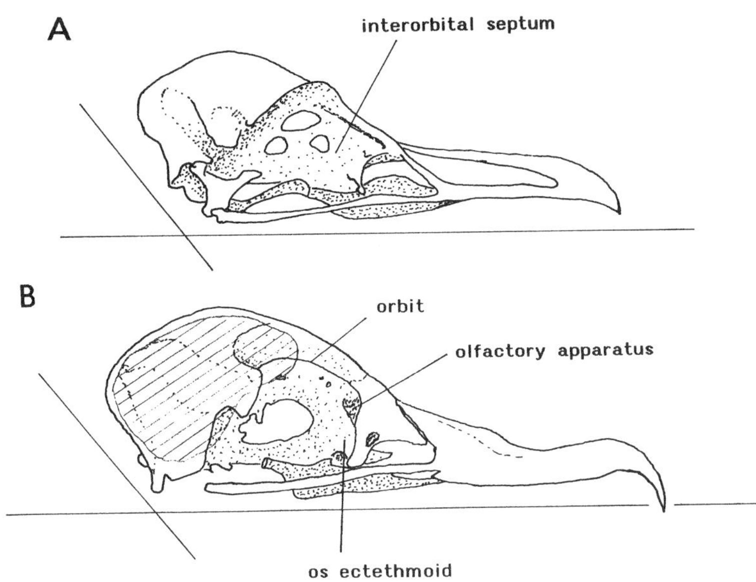

Figure 4.2. (opposite page) Juvenile skull showing constituent bones. (A) Photographs of a juvenile Dinornis *(MNZ S37876). (B) Drawings interpreting structure.*
ali = alisphenoid;
bas = basisphenoid;
bas proc = basipterygoid process;
bo = basioccipital;
exo = exoccipital;
fr = frontal;
l = lachrymal;
mt = mamillar tuberosities;
n = nasal;
oc = olfactory chamber;
pmx = premaxilla;
par = parietal; pro = prootic;
r = basisphenoid rostrum;
sq = squamosal;
so = supraoccipital;
v = vomer.
Scale in millimeters.
Drawings by J. Ogier.
Photographs by J. Palmer.

*Figure 4.3. Two neognath skulls to show the markedly different configuration in their skulls from those of moa. (A) Gull (*Larus *sp.). (B) Petrel (*Procellaria *sp.). The lines show that the occipital plane is at an acute angle to the skull. The brain (hatched area in B) lies behind the orbits, which are separated by a perforate interorbital septum (os mesethmoidale). An anterior extension of the septum (os ectethmoid) forms the rear wall of the olfactory chamber, which is small in most birds and widely separated from the brain.*

mandibular musculature, is formed by the fusion of many discrete elements, none of which is discernible in the adult.

The moa cranium differs substantially from those of all other birds because it lacks an interorbital septum or plate (Fig. 4.3). In other birds, the braincase is aligned at an angle to the axis of the skull to meet the ventrocaudal position of the vertebral column and is separated widely from the olfactory apparatus, which is usually poorly developed. The unique form of moa is achieved by rotating the brain box forward so it is parallel to the axis of the skull and enlarging the olfactory apparatus massively so that they abut each other, as can clearly be seen in the juvenile skull (Fig. 4.2).

Cranium

In the hatchling moa, all constituent bones of the cranium were unfused. Earlier researchers, including Parker, Archey, and Oliver, did not have access to immature skulls in which the fusion of overlying elements had not already occurred, so they could not observe where the symphyses were, and thus the following descriptions differ somewhat from their interpretations.

The top and front of the skull is formed by the paired nasals, which, posteriorly, have an overlapping articulation with the frontals (Fig. 4.2). The premaxilla articulates with and overlies the paired nasals but is never fused to them. The fusion of the nasals to the frontals—and to each other—is one of the last events in the ontogeny of the cranium. Descending laterally from the nasals are the lachrymals, which provide protection to the ethmoid, or olfactory capsule.

The roof of the skull is formed mainly by the paired frontals and parietals, which abut in the midline and extend down the sides and under the cranial vault. The frontals may connect ventrally immediately posterior to the olfactory capsule. Ventrolaterally, the alisphenoid overlies the junction of the parietal and the frontal. The optic nerve exits the braincase in front of the alisphenoid and between the frontal and basisphenoid. At the posterior border of the alisphenoid and below the rear end of the parietal is the prootic. The trigeminal foramen (foramen ovale) is found at the junction of the prootic and the alisphenoid. The exoccipital bounds the prootic caudally. The squamosal overlies the top half of the prootic and articulates with the lower posterior corner of the parietal, the upper posterior corner of the alisphenoid, and the exoccipital as well.

Posteriorly, the cranium is completed by the four occipital bones. Centrally, above the foramen magnum, the supraoccipital is bounded on either side by an exoccipital. The exoccipitals extend medially under the foramen magnum and contribute the upper third of the occipital condyle. The exoccipital and supraoccipitals seem to be the first bones to fuse together in the animal's ontogeny. Ventrally, the occipitum is completed by the basioccipital, which contributes the medioventral third of the condyle. Some authors describe the broad descending lateral process of the exoccipital as the paroccipital.

The base of the cranium, or the basitemporal platform, is formed mainly from two bones. The basioccipital contributes about the rear third and articulates anteriorly with the basisphenoid. The basisphenoid also articulates with the prootic posterodorsally, and the alisphenoid does so dorsally in the region above the basipterygoid processes. The anterior part of the basisphenoid is produced as an elongate rostrum that lies below and articulates with the olfactory capsule. Even in very juvenile skulls, the anterior part of the basitemporal platform has a pair of anteriorly directed, ventrolateral projections (basipterygoid processes), which articulate with the pterygoids. In the adult cranium, the basioccipital part of the basitemporal platform may develop a pair of tuberosities or prominences (mamillar tuberosities), which may be poorly developed or, in some taxa, well developed.

During maturation, the anterior end of the rostrum develops triangular processes that diverge on the horizontal plane in front of the antorbital. These form the base of what becomes the anterior section of the olfactory chamber, which houses the turbinals. Parker (1895) interpreted them to be the base of the mesethmoid. In emeids, the rostrum is in the same plane as the basitemporal platform, but it forms an obtuse angle of about 150° with it in dinornithids (Parker 1895: 385).

The olfactory chamber houses the olfactory capsule (ethmoid), a complex bony structure completely enclosed by other bones of the skull, which seems to be a feature unique to moa. In the most juvenile specimens, it is a single structure with no visible sutures and is a separate entity unconnected to the rest of the cranium. None of the early workers had the juvenile specimens that are now available at their disposal, and so it was interpreted as the fusion of a complex of several bones. Parker (1895) called the ventromedial part that articulates with the rostrum the mesethmoid. The posterolateral parts of the capsule are what Parker calls the inferior aliethmoid, and the roof, which lies under the nasals, is the superior aliethmoid. The posterior part of the olfactory capsule does not articulate with the frontals, as Parker suggested, but rather is enveloped by them; fusion with them occurs later.

The capsule has a medial septum (entire in dinornithids and partial in emeids) that divides it into two lateral halves. Each was connected posteriorly to the brain by a foramen. There is no separate presphenoid or orbitosphenoid; rather, there are ankylosed (fused) regions at the base of the frontals and to the rear of the olfactory capsule. The uniquely enlarged form of the olfactory chambers in moa is achieved by its posterior expansion, which thus occupies the area where the interorbital septum of other birds lies. The interorbital septum is also absent because the axis of the braincase is about horizontal, which rotates it forward between the eyes from the rear.

In skulls of adult individuals, a thin antorbital plate diverges from the inner wall of the orbit and connects to the descending lachrymal, forming the anterior wall of the orbit. It is thus the modified anterolateral wall of the olfactory chamber. A spiral ingrowth from the posterior end of the antorbital becomes the turbinal (Fig. 4.4). The number of spirals varies from 1 to 3 between taxa. The ventral side of the antorbital is delicate and curves forward at an angle of 45–60° to the medial line. It does not connect to the rostrum in emeids, as seen for *Euryap-*

Figure 4.4. Anterior view of a Dinornis *skull to illustrate the well-developed turbinals in the nasal cavity. The deeply grooved dorsal surface of the mandible can clearly be seen. The cranium is about 100 mm wide. Photograph of unidentified specimen by Augustus Hamilton, Museum of New Zealand Te Papa Tongarewa, B4974.*

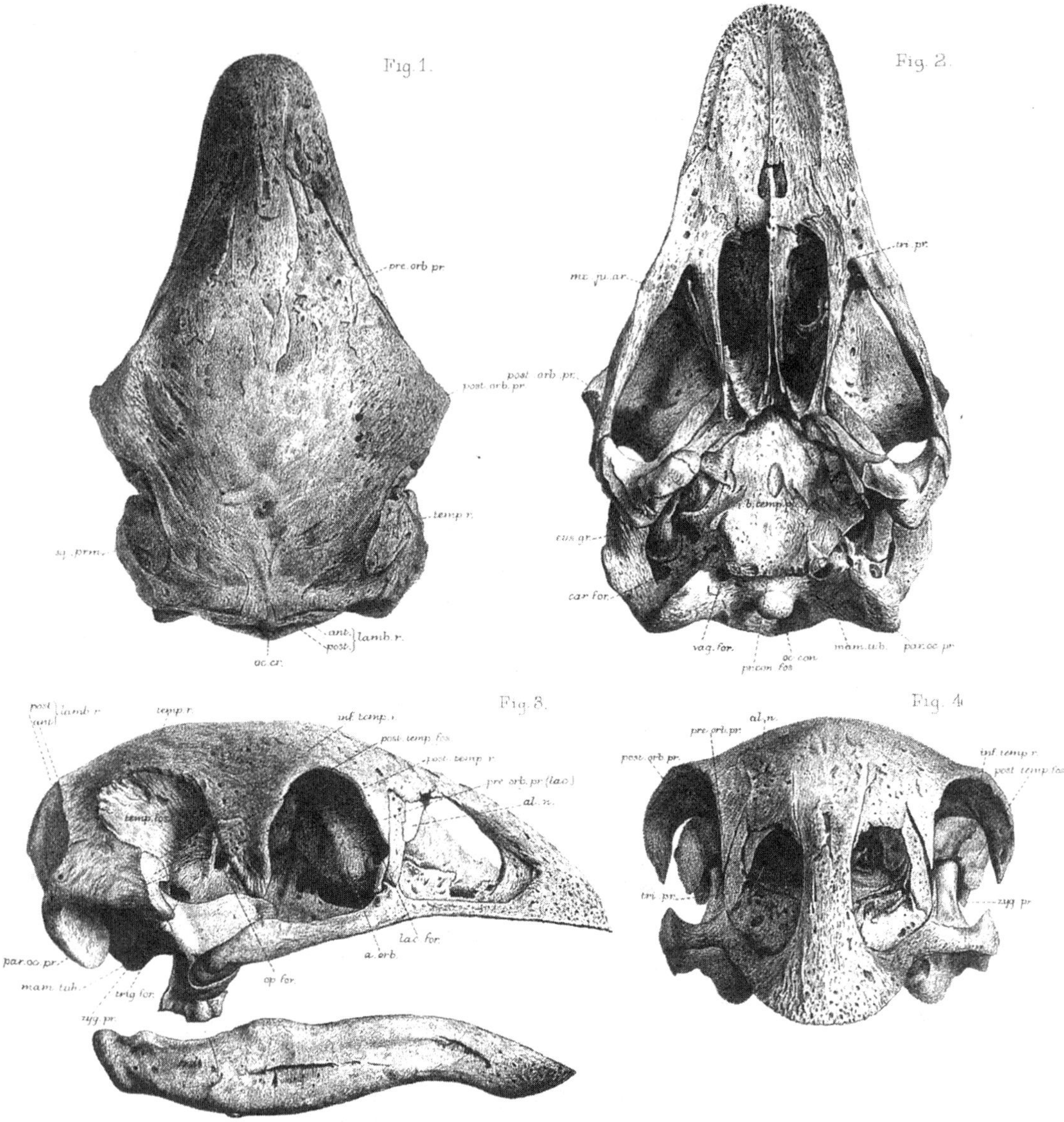

Figure 4.5. Skull of an emein moa Euryapteryx geranoides *illustrating the broadly rounded bill (after Parker 1895: pl. 56). The cranium is about 80 mm wide (cf. dinornithid skulls shown in Fig. 4.1). Abbreviations for Parker's annotations: a. orb. = antorbital plate; al. n. = alinasal or turbinal; b. temp. pl. = basitemporal platform; car. for. = carotid foramen; eus. gr. = groove for eustachian tube; inf. temp. r. = inferior temporal ridge; lac. for. = lachrymal foramen; lamb. r. = lambdoidal ridge; mam. tub. = mamillar tuberosity; mx. ju. ar. = maxillojugal arch; oc. con. = occipital condyle; oc. cr. = occipital crest; op. for. = optic foramen; par. oc. pr. = paroccipital process; post orb. pr. = postorbital process; post temp. foss. = posttemporal fossa; pr. con. fos. = precondylar fossa; pre. orb. pr. = preorbital process; sq. prm. = squamosal prominence; temp. fos. = temporal fossa; temp. r. = temporal ridge; tri. pr. = triangular process of rostrum; trig. for. = trigeminal foramen; vag. for. = vagus foramen; zyg. pr. = zygomatic process. The antorbital plate ("ant" in Fig. 4.1) is shown here as "a. orb."*

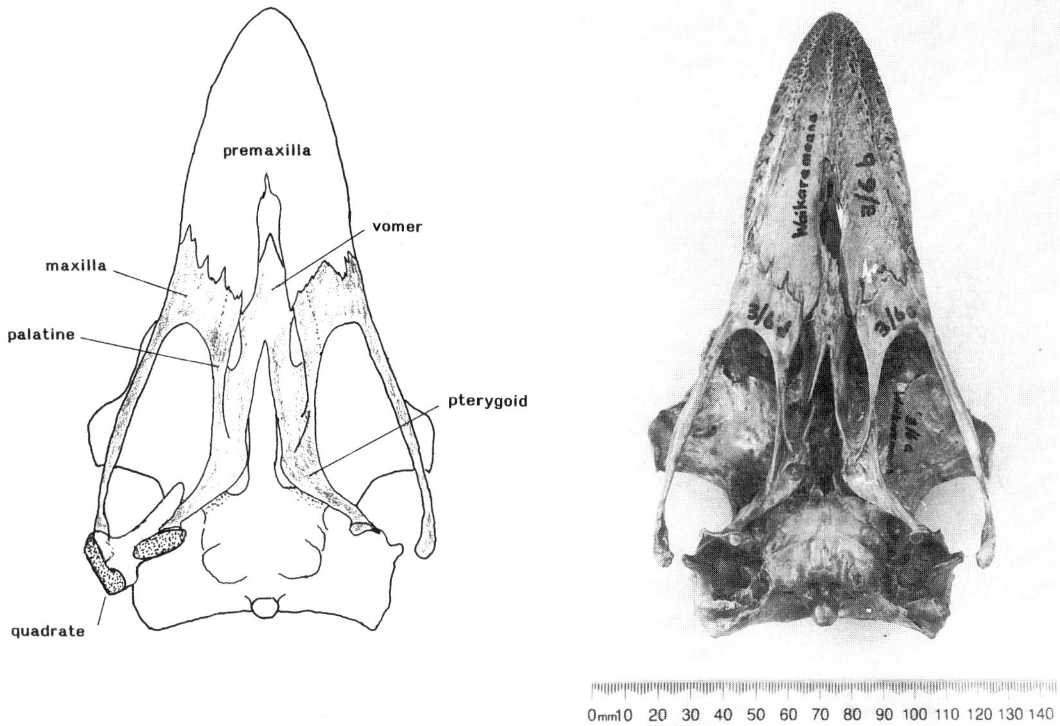

Figure 4.6. Ventral view of Dinornis *skull showing the arrangement of the bones of the palate. The palate structure varies little among moa species. All adults have a fused pterygoid–palatine–vomer, with the vomer in turn fused to the premaxilla anteriorly and the palatine to the maxilla. Posteriorly, the pterygoid articulates medially with the basipterygoid process and sleeves the ventromedial part of the orbital process of the quadrate, thus preventing a quadrate–basipterygoid process junction.*

teryx geranoides in Figure 3.8 (postchoanal bar, labeled "p.bar" in the figure) or Figure 4.5, where it is labeled "a. orb." It forms a stout bar connected to the rostrum at right angles in dinornithids.

Palate

The palate is a complex of several bones that articulate with the cranium, quadrates, and premaxilla to form the roof of the mouth (Fig. 4.6). The structure of the palates of moa is pivotal in grouping them with other ratites as paleognathous birds (e.g., Bock 1963; Cracraft 1974b; Huxley 1867). This complex structure and a few other skeletal features are not shared with other birds and so form the phylogenetic basis of uniting ratites as a monophyletic group.

The paleognathous palate is defined by the combined presence of the following features. (1) A large vomer articulating by a tight suture or fusion anteriorly with the premaxillae, the maxillopalatines, or both, and posteriorly fused to the ventral surface of the pterygoid. The palatines fuse to the ventral surface of the pterygovomer articulation. (2) The pterygoid prevents the palatine from articulating medially with the basisphenoid rostrum. (3) The pterygoid and the palatine fuse along a suture in a nonmoveable joint. (4) The articulation on the pterygoid for the basipterygoid process of the basicranium is near the pterygoid–quadrate articulation. (5) The pterygoid–quadrate articulation is complex and involves portions of the orbital process of the quadrate. All other birds (neognaths) have a short vomer, or it is lost altogether. If present, the prevomer rarely contacts the premaxilla. The palatines extend medially to the rostrum, thereby separating the vomer (if present) from the pterygoids. Most joints in the neognathous palate are of the ball-and-socket type rather than sutures. The neognath structure is lighter and allows more flexibility than the paleognathous palate.

In all moa, the two sides of the delicate vomer in older birds are fused anteriorly, form a sleeve around the rostrum, and articulate along a long suture with the ventromedial processes of the premaxilla. All the features of the paleognathous palate are found in moa. In fact, the rigidity of the palate is enhanced by the maxilla, palatines, and pterygoids being fused together in adults.

The pterygoids are short, complex bones that articulate with the quadrate, the basipterygoid processes, and the palatines. At their midpoint, they lie between and articulate with the basipterygoid process of the basicranium and the orbital process of the quadrate. Caudally, they partially sleeve the orbital process of the quadrate, and anteriorly, they fuse with the vomer medially and the palatines laterally. The palatines are delicate bones that join the ventral part of the maxilla anteriorly. The maxilla is an odd-shaped bone that in moa usually encloses a chamber (antrum) that opens posteriorly. The shape of the maxilla varies greatly among taxa. A lateral process extends posteriorly from the antrum to join the jugal, which articulates with the quadratojugal. All three bones fuse early in development and form a single bone in adults, which articulates with the quadrate posteriorly.

Mandible

The mandibles of the various moa differ in parallel with their differing premaxillae. All moa have a mandible that is somewhat sigmoid in lateral view, with the ventral surface convex in the posterior half and concave in the front section, following the down-curved tip. The degree of downward curvature is greatest in *Dinornis*. In dorsal view, there is marked variation in shape. Emeid mandibles have straight sides, or rami, that diverge to differing degrees. Some have pointed tips; others have blunt, rounded tips (Fig. 4.7). Mandibles of *Dinornis* differ from those of emeids in that the rami curve outward and converge on a blunt tip.

Figure 4.7. (opposite page) Dorsal views of moa mandibles from representative species of each genus. The cross-sectional shape of the dorsal surface of the mandibles is shown as offsets. The figures have been made to roughly equal length to better show differences in shape because species in a genus differ little and the total size range is large. c.l. = cotyla lateralis; pmm = processus mandibulae medialis.

The mandible articulates with the quadrate at a cuplike expansion at the end of the rami. Two articular surfaces lie within this for the articulation with the condyles of the quadrate. The outer edge (cotyla lateralis) has a prominent posterior process in all emeids, but in *Dinornis* (and all other ratites), there is no equivalent process, and the cotyla lateralis is rounded. Another process produced posteromedially from the cup (processus mandibulae medialis) houses a circular pneumatic foramen in all moa genera except *Emeus*.

Premaxilla

The nasal process of the premaxilla of paleognathous birds is different from that of other birds in that it is a single median bone, rather than paired (Pycraft 1900). The premaxillae of paleognaths are further

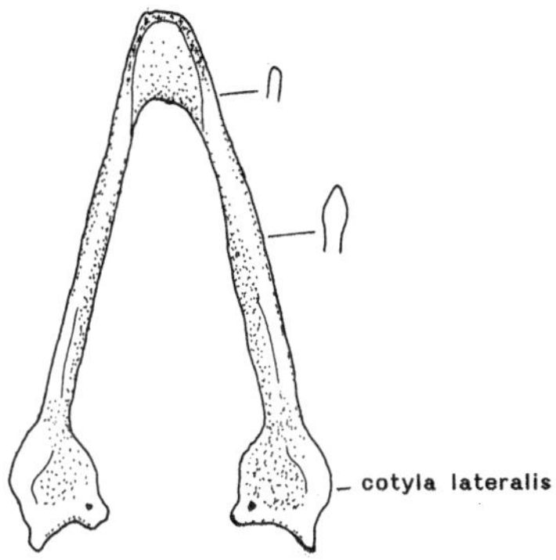

ANOMALOPTERYX DIDIFORMIS

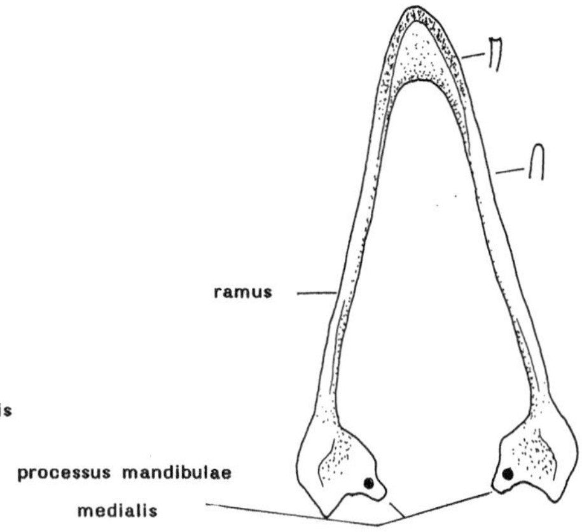

MEGALAPTERYX DIDINUS

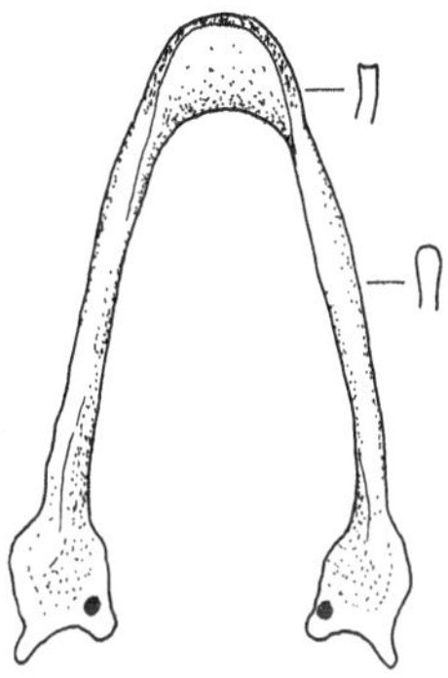

EURYAPTERYX GERANOIDES

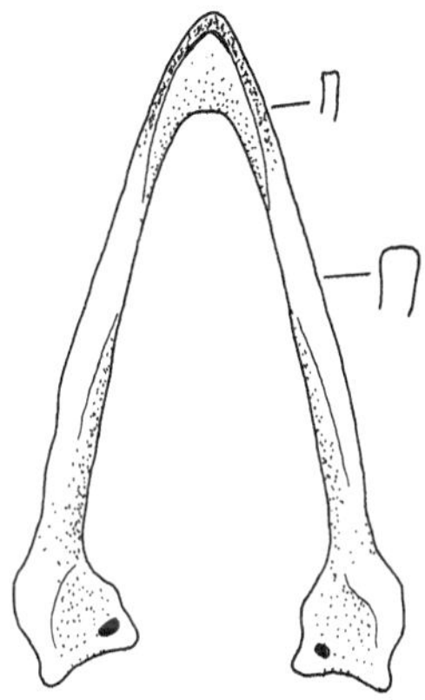

PACHYORNIS MAPPINI

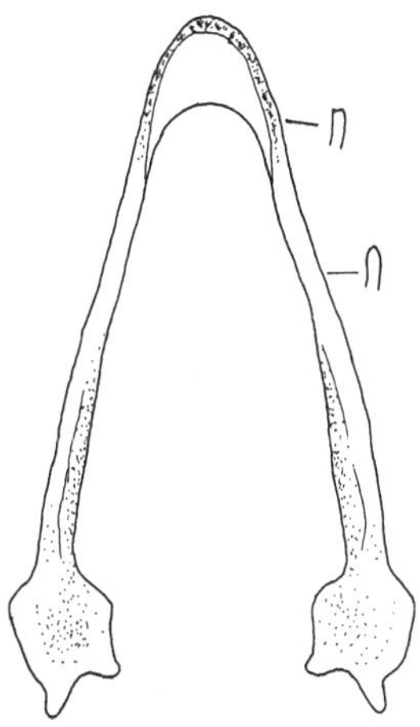

EMEUS CRASSUS

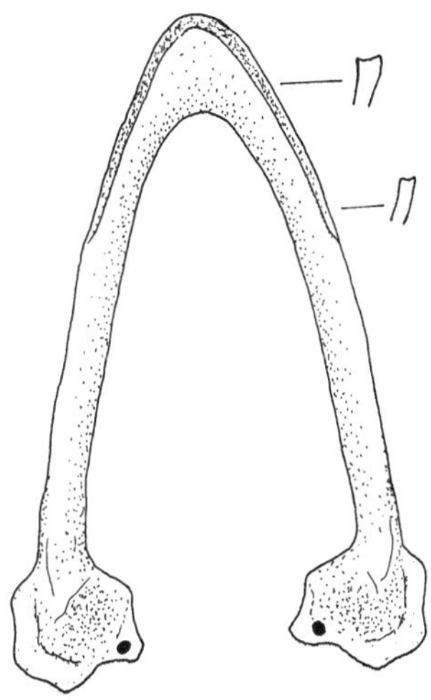

DINORNIS STRUTHOIDES

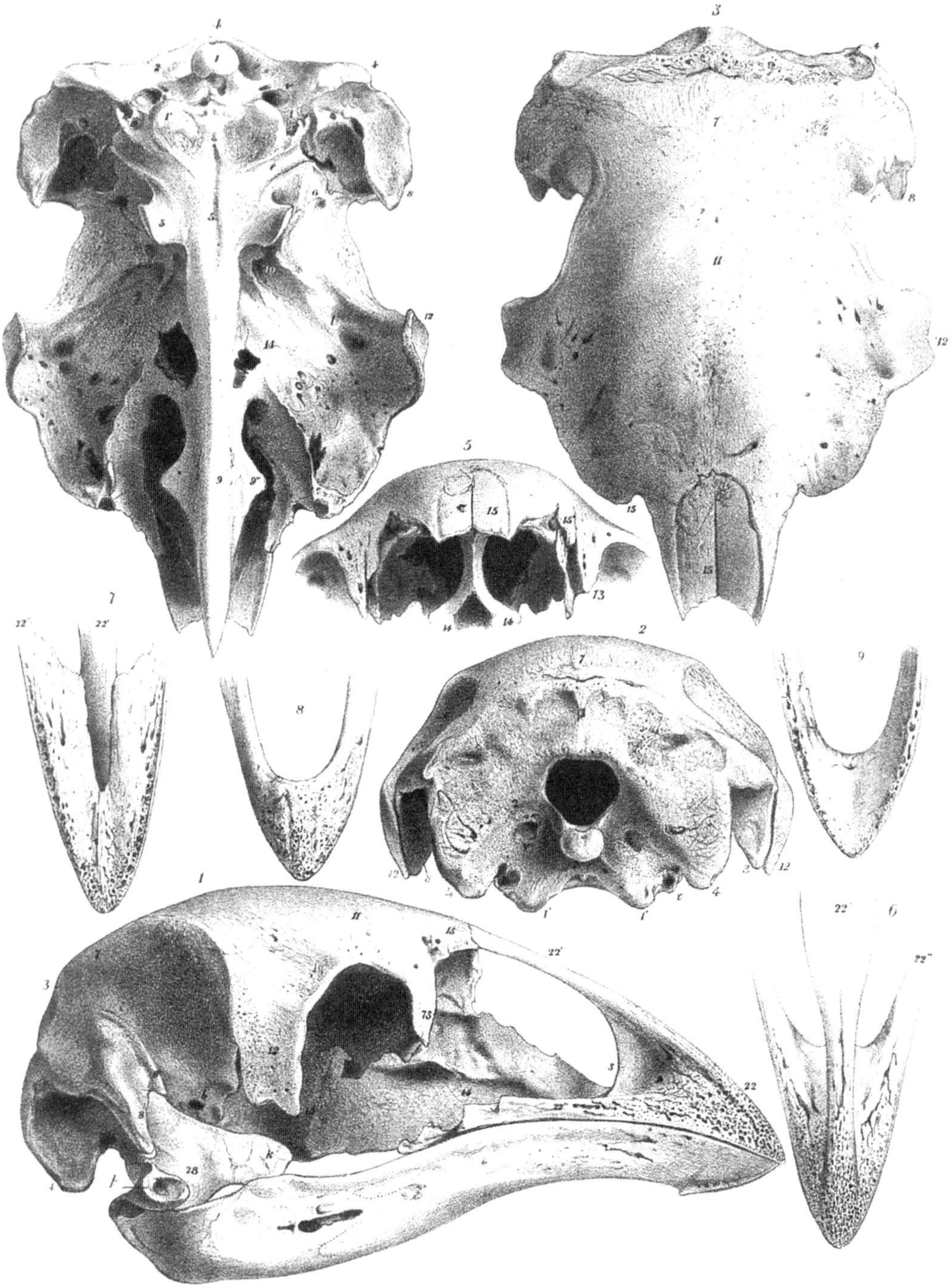

Figure 4.8. Skull of an anomalopterygin moa Pachyornis elephantopus *to show the typically pointed bills of this subfamily. The cranium is about 80 mm wide. From Owen (1879: pl. 77 [wrongly labeled plate 76]).*

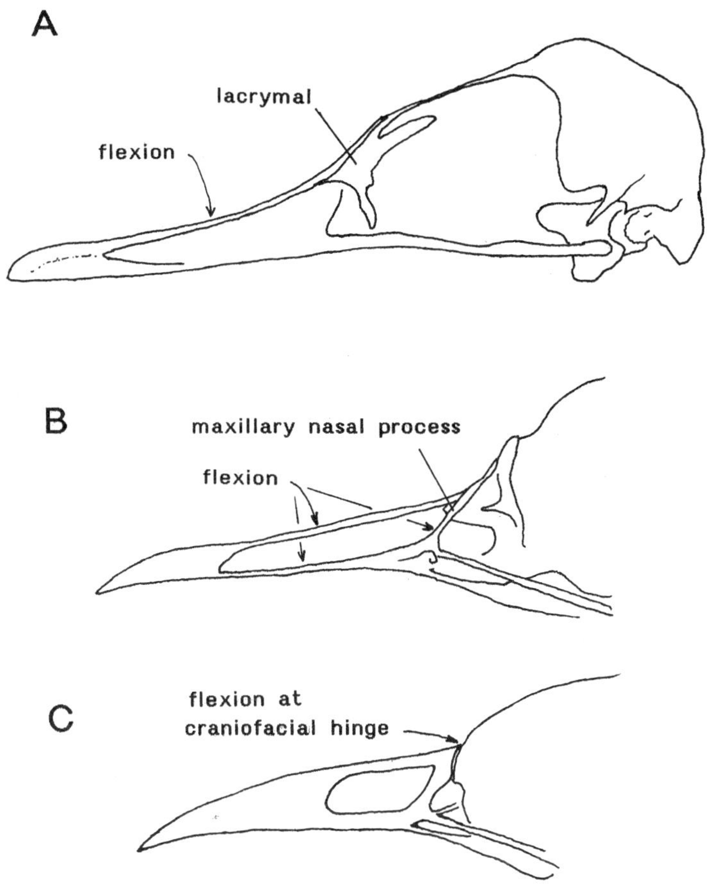

*Figure 4.9. Different types of rhynchokinesis in birds. (A) The ratite rhynchokinetic condition shown by a rhea, with flexion in the middle of the bill. The premaxilla overlaps the nasal between the lachrymals, so there is no craniofacial hinge. It lacks a maxillary nasal process, and the lachrymal does not conjoin with the quadratojugal, allowing movement along the ventral border of the skull. (B) Typical neognath rhynchokinetic condition shown by a plover (*Pluvialis squatarola*), with flexion in at least three places. (C) The prokinetic skull of a jay (*Cyanocitta stelleri*) showing flexion at the craniofacial hinge and a solid upper bill and maxillary nasal process. B and C drawn after Baumel and Witmer (1993: fig. 5.2).*

characterized by the culmen (Fig. 4.1) being a raised, broad, flattened, ridge anterior to the nasal process on the tip (rostrum maxillae), which is defined by a lateral groove. A ridge of equal width parallels this structure on the ventral surface of the mandible symphysis (rostrum mandibulae). This bony structure of the premaxilla and mandible supports a unique similarly ridged rhamphothecal structure—that is, the horny outer sheath of the bill. The upper bill is essentially composed of three parts; an elevated, flattened central ridge (the culmen) is laterally expanded anteriorly and separated from lateral segments by distinct grooves that rise from each side of the base of the culmen and pass forward above the nostrils (Parkes and Clark 1966). In moa, where this feature is perhaps the most strongly developed of the ratites, the central ridge is obvious. Parkes and Clark (1966) found that this rhamphothecal pattern was most easily discerned in juvenile birds but was uniquely discernible in all genera of ratites, and thus is a significant feature supporting monophyly of this group.

The premaxilla of all moa species are relatively shorter than in any other ratite. The sides of the premaxilla extend posteriorly in a flattened, toothed, maxillary process, which articulates with the maxilla; the latter overlaps it dorsally.

As first described above in the section on mandibles, the bills of moa species vary greatly. The premaxillae of emeids have straight sides and are higher at the caudal end of the nasal septum than wide. Of these, emeins have broadly rounded premaxillae (Fig. 4.5) and anomalopterygins pointed tips (Fig. 4.8). Dinornithids differ from the emeids in the marked dorsoventral compression of the premaxilla, which is wider than high at the nasal septum (Figs. 4.2, 4.4). The ridge from the tip to the nasal process is prominent and forms a gentle curve in lateral profile. In dorsal view, the sides are convex and converge on a blunt tip.

Most neognaths have a nasofrontal hinge that allows bending or rhynchokinesis of the upper jaw (Fig. 4.9). Bock (1963) considered that all ratites had or have a kinetic (i.e., allows bending) skull, although it was noted that some authors considered aepyornithids to have too thick an upper premaxilla to allow this. In prokinetic birds (those with a functional nasofrontal hinge), the entire upper jaw moves upward. The overlapped nasal–premaxilla junction of ratites differs from that seen in many neognaths and impedes bending at this point. Most ratites do not have the maxillary process of the nasal fused to the maxilla (forming a solid strut), as do most neognaths, which, in combination with the slender, usually elongate, premaxilla (cf. rheas, ostriches, cassowaries, and emus), allows the ratite upper jaw to bend upward about the nasal process of the premaxilla (rhynchokinesis). However, moa have a relatively short, robust premaxilla, which is usually associated with a maxillary–nasal process that is articulated with the maxilla. It thus seems unlikely that much bending was possible in the moa upper jaw.

Vertebrae and Ribs

Moa in the family Emeidae had 27 presacral vertebrae, although occasionally the first sacral vertebra (vertebra 28) was not fused to the sacrum (fused vertebrae of the pelvis). There were eight or nine pairs of ribs, and of these, the last two or three articulated with pelvic vertebrae. Therefore, vertebrae 22–27 carried ribs, and they have been called "thoracic vertebrae" by workers such as Owen and Archey. However, because the ribs associated with vertebrae 22–24 did not articulate with sternal ribs, vertebrae 22–24 are more properly considered to be cervicothoracic vertebrae. Here, thoracic vertebrae are defined as the series from the first vertebra, the rib of which articulates with a sternal rib, to the last presacral vertebra. In emeids, it was usual for vertebrae 25 and 26 to articulate with the first and second sternal ribs, respectively. The sixth pair of ribs (on vertebra 27) usually did not articulate with their sternal ribs. *Megalapteryx* differs from other emeids in that vertebrae 22 and 23 were cervicothoracic and in that there were four thoracic vertebrae (vertebrae 24–27). Only the first two pelvic vertebrae had associated ribs. *Dinornis* was different again, as it had seven rib-bearing vertebrae anterior to the pelvis (vertebrae 24–30). Of these, verte-

TABLE 4.1.
Vertebral formulae in moa compared with modern ratites.[a]

Ratite	Cervical and cervicothoracic	Thoracic vertebrae	Sacral vertebrae	Caudal vertebrae
Rhea	14 + 3	18–22	23–44	45–50
Struthio	17 + 3	21–25	26–46	47–56
Dromaius	17–18 +3	22–26	27–47	48–54
Casuarius	15–16 + 3	20–26	27–50	51–59
Apteryx	15 + 1	17–24	25–37	38–45/47
Emeidae	21 + 2	24–27	28–45	46–57
Dinornis	23 + 2	26–30	31–48	49–60

[a] Data for modern ratites are from Mivart (1874, 1877). The first thoracic vertebra is that which articulates with the first sternal rib. In the cervical and cervicothoracic column, the first value is the number of cervical vertebrae; the value after the plus sign is the number of cervicothoracic vertebrae. In the other columns, the numbers refer to the place in the vertebral series, where the numeral 1 is the first (= atlas) vertebra.

---

brae 24 and 25 had free ribs, and the remaining five were true thoracic vertebrae. The first three thoracic vertebrae (vertebrae 26–28) all articulated with sternal ribs (Table 4.1).

All moa had very short midcervical vertebrae. In most species, these vertebrae were about as long as they were wide. Such short vertebrae are in marked contrast to those in other ratites, which in a series culminating in *Rhea* have progressively elongated midcervicals (Figs. 4.10, 4.11).

Dinornithids had 30 presacral vertebrae, which differed from those of emeids in details of vertebral structure. For example, the upper cervicals (vertebrae 3–7) had a single, small dorsal spine, whereas emeids had a bifid spine on these vertebrae.

Moa had a reduced set of caudal vertebrae. Up to 11 are discernible, though the final two to three were ankylosed into one unit. None had lateral processes, and they declined quickly in size caudally. There was no true pygostyle, so as in other ratites, there were probably no well-formed tail feathers.

The thoracic ribs that articulated with the sternal ribs had uncinate processes, which were fused to the ribs in *Dinornis* but usually not in emeids.

The number and arrangements of ribs vary among other ratites. Emus have nine pairs (three cervicothoracic, 19–21; five thoracic, 22–26; and one pelvic, 27); cassowaries have eleven pairs (three cervicothoracic, 17–19; seven thoracic, 20–26; and one pelvic); rheas have eight pairs (three cervicothoracic, 15–17; five thoracic, 18–22; one pelvic); and ostriches have ten pairs (three cervicothoracic, 18–20; five thoracic, 21–25; and two pelvic).

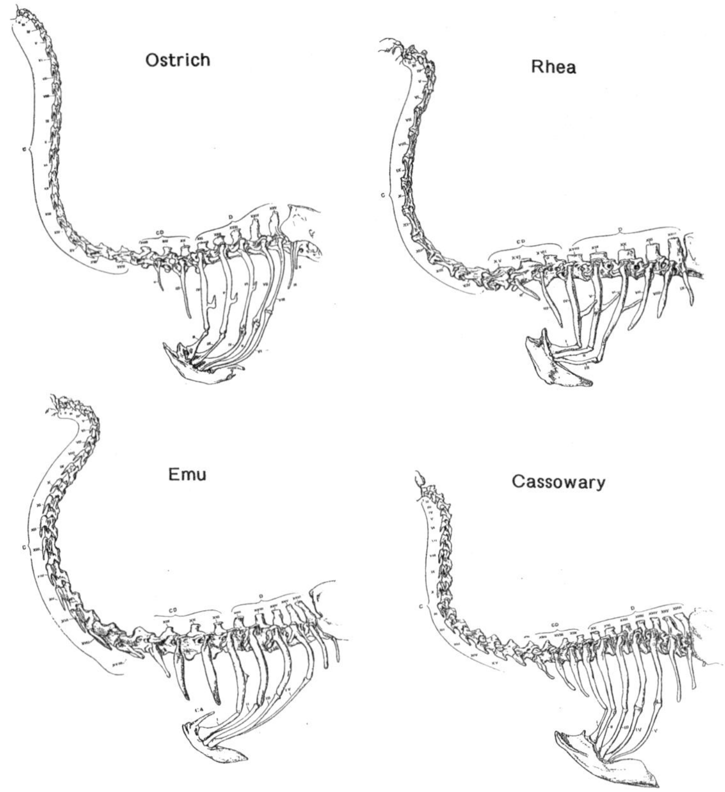

Figure 4.10. Vertebral column of large, modern ratites drawn to equal size after Mivart (1874, 1877). The length of the neck of cassowary < emu < rhea < ostrich.

Sternum

The sterna of all moa (Fig. 4.12) are typical of ratites in that there is no trace of the central keel, or carina, and that their general form is a flattish, dish-shaped plate with a raised ridge across the anterior border. Unlike rheas, elephantbirds, emus, or cassowaries, all moa had sterna with well-developed, diverging lateral processes bounding a lateral notch and a single central xiphoid process (bifid in dinornithids and single in emeids). The caudal margin of the sterna of ostriches have two weakly developed lateral processes—the external xiphoid processes of Mivart (1874)—but these are quite different from the lateral processes of moa sterna. The structure in ostriches is more like one developed from a single, subrectangular bone, such as is present in emu or rhea, by the nonossification, or erosion, of parts of the caudal margin. Kiwi sterna also have lateral processes. The sternal ribs articulated at the anterior end of each side just behind a short, blunt costal process. Well-formed pits for the coracoid articulation are present in *Dinornis, Anomalopteryx,* and *Pachyornis* spp., but these are lacking or very small in *Megalapteryx, Euryapteryx,* and *Emeus.*

Pectoral Girdle

The pectoral girdle of birds typically has three bones that support the forelimb. The strut from the sternum is the coracoid, which articulates at its cranial end with the scapula, the blade of which extends

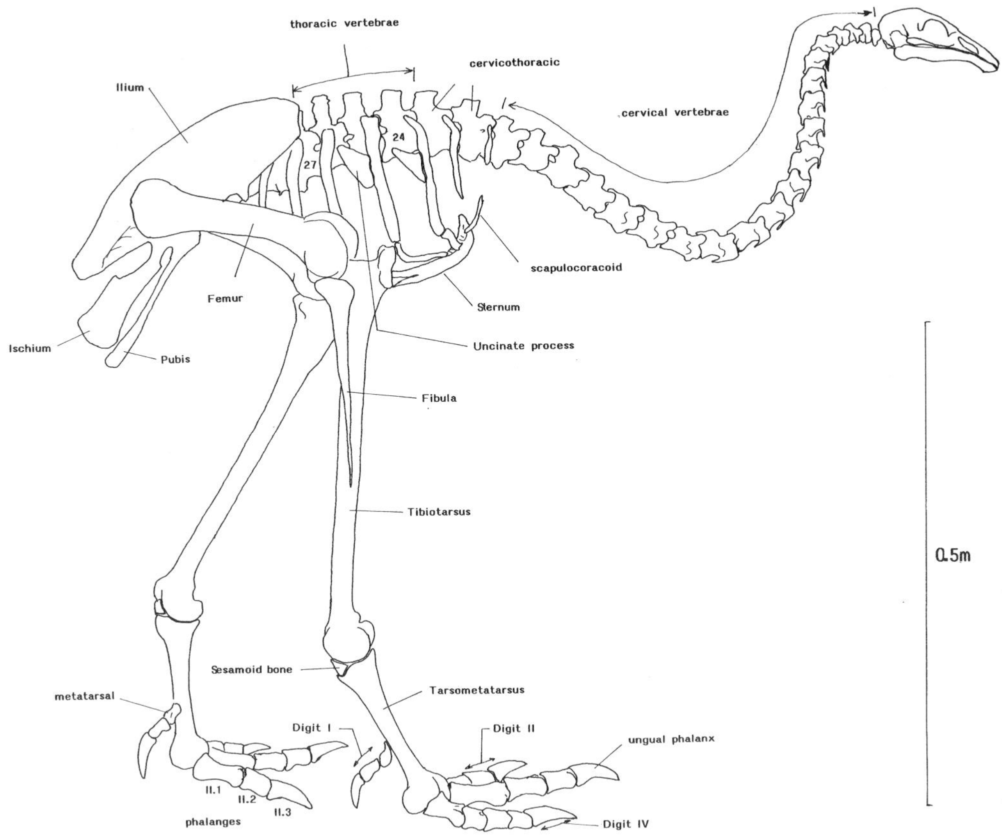

Figure 4.11. Skeletal reconstruction of Megalapteryx didinus. *As in all moa, there is a deeper ventral loop in the vertebral column than in other large ratites, and the vertical part of the neck is short.*

caudally over the ribs. The third bone of the pectoral girdle is the furcula, or wishbone, which in birds is usually a single bone assumed to be homologous with the fused clavicles of other animals. The pectoral girdle bones are supported by ligaments and muscles and are not ankylosed to the vertebral column. All three form the pivot point (glenoid) for the humerus.

Moa are unique among birds in that they have lost all traces of the wing. There is not even a humerus, let alone more distal bones, and no wishbone. The elephantbirds had exceedingly reduced wings, with fusion of several distal elements, but they still had wing skeletons. Kiwi, too, under their hairlike feathers, have a complete wing, albeit a tiny one. But moa had none. The sole clue that their ancestors once had forelimbs is that they had a single, small bone articulating to the front of the sternum. This is the scapulocoracoid, the result of fusion of these two elements (Fig. 4.13). It is longest and most easily seen to consist of a short coracoid and a longer, tapering scapula in *Dinornis, Anomalopteryx,* and *Pachyornis* spp. In *Megalapteryx, Euryapteryx,* and *Emeus,* it is very small, which corresponds to the loss of a coracoidal fossa in these genera. In no species of moa is there a glenoid fossa for the articulation of the humerus, as there is in kiwi and all other ratites that have wings.

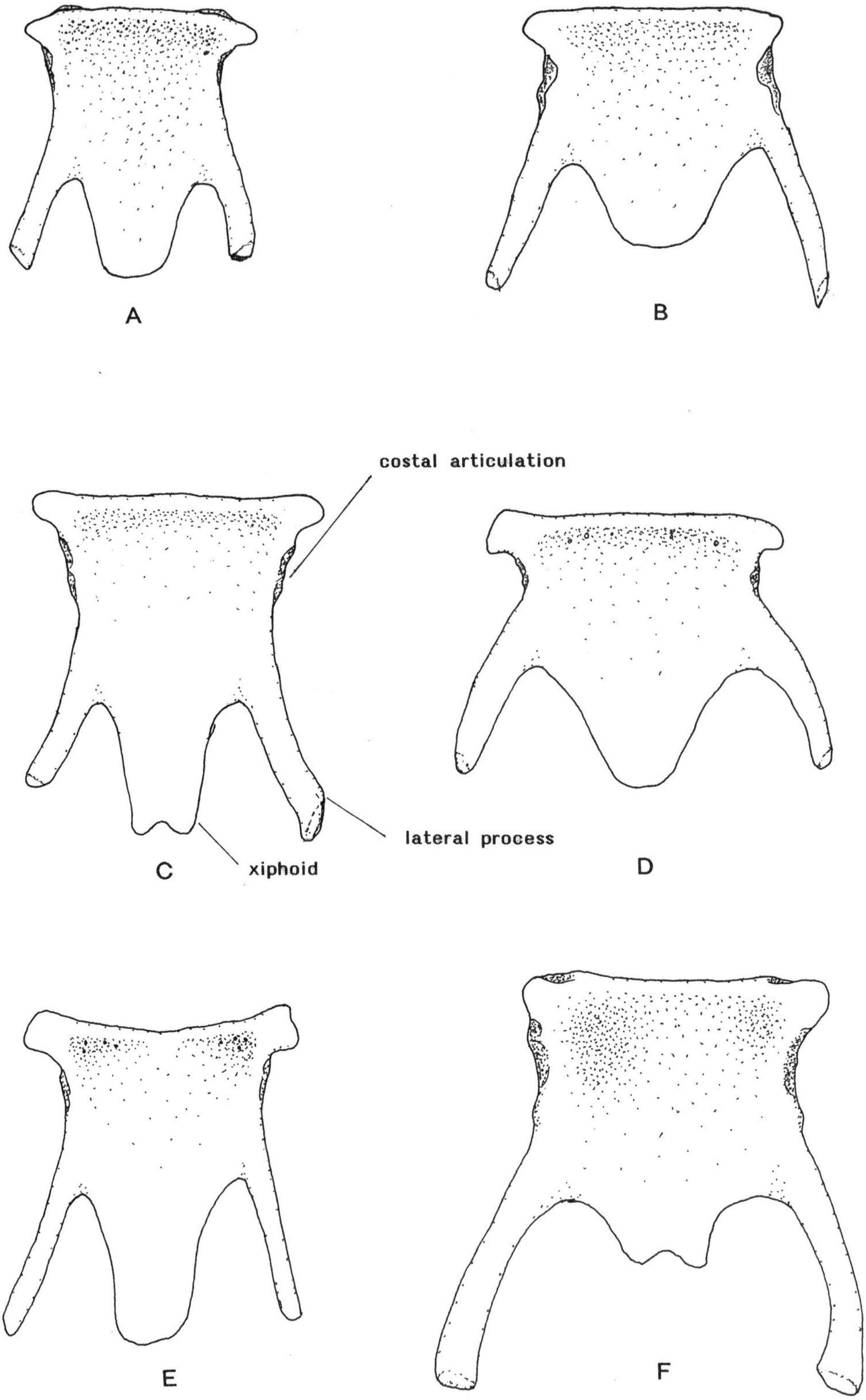
A
B
costal articulation
C
xiphoid
lateral process
D
E
F

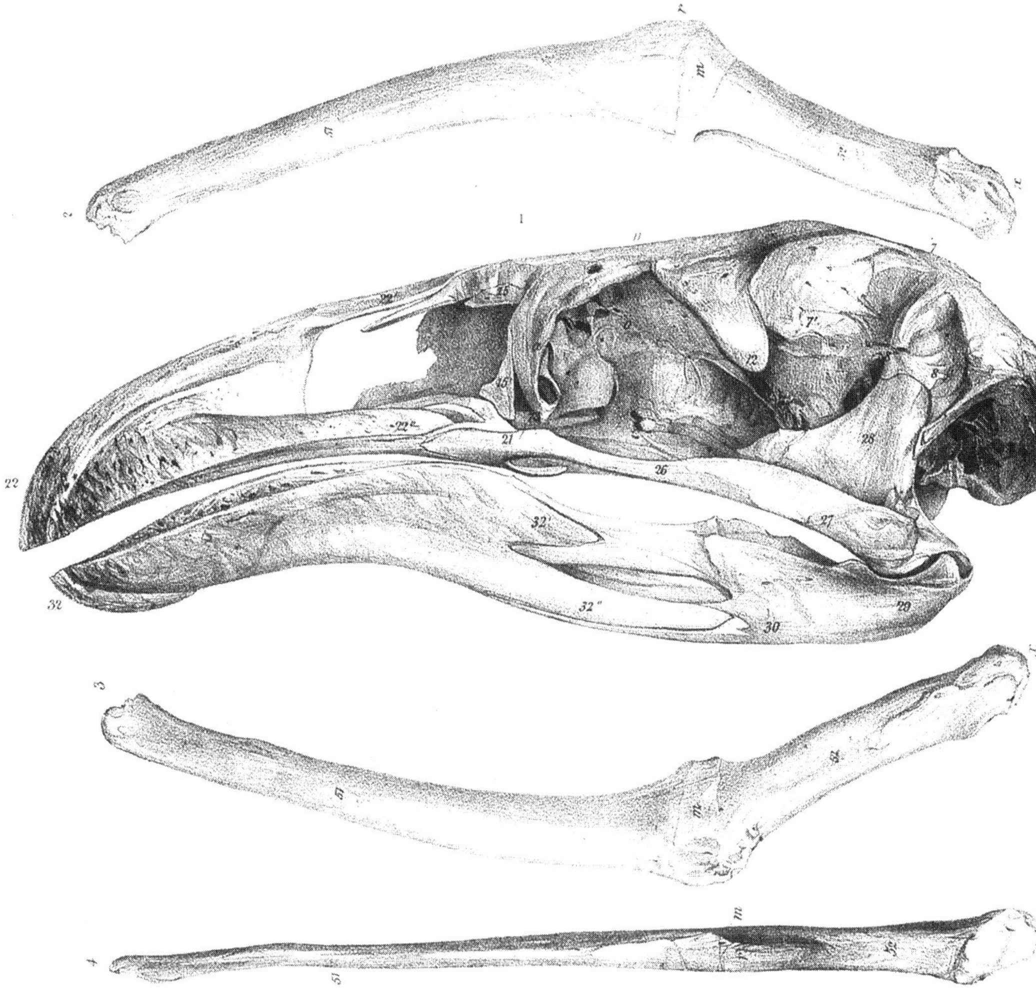

Figure 4.12. (opposite page) Moa sterna drawn to approximately the same size. (A) Anomalopteryx. *(B)* Megalapteryx. *(C)* Euryapteryx. *(D)* Pachyornis. *(E)* Emeus. *(F)* Dinornis. *Within a species, there is a marked size range, and interspecific variation within a genus contributes to further size variation. Shape, however, is constant within a genus.*

Figure 4.13. (above) Scapulocoracoid of Dinornis novaezealandiae *with the skull of the same individual, after Owen (1879: pl. 64). Moa scapulocoracoids have no articular facet for the sternum and no glenoid fossa for the humerus. The scapulocoracoid shown here is probably about 180 mm long.*

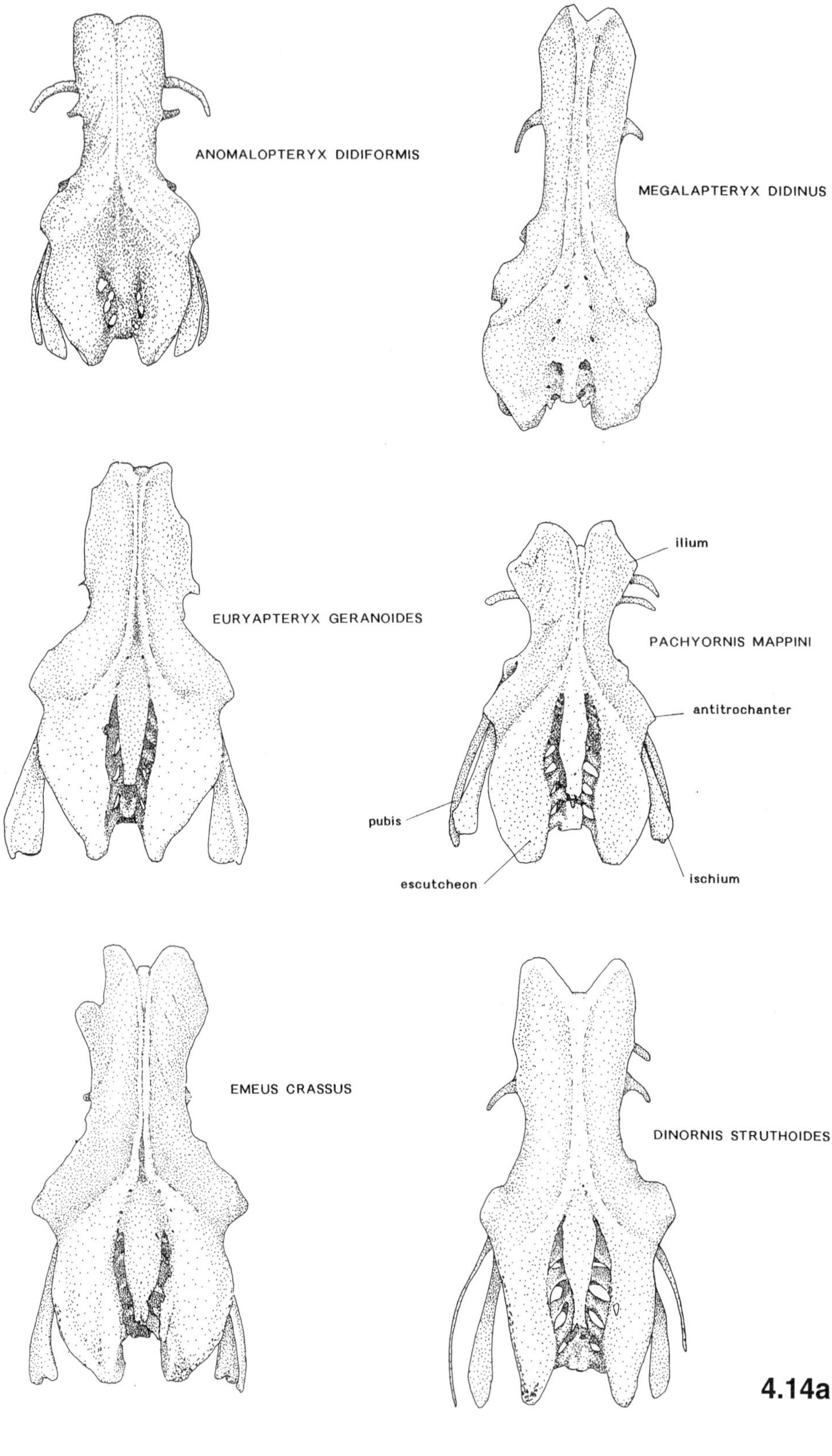
ANOMALOPTERYX DIDIFORMIS
MEGALAPTERYX DIDINUS
EURYAPTERYX GERANOIDES
ilium
PACHYORNIS MAPPINI
antitrochanter
pubis
escutcheon
ischium
EMEUS CRASSUS
DINORNIS STRUTHOIDES

4.14a

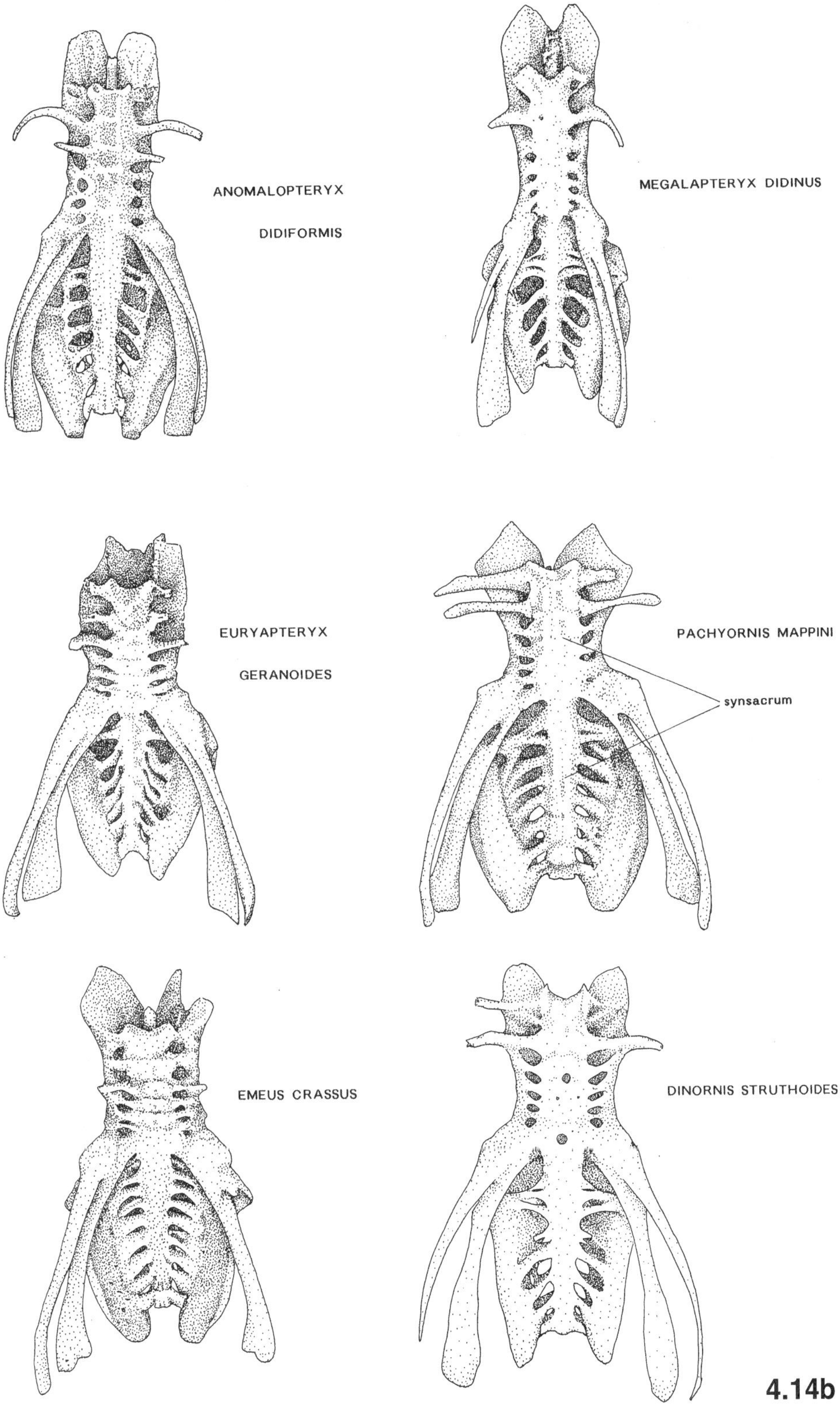
ANOMALOPTERYX
DIDIFORMIS
MEGALAPTERYX DIDINUS
EURYAPTERYX
GERANOIDES
PACHYORNIS MAPPINI
synsacrum
EMEUS CRASSUS
DINORNIS STRUTHOIDES

4.14b

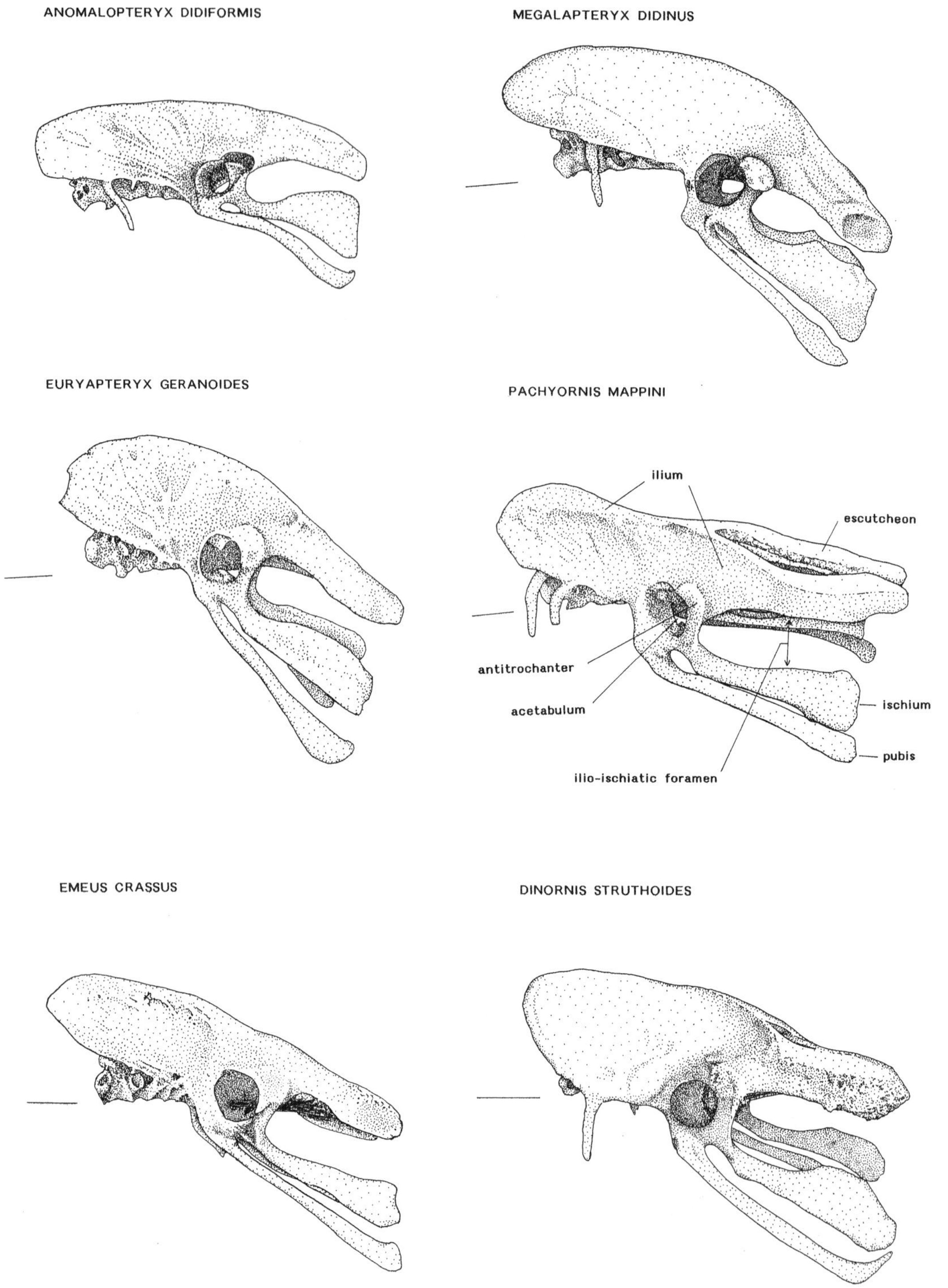
ANOMALOPTERYX DIDIFORMIS
MEGALAPTERYX DIDINUS
EURYAPTERYX GERANOIDES
PACHYORNIS MAPPINI
ilium
escutcheon
antitrochanter
acetabulum
ischium
pubis
ilio-ischiatic foramen
EMEUS CRASSUS
DINORNIS STRUTHOIDES

4.14c

Pelvis

The moa pelvis is a large, robust compound bone. A row of fused sacral vertebrae form the synsacrum, which is enclosed dorsally by a pair of concave ilia that fuse along the midline anteriorly (Fig. 4.14). The femur articulated in the socket joint of the acetabulum. There are normally 18 vertebra in the pelvic series: 7 preacetabular, 4 acetabular, and 7 postacetabular. The first sacral vertebra usually has freely articulating ribs, and the second sometimes does. The third sacral vertebra (vertebra 30 in emeids) has a fused rib. Approximately midway along each side is the open acetabulum, which is a circular fossa fenestrated medially, in which the ball of the femur articulates. The prominent antitrochanter bounds the posterodorsal part of the acetabulum. It articulates with the antitrochanteric facies of the femur and thus constrains the angle of the femur–pelvis articulation. Its position restricts the femur to movement through a 15° arc, about a plane at about 30° below the line of the sacral vertebrae. Moa femora were not vertical.

Posterior to the acetabulum, the ilia diverge and are connected by the costal processes of the sacral vertebrae, which form a broad, flat platform known as the escutcheon. Extending caudally from the ventral rim of the acetabulum are two long processes. The dorsal one is the ischium, which lies slightly outside of and below the ilium and ends in a slightly down-curved expansion. It never fuses or articulates with the ilium. Below the ischium is the thinner pubis, which usually extends slightly more distally than the ischium. The two processes usually meet, but they do not fuse at their distal extremities.

That the ischium in moa, as in all other ratites, does not fuse with the ilium—or if it does, then it does so only over a short distance, thus creating a very large ilioischiatic foramen—is one of the characters that indicates the monophyly of ratites (Cracraft 1974b). It is usual to consider a broad postacetabular portion to the pelvis as the primitive condition and a narrow one, as the derived state (Cracraft 1974b). Tinamous have widely separated ilia posteriorly, and like moa, they are considered to be primitive among the ratites. Emus, rheas, ostriches, and cassowaries all have posteriorly convergent ilia but have a wide, flat area above the acetabulae. Elephantbirds had a broad acetabular region, like moa. In contrast, kiwi ilia are closely applied, although not fused, along their entire length, and there is no broad, flat escutcheon. Unlike moa, the ilia of kiwi are poorly, if at all, ankylosed to the synsacrum in adult birds.

The pelves in various species and genera of moa vary in lateral profile, relative width, the angle of the costal processes in the escutcheon, and the relative position of the acetabulum. However, identification of pelves is sometimes difficult.

Figure 4.14. (pages 98–100) Moa pelves in (a) dorsal, (b) ventral, and (c) lateral views, all drawn to a similar size. All species within a genus have pelves that are similar in shape, but there is marked intraspecific and intrageneric size variation; hence, the figure is designed to demonstrate shape differences. For example, a small Dinornis *pelvis, such as the type of* Dinornis gazella *Oliver (=* Dinornis struthoides*) is the same size as an average* Anomalopteryx didiformis *pelvis but the same shape as that of* Dinornis giganteus, *whose largest examples may be over 0.5 m long. Drawn by J. Ogier.*

Hindlimbs

In the hindlimb of moa, the femur was short and stout, the tibiotarsus long, and the tarsometatarsus short to very short. Alongside the tibiotarsus was a stilettolike fibula, the distal end of which was fused to the tibiotarsus. Between the tibiotarsus and the tarsometatarsus was a

PL. LXI.

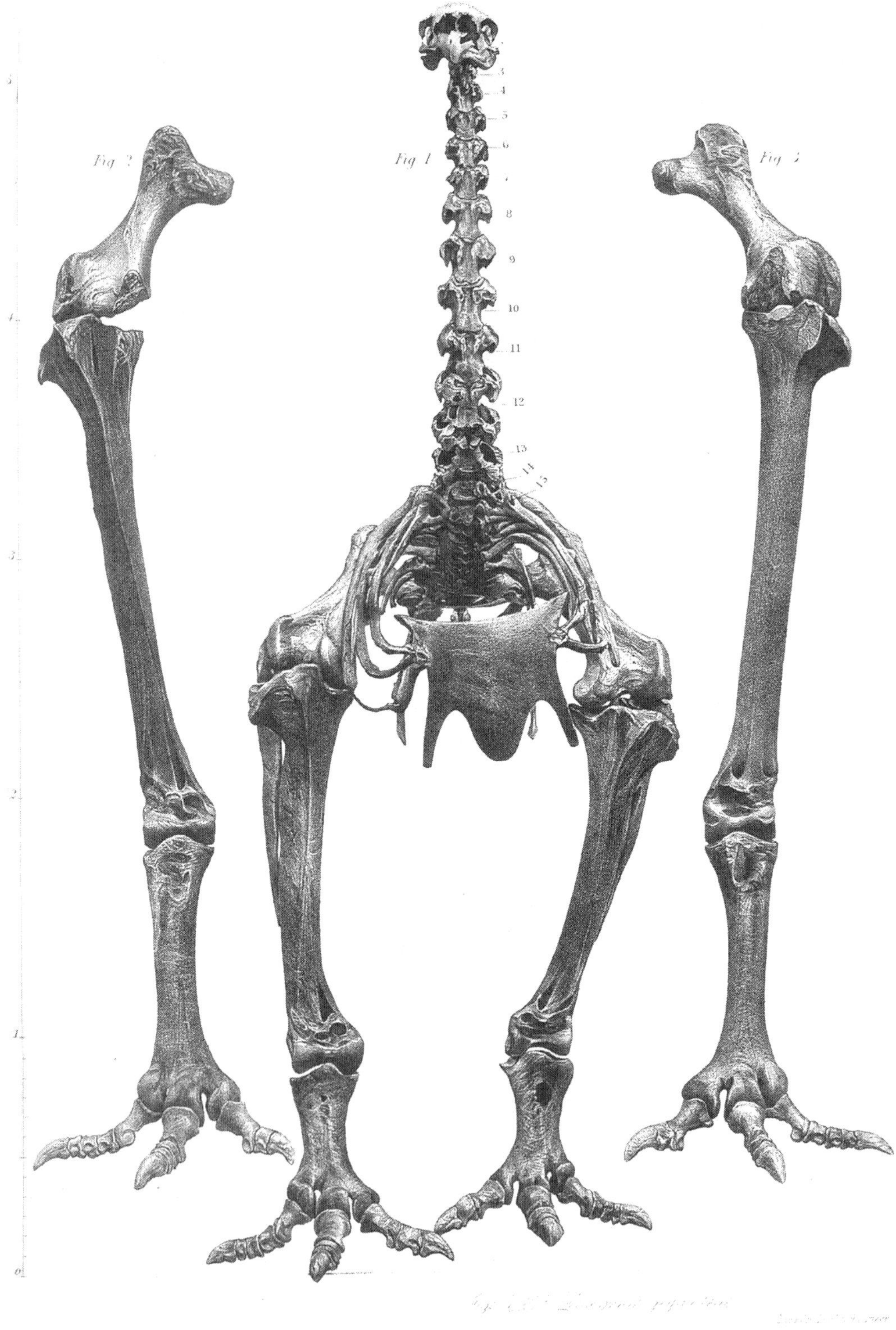

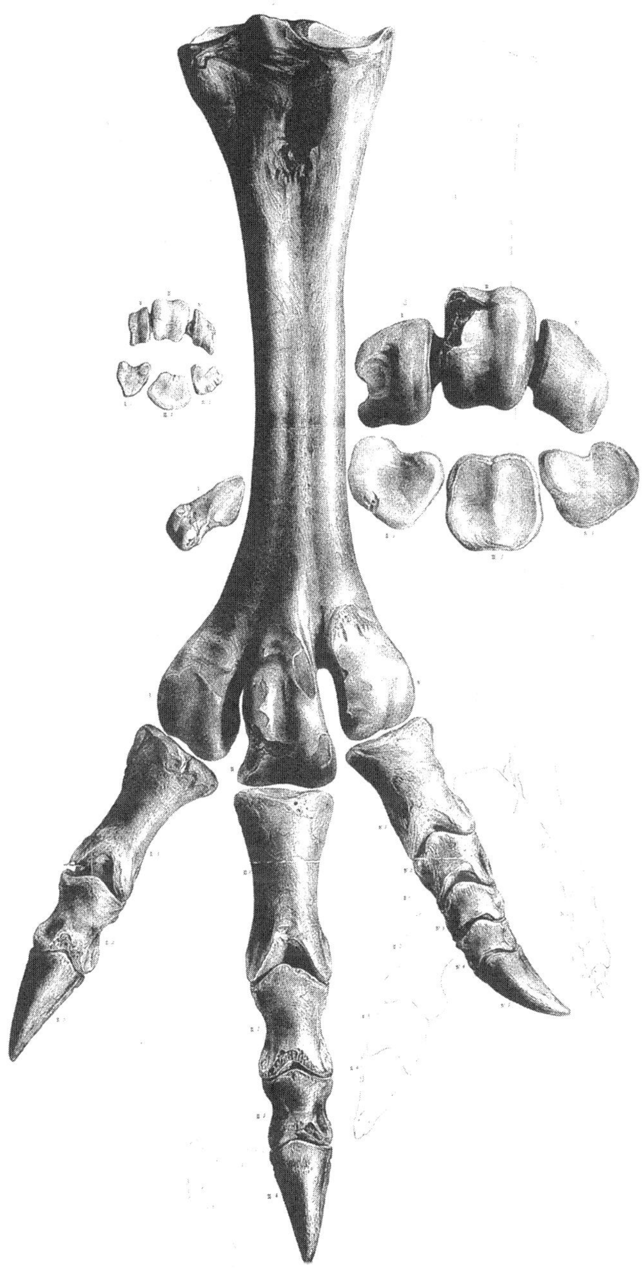

Figure 4.15. (opposite page) Pachyornis elephantopus *skeleton shown with* Dinornis giganteus *legs (from Owen 1879: pl. 61). The* Dinornis *tibiotarsi are about 800 mm long.*

Figure 4.16. Dinornis *left foot skeleton, showing tarsometatarsus and phalanges (after Owen 1879: pl. 49). The tarsometatarsus is ~400 mm long.*

small, triangular sesamoid bone. In adults, the patella was always fused to the tibiotarsus to form part of the procnemial ridge. All moa had four toes, in which the usual phalangeric formula was 2:3:4:5, except *Euryapteryx* and *Emeus*, in which it was 2:3:4:4. The hind toe (digit I) had two very reduced phalanges articulated to a reduced metatarsal. The reduction in this hind toe was associated with the frequent absence of

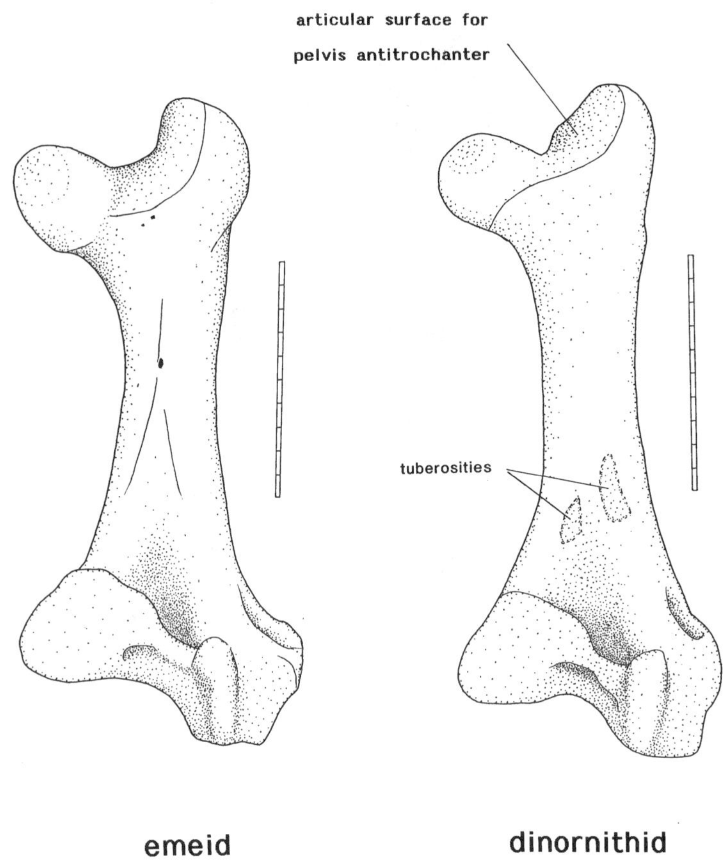

Figure 4.17. Ventral views of femora showing marked differences between the emeid femur (left) and dinornithid (right). Emeids have markedly constricted neck and a ventral ridge that results in a triangular raised feature. The femur neck in dinornithids is not constricted; there are always two raised tuberosities on the shaft, separated by a depressed smooth zone aligned along the shaft. Scale bar is in centimeters.

any discernible metatarsal scar on the tarsometatarsus. It is probable that of the hind toe (digit I), only phalanx I.2 (the ungual) projected from the rear of the leg as a small spur. Digit III (the middle toe) was always longest, followed by digit II (the inner toe), then digit IV (the outer toe). All terminal phalanges, or unguals, are pointed, slightly down-curved claws. None of the toes had large spurs, as, for example, cassowaries have.

The various genera of moa differed slightly in the proportions of their major leg elements, but all had relatively shorter tarsometatarsi than other ratites (Figs. 4.15, 4.16). All genera and species can be diagnosed by use of specific shape characters on each main leg element, independent of lengths and widths (see Appendix 3).

Moa are distinguishable from other ratites on several features of the hindlimb. Apart from *Dinornis,* the ball of the femur (Fig. 4.17, 4.18) is separated from the trochanter by a well-defined neck, the antitrochanteric facies (which articulates with the trochanter on the pelvis) is concave (convex in all other ratites except *Aepyornis*), and the distal femur condyles are broadly separated. On the tibiotarsus (Fig. 4.19), the cnemial ridges or crests are widely separated at their base (constricted in *Struthio* and *Rhea*), distally the groove between the

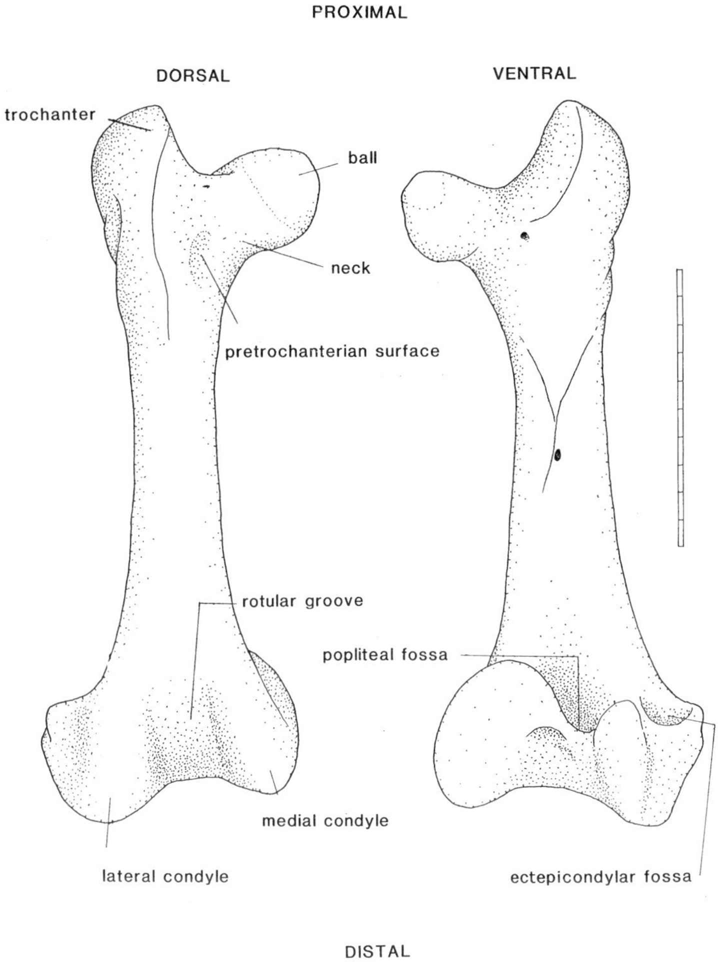

*Figure 4.18. Moa femur (*Anomalopteryx didiformis*), showing features identified in the text. Scale bar is in centimeters.*

anterior points of the condyles is wide and shallow, and the tendinal groove has a small tendinal bridge over it. The tarsometatarsus (Fig. 4.20) has two distinct, although short, hypotarsal ridges, the outer one longer proximodistally, and there are three trochleae. The middle trochlea always projects farthest distally and appreciably forward of the shaft. There is no distal foramen. The tarsometatarsus is relatively short, being shorter than the femur in all emeids, and is only slightly longer in dinornithids. In all other large ratites, the tarsometatarsus is considerably longer than the femur, the middle trochlea is not produced forward of the shaft to any extent (slightly in aepyornithids), and there is only one (the outer) hypotarsal ridge, which is high and long.

Larynx

The moa larynx consisted of several ossified elements. A broad thyroid or ossified cricoid cartilage enclosed two ossified arytenoids dorsally on its rear, as shown by Oliver (1949: fig. 84). Scarlett (1975b)

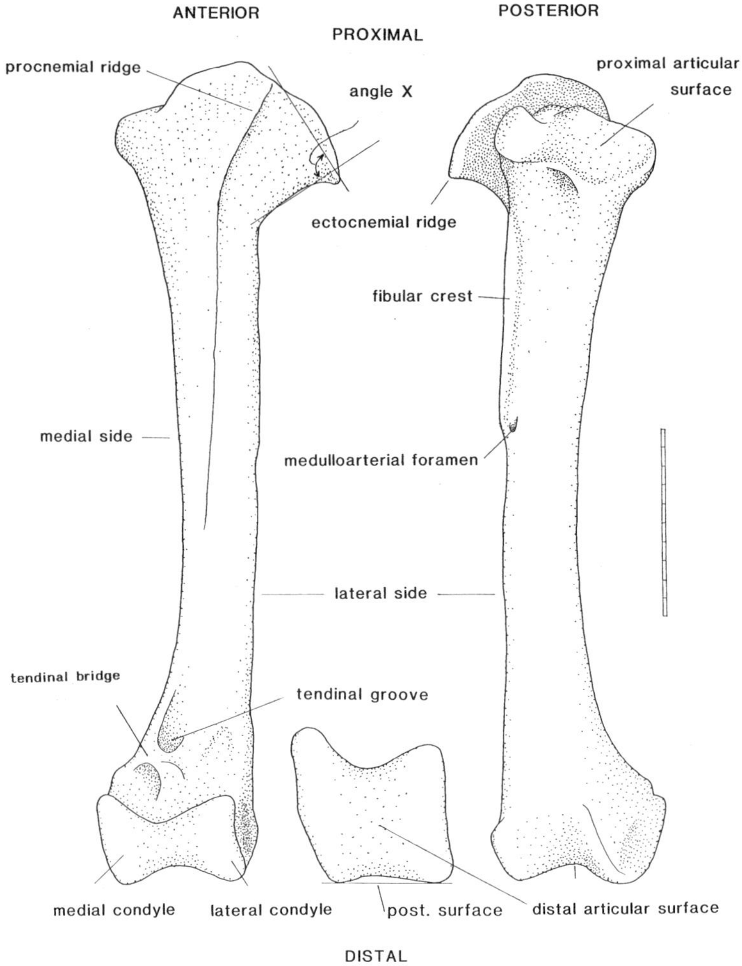

*Figure 4.19. Moa tibiotarsus (*Emeus crassus*), showing features described in the text. Scale bar is in centimeters.*

described the thyroid bones of most genera. All have the form of a broad, triangular basin, concave in dorsal aspect, with the point anteriorly. Posteriorly, one to three tracheal rings are fused to it. Scarlett (1975b) reported that *Dinornis* differed from all other taxa in having a strong ridge dividing the basin on the dorsal surface, but specimen MNZ S400, extracted from the mummified head of *Megalapteryx didinus* from Cromwell, also shows this feature. Such a ridge appears primitive because anatids and galliforms have it. One adult *D. giganteus* thyroid was 36 mm long, 41 mm wide, and 14 mm deep. On its ventral surface, the basin of the moa thyroid has a pair of rounded ridges, which are poorly developed in *Pachyornis elephantopus* but strongly developed in *Euryapteryx geranoides* (Scarlett 1975b). Thyroids of *Pachyornis elephantopus* have a deeper basin than other emeids; measurements of one adult were 33 × 30 × 15 mm. *Emeus crassus* has thyroids similar to *Euryapteryx geranoides,* but smaller, and the ventral ridges are not as well developed. *Euryapteryx geranoides* thyroids ranged

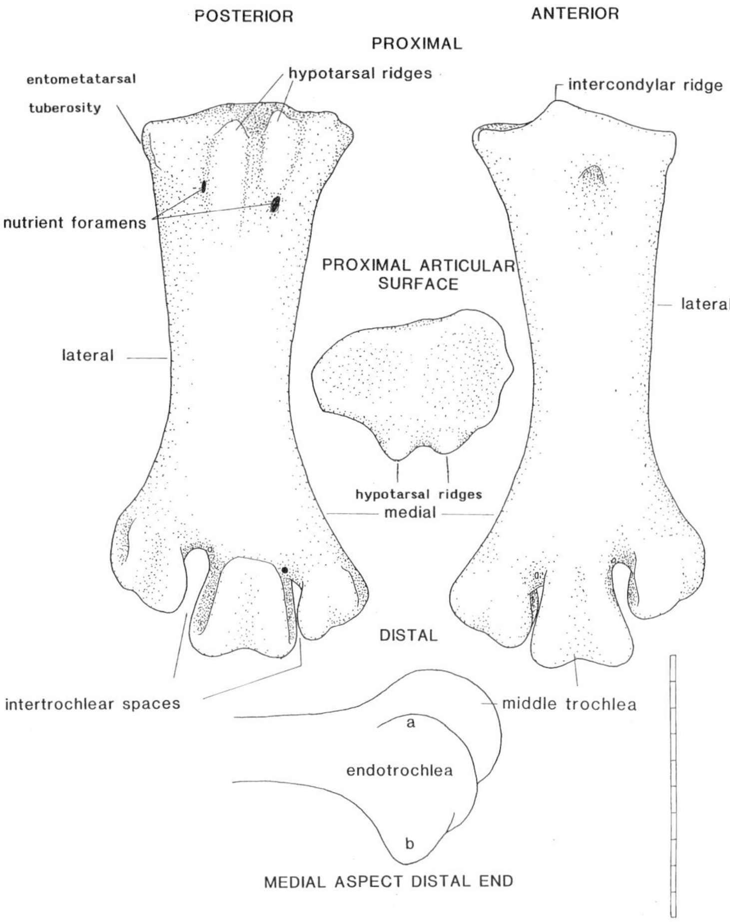

*Figure 4.20. Moa tarsometatarsus (*Emeus crassus*) illustrating features referred to in the text. The points labeled "a" and "b" on the medial view of the distal end are useful landmarks. In* Emeus, Euryapteryx, *and* Pachyornis, *"a" is above "b" as shown, but* Megalapteryx *and* Anomalopteryx *differ, with point "a" rotated markedly distally of "b." Scale bar is in centimeters.*

from 21–27 mm long by 23–30 mm wide by 8–13 mm deep and were generally wider than long. *Emeus crassus* thyroids ranged from 24–25 mm long and were about as wide. *Anomalopteryx* thyroids (~22–23 mm long and wide) are characterized by well-developed posterior extensions to the sides of the basin. That of *Megalapteryx* is of similar size to *Anomalopteryx* but lacks the posterior projections.

Continuing from the fused tracheal rings of the thyroid, the rings of the trachea extended in a long series back to the lungs. The tracheal rings from the neck region of all moa were short (1–2 mm long) and varied from circular to elliptic in shape. At least some moa had an ossified syrinx. Oliver (1949) described a compressed elliptic bone formed from five to six fused rings (Fig. 4.21D, E) to which another ring, open on one side, was probably attached, and which marked the divergence of the bronchia.

In several specimens, several half-rings have been recovered that are from the separate bronchial tubes themselves. There is no evidence of an ossified syringeal bulla in any moa. The bone Owen (1871b: 389, pl. 93) considered to be a complex syrinx is most likely to be the

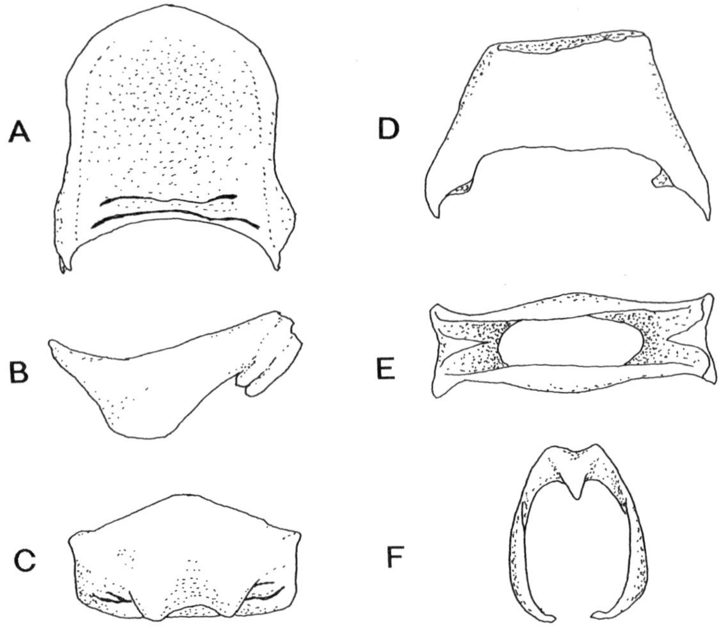

Figure 4.21. Thyroid bone of a moa attributed to Euryapteryx geranoides *by Scarlett (1975b). (A) Dorsal, (B) lateral, and (C) front view, and the (D, E) syrinx and the (F) tracheal ring that articulate with it. Redrawn from Oliver (1949: figs. 25–27).*

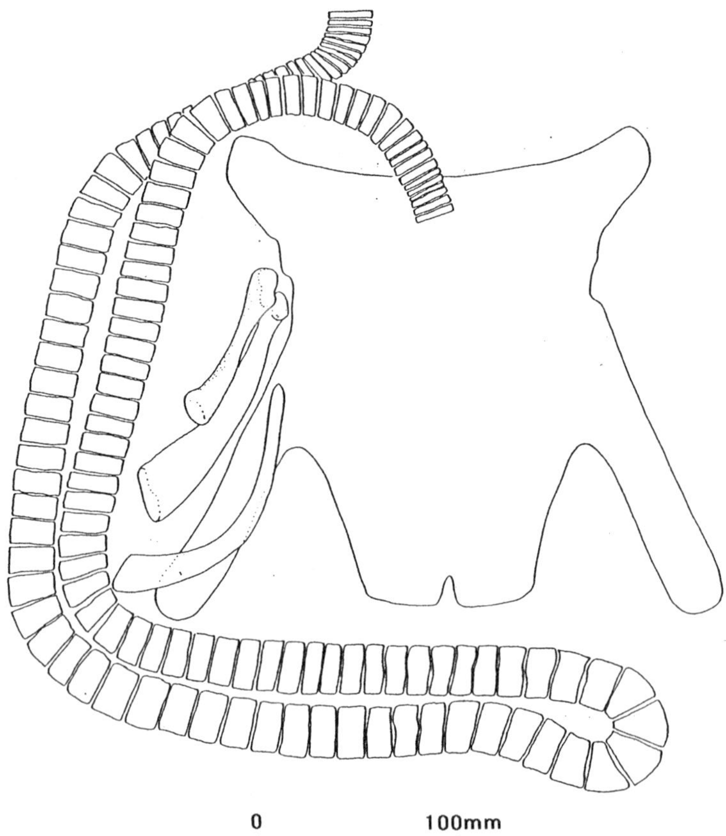

Figure 4.22. Body loop of the trachea in Euryapteryx geranoides, *as excavated with an articulated skeleton in Honeycomb Hill Cave. From Worthy (1989c), published courtesy of the* New Zealand Journal of Archaeology.

isolated olfactory capsule of a juvenile moa. In *Euryapteryx* and *Emeus*, the tracheal series was remarkably elongated in the thoracic cavity. The trachea extended in a long loop down the left side of the body cavity, then across it, before doubling back on itself to the breast and then into the lungs. This loop in *E. geranoides* was about 1 m long (Fig. 4.22) and so probably doubled the length of the trachea. It consists of elongated segments, each 10–20 mm long, with very robust walls. They were round in section and about twice the diameter of the neck rings (Worthy 1989c).

Tracheal elongation has been documented in 60 species of birds in 11 families in 6 orders: 29 species of cracids, 13 cranes, 5 manucodes in the bird of paradise family, 4 swans, 2 guinea fowl, and single species in several other groups (Fitch 1999). *Emeus* and *Euryapteryx* are therefore the only ratites with such a feature. In most species showing the elongate tracheae (39 species), the elongation is of a subdermal type, where it is found between the skin and the pectoral muscles. Alternatively, the tracheal elongation is housed within the sternum, as in swans and cranes (17 species). In the guinea fowl, it occurs in the clavicles, and only in two species (European spoonbill and African wood ibis) does it occur within the thoracic cavity. *Euryapteryx geranoides* and *Emeus crassus* can thus be added to this select group where the tracheal elongation is of the intrathoracic type. Fitch (1999) found that the only things birds with tracheal elongation had in common was that most were relatively large; made loud, far-reaching calls; and lived in (or at least nested in) dense habitats with restricted visibility.

Usually, these modifications are interpreted as enhancing acoustic ability. Hypotheses that the structure serves to lower the pitch of the call or increase its loudness are not supported by evidence; instead, these effects are achieved by a larger syrinx (Fitch 1999). Rather, it seems that the elongated trachea enables sounds with closely spaced formant frequencies, which results in a deeper or more resonant sound that travels far. This appears to enhance acoustic communication in dense habitats where sight is of limited use. Fitch (1999) hypothesized that the tracheal elongation was a mechanism to exaggerate the size of the bird. This may be expressed as females being attracted to the "largest" male or the "larger" male that may better defend its territory. Whatever the specific reason, it seems that at least one sex of *Emeus crassus* and *Euryapteryx geranoides* had the ability to make loud, far-carrying sounds, whereas other moa did not.

Stapes

The stapes (columella) is the only ear bone in birds, as well as the smallest bone in a bird. Among moa, only the stapes of *Megalapteryx didinus* has been described to date (Vickers-Rich et al. 1995), although that of *Dinornis* is now known. The stapes of *Megalapteryx* is about 6 mm long and has a distinctive twisted, sigmoid curve to the shaft. The distal (or lateral) end is flattened anteroposteriorly. The shaft merges with the footplate on the anterior end around its outer edge, except where there is a large basal fossa. This is unlike *Apteryx*, in which the

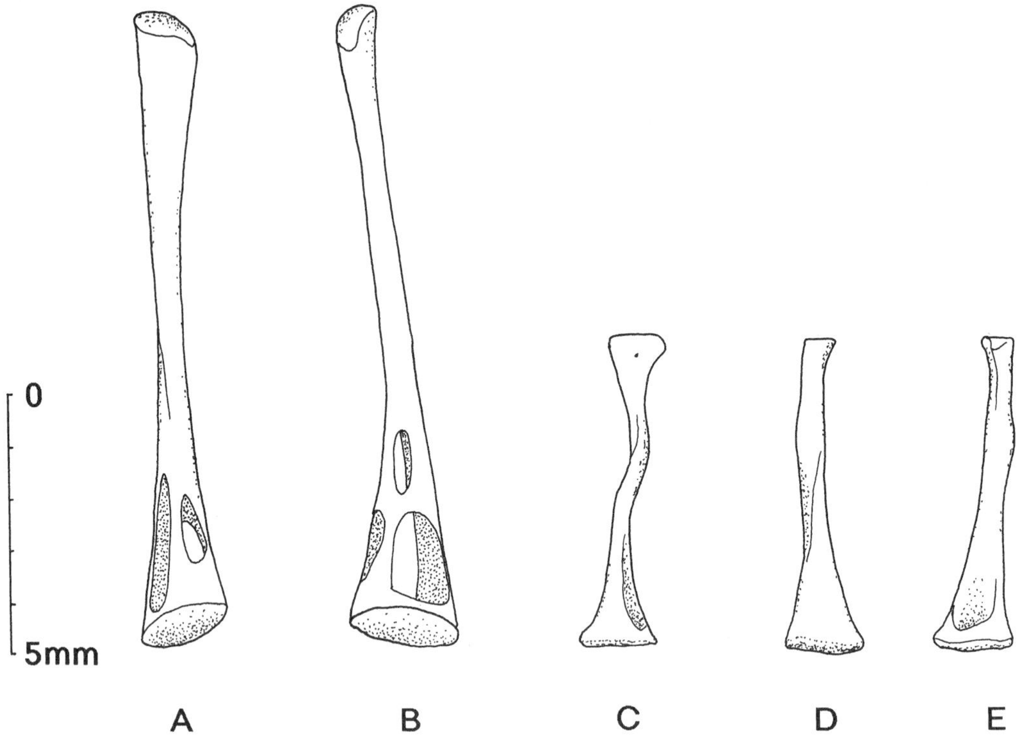

Figure 4.23. Moa stapes rotated to show several views. (A, B) Left stapes from Dinornis novaezealandiae *MNZ S32677. (C–E) Right stapes from* Megalapteryx didinus *(redrawn from Rich et al. 1995). The distal end of the bone is at the top of the figure, and the anterior end is at the bottom of the figure.*

shaft joins to the center of the footplate, but somewhat similar to the condition in *Dromaius* and *Struthio,* although in these taxa the bone is thin and pieced by several fossae. In rheas (e.g., *Pterocnemia*) and tinamous (e.g., *Rhynchotus*), the shaft joins the footplate as a series of struts. In other words, there are more fossae than in moa, both around the margin and in the center in rheas, but restricted to the center in tinamous. In *Megalapteryx,* both ends have a similar degree of expansion, in contrast to other ratites, in which the basal plate is considerably larger.

Not surprisingly, the stapes of *Dinornis novaezealandiae* (part specimen MNZ S32677; Fig. 4.23) is more similar to that in *Megalapteryx* than to the stapes of other ratites. The single specimen is 12.2 mm long, has a slight twist in the shaft, and anteriorly, the shaft is hollow, has four fenestrae, and merges with the margin of the footplate.

Hyoids

All moa had two ossified basal hyoid bones, the ceratobranchials, which were elongated, rodlike structures up to 80 mm long. The second, more caudal, pair of hyoid bones, the epibranchials, was probably usually cartilaginous, as is normal in paleognathous birds (Pycraft 1900: 195). However, ossified epibranchials are known from at least one specimen of *Megalapteryx.* The basihyal, to which the ceratobranchials articulated, was also probably cartilaginous, as in other

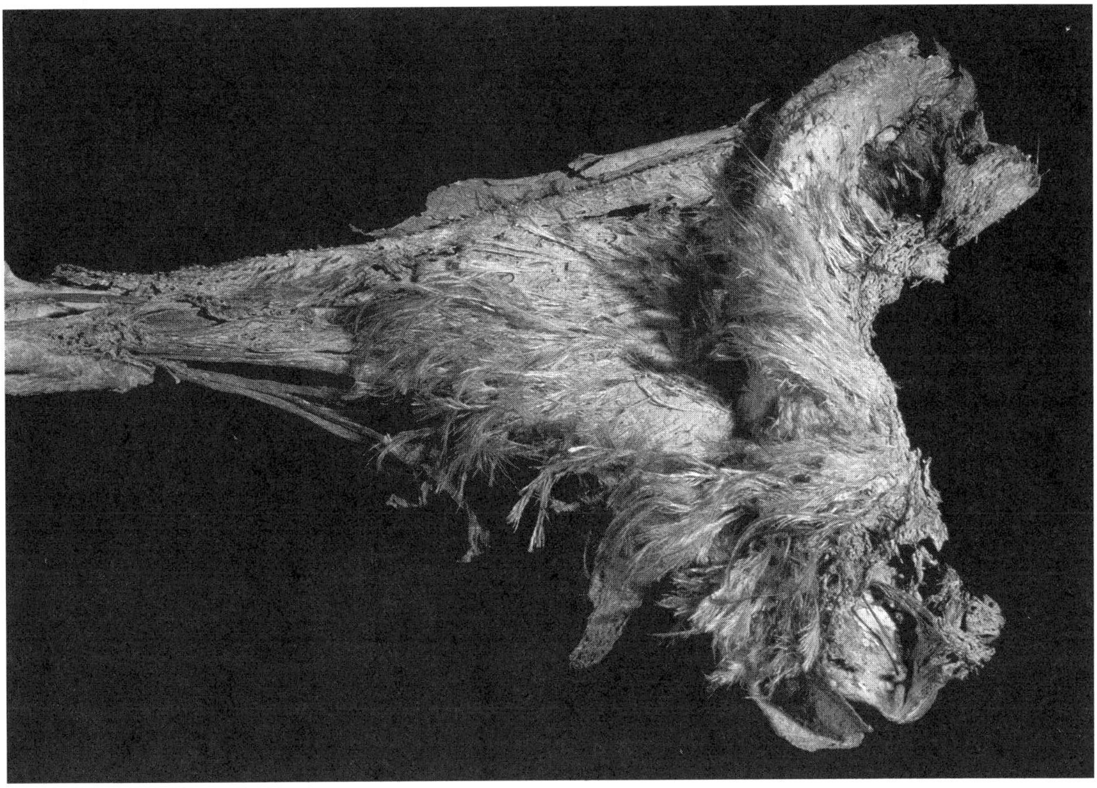

Figure 4.24. The dried leg of Megalapteryx didinus *from the Waikaia catchment, Otago. Photograph by R. Morris.*

ratites, but none has been found. Similarly, the glossohyal was probably cartilaginous.

"Soft" Parts

There are nine remarkable specimens that have preserved dried skin and other tissues. Of these, the type of *M. didinus* (see Fig. 3.10) is the most complete. The tongue is preserved, and it is the only one in which the horny claw on the toes is present. Three other mummies of this species are in New Zealand collections: an articulated feathered leg from Waikaia, Otago (Fig. 4.24); the Cromwell specimen (Vickers-Rich et al. 1995), which includes a dried head (including the eye) and neck (Fig. 4.25); and parts of a skeleton, including an articulated foot from Mount Owen (Worthy 1989g; Fig. 4.26). For *Dinornis,* there is a lower leg with preserved scutes from Knobby Range in Otago (Fig. 4.27; Hutton and Coughtrey 1875a) and the Tiger Hill skeleton, now in York Museum, England (Hector 1872a). For *Emeus,* there is a neck and odd bones with fibrous tissue (possibly ligament) on them from Earnscleugh Cave (Hutton and Coughtrey 1875b). A remarkable recent discovery was of a whole skeleton of *Anomalopteryx didiformis* with much skin and tissue on it (Figs. 4.28, 4.29; Forrest 1987). The Tiger Hill skeleton shows the remains of footpads (Fig. 4.30). A tarsometatarsus in the zoology department of the Cambridge University Museum, England, has a small patch of skin and scutes (Anderson 1989c). Apart from

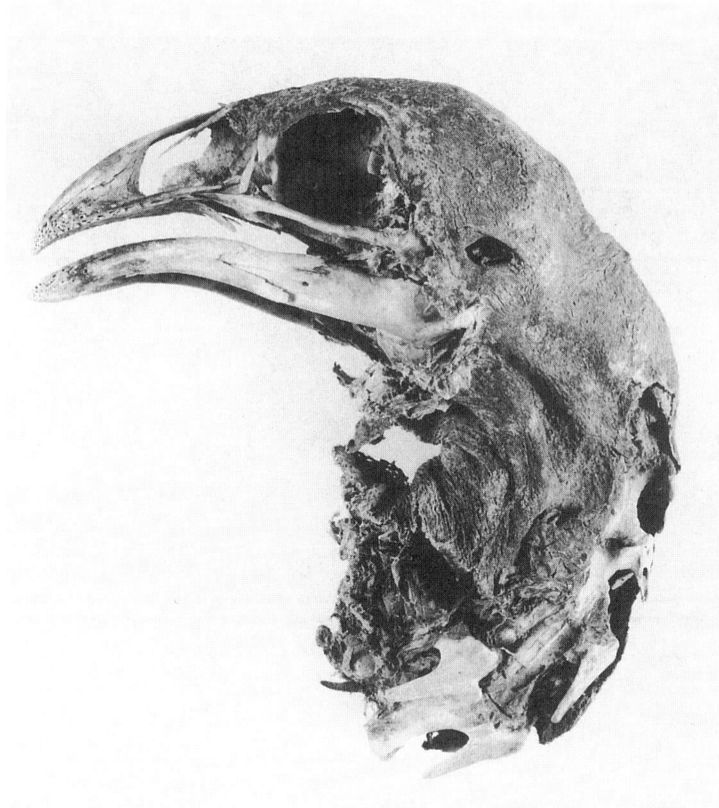

Figure 4.25. Dried head of Megalapteryx didinus *found at Cromwell, Otago. Photograph by Warwick Wilson, Museum of New Zealand Te Papa Tongarewa, B16298.*

Figure 4.26. (below) The dried foot of the Mount Owen Megalapteryx didinus. *Photograph by R. Morris.*

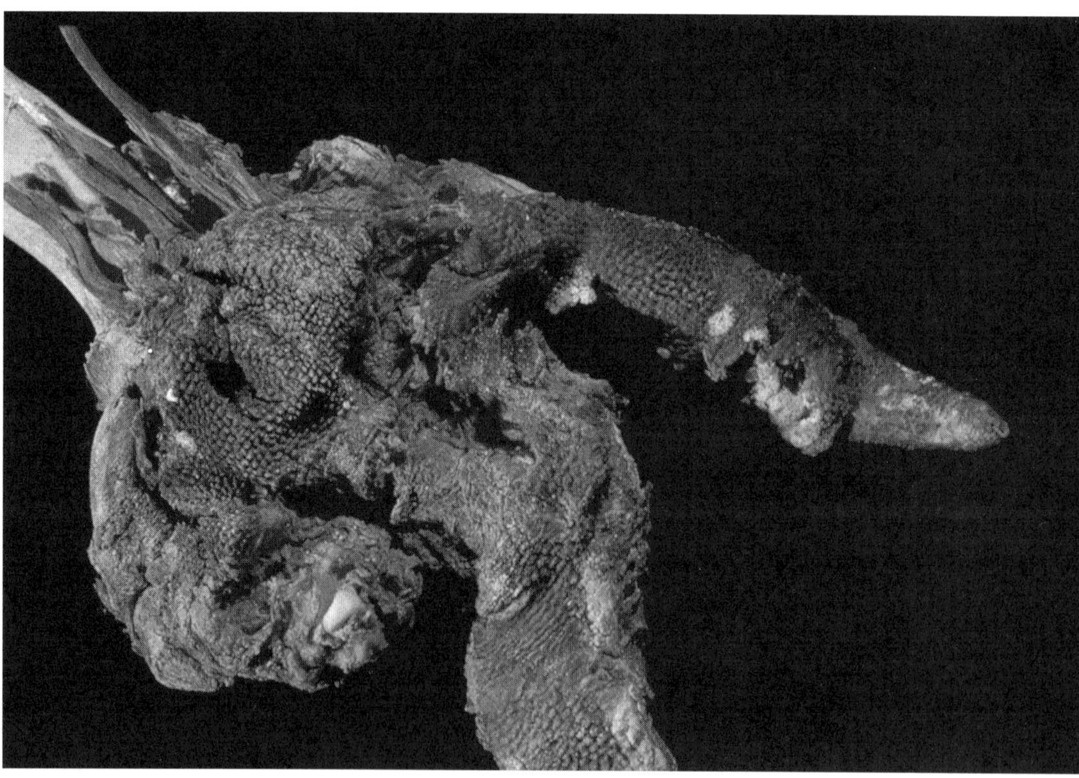

Figure 4.27. Dried foot of Dinornis novaezealandiae found in Knobby Range, Otago. Photograph by Augustus Hamilton, Museum of New Zealand Te Papa Tongarewa, C1079.

these more significant specimens, a few pieces of skin have been recovered, and there are several unsubstantiated records listed by Anderson (1989c: table 5.2).

The ages of the mummified specimens have been the subject of much speculation in the past, and the apparently fresh remains have often been used to suggest that moa must have survived into recent centuries. However, as Hutton (1892c) foresaw, the mummies are not necessarily recent. Their survival is related to the dry preservation conditions they were subjected to. Most mummies have been found in the dry interior of Otago and northern Southland, where annual rainfall is less than 500 mm. The exception is the Mount Owen moa, which was found in the throat of Blowhole Cave in the narrow connecting fissure that separates this cavern from the very large Whalesmouth Cavern. The latter cave opens to the outside world at a much lower altitude, and the resultant perpetual wind through the fissure to Blowhole, combined with the location of the cave near the tree line, resulted in the effective freeze-drying of this specimen. The Cromwell moa (*Megalapteryx*) gave two dates of 646±95 radiocarbon years before the present (B.P.); 1284–1408 cal (calendar years) B.P. (68%) and 690±120 years B.P.; and 1246–

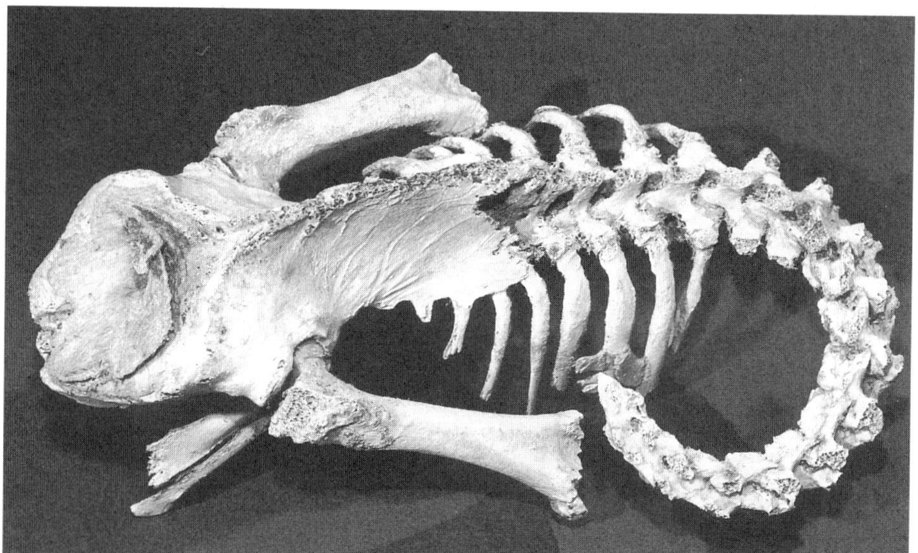

Figure 4.28. (above) Mummified remains of Anomalopteryx didiformis *from Southland. A lateral view of the head. The bill and eye are relatively small. Mice had gnawed prominent tissue, and the rhamphotheca is lost except between the mandible and premaxilla. Photographs reproduced courtesy of the Southland Museum and Art Gallery.*

Figure 4.29. (below) Mummified remains of Anomalopteryx didiformis *from Southland. A dorsal view of the body. The femora are preserved in their natural position, showing divergence from the center line and only slight declination from the line of the thoracic vertebrae. Photographs reproduced courtesy of the Southland Museum and Art Gallery.*

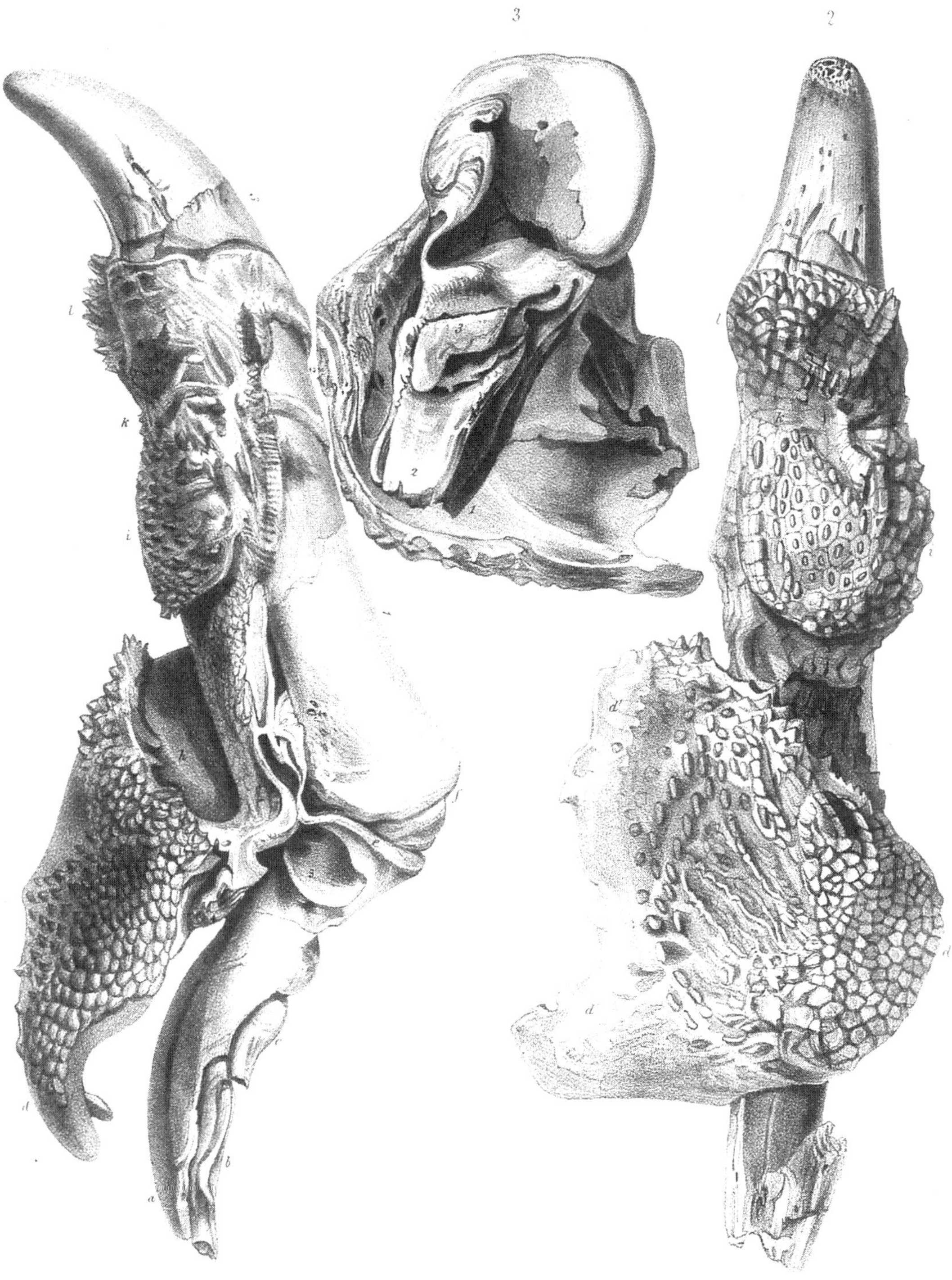

Figure 4.30. Digit II of the Tiger Hill Dinornis novaezealandiae *showing the footpads. After Owen (1879: pl. 71).*

Figure 4.31. Moa feathers, probably from Monck's Cave, Sumner, Canterbury, that were recently obtained from a private estate. Photograph reproduced courtesy of the Southland Museum and Art Gallery.

1403 cal B.P. (68%) (Vickers-Rich et al. 1995), which are not statistically different. The Mount Owen moa gave a result of 3350±70 cal B.P.; 3471–3638 cal B.P. (68%) (Worthy 1989g). The Cromwell moa may well have seen the early Polynesian inhabitants of New Zealand, but the Mount Owen moa lived 2500 years before people arrived.

Apart from the Waikaia leg, none of the material preserves much information about the plumage. Most specimens have only feather bases intact, so the Waikaia leg, with its numerous feathers, provides the most information. As a result, we know that the legs of *M. didinus* were feathered to the toes, as are those of Darwin's rhea (*Pterocnemia pennata*) (Pycraft 1900: 155), which is also an inhabitant of upland areas that are periodically covered in snow. Moa, like *Dromaius* and *Casuarius* but not *Rhea, Struthio*, or *Apteryx*, had at least some feathers with an aftershaft. Whether an aftershaft was present on all feathers is not certain because, as noted by Oliver (1949), many feathers may have lost theirs as a result of poor preservation. *Megalapteryx* feathers from the Waikaia site and Takahe Valley have grayish-white down toward the base and are dark—almost black—in the distal half or third of the length; the tips of most are white. Most of these feathers are less than 18 cm long and are in two forms: those that are widest near the base and taper toward their end, and those that are broadest close to their end. Feathers found in Monck's Cave, Sumner, Christchurch, South Island, are up to 23 cm long. These were brownish red and characteristically tapered to a point in the distal half (Fig. 4.31). Feathers from several sites, but of uncertain specific identity, have been described with a range of colors. One was reported to be white; some were reddish brown, grading distally to black with a white tip; others had a central yellow stripe bordered by purplish brown (Oliver 1949).

Special Features of the Skull

Many features of moa skulls set the group apart from other ratites. The posterior articulation of the vertebral column with the cranium is one with considerable implications in defining the stance of the birds. Rheas, ostriches, and emus all live in savanna-type habitats, where the height of the eyes above ground is paramount for the detection of predators. They have elongated tarsi, elongated cervical vertebrae, and a cranium articulated below the skull, which are all features that serve to raise the height of the bird's head. The long legs are of course adapted for fast running, and the elongated neck is necessary to enable the bird to reach the ground for feeding or drinking. In their structure, moa were more like cassowaries. These species live in closed forest, have short cervical vertebrae, and walk with their head at about the level of the back. All moa had relatively short vertebrae, short tarsometatarsi, and more of their neck behind their head. The neck would have been in line with the back of the head, which was normally carried at the level of, or slightly above, the back.

Moa were unique among birds in the huge development of the olfactory chambers, which extended posteriorly, impinging on the area where the interorbital septum usually is in birds. *Apteryx* does not have a single interorbital septum, as in other birds, either. Although the septum is present, the huge olfactory capsules are closely applied on either side of it and extend rearward to near the basisphenoid processes; they actually partially lie under the brain cavity. The structure of the olfactory lobes is thus totally different from that of moa.

The preorbital structure of moa species differs considerably from that in other ratites. The lachrymal does not have a posterior process lying beside the frontal along the dorsal surface of the cranium, and the maxillary process of the nasal is usually well developed, extending to and articulating with the maxilla, not small and unconnected. The orbit is separated from the anterior part of the olfactory chamber that houses a spiraled turbinal process by the large antorbital plate. The rostrum has wide, triangular processes developed from it, which form the floor of the large anterior part of the olfactory chambers. None of the other ratites has a similar structure.

The orbit was smaller in proportion to the size of the skull than in most birds and all ratites. Presumably olfaction was very important in moa. The bills of all moa were short and stoutly built, despite having variation in sharpness, height, and steepness. In most ratites, the premaxillary septum is small and the nasal process of the premaxilla is a single wide, thin, flattened, and more elongate bone, which allows the upper mandible to be raised or bent, which means there is rhynchokinesis in these ratites. This facility is further made possible by the lack of connection between the maxilla and the nasal. However, in moa, the septum was large, the premaxilla short and very stout, the nasal connected to the maxilla with a strut, and the various components of the paleognathous palate robust, so it seems unlikely that much, if any, bending could have occurred in the upper jaw.

Systematic Diagnoses

Order Dinornithiformes

Moa shared with other ratites the following derived characters. They had a paleognathous palate in which the thin vomer encloses the rostrum anteriorly, with paired caudal processes ankylosed to the palatine and pterygoid. The premaxilla and mandible had a pronounced tripartite rhamphothecal structure with a central ridge that was flattened and that widened anteriorly and was separated from lateral plates by distinct grooves. The nasal process of the premaxilla was unpaired and was not fused to the nasals. The pelvis had long, posteriorly open, ilioischiatic foramina. The sternum was broad and flat, without trace of a keel.

Moa differed from all other ratites, and hence can be placed in a separate order, in the following characters. The skull has small orbits, large olfactory cavities, and is broad across the occipital region. The basitemporal platform has prominent basipterygoid processes. The basiparasphenoidal rostrum is long and has triangular expansions extending laterally beneath the antorbital plate. The maxilla has a large, dorsal inflation enclosing a cavity—the maxillary antrum—although it is secondarily collapsed in two genera. The beak is short and may be pointed or rounded, and the mandible is curved down at the tip, not flat. Moa have 21–23 cervical, 6 dorsal or rib-bearing presacral, 18 sacral, and 11 caudal vertebrae with no pygostyle (coalesced caudal vertebrae modified for the insertion of tail feathers). The pelvis is broad at and posterior to the acetabulum. The wing is entirely absent, and the pectoral girdle is represented only by a vestigial single scapulocoracoid bone, without a trace of a glenoid facet (usually a fossa for the attachment of the humerus). The sternum has well-developed lateral processes. The femur has a concave articular surface for the antitrochanter on the pelvis (convex in other ratites), the tibiotarsus has an osseous bridge over the tendinal groove, and the tarsometatarsus has two hypotarsal ridges (one in other ratites except *Apteryx*).

Family Dinornithidae

Genus Dinornis

The bill is broad and flattened, wider than high at the caudal margin of the nasal septum of the premaxilla, and its lateral margins convex. The mandible has outwardly curved rami, in which, caudally, the cotyla lateralis is rounded and lacks a posterior process. The cranium is more than twice as broad as it is high. The preorbital plate is connected to the rostrum by a robust ventral bar that rises at right angles to the rostrum. The maxillary process of the nasal is a slender bone firmly connected to the nasal that lies anterior to the lachrymal and expands ventrally and articulates with the maxilla. The turbinal in the olfactory chamber has two complete spirals. The planes of the rostrum and basitemporal plate meet at about 150°. There are 29–30 presacral vertebrae, of which those at the nape have a single neural spine. The cervical vertebrae have laterally compressed vertebrarterial canals. The uncinate processes are firmly ankylosed to the ribs. The femur, tibiotarsus and tarsometatarsus have relative lengths of 1:2:1.1. The ventral surface of the femur shaft has two distinct tuberosities separated by a groove; the ventral surface of the proximal shaft is flat and forms a right angle with the lateral surface of the trochanter (Fig. 4.17). The ball is separated from the shaft by an indistinct neck that is not narrower than the ball. On the tibiotarsus, the angle formed by the dorsal surface of the ectocnemial ridge and its ventral buttress (see Fig. 4.19) is distinctly greater than 90°. The length of the tarsometatarsus is 2.5–3.0 times its maximum width, and the shaft has distinctly parallel sides. In dorsal view, the lateral edges of the pelvis escutcheon are straight and slightly convergent. The sternum has the left anterior depression markedly deeper than the right one, the lateral processes are curved outward, and the xiphoid process is bifid (Worthy 1994a).

Dinornis struthoides

The cranium–postorbital width is 101–127 mm (mean, 112 mm); temporal fossae/postorbital width ratio ranges 0.50–0.59 (mean, 0.55; Fig. 4.32). The occipital region has strongly developed protuberances lateral to the occipital crest, and the mamillar tuberosities are strongly developed ventrally and protrude variably posteriorly and usually have a marked, flattish hollow with a low medial ridge between them. The occipital condyle has no neck between the articular surface and the cranium, and the articular surface is spread ventrally so in ventral view the condyle appears circular. The rostrum is compressed ventrally. The medial flange of the pterygoid is elongate posteriorly and overlaps the articulation with the basipterygoid. Leg bone lengths are as follows: femora 261–307 mm, tibiotarsi 471–609 mm, and tarsometatarsi 247–319 mm.

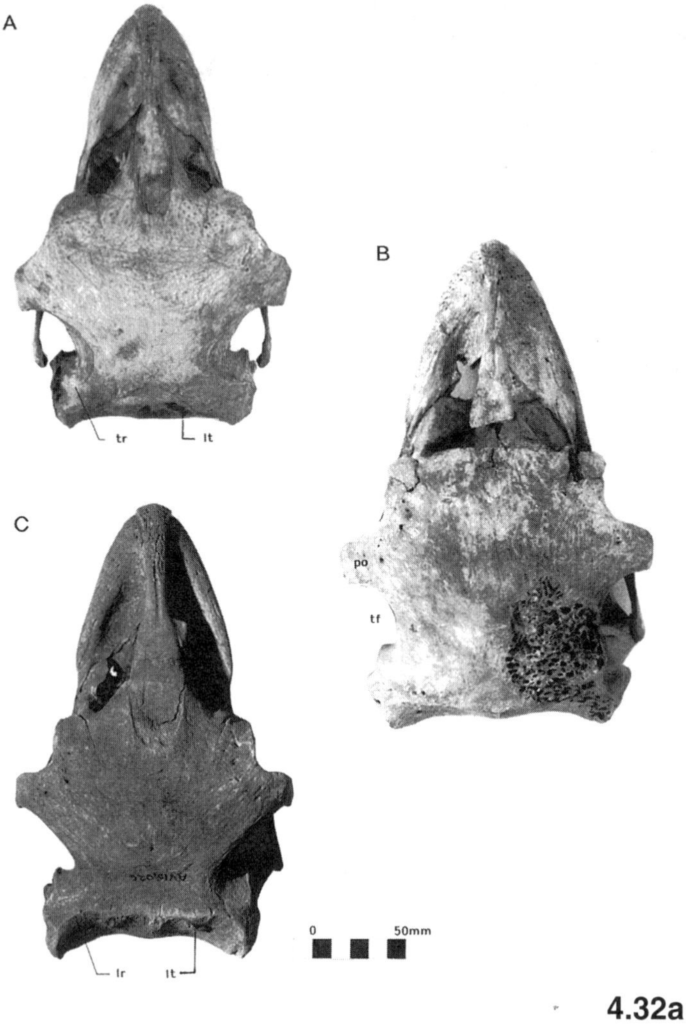

Figure 4.32. Dorsal (3.32a) and ventral (3.32b) views of Dinornis *crania. (A)* Dinornis struthoides *from Brian Reeve's collection. (B)* Dinornis novaezealandiae *MNZ S32667. (C)* Dinornis giganteus *CM Av15026. b = basisphenoid process; lr = lambdoidal ridge; lt = lateral tuberosity on occipital crest; mpt = medial flange of pterygoid; mt = mamillar tuberosity; oc = occipital condyle; po = postorbital process; tf = temporal fossae; tr = temporal ridge. Figure reproduced with permission of the Royal Society of New Zealand from Worthy (1994a: figs. 1, 2).*

Dinornis novaezealandiae

The cranium–postorbital width is 104–125 mm (mean, 114 mm); temporal fossae/postorbital width ratio ranges 0.53–0.65 (mean, 0.58; i.e., postorbital width is relatively less than in *D. struthoides* or *D. giganteus*). The occipital region is flat or has only weakly developed protuberances laterad of the occipital crest (prominent in *D. struthoides* or *D. giganteus*), and the mamillar tuberosities are weakly developed ventrally (prominent in *D. struthoides* or *D. giganteus*) and usually have a shallow smooth hollow between them. The occipital condyle has no neck between the articular surface and the cranium (as in *D. struthoides,* but unlike in *D. giganteus*), and the articular surface is caudally oriented so in ventral view the condyle appears crescent-shaped. The rostrum is rounded ventrally (compressed in *D. struthoides* or variable in *D. giganteus*). The pterygoid medial flange is short pos-

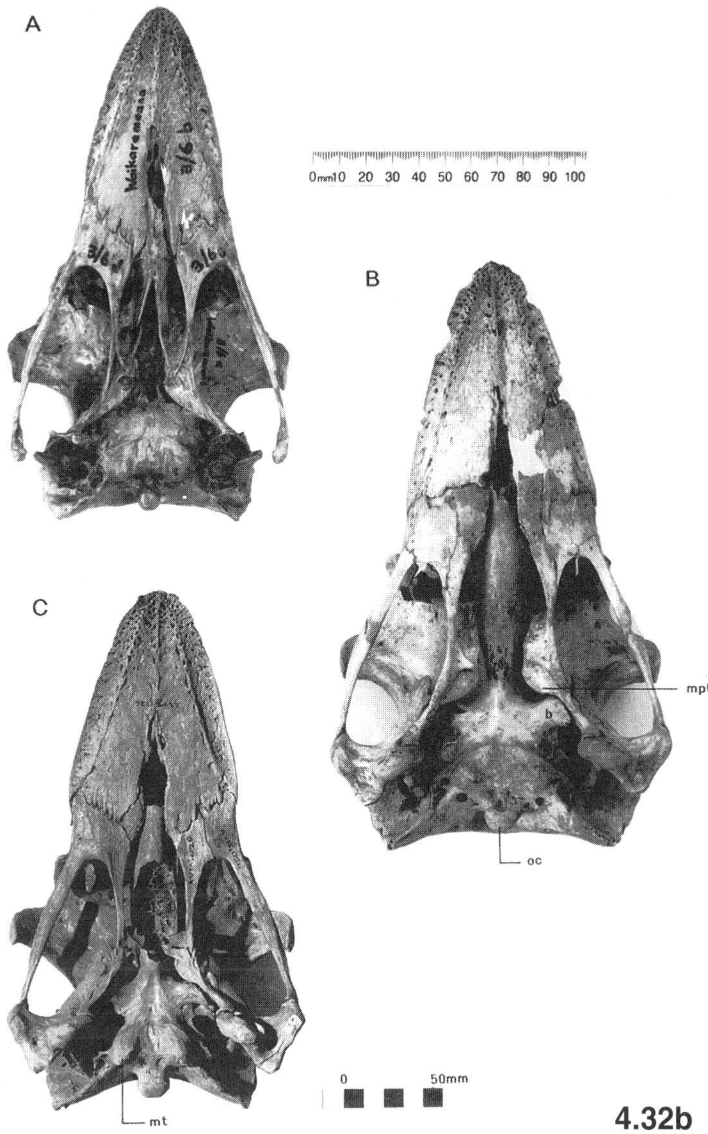

4.32b

teriorly and does not overlap the articulation with the basipterygoid (long and overlaps articulation with basipterygoid in *D. struthoides*). Leg bone lengths are as follows: femora 295—356 mm, tibiotarsi 605—685 mm, and tarsometatarsi 304–383 mm.

Dinornis giganteus

The cranium–postorbital width is 120–145 mm (mean, 134 mm), and is therefore larger than in other species; temporal fossae/postorbital width ratio ranges 0.5–0.6 (mean, 0.54). The occipital region has strongly developed protuberances laterad of the occipital crest, and the mamillar tuberosities are strongly developed ventrally and protrude variably posteriorly; they usually have a marked hollow between them (features that distinguish it from *D. novaezealandiae* and ally it to *D. struthoides*). The occipital condyle has a pronounced neck between the articular surface and the cranium (unlike *D. struthoides* and *D. novaezealandiae*), and the articular surface is caudally oriented so that in ventral view, the condyle appears crescent-shaped (in which it agrees with *D. novaezealandiae* and differs from *D. struthoides*). The rostrum may or may not be compressed ventrally. The pterygoid medial flange is short posteriorly and does not overlap the articulation with the basipterygoid (as in *D. novaezealandiae*). Leg bone lengths are as follows: femora 328–447 mm, tibiotarsi 675–992 mm, and tarsometatarsi 345–531 mm (Holdaway and Worthy 1997; Worthy 1994a).

Family Emeidae

The bill is either broad or narrow at the tip but is not flattened. It is higher than wide at the caudal margin of the nasal septum of the premaxilla, which has straight sides. The mandible also has straight rami, which have a marked projection on the lateral cotyla. The cranium is less than twice as broad as high. The delicate preorbital plate is not connected to the rostrum and is angled forward. The planes of the rostrum and basitemporal plate coincide, with no angle between them. There are 27 presacral vertebrae; those at the nape have a bifid neural spine. The vertebrarterial canals of the cervical vertebrae are not laterally compressed. The uncinate processes are either not ankylosed or not well ankylosed to the ribs. The femur, tibiotarsus, and tarsometatarsus have relative proportions 1:1.6–1.8:0.7–0.8. The femur has one complex tuberosity on the ventral side of the shaft, the neck separating the ball from shaft is distinct and narrower than the ball, and the ventral surface of the proximal shaft is convex and forms a continuous curve to the lateral surface of the trochanter (Fig. 4.18). The angle formed by the dorsal surface of the ectocnemial ridge and its ventral support on the tibiotarsus is less than or equal to 90°. The length of the tarsometatarsus is less than 2.5 times its maximum width, and the sides of the shaft are not parallel. The sides of the pelvic escutcheon are curved in dorsal view. The sternum has symmetric depressions within its body and has straight lateral processes; the xiphoid process is not bifid.

Subfamily Anomalopteryginae (Genera: Pachyornis, Anomalopteryx, Megalapteryx*)*

Digit IV with five phalanges; trachea with no body loop; bill with pointed tip; olfactory chamber expanded.

Genus Pachyornis

The bill tip is fine and pointed and in dorsal view has straight sides that diverge caudally (not outwardly curved or becoming parallel caudally). The mandible is robust. In lateral view, the premaxilla tip is sharp and the culmen straight. The maxillary process of the nasal is a slender bone that lies anterior to the lachrymal, expands ventrally, and articulates with the maxilla. The turbinal in the olfactory chamber has one complete spiral. The maxillary antrum is expanded, with the chamber about as wide as it is long. The skull has wide temporal fossae; whether the temporal and lambdoidal ridges are in contact depends on the species. The sternum has widely divergent lateral processes, and the body is much shorter than total width. The pelvis is wide with a flattened dorsal profile, the base of the acetabulum meets the synsacrum in a curved junction, and the costal processes of the escutcheon diverge widely. The scapulocoracoid is relatively long, with the scapular segment being longer. The uncinate processes are stout and not fused to the ribs. The distal end of the femur is expanded markedly and the shaft is short and has nonparallel sides (in dorsal view); the shaft width is more than 14.5% of total length. The rotular groove on the femur is broad. The anterior tibiotarsus has a prominent ridge extending from the buttress of the ectocnemial ridge between the procnemial ridge and fibular crest, which results in the shaft in this region being markedly convex. The procnemial ridge is bent laterally from the axis of the shaft in its upper section, and the distal end of the bone is greatly expanded medially. In anterior view the shaft is "stepped" above the medial condyle. The medial or endotrochlea of the tarsometatarsus is angular medially; its widest point is level with the proximal end of the adjacent intertrochlear space. The middle trochlea arises abruptly from the anterior face of the shaft, and the lateral hypotarsal ridge is longer and higher than the medial one.

Pachyornis elephantopus

This species was massive for its height, being the most thick-set moa. On the skull, the temporal fossae spread over the dorsal surface of the cranium (Fig. 4.33). There is little or no gap between the temporal and lambdoidal ridges. The zygomatic and postorbital widths are about equal, which makes the skull rather rectangular in dorsal view. The sternum lacks coracoidal sulci. Size varies according to sex and geologic age. The femora are 265–335 mm long, and the popliteal and ectepicondylar fossae are about equidistant from the proximal end. The tibiotarsi have a short fibular crest (~20% total length), and the nutrient foramen (medulloarterial foramen) is distal to the end of the fibular crest. Lengths range from 500 to 620 mm. The tarsometatarsi are 200–270 mm long and very wide in proportion to their length (shaft width about 26% total length; distal width about 59% length).

Pachyornis australis

This was also a massive, squat bird but was more lightly built than *Pachyornis elephantopus.* In the skull, the temporal fossae do not spread dorsally over the cranium. They are instead sharply defined medially. Unlike *P. elephantopus,* there is a wide gap between the temporal and lambdoidal ridges, and the postorbital width is markedly greater than the zygomatic width, so the cranium is not rectangular in plan view. The dorsal surface of the cranium, especially on the frontals, is covered in small pits up to 2 mm wide and about 0.5 mm deep. These are interpreted as feather pits. They are only variably present on crania of *Dinornis* spp., *Anomalopteryx,* and other *Pachyornis* spp. Equivalent pits are absent on crania of emeids. Uniquely among moa, the quadrate has a pneumatic fossa behind the head of the otic process. The

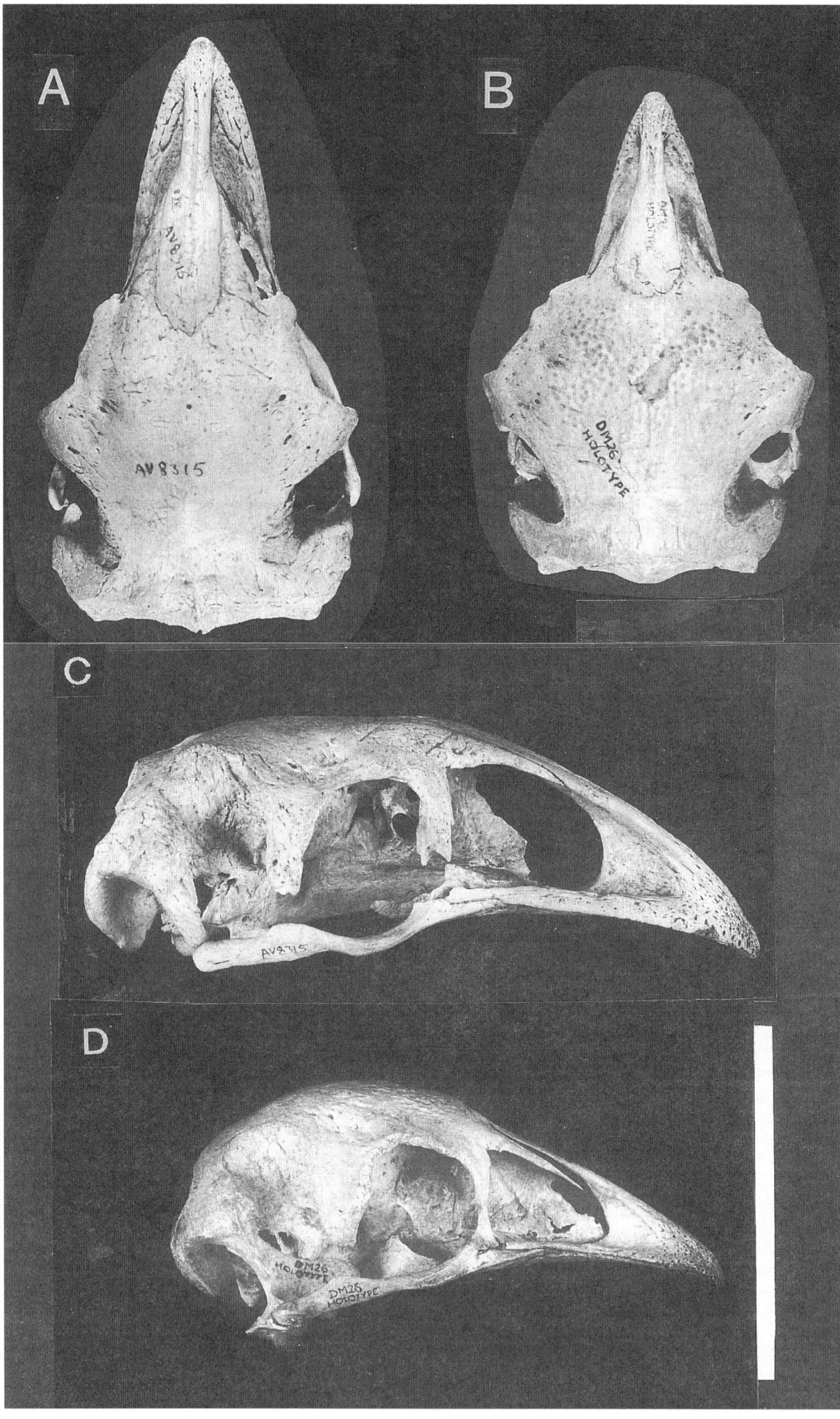
A
B
AV8315
DM26
HOLOTYPE
C
D

sternum has deep sulci, or hollows, for the articulation of the coracoid. The femur has a large, pneumatic foramen on the dorsal pretrochanteric surface, and the popliteal fossa is more distal than the ectepicondylar fossa. Femora are 240–310 mm long. Tibiotarsi are 430–500 mm long and have a long fibular crest (~25% total length; Fig. 4.19). The nutrient foramen is at the distal end of the fibular crest. Tarsometatarsi are 190–255 mm long and not as stout as those of *P. elephantopus*, with a shaft width of about 23% total length; the distal width is about 52% of the length).

Pachyornis mappini

This is a stoutly built but relatively small species. The temporal fossae do not spread dorsally over the cranium but are sharply defined medially. There is a wide gap between the temporal and lambdoidal ridges, and the postorbital width is markedly greater than the zygomatic width. The sternum has deep coracoid sulci. The femur has a large pneumatic foramen on its ventral surface at the proximal end. The popliteal fossa is more distal than the ectepicondylar fossa. Femora are 140–230 mm long (for Holocene specimens) or 199–274 mm long (from Pleistocene deposits). Tibiotarsi have a short fibular crest (~22% total length), and the nutrient foramen is usually distal to the end of the fibular crest. Tibiotarsi from Holocene specimens are 240–372 mm long but are 345–492 mm long in Pleistocene specimens. The tarsometatarsi are 115–170 mm long (Holocene) or 150–210 mm (Pleistocene).

Genus Anomalopteryx

The premaxilla tip is robust, and after initially diverging (proximal to the tip), the sides diverge only slightly. The ridge of the culmen rises steeply at first, then flattens out toward the nasals. The mandible is robust, with straight, divergent rami. The maxillary process of the nasal is slender and poorly connected to the nasal and so is usually lost; it lies anterior to the lachrymal, expands ventrally, and articulates with the maxilla. The maxillary antrum is expanded—the chamber is longer than it is wide. The turbinal in the olfactory chamber has one complete spiral. The cranium has wide temporal fossae spreading over the dorsal surface, causing the temporal and lambdoidal ridges to be in contact. The orbital margins are parallel rostrally, and the postorbital width is equal to the zygomatic width. The sternum has narrowly divergent lateral processes that are shorter than or equal to the xiphoid. The overall width of the sternum is about the same as the length, and the coracoid sulci are deep and obvious. The pelvis is narrow with a curved dorsal profile in lateral view, and the base of the acetabulum meets the synsacrum in a curved junction. The scapulocoracoid is relatively long (about as long as the sternum is wide) with the scapular segment longest.

The uncinate processes are robust and are usually ankylosed to the ribs. The distal end of the femur is not markedly expanded, and the shaft is more elongate (than all moa except *Megalapteryx*) with parallel sides (shaft width <14.5% of total length). The lateral condyle in dorsal view is semiparallel to the shaft (unlike all other moa), there is a broad

Figure 4.33. (opposite page) Crania of the sympatric Pachyornis elephantopus *and* Pachyornis australis *in (A, B) dorsal and (C, D) right lateral views. The smaller* P. mappini *has a cranium similar to that of* Pachyornis australis *but is found only in the North Island. The anterior cranium of* P. australis *has strongly indented pits. (A, C)* P. elephantopus *CM Av8315. (B, D)* P. australis *holotype MNZ S26. Scale bar = 100 mm. Figure modified from Worthy (1989b).*

rotular groove, and the popliteal fossa is the same distance as the ectepicondylar fossa from the proximal end. The anterior tibiotarsus shaft is flat between the ridge from procnemial ridge and the fibular crest. The procnemial ridge is bent laterally from line of shaft in its upper section, and the distal end is not greatly expanded medially. The tarsometatarsus is relatively elongate, with straight sides that converge slightly distally. In anterior view, the lateral side of the outer trochlea and the medial side of the inner trochlea are subparallel. In medial aspect, the anterior extreme of the medial trochlea (point "a" in Fig. 4.20) is markedly more distal to the caudal extreme (point "b" in Fig. 4.20), not opposite it, as in *Emeus* or *Euryapteryx*.

Anomalopteryx didiformis

Same characters as for genus. Femora 210–260 mm long, tibiotarsi 315–430 mm long, and tarsometatarsi 145–200 mm long.

Genus Megalapteryx

The bill tip is pointed, delicate, and has divergent sides. A much reduced maxillary process of the nasal lies medial to the lachrymal and is articulated ventrally with the maxilla. Its poor connection with the nasal means it is usually lost (but see Oliver 1949: figs. 6, 121). The maxillary antrum is nearly collapsed, leaving a small remnant chamber. The turbinal in the olfactory chamber does not have one complete spiral. The cranium has small temporal fossae; the temporal and lambdoidal ridges are not in contact. The orbital margins converge anteriorly, and the postorbital width is greater than the zygomatic width. The sternum has narrowly divergent lateral processes that are not longer than the xiphoid. Its overall width is about the same as its length, and the coracoidal sulci are indistinct or absent. The pelvis is narrow with a curved dorsal profile in lateral view, and the base of the acetabulum meets the synsacrum at right angles. The scapulocoracoid is relatively short (much shorter than the anterior width of the sternum), and the scapular segment is shortest.

The distal end of the femur is not markedly expanded. The femur has a narrow rotular groove and the shaft is more elongate (than all other moa) with parallel sides (shaft width <14.5% of total length); the popliteal fossa is distal to the ectepicondylar fossa. The anterior tibiotarsus has a prominent ridge extending from the ectocnemial ridge to between the ridges from the procnemial ridge and the fibular crest, so that the shaft surface in this region is markedly convex. The procnemial ridge is straight and not bent laterally where the proximal end expands; the distal end is not greatly expanded medially. On the tarsometatarsus, a prominent ridge bounds the nutrient foramen beneath the medial hypotarsal ridge anteriorly, and it is not set on a flat surface as in other species. In anterior view, the outer sides of the lateral and medial trochlea are subparallel.

Megalapteryx didinus

This rather gracile moa, with long, supple toes, has the characters of the genus. Its femora are 200–330 mm long, the tibiotarsi 270–490 mm long, and the tarsometatarsi 135–220 mm long. Pleistocene specimens are much bigger than Holocene-age specimens, which contributes to the large size range.

Subfamily Emeinae (Genera: Euryapteryx, Emeus*)*

Digit IV has four phalanges, and the ventral surfaces of the unguals are nearly flat. The trachea includes a body loop (Fig. 4.22). The bill has a rounded tip, and the mandibles are weakly constructed. The olfactory chamber is relatively small. The turbinal in the olfactory chamber has one complete spiral. The temporal fossae are small. The sterna are longer than wide and have no coracoidal pits. Scapulocoracoids are not known. The uncinate processes are thin and not fused to the ribs. The anterior shaft of the tibiotarsus is flat between the procnemial ridge and the fibular crest, and the procnemial ridge is bent laterally from the line of the shaft in its upper section. In anterior view, the medial side of the tibiotarsus shaft is not stepped, as in *Pachyornis,* but rather meets the medial condyle in a smooth, shallowly concave curve. The tarsometatarsus is narrowest in its lower third rather than at midlength. The middle trochlea arises from the shaft at a shallow angle, not abruptly (Worthy 1992b).

Genus Euryapteryx (Euryapteryx curtus, E. geranoides)

The skull lacks distinct depressions above the orbits, has no marked depressions mesial to the paroccipitals, and is not always domed between the postorbital processes; the dorsal margin of the occipital foramen is thin. The premaxilla and mandible are broadly rounded, so the rami do not diverge widely. The maxillary antrum is collapsed, with no remnant chamber. The maxillary process of the nasal is either absent or reduced to a thin, unconnected sliver of bone lying anterior to the lachrymal, as in the type of *Euryapteryx boothi* (now *Euryapteryx geranoides*) Oliver (1949: fig. 82; Parker 1895: pl. LVI). The sternum has a single hollow behind the raised anterior margin in which there are a few scattered foramina, and the lateral processes are shorter than the basin length anterior to the lateral notch.

The pelvis is wide with a curved dorsal profile in lateral view, the base of acetabulum meets the synsacrum in a curved junction, and the costal processes of the escutcheon do not diverge widely from the sacrum. The iliac plates do diverge widely from the center line above the midpoint of the acetabulum. The femur shaft is straight, relatively short, and robust (width/total length = 0.145–0.220). There are no large anterior fossae, and the crest of the trochanter does not extend along the neck to the ball, so the neck is rounded ventrally. The popliteal fossa is distal to the ectepicondylar fossa. In the tibiotarsus, the procnemial ridge has a flattened top, and the fibular crest is long, with the nutrient foramen usually proximal to its distal end. The medial extreme of the inner trochlea of the tarsometatarsus is angular. In dorsal view, the anterior side of the articular surface is flat so that the anteroposterior depth is narrower than in the similarly stout *Pachyornis* or *Emeus.*

Euryapteryx geranoides

This is the larger of the two species now recognized in the genus. The skull has well-developed mamillar tuberosities, so that the basitemporal platform appears to be concave in posterior view. The basitemporal platform and the rostrum form an angle of about 100° with the plane of the occipital foramen. The paroccipitals are prominent so that a line along their posterior margin clears the occipital condyle. The femur is narrowest at midlength. The tarsometatarsus is less than 2.1 times longer than its distal width. The extent of the distal projection of the hypotarsal ridges varies in this stout species, but the entometatarsal tuberosity is prominent. Femora are 215–340 mm long, tibiotarsi 405–600 mm long, and tarsometatarsi 150–270 mm long.

Euryapteryx curtus

The skull has weakly developed mamillar tuberosities, so that in posterior view the basitemporal platform appears flat or there is only a very shallow groove between them. The basitemporal platform and rostrum form an angle of about 120° with the plane of the occipital foramen. The paroccipitals are not prominent so that in dorsal view, a line along their posterior margin intersects the occipital condyle. The dorsal border of the tympanic cavity forms a ridge that is continuous anteriorly over the zygomatic process. Crania may have a prominent interorbital elevation, in addition to the dome between the temporal fossae. The femur is narrowest in its proximal third. The posterior margin between the distal condyles of the tibiotarsus is flat and not deeply concave, as in the similar-size *Pachyornis mappini*. The tarsometatarsal hypotarsal ridges project distally to equal extents. Femora are 150–225 mm long, tibiotarsi 240–380 mm long, and tarsometatarsi 110–165 mm long.

Genus Emeus

The skull has distinct depressions above the orbits within which are distinct foramina, there are marked depressions mesad of the paroccipitals, it is always domed between the postorbital processes, and the dorsal margin of the occipital foramen is swollen to form a prominent supraforaminal ridge. The basitemporal platform is flat in posterior view because the mamillar tuberosities are very weakly developed. The maxillary antrum is expanded into a small remnant chamber. The maxillary process of the nasal appears to be fused to the anterior edge of the lachrymal and is not expanded ventrally. The medial process of the mandible lacks a pneumatic foramen, unlike in all other genera. The sternum has a two distinct hollows behind the raised anterior margin in which there are two groups of foramina, and the lateral processes extend caudally of the lateral notches a greater length than that of the basin anterior of the notch.

In the pelvis, the iliac plates diverge gradually from the center line at a point adjacent to the anterior margin of the acetabulum (not as abruptly or as caudally as in *Euryapteryx*). The femur is deeply concave on the pretrochanteric surface and has a well-developed scar (often a pit) for the insertion of the ?ligament iliofemorale. The femur shaft is usually laterally compressed and has only one ventral ridge; the trochanteric ridge is continuous across the neck to the ball. The popliteal and ectepicondylar fossae are equidistant from the proximal end. In the tibiotarsus, the nutrient foramen is at the distal end of the fibular crest, and the procnemial ridge in anterior view slopes up to its greatest height rather than being flattened across the top, as in *Euryapteryx*. On the tarsometatarsus, the medial extreme of the inner trochlea is angular but is opposite the greatest depth of the adjacent intertrochlear notch. The tarsometatarsus is relatively elongated, with straight, slightly convergent (distally) sides, and the shaft is narrowest in its distal third. In medial aspect, the anterior extreme of the medial trochlea is equidistant with the caudal extreme of the trochlea from the proximal end, which distinguishes it from the similarly proportioned tarsometatarsi of *Anomalopteryx didiformis*. The anterior edge of the proximal end is not flat above and lateral to the intercondylar ridge, and the entometatarsal tuberosity is weak or absent.

Emeus crassus

Same characters as the genus. Femora are 215–300 mm long, tibiotarsi 340–495 mm long, and tarsometatarsi 160–227 mm long.

Identification of Moa Leg Bones

An illustrated key to the identification of moa femora, tibiotarsi, and tarsometatarsi by use of the characters identified above is given in Appendix 2.

A New Look at Moa Species Interrelationships

The foregoing discussion has largely been aimed at diagnosing the individual species within the moa family. This was done within the bounds of the current taxonomy as given above and as advocated by Worthy (1988c) and Turbott (1990a). However, in this earlier work where Worthy better defined the various species (e.g., Worthy 1988c, 1988d, 1989a, 1992b, 1994a), the higher taxonomy of moa was not addressed. Here we present a cladistic analysis of the morphologic characters used in the foregoing species diagnoses in an attempt to determine these higher relationships.

The characters used are described in Appendix 3. They were analyzed with the computer program PAUP version 3.1.1 (Swofford 1993). This program assesses the distribution of characters among the compared taxa and constructs a tree where taxa are united by the shared presence of derived characters. The tree uniting the taxa with the least number of branches is the preferred tree. This is said to be the most parsimonious arrangement, which involves the fewest changes in characters. Although evolution does not necessarily always proceed parsimoniously, the method does provide the least number of evolutionary steps. The analysis was carried out in the simplest way, with no subjective weighting of characters, treating them as unordered (i.e., the direction of evolutionary change is not assumed), and variable character states were coded separately. The coding of species is shown in Appendix 3, Table 1.

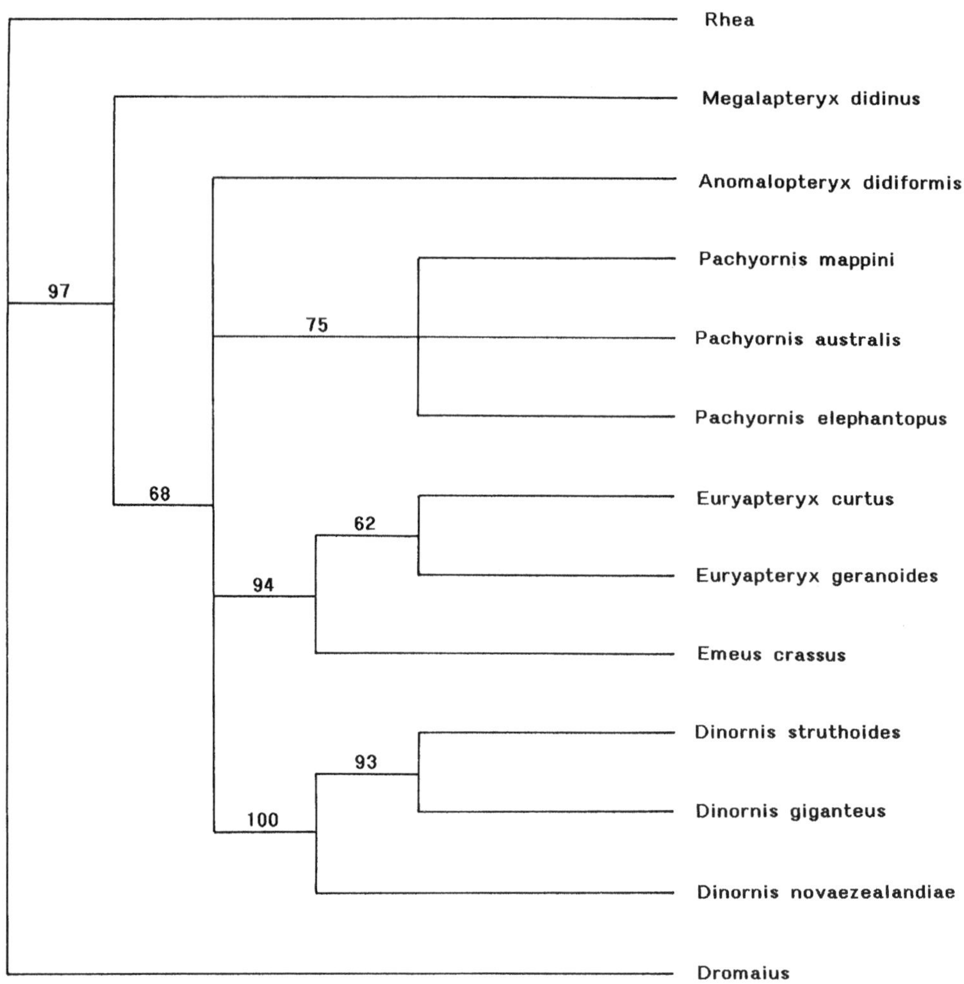

Figure 4.34. Phylogenetic tree of moa, with emu and rhea as outgroups, based on cladistic analysis of 82 morphologic characters. The characters are unordered and unweighted. This consensus parsimony tree shows bootstrap values based on an unweighted analysis (100 replicates, PAUP version 3.1.1; Swofford 1993) that indicate the percentage of trees that share that typology. For example, 100% of trees had all Dinornis *species together and 75% had* Pachyornis *species associated, but only 68% of trees had* Megalapteryx didinus *separated from the rest of moa.*

A test to show whether the tree produced by PAUP has any biologic significance, such as reflecting phylogeny, is available. The pattern of the frequency distribution of tree lengths from 1000 trees produced at random indicates whether the data are structured, and structure is considered to mean that the data reflect a phylogenetic signal (Hillis and Huelsenbeck 1992). The tree-length distribution of 1000 random trees generated from the moa data set in Appendix 3 had a g_1 value of −1.132978 ($P < 0.01$), indicating a highly significant skew (Hillis and Huelsenbeck 1992), showing the data set has structure. The provisional cladistic analysis revealed here resulted in two shortest trees of 211 steps (consistency index [CI] = 0.602). These differed in whether *Pachyornis elephantopus* was paired with *P. mappini* or *P. australis.* However it was noted that at 212 steps, 5 trees were obtained, and at 213 steps, 16. With emu and rhea as the designated outgroups, a consensus parsimony tree was obtained by use of the bootstrap method

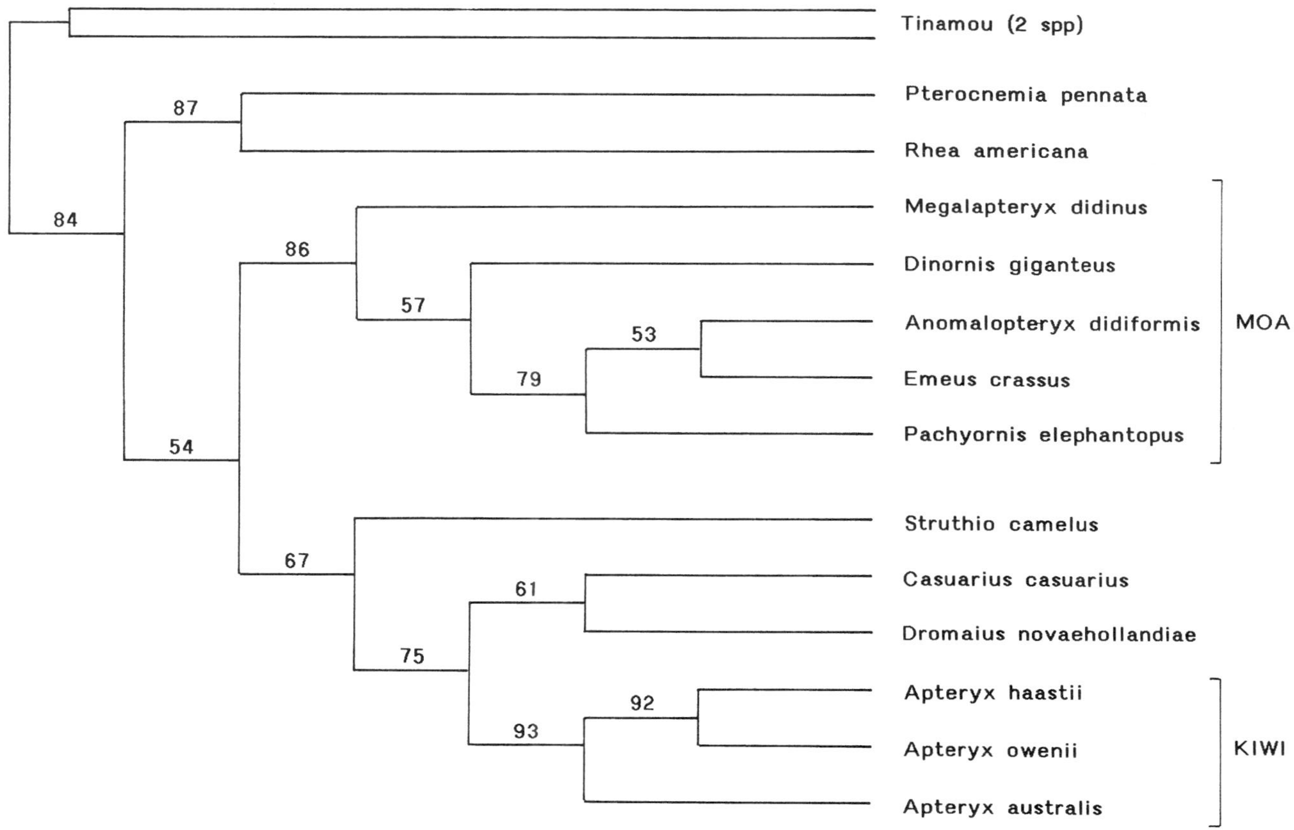

Figure 4.35. A phylogenetic tree of moa taxa derived from mitochondrial DNA analyses. Branches show relationships only, not genetic distance. After Cooper (1997).

with a heuristic search (Fig. 4.34). The method involves resampling with random changes in the data set to see what the possible distribution of the data could be. All moa taxa form a single clade. Surprisingly, and against predictions from the current taxonomy, *Megalapteryx didinus* was the basal species within moa. However, as predicted from the number of trees obtained that were only 1–2 steps longer, the major feature of the tree was an unresolved polychotomy of *Anomalopteryx*, *Pachyornis* spp., *Dinornis* spp., and the emeins (*Euryapteryx* and *Emeus*). Within these clades, the *Pachyornis* species formed a further polychotomy, but *Dinornis* species and the emeins grouped as expected on the basis of the other information.

We then constrained the analysis so that *Megalapteryx didinus* was a sister group of *Anomalopteryx didiformis* and found a tree of only 215 steps (CI = 0.591). Constraining the analysis to the typology predicted from the traditional taxonomy required a tree only one step longer than that at 216 steps (CI = 0.588). These analyses indicate that the genera of moa as now defined are well supported, but that the present data are insufficient to resolve the intergeneric relationships. There is a weak signal that *Megalapteryx didinus* is not the sister taxon of *Anomalopteryx didiformis*. Generally, these data do not support the presence of two moa families (Emeidae and Dinornithidae) with two subfamilies of emeids.

DNA and Interspecific Moa Relationships

The first genetic data that compared several moa species was that by Cooper et al. (1992). In this and subsequent studies (e.g., Cooper 1997), a surprising feature has been obvious. *Megalapteryx didinus* was found to be the basal group within moa, branching off below *Dinornis*, with other genera forming a more recent clade (Fig. 4.35). This result is inconsistent with the traditional taxonomy made on the basis of morphologic analyses, in which the three species of *Dinornis* comprise the family Dinornithidae and the other genera together comprise the Emeidae. Within Emeidae, *Megalapteryx* is aligned near *Anomalopteryx* with *Pachyornis* in the Anomalopteryginae, and *Emeus* and *Euryapteryx* form the Emeinae. However, these DNA analyses do support the cladistic analysis of morphologic characters presented here (Fig. 4.34) that suggested *Megalapteryx didinus* is basal in the moa family and widely separate from *Anomalopteryx didiformis*.

Further genetic studies comparing all moa taxa with more sequence data are in progress, and their analyses will incorporate the morphologic data. It will be interesting to see whether these initial results are borne out.

An unexpected outcome of these intraspecific comparisons of moa was that their overall genetic diversity is low. This has been related to a reduction in actual diversity of moa ancestors, and it has been suggested that the present diversity had its origin after the Oligocene (Cooper et al. 1992; Cooper and Cooper 1995). The Oligocene saw much of New Zealand submerged—only 18% of the present land area remained above sea level and thus had low relief—so possibly only one ancestor survived this period. Similar degrees of genetic diversity are seen in kiwi and wrens, and so both these groups may each have a single ancestor giving rise to them since the Miocene.

Relationships among Ratite Families: Morphology

The above discussion has clearly shown that moa are ratites. Definition of this group has been a huge problem in avian taxonomy and will be reviewed briefly below. Resolution of the problem is confounded by the apparent primitiveness of all ratites, and modern taxonomy requires derived characters to define a group. Second, the extreme modifications accompanying their development as huge cursorial birds are dominant and obscure the features of historical phylogenetic significance. Currently, ratites include the flighted tinamous of South America (nine genera and 45–50 species), ostrich (*Struthio camelus*) of Africa and formerly in Eurasia, rheas (*Rhea americana, Pterocnemia pennata*) of South America, elephantbirds or aepyornithids of Madagascar (*Mullerornis,* three species; *Aepyornis,* four species), cassowary (*Casuarius,* three species), emus (a single living member *Dromaius novaehollandiae,* and several extinct forms), kiwi (*Apteryx,* four species and an extinct form), and moa, Dinornithiformes (six genera and 11 species).

The large extinct Dromornithids or mihirung birds of Australia were considered to be ratites most closely related to cassowaries by Vickers-Rich (1980). This group ranged from the Miocene to the Pleistocene and has eight species in five genera, including one species, *Dromornis stirtoni,* that may have rivaled or exceeded in weight *Aepyornis maximus,* often considered bird heavyweight of the world. However, Olson (1985) noted that in *Genyornis,* the deep, troughlike mandible, which lacked traces of rhamphothecal grooves, and the quadrate were unlike those of any ratite. He also noted that the ilioischiatic foramen of the pelvis was closed, so this genus at least lacked the diagnostic characters of ratites. Bledsoe (1988) included the dromornithids in his phylogenetic analysis of the ratites and found them to be most similar to the cassowaries and emus, but because he lacked material sufficiently complete to diagnose dromornithids as ratites in the first place, he considered their relationships to still be unresolved. Vickers-Rich (1990) has commented that the cranial morphology discernible in some recently prepared specimens indicates a structure so highly derived that it is difficult to associate mihirung birds with any known group. A detailed description of these and newly found cranial specimens by Murray and Megirian (1998) has convincingly shown that dromornithids were not ratites and indicated instead that they had a close relationship with anseriform birds. Their similar bill morphology suggests dromornithids were the ecologic equivalents of *Diatryma,* an equivalent-size Eocene bird of the Northern Hemisphere, and Andors (1992) showed that *Diatryma* also had anseriform affinities. Both appear to be best interpreted as giant herbivores (Andors 1992; Murray and Megirian 1998).

Around the end of the 19th century, most investigators (e.g., Fürbringer 1888; Pycraft 1900) accepted a polyphyletic origin (the members of a polyphyletic group do not share a single common ancestor) for

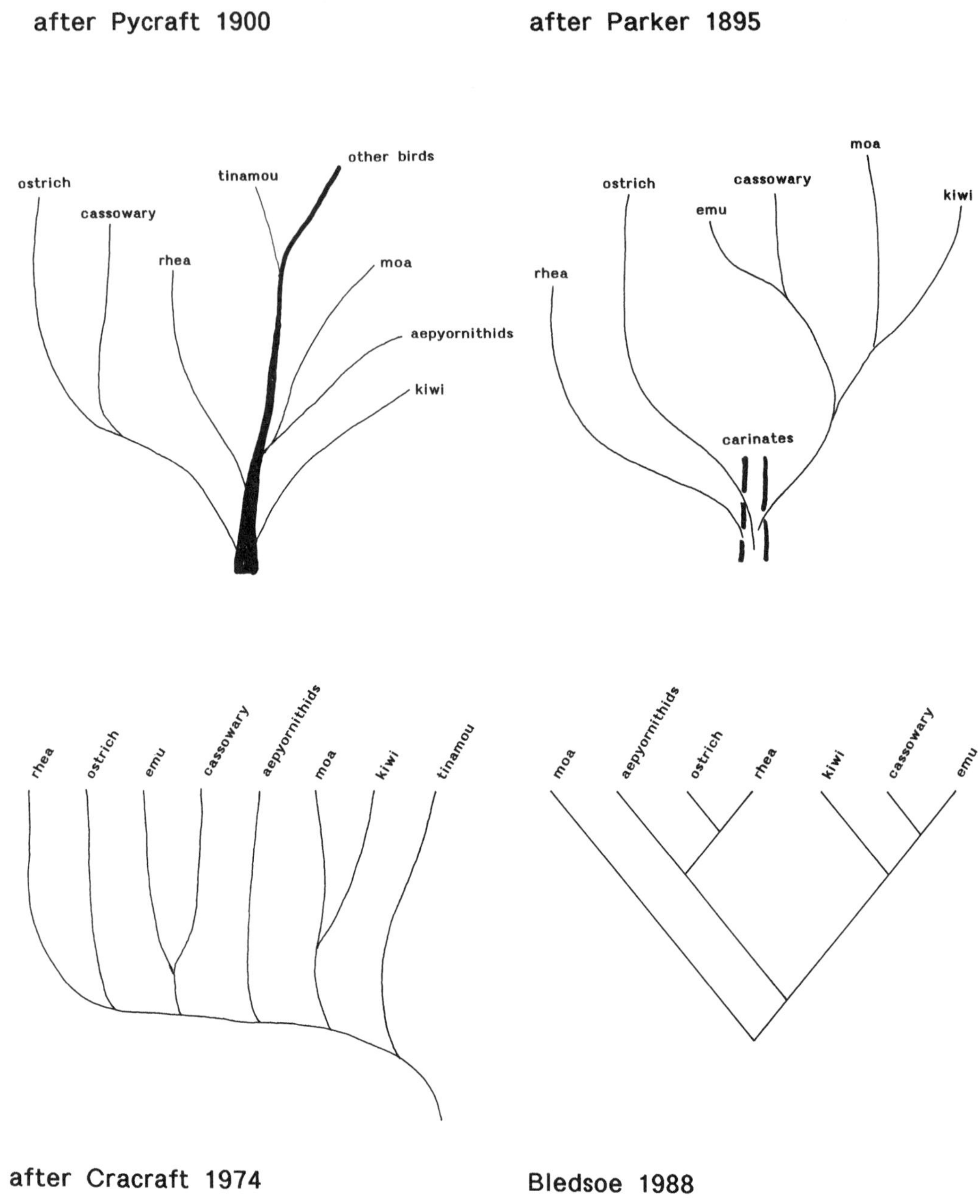

Figure 4.36. Four different hypotheses of ratite relationships spanning a century of work. All studies were based on morphologic analyses.

the ratites, a view no doubt arrived at in part as a result of the perceived fixity of continents—the theory of plate tectonics and continental drift had yet to be invented and accepted. Perhaps one of the most influential works was that of Pycraft (1900). He considered ratites to be polyphyletic and primitive in relation to other birds. Within ratites, he allied moa most closely to aepyornithids, swayed in large part by the shared presence of an escutcheon to the pelvis in these taxa. He also noted some similarities with *Rhea*, especially the structure of the palate. Significantly, he placed *Casuarius* and *Dromaius* distant from moa on his tree (Fig. 4.36).

Parker (1895) also advocated a polyphyletic origin for the ratites (Fig. 4.36). He considered *Apteryx* to be the closest ally of moa; he considered the marked differences between these taxa as being of an adaptive nature. He placed the four Australasian families as being most closely related, but he saw no close relationship between these and rheas or ostriches. In this he more or less agreed with Fürbringer (1888), who derived Casuariformes (emu and cassowary) and Apterygiformes (moa and kiwi) from separate primitive stock.

However, from the earliest investigations, some researchers concluded that ratites were monophyletic. From his studies of the axial skeleton of various ratites, Mivart (1877) considered all ratites as monophyletic, with *Dromaius* and *Casuarius* as sister taxa, as were dinornithids and *Apteryx*. Together, these two pairs were closer to each other than to *Rhea* and *Struthio*. One of the most notable ornithologists of the last century was the English ornithologist A. Newton, and he clearly stated that all ratites were related most closely to each other than to any other birds (Newton 1885).

One of the most influential articles dealing with ratite origins was that of Cracraft (1974b). He performed a cladistic analysis on 25 skeletal characters of ratites and found they formed a monophyletic group (as in Fig. 4.36). Significantly, his results placed dinornithids and *Apteryx* as sister taxa emerging as a somewhat basal limb from the tree, and not very close to *Dromaius* and *Casuarius*, which formed a species pair higher on the tree. His results concurred with the evolution of ratites on Gondwana. The relatively new knowledge of plate tectonics and the subsequent new history of continental relations provided the mechanism (vicariance) by which populations of a protoratite could be dispersed around the world and evolve their differences. However, this result immediately drew criticism and stimulated further studies. Particularly, the development of DNA analytic techniques offered a new raft of tools to test the hypotheses.

However, Bledsoe (1988) challenged Cracraft's (1974b) conclusions on the same morphologic ground. He reexamined the question of ratite phylogeny by use of a cladistic analysis of 88 postcranial characters and concluded that moa were basal on the tree, that *Aepyornis*, *Rhea*, and *Struthio* formed a clade, and that *Apteryx*, *Casuarius*, and *Dromaius* formed a clade. Dissenting considerably from most previous workers, he found that moa were not the sister taxon of any of these groups and placed them basally on the tree; however, 28 of his characters pertained to the pectoral girdle elements, which are missing in moa. Such a large number of missing characters could affect the analysis.

Although Bledsoe (1988) and Cracraft (1974b) differed on interpretations of the interrelationships of the ratites, they both supported monophyly of the ratites. But as reviewed by, for example, Houde (1988), the debate on ratite monophyly versus polyphyly has continued into recent times. When Sibley and Ahlquist (1990) reviewed the work on ratite relationships, they listed no fewer than 75 individual workers or groups who had considered the matter, indicating the huge amount of effort put into this problem. To a certain extent, it has depended on whether authors have focused on differences among the taxa or sought their common features. But as noted above, the skeletal features that ratites have in common are few—paleognathous palate, open ilioischiatic foramen, and rhamphothecal grooves (Bock 1963; Cracraft 1974b). Some have argued that all could be derived through neoteny (e.g., Olson 1985).

However, the evidence is stacked in favor of a common origin for ratites. Apart from the above-mentioned osteologic features, similarities have been demonstrated between ratites in other morphologic features, particularly the structure of their eustachian tubes, middle ear morphology, coelomic structure, jaw musculature structure, carotid arteries, pattern of bone vascularization, parasite complements, chick plumage, pterylography, behavior, and spermatozoa (Houde 1988).

However, since the 1980s, reexamination of some fossils and discoveries of others in the Northern Hemisphere have put a new slant on the question of ratite relationships. *Palaeotis weigelti,* formerly called a bustard, of the middle Eocene from Germany, is actually a paleognathus bird, properly included in the Struthionidae (Houde and Haubold 1987). Studies of the smaller volant birds known as lithornithids by Houde (1986, 1988) showed that this group of species are also paleognathous carinates. However, their relationships to surviving ratites remain unclear—they have a combination of tinamoulike and ratitelike characters and thus could be ancestral tinamous or sister taxa to either tinamous or some ratites (Houde 1988). They date from Paleocene and Eocene deposits of North America and thus extend the geographic range of ratites.

Figure 4.37. (opposite page) Two recent hypotheses of ratite relationships. The relationship shown here from Lee et al. (1997) was based on morphology for ratites, including moa, but was the same when a total-evidence solution was attempted in which the morphologic data were combined with a data set for mitochondrial DNA (mtDNA) that excluded moa. Cooper's (1997) hypothesis was based on mtDNA data for all included taxa and included several species in each group —for example, three kiwi, two tinamous, and five moa.

The similarity of the lithornithid *Paracathartes* to kiwi is striking and may reflect a recent evolution of typical ratite morphology convergent with other ratites. Indeed, there is support for this in DNA evidence, as seen below. However, there are mid-Paleocene rhea fossils (Tambussi 1995; Tambussi et al. 1994), so *Lithornis* and other Eocene lithornithids cannot be ancestral to *Rhea,* but perhaps their ancestors were. The possibility remains that some Gondwanan flightless forms now surviving had a flighted ancestor and evolved independently of other ratites that were distributed by vicariance.

A Variety of Methods Support Ratite Monophyly

The question of ratite phylogeny has been addressed by many other methods at the cellular and molecular level (reviewed by Sibley and Ahlquist 1990). A sample to show the breadth of these studies is given below.

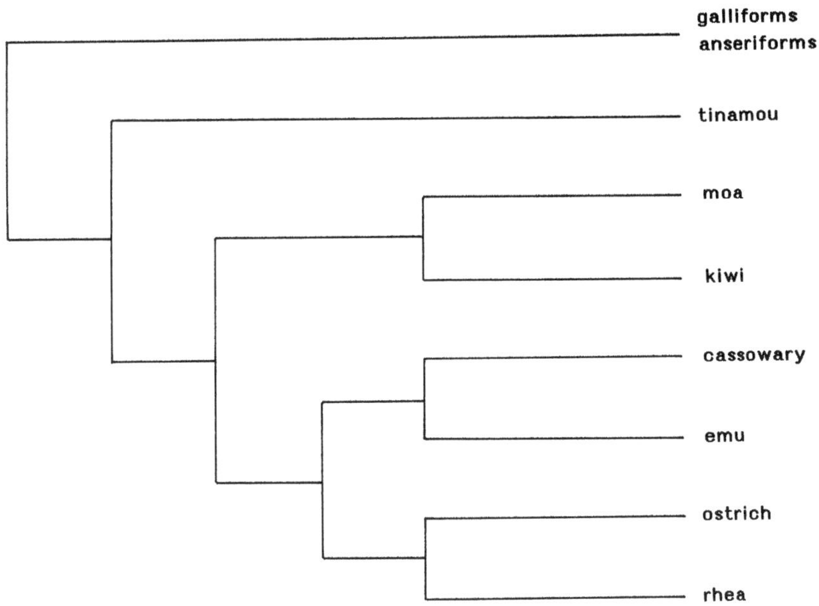

after Lee et al. 1997

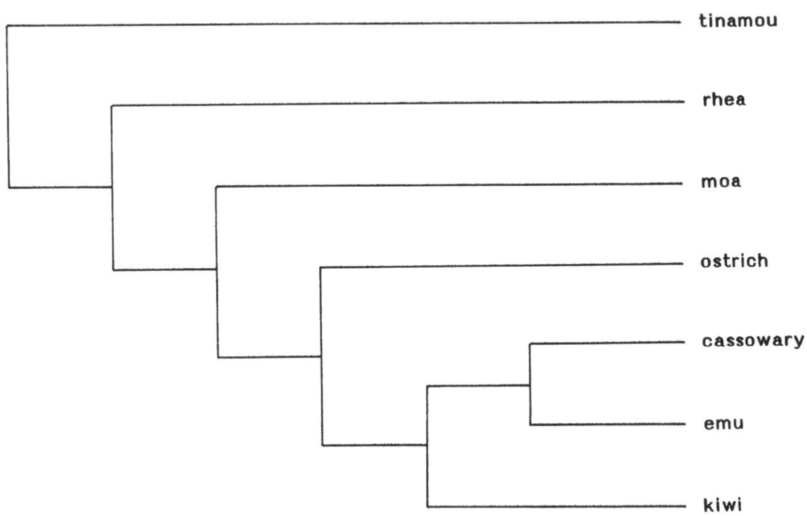

after Cooper 1997

De Boer (1980) addressed the question of ratite monophyly by use of chromosome analysis and found those of kiwi to be similar to other large ratites and to have an ancient character. They share a unique sex-determining system in that they do not have a clearly heteromorphic sex chromosome pair, as in other birds; the fourth chromosome pair differs only slightly in size.

The structure of the eye protein α-crystalline A indicated that ratites were the sister group of other birds (Stapel et al. 1984). Several studies demonstrated similarities in egg-white proteins (Sibley and Ahlquist 1990). But perhaps the most powerful techniques for demonstrating relationships are those that use DNA. After all, DNA forms the genes that define the animal, so relationships in the DNA should mirror phylogeny.

The technological innovation of DNA-DNA hybridization and analysis was first used on the ratite phylogeny problem by Sibley and Ahlquist (1981). Their results contrasted markedly with Cracraft's (1974b) in that *Apteryx* was found to be the sister group of *Dromaius* and *Casuarius*, but how this combined clade related to *Struthio* and *Rhea* was unresolved. Later, Sibley and Ahlquist (1990) published a new DNA-DNA hybridization study that retained the *Casuarius, Dromaius* and *Apteryx* clade, but they found two alternatives for the relationships of *Rhea* and *Struthio*.

Then in 1988, the discovery of the polymerase chain reaction allowed sequencing of DNA and RNA, enabling access to a whole new range of data that could be used in phylogeny. In 1992, the first results of mitochondrial DNA (mtDNA) amplification from moa tissue obtained from four taxa showed that dinornithiformids were not closely related to either *Apteryx* or the *Dromaius* and *Casuarius* clade (Cooper et al. 1992). This initial analysis was made on a relatively small sequence of 390 base pairs of the mitochondrial 12S genome and showed *Dromaius* and *Casuarius* as sister taxa. This pair was also shown to be the sister group to *Apteryx*, thus supporting Sibley and Ahlquist's DNA work. Cooper (1997) expanded upon this work with new sequences from other genes (ND6/transfer RNA–proline intragenic region; nuclear *c-mos* proto-oncogene) to produce a combined data set of 1023 base pairs (Cooper 1997). Again, a sister-group relationship of kiwi to the *Casuarius* and *Dromaius* clade was obtained. A new analysis of the 12S data set (enlarged by the addition of several different moa species) produced a tree (Fig. 4.37). Moa are near the base of the tree and not close to kiwi at all (Cooper 1997).

Significantly, Cooper (1997) showed that the inclusion of moa in the analyses greatly improved the resolution of the resultant phylogeny. In the same volume, Lee et al. (1997) used a large data set (5444 base pairs) of nucleotide sequences from a variety of genes in the mitochondrial genome for the extant ratites. They obtained a tree with the same typology as Cooper (lower tree in Fig. 4.37). But they noted that if tinamous were used to root the tree, although the same typology was obtained, the second most parsimonious tree was only five steps longer and produced a tree with *Apteryx* given a more basal origin and separated from the *Casuarius* and *Dromaius* clade.

A new morphologic analysis of 58 skeletal characters by Lee et al. (1997) resulted in a tree that was similar to the second most parsimonious tree obtained above (upper tree in Fig. 4.37). When they combined the nucleotide and morphologic data sets to produce a "total evidence solution," a single most parsimonious tree with the same typology was obtained. However, the next most parsimonious tree was only two steps longer and gave the typology obtained by the nucleotide data alone—that is, that obtained by Cooper (1997). It therefore seems that the application of the data set of Lee et al. (1997) was unable to satisfactorily discriminate between placing *Apteryx* with the *Casuarius* and *Dromaius* clade, or basal on the tree after tinamous. Cooper, with the advantage of the moa data set, was able to overcome some of the analytic problems caused by the great divergence between these few taxa (the so-called long branch attractant problem) and obtain greater resolution of the phylogeny. Significantly, his data do not place the moa near *Apteryx,* and although his data place moa basally on the tree, they put *Apteryx* with the *Casuarius* and *Dromaius* clade.

In the latest development in the search for ratite origins, Cooper's team has sequenced the entire mtDNA genome for two moa—*Emeus crassus* and *Dinornis giganteus*—and placed these in a framework of a greatly enlarged database for extant ratites and included aepyornithids for the first time (Cooper et al. 2001). The study supports a late Cretaceous origin of ratites and their subsequent evolution by vicariance as Gondwana broke up. Aepyornithids are shown to have an ancient lineage and are not clearly related to any modern group. In a radical departure from the convention of the split of Africa and South America facilitating the ostrich–rhea split, Cooper et al. (2001) suggest their data better support a Eurasian route for ostriches. Late Cretaceous connections between Australia/Antarctica and Indo-Madagascar via the Kerguelen Plateau would have allowed the ancestors of elephantbirds and ostriches to enter Indo-Madagascar. The subsequent northward movement of India would carry the ostrich to Eurasia, from where it dispersed later to Africa. Significantly, kiwi are again allied with the emu–cassowary clade and are well separated from moa. The estimated divergence times suggest kiwi separated from the emu–cassowary clade some 68 Ma, which is long after New Zealand ruptured from Australia (82 Ma). This means that ancestral kiwi dispersed to New Zealand over water, probably using the then-emergent sections of the Lord Howe and Norfolk Ridges. The two moa taxa indicate the familial split within moa occurred about 13.2 Ma and is thus coincident with other radiations of taxa as New Zealand emerged from a period of extensive land submergence during the Oligocene (Cooper and Cooper 1995).

In summary, more than 100 years of phylogenetic endeavor has largely reached consensus on the monophyly of ratites. But the substantial divergences within ratites present continuing problems to resolving details of the relationships—in particular, the placement of kiwi on the family tree remains contentious. Most analyses agree that moa are relatively primitive and occupy a basal position on the tree. What is sorely needed now are fossils from the periods of early ratite evolution to test some of the new hypotheses that are based on DNA.

5 • The Living Moa

Post-Pleistocene Dwarfing of Moa

The reduction in mean individual size of species between the last Glacial maximum and the Holocene is a well-known phenomenon in mammals. For example, in Australia, there were giant morphs of such recent taxa as Tasmanian devils (*Sarcophilus*), long-beaked echidnas (*Zaglossus*), and grey kangaroos (*Macropus*) (Murray 1991). *Macropus giganteus* and *M. robustus* are morphologically identical but smaller equivalents of, respectively, *M. titan* and *M. altus* from the Pleistocene (Dawson and Augee 1997). Giant morphs in birds over this time period are a much rarer phenomenon—if indeed any have been described—but several species of moa had larger morphs during the last Glacial period. Giant morphs can only be recognized if a species is identified by morphologic characters rather than by dimensions. Such characters are now known for all moa species.

The first moa in which temporal size variation was studied is *Megalapteryx didinus*. At one time, as many as three species of *Megalapteryx* were recognized, separated by size. When this genus was reexamined, only one species could be accepted because there were no observable differences in form. The coefficients of variation (CVs) of the dimensions of any single population were within limits acceptable for a sexually dimorphic population (Worthy 1988d). The size variation was found to be related to age. Dates given by Worthy (1988d) and subsequently by Worthy and Holdaway (1993, 1994b) show that none of the smallest specimens and all of the larger individuals (previously called *M. benhami*) are of the Pleistocene (Otiran Glacial) age.

The smallest individuals of *M. didinus* are from Late Holocene deposits in the Punakaiki karst (100–200 m) on the West Coast. They are much smaller than individuals from contemporary populations in the subalpine zone above 1200 m in northwest Nelson, only 100 km to the north. Fossil sites from the subalpine zone are of very local occurrence in New Zealand: significant faunas are known only from northwest Nelson (>1200 m) and Fiordland (>900 m). In both areas, the moa fauna is dominated by *M. didinus,* which was also common in the high country of Otago (>500 m) during the Holocene. *M. didinus* therefore seems to have been the dominant moa in high-country environments along the length of the South Island. There was little size variation in these higher altitude populations, so presumably those living in the mountains inland of the West Coast study area were of similar size to other high-country individuals.

The exceedingly small individuals comprising the Late Holocene lowland populations suggest that there was clinal variation in size with altitude. The absence of a similar cline with latitude in the high-altitude population may be because the species lived primarily in the subalpine zone, the lower boundary of which is depressed with increasing latitude but within which the temperatures are similar. So although *M. didinus* became smaller between the Late Otiran and the Late Holocene, there is also evidence that it was smaller down an altitudinal gradient. Both trends indicate that size in this species was inversely correlated with temperature.

The mean size of individuals of *Pachyornis elephantopus* also varied through time (Fig. 5.1). Bones of the mid-Otiran (30,000–40,000 years before the present [B.P.]) population on Mount Cookson in northeastern South Island were significantly smaller than those in the population there during the coldest part of the last Glacial period 14,000–20,000 years ago (Worthy and Holdaway 1995). About 60 km south, in the Waikari district, the same species trapped 10,000–12,000 years ago in the Glencrieff site had significantly larger bones than the population trapped 1.5 km away in Pyramid Valley 2000–4000 years ago (Worthy and Holdaway 1996b). Birds in the Albury Park site in South Canterbury, which has a similar age, had similar mean sizes to those in Glencrieff. The Cheviot site in North Canterbury is younger than Pyramid Valley, judging by a radiocarbon age on the enclosing peat, and the mean size of *P. elephantopus* bones is a little smaller again than at Pyramid Valley. It therefore seems that the mean size of *P. elephantopus* increased with the cooling temperatures until the Late Glacial maximum and thereafter declined.

In the North Island, *Pachyornis mappini* and *Euryapteryx curtus* were also larger during the Otiran than they were in the Holocene (Worthy 1987a: Fig. 5.2). What all these species have in common is that they inhabited the most open environments available to moa. Species that frequented closed forest have the same size range in deposits of different ages and in different areas of New Zealand. Therefore, the mean individual size of moa most exposed to cold temperatures—those that preferred open shrubland or grassland habitats—changed in accordance with Bergmann's Rule.[1] Mean size in these species probably increased and decreased with every cold–warm cycle of the climate. The trends emphasize the fact that size is a poor taxonomic feature in moa species.

1. Bergmann's Rule states that as one moves down a temperature gradient (e.g., toward higher latitudes), the mean size of warm-blooded animals increases.

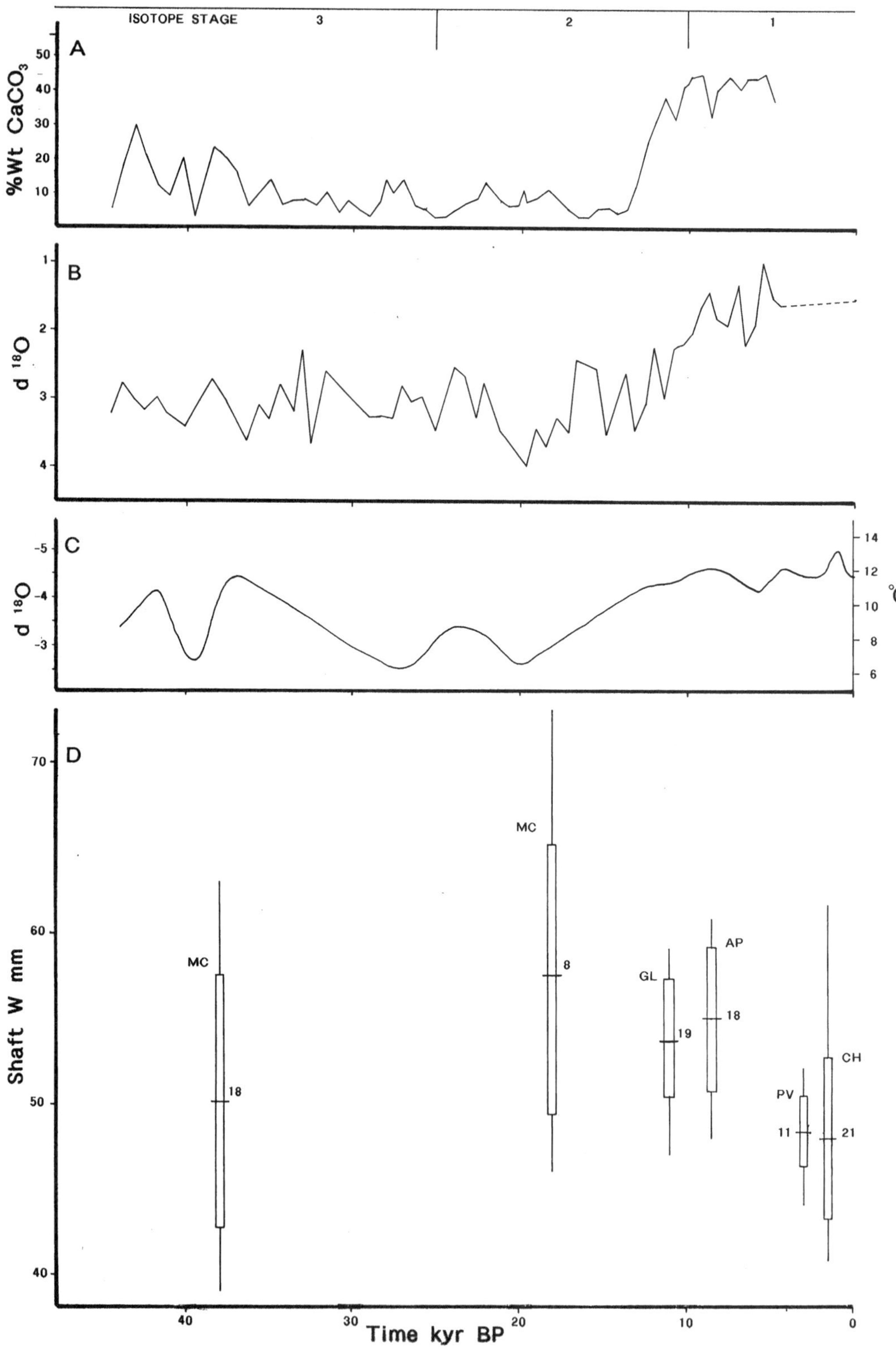
ISOTOPE STAGE
3
2
1
A
%Wt $CaCO_3$
50
40
30
20
10
B
d ^{18}O
1
2
3
4
C
d ^{18}O
-5
-4
-3
14
12
10
8
6
°C
D
Shaft W mm
70
60
50
40
MC
18
MC
8
GL
19
AP
18
PV
11
CH
21
40
30
20
10
0
Time kyr BP

Changed Proportions in Element Lengths

A much rarer phenomenon than post-Pleistocene dwarfing is altered proportions of different skeletal elements in a similarly short period of time. Over longer geologic time periods, this is a common trend and is well documented for evolution of deer, elephants, and bovids on islands, where shortening of the distal leg is a common feature (Sondaar 1977). In birds, altered proportions of skeletal elements are associated with the formation of many island species by the degeneration of wings and hence loss of powered flight (e.g., Livezey 1990; Olson and Jouventin 1996). Rails in particular have speciated prolifically in this manner (Olson 1990b; Steadman 1989a). However, in the short term of the last 10,000–20,000 years, few species are known to have undergone significant shifts in body proportions. This may be because detection of such morphologic trends requires large sample sizes, which are rarely available, especially from many sites that each span a limited time period. These conditions were met by the ptarmigan (*Lagopus mutus*), one of the most common species in palearctic Pleistocene avifaunas (Tyrberg 1995). Modern specimens of *L. mutus helveticus* have longer tarsometatarsi and shorter carpometacarpi than fossil specimens from the last Glacial period (Bochenski and Tomek 1994).

A common member of the communities inhabiting New Zealand's dry vegetation mosaics was Finsch's duck (*Euryanas finschi*). Populations of Glacial and Late Glacial age had wings that were about 10% longer than their Late Holocene successors (Worthy 1988b). Absolute size, as indicated by femur length, did not change in the same period. Since that study was published, populations of Glacial age on Takaka Hill, as well as in North and South Canterbury, have all been found to have relatively long wings, confirming that the wing–leg proportions are a useful chronologic indicator in this species.

The apparently rapid regression in flight ability indicated by these observations suggests that *Euryanas* was responding to a relatively new release from the selection pressures that maintained flight ability in its ancestors. Against Worthy (1988b) and Livezey (1989b), but in concurrence with Oliver (1955) and Olson (1991), we now believe that *Euryanas* was probably derived from the Australian wood duck (*Chenonetta jubata*), which favors grazing terrestrial grasslands. If so, a relatively recent arrival of this duck in New Zealand is indicated because exploitation of the terrestrial grazing niche it was preadapted to—and its concurrent release from predation—would allow flightlessness to evolve. Such a colonization was probably in one of the preceding glacial periods when open habitat was widely available. This hypothesis predicts that *E. finschi* would not be found in early Pleistocene faunas of New Zealand. *Euryanas finschi* is the only species in New Zealand where the morphologic changes toward flightlessness have been measured, despite the presence of many flightless species.

Fossil sites that provide samples of terrestrial avifaunas more than 50,000 years old are virtually unknown in New Zealand, so there is no evidence to show when flightlessness evolved in most species. The few

Figure 5.1. (opposite page) Various environmental parameters plotted against time (radiocarbon years ago). (A) The $CaCO_3$ content of the Deep Sea Core DSDP 594 hole (from Nelson et al. 1993). Low $CaCO_3$ content is the result of dilution by terrigenous silt supplied by rivers and wind. (B) The $\delta^{18}O$ values of the planktonic foraminifera Globigerina bulloides *from DSDP 594 (from Nelson et al. 1993). The $\delta^{18}O$ values are generally held to have an inverse relationship with temperature—that is, more negative values relate to higher temperatures. (C) The mean annual temperature determined from $\delta^{18}O$ values of speleothems (after Hendy 1969). (D) Pictorial representation of summary statistics of shaft widths for several populations of* Pachyornis elephantopus *plotted at the modal age of the deposit. AP = Albury Park; Ch = Cheviot Swamp; MC = Merino Cave, Mount Cookson; GL = Glencrieff Swamp; PV = Pyramid Valley. The boxes enclose one standard deviation, the vertical bar is the range, the horizontal bar the mean, and* n *is the sample size. Dates for sites: Mount Cookson, Worthy and Holdaway (1995); Pyramid Valley, Glencrieff, Worthy and Holdaway (1996b); Albury Park, Worthy (1997b); Cheviot Swamp, bones from 0.5–1.5 m deep in peat. The peat was radiocarbon dated as 580 ± 70 years (Wk1671) at 0.5 m and 5230 ± 80 years (Wk1670) at 4.0 m, so the assumed age of bones is 1–2000 years ago.*

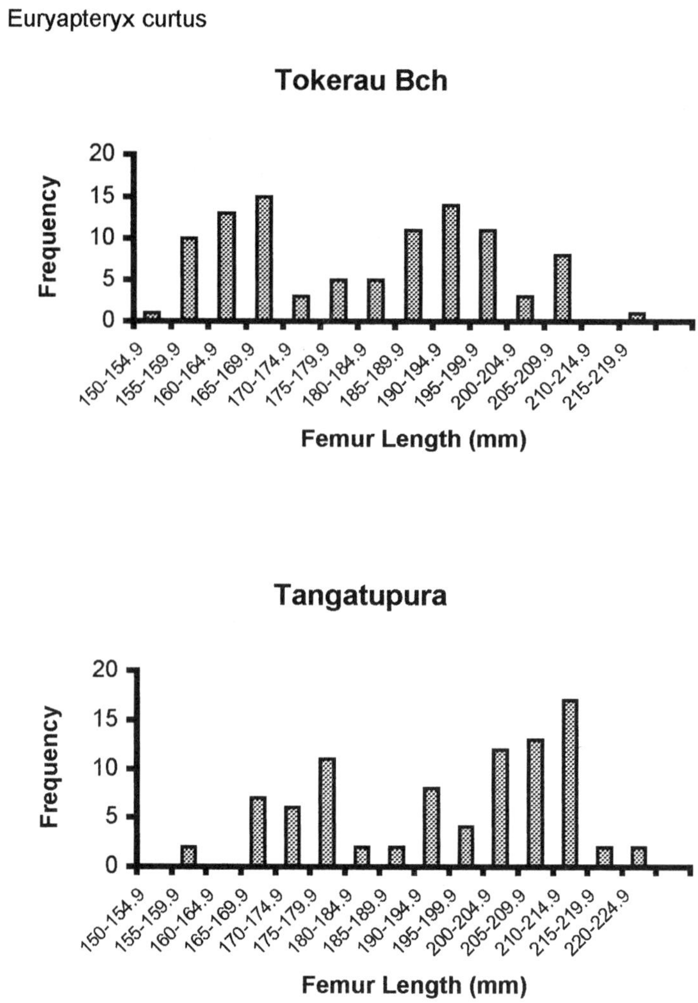

Figure 5.2. Size frequency distributions for Euryapteryx curtus *from Tokerau Beach (Holocene) and Tangatupura (Otiran Glacial) deposits, on the basis of femur length. Length is clearly bimodal in each site, and the median points in Tokerau Beach are clearly less than in Tangatupura, reflecting postglacial dwarfing in this species.*

moa fossils older than 100,000 years show that all species had their Late Holocene form 1 million years ago (Ma), and for some, 2.5 Ma (Worthy et al. 1991). So in this group, flightlessness was not a recent phenomenon.

Biology of Moa

How Heavy Were Moa?

It is generally held that moa were among the largest birds known but that elephantbirds were larger (Alexander 1983a). One of the earliest estimates of moa mass, and perhaps that most widely quoted, is that by Amadon (1947). He reported that *D. maximus* weighed 230 kg and *D. giganteus* 242 kg. As it happens, these two taxa are the South

TABLE 5.1.
Estimates of moa body mass (kg) by Alexander (1983a).

Taxon and catalog number[b]	Method[a] 1	2	3	4
Anomalopteryx didiformis BMNH A3	29		30	
Anomalopteryx didiformis BMNH A129		46		42
Euryapteryx geranoides BMNH A95		66		75
Pachyornis elephantopus BMNH A3620		163		
Pachyornis elephantopus BMNH unreg		119	115	
Pachyornis elephantopus LCM	129		127	170
Dinornis giganteus BMNH A608	242		199	
Dinornis giganteus BMNH 32039-42		179		
Dinornis giganteus BMNH unreg		304		

[a] Methods were as follows: 1, from scale models; 2, assumes mass was proportional to the cube of leg length; 3, assuming trunk mass [trunk length (thoracic vertebrae to end of ilium) × trunk width (mean of maximum width of rib cage and pelvic girdle) × trunk depth (mean of femur length and depth from the posterior end of the sternum) in mounted skeletons] is a constant proportion of body mass determined volumetrically; 4, assuming skeletal mass is a constant proportion of body mass.
[b] BMNH = British Museum (Natural History), London; LCM = Leeds City Museum specimen, Leeds, England; unreg = unregistered specimen.

TABLE 5.2.
Body masses of moa (kg) calculated using the femur circumference method of Atkinson and Greenwood (1989) compared with values arrived at using this and a leg length method by Anderson (1989c).

Moa species	Atkinson and Greenwood 1989				Anderson 1989c	
	n	Mean	Minimum	Maximum	Leg length	Femur circumference
Megalapteryx didinus	10	23.7	16.9	34.3	53	25
Anomalopteryx didiformis	16	41.3	31.7	58.7	46,49	32
Emeus crassus	24	74.9	45.6	120.1	61	51
Euryapteryx curtus					26	22
Euryapteryx geranoides	19	95.7	48.7	139.9	57, 88	61, 61
Pachyornis mappini					30	28
Pachyornis elephantopus	20	146	96.7	247.4	123	105
Dinornis struthoides	12	96.2	82.0	114.8	106	117, 125
Dinornis novaezealandiae	8	143.7	100.0	199.7	169, 280	167
Dinornis giganteus	10	177.9	133.6	272.7	242	NA

Abbreviation: NA = not available.

TABLE 5.3.

Body masses (kg) of moa species calculated using algorithms that are based on femur lengths (Prange et al. 1979), femur shaft width (Campbell and Tonni 1983), and femur circumference (Anderson et al. 1985).[a]

Species and age/site	Reference	Length (mm)	Shaft (mm)	Mass (kg) based on length	Mass (kg) based on circumference	Mass (kg) based on diameter
Dinornis struthoides	Worthy 1994a	284.4 (261–307); n = 8	41.3 (38–46.2); n = 8	71 (56–88)	$\bar{x}$ = 71 (43–115) Min 59 (36–95) Max 92 (56–151)	$\bar{x}$ = 68 (52–79) Min 59 (43–71) Max 81 (67–92)
Dinornis novaezealandiae	Worthy 1994a	319.6 (295–356); n = 12	51.5 (47.1–59); n = 12	98 (78–132)	$\bar{x}$ = 117 (71–195) Min 58 (36–93) Max 160 (95–270)	$\bar{x}$ = 98 (85–106) Min 84 (70–94) Max 122 (115–127)
Dinornis giganteus	Worthy 1994a	381.6 (332–447); n = 13	57.5 (45–64); n = 12	160 (109–249)	$\bar{x}$ = 151 (90–252) Min 86 (53–142) Max 93 (113–328)	$\bar{x}$ = 117 (109–123) Min 78 (62–89) Max 140 (138–142)
Anomalopteryx didiformis Makirikiri	Worthy 1989a	232.8 (199–259); n = 207	NA	40 (26–54)	NA	NA
Anomalopteryx didiformis	Archey 1941	240.25 (200–275); n = 32	32.4 (26–39); n = 32	44 (26–64)	$\bar{x}$ = 41 (26–65) Min 25 (16–38) Max 62 (38–101)	$\bar{x}$ = 35 (22–47) Min 31 (19–42) Max 61 (46–73)
Megalapteryx didinus Holocene	Unpublished data	231.3 (203–268); n = 21	27.8 (23.2–35); n = 27	40 (28–60)	$\bar{x}$ = 29 (18–45) Min 19 (12–29) Max 49 (30–78)	$\bar{x}$ = 35 (26–51) Min 26 (14–37) Max 51 (36–63)
Megalapteryx didinus Glacial	Unpublished data	262.2 (200–297); n = 21	34.2 (28–45); n = 35	56 (26–80)	$\bar{x}$ = 46 (29–74) Min 29 (19–46) Max 86 (53–142)	$\bar{x}$ = 49 (34–61) Min 35 (22–47) Max 78 (63–89)
Pachyornis mappini Holocene, Makirikiri	Worthy 1989a	201 (172–224); n = 22	NA	27 (17–36)	NA	NA

Species and age/site	Reference	Length (mm)	Shaft (mm)	Mass (kg) based on length	Mass (kg) based on circumference	Mass (kg) based on diameter
Pachyornis mappini Glacial, Tangatupura	Worthy 1987a and unpublished data	233.6 (199–274); *n* = 52	26.7 (28.9–44.3); *n* = 52	41 (26–64)	$\bar{x}$ = 26 (17–41) Min 31 (20–49) Max 83 (51–136)	$\bar{x}$ = 33 (20–44) Min 37 (24–49) Max 76 (61–87)
Pachyornis australis	Worthy 1989b	(240–310); *n* = 36	22.9; *n* = 11	44–90		
Pachyornis elephantopus		See Table 5.6	See Table 5.6			
Emeus crassus Pyramid Valley	Worthy and Holdaway 1996b	265.4 (224–296); *n* = 56	NA	58 (36–79)	NA	NA
Euryapteryx curtus Doubtless Bay, Far North	Unpublished data	181.6 (152–218); *n* = 100	NA	20 (12–34)	NA	NA
Euryapteryx curtus Glacial, Tangatupura	Worthy 1987a and unpublished data	194.4 (156.8–223.0); *n* = 88	27.1 (20.3–32.5); *n* = 88	24 (13–36)	$\bar{x}$ = 27 (17–42) Min 14 (9–21) Max 41 (26–65)	$\bar{x}$ = 33 (20–45) Min 21 (11–31) Max 45 (31–57)
Euryapteryx geranoides Southern North Island	Worthy 1992b	255.7 (225–282); *n* = 10	NA	53 (37–69)	NA	NA
Euryapteryx geranoides Paerau, Otago	Worthy 1992b	300 (254–332); *n* = 21	NA	82 (52–109)	NA	NA
Euryapteryx geranoides Pyramid Valley	Unpublished data	284.3 (265–305); *n* = 7	45.1 (40–51); *n* = 7	71 (58–86)	$\bar{x}$ = 87 (53–142) Min 66 (41–107) Max 115 (69– 191)	$\bar{x}$ = 78 (63–89) Min 64 (49–76) Max 96 (83– 105)

[a]Measurement values are given as mean (minimum–maximum) values. Calculated masses using femur length give single-figure estimates, so a mass based on the mean is followed by parenthetical values for the minimum and maximum femur lengths. The algorithms for estimating mass from femur circumference (computed assuming round shafts) and femur diameter generate estimate ranges; three sets of values are therefore presented for each species: the calculated mass and estimate range for the mean ($\bar{x}$), minimum (Min), and maximum (Max) values. NA = not available.

TABLE 5.4.

Body masses and heights of modern large ratites from cited sources. In *Struthio* and *Rhea*, males (M) are the larger sex, whereas in other genera, females (F) are largest.

Species	Mass (kg)[a]	Height (m) and mass (kg)[b]	Height (m) at head[a]	Height (m) at back[c]
Struthio camelus	M <150	M <2.5 m, <135 kg	M 2.10–2.75; F 1.75–1.90	111
Rhea americana	M <25	M <1.6 m, <25 kg	M 1.60	87
Pterocnemia pennata	NA	Average 0.9 m	NA	NA
Casuarius casuarius	M 34.1, 29.2 (n = 2); F 58.5 (n = 1)	1.8 m	1.5–1.7	93–102
C. unappendiculatus	NA	1.6 m	1.6	NA
C. bennetti	NA	1.1 m	1.1	NA
Dromaius novaehollandiae	25–48; mean F 36.9; mean M 31.5	1.5–1.8 m; F 41 kg, M 36 kg	1.5–1.9	102–112

Abbreviation: NA, not available.
[a] Marchant and Higgins (1990).
[b] Harrison (1978).
[c] Anderson (1989c).

Island and North Island representatives of the species *Dinornis giganteus* as accepted here. In calculating the body masses, Amadon used a different algorithm for each taxon. For *D. maximus,* he used the proportion of femoral cross-sectional area to mass in the cassowary (*Casuarius* sp.) as his model and for *D. giganteus,* the proportion of body length (anterior border of thorax to acetabulum) to mass in the cassowary. The resulting values are not therefore directly comparable, but they did suggest that these birds were massive, weighing about as much as a cow.

Alexander (1983a) was next to estimate the body mass of moa. However, because he wanted to determine the factors of safety in the leg bones, he derived masses for only four species of moa. Four methods were used (Table 5.1).

Of these, method 4 can be disregarded because bone mass in fossils depends on preservation, and the mass of equal-size bones of the same species from a single site can vary widely. Method 2, which assumes that mass was proportional to leg length, is likely to be in error because moa have relatively shorter femora than other birds. The two volumetric methods (1 and 3) gave comparable results and suggested that different moa species varied widely in their body masses. The variability within species was not assessed.

At about the same time, Anderson (1984) listed estimated body masses for five species of moa. They were calculated by Ian Smith of the anthropology department of Otago University from femur cross sections; the body masses were for *Euryapteryx geranoides, Emeus crassus, Megalapteryx didinus, Dinornis struthoides,* and *D. giganteus.* Anderson (1989c) expanded the list to include all species for the first time (Anderson 1989c; Table 5.1) by use of Alexander's method of equating mass to the cube of leg length (Table 5.2).

However, these early estimates of moa weight failed to acknowledge the biologic reality of moa as a species, rather than individuals, and so came in a range of sizes for each. As all students of moa systematics have noted, the variation in size is quite high. In fact, it is obvious because of the great size of the birds. When the variation is reduced to a size-independent value such as a CV, then the variation is seen to be about the same as in many species (see e.g., Cracraft 1976b). Nevertheless, there are within a species small individuals and others considerably larger—in some cases, nearly twice as large as the small ones. The first authors to attempt to document this range were Atkinson and Greenwood (1989).

Atkinson and Greenwood (1989) used femur circumferences and applied the allometric relationship $W = 1.08 \times C_f^{2.28 \pm 0.1}$ (where W is weight and C_f is femur circumference) that was derived from the relationship between femur circumference and body weight in 72 species of flying birds by Anderson et al. (1985). As the femur was usually oriented in life at about 30° downward from the horizontal, it takes the load of the animal as a bending moment on its shaft, so the diameter or circumference of the shaft should directly reflect a bird's mass. Atkinson and Greenwood's analysis substantially improved the picture of the relative sizes of the various moa, but there was still no information on the size range of the two small North Island species, and the estimates for *Megalapteryx didinus* were based only on small specimens of that species.

An added complication is that several species varied considerably in size between the Pleistocene Glacial period and the Holocene. To illustrate this, we recalculated body masses for all species using measurement data from various sources (Table 5.3), including published sources as indicated, but some of it new. We used published algorithms that were based on femur length (Prange et al. 1979), femur circumference (Anderson et al. 1985), and greatest femur shaft width (Campbell and Tonni 1983). We prefer the results given by femur shaft width because it most closely reflects the fact that body mass is transmitted to the feet by a bending moment in the femur.

Moa obviously varied widely in size. Some were not much larger than a jumbo-size turkey; others were truly giants of the bird world. As the data given by Atkinson and Greenwood (1989) showed, *Dinornis giganteus* was not as heavy as suggested by earlier estimates. It had a relatively slender pelvis, and improbable as it may seem, *Pachyornis elephantopus*, although considerably shorter in stature, was often as heavy as the giant moa.

How Do These Body Masses Compare with Those of Living Ratites?

Body masses for modern ratites are given in Table 5.4. As their slender proportions would indicate, all living ratites are lighter than moa of the same height. Emeid moa of 25–40 kg would be no more than 50–60 cm high at their back.

TABLE 5.5.
Body masses for aepyornithids calculated from algorithms for femur length, femur circumference, and greatest femur shaft diameter (preferred); all measurements are in millimeters.[a]

Species	Reference or catalog no.	Element[b]	Length	Circumference, shaft diameter	Mass by length	Mass by circumference (estimate, range)	Mass by diameter (estimate, range)
Aepyornis gracilis	Monnier 1913	Fem	322	(170) 60	100	131, 78–219	125, 119–129
Aepyornis hildebrandti	Monnier 1913	Fem Max	320	(195) 70	98	179, 106–304	162, 159–168
Aepyornis hildebrandti	MNZ S38301	Fem L	320+	59.5 × 52	98+	140, 83–235	124, 117–129
Aepyornis ?hildebrandti	UA4902	Fem L	306	58.7	86	158, 94–267	121, 114–126
Aepyornis ?hildebrandti	UA8647	Fem R	325	63.5	102	189, 112–322	140, 137–141
Aepyornis medius	Monnier 1913	Fem Min	330	(210) 72	107	212, 124–363	170, 165–179
Aepyornis medius	Monnier 1913	Fem Max	368	(240) 82	145	288, 166–499	212, 197–239
Aepyornis medius (as *hildebrandti*)	L'Academie Malgache	Fem R	350	68.6	126	226, 132–386	157.5, 155–161
Aepyornis medius (as *hildebrandti*)	L'Academie Malgache	Fem L	355	66	131	207, 121–352	147, 147.8–147.6
Aepyornis maximus	Monnier 1913	Fem Min	410	(270) 90	196	377, 215–660	247, 223–294
Aepyornis maximus	Monnier 1913	Fem Max	465	(285) 96	278	427, 242–751	275, 243–339
Aepyornis maximus	L'Academie Malgache	Fem R	430	92 × 67	223	316, 182–550	256, 229–309
Aepyornis maximus	L'Academie Malgache	Fem L	420	90 × 68	210	311, 179–539	247, 223–294
Aepyornis maximus	UA8648	Fem R	462	90	273	419, 238–737	247, 223–294

Species	Reference or catalog no.	Element[b]	Length	Circumference, shaft diameter	Mass by length	Mass by circumference (estimate, range)	Mass by diameter (estimate, range)
Aepyornis maximus	MNZ S38302	Fem L	435	85.4 × 65.3	231	280, 162–484	226, 208–262
Aepyornis ?maximus	AMNH 3124	Fem LR	392	84 × 62	173	259, 150–446	220, 203–252
Aepyornis ?maximus	AMNH 3124	Fem L	382	75 × 45	161	165, 98–279	182, 175–196
Mullerornis agilis	Lamberton 1934	Fem Min	247	(105)	48	43, 27–69	NA
Mullerornis agilis	Lamberton 1934	Fem Max	275	(135)	64.4	77, 47–127	NA
Mullerornis agilis	Lamberton 1934	Fem Mean	257	(126)	53.4	66, 41–107	NA
Mullerornis agilis	MNZ S38300	Fem L	265	38 × 35.3	58	55, 34–88	59, 43–70
Mullerornis grandis	Lamberton 1934	Fem Min	270	(125)	61.2	65, 40–105	NA
Mullerornis grandis	Lamberton 1934	Fem Max	295	(145)	78.3	91, 55–150	NA
Mullerornis grandis	Lamberton 1934	Fem Mean	283	(134)	69.8	76, 46–124	NA
Mullerornis sp.	AMNH 3123	Fem		39.5 × 36.6	NA	58, 36–94	62, 47–74
Mullerornis betsilei	Lamberton 1934	Fem	235	NA	41	NA	NA

Abbreviations: AMNH = American Museum of Natural History; dR = distal right; fem = femur; L = left; Max = maximum; Min = minimum; MNZ = Museum of New Zealand Te Papa Tongarewa; NA = not available; R = right; sR = shaft right; UA = Smithsonian Institution.

[a] Data are from cited references or given specimens.

[b] Measurements given are femur length (mm), femur circumference (mm), and estimated shaft width assuming asymmetry, or where format is $x \times y$, as width times depth (mm). Mass estimates for femur length are single values; estimates based on circumference and diameter have a range and thus are given as estimate (range).

Extinct Ratites

Elephantbirds (Aepyornithidae), especially the heavyweight of the family, *Aepyornis maximus,* have repeatedly been claimed to have been the heaviest birds to have lived. There are, in fact, few estimates of their size. Most authors cite Amadon's (1947) value of 438 kg. Amadon used a femur cross-sectional area of 60 cm^2, obtained from the modal value of the range of shaft widths of 78–92 mm from Monnier (1913). However, the femora of *Aepyornis maximus,* unlike those of *Dinornis,* are markedly asymmetric in section, with different sagittal and transverse widths. As Table 5.5 shows, the shaft of the medium-size specimen in the Museum of New Zealand measured 85.4 by 65.3 mm in width. Taking an average diameter of 75.35 mm, and hence a cross-sectional area of 44.6 cm^2, the weight calculated from Amadon's data for the cassowary and the proportion of femur area to weight would be 310.7 kg. The often-published value of 438 kg is therefore a considerable overestimate.

More importantly, *Aepyornis maximus* is the largest of a group of elephantbird species, which forms a neat analogy to the moa species in New Zealand. The seven or eight species of elephantbirds presently listed differed markedly in size. We have taken mean and ranges of femur sizes given for large samples in the literature and used the same methods that we applied to the moa to obtain comparable estimates of live mass (Table 5.5). We have augmented these data with values from actual specimens (catalog numbers cited).

Mass estimates for *Aepyornis* spp. using femur length and femur diameter broadly agree, but those using femur circumference are considerably higher and become more so as femur size increases. We think it probable that the algorithm for femur circumference markedly overestimates mass in the largest individuals. The marked cross-sectional asymmetry of *Aepyornis* femora, but not those of *Mullerornis,* may relate to the wide body of these birds. To bring the feet under the body, the distal end of the femur is modified so that the lateral condyle is markedly distal to the medial condyle and is somewhat rotated ventromedially so as to direct the tibiotarsus medially. The lateral and rotational stresses associated with the outwardly directed and rotated femur have probably necessitated the unusual thickening seen in *Aepyornis* femora. If so, the maximum shaft diameter value would be expected to overestimate body mass. Whether or not this is true, the range of femur diameters for species as given by Monnier (1913; 72–82 mm and 90–96 mm) suggests that both *A. maximus* and *A. medius* were markedly heavier than *D. giganteus.* Estimates using femur length, however, indicate that *A. medius* and *A. maximus* combined had a size range similar to that of *D. giganteus. A. hildebrandti* was the smallest *Aepyornis* species, and its femur length suggests it was similar in size to *Euryapteryx geranoides.*

The three or four species of *Mullerornis* differed in size and were equivalent to various emeids in New Zealand. *Mullerornis agilis* may have been similar in size to *Emeus crassus.* No femur measurements were available for *M. rudis,* but the tibiotarsus and tarsometatarsus were of similar size to that of *M. betsilei* (Lamberton 1934: 127). Lamberton (1934) gives the femur length of *M. betsilei* as 235 mm, so this species is much smaller than *M. agilis. Mullerornis betsilei* would have been of similar weight to *Anomalopteryx didiformis* at about 40–45 kg.

The elephantbirds consist of two genera within which species are separated by size, just as are several of the moa. Two species, as currently defined, of *Aepyornis* are here confirmed as having been much heavier than any moa, but *A. maximus* probably reached only about 275 kg. However, as *A. medius* and *A. maximus* are distinguished mainly by size, with only minor cranial differences proposed (Wiman and Edinger 1941), it is likely that these named forms belong to a single species. Certainly the observed range in size of the bones of the two species combined would not be out of keeping with that seen in moa.

Mass Estimates and Size Change with Time

It has been shown above that several moa species were considerably larger during the last Glacial period than they were in the latest Holocene. The mass estimates generated here, which are based on femur length (Tables 5.3, 5.6), suggest that the mean body mass of *P. mappini* decreased from 41 to 27 kg between the height of the last Glacial period 18 ky B.P. and the Holocene. The size decrease for *Euryapteryx curtus* was less spectacular: mean mass (as estimated from femur length) went from 24.5 to 20 kg in the same period. Analysis of data for *Megalapteryx didinus* shows a similar story, with Glacial populations shrinking from a mean mass of 56 to 40 kg (Table 5.3). The data for *Pachyornis elephantopus* are particularly good, with well-dated sites in close proximity that show that *P. elephantopus* increased in size as the coldest part of the glaciation approached, peaked during the coldest period (20–18 ky B.P.), and thereafter declined in size (Fig. 5.1). The Glacial population was 34% heavier than the latest Holocene population.

These changes show the plasticity of size in moa through time, making it inadvisable to use size alone to distinguish species. In this regard, Lamberton's (1934) naming of *Mullerornis grandis* as distinct from *M. agilis* because of an average size increase of about 10% could be questioned. It is likely that the Ampasambazimba specimens on which Lamberton based the name *M. grandis* represent a geologically older population of *M. agilis*. Recently, Goodman (1999) has obtained accelerator mass spectrometry radiocarbon dates for some of the Madagascar sites, and one of the Ampasambazimba bones was 22,550 ± 170 years B.P., indicating that some bones in the site were indeed of Glacial age. The only other date for the site (~1073 years B.P.) was obtained in 1973, before effective purification procedures for bone material were developed and used, and so is suspect. As in moa, the size variation used to define species in *Aepyornis* species may reflect size variations through time.

TABLE 5.6.

Variation in body masses (kg) of individuals calculated from femur measurements (mm) in different populations of *Pachyornis elephantopus* illustrating changes in size over time.[a]

Site and age	Reference	Length	Shaft width	Mass by length	Mass by circumference	Mass by diameter
Merino Cave east deposit, North Canterbury, 38 ky	Worthy and Holdaway 1995	NA	50.1 (39–63); n = 18	NA	$\bar{x}$ = 110 (66–183) Min 62 (38–101) Max 186 (110–315)	$\bar{x}$ = 93 (80–102) Min 61 (46–73) Max 137 (133–139)
Merino Cave west deposit, North Canterbury, 14–20 ky	Worthy and Holdaway 1995	NA	57.5 (46–73); n = 8	NA	$\bar{x}$ = 151 (90–254) Min 91 (55–149) Max 260 (151–447)	$\bar{x}$ = 117 (109–123) Min 81 (66–91) Max 175 (163– 84)
Honeycomb Hill, 18–22 ky	Unpublished data; MNZ specimens	340 341 338 325	57.7 × 63 58 × 64 58 × 69 50.5 × 60.6	116 117 114 103	168 (100–285) 173 (102–293) 189 (111–321) 139 (83–234)	137 (133–139) 140 (138–142) 159 (157–163) 128 (122–132)
Glencrieff, North Canterbury, 10–12 ky	Worthy and Holdaway 1996b	312.6 (265–337); n = 19	53.7 (47–59); n = 19	92 (58–113)	$\bar{x}$ = 129 (72–216) Min 95 (58–157) Max 160 (95–270)	$\bar{x}$ = 105 (93–112) Min 84 (69–94) Max 122 (115–127)
Albury Park, South Canterbury, 10–12 ky	Unpublished data; CM specimens	310 (288–329); n = 9	55.1 (48–60.8); n = 18	90 (73–106)	$\bar{x}$ = 137 (82–229) Min 100 (61–165) Max 171 (101–290)	$\bar{x}$ = 109 (99–116) Min 87 (73–96) Max 129 (123–132)
Pyramid Valley, North Canterbury, 2–4 ky	Worthy and Holdaway 1996b	305.2 (292–320); n = 11	48.4 (44–52); n = 11	86 (76–98)	$\bar{x}$ = 102 (62–169) Min 82 (50–134) Max 120 (72–200)	$\bar{x}$ = 88 (74–98) Min 75 (60–86) Max 99 (87–107)
Cheviot, North Canterbury, 1–2 ky	Unpublished data; CM specimens	289 (250–327); n = 22	48.0 (40.8–61.3); n = 21	74 (49–104)	$\bar{x}$ = 100 (61–165) Min 69 (42–112) Max 175 (103–296)	$\bar{x}$ = 87 (73–96) Min 66 (51–78) Max 151 (125–134)

[a] $\bar{x}$ = mean; Min = minimum; Max, maximum; NA = not available.

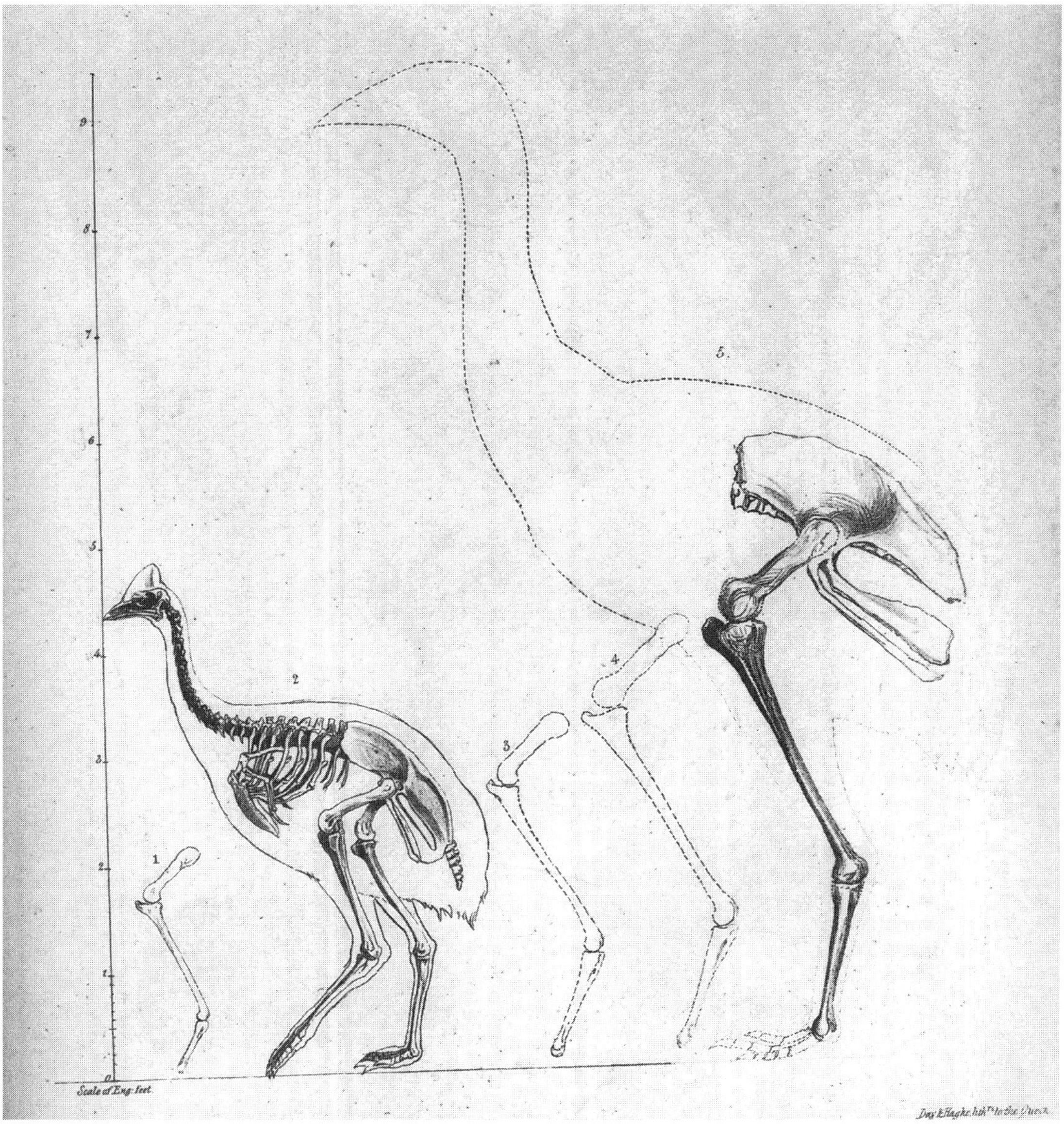

Figure 5.3. Richard Owen's first reconstruction of a moa (Owen 1844a: pl. 30).

Posture and Body Proportions

From virtually the first published descriptions of moa, there has been speculation on their posture. In his landmark article of 1844, Owen portrayed *Dinornis* as an enlarged cassowary (1844a: pl. 30). In doing so, he had considered all extant ratites, concluding that *Dinornis* was more similar to cassowaries than the other ratites (kiwi, ostrich, emu, and rhea):

> But since the Cassowary and Apteryx, as compared to the Ostrich and Emeu, combine shorter tarso-metatarsals with their shorter necks, the *Dinornis* is much more likely to have resembled these birds than the Ostrich in the proportionate length of its neck, and we know that it resembled the *Apteryx* much more than the Os-

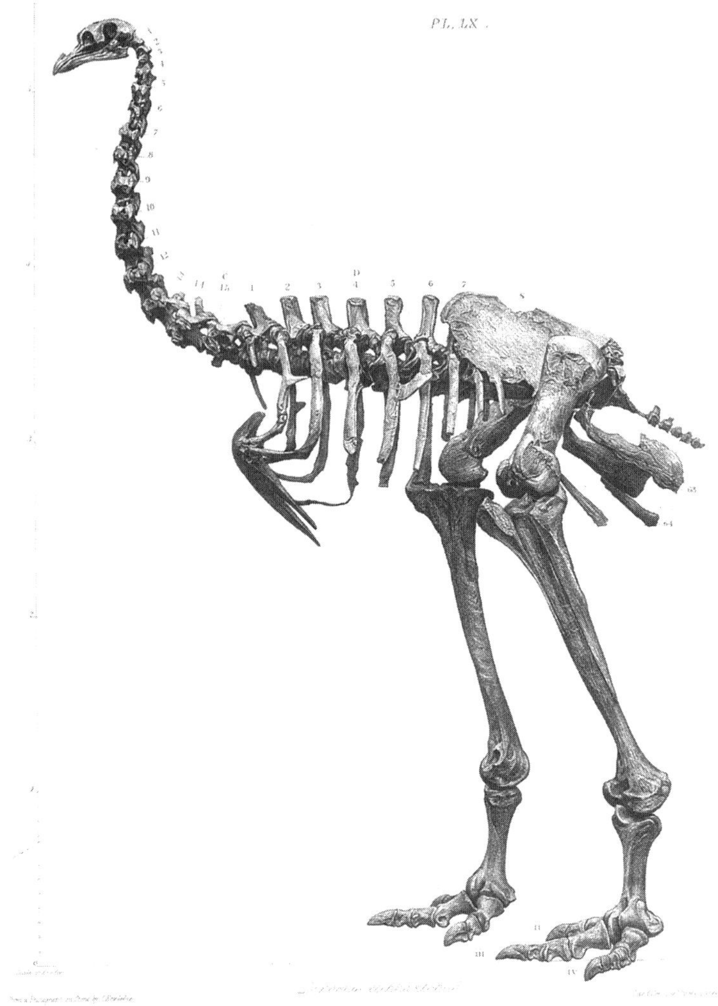

Figure 5.4. Richard Owen's reconstruction of Pachyornis elephantopus *(Owen 1879: pl. 60).*

> trich in the robust proportions of the cervical vertebrae. (Owen 1844a: 264).

Using the cassowary analog, Owen estimated the height of *D. giganteus* at no more than 3 m (10 feet). However, as Owen's plate 30 (Fig. 5.3) clearly shows, he was obviously unfamiliar with the normal posture of a cassowary because he gave it a markedly posteriorly down-curved pelvis, and the thoracic vertebrae sloped upward. His estimate for the height would have been much less if the cassowary had been articulated as in life.

In 1858, Owen published a reconstruction of *Pachyornis elephantopus* (Fig. 5.4) on the basis of a composite skeleton from bones obtained at Ruamoa in Otago (Owen 1858b: pls. 46, 47). He inferred that there were only 21 presacral vertebra (not the 27 now known) and maintained the slightly upward-sloping posture he had advocated in 1844. At about the same time, Owen reconstructed *Dinornis novae-*

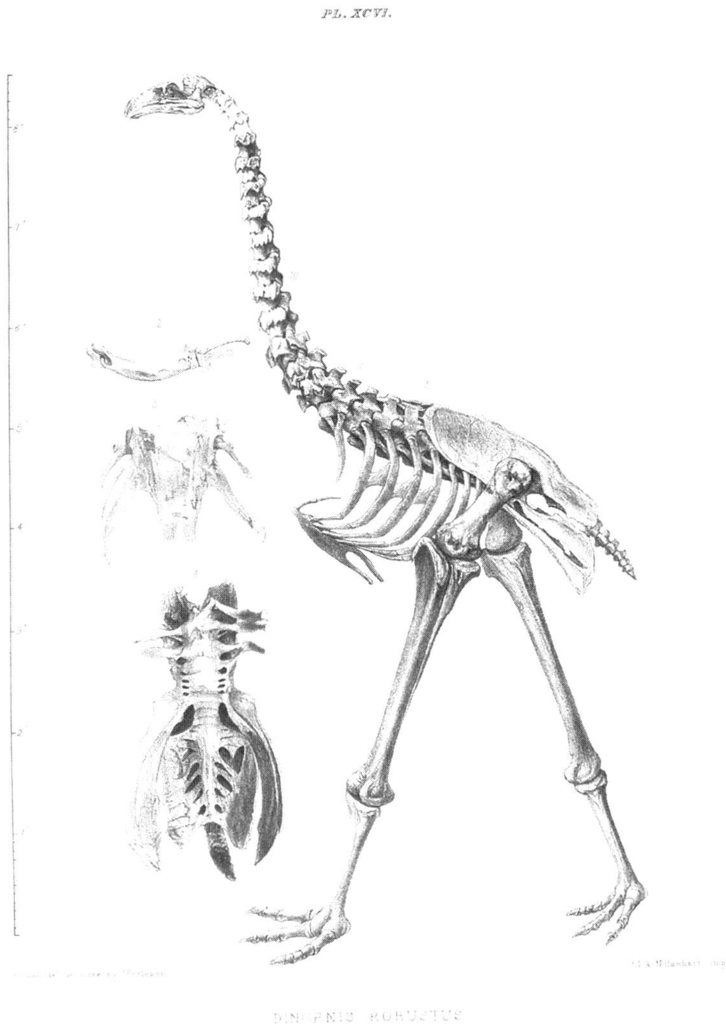

Figure 5.5. Richard Owen's reconstruction of Dinornis robustus *(=* novaezealandiae*), the Tiger Hill, Manuherikia, specimen (Owen 1879: pl. 96).*

zealandiae in a rather erect posture (Fig. 5.5), the more so now that we know the reconstruction lacked a third of its real number of vertebrae. It was not until the 1880s that Owen had a complete vertebral column from which to reconstruct the true posture of moa. He described the complete skeleton of *Dinornis parvus* (now *Anomalopteryx didiformis*) in 1883 (Owen 1883a: pls. 51, 58), and depicted in the mount what we now know to be the characteristic loop of the lower cervicals, seen in all large ratites (Fig. 5.6). However, although he had the thoracic vertebrae angled more horizontally than previously, they were still made to slope upward anteriorly.

By this time, the influential Julius Haast, working in New Zealand, had constructed many composite skeletons from the mixed bones excavated from the rich Glenmark Swamp deposit. The skeletons were sent to sites worldwide in exchange for specimens to fill Haast's new Canterbury Museum, and with them went their incorrect posture.

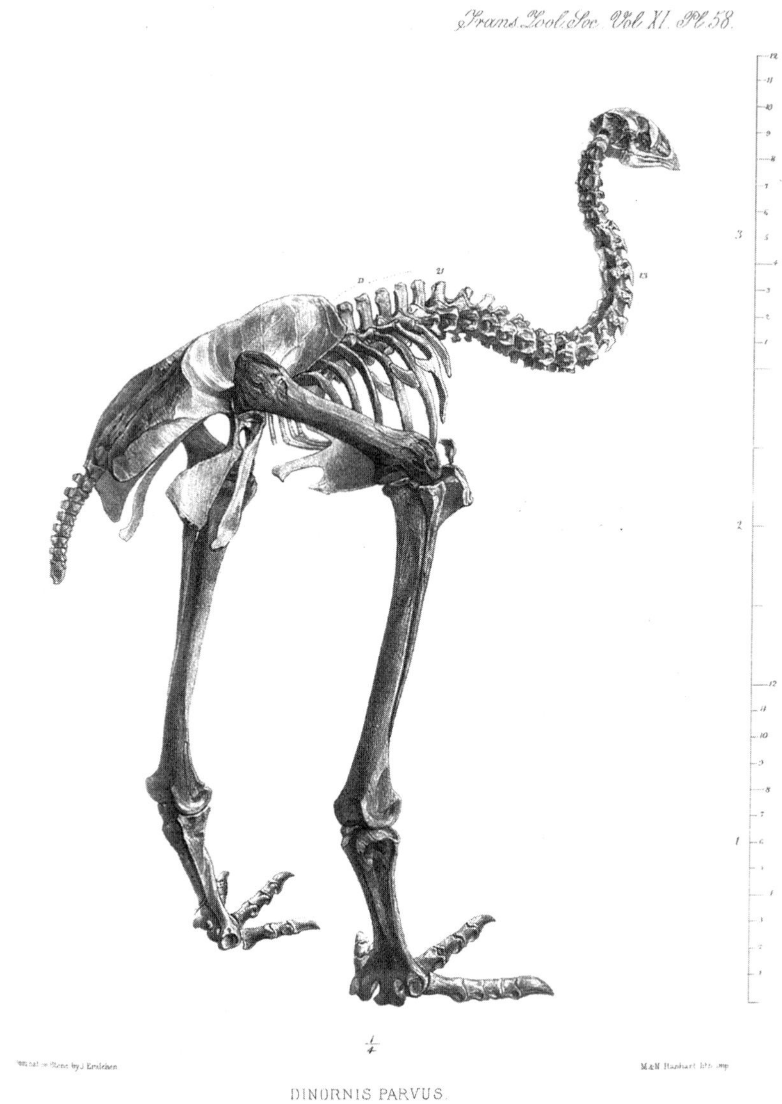

Figure 5.6. Richard Owen's reconstruction of Dinornis parvus *(=* Anomalopteryx didiformis*), from a specimen from Takaka Hill (Owen 1883a: pl. 58).*

All were mounted in the posture first advocated by Owen, with its upward-sloping thoracic vertebrae (Fig. 5.7). With a thirst for the tallest mount—but perhaps with scant regard for reality—often too many vertebrae, especially of the lower cervical series, were strung together, giving very tall constructions indeed. Nothing explicit has been written about the technique of articulating a moa skeleton, so many such constructions followed the perceptions of those who did them first. From the 19th century into the mid-20th century, it seems that most exhibiters have favored the ostrich or emu as an analog. For example, the *Dinornis novaezealandiae* skeleton from Takaka (as *D. torosus,* Oliver 1949: fig. 127) was articulated in a very erect attitude, rather like an ostrich. Despite all the leg bones being placed in a vertical posture, the back was kept horizontal and the neck articulated under the cranium as

Figure 5.7. Moa skeletons displayed in the Mammal Room, Canterbury Museum, June 1906. Photograph courtesy of Canterbury Museum.

Figure 5.8. An articulated skeleton of Dinornis struthoides *in the Wanganui Regional Museum. Photograph courtesy of the Wanganui Regional Museum.*

in an ostrich. Several skeletons in the Wanganui Museum were erected this way too (Figs. 5.8, 5.9).

So what is wrong with these and previous articulations? First, preconceptions developed from observations of other birds should be ignored. Each species is unique, and there is great variety in form. But if the simple rule of articulating opposing facets in all joints is followed, then the average posture of that animal must result. The outward form of a kiwi does not hint at the amazing anterior loop in its lower cervicals (Fig. 5.10), yet articulating the vertebrae by this simple rule will produce this loop.

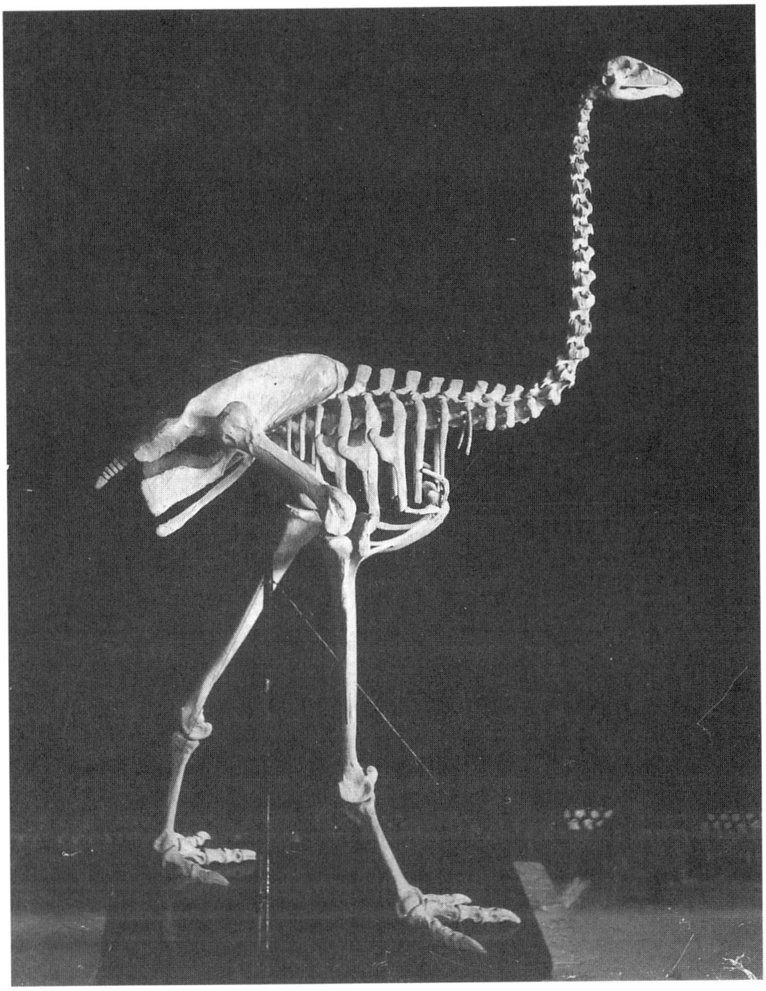

Figure 5.9. *An articulated skeleton of* Anomalopteryx didiformis *in the Wanganui Regional Museum. This style of articulation where moa were made to look like ostriches was favored from about 1900 to the 1970s. Photograph courtesy of the Wanganui Regional Museum.*

We can be reasonably confident that when a moa stood still, its tarsometatarsi were vertical or sloping slightly (~10° forward to the ground from vertical) so the rear of the toe was slightly forward of any part of the tarsometatarsus as in ostriches, cassowaries, and emus—in fact, most large birds. The distal tibiotarsus is then articulated with the top of the tarsometatarsus in the middle of the possible range of movement at that joint (to form the avian ankle). From the tarsometatarsus, the tibiotarsus slopes upward and forward to the knee, usually at an angle of 10–20° to the vertical. The femur then articulates with the tibiotarsus, passing rearward to the pelvis, forming a caudal femur–tibiotarsus angle of about 90°. The knee and ankle joints have a wide range of possible movement to accommodate walking and sitting. However, the femur differs from the distal bones in being wholly buried within a muscle mass and it lies close to the pelvis with the femoral trochanter next to the antitrochanter. Articulated in this position, the femur forms an angle of about 30° declination to the anterior part of the synsacrum. It was not possible for the femur to move more than

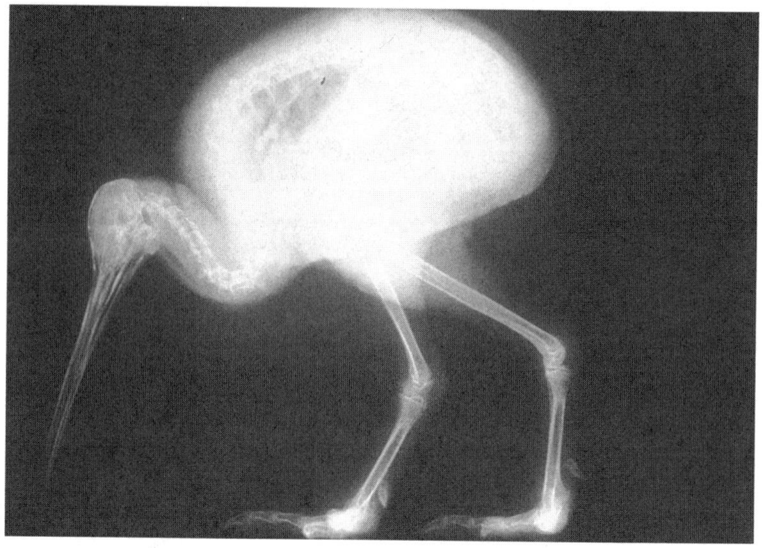

Figure 5.10. Radiograph of a living kiwi, from Beale (1991). Photograph courtesy of Auckland Museum and republished with permission of Royal Society of New Zealand.

Figure 5.11. Drawings of emus traced from photographs to illustrate several postures. (A, B) Normal, relaxed postures. (C) The bird is demonstrating interest in something. (D) Erect posture affected by the female during sexual display. (E) Erect posture taken during a period of intense interest in a relatively distant object. Note that in A–D, the back slopes down anteriorly but is flat in E. An upward-sloping back is only rarely achieved, and then only momentarily, during interactions between individuals.

about ± 15° about the median point of its articulation without rupturing the joint. In all moa species, as in all ratites, the anterior synsacrum must then lie parallel to the ground; in turn, the thoracic vertebrae must also be parallel to the ground. This should not be too surprising because it is the normal posture of emus, cassowaries, ostriches, and rheas (Figs. 5.11, 5.12).

Figure 5.12. Drawing of a cassowary in a normal, relaxed posture. Among living ratites, cassowaries are most like moa, which generally are built even more heavily and have a relatively shorter neck and shorter tarsi. Drawn by A. Worthy.

However, from the body forward, the posture is unique to moa. Immediately forward of the thoracic vertebrae, the vertebral column is bent downward into a loop, the lowest point of which is between vertebrae 16 and 12 (numbered from the cranium). The upward rising set of vertebrae are short; in fact, they are the shortest of all large ratites, as described above. So this rising set of vertebrae hardly reaches above the line of the back. The same number of vertebrae in emus rises twice as far because they are each relatively twice as long as the corresponding element in a moa (see Fig. 4.10). Also, it must be remembered that in moa, the vertebrae articulate with the back of the skull, not its base, as in ratites of savanna habitats. Vertebrae 1–3 are therefore parallel to the ground and the downward bend starts at 4–5. The result is the average posture of a moa—not its possibilities.

Moa were very long birds, not tall ones. The moa head was held only slightly above the level of the back. Its stance was, ironically, very like that of a cassowary, except that Owen (1844a) constructed the cassowary incorrectly and followed suit with the moa. *Dinornis giganteus* was huge—nearly 2 m high at the back—but normally its head would not have been far above it. Yet obviously, by tilting its pelvis back and reaching up, it could undoubtedly have attained 3 m. We do not usually visualize a sheep or a cow jumping over a fence, but they still—on rare occasions—do just that.

From the formerly widespread perception based on analogy with the ostrich, a general consensus developed in the 1980s that moa held their heads in a low-slung posture. However, the change was not the result of an appreciation of the morphologic constraints but because of ecologic enlightenment. The moa had been assumed to be grassland birds, but as knowledge of the prehuman New Zealand environment grew, especially from palynologic studies, it became obvious that many of the moa, if not all, were forest birds. An ostrich, for example, could not pass through the tangled, dense forests then characteristic of much of New Zealand. Its vertical neck and high head would have been a positive hindrance. So people turned again to the cassowary for an analog. Now, most modern reconstructions show moa with a flat back that slopes downward to the base of a neck before rising in a sigmoid curve to articulate with the back of the skull (Fig. 5.13; see Fig. 4.11). In Figures 5.14 to 5.18, we incorporate all the above data to reconstruct the fleshed-out moa.

Body Proportions

The range in body form of moa species is similar to that seen in aepyornithids. The smaller ones such as *Anomalopteryx didiformis* or *Emeus crassus* were built much like the similar-size *Mullerornis* (Fig. 5.19). The heaviest, *Pachyornis elephantopus,* was somewhat similar to *Aepyornis maximus,* although the amazingly short tarsometatarsus of the former accentuated its graviportal qualities.

The overriding fact about moa is that they were huge graviportal

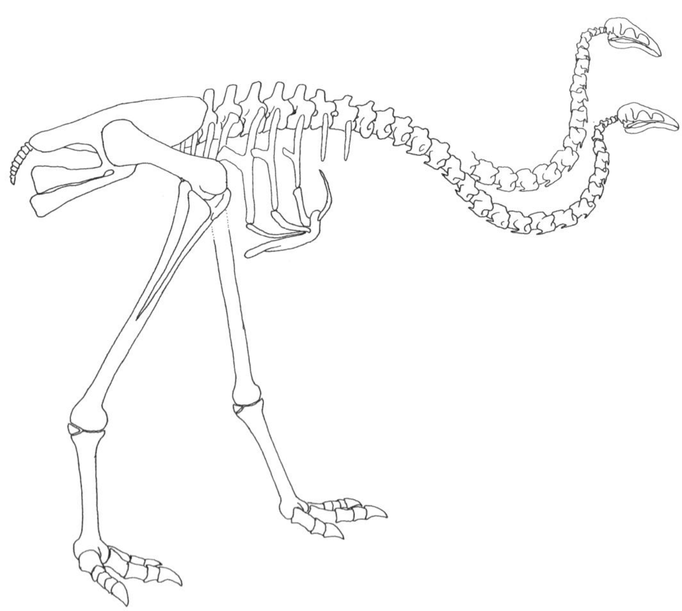

Figure 5.13. (this page) A reconstruction of the skeleton of Dinornis novaezealandiae *in a feeding pose (lower neck) or alert pose (upper neck position).*

Figure 5.14. (opposite page, upper left) Members of the genus Pachyornis, *exemplified here with* Pachyornis mappini, *were extraordinarily squat and solid birds. Note the robust sharp bill. The takahe (for scale) is about the size of a large chicken and stands about 40 cm high in this and the following reconstructions. Drawing by J. Winn.*

Figure 5.15. (opposite page, upper right) Anomalopteryx didiformis *is a relatively small, slender species with a stout bill. Drawing by J. Winn.*

Figure 5.16. (opposite page, lower left) Megalapteryx didinus *is the most slender and presumably athletic of all moa. Its large, slender toes and feathered tarsi are some of its adaptations to its mountain habitat. Drawing by J. Winn.*

Figure 5.17. (opposite page, lower right) The genera Euryapteryx *and* Emeus *are similar stout, short-legged birds with a broad, blunt bill, here exemplified by* Euryapteryx geranoides, *which is the most stout of the three species. Drawing by J. Winn.*

Figure 5.18. (top) Dinornis, *here shown by* Dinornis novaezealandiae, *were relatively slender and long birds characterized by broad, flattened heads. Drawing by J. Winn.*

Figure 5.19. (bottom and opposite page) Skeletons of elephantbirds as articulated by Lamberton in L'Academie Malgache, Madagascar. (right) Mullerornis agilis. *(opposite) The display in L'Academie with* Aepyornis maximus *in the center. From Lamberton (1934).*

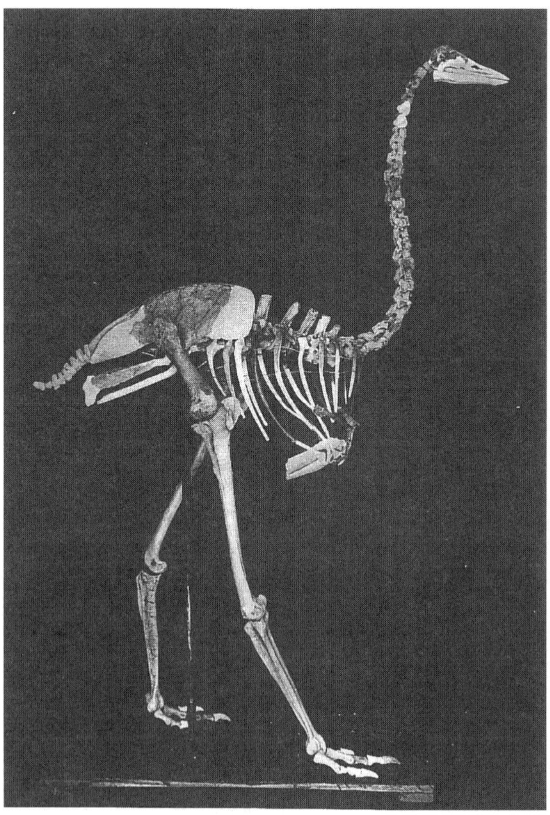

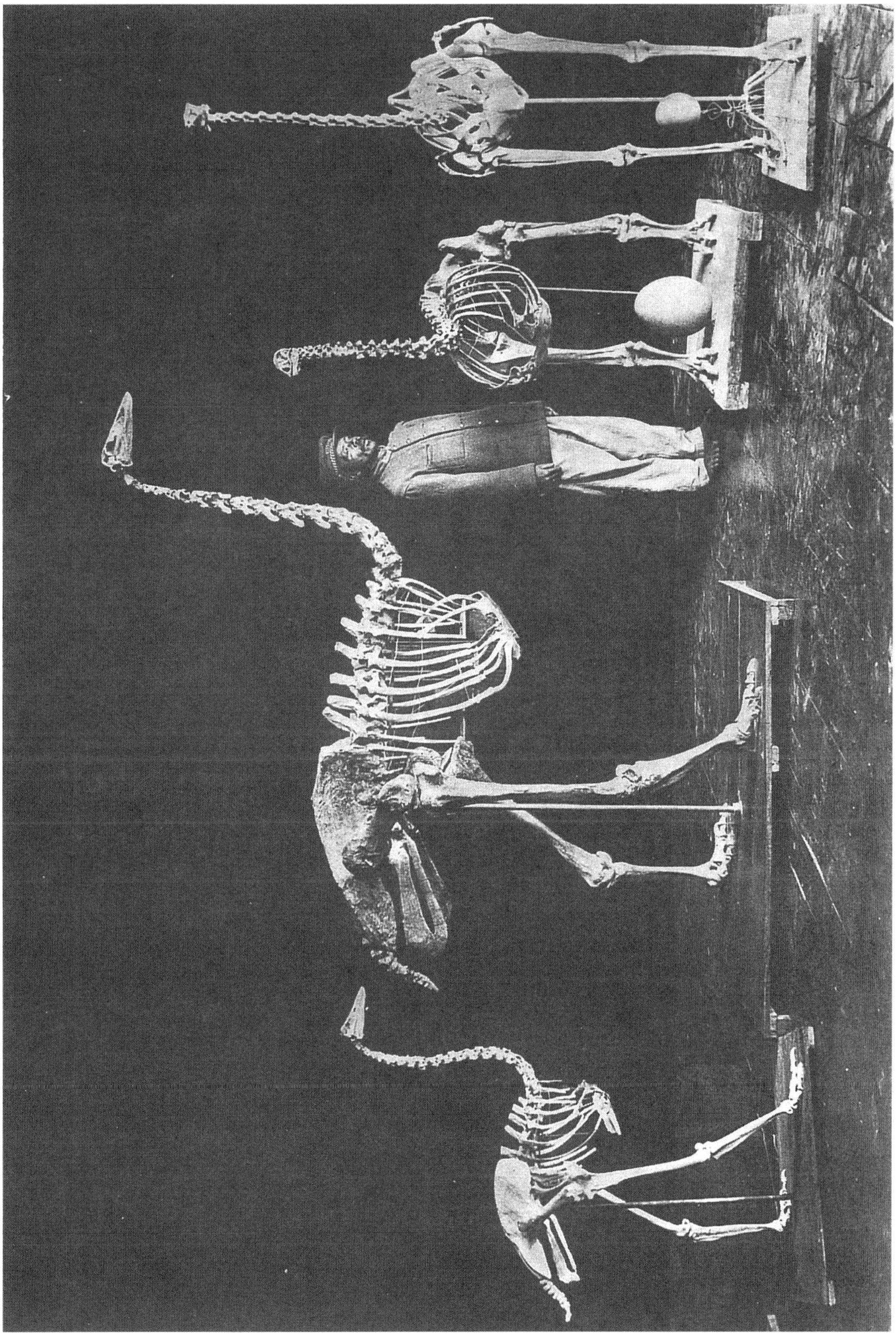

birds. It follows that they had large legs and no wings, and hence no appreciable breast muscles. Even considering that they were low-slung birds of the forest, they did have, seemingly, quite stout legs, even for large ground birds. Or did they? Their legs were certainly stouter than those of emus, cassowaries, and ostriches. Furthermore, all of these living species differ from moa in having a much longer tarsometatarsus—indeed, much longer than the femur. The femoral/tarsometatarsal ratio of moa is 1:1 for emeids, or 1:1.1 for dinornithids, which means that moa had very short tarsometatarsi. Elephantbirds had relatively longer tarsometatarsi than moa, but they too had much shorter tarsometatarsi than in their nearest relative, the ostrich (Fig. 5.19). Both groups are overall more stoutly built than other ratites, and both have pelves that are broad posteriorly.

The proportions of the lengths of moa leg bones led Alexander (1983a) to observe that species within the "Anomalopterygidae" (= Emeidae here) had short femora and particularly short tarsometatarsi compared to their tibiotarsi. As a result, he postulated that because of the loss of wings and their associated musculature, the center of mass of moa may have been further posterior than in most modern birds. This would in turn have meant that they may have stood with their femora aligned more vertically than in most modern birds. Furthermore, he suggested that the short tarsometatarsus in moa was an adaptation to maintain the balance of the bird on its feet, given that they had retained broad pelves. Other ratites with long tarsometatarsi had narrow pelves. This is an interesting hypothesis that has not been further explored. But whatever explanation is adopted, it must take into account the morphologic constraints presented by the bones themselves. The femora were not held vertically; this would have been a morphologic impossibility in all moa. They were held as shown in the modern skeletal reconstruction (Fig. 5.20). Owen had these articulated more or less correctly from the beginning. So whatever the reason may have been for

Figure 5.20. A modern museum articulation of the moa Anomalopteryx didiformis. *"Humpty" was articulated from bones excavated from a cave called "The hole the bullock fell down" in the Puketoi Range, southern North Island, on August 13, 1983, by Richard Cassels of the Manawatu Museum, members of the Manawatu Spelelological Group, C. M. Worthy, and T.H.W. The articulation by Blake Abernethy was completed in February 1986 for the Science Centre and Manawatu Museum. Photograph by R. Morris, courtesy of the Manawatu Museum Society and the Science Centre and Manawatu Museum.*

short tarsometatarsi, it would have to have evolved within the possible postures allowed by the joints themselves.

Alexander (1983b) also examined the apparently extraordinarily robust bones of *Pachyornis elephantopus*. By estimating the body weight of the bird and knowing the cross-sectional area of the limb bones, he was able to calculate the stress on the skeleton at a slow running speed of 3 m/s and a fast one of 10 m/s. He found that although the tibiotarsus had great absolute strength, such strength was, as a proportion of body weight, more or less similar to that in ostriches. The femur and tarsometatarsus were stronger, or had higher factors of safety (calculated by dividing strength by stress), but this was explained by their relative shortness. However, the stresses likely to be experienced by all elements were similar to other animals. Therefore, the bones were not disproportionately strong in comparison with the body mass. In other words, the bones had to be that big to support the bird. Nevertheless, he explained the high factors of safety he had calculated as resulting from reduced costs of strengthening bones because the moa was a browser without large predators to contend with.

Reif and Silyn-Roberts (1985) challenged this conclusion, citing marked differences in the ecology of ostriches and moa, which preclude their direct comparison. Instead, they suggested that the dense shrublands inhabited by moa required the perceived greater strength to effectively force a passage for the birds. Their main point was that until the ecology of an animal's habitat is known, then neither are its strength requirements, and thus factors of safety cannot be extrapolated from one animal to another. Taking Reif and Silyn-Roberts's warning one step further, and with the benefit of a little knowledge of the paleoecology of *Pachyornis elephantopus,* we could speculate that this inhabitant of open grasslands may have varied widely in body weight through the course of a year. Food would be seasonally abundant or sparse, and it seems likely that this species could have used a fattening strategy so as to live through the hard times of winters, when food was sparse and temperatures coldest. Seasonal fattening cycles are known for many birds and have been inferred for other large graviportal birds such as the dodo (Kitchener 1993). Undoubtedly, *Pachyornis elephantopus* was one of the heavyweights of the avian world.

Locomotion and Footprints

There are only seven or eight known occurrences of moa trackways. The first, and best known, was recorded by Williams (1872), Cockburn-Hood (1874), and Plant (1877) from the banks of the Taruheru River (also known as Waikanae Creek), near Turanganui in Poverty Bay (northeastern North Island). An area 4.3 m long by 1.5 m wide (14 by 5 feet) had many individual prints on it. Among these, two sets of tracks could be discerned (Williams 1872). These were sketched, and they show the passage of two individuals. Owen (1879a) reproduced William's plate as his tracks 3 and 4 in plate 116 (Fig. 5.21). The largest was 200 mm (7⅞ inches) from "heel" to tip of the middle toe, and 153 mm (6 inches) for both the inner and outer toe lengths. They were 178 mm (7 inches) across the toes. The heel-to-heel distances given by Wil-

Figure 5.21. Moa footprints from Owen's Memoirs on the Extinct Wingless Birds of New Zealand *(1879: pl. 116). The pair of trackways on the right were copied from Williams's (1872) sketch. They indicate that three species walked on the Turanganui site.*

Figure 5.22. (opposite page) Moa footprints revealed on the bank of the Manawatu River in 1912. Photograph by E. R. Whalley. Photograph reproduced courtesy of the Science Centre and Manawatu Museum.

liams (1872) were 19.5, 20.25, 21.75, 19.5, 20.5, 19.5, and 19 inches (mean, 20 inches, or 508 mm) for the larger series and 13.5, 12.5, 13, and 12.25 inches (mean, 12.8 inches, or 325 mm) for the smaller prints. It is also observed that in trackway 3, the upper and lower prints must be the left foot and the middle one the right because it is offset to the right from the other two; hence, the digits on the lower two prints are incorrectly labeled (J. O. Farlow, personal communication).

Both the Archdeacon W. L. Williams and His Honor T. B. Gillies collected slabs with prints from this site. Gillies (1872) recorded having obtained blocks with a total of 8 or 10 prints in them. Voy (1880) recorded that he excavated a slab containing four prints and part of another, which was given to the museum of the University of California, Berkeley. Later, Hill (1895) reported having several of the prints in his possession. In May 1912, several more prints were exposed and collected. Two are in the Gisborne Museum, and others are in the Museum of New Zealand. Of the original series, two (43554, presented by T. H. Cockburn-Hood, and 46988, presented by Reverend R. H. Davies) and some casts are now in the British Museum(Natural History), London.

PALMERSTON NORTH
MOA'S FOOTPRINTS

Figure 5.23. Excavation of the moa footprints revealed on the bank of the Manawatu River in 1912. The discoverer, Henry Coles, is on the left. Photograph reproduced courtesy of the Manawatu Museum Society and the Science Centre and Manawatu Museum.

The Auckland Institute and Museum, Auckland, New Zealand, has five different slabs with moa footprints or casts, including two slabs totaling six footprints, which are probably part of the Gillies (1872) Turanganui series. One print is in each of the Napier and Otago Museums, and several are in the Museum of New Zealand.

Owen (1879a) referred these tracks to three species, the largest, series 2 in Figure 5.21, to *Dinornis ingens*, which would equate to *Dinornis novaezealandiae*. Owen noted the impression of a hind toe in these prints, which would accord with the well-developed digit I of this species. The next set of three prints Owen referred to *Dinornis struthoides* and the smallest ones to *Dinornis dromioides* (= *Anomalopteryx didiformis*).

Moa footprints were seen at the reservoir at Lighthouse Road, Napier, by R. Lamb and H. Hill in 1887 (Hill 1914: 350). No specimens are known. A few years after the first discovery at Gisborne, Hill (1895) reported the discovery of prints in the bed of the Manawatu River. These were much larger—one was 381 mm (15 inches) from the heel to the tip of each toe. Another was 210 mm (8.25 inches) from the heel to the tip of the middle toe but 254 mm (10 inches) to the outer toes, and the stride length was measured as 660 mm (26 inches). Hill reported smaller prints as well, but he only sketched three of the larger prints and did not collect them.

On August 13, 1912, a flood exposed a further set of tracks in the Manawatu River within the city of Palmerston North (Wilson 1913). Four were particularly well preserved and were cut out and kept, but before this, they were photographed and measured (Figs. 5.22, 5.23). They were large: 457 mm (18 inches) across the toes, with the end of middle toe to heel measuring 305 mm (12 inches) and a heel-to-heel length of 762 mm (30 inches). From the size of these tracks, it is likely a large *Dinornis* made them. It may be the same site that Hill saw in 1895.

Another site was apparently never reported in the literature but is recorded by a footprint preserved in the Museum of New Zealand. The site was described as "On a sand layer, Marton, on bank of Tutaenui in washout 15 feet below surface," and the print was presented by D. Matheson in May 1896. Marton is just east of Whanganui in the lower North Island.

In March 1939, mudstone rocks were exposed in the bed of the Rangitikei River at Tangimoana (western lower North Island) that contained the footprints of several moa (Oliver 1949: 24–25). There is apparently no pictorial record of these, but a cast (MNZ S462) was made and collected by J. C. Young; it was presented to the museum on April 5, 1948.

In 1946, R. H. D. Stidolph recorded the following:

> An interesting discovery, three imprints of a moa's foot or claw, was made recently near the mouth of the Waiwakaiho Stream, Fitzroy, New Plymouth, by Mr Joseph Iorns. After a flood and a spring tide, which washed off sand and timber from a shelf of brown swamp sludge and sand formation, three good footprints were laid bare, all the same size, twelve inches by 8½ inches, in a line, about six feet apart. The ground was too soft to allow the footprints to be cut out, so Mr Iorns mixed some fine concrete and filled in the imprints. The next tide covered them again, and more recently one of the footprints was exposed once more and was cut out by Mr Iorns. He presented it to the museum at New Plymouth. The concrete saved the imprint from scouring out, and it is the intention of the museum people to lift it out to show the original impression. (Stidolph 1946)

One of the molds is probably MNZ S24314 in the Museum of New Zealand collection, but the specimen lacks collection data, so this is not certain. From their size, it is likely a *Dinornis* made the prints, because only large *Pachyornis* species had similarly splayed toes and similar-size feet, and there were no large *Pachyornis* in the North Island.

A further unpublished record is a photograph taken in 1973 of

TABLE 5.7.
Summary of stride length and inferred speed of moa on the basis of trackways.[a]

Source	Mean print interval (inches)	Stride length (mm)	Hip height (mm)	Speed (km/h)[a]	Speed (km/h)
Williams 1872, larger series	19.9	1012	704[b]	5.04	4.34[b]
Williams 1872, smaller prints	12.8	650	—	—	—
Hill 1895	26	1321	1094[b]	2.88	4.03[b]
Wilson 1913	30	1524	1400[c]	3.6	3.84[c]

[a] Speed is derived from Alexander (1983b).
[b] Values of hip height and speed from Duncan and Holdaway (1989).
[c] Calculated here. Differences in speed derive from differences in assumed hip height. Alexander (1983b) assumed a hip height of 1.5 m for the bird that made the prints described by Hill (1895) and Wilson (1913), but we consider this too high and modify them as shown.

prints underwater, on the bed of Lake Taupo, the largest volcanic lake in the central North Island (B. Gill, personal communication).

Speed of Moa

The speed of an animal can be calculated from its tracks if its hip height is known. Alexander (1983b) and Duncan and Holdaway (1989) used the few trackways available to determine the speeds of the track-makers (Table 5.7). Their calculations show that the tracks were made by walking moa.

Life History Strategies

We can only guess at the lifespan of a moa. Some birds and dinosaurs have microlaminations laid down annually on the internal surface of long bones (Horner et al. 2000 and references therein), but these have not so far been looked for in moa. Similarly, how many eggs they laid and most features of their biology are largely guesswork. Moa were, however, very large birds, and large birds generally live a long time. Many bird species in New Zealand live to comparatively old ages compared with relatives elsewhere. Long life—or low adult mortality—is a feature of what are known as K-selected species. This is the ecologic term for species that generally live a long time and thus mature late and have a low rate of reproduction; these species also live at or near the carrying capacity of stable habitats. Most have few or no predators as adults. Many New Zealand birds fit this description—and New Zealand had no mammalian predators, climatic seasonality is minimal, and the vegetation is evergreen. Thus food supplies are generally constant and predictable. We could therefore look to the cassowaries of New Guinea and Australia as the nearest living analog to moa. But they evolved in the presence of predators such as the thylacine, crocodilians, giant lizards (the 7-m-long *Megalania*), and giant snakes (*Wonambi*), so moa may have been even more K-selected. The forest species therefore probably lived either singly or in pairs and probably lived to be 50 or more years old.

Sex Ratio

The size variation in moa is known to have been large. For some species, it was remarkable, and in these, it was attributed to sexual dimorphism, first by Cracraft (1976b). Cracraft found that the CV in bone lengths was often up to 6 and 7%, but in *Pachyornis mappini* it reached 10–13%, and 8–10% in *Emeus crassus*. Species with little sexual dimorphism have CVs in the range 2–4%. The species with smaller CVs exhibited unimodal size distributions. This does not mean that the sexes were not significantly different in size—the lengths of sexed kiwi bones are significantly different in segregated samples, but the combined samples still give a unimodal distribution (Worthy 1987a).

However, distributions of leg bone length in the species with high CVs were bimodal in large samples. If the size classes are equivalent to sexes, they provide an opportunity to investigate sex ratios. Large samples of *Euryapteryx curtus* and *Pachyornis mappini* from North Island sites (Worthy 1987a) allow some insight into relative proportions of the sexes, if both sexes were trapped in the same proportions. In the dunes at Tokerau Beach, where entrapment is not relevant, the smaller and larger sexes of *Euryapteryx curtus* are roughly in equal proportions, whereas in the swamp miring site of Tangatupura, the larger sex is more numerous (Fig. 5.24). In the same deposit, the proportions of *Pachyornis mappini* seem to be about equal. Assuming the lighter sex of *E. curtus* was less likely to be mired, we can infer that in both these species, the sex ratios were about equal.

Measurements of the large *Dinornis giganteus* sample from Pyramid Valley also show a bimodal component for this species, with a bimodal length frequency histogram for tibiotarsi (Fig. 5.25). If this distribution reflects sexual dimorphism, it suggests that the sexes had overlapping sizes. In the large sample of *Emeus crassus* from Pyramid Valley, length–frequency histograms of long bones (Fig. 5.26) do not suggest any bimodality in the population, but plots of ratios do. The femur/tarsometatarsi ratio plotted against the femur/tibiotarsi ratio for *Emeus* is strongly grouped into two clusters (Fig. 5.27). The known female (an egg was associated with the articulated skeleton) was a larger individual among the group with relatively shorter tibiotarsi. A relatively shorter tibiotarsus equates to a relatively longer femur, which may relate to females having a broader pelvis, necessary for the passage of the egg. Interestingly, this group accounts for 17 individuals, whereas there are 35 in the grouping of birds with relatively shorter tibiotarsi, which suggests a sex ratio of 2:1 in favor of males in this species.

Age Structure

Generally, we have not been able to examine the population structure of any moa populations, but it has been possible to estimate adult–juvenile proportions in the large samples of bones from some sites. Osteologic maturity in moa was progressive, reaching completion in different elements at different times. In the growing bird, the astragalus was not fused to the distal end of the tibiotarsus, but shortly after the adult length of the bone was reached, it fused. Some time later, the

TABLE 5.8.
The proportion (%) of osteologically immature moa for each species in the given sites.

	Site			
Species	Makirikiri	Pyramid Valley	The Graveyard	Cheviot
Dinornis struthoides	40			
Dinornis novaezealandiae	13			
Dinornis giganteus	9	16		
Anomalopteryx didiformis	14			
Megalapteryx didinus			56	
Emeus crassus		18		30
Pachyornis elephantopus				12
Pachyornis australis			30	
Pachyornis mappini	29			

patella fused to the top. In kiwi, the astragalus–tibiotarsal fusion occurs after one year, but the patella does not fuse until the bird is four years old (Beale 1991).

At Pyramid Valley, 67 *Emeus* were assessed for age by these characters (Table 5.8). There were no small chicks, but nine were juvenile with unfused patella, and a further three were subadult because symphyses were still clearly visible. Therefore, 82% were adult, or more advanced in ontogenetic age than a four-year-old kiwi. Similarly, in the large *Dinornis giganteus* sample from Pyramid Valley, 55 skeletons could be assessed for age in this way. Forty-six (84%) were adult, one (2%) was subadult, and eight (14.5%) were juvenile. At Cheviot in North Canterbury, where small samples only are represented, *Emeus crassus* juveniles were 30% of 10 individuals and *Pachyornis elephantopus* juveniles 12% of 17 individuals. Only one other site has a surviving large enough sample size for *Dinornis,* and this is the swamp deposit at Makirikiri near Whanganui in the North Island. There, *D. giganteus* juveniles represented 9%, *D. novaezealandiae* 13%, and *D. struthoides* 40% of the populations of each species. Also at Makirikiri, *Anomalopteryx didiformis* juveniles represented 14% of 181 birds and *Pachyornis mappini* juveniles 29% of 24 birds. In the large sample of moa from Layer 3 in The Graveyard at Honeycomb Hill Cave, northwestern South Island, immature birds represented 56% of 73 individuals for *Megalapteryx didinus* and 30% of 30 individuals for *Pachyornis australis* (Worthy and Mildenhall 1989).

The proportions of young in deposits may be affected by site-specific factors—for example, a cave pothole might preferentially trap the naïve young moa. But all species in that trap experience the same set of conditions, so single sites provide valid interspecific comparisons. There was a marked difference in proportions of juveniles between the two common species in The Graveyard deposit. Similar differences were observed between taxa at Makirikiri. Notably, the two similar-size emeids *Anomalopteryx didiformis* and *Pachyornis mappini* had markedly different proportions of young. If these proportions reflect

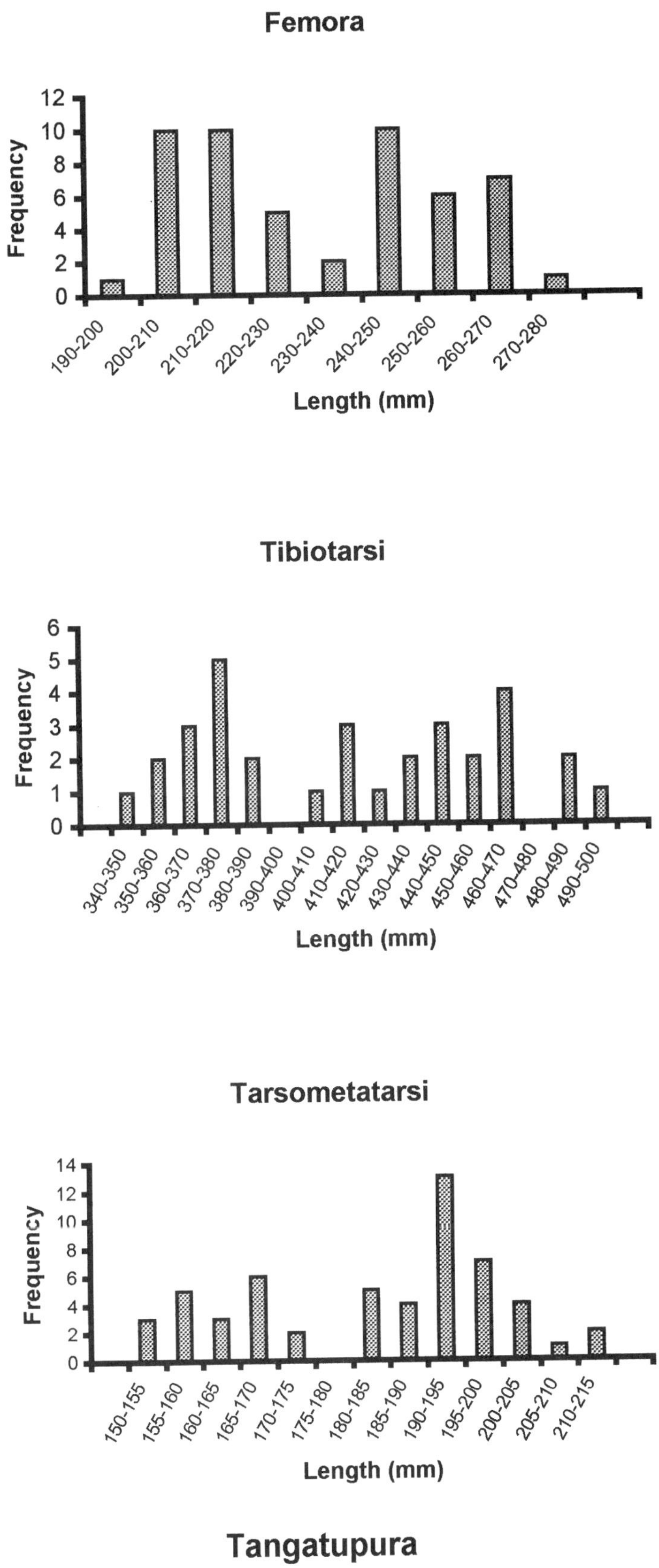

Figure 5.24. Length frequency distribution of long bones of Pachyornis mappini *from Tangatupura show a pronounced bimodality, which has been interpreted as sexual dimorphism.*

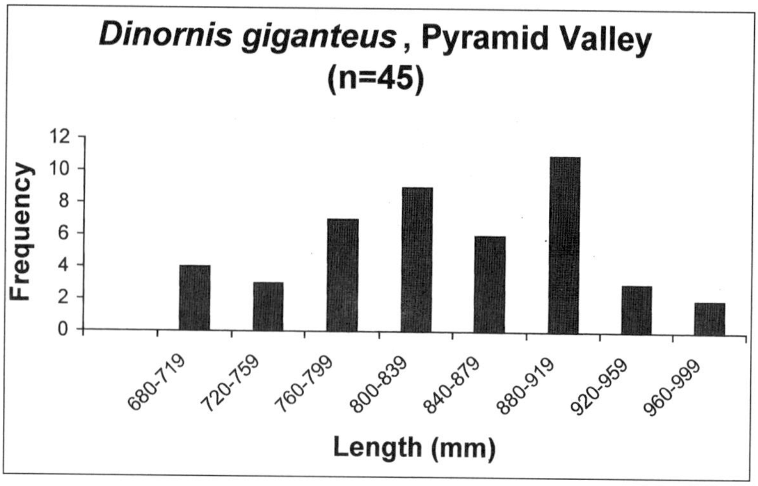

Figure 5.25. Length frequency histogram for tibiotarsi of Dinornis giganteus *from Pyramid Valley. Modified from Holdaway and Worthy (1997), with the permission of the Royal Society of New Zealand.*

actual population structure, then the marked differences may reflect reproduction abilities. The upland, shrubland *Megalapteryx didinus* seems to have had markedly more young in its population than its lowland counterpart *Anomalopteryx didiformis. Pachyornis* species had similar proportions in the two sites with a large sample size. The data for emeids appear to show a relationship between proportions of young and the degree of seasonality in the preferred habitat of each species. The moa in the most strongly seasonal environment, the subalpine zone, appeared to have produced more young than those in more equable environments, so the species in the most stable environment—closed-canopy lowland forest—had the fewest young. *Dinornis struthoides* is the smallest dinornithid and preferred shrublands. Analysis of the few data available suggests that it produced more young than the other dinornithids. However, analysis of data from Makirikiri suggests that the proportions of young in *Dinornis* spp. may be inversely proportional to the size of the species.

Maturity would have come late: possibly moa were 5–10 years old before they could hold a territory and breed successfully. But from the above data, we could infer that the larger dinornithids matured later than the smaller ones, and perhaps the moa of the deep forest matured later than those of the shrublands. The forest species had a lower productivity to contend with, which meant that territories were larger, and hence fewer and more sought after.

Breeding and Eggs

There are few data on moa clutch size, but analysis of existing data suggests that most species had a clutch of only one or two eggs. Excavation of rockshelters in Hawke's Bay has revealed eggshell, and no more than one or two eggs are represented by any one collection of eggshell from each discrete event (Hartree 1999; data not shown).

Similarly, in the dunes at Tokerau Beach, eggshell is commonly found as discrete scatters on the dune surface. Falla, in Golson (1957: 272), interpreted reports by Frost that at Tokerau Beach, skeletons

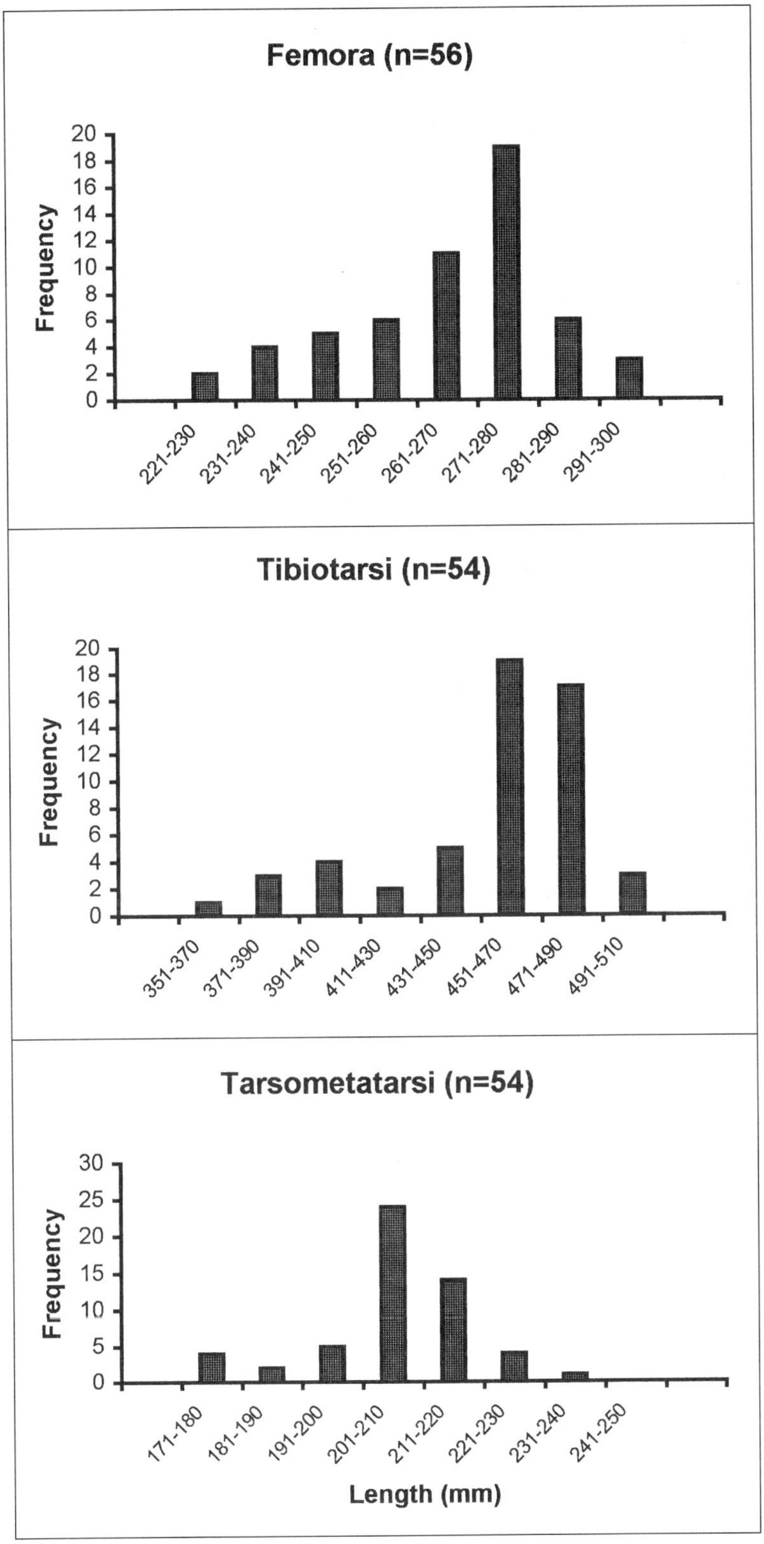

Figure 5.26. Length frequency distributions for Emeus crassus *long bones from Pyramid Valley. Modified from Holdaway and Worthy (1997), with the permission of the Royal Society of New Zealand.*

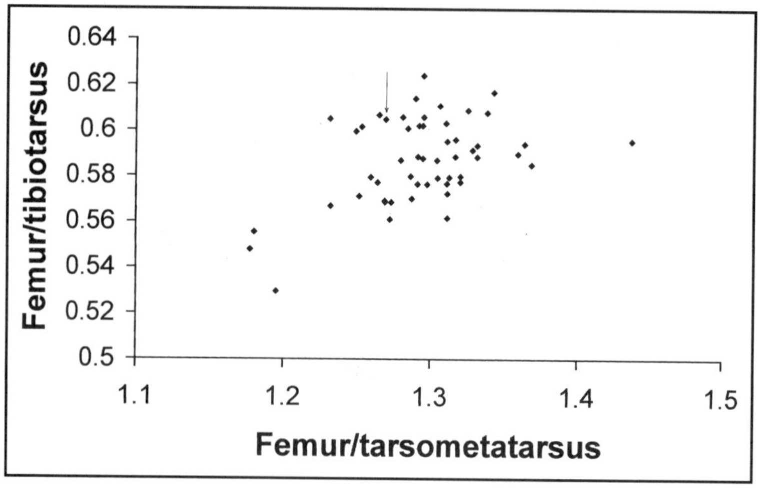

Figure 5.27. Ratio plots for Emeus crassus *from Pyramid Valley using the same data as in Figure 5.26. Two distinct clusters are apparent and are distinguished by the relative size of the tibiotarsus. The known female (arrow)—it was gravid when trapped in the swamp—is a large individual and is in the grouping with relatively shorter tibiotarsi. Two of the outliers in the lower left of the plot are subadult individuals, known because the tibiotarsus–patella synostosis was incompletely fused; this is one of the last junctions to become indistinct in moa growth. Modified from Holdaway and Worthy (1997), with the permission of the Royal Society of New Zealand.*

were closely spaced and each covered one or more eggs, to result from "seasonal corraling [*sic*] of the birds by man, a kind of semi or temporary domestication." Such fanciful interpretation is not supported by the facts. First, only one or two of the more than 100 partial skeletons from this locality in museum collections have eggshell with them, and even then direct association is equivocal. Second, the moa remains come from prehuman horizons. Although eggshell is abundant at this and other Far North dune localities, carefully made collections of discrete scatters provide eggshell sufficient for usually only one or two eggs.

There are relatively few more or less complete eggs in existence, and of these, even fewer can be referred to a species with any confidence. One is an egg of *Emeus crassus* that was found with a skeleton in Pyramid Valley swamp, as mentioned above. Presumably this was an unlaid egg. A green egg from Chatto Creek, discussed in more detail below, is referred to *Megalapteryx didinus* because when moa bones are found in intimate association with green eggshell, they belong solely to this species, or it is always present in the assemblage. And the large white egg from Cromwell, which was found in the bank of a river, contained an unhatched embryo. This was described by Hector (1872a) (Fig. 5.28) and is clearly referable to *Pachyornis* by the form of the cranial bones, specifically the sharp pointed premaxilla, which has a straight culmen. It was probably *Pachyornis elephantopus,* which is the principal species in this genus known from Central Otago.

Oliver (1949) listed 19 eggs sufficiently complete for one or other diameter to be measured. He allocated them to species by relative size from the known ratio of the *Emeus* egg: the *Emeus* egg was $^2/_3$ the length of the femur, $^3/_8$ that of the tibiotarsus, $^6/_7$ the length of the tarsometatarsus, and $^9/_{20}$ that of the pelvis. The largest egg he referred to *Dinornis maximus* (here *Dinornis giganteus*).

In our analysis, we note that Oliver's method incorrectly identified the *Pachyornis elephantopus* egg (with the juvenile) as a medium-size *Dinornis.* It seems that *Pachyornis,* with its relatively broad pelvis, had a relatively large egg. If this egg is attributed to *P. elephantopus,* the

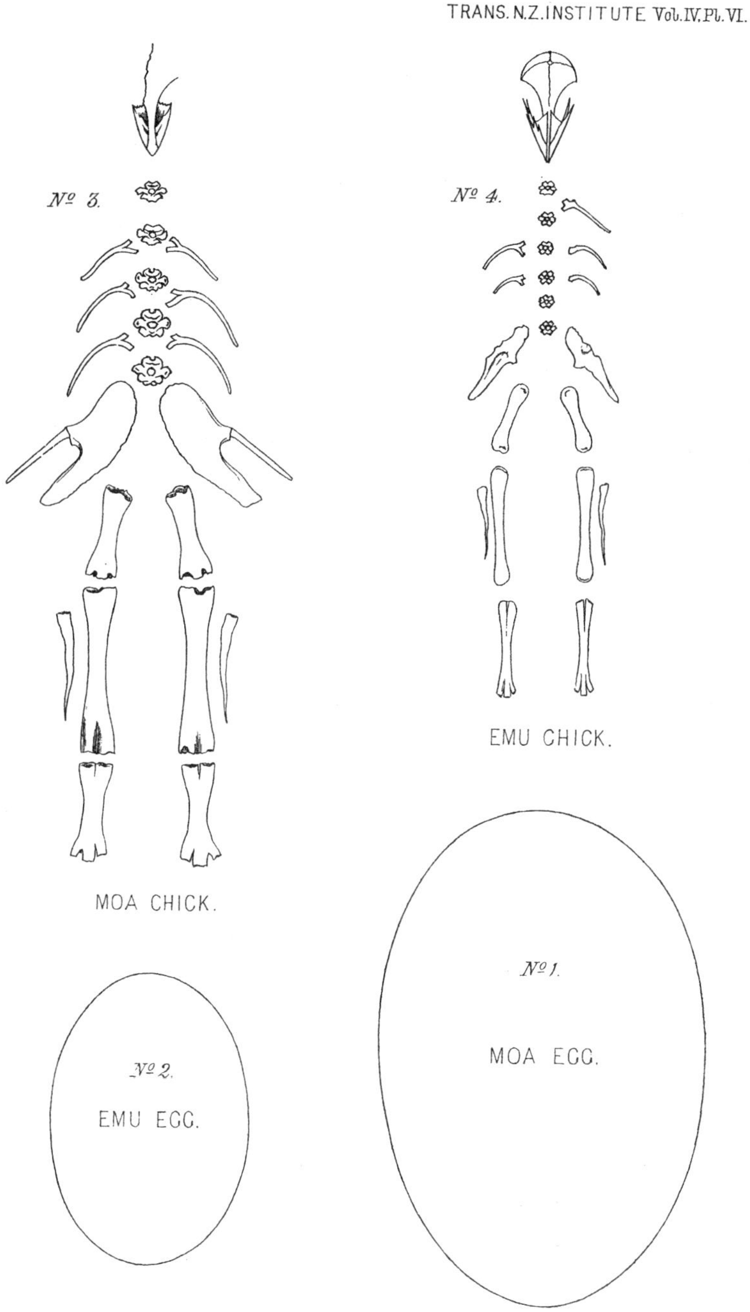

Figure 5.28. Embryonic skeleton of a moa (MNZ S4), identifiable as a species of Pachyornis *by the distinctive premaxilla. It was found within an egg in Central Otago. After Hector (1872a).*

Shag River egg can be referred to the same species because it is of similar size. Then the large group of somewhat smaller eggs (194–215 mm in length, attributed by Oliver 1949 to *P. elephantopus*) probably all belong to *Euryapteryx geranoides*. This would agree with the relative frequency of species represented by bones in the same deposits. At Wairau bar and Awamoa, *E. geranoides* dominates and *Pachyornis elephantopus* is much less common. The Waitomo, Waikaremoana,

TABLE 5.9.

Moa eggs for which length and width measurements can be obtained.[a]

Locality	Collector	Date	Size (mm)	Oliver's species	Inferred species	State of egg	Museum	Site type
Kaikoura	Fyffe	1859	240 × 178	*Dinornis giganteus*	*Dinornis giganteus*	P	MNZ	Archaeologic, burial
Mangawhitikau Valley, Waitomo	C. Matthew	ca. 1970	197 × 151	NA	*Dinornis novaezealandiae*	R	Waitomo	Rockshelter
Awamoa[b]	W. B. D. Mantell	1852	200 × 150	*Euryapteryx geranoides*	*E. geranoides*	R	BMNH A222	Archaeologic, midden
Awamoa	W. B. D. Mantell	1852	200 × 140	NA	*E. geranoides*	R	BMNH A222a	Archaeologic, midden
Awamoa	W. B. D. Mantell	1852	215 × 150	NA	*E. geranoides*	R	BMNH A222b	Archaeologic, midden
Clutha River	S. Coard	1899	195 × 135	*Pachyornis elephantopus*	*E. geranoides*	P	OM Av7475	Alluvial
Clutha River	P. Weaver	1900	201 × 138	*P. elephantopus*	*E. geranoides*	P	BMNH	Alluvial
Clutha River	P. Weaver	1900	— × 142	*P. elephantopus*	*E. geranoides*	I	OM Av9033	Alluvial
Ettrick	C. Gibson	1911	200 × 138	*P. elephantopus*	*E. geranoides*	P	OM Av7479	Alluvial
Wairau Bar	J. Eyles	1942	205 × 158	*E. geranoides*	*E. geranoides*	R	CM	Archaeologic, burial
Wairau Bar	J. Eyles	1939	194 × 139	*P. elephantopus*	*E. geranoides*	P	MNZ	Archaeologic, burial
Wairau Bar	J. Eyles	1942	207 × 139	*P. elephantopus*	*E. geranoides*	P	CM	Archaeologic, burial
Wairau Bar	J. Eyles	1942	207 × 145	*P. elephantopus*	*E. geranoides*	R	CM	Archaeologic, burial
Wairau Bar	J. Eyles	1942	215 × 146	*P. elephantopus*	*E. geranoides*	R	CM	Archaeologic, burial
Doubtless Bay	Matthews	1900	120 × 91	*Euryapteryx curtus*	*E. curtus*	P	AIM	Dune
Doubtless Bay	Matthews	1900	120 × 97	*E. curtus*	*E. curtus*	I	AIM	Dune
Doubtless Bay	Deeming	1940	125 × ?	*E. curtus*	*E. curtus*	R	?	Dune
Pyramid Valley	R. A. Falla	1939	179 × 134	*Emeus crassus*	*E. crassus*	R	CM	Swamp
Chatto Creek	G. Hamel	1981	160 × 120	NA	?*E. crassus*	I	OM Av7400	Mud flow
Waipiata, Taieri River	J. Darby, E. Barker	1994	180 × 129	NA	?*E. crassus*	R	OM Av7840	Alluvial

Locality	Collector	Date	Size (mm)	Oliver's species	Inferred species	State of egg	Museum	Site type
Cromwell	C. O'Neal	1866	226 × 155	*Dinornis robustus*	*P. elephantopus*	R	MNZ S4	Alluvial
Shag River	G. Griffiths	1941	221 × 150	*D. robustus*	*P. elephantopus*	P	OM Av9032	Archaeologic
Chatto Creek	C. I. Irwin, OM party	1953	160 × 107.5	NA	*Megalapteryx didinus*	R	OM Av7469	Mud flow
Kai iwi	A. E. Clutterbuck	1931	159 × 114	*Anomalopteryx didiformis*	*A. didiformis*	P	Wanganui	Alluvial
Tukituki River	J. C. Jorgensen	1936	152 × 121	*A. didiformis*	*A. didiformis*	R	HBCT	Rockshelter
Mangawhitikau Valley, Waitomo	C. Matthew	ca. 1970	168 × 120	NA	*A. didiformis*	R	Waitomo	Rockshelter
Rosie Bay, Lake Waikaremoana	W. H. Hartree,	Before 1957	—	NA	*A. didiformis*	R	MNZ S473	Rockshelter
Hukanui 7b, Hawke's Bay	W. H. Hartree, J. C. Yaldwyn	Before 1957	180 × 115	NA	*A. didiformis*	R	HBCT	Rockshelter
Martinborough 5, Haurangi Hole	W. H. Hartree	1952	165 × 120	NA	*A. didiformis*	R	MNZ S457	Rockshelter
Martinborough 5, Haurangi Hole	W. H. Hartree	1952	— × 120	NA	*A. didiformis*	R	MNZ S458	Rockshelter

Abbreviations: AIM = Auckland Institute and Museum; BMNH = British Museum (Natural History), London; CM = Canterbury Museum; HBCT = Hawke's Bay Cultural Trust, Napier; OM = Otago Museum; I = imperfect; MNZ = Museum of New Zealand Te Papa Tongarewa; P = perfect; R = restored.

[a] Data from Oliver (1949), Simpson (1955), Anderson (1989c), Benham (1902), and our observations. Oliver's species are those attributed by Oliver. Under the inferred species column, we list our choice of species attributed to that egg. The measurements for the *Emeus* given by Oliver incorrectly quoted the egg length as 129 mm instead of 179 mm. Oliver gave the measurement of the Kaikoura egg as 253 × 178 mm, which is the measurement of a plaster model in the Museum of New Zealand, but Dell and Falla (1972) give the measurement of the actual egg, and that is used here. The larger Waitomo egg is referred to *Dinornis novaezealandiae* because this is the common large dinornithid of the region.

[b] The egg measured by Owen is one of the following three partly restored eggs from the Mantell collection, for which measurements were given by Anderson (1989c).

and Martinborough smaller eggs are referred here to *Anomalopteryx didiformis* because this was the most common small moa in the sites of surrounding areas. However, *Pachyornis mappini* was relatively common and included juveniles in the latter two sites, so may also be represented among these smaller eggs. Similarly, the eggs from Tokerau Beach in Doubtless Bay can confidently be ascribed to *Euryapteryx curtus*, because about 95% of more than 100 individuals represented by bones in that area are of this species. At least 30 eggs have length and width dimensions that are measurable (Table 5.9).

Eggs have been found in several situations. Most surprising, perhaps, are those that were washed from alluvial deposits during gold-dredging operations in the 1900s. Two were even found floating in the dredging pond. At least seven are from alluvial deposits. One (the *Emeus* egg) was reconstructed from crushed remains found in a swamp. Several have been obtained from archaeologic deposits. Those from Wairau Bar and Kaikoura were associated with burials. Those from Awamoa had been cooked and eaten, and others, from Tukituki River and Shag Mouth, are of archaeologic origin but from an uncertain context. A few, including the Patoka and Martinborough eggs, have been obtained from shallow rockshelter deposits, which had been used for nesting. Collections of broken shell from similar Hawke's Bay sites and the dune fields of Tokerau Beach would yield many more measurable eggs if they were reconstructed.

An interesting feature of all moa eggs is that they are relatively big compared with those of living large ratites (Fig. 5.29). Even the small eggs attributed to the small (~20 kg) *Euryapteryx curtus* are about the same size as those from the much bigger emu (average female weight about 37 kg). Emu eggs vary considerably in size; for example, Marchant and Higgins (1990) cite data for 17 eggs with a mean of 130.3 mm long (range, 125.5–138.5 mm) by 89.6 mm wide (range, 81.9–93.4 mm). Eggs laid later in a clutch are invariably smaller than the first ones. Cassowary *Casuarius casuarius* eggs are little bigger than emu eggs: 135 mm long (range, 118–158 mm) by 95 mm wide (range, 89–100 mm; n = 37; Marchant and Higgins 1990), and the female may weigh about 60 kg. The cassowary therefore weighs about the same as *Emeus crassus*, and as for the emu, its egg is considerably smaller than that of moa of similar size.

Similarly, the ostrich is the biggest of the living ratites, but its eggs are small compared with those of most moa. The kiwi has a very large egg compared with its body size, and its chicks at hatching are precocious. The large eggs of the moa suggest that moa chicks too would be well developed upon hatching and probably would be able to feed by themselves almost immediately, as can a kiwi. The necessary corollary of this is that moa would require a protracted incubation period to facilitate this development. Thus we arrive at a picture of breeding in moa involving a nest with only one or two eggs in it, a long incubation period, probably by the male, as in most extant ratites, and precocious chicks able to feed on their own shortly after hatching.

The structure of moa eggshell has received very little study. Nathusius (1870, in Tyler 1964) and Hutton (1872c) briefly reported the

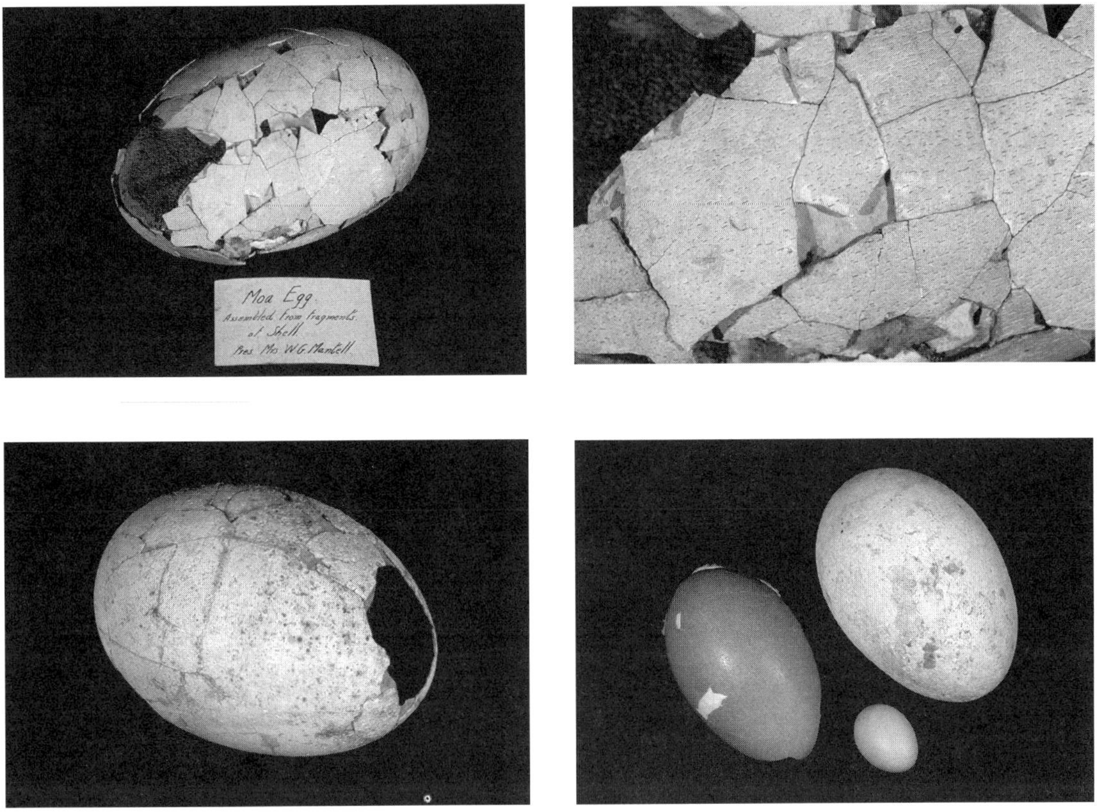

*Figure 5.29. A montage of moa eggs. (A) One of Mantell's original reconstructions (MNZ S00002) of ?*Pachyornis elephantopus *from Awamoa, Otago. (B) Detail of same. (C)* Anomalopteryx didiformis *MNZ S457, from Martinborough. (D) Moa eggs referred to* Euryapteryx geranoides *(OM Av7479) from Ettrick Burn, Otago (largest) and* Megalapteryx didinus *(OM Av7469) from Chatto Creek, Otago (left), compared with a chicken egg. Photographs by R. Morris.*

physical structure of moa shell. Liversidge (1881) reported the chemical composition of moa shell. Tyler (1957) undertook more detailed study of moa shell, particularly the pore structure, but much more remains to be done, particularly in describing interspecific variation. Most recently, Gill (2000) presented data on thicknesses of eggshell from dune deposits on Karikari Peninsula, Northland.

Moa eggs were smooth and pitted with either circular or linear markings. The polar regions of the egg have only circular pits, whereas equatorial parts of the egg commonly have linear pits. Within these linear depressions, a pore may be branched and have up to four pore mouths opening to the pore. The outer two thirds of the thickness is termed the spongy layer, and the lower third is called the mammillary layer (Tyler 1957).

Moa eggshell is thin. Hutton (1872c) said moa eggshell was 0.07 inches (1.78 mm) thick, and eggs from Tyler's (1957) small sample taken from Otago locations were all about 1.4 mm thick. Oliver (1949) recorded a maximum thickness of 2 mm from the *Pachyornis* egg from the Clutha (here listed as from Cromwell, a town on the Clutha) and the Shag River egg here assigned also to *Pachyornis*. Smaller moa laid eggs with much thinner shell, with most from Tokerau Beach being about 1 mm thick. The eggshell there was most likely laid by *Euryapteryx curtus* because this is the dominant moa in assemblages in this site complex. McCulloch (1992b) reported dark olive-green eggshell from 1000-m elevation that was 1.16 mm thick and referred it to *Megalapteryx*. McCulloch noted that other shell in the site was a much paler green. It is pertinent to note that emu eggs vary markedly from nearly black to pale green, and all fade with exposure to sun. Thus it is likely that the different moa species had shells of different thicknesses to match their size, but all were thin compared with ostrich eggs, which range from 1.7 to 2.1 mm thick (Tyler and Simkiss 1960), yet which in terms of length and breadth are smaller than eggs of *Anomalopteryx didiformis*.

Hutton concurred with Nathusius's (1870, in Tyler 1964) conclusion that moa eggshell was more like that of the South American rhea than any other ratite. Tyler and Simkiss (1960) noted that pore structures of moa were most similar to either rhea or *Aepyornis* but quite different from ostrich, emu and cassowary, or kiwi.

Pathology in Moa

Very little work has been done on the pathologic conditions found in moa bones. Unlike the case in many small birds where there is often found evidence of a fracture that has healed, even in such necessary elements as the humerus of a pigeon, we have seen no evidence that a moa ever survived a broken leg. The leg bones of moa were strong, and if a moa had ever had an accident that broke one of the three major bones, it is doubtful that it would have survived. Even the numerous instances of moa having fallen down vertical drops into caves rarely resulted in a fractured leg bone. We know of only one skeleton where the femur broke; in this individual, the bird had fallen about 15 m down a shaft into a cave. We know this because the pieces of the femur bear the characteristic marks of a greenstick fracture. A greenstick fracture happens when the bone is broken when the animal is alive or freshly dead; the fracture is smooth and has a shallow, often curved plane through the shaft, rather than at right angles to the surface, and with a surface like broken chalk. Such greenstick fractures are of course common in archaeologic sites.

Fractures of minor bones such as ribs and fibulae were more common. Worthy (1998b: fig. 9) published a photograph showing the articulated legs of an *Emeus crassus* in situ in which the fibula had broken in life and twisted outward at 90° to the tibiotarsus. That this was not related to the immediate death of the animal is shown by the fact that the break had partly healed in a deformed fashion.

No survey of disease has been made in moa. However, a single spectacular example was recorded by Anderson and Sorenson (1944) when they described a case of chronic osteomyelitis in the right tarsometatarsus of a subadult *Emeus crassus*. The huge abscess that resulted completely deformed the bone (Fig. 5.30). A similar abscess is seen in the mandible of an *Anomalopteryx didiformis* (Fig. 5.31).

Many examples occur in the main collections in New Zealand of less serious cases of arthritis, mainly in the toe joints of various moa. These have also been reported by several authors (Booth 1875; Anderson and Sorenson 1944).

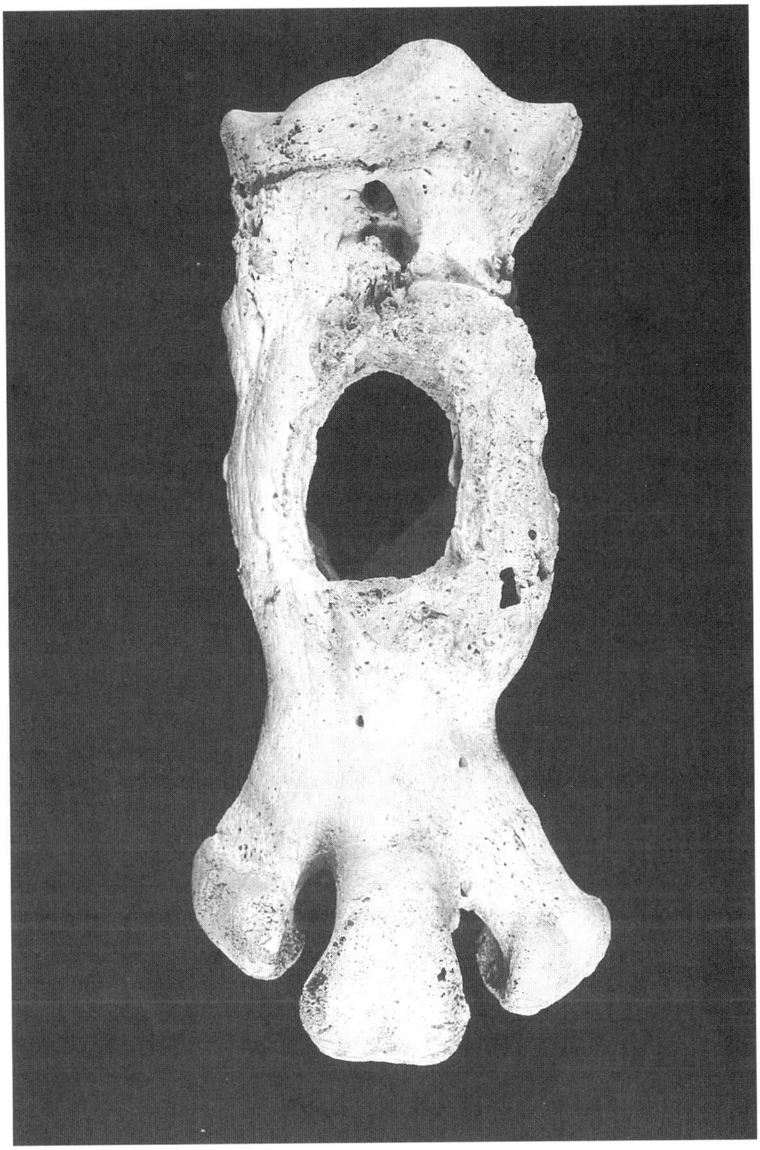

Figure 5.30. Case of chronic osteomyelitis seen in a tarsometatarsus of Emeus crassus *(Southland Museum A42.15). Photograph reproduced courtesy of the Southland Museum and Art Gallery Collection.*

Ecology of Moa: Habitat and Habitat Use

Historical Review of Perceptions

The habitats of moa did not greatly interest most workers during the 1800s. They were concerned primarily with defining the taxonomy of the group. The first statement about moa in a specific environment was by Julius Haast. His general theme dominated popular and scientific perception of the group for nearly a century:

> Judging from the structural character of the different species of *Dinornis,* they must have inhabited the open country where such existed, and not the forested regions, where not only innumerable

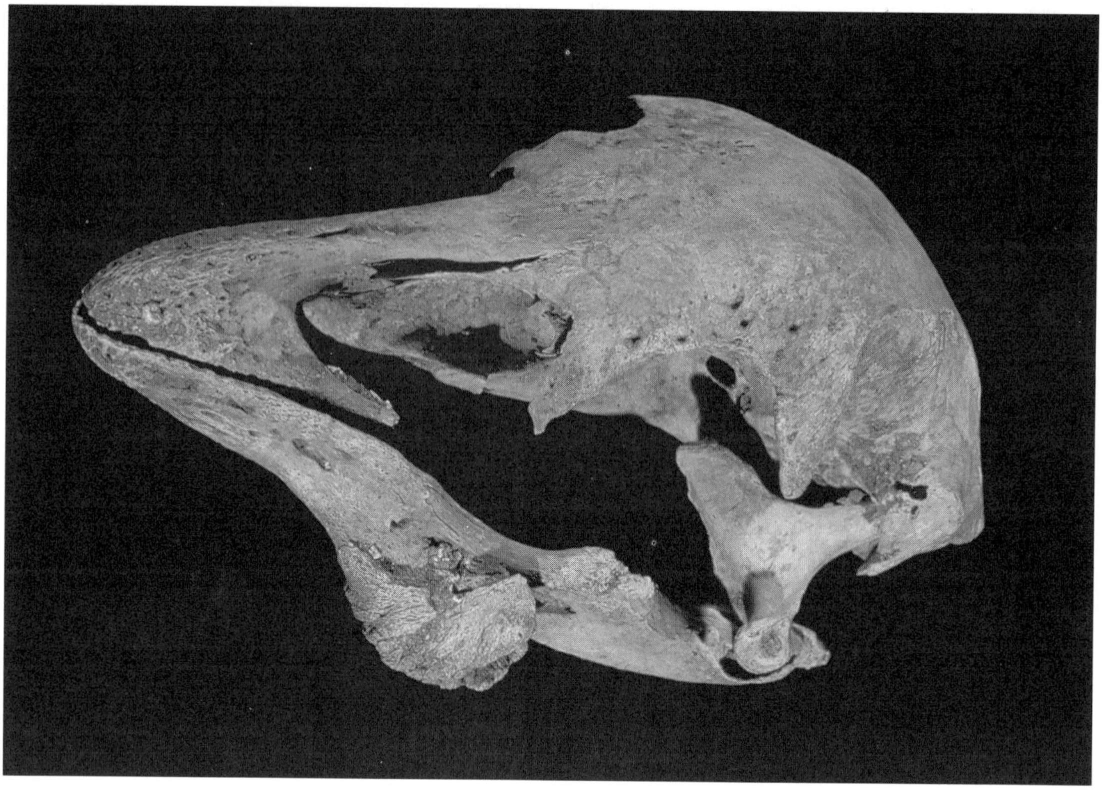

Figure 5.31. Skull of Anomalopteryx didiformis *from Hukanui, Hawke's Bay, with evidence of an abscess or cancerous growth on its mandible. Photograph by R. Morris.*

impediments to locomotion would have stood in their way, but where they also would probably have found little food suitable to them. In the term "open" I include plains and hillsides in the low lands covered with grass, fern, tutu (*Coriaria ruscifolia*), flax (*Phormium tenax*), and cabbage trees (*Cordyline australis*), and the subalpine regions. . . . Moreover, I have no doubt that the different species of *Dinornis,* like those of the *Apteryx,* were omnivorous. (Haast 1872b: 73)

Haast's view was dominated by his perception of what is possible for a big bird; he was not familiar with the big cassowaries of the lowland rainforests in Papua New Guinea. Mainly he was misled by misconceptions of the antiquity of the environment that greeted the Europeans who came to New Zealand. He, like everyone else until the mid-20th century, thought that the grasslands and vast areas of fern and shrublands that European settlers found in all drier parts of the country, but particularly in Canterbury and Otago, where he worked most, were natural. They were not.

Thereafter, scientists paid little attention to moa habitat. Popularists, such as Buick (1931: 213), continued to promote Haast's suggestion: "It [moa] browsed upon the hillside grasses as the cattle do, and devoured leaves and berries when it could reach them. At the season of the year when tawa, the karaka, and the matai trees were in fruit, the birds would penetrate into some of the tighter bush, but at all other times of the year it preferred the open spaces."

Archey, who led in developing a modern taxonomy of moa, favored an open habitat. He was swayed in part by the observations that other large ratites favored mesic habitats:

> It was in the latter extensive open grasslands of the South that the moa abounded, particularly the massive forms whose ungainly bulk would probably have caused them to favor the lowlands, leaving the higher tussock and scrub country to the more slender species, e.g., of *Anomalopteryx* and *Megalapteryx*, and probably of the tall Dinornis also, despite the latter's frequent occurrence in swamp deposits. (Archey 1941: 81)

Despite its emphasis on nonexistent grasslands, this statement was significant because for the first time, someone acknowledged that different species might have lived in different habitats. But Archey still could not conceive of any moa being primarily a forest bird: "Another adverse factor [relating to their extinction] may have been the extension of the forest during the pluvial period. It is well known that the moa penetrated into the forest—they were trapped in caves far into the bush; . . . possibly they wandered in in search of food, but bred in more open situations" (Archey 1941: 91).

Oliver (1949, 1955) also believed that moa lived mainly in open country, leaning on Māori tradition and the evidence of the location of moa bone discoveries. At the same time, Duff (1949a, 1949c, 1951) brought the idea of moa as grassland birds to the general reader, and a whole generation of New Zealanders grew up reading this material. It is ironic that the very moa that provoked Duff's publications provided data on moa diet that Duff chose to ignore. Many of the skeletons were associated with gizzard contents. Along with the gizzard stones was well-preserved plant matter. The first studies of these gizzards were actually published nearly a decade before Duff's bulletin (Allan et al. 1941) and showed that the gizzards were full of seeds and twigs, not grass.

Falla (1962), although repeating Archey's suggestion that *Anomalopteryx* and *Megalapteryx* were presumably small bush moa, noted that nothing much was known about the habitat or food preferences of *Pachyornis*. Echoing Haast, he believed that the structure of *Euryapteryx* meant that it was almost certainly a low country and probably a grassland and forest edge species.

It was more than 25 years after the recovery of abundant gizzard contents in Pyramid Valley before there was any appreciation that they revealed moa diet. An improved understanding of moa habitat preferences was left to an archaeologist. Simmons (1968) drew on the evidence offered by palynology (Cranwell and von Post 1936; Wardle 1963) to point out evidence that in both old and young fossil deposits, including those dating to the time of Māori arrival, New Zealand was primarily forested. Therefore, "it is likely that the moa at the earlier stages of their history, at the time of the earliest known moa fossils and in recent deposits, were forest or forest fringe birds." He used the association of forest birds and moa in Pyramid Valley and in caves in the Wairarapa to strengthen the case that "the moa was associated with forest" (Simmons 1968: 116).

Simmons was instrumental in establishing a new perception of the preferred habitat of most moa. Nevertheless, the idea of forest-dwelling moa was at first not generally accepted. For example, Stevens (1980a: 252) stated,

> Most of them probably grazed grassland plains, shearing off the grasses with their massive beaks as neatly as an old-time harvester with his scythe. *Dinornis maximus* probably required as much grass per day as a bullock. . . . In ancient New Zealand the grassland moas thus filled the role now occupied by grazing mammals (sheep, goats, cattle etc).

However, he did allow that some of the lesser moa (Anomalopterygidae, or Emeidae as herein used) lived in forest. But well before 1980, abundant evidence from pollen analyses had shown that grasslands were rather limited in distribution during the Holocene (Cranwell and von Post 1936; Lintott and Burrows 1973; McGlone and Moar 1977; Moar 1971, 1973), yet the fossils of the larger moa were widespread. Stevens had either misrepresented or ignored published information on moa diet dating back to 1941.

About the time Stevens's influential book appeared, Colin Burrows of Canterbury University had finally taken up the challenge of seeing just what was in all those gizzards that had been sitting in the Canterbury Museum for 30 or 40 years. His detailed research into moa diet was based on the examination of 14 *Dinornis giganteus* gizzards from Pyramid Valley and Scaifes Lagoon (Burrows 1980a, 1980b; Burrows et al. 1981). It showed conclusively that the dominant food eaten by *Dinornis* was twigs of a range of shrubs and trees, augmented by a few leaves and fruit.

The results from gizzard analyses were convincing and helped to sway the perception of where moa lived—but no more so than the greatly increased data from pollen analysis, which showed that grasslands were of limited extent, except in the subalpine zone, during most of the Holocene. Nevertheless, as McCulloch (1992a: 49) said, "the myth of grazing, grassland moas continues to die hard."

By the time of a seminal symposium on moa entitled "Moas, Mammals and Climate in the Ecological History of New Zealand," held in 1986, there was a general acceptance that moa were forest birds. For example, Clout and Hay (1989: 27) state that "the extinct moas (Dinornithidae) were forest-dwelling browsers and frugivores." But Caughley (1989: 6) was more circumspect. He noted that although it was certain moa lived in forest environments, he "[could] think of no ecological reason why moas would avoid alpine grassland and sub-alpine scrub." But he noted that the evidence was thin; only a few bones were known from loess deposits or from the subalpine zone (see below).

In an astute work, Flux (1989) made use of biogeographic theory to predict that if forest was the only habitat type, then we would only expect about three species in each main island, or a maximum of six species in total. The fact that there were 12 (and that figure did not seem likely to be reduced much) meant that "several species of moa did not live in forest but in more open habitats—swamps, coastal or riverine areas, and alpine grassland" (Flux 1989: 36). The flaw in this analysis is that it assumes forest presented a single habitat, yet the casual observer can see great differences between the almost monotypic and single-storied beech (*Nothofagus*) forests compared with the multi-tiered, podocarp forests or hardwood forests.

Batcheler (1989) also favored a mix of habitats. He disregarded the evidence from gizzards obtained from moa in swamps as being unrepresentative of any favored habitat. On the basis of the observation that more open habitats support greater mammalian biomass, he advocated scrub and grassland communities for the larger moa and forest for the smaller ones. Moreover, because he viewed moa as being clumsy, he said that "moas were not habitual occupants of the alpine zone" (Batcheler 1989: 63).

Atkinson and Greenwood (1980, 1989) developed a novel hypothesis that moa and plants coevolved in New Zealand to produce an abnormally large number of divaricating shrubs and other growth forms that gave protection against browsing. In it, they assumed that "all moa species were probably adapted to feeding in the forest and scrub even if a few were specialised to make greater use of more open habitats" (Atkinson and Greenwood 1989: 70).

Where They Actually Were

By the end of the 1980s, it was generally accepted that most moa lived in forest or at forest margins, but that some may have used the available (mainly subalpine) grasslands. Some thought that certain species visited the subalpine zone, but most thought this would have been unusual. In all this, however, the distribution of the fossils themselves was largely being ignored. Worthy had explored caves in the subalpine zone at up to 1800 m and knew that moa bones were abundant at these altitudes. Scientists had not been active there and so could not know. So a short article (Worthy 1989e) reviewing the distribution of moa in the subalpine zone was the first to reveal the frequency of their remains above the tree line. It showed that the small upland moa *Megalapteryx didinus* was most common but that *Pachyornis australis, Dinornis struthoides,* and *Dinornis novaezealandiae* were also found in sites above the tree line.

Similarly, there had been no review of specimens from the extensive South Island Pleistocene loess deposits, which again led to the assumption that moa remains were rare in loess. Perhaps the neglect had been caused by the loess bones not being "pretty" specimens and so were often overlooked by curators, who, with a stamp-collector viewpoint, favored better-looking, complete specimens. A survey of specimens in Canterbury Museum showed that 65 loess sites had been cataloged. *Pachyornis elephantopus* accounted for 62% of finds, but *Euryapteryx geranoides, Emeus crassus, Dinornis struthoides,* and *Dinornis giganteus* were also present. Apart from moa, the loess fauna contained *Aptornis defossor, Cnemiornis calcitrans,* and *Tadorna variegata* (Worthy 1993a).

But there were still unrecognized patterns in the distribution of moa remains over the whole of New Zealand. It turned out there was not as much significance in which species were present in a site as there was in which were the most common. The New Zealand fossil record is so extensive that representation by numbers is significant. If enough bones are collected in an area and no cognizance is taken of geologic age, the species list will include most, if not all, moa species known from the island. Only a few of these, however, will contribute most individuals. With the advantage of having examined collections of moa from several different areas in the country T.H.W. saw a pattern emerging. In most places, only three or four moa species were common, but the dominant one varied from place to place. Data from natural and archaeologic sites were combined, taking into account geologic age, and a general picture of moa distributions was revealed in which each species could be shown to have a preferred habitat (Worthy 1990). In that article, three principal habitats were identified: upland zone, lowland wet forest zone, and lowland dry climate zone.

Upland Zone

The lower limit of the upland zone or subalpine zone is the tree line, at which temperatures prohibit forest growth, which in New Zealand is at about 1200 m in the north but lowers to about 900 m in Fiordland in the southwest of the South Island. The zone supports a wide variety of herbs and tussock (clump) grasses, many of which are palatable to mammalian herbivores. Although snow covers these habitats for part of each winter, most are accessible for at least eight months of the year. Fossils can only be found where suitable fossil sites exist, which in New Zealand normally means where there is limestone. Fortunately, in northwest Nelson and in Fiordland, there are high-altitude limestone karst fields with numerous caves. The fossil faunas in the caves show that *Megalapteryx didinus* was the most common moa in these places. Its remains are, in fact, numerous. Three other species (*Pachyornis australis, Dinornis struthoides,* and *D. novaezealandiae*) are also found, along with a range of other birds, including Finsch's duck (*Euryanas finschi*), New Zealand coot (*Fulica prisca*), kiwi (*Apteryx* spp.), Haast's eagle (*Harpagornis moorei*), Eyles's harrier (*Circus eylesi*), and weka (*Gallirallus australis*).

The members of this high-altitude fauna would have had to move downslope during winter, at which time they would have overlapped with the altitude range of members of the adjacent forest communities. This is reflected in faunas at successively lower altitudes, where the proportion of *M. didinus* steadily declines in favor of another small emeid *Anomalopteryx didiformis,* which makes its appearance at about 1000 m and by 600 m is the only one of these two moa present. However, there may have been temporal separation because the lower altitude species may well have moved down and up in unison with the subalpine birds.

Lowland Wet Forest Zone

The lowland wet forests are typical of the western and northern parts of both main islands and of Stewart Island, where the annual rainfall was greater than 1200 mm. For most of the Holocene, these areas had a cover of tall forests with a continuous canopy. Most forests were dominated by Podocarpaceae, but types ranged from those dominated by rimu (*Dacrydium cupressinum*) to mixed hardwood–podocarp forests of Northland to beech forests of higher altitudes in the lower North Island and the South Island. Forest structure rather than the dominant tree seems to have been the important factor. So although a *Nothofagus* beech forest differs greatly from a tall podocarp forest, so long as it had a continuous canopy and was in the wetter regions, *A. didiformis* was the main small emeid present.

The other moa typical of wet forests was *Dinornis novaezealandiae,* but depending on the main island (North Island or South Island), one or two other moa species were usually present. In the North Island, *Pachyornis mappini* often lived alongside *A. didiformis,* and *Dinornis struthoides* was also present. In the South Island, closed canopy rain forests usually abutted the subalpine zone, and in the ecotone between them at 500–1000 m, both *M. didinus* and *A. didiformis* were often present, the latter being more abundant at lower altitudes (Fig. 5.32).

Lowland Dry Climate Zone

In this zone are included all areas where there was abundant shrubland as well as open-canopy forest and some grassland. Thus in the Holocene, most such regions would have held a mosaic of forest margins, shrublands, and grasslands. Such habitats were most common in the rain-shadow zone to the east of the Southern Alps, where seasonal water deficits restricted forest growth. In these same areas, large braided rivers augmented the patchiness of the environment by inducing seral habitats as the rivers periodically changed course. On the coasts, developing dune fields also featured similar habitats.

In the North Island, the rain-shadow effects were restricted to the Hawke's Bay area in the central East Coast. However, north of Auckland, there were dune fields in the Far North, from Kaitaia to North Cape, and in smaller areas such as Doubtless Bay. The vegetation mosaics in all these areas of both islands were critical to the third group of moa, the emeins. In the dune lands of the Far North, the little coastal moa *Euryapteryx curtus* was common, where it was associated with occasional *Dinornis giganteus, Dinornis struthoides,* and *Pachyornis mappini.* In the dunes along the southern North Island's East Coast, these species were joined by *Euryapteryx geranoides.* The transition to the closed canopy forest inland was abrupt. Sites only a kilometer or so inland are dominated by *Anomalopteryx didiformis.*

Central North Island is dominated by active volcanism. The andesite volcanoes of Ruapehu, Ngauruhoe, and Tongariro dominate the terrain at present, but important volcanic events with regional effects have been associated with the Taupo volcanic zone. Most recent vents have been in or near Lake Taupo itself, and a series of destructive rhyolite ash falls and ignimbrite flows have remodeled the terrain. The rarer ignimbrites have extended westward as far as the coast (100 km) and on occasion have overrun the axial ranges to the east and nearly reached the coast (100 km) in that direction also. In the later Holocene,

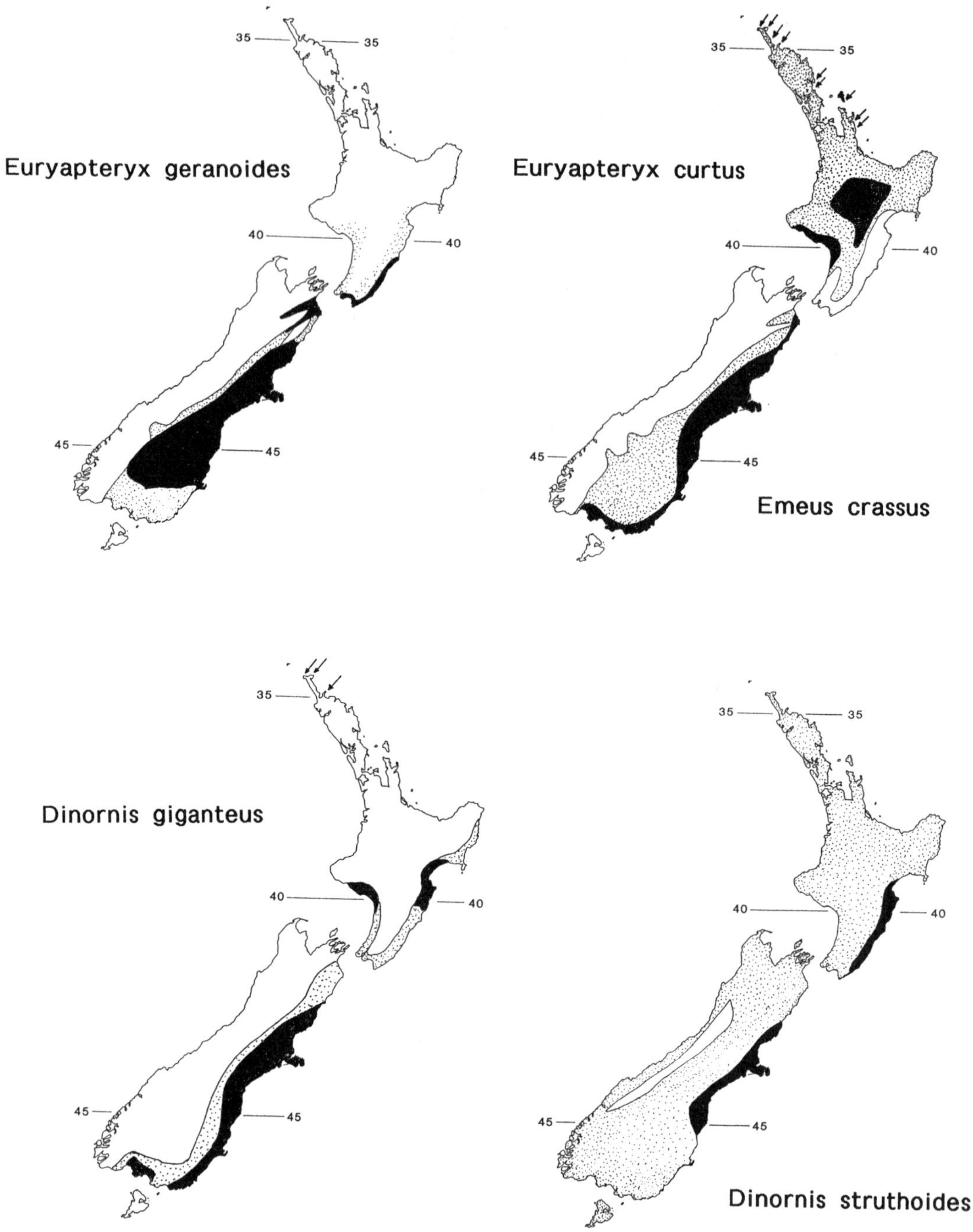

Figure 5.32. Maps of the distributions of the different moa species during the Holocene. Black areas indicate where the species was prominent in the fauna. Stippled areas indicate areas outside of the main distribution where the species was regularly present, and areas not shaded are where the species was rarely, if ever, present. Arrows indicate localized areas where the species has been found outside of its major range. Latitude is indicated by the numbered (degrees) horizontal lines.

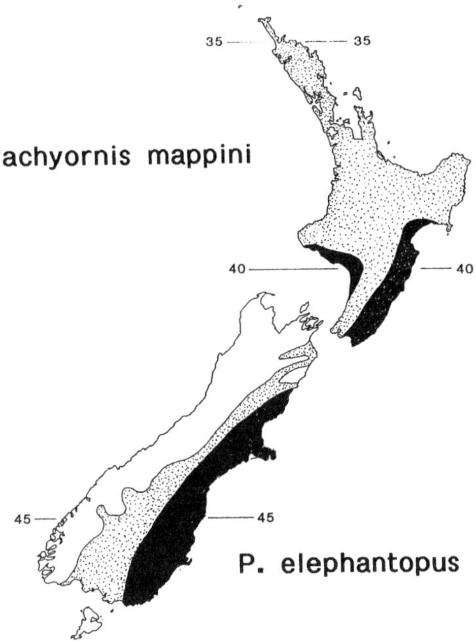

Pachyornis mappini

P. elephantopus

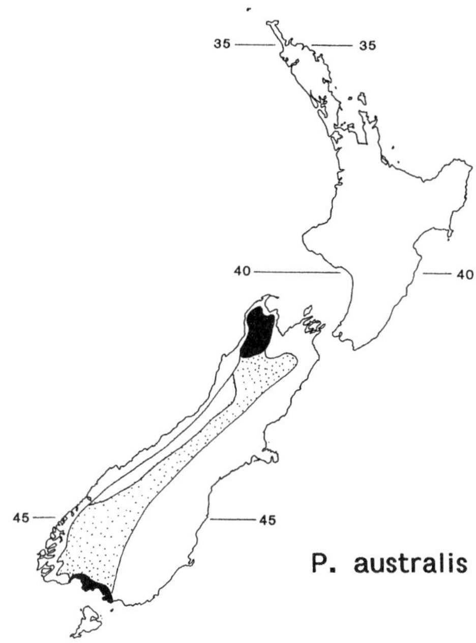

P. australis

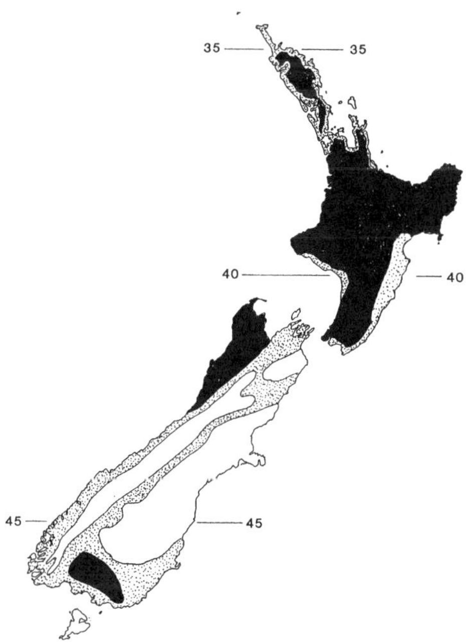

Anomalopteryx didiformis

Dinornis novaezealandiae

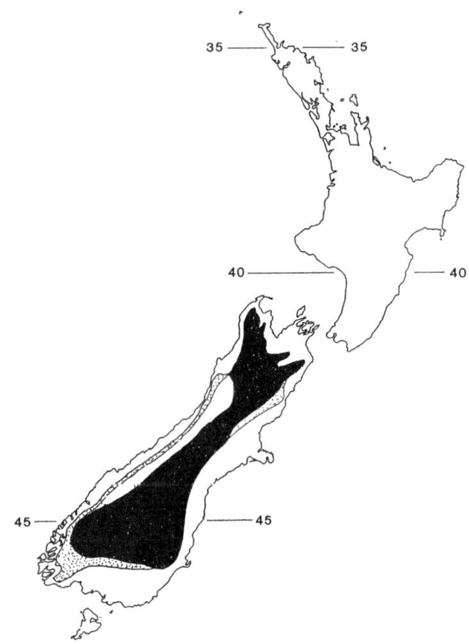

Megalapteryx didinus

particularly destructive eruptions took place 3300 years B.P. (Waimihia event), and notably the Taupo tephra eruption of about 1850 years B.P.

A major volcanic event has occurred every 1000–2000 years from somewhere in the central North Island volcanic zone during most of the last 100,000 years, and all eruptions have resulted in varying degrees of forest destruction. The Taupo eruption of 1850 years B.P. caused complete destruction of forest over a 80-km radius and caused substantial destruction for several further kilometers. Inundation of low-lying areas by vast floods of outwash pumice destroyed valley and floodplain vegetation in all the major river systems of the central North Island. Periodic forest destruction by volcanoes, along with a relatively high altitude in parts of the area, has resulted in a vegetation pattern that has shifted for thousands of years. Throughout the area over that time, there would have been substantial amounts of shrubland on or around the central plateau. Although there are few fossil sites, some on the periphery suggest that *Euryapteryx curtus* was common. The only moa-hunting sites known from this area (Tokoroa and one on Lake Taupo) contain only *E. curtus*.

There is some indication from the fossil record that not all dune lands supported the same fauna. Although in most dune sites throughout the North Island *Euryapteryx curtus* dominated the moa faunas, in one area they did not. At Waikuku Beach, just southeast of North Cape, *Pachyornis mappini* dominated (Table 5.10). The high proportion of *P. mappini* at Waikuku coincides with the presence of a wetland behind the dunes, which is unknown elsewhere in North Island dune sites. However, in the Holocene deposits of Lake Poukawa in central Hawke's Bay, the moa fauna is dominated by *Pachyornis mappini*. The species is also frequent in the midden deposits of South Taranaki, where the coast is backed by a series of parallel dunes with intervening swale wetlands. Despite being in seasonally dry regions, these places all have wetlands nearby. It appears that *Pachyornis mappini* preferred wetland vegetation.

The largest areas of xeric vegetation in New Zealand are in the South Island. The Southern Alps and other major ranges stretch for 400 km in a northeast–southwest chain, creating a rain-shadow area to their east, in Canterbury and Otago. Here, the average annual rainfall is as low as 400 mm in the driest areas and rarely is over 1200 mm. Seasonal (summer) drought is common. Large rivers rise in the mountains along the western margin, which, because of the rapid tectonic uplift of the ranges, carry huge bed loads of gravels. The rivers are usually braided channels in beds up to 1 km wide, and the whole systems meander across the plains and intermontane basins, periodically altering their course.

The combination of dry climate and physical structure meant that throughout the Holocene, the landscape supported a mosaic of vegetation. Tall forest was established in favorable places, but poorer soils and ecotonal zones of riverbeds supported shrublands and grassland. The drier and higher areas were mostly grassland or dense low shrubs. In this habitat, which was seasonally cold with winter snow but had

TABLE 5.10.
Comparison of the moa faunas from Waikuku and Tokerau beaches in Northland.[a]

Species	Waikuku Beach (MNI)	Frequency of Waikuku moa (%)	Tokerau Beach (MNI)	Frequency of Tokerau moa (%)
Pachyornis mappini	25	55.6	15	8.2
Euryapteryx curtus	11	24.5	157	86.3
Euryapteryx geranoides	0	0	2	1.1
Anomalopteryx didiformis	1	2.2	0	0
Dinornis giganteus	1	2.2	2	1.1
Dinornis novaezealandiae	2	4.4	2	1.1
Dinornis struthoides	5	11.1	4	2.2
Total	45	100	182	100

Abbreviations: CM = Canterbury Museum; MNZ = Museum of New Zealand Te Papa Tongarewa; WO = Waitomo Museum.
[a] Data from sites 7–9 for Waikuku Beach and from sites 33–39 for Tokerau Beach are from Millener (1981a).

summer temperatures exceeding 30°C, a distinctive suite of moa was found. The eastern moa *Emeus crassus* was restricted to the eastern lowlands of the South Island and dominated areas below 200 m. *E. crassus* was always associated with the larger emeids *Euryapteryx geranoides* and *Pachyornis elephantopus* and the dinornithids *Dinornis giganteus* and *D. struthoides*. These species displayed some niche partitioning along an altitude gradient because *Euryapteryx geranoides* gradually replaced *Emeus* upslope above 200 m. *Euryapteryx* may have preferred vegetation on the better-drained soils of the hills and *Emeus* that of the wetter, swamp-forests in the flat, lower-lying interfluves. The coasts of the South Island had dune fields in which the natural deposits are dominated by *E. crassus*, which may have been an analog of *Euryapteryx curtus* of the North Island.

Faunal Composition: An Insight to Site Age

The identification of these discrete patterns of moa distribution in the Holocene enabled the identification of Quaternary chronofaunas in New Zealand for the first time. It provided a hypothesis that could be tested by analysis of cave and other faunas. Dominant representation of a species or a particular suite of species in a deposit indicates the presence of a particular habitat type; hence the deposit could be dated from knowledge from palynology as to the time of occurrence of that habitat (e.g., shrubland) in that area. For example, abundant *Euryapteryx geranoides* in a deposit in a wet western forested area indicates that the deposit was laid down when shrubland predominated. From that, we know that *Anomalopteryx didiformis* dominated Holocene deposits of this area. Palynologic data show that shrubland was last present in this area before 12,000 years B.P. (in the Otiran Glacial). The prediction that the deposit containing *E. geranoides* was of the last Glacial age could then be tested by radiocarbon dating. None of the many sites dated to present has contradicted the rule that the dominant emeids in a deposit clearly reflect a particular habitat at the time of deposition.

Moa Diet

The diet of extinct species can rarely be observed directly, but several lines of evidence can be used to reconstruct the general diet. First, the structure of the feeding apparatus places physical constraints on what can be eaten. A comparison of the shape and relative strength of the moa bill can therefore be informative.

Second, the gizzard, the function of which to some extent parallels that of teeth in dentate animals, also varied among moa species. The gizzard is the stomach chamber in which food is ground up by stones the bird swallows to aid digestion. Powerful muscles squeeze the chamber and so grind the stones and food together. Although in all moa we can no longer dissect the tissues of the gizzard, complete sets of gizzard stones have been recovered with skeletons. The size of the individual stones and their total mass can be related to the ability to break down fibrous plant matter. A large set of stones, including some that were big, probably enabled much fibrous food to be consumed. Small gizzards with small stones suggest that ingested foods were soft. Furthermore, some of the small gizzards are in relatively big species and the bigger sets in small species.

Third, actual gizzard contents have been recovered in some sites. Analysis of the plant fragments they contain provides the best evidence about food species. A further indirect method is the analysis of stable isotopes such as $^{13}C/^{12}C$ and $^{15}N/^{14}N$ in moa bones (discussed in more detail below). Data on the carbon isotopes are routinely obtained as a by-product of radiocarbon dating.

Moa Bills

The bills of the various genera of moa are distinctly different. Unfortunately, the rhamphotheca is known for only one specimen (the type of *Megalapteryx didinus*), so the cutting edge of the bill is unknown in most species. However, the edges of the maxilla and mandible give information on the cutting apparatus and these do differ among taxa (see Fig. 4.7). Some moa (for example, *Dinornis* spp.) have a distinct U-shaped groove in the upper surface of the mandible along the anterior half. The groove sits wholly within a wide, shallow groove in the premaxilla when the two are opposed. Atkinson and Greenwood (1989) postulated that the mandibular groove supported a sharp-edged double rhamphothecal ridge, which conferred great cutting power when feeding on plants.

However, other birds that cut fibrous plant matter do not have this bill structure. For example, the rails *Porphyrio porphyrio* and *P. hochstetteri* routinely clip tough grass tillers. Their bills have single-ridged, heavily built, opposing, and sharp upper and lower edges. The endemic New Zealand kakapo parrot (*Strigops habroptilus*) eats a wide range of plant leaves and fruit, including fibrous material, yet its bill is typically parrotlike and lacks the grooves seen in moa. The grooved structure of the moa bill is not found in rails, ducks, procellariids, larids, or parrots but is seen in shags and cormorants (Phalacrocoracidae).

Bills of *Phalacrocorax* and *Leucocarbo* species have grooves similar to those of moa on the opposing upper and lower surfaces of the mandible and premaxilla, respectively. The shags hold slippery fish in their bills while orienting them for swallowing. The double-ridged structure is not used to cut up the fish but allows a good grip. The ridges on the moa bill may have had a parallel function, enabling the moa to gain a better hold on a plant twig. Possibly the structure allowed a moa to orient a twig in a particular way to be able to cut it: actual gizzard contents show that *Dinornis* clipped the twigs off plants rather than pulling and breaking them off. The bill structure may have enabled leaves and fruit to be stripped from a stem as well.

The surface of the mandible that opposes the premaxilla is not the same in all moa. The deep groove seen in *Dinornis,* on a mandible whose dorsal surface is narrow posteriorly and widens toward the tip, is unique to that genus. The mandibles of all *Pachyornis* spp. are different. The dorsal surface is wide and flat or very shallowly grooved at the base of the gape, and the surface narrows toward the tip without a deep groove ever being present. In *Megalapteryx,* the mandible is thin and carries only a shallow groove anteriorly. In *Emeus* and *Euryapteryx,* the mandible is convex dorsally at the gape and narrows anteriorly, forming a shallow groove. *Anomalopteryx* differs from all other species in having a wide mandible, which carries a sharp dorsal crest at the gape that narrows toward the tip, with no sign of a groove.

The mandibles of the various genera differ in their relative robustness. When depth and width over their length are considered, it is clear that the blunt-billed taxa (*Euryapteryx*) have weakly constructed mandibles. The small species with pointed and low-angled culmens (dorsal part of premaxilla), such as *Megalapteryx* and *Emeus,* also have weak mandibles. Those with steep culmens and pointed bills (*Pachyornis* and *Anomalopteryx*) have robust mandibles. The degree of this robustness is directly related to the size of the temporal fossa (the insertion area for the retractor muscles of the mandible). Species with large, robust mandibles have large temporal fossa that spread over the crania medially and posteriorly: *Pachyornis elephantopus* and *Anomalopteryx didiformis* have the largest of all moa. The broad, flat surface opposing the premaxilla in *Pachyornis* species would have allowed a proportionally great force to be applied by the large jaw musculature. The bill of *Anomalopteryx* was even better constructed for cutting because the sharp edge of the mandible at the base of the gape would have overlapped the opposing premaxilla. Their bills probably functioned like secateurs and would have been able to give a nasty bite.

Gizzard Size

The mass of complete gizzard sets varies with species (Table 5.11). In *Dinornis,* the largest species have huge sets exceeding 5 kg, in which stones can exceed 4 cm in one axis. The size of gizzards is smaller in

TABLE 5.11.
Mass of gizzard stone sets for individuals of some moa species.[a]

Species	Mass (g)	Body mass (kg)	Specimen and reference
Euryapteryx geranoides	253	?	CM Av20113, Gregg 1972
Euryapteryx geranoides, large individual	630	129, 124–133	MNZ S25656, Worthy 1989c
Euryapteryx geranoides, large individual	330	?	MNZ S27794, Worthy 1989c
Euryapteryx geranoides, small individual	92	~40	MNZ S27895, Worthy 1989c
Euryapteryx curtus	140	67, 51–78	WO555, Worthy, unpublished data
Euryapteryx curtus	80	32, 19–43	WO580, Worthy, unpublished data
Euryapteryx curtus	55	33, 20–45	WO581, Worthy, unpublished data
Emeus crassus	35.5	?	CM Av20119, Gregg 1972
Anomalopteryx didiformis	1105	~41	Worthy 1989c
Dinornis giganteus	5600	?	CM Av28434, Burrows et al. 1981
Dinornis giganteus	4380	99, 87–107	CM Av20118, Gregg 1972
Dinornis novaezealandiae	2450	93, 81–103	MNZ S25765, Worthy, unpublished data
Dinornis novaezealandiae	2310	88, 74–92	MNZ S23654, Worthy, unpublished data
Dinornis novaezealandiae	3080	102, 91–110	MNZ S32678, Worthy, unpublished data
Dinornis struthoides	2090	61, 46–73	MNZ S28225, Worthy, unpublished data
Dinornis struthoides	2600	?	WO493, Worthy, unpublished data

Abbreviations: CM = Canterbury Museum; MNZ = Museum of New Zealand Te Papa Tongarewa; WO = Waitomo Museum.
[a] Body masses are estimates, or if given as *x*, *y–z*, *x* is body mass and *y–z* is the standard error margin of estimated mass, calculated from the femur diameter. In each case, the set was excavated under conditions that give confidence the set is more or less complete. Most were excavated by T.H.W., as indicated in the references.

smaller species. This relationship is not consistent across genera: *Euryapteryx* is considerably larger than *Anomalopteryx,* yet the latter has the larger gizzard. At the same time, *Anomalopteryx* gizzards regularly contain much larger stones than are ever found in a *Euryapteryx* or *Emeus* gizzard.

Analysis of these data suggests that species with weakly constructed mandibles—the blunt-billed *Euryapteryx* and *Emeus* with its finer, somewhat more pointed bill—had small gizzards with small stones. They did not need to deal with fibrous vegetable matter to the extent that *Dinornis* spp. and *Anomalopteryx* did.

Food Remains from Preserved Gizzards

Hamilton (1892a) was the first to observe what the contents of moa gizzards comprised. He observed clumps of gizzard stones eroding from the peat on the peat-covered surface of Swampy Hill, near Dunedin. No bones were present because all had been dissolved by the acidic peat. One gizzard was described thus: "The stones were found to be closely compacted together in a ball-shaped mass, (as large as a man's head) and were of all sizes, from those of less than a pea to those

TABLE 5.12.
A list of plant food species from gizzards of non-*Dinornis* species of moa in Pyramid Valley.

Moa specimen	Food plant species	Seeds or fruit (*n*)	Leaves and petioles
Emeus crassus XVIb	*Prumnopitys spicata*	58	
(Falla 1941)	*Myoporum laetum*	8	
	Nertera sp.	1	
	?*Carmichaelia* and *Gaimardia*		
Emeus crassus CM Av20119	*Prumnopitys spicata*	Y	Y
(Gregg 1972)	*Rubus* sp.	Y	
	Coprosma rotundifolia	Y	
	Coprosma sp.	Y	
	Pittosporum sp.	Y	
	Scirpus sp.	1	
	Indet sp.		Y
Euryapteryx geranoides	*Prumnopitys spicata*	Y	Y
CM Av20113 (Gregg 1972)	*Rubus* sp.	Y	
	Muehlenbeckia australis	Y	
	Cladium sp.	Y	

Abbreviation: Y = species present.

larger than a pigeon's egg." The washed and dried stones weighed 2 kg (4.5 pounds). A second example, whose stone complement weighed 2.8 kg (6 pounds), had the same comminuted vegetable matter, including large numbers of seeds of *Leucopogon* and *Coprosma,* as well as short twigs and branches.

More than 40 years elapsed before further observations were made. During the excavations since 1939 at the remarkable Pyramid Valley moa swamp, from which no fewer than 183 partial moa skeletons have been excavated, gizzards were often recovered with the skeletons. Studies on some of these gizzards were published by Falla (1941), Gregg (1972), and Burrows et al. (1981), but many were unfortunately discarded before analysis. A summary of the food contents is given in Tables 5.12 and 5.13.

The data in Tables 5.12 and 5.13 are heavily biased toward the diet of *Dinornis giganteus.* The contents of only three gizzards of other species have been described. The gizzard from *D. giganteus* 122B (see Table 5.13) was complete and analyzed in full. More than 50% of the plant matter consisted of short pieces of *Olearia* twigs (mainly 2–3 cm long, but with a range of 1–6 cm) up to 6 mm in diameter. *Dinornis giganteus* clearly browsed on the leaves and terminal branchlets of shrubs and trees, as well as their covering of vines. They also ate berries from these same shrubs. There were no ferns and very few herbs or grasses in the samples. If the Scaifes lagoon samples were all from *Dinornis struthoides,* which is probable, then this smaller species ate much the same range of taxa. There, at least two specimens contained

TABLE 5.13.
The food plant composition of *Dinornis* gizzards from Burrows (1980b) and Burrows et al. (1981).[a]

	Specimen of *giganteus* sp.											Specimen of *struthoides* sp.		
Food plant	122B CM Av28434	76D CM Av8468	76K CM Av8471	76M CM Av8473	108D CM Av15025	108E CM Av15024	AA ?	BB ?	89B CM Av8462	XA CM Av13899	121D CM Av20118	I	II	III
Trees														
Carpodetus serratus													1s	
Cordyline australis	1s									2s		5s	3s	5s
Elaeocarpus hookerianus										1s		1s		
Pennantia corymbosa	1s													
Plagianthus betulinus	1s, >10t	>30t	∞t	∞t			>3t		1t	1s, ∞t, >10b	>3t	1s		
Podocarpus spicatus	2s, 230p	1s, >35p	3s, >10p	7p	4p			3p	11s	2s, 4p	48s, >3000p	5s	15s	4s
P. dacrydioides													3t	
P. hallii												1p	7p	
Pseudopanex ferox	16s									4s				
Pseudopanex sp.	1p					1s						1s		
Shrubs														
Coprosma sp.	>700s	>65s	>100s	>400s	>400s	>30s		19s	6s, 1f	>150s, 8f	6s	64s, 2f	152s	140s, 3f
Coprosma rhamnoides	>30s		>1000s											
C. rotundifolia	32s		1s	10s	2s	5s				19s	4s	1s	10s	1s
C. sp. cf. *microcarpa*	>10s													
C. sp. cf. *robusta*	>10s													
Corokia cotoneaster			11s	2s	11s		10s			>1750s			>600s	
Leptospermum scoparium	1c						33p			2c, 1p				
Lophomyrtus obcordata	1s					1s								

	Specimen of *giganteus* sp.											Specimen of *struthoides* sp.		
Food plant	122B CM Av28434	76D CM Av8468	76K CM Av8471	76M CM Av8473	108D CM Av15025	108E CM Av15024	AA ?	BB ?	89B CM Av8462	XA CM Av13899	121D CM Av20118	I	II	III
Myrsine divaricata	43p	6s, >75p	1s, >20p	4s, >50p			1s, 4p	1s, >8p	∞p	35s, >110p	>10p			
Olearia virgata	∞t	>16t	5t	>8t	>5t	>2t	>3t	>8t	>1t	>30t		>7t	11t	>5t
Pimelea sp.													1s	
Teucridium parvifolium			12s							13s				
Phyllocladus alpinus											1p			
Vines														
Muehlenbeckia australis	4s	2s										1s	6s	1s
M. complexa	1s			1s				1s		>200s				
Rubus sp. cf. *squarrosus*	∞p					>20p		∞p	∞p					
Rubus sp. cf. *schmidelioides*		∞p												
Rubus sp.	14s, ∞t	3s, ∞t	3p, 1t	14s, >10p	>5t	7s, >20p	1t	∞t	1s, ∞t	1t	5s, ∞p, >5t	61s, ∞t	184s, ∞t	48s, ∞t
Tetrapathaea tetrandra	2s													
Clematis sp.		1s												
Herbs														
Carex secta	12s	4s	1s	2s	17s		1s	1s		12s		>50s	27s	9s
Chenopodium sp. cf. *allanii*	48s					1s						1s	5s	1s
Cyperaceae	6s				1s					1s		6s	60s	22s
Phormium tenax		1s	2s									>18s, ∞p fragment	>26s, ∞p fragment	>10s
Unidentified grass/sedge	3p	2p	1p							3p				

Abbreviations: c = capsule; f = fruit; p = leaf, petiole, or cladode; s = seed; t = twig or short shoot.

[a] Most samples are only partial gizzards, the exception being 122B; 76K had been previously reported on by Falla (1941), XA was mostly complete, and 121D was nearly complete and partly reported on by Gregg (1972). Weights of stones are only given if the gizzard was near complete. All *D. giganteus* were from Pyramid Valley. The *Dinornis struthoides* numbers I, II, and III were from Scaifes Lagoon near Wanaka in Otago. They were each associated with one of the following skeletons: CM Av20587 *D. giganteus*, and CM Av20588, CM Av20589, CM Av20590, and CM Av20591, four juvenile *D. struthoides*.

significant quantities of flax (*Phormium tenax*), which is a fibrous monocotyledon living in wetlands but which, according to analysis of pollen data, was absent from Pyramid Valley.

The preserved ends of some of the twigs show that they were sheared off, rather than torn, indicating that the bills must have had sharp edges. The contents of the three gizzards from *Emeus* and *Euryapteryx* did not include twigs, only fruit and leaves. If these are representative, then these emeids preferred a far less fibrous diet than dinornithids.

In summary, bill structure, gizzard stones, and gizzard contents all agree and show that *Dinornis* preferred a very fibrous diet dominated by twigs snipped from low trees and shrubs. *Emeus* and *Euryapteryx* spp. had a less fibrous diet probably dominated by leaves and fruit. Extrapolating from the relationship between bill structure and known diet in other genera, the diet of *Pachyornis* and *Anomalopteryx* probably also included much fiber. Their bill and gizzard structure would allow the cutting and processing of such material. Unpublished observations of *Pachyornis elephantopus* gizzard contents from a swamp in Canterbury (B. McCulloch, personal communication) indicate that this bird could snip pieces from the extremely fibrous leaves of New Zealand flax. These leaves are resistant to tearing and must be cut.

Did Moa Species and Plants Coevolve?

Botanists have long been aware that the New Zealand flora has an unusually high number of divaricating plants, those where branching of shoots occurs at a wide angle and the twigs tend to interlace as a result. Often this growth form is present in juveniles, but above a certain height a quite different, more open structure is assumed. Some plants are spiny, and others coppice vigorously from a browsed or cut-off stump. Yet it was tacitly assumed by most that this flora evolved in the absence of browsing pressure: there were no browsing mammals or marsupials.

Although some authorities had allowed that moa would have browsed plants, it was Greenwood and Atkinson (1977) who were the first to explicitly state that the growth forms of many New Zealand plants had evolved in response to moa browsing. They considered that the principal growth form coming out of this browsing pressure was the divaricating structure. It was a defense against the perceived feeding mechanism of clamping, pulling, and breaking (as opposed to the clipping of ungulates) and yet was abandoned when the plant exceeded about 3 m in height, which coincides with the reach of the largest moa. Small leaves were protected in the interior of the bush by matted, tough, springy branches. They also noted that some plants are spiny, particularly the many members of the genus *Aciphylla,* and that other, softer species mimic them (e.g., the daisy *Celmisia lyalli*). Other features of the flora that may relate to browsing were toxic compounds in leaves (e.g., in *Pseudowintera colorata*) and tough leaves such as in *Phormium* and *Cordyline.*

However, climate-based explanations have also been offered for these structures. Early botanists, for example Cockayne (1912), interpreted the divaricating structure as an adaptation to windy and xeric climates because most species that attained this plant form lived in windy and dry habitats. McGlone and Webb (1981: 21) challenged the moa-browsing hypothesis by resubmitting that climate was the best explanation of the divaricating growth form. They stated, "the majority of divaricating plants and heteroblastic juveniles in the New Zealand flora have evolved in response to lowland and montane environments, which are unsuitable for normal tree growth because of the high and unpredictable incidence of frost, wind and drought." In support of this hypothesis, they noted that although an individual of a species is often divaricating in open or forest margin habitats, the same species just inside a forest is considerably more loose in form and has larger leaves. Similarly, divaricating forms dominate shrublands of frost-prone river terraces, and there was no evidence of their replacement by more vigorous, less browsing-resistant species in the centuries in which moa had been extinct.

In support of their rebuttal, McGlone and Webb (1981) noted that Greenwood and Atkinson's (1977) hypothesis was challenged by the analysis of gizzard contents (Burrows 1980b), which revealed that the diet of *Dinornis giganteus* was mainly twigs, often of divaricating species, which had been clipped off, not broken. *Aciphylla* was often browsed by mammals, and given the cutting ability that moa had, it seemed unlikely that moa could not have eaten them. Further, they showed that the plants listed by Greenwood and Atkinson (1977) with tough leaves or toxins within them were representative of genera dispersed widely outside of New Zealand, so such traits were most likely inherited rather than local responses.

In succeeding papers (Atkinson and Greenwood 1980, 1989; Lowry 1980), the moa browsing hypothesis was further developed. In their 1989 article, Atkinson and Greenwood clearly showed that several moa species lived in sympatry facilitated by a range of sizes, using a number of feeding strategies revealed by different bill morphologies. The hypothesis changed from one where plants had defenses against browsing to one where such structures conferred a selective advantage on the plant because of reduced moa browsing. It focused on the forms, spiny tussocks, mimicry, reduced visual apparency, and divarication.

Spiny Forms

Sixteen of more than 30 species of *Aciphylla* are spiny, and the largest, most spiny of these live in lowlands, along watercourses or on terraces (Fig. 5.33). The small, nonspiny forms are at high altitudes, often on cliffs, and the Chatham and Australian forms of the genus are not spiny. This distribution correlates with a presumed once-greater browsing pressure at sites where the large spiny forms are found and not with the distribution of xeric habitats or wind exposure.

Mimicry

Atkinson and Greenwood (1989) postulated that moa had acute vision and a poor sense of smell; hence, color and form were significant

Figure 5.33. Aciphylla ferox, *one of the numerous spiny, rosette herbs found in open habitats in New Zealand. Photograph by T.H.W.*

in food choice. Some palatable plants look similar to toxic ones; for example, *Alseuosmia pusilla* is similar to the toxic *Pseudowintera colorata.* Others look like unpalatable dead twigs—for example, *Muehlenbeckia ephedroides,* which is a prostrate vine usually found on unconsolidated river gravels. The significance of this advantage is challenged by the observation that moa had relatively small eyes and, if their extraordinarily large olfactory chambers reflected their olfactory ability, a good sense of smell (Worthy 1989c).

Reduced Visual Apparency

Atkinson and Greenwood (1989) noted that in other ratites, colors at the indigo–violet end of the spectrum are not perceived well, so moa may have had similar limitations. Such coloration would make plants less obvious and so escape browsing. This might explain why purplish-black or dark bronze coloration was common in juveniles of many New Zealand woody plants within the forests, rather than the dark green that is more common elsewhere. Other plants look like insubstantial, apparently leafless twigs.

Divaricating Form

Atkinson and Greenwood (1989) listed 53 species characterized by interlaced branches with tough stems and reduced size and numbers of leaves on the outer branches. Twenty-two of the 53 were said to be

primarily forest trees and shrubs that were widespread and abundant, although many occurred on forest margins and in scrub as well (Figs. 5.34, 5.35). Most were strongly divaricating, even in shaded locations. But other divaricating species were found only in xeric conditions—some in wet, swampy conditions, others where frost is common, and yet others where it is rare. According to Atkinson and Greenwood (1989: 87), to account for the distribution of this form

*Figure 5.34. Example of a divaricating plant (*Myrsine divaricata*), a small shrub. Photograph by T.H.W.*

> requires a selective factor capable of influencing plants growing in this wide range of latitude, altitude, topographic and geological conditions. Moa meet this criterion of wide distribution and provide the most satisfactory selective factor that can explain the

Figure 5.35. Detail of the interlocking branching structure of the divaricating shrub Myrsine divaricata. *Photograph by T.H.W.*

characteristics of divaricating plants, namely, multiple growing points, interlaced branching, tough stems and the reduced size and number of leaves on the exterior of the plant.

If there was coevolution of plants and moa, some adaptations need to be demonstrated in moa. Atkinson and Greenwood (1989) suggested that some moa may have increased in height and others obtained larger gizzards. However, the fossil record is silent on this aspect. There is no evidence for the course of moa evolution: we have only the final chapter, as it were. The physical differences observed between the moa are easily explained as necessary for niche partitioning in the sympatric species.

In summary, moa browsing seems to provide a plausible hypothesis to explain the unusual frequency of particularly divaricating plants in New Zealand. But as in most biologic–environmental interactions, the relationship is probably not simple, and climatic factors are likely to have been significant in some situations.

Are the Browsing Effects of Moa Replaced by Introduced Ungulates?

As moa have only recently been removed from the environment, it has been argued by some that introduced ungulates such as deer and goats are just replacing them. It is true that all are browsers, but their effects are rather different. Deer and goats strip bark off trees, a deleterious effect that is unlikely to have occurred in moa browsing. Moreover, ungulates have a more sophisticated mechanism for obtaining browse: a prehensile tongue and teeth that they use to grasp and break down fibrous matter. In addition, the cellulose-digesting system is a feature of ruminants that enables better use of plant matter and doubtless contributes to the greater density of ungulates in forests. They travel in herds. In contrast, it is likely (by analogy with cassowaries) that moa moved about either singly or in small family groups, so the browsing impact is likely to have been much lower.

The observation that several species of tree—for example, *Pseudopanex* sp., *Griselinia littoralis, Brachyglottis rotundifolia, Aristotelia serrata,* and *Fuchsia excorticata*—are greatly depleted in areas suffering high ungulate browsing pressure indicates that such browsing is not the same as that formerly made by moa. In the subalpine herb fields, many species are severely depleted where browsing pressures are high, yet moa lived in these zones. But the question arises, what are they depleted with respect to? Have not the centuries since moa were made extinct been a long enough period for a rearrangement of the plant distributions? As Caughley (1989: 7) stated,

> First principles of range management inform us that, during the interregnum between the moas departing and the deer arriving, the "decreaser species" would have out-competed the "increaser species." They would have proliferated at the expense of the latter, and would have banished "invader species" from many plant communities. But we have scant information on which plant species fitted into each of those categories under the regime of defoliation enforced by the moas.

So the deer are having an effect, but it would not be the same as that made by moa: no two herbivores have an equivalent effect on vegetation (Caughley 1989). Moreover, this effect is not on the natural forest of New Zealand; it was natural when the browsing guild of moa were present. So removing deer and other ungulates does not return it to its "natural" state. But it is most likely that New Zealand forests never experienced the severe degradation that is seen when introduced ungulates are present in high densities.

The situation is summed up by Atkinson and Greenwood (1989: 92): "For the most part, introduced browsing mammals are acting as a new influence on vegetation and, in many places, are depleting its structure and composition compared with that of pre-Polynesian New Zealand."

Vocalization

Moa can be heard no more. There were almost certainly no more moa to make any sound by the time Captain James Cook landed in New Zealand in 1769. If any were to be heard at all, Cook was probably in the right place in Dusky Sound, Fiordland, as one of the remote, wet upland valleys of that region is where the last moa probably survived. No believable descriptions of moa have been handed down in Māori mythology. Certainly, none was recorded until after Owen described the first bones. Moa had disappeared from the menu between 400 and 300 years ago (Anderson 1989c), and so it seems that in the succeeding 10–15 generations, moa disappeared from the oral traditions of the surviving people as well. In short, we have no record of the sound of any moa.

However, all moa species, as in all birds, had a syrinx at the base of the trachea where the bronchi branch off, and the emeins (*Euryapteryx geranoides, Euryapteryx curtus,* and *Emeus crassus*) and possibly *Megalapteryx didinus* all had an elongated body loop in the trachea. The extent of this structure has only been documented once, when T.H.W. was fortunate enough to find and excavate an articulated skeleton of *Euryapteryx geranoides* in Honeycomb Hill Cave. The bird had died as though sitting on its nest and was on a flat silt surface, so that the bones and other structures were undisturbed. The lateral symmetry of the skeleton was preserved. This good fortune was unfortunately offset by extreme bad luck in that all the photographs were lost in transit to or from the processing laboratory. However, a drawing was made at the time of the structure as it was in the cave, and this is reproduced in Figure 4.22.

The tracheal rings of the windpipe, or upper trachea in the neck, are typical in that they were short, oval, bony structures, similar to those in other moa and merely huge versions of those in other birds. However, just within the body cavity itself, where the tracheal rings normally might be expected to disappear as the bronchi enter the lungs, a remarkable intrathoracic loop was present. The loop was 1.2 m long, consisted of over 125 rings, and extended down the left side of the body cavity and around to the posterior right side of the sternum before looping back on itself to the anterior of the sternum and then, presumably, the lungs. In the region where the loop bent over itself, the rings were fused. This section had been found before (Archey 1941: fig. 13). The rings of the entire body loop were 1–2 cm long and comparatively thick (1–2 mm), as against 1–2 mm long and less than 1 mm thick at the cranial end.

The structure was far more robust than required simply for breathing, and as in all other species with elongate trachea (see above), it probably was used in some way during vocalization. Some birds, such as the trumpeter swan *Cygnus buccinator* and whooping crane *Grus americana,* which are capable of loud, far-carrying calls, have extended tracheal loops in the body cavity. So these moa too could likely generate calls with a low pitch that would ensure they could be heard at a great distance. The specialized structures in the shrubland taxa of moa (*Euryapteryx, Emeus,* and *Megalapteryx*) suggest that they had the greatest vocal abilities. Because these moa also had the least olfactory ability (as determined by relative size of their olfactory chambers), it is possible that sound rather than olfaction was somehow important in social interaction for them.

Within their range, these species were also seemingly quite abundant—a conclusion based not on natural fossil deposits, which offer little information on abundance, but from archaeologic sites, in which *Emeus* and *Euryapteryx* in particular were present in huge numbers. Perhaps the emein moa needed loud calls in their dense grassland–shrubland–forest mosaic habitats. Alternatively, the males may have had lek behavior. After all, New Zealand has singing bats and booming kakapo parrots that lek nocturnally. In the presence of the giant diurnal avian predators, nocturnal activity patterns may have been common or even the rule. Imagine moa spaced out over the downlands, vigorously defending some desirable waterhole or prime ridge top, and drumming up a harem.

STABLE ISOTOPES AND NEW ZEALAND PALEOBIOLOGY

Studies using stable isotopes are comparatively new to the paleobiology of New Zealand vertebrates and have been largely by-products of the radiocarbon dating program. The amount of the less common stable carbon isotope (^{13}C) in a sample is measured as part of the dating process. For paleodietary studies, the stable nitrogen (^{15}N) composition is important, but there are still very few data on stable nitrogen ratios from fossil bones in New Zealand. Results so far show that stable isotopes have great promise to provide important information on, for example, the paleoecology of the extinct fauna, the effects of introduced predators, and changes in vegetation structure.

Here we provide only a brief summary of some results that are based on data from the present radiocarbon database and discuss the potential for future studies. Almost all the research has concentrated on the stable isotopes in purified gelatin derived from the collagen fraction of the bone proteins. Rates of carbon turnover are very slow in collagen, and the isotopic composition reflects the average diet integrated over the lifetime of the individual (Tieszen et al. 1983). Bone proteins have a much longer half-life than those in muscle or other tissues.

Paleodiet

One long-standing problem in New Zealand Quaternary paleobiology has been the role played by the two species of unusual gruiforms, the North and South Island adzebills (*Aptornis*). The diet of these birds had been debated since Richard Owen described them in the mid-19th century (Owen 1844a, 1848a). Opinions have been divided as to whether they were herbivores or carnivores (Holdaway 1989). In the absence of gizzard contents, and given the ambiguity of information inferred from the structure of the beak and jaw, stable isotopic analysis seemed to provide the best method of determining the trophic level of both species.

As it progresses up the food chain, the nitrogen content of protein in the tissues becomes more enriched in the heavier stable isotope, ^{15}N, because the heavier isotope is favored in transmembrane reactions. It follows that the degree of enrichment of ^{15}N in bone collagen should indicate the trophic level of the animal (Minagawa and Wada 1984; Schoeninger and De Niro 1984). Similarly, the proportion of ^{13}C, the heavier of the two stable carbon isotopes in tissue, increases with each trophic level, from plants to herbivores and then to carnivores (De Niro and Epstein 1978b). Actually, it is the degree of *depletion* of the heavier isotopes that declines, plants typically being from 24 to 30 parts per thousand (‰) depleted in ^{13}C relative to the amount in atmospheric CO_2. If the levels of enrichment or depletion of ^{15}N and ^{13}C are plotted against each other for different taxa, the trophic level of each taxon in the assemblage can be determined (Bocherens et al. 1997; Schoeninger and De Niro 1984). Although there are difficulties in the interpretation of isotopic data as applied to paleodiets, especially in marine or boreal ecosystems, particularly with humans (Hancock et al. 1989), the technique has been used with some success to infer paleodiet (e.g., Bocherens et al. 1997; Matheus 1995) and to explore aspects of the diet of living species that are difficult to sample otherwise (e.g., Ames et al. 1996; De Niro and Epstein 1978a; Fleming 1995; Hobson and Sealy 1991; Hobson et al. 1997; Mizutani et al. 1990; Ramsay and Hobson 1991; Rau et al. 1983; Tieszen et al. 1979).

The ^{13}C and ^{15}N compositions of bone gelatin samples from individuals of *A. defossor* and *A. otidiformis* showed that both species were predators. Levels of enrichment in ^{13}C and ^{15}N for two specimens of *Aptornis otidiformis* are shown in Figure 1, in comparison with values for the owlet-nightjar (*Aegotheles novaezealandiae*), a moa (*Euryapteryx curtus*), and Finsch's duck (*Euryanas finschi*). The adzebill samples were more enriched in ^{15}N than are those of the insectivorous owlet-nightjar, which suggests that the adzebill ate species higher in the food chain than insectivores—and hence were predators. Substantial differences in the level of enrichment in ^{15}N between the North and South Island species appear to result from environmental fac-

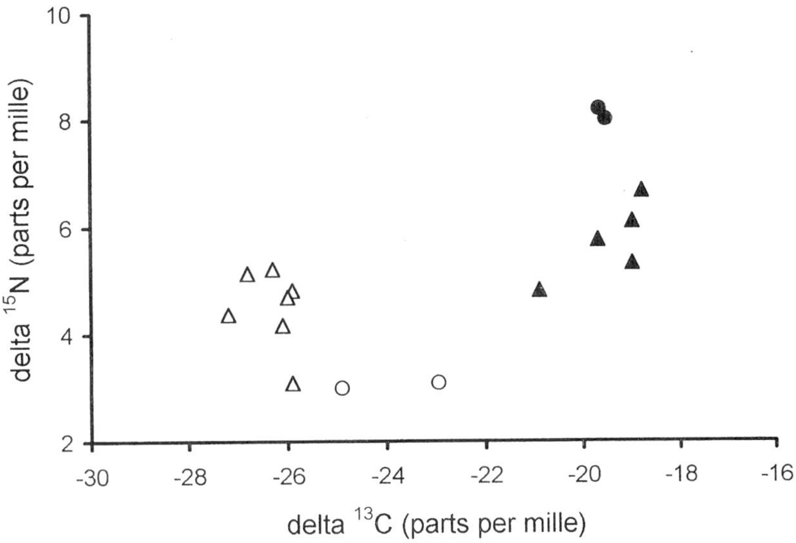

*Figure 1. Stable isotope enrichment levels (δ^{15}N ‰, δ^{13}C ‰) of bone gelatin from the North Island adzebill (*Aptornis otidiformis*; filled circles) in comparison with values for an insectivore (*Aegotheles novaezealandiae*; filled triangles) and two herbivores, a moa (*Euryapteryx curtus*; open circles) and Finsch's duck (*Euryanas finschi*; open triangles) from comparable environments. The adzebills (from sites 255 km apart) had almost the same isotopic composition, which was enriched in ^{15}N in comparison to that of the series of owlet-nightjars, showing its higher position in the food chain. The greater depletion in ^{13}C exhibited by the duck than the moa suggests that it ate vegetation that grew beneath a dense canopy.*

tors that are still being investigated (Holdaway et al., in preparation).

Herbivores: Useful Anomalies?

The isotopic composition of carbon in bone proteins depends on the composition of the diet (different plants fractionate the carbon isotopes to different degrees). The isotopic signal in the diet of herbivores is derived from the protein in the plants they eat. In turn, the isotopic composition of the plant is affected by various factors in the environment, including the amount of moisture in the environment, the altitude, and whether the plant was growing beneath a closed canopy forest or in open vegetation (Lajtha and Marshall 1994). Because the isotopic composition of the collagen in an animal is similar to that of the plants in its environment (Koch et al. 1994), measuring the proportion of ^{13}C in an animal's bone protein can potentially provide information on its microhabitat and, under certain circumstances, its feeding ecology (Koch et al. 1994). The information can also be used to infer climatic change over longer periods of several thousand years by inferring changes in the structure of the vegetation at individual sites.

The situation is complicated by the presence of different habitats within the "collection" range of a fossil site. For example, if two habitats meet near a site, individuals of species that prefer either one of the two may be preserved there. And species that prefer the vegetation at the ecotone between the two larger habitats will also be present. Where the habitats are known from other evidence, it may be possible to explain the differences. If no additional information is available, the isotopic data may, with care, be used to reconstruct the vegetation at one time or through time, and the hypotheses may be tested later when more data are available.

Data for a single species at one site may reflect changes in the habitat or diet of that species through time. One species of moa does seem to be providing such paleoclimatic information. The δ^{13}C values for specimens of the upland moa *Megalapteryx didinus* from Te Ana Titi Cave on the West Coast of the South Island varied between –20.5‰ and –30.1‰ for bones whose ages ranged from 19,000 years B.P. to just under 4000 years B.P. During the Holocene, the area was inhabited by a very small morph of the species. Interpretation of the δ^{13}C values ranged from random variation to possible systematic differences between the temporal populations. However, when plotted against the conventional radiocarbon age, the δ^{13}C values show a definite pattern, with a significant depletion of ^{13}C starting at about 12,000 years B.P.; the same pattern is seen in dates on the same species from Honeycomb Hill Cave, to the north of Te Ana Titi (Fig. 2). The timing matches that of the change from glacial to interglacial climatic conditions.

When data from a range of moa species and sites within the South Island are plotted, a similar pattern emerges (Fig. 3). Here, all the species sampled from more than 15,000 radiocarbon years B.P. had δ^{13}C values indicating open vegetation structure—none was less than –24‰ regardless of species. Even a specimen of *Dinornis novaezealandiae*, which in the Holocene was invariably associated with closed canopy wet forests and whose δ^{13}C value in deposits of Holocene age was –28‰ to –30‰, had a δ^{13}C value about as enriched as any of the open vegetation moa. It was living in open-structured vegetation, probably (on other evidence) a shrubland–grassland mosaic with patches of forest (Worthy and Mildenhall 1989).

Within the past 10,000 radiocarbon years B.P., the band widens dramatically. The increasing depletion (increasingly negative values of δ^{13}C (Fig. 3) shows that these birds were now living in closed-canopy forests that developed with the ameliorating climate of Late Glacial and Holocene time. Insufficient data are available yet for North Island sites. The stable isotope composition of the moa bone gelatin in this instance promises to provide a proxy indicator for the timing of the spread of forest and hence of climatic change.

The vegetation changes mirrored by the

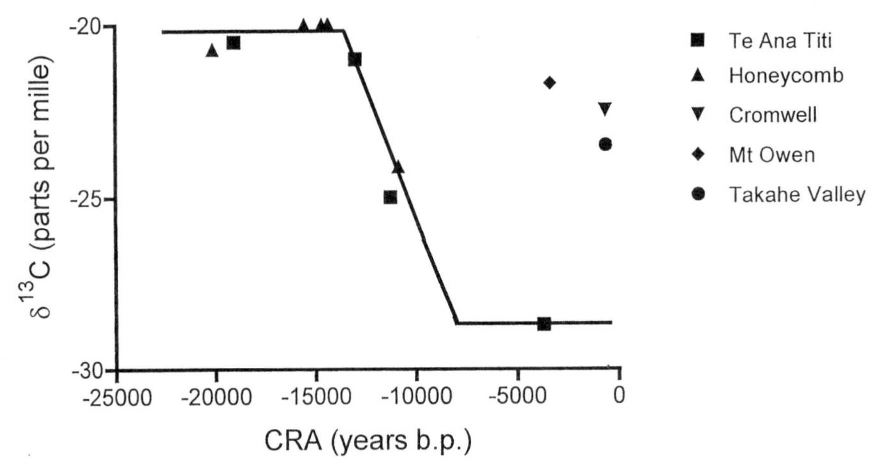

*Figure 2. Return of tall forest on the West Coast of the South Island of New Zealand at the end of the Pleistocene, as indicated by depletion of [13]C (δ^{13}C ‰) in bone gelatin of the upland moa (*Megalapteryx didinus*) from Te Ana Titi [cave] (filled square) and Honeycomb Hill caves (filled triangle) against conventional radiocarbon age (radiocarbon years) for each specimen. Specimens from sites with open vegetation (diamond, Mount Owen, northwest Nelson; inverted triangle, Cromwell, central Otago; filled circle, Takahe Valley) shown for comparison. The sudden increase in the depletion of ^{13}C in the gelatin evident in specimens from Te Ana Titi and Honeycomb Hill Cave after 14,000 years before the present probably results from consumption of ^{13}C-depleted vegetation beneath a closed canopy.*

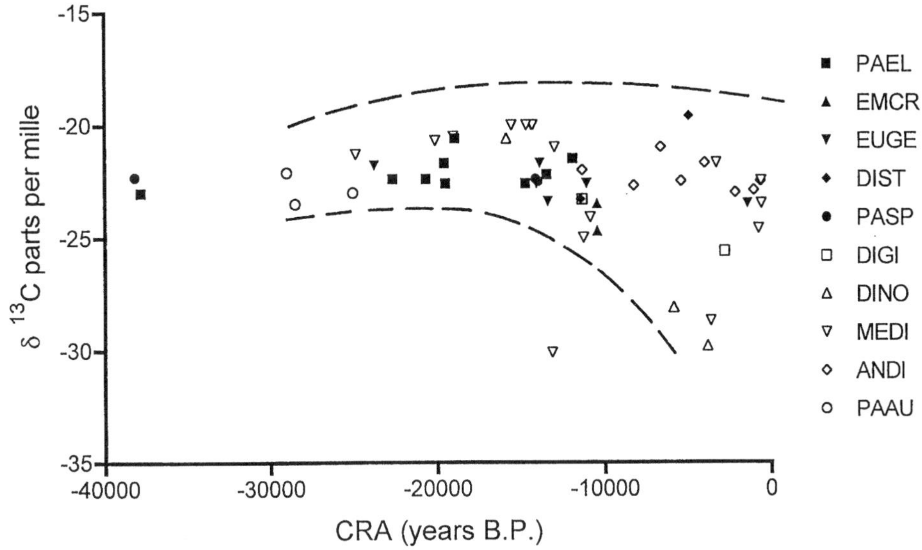

Figure 3. Return of tall forest in the South Island, New Zealand, as indicated by depletion of ^{13}C *(*$\delta^{13}C$ *‰) in bone gelatin of nine species of moa (plus values for indeterminate species of* Pachyornis*) against conventional radiocarbon age for the samples. Most taxa fed on vegetation in open situations, but two (*Megalapteryx didinus *on the West Coast and* Dinornis novaezealandiae*) appear to have consumed vegetation beneath a dense canopy when that was available.* Anomalopteryx didiformis *(open diamond) is characteristic of deposits from areas clothed in dense forest, but the moderate levels of depletion of* ^{13}C *in its bone gelatin suggest that it ate vegetation at forest margins.*

Moa taxa in the key are acronyms with the first two letters of the genus and the first two letters of the species: Pachyornis elephantopus, Emeus crassus, Euryapteryx geranoides, Dinornis struthoides, Pachyornis *species,* Dinornis giganteus, Dinornis novaezealandiae, Megalapteryx didinus, Anomalopteryx didiformis, Pachyornis australis.

changing levels of $\delta^{13}C$ in moa bone gelatin were natural and were forced by the changing climate at the end of the Pleistocene. More recently, the vegetation pattern of New Zealand was drastically altered by the effects of settlement by first Polynesians and then Europeans. Unfortunately, moa and the other extinct birds cannot provide information on the Polynesian deforestation of the eastern South Island because the mosaic vegetation that was replaced was not a full multitier forest, and so the signal that appears to be present on the West Coast should not be apparent in the east. Notwithstanding limitations such as this, studies of stable isotopes of bone gelatin from vertebrate fossils are beginning to provide useful new information on the paleoecology of New Zealand birds. As with the application of any new technique to a different situation, thus far, more questions have been raised than answered. That a limited application of stable isotope research has resulted in a solution to the puzzle of the adzebills' diet and appears to provide information on former vegetation and climate is an impressive demonstration of the potential of stable isotope studies to New Zealand paleobiology.

6 • Kiwi—Ratites of the Night

Family Apterygidae

This family of ratites is confined to New Zealand. An interesting possible record of this family comes from Australia. De Vis named a tarsometatarsus from the Pliocene as *Metapteryx bifrons*; however, this is almost certainly from an immature emu (Brodkorb 1978: 215).

Genus *Apteryx* Shaw and Nodder

The various forms of kiwi are readily divisible into two distinctive groups: the brown kiwi and the spotted kiwi. Brown kiwi are presently widespread in the North Island and have a disjunctive distribution in the South Island. All populations are contracting. Formerly, populations had densities of 40–100 birds per square kilometer and still do on predator-free islands, but in mainland situations, density seldom exceeds four birds per square kilometer (McLennan et al. 1996). There is a small population at sea level at Okarito, one 150 km to the south between the Haast and Cascade Rivers, and a larger one in Fiordland. Brown kiwi are also present on Stewart Island. Conventionally, the taxonomy of brown kiwi is geographically based, with North Island, South Island, and Stewart Island birds each given subspecific status (Marchant and Higgins 1990; Turbott 1990a).

The little spotted kiwi *A. owenii* was once widespread over North and South Islands but is now confined to managed predator-free near-shore islands. The great spotted kiwi *Apteryx haastii* has only ever been recorded from the South Island, in areas mainly west of the Alps. Densities were highest in subalpine zones in northwest Nelson (six pairs per square kilometer) in the 1980s (Marchant and Higgins 1990: Fig. 6.1). By the 1980s, *A. haastii* was reduced to three discrete populations, one in northwest South Island, one centered on Arthur's Pass, and one in Westland. Its range does not presently overlap that of brown kiwi.

Brown kiwi live to be 25–30 years old in captivity. They are unique

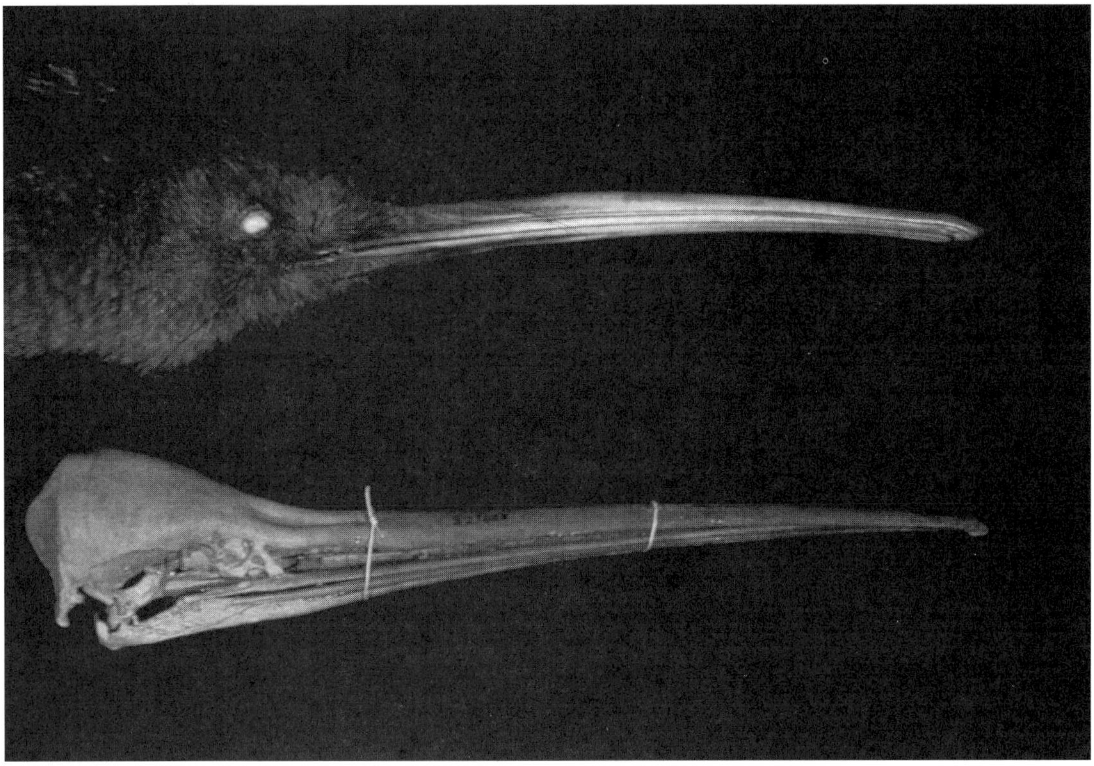

Figure 6.1. Kiwi skull (MNZ S21400) (bottom) compared to the head in a study skin of a female Apteryx haastii. *Photograph by R. Morris.*

birds for their many mammalian convergences. Their sense of smell is highly developed and is used to locate food, which ranges from worms to small arthropods to freshwater crayfish, but does include some plant material such as fruit and leaves. Their nostrils are uniquely situated in the tip of their long bill, which can be deeply probed into the ground in search of prey. Females are larger than males and have relatively longer bills than males. They are nocturnal over most of their range, but on Stewart Island, they often forage through the daylight hours as well. They live in a variety of habitats, from the seashore, where they probe in decaying kelp for sand hoppers, to the subalpine zone in mountains, where snow frequently lies on the ground. They can be found in open marram grass dunes, scrub habitats, tall forests, and even occupy (at least temporarily) pine forests. They are extreme K-selected species, laying an average of two eggs in a burrow where they are incubated for up to 90 days before hatching, usually by the males.

In recent years, research has shown conclusively that the single most significant threat to the survival of kiwi on the mainland are stoats, but dogs and accidental deaths as a by-product of possum trapping are also significant. Predation, mainly by stoats, results in over 94% of young being killed by the time they are 100 days old, and they do not reach adult size for a further 300 days, whereupon they are of sufficient size to be rarely killed (McLennan et al. 1996).

Recently, there has been much research into morphologic and genetic variation in kiwi. This has resulted in the discovery that the morphologic conservatism of kiwi is possibly masking several cryptic species (Baker et al. 1995; Szabo 1993), and now management aimed at conserving genetic diversity is treating brown kiwi as three distinct species. They differ in allozymes and in mitochondrial DNA more than their morphology would indicate. In April 1993, the New Zealand Department of Conservation announced that a new kiwi had been discovered. The tokoeka was reportedly the first new bird species to be discovered since takahe in 1948. Unfortunately, someone forgot to tell them that the first kiwi named from New Zealand was a tokoeka (i.e., *Apteryx australis* Shaw and Nodder 1813) from Dusky Sound in Fiordland (Buller 1869: 226).

Nevertheless, the research based on DNA analyses indicated that brown kiwi from Haast, Fiordland, and Stewart Island were most similar to each other and probably one species. Some long-term Stewart Island residents believe the kiwi was introduced historically at Port Pegasus on the island and have documented a northward spread over the 20th century, which would fit with the form being indistinguishable from Fiordland birds. The Okarito population about halfway up the West Coast was, surprisingly, found to be most similar to North Island brown kiwi. Research by Charlie Daugherty (Victoria University, Wellington) and Alan Baker (Baker et al. 1995) found that allozyme variations were matched by marked differences in the mitochondrial DNA. However, each population of brown kiwi could be distinguished genetically. These results can be interpreted in two ways: either the various kiwi populations have gone their separate ways for hundreds of thousands of years, or they have much greater genetic diversity than other bird species.

The skeletal morphology of kiwi is as conservative as their plumage. *Apteryx owenii* can be distinguished by their smaller size, but all forms of brown kiwi and *A. haastii* have considerable overlap in the size of their bones. *Apteryx haastii* have the biggest bones, but at least half of the size range is shared with *A. australis*. Although it is often said that North Island kiwi bones are smaller than those of South Island brown kiwi, there is considerable overlap, and the biggest North Island bones are only slightly smaller than the biggest South Island bird (Worthy 1997b). No qualitative differences in form are apparent between any of the larger kiwi. Leg bones of *Apteryx haastii* tend to have narrower shafts, but again, there is considerable overlap in measurements with brown kiwi. For these reasons, kiwi bones are usually only identified to one of two taxa: *A. owenii* and *A. australis/haastii*. Recently, a third taxon, the Eastern kiwi, has been identified from fossil bones only.

Apteryx mantelli Bartlett, 1852 (North Island Brown Kiwi)

Bartlett (1852)[1] distinguished this species of kiwi from the first-named *A. australis* on account of several features: it has a scutulated

1. Bartlett published in the *Proceedings of the Zoological Society* for 1850, but the section was not published until 1852 (Duncan 1937).

rather than a reticulated tarsus; the claw of the middle toe was longest, rather than all claws being of equal length; the claws were a dark horn color, rather than nearly white; the wings had strong, thick quills, rather than soft, slender quills; the face had long, straggling hairs about the base of the bill, rather than short hairs; and the plumage was rufous brown rather than pale grayish brown. The plumage of this species varies greatly between individuals but is always variously darkly streaked lengthwise with rufous brown and blackish brown. The plumage has a shaggy texture because of widely spaced barbs and few or no barbules. Rachides protrude beyond all rami, giving the feathers a harsh and prickly feeling if stroked backward (Marchant and Higgins 1990). Males range in weight 1.8–2.5 kg and females 2.0–3.2 kg. These animals inhabit North Island only.

Fossil bones of big kiwi in North Island sites have been attributed to *A. mantelli* because *A. haastii* was not recorded historically from the North Island. Their bones are present in deposits from North Cape to Wellington but are most common in cave deposits in wetter western regions. Bones of *A. owenii* have a similar distribution in North Island.

Apteryx sp. Okarito Form

The brown kiwi at Okarito on the South Island West Coast are genetically closer to *A. mantelli* from the North Island than to *A. australis* immediately to their south. As yet, no specimen of this form has been collected or has been identified in museum collections, and nothing is known about its skeleton. It is said to be smaller, and its plumage is undescribed.

Apteryx australis Shaw and Nodder, 1813 (South Island Brown Kiwi, Tokoeka)

Synonymous with *Apteryx australis* is *Apteryx lawyri* Rothschild, 1893 (Stewart Island brown kiwi).

Compared with *A. mantelli,* the plumage of this species is noticeably lighter, lacking the rufous coloring to the back, and is softer, lacking the harsh, prickly feeling (Marchant and Higgins 1990). However, Oliver (1955) says that *A. australis* is slightly rufous toward the feather tips and that *A. lawryi* is heavily striped with rufous brown to black, although it is most similar to *A. australis.* Yet he also says *A. mantelli* is darker and more rufous than either southern form. Obviously, there is some subjectivity here, but North Island birds are generally considered to be darkest. *A. australis* females weigh 2.0–3.4 kg; *A. lawryi* males weigh 2.3–3.0 kg and females 2.7–3.5 kg. These animals inhabit South Island from Haast to Fiordland, and Stewart Island (Fig. 6.2).

It was obviously once more widespread, as Oliver (1955) records it from the forests surrounding Otago Harbour in 1873, and it was seen in Marlborough Sounds in 1931. However, the identity of the kiwi in these latter records may have been assumed, as specimens documenting them are unknown.

In the caves of western and northern regions of the South Island,

*Figure 6.2. Male living brown kiwi (*Apteryx australis*) from Stewart Island in typical foraging pose. Photograph by R. Morris.*

large kiwi bones are common, especially in pitfall traps (Worthy 1993b; Worthy and Holdaway 1993, 1994b). Because of the aforementioned problems in identifying kiwi bones, these could be either *A. haastii* or *A. australis*. Some of the bones from Honeycomb Hill and all those from the subalpine sites on Mount Arthur and Mount Owen in northwest Nelson are of very large kiwi, outside the known size range for brown kiwi, and thus are probably *A. haastii*. These sites are from within or on the fringes of the range of *A. haastii* as now known.

Apteryx sp., Eastern Kiwi

On the eastern side of South Island, "large" kiwi bones are locally common. In 1997, Worthy used the large sample of kiwi bones from Kings Cave in South Canterbury to demonstrate that two species with nonoverlapping size ranges were present. One was indistinguishable from *A. owenii*, but the larger one had leg bones that were relatively stouter than either *A. australis* or *A. mantelli*, yet were shorter than either of these species. The form has since been recognized as the large kiwi in Pyramid Valley (Holdaway and Worthy 1997), Otago (Worthy 1998b), and Castle Rocks, Southland (Worthy 1998d). The leg bones in the large sample of fossils from Castle Rocks were found to be significantly shorter than *A. australis* but of equivalent stoutness. They did not differ from South Canterbury bones (Worthy 1998d). Thus, throughout the eastern region of the South Island, there was a large kiwi present in the Holocene that was significantly smaller than both *A. australis* and *A. mantelli,* yet tended to be stouter than these species. The differences suggest that this form is a distinct species, but as yet, no single skeleton is available.

Apteryx owenii Gould, 1847 (Little Spotted Kiwi)

The soft plumage of this small kiwi is brownish gray and mottled or banded horizontally, rather than streaked lengthways, as in brown kiwi. It is small, with males weighing about 1.1 kg and females 1.3 kg

The historic distribution of this kiwi was Marlborough (including D'Urville Island), Nelson, and the West Coast in the South Island. However, Morgan Carkeek, the same surveyor who caught a probable takahe in the North Island, is recorded as having caught a little spotted kiwi in the Tararua Ranges in 1875. The last little spotted kiwi were removed from D'Urville in the 1980s, and the last mainland specimen was an adult female from Westhaven Inlet in northwest Nelson on July 25, 1978 (MNZ 23038; Worthy and Holdaway 1994b: 307). The species survives now in managed populations on Kapiti Island and a few other, smaller populations.

Fossil bones of this species have been found in widespread localities throughout North and South Islands. Their bones have not been recorded from the subalpine zone, where bones of apparent *A. haastii* are common.

Only one fossil kiwi has been named, and it is now in the synonymy of *Apteryx owenii.* Lydekker (1891) founded the genus and species *Pseudapteryx gracilis* on a solitary left tarsometatarsus (BMNH 32237a) from Mantell's collection. It is primarily distinct from other kiwi in that the outer distal nutrient foramen is not enclosed and that the proximal foramina were offset from one another. Hamilton (1893: 100–101) used the series of *Apteryx* bones from Castle Rocks to show that these principal distinguishing characters were attributable to individual variation. Millener (1987), apparently unaware of this work, came to the same conclusion.

Apteryx haastii Potts, 1872 (Great Spotted Kiwi, Roa)

Like its smaller namesake, the great spotted kiwi's plumage is also mottled or banded horizontally. Adult females of this species attain the largest size of any kiwi—some were 4.5 kg (McLennan and McCann 1991). Upland populations in the subalpine zone comprise significantly larger individuals than nearby lowland populations, conforming with Bergmann's Rule. In the upland population studied in Gouland Downs by McClennan and McCann (1991), males range from 2.1–3.0 kg (mean, 2.6 kg) and females 3.0–4.3 kg (mean, 3.4 kg). In the near–sea level site at Kahurangi Point, males range 1.8–2.3 kg (mean, 2.1 kg) and females 2.9–3.2 kg (mean, 3.1 kg). Only a single egg is laid, not two or more as in brown kiwi, and this egg (117–132 mm long, 71–84 mm wide, $n = 7$) is, relative to female size, 10%–20% smaller than those of brown kiwi.

The habitats occupied vary from tussock shrub mosaics of the subalpine zone, where winter snow and frosts are normal, to dense, subtropical rainforest at coastal locations, which never experience frost and snow. Their diet is dominated by worms and beetles but includes centipedes and spiders, as well as rare other invertebrates such as large snails and freshwater crayfish. They often eat seeds and leaves (McLennan and McCann 1991).

The fossil record of *A. haastii* has not been reviewed. However, the difficulties of identifying fossil kiwi bones mean that only the specimens outside of the size range of brown kiwi might be correctly identified. No such large specimens have been found on the eastern side of the Alps. The numerous kiwi bones on Takaka Hill in the north of the South Island include no large specimens, whereas those from Mount Arthur and Mount Owen, 30 and 80 km, respectively, to the south, do. It seems possible that Takaka specimens could be brown kiwi, but because lowland populations of *A. haastii* are significantly smaller than their upland counterparts, the possibility that the Takaka bones are just from smaller birds than the subalpine northwest Nelson birds cannot be discounted.

7 • Waterbirds of Many Kinds

Anatid Fauna

The duck fauna of New Zealand is relatively diverse. Of the 140 breeding species of bird in the North Island, South Island, and their nearshore islands, 93 were land species, freshwater species, or both, and 13 of these were anatids. Diversity was high, with 10 genera, but as with other groups, there are or were few species within genera. There was one more endemic species on the Chatham Islands.

In faunas of the outer islands, there are equivalents of mainland species, but no new groups. For example, Millener (1996) suggested that on the Chatham group, *Cygnus, Tadorna, Mergus,* and teal were distinct species from those on the mainland, but this remains to be investigated. All are now extinct on the Chatham Islands, and two, *Tadorna* and *Cygnus,* have been replaced by new populations self-colonizing from the mainland. Only one distinct anatid has been described from the Chatham Islands, the Chatham Island duck *Pachyanas chathamica.*

Elsewhere in the New Zealand archipelago, ducks are or were present on the subantarctic Auckland, Campbell, and Macquarie Islands. They included small, flightless teal and the merganser *Mergus australis* (Figs. 7.1, 7.2). The Macquarie Island *Anas* species and *M. australis* became extinct in the early 20th century. The teal have often been considered as only subspecifically distinct from the mainland brown teal (e.g., Turbott 1990a), but this ignores obvious morphologic traits that have been corroborated recently by DNA analyses. Each is now recognized to be a distinct species (Livezey 1990; Marchant and Higgins 1990; Oliver 1955). Johnson and Sorenson (1999) and Kennedy and Spencer (2000) showed that the New Zealand radiation of teal is separate from and possibly earlier than the separation of the chestnut teal (*A. castanea*) and grey teal (*A. gracilis*) in Australia. *Anas nesiotis* of Campbell Island and *A. aucklandica* of the Auckland Islands inhabit

Figure 7.1. Head of Mergus australis *(OM Av1110), one of the few specimens collected from the Auckland Islands in the 19th century. Photograph by R. Morris.*

tussock grass areas and the kelp beds of the islands' littoral zones. The New Zealand mainland form *A. chlorotis,* now critically endangered, was formerly widespread in habitats as diverse as coastal lagoons, lakes, and tall rainforest.

Of the 13 resident species in the North Island and South Island, 7 have become extinct, though 1 (the black swan) has since recolonized. Most of New Zealand's waterfowl—the main exception being the enigmatic blue duck *Hymenolaimus malacorhynchos*—have affinities with Australian taxa and can be assumed to have been derived from ancestral species there. An interesting feature of the New Zealand anatid fauna is that the most derived species have become terrestrial or spend a lot of time away from water. The two extinct geese *Cnemiornis* spp. and Finsch's duck *Euryanas finschi* were entirely terrestrial. Both the blue duck *H. malacorhynchos* and brown teal *A. chlorotis* forage at considerable distances from water at night.

Extinct Taxa

Cygnus sumnerensis/atratus

Genus Cygnus *Bechstein*
Cygnus atratus *(Latham, 1790) (Australasian Black Swan)*
Cygnus sumnerensis *(Forbes, 1890) (New Zealand Black Swan)*
Cygnus chathamensis *Oliver, 1955 (Chatham Island Black Swan)*

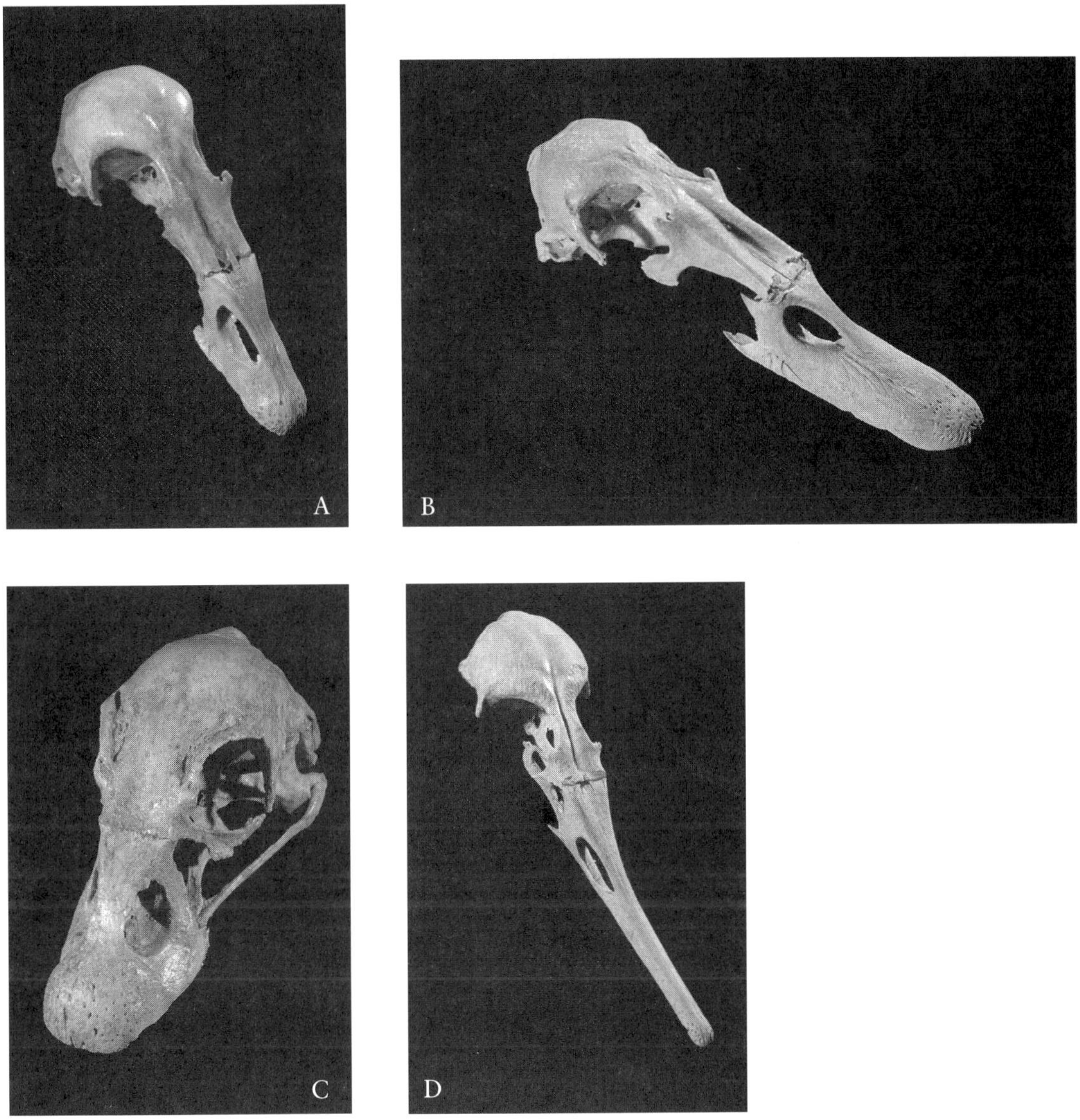

Figure 7.2. Photographs of skulls of selected anatids. (A) Euryanas finschi *(T.H.W., private collection). (B)* Pachyanas chathamica *(MNZ S29475). (C)* Cnemiornis calcitrans *(MNZ S35266. (D)* Mergus australis *Te Ana-a-moe [cave], Chatham Island (MNZ, uncataloged). Photography by R. Morris.*

Swan bones were first described from New Zealand by Forbes (1890a), when he named material (three coracoids, one distal, and one proximal humeri) found in Monck's Cave, Sumner, Christchurch, as *Cygnus sumnerensis*. In a typically brief Forbesian diagnosis, the only character given was that the bones were bigger than those of the black swan (*Cygnus atratus*) of Australia. Forbes was influenced by the observation that swans were not recorded historically in New Zealand and that the fossils antedated the introduction of *Cygnus atratus* to New Zealand. About 100 *Cygnus atratus* were introduced to New Zealand between 1864 and 1868 from Australia, but the population expanded so rapidly within this time that self-introduction from Australia seems highly probable (Heather and Robertson 1996; Turbott 1990a).

Forbes did not consider the possibility that *C. atratus* may have lived in prehistoric times in New Zealand and then been exterminated by Māori hunting. Oliver (1955: 603), under the impression that Forbes's type material of *C. sumnerensis* was lost and the name a *nomen nudum*, and that New Zealand and the Chatham Islands shared a single species of swan, erected *Cygnus chathamensis* on the basis of swan bones from the Chatham Islands. Since the rediscovery of Forbes's type material (Dawson 1958b), Oliver's name has been synonymized with *C. sumnerensis*. All fossil and archaeologic bones from New Zealand have been referred to *C. sumnerensis* because *C. atratus* was introduced in the 1860s, and therefore it was thought that fossil material could not belong to it.

The status of *Cygnus sumnerensis* seems to never have been seriously questioned. Scarlett (1972b: 10) noted that "apart from the greater average size, and slight differences in the pelvis and sternum, the bones are indistinguishable from those of a Black Swan, of which it was probably only a sub-species." The size difference mentioned by Forbes, Oliver, and Scarlett has never been quantified, and the "slight differences" in the pelvis and sternum remain undescribed. Worthy (1998a) tested the assumption that the fossil material was larger than recent bones by use of the large sample of fossil bones from Marfells Beach in the South Island. The sample had about the same size range as modern specimens, and no significant differences in form were apparent. Mean lengths of fossil leg bones tended to be longer than in recent birds, but the differences were small (femora 91 versus 87 mm; tibiotarsi 187 versus 175 mm; tarsometatarsi 102 versus 95 mm). These apparent differences were possibly influenced by sample size, and the dune ta-

Figure 7.3. Plots of selected skeletal elements against femur length for individuals of Cygnus atratus. *The linked points are three fossil specimens, which are neither bigger nor indicative of different body proportions to the rest of the sample, which were modern specimens. After Worthy (1998a).*

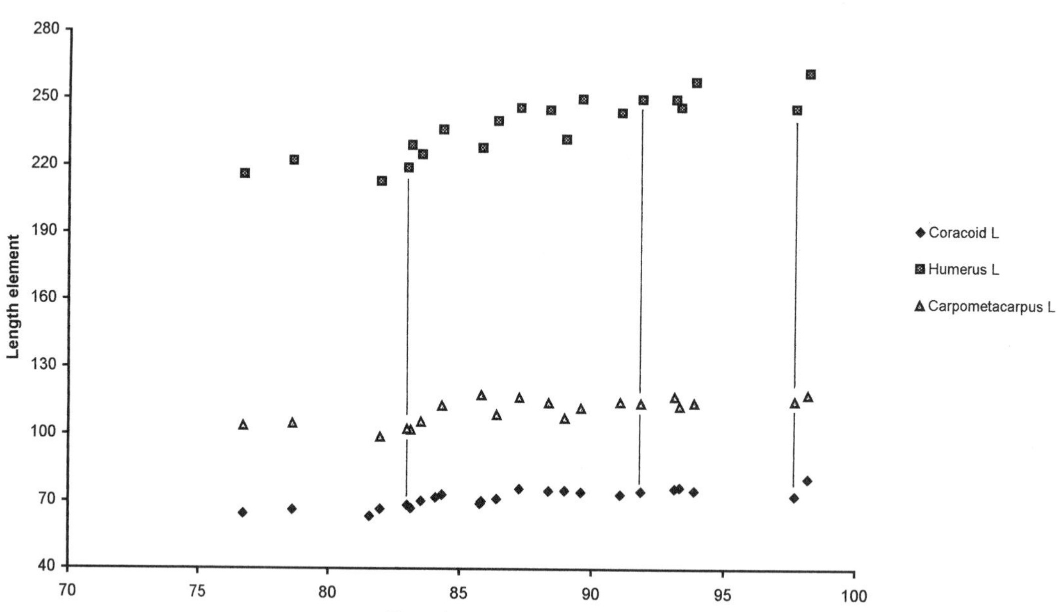

phonomy would have favored preservation of bigger, stouter bones. Measurements for wing bones and the coracoid did not differ significantly.

Proportional differences in bone lengths can be assessed properly only by comparing associated skeletons. The few associated fossil skeletons had wing–leg proportions similar to those of modern skeletons (Fig. 7.3). The fossil bones of swan in mainland New Zealand are now referred to the Australian black swan (*Cygnus atratus*).

The Chatham Island fossil swan may in fact have been a distinct species. Although unstudied, it is said to have had a larger and stouter bill and relatively shorter wings than the mainland form (Millener 1996, 1999). Complicating things somewhat is the fact that the type of *C. chathamensis* Oliver is now lost or not identifiable as such in the Museum of New Zealand, so the status of fossil swans from the Chatham Islands awaits a full review.

Cnemiornis calcitrans

Genus Cnemiornis *Latham*
Cnemiornis calcitrans *Owen, 1865 (South Island Goose) and*
Cnemiornis gracilis *Forbes, 1892 (North Island Goose)*

Among moa bones sent to Sir Richard Owen from a fissure near Timaru in the South Island were many smaller specimens, which he recognized as belonging to an anatid. He described them as the new genus and species *Cnemiornis calcitrans* (Owen 1866b). Owen thought that *Cnemiornis* was most similar to *Cereopsis novaehollandiae,* the Australian Cape Barren goose (Figs. 7.4–7.7). *Cnemiornis* was rare in the collections in the 19th century, so the discovery of an almost complete skeleton of *Cnemiornis* from Earnscleugh Cave in Central Otago was noteworthy (Hector 1874).

A smaller fossil species of *Cnemiornis* was later founded on the basis of a slender tibiotarsus from the North Island. Forbes (1892d: 187) erected the name *Cnemiornis gracilis* for "a most elegantly moulded goose from the North Island." This was the description, which lacks information regarding what element or specimen was at hand, where it was from, or how it differed. As with the swan, the type of *C. gracilis* referred to by Forbes was long thought to have been lost, and Oliver (1955) described *Cnemiornis septentrionalis* on the basis of bones from Hunterville. However, along with others, the lost type of *C. gracilis* was rediscovered in the Palaeontology Department of the British Museum (Natural History), London, by Elliot Dawson, and the name was reinstated (Dawson 1958b). The specimen, labeled as being from Te Aute, is a left tibiotarsus (NHM A2074) and is cataloged as the type of *Cnemiornis gracilis* Forbes. Apart from its being considerably smaller (Table 7.1), the North Island goose differs from the South Island bird in some osteologic features, mainly in the greater pneumaticity of the coracoid, scapula, and furcula (Worthy et al. 1997).

In the same publication in which Forbes erected *C. gracilis,* he founded *Cnemiornis minor,* "founded on tibiae now in the Canterbury Museum, which for many years have passed as belonging to the long-known calcitrans, but which require only to be compared with the type

Figure 7.4. Cape Barren goose Cereopsis novaehollandiae *of Australia, the closest relative of the New Zealand goose* Cnemiornis. *Photograph by R. Morris.*

species to disclose their distinctness" (Forbes 1892d: 187–188). The name is a *nomen nudem* because the specimens were not identified; it was not said how they differed; and they are not now identifiable. Last, the synonymy of *Cnemiornis* must also include *Cereopsis novaezealandiae* Forbes. Forbes (1892d: 188) reported, "It is with much satisfaction that the author has to announce the addition to the New Zealand fauna of a species of *Cereopsis* itself. The species is founded on a portion of the cranium, which, except that it is slightly larger, is almost indistinguishable from *Cereopsis novaehollandiae.* The species has been designated *Cereopsis novaezealandiae.*" Forbes (1892b) made it clear that this specimen was from Enfield, in South Island. Dawson (1958b) claimed to have found the type skull in the British Museum (Natural History) and referred it to *Cnemiornis calcitrans,* but he did not indicate where in the museum the specimen was kept or give the specimen's catalog number. To further complicate matters, Oliver (1955) described a cranium in the Museum of New Zealand under the same name: *Cereopsis novaezealandiae* Oliver. However, this is in fact a recent cranium of *Cereopsis novaehollandiae,* so in this instance, the name falls into the synonymy of the Australian species.

Most researchers have agreed with Owen that *Cnemiornis* was closely related to *Cereopsis* (Brodkorb 1964; Howard 1964; Oliver 1955; Woolfenden 1961). More recently, Livezey (1989b) studied the phylogenetic relationships of *Cnemiornis* on the basis of cladistic analyses of skeletal characters and concluded that *Cnemiornis* was not a goose and was not closely related to *Cereopsis,* but instead represent-

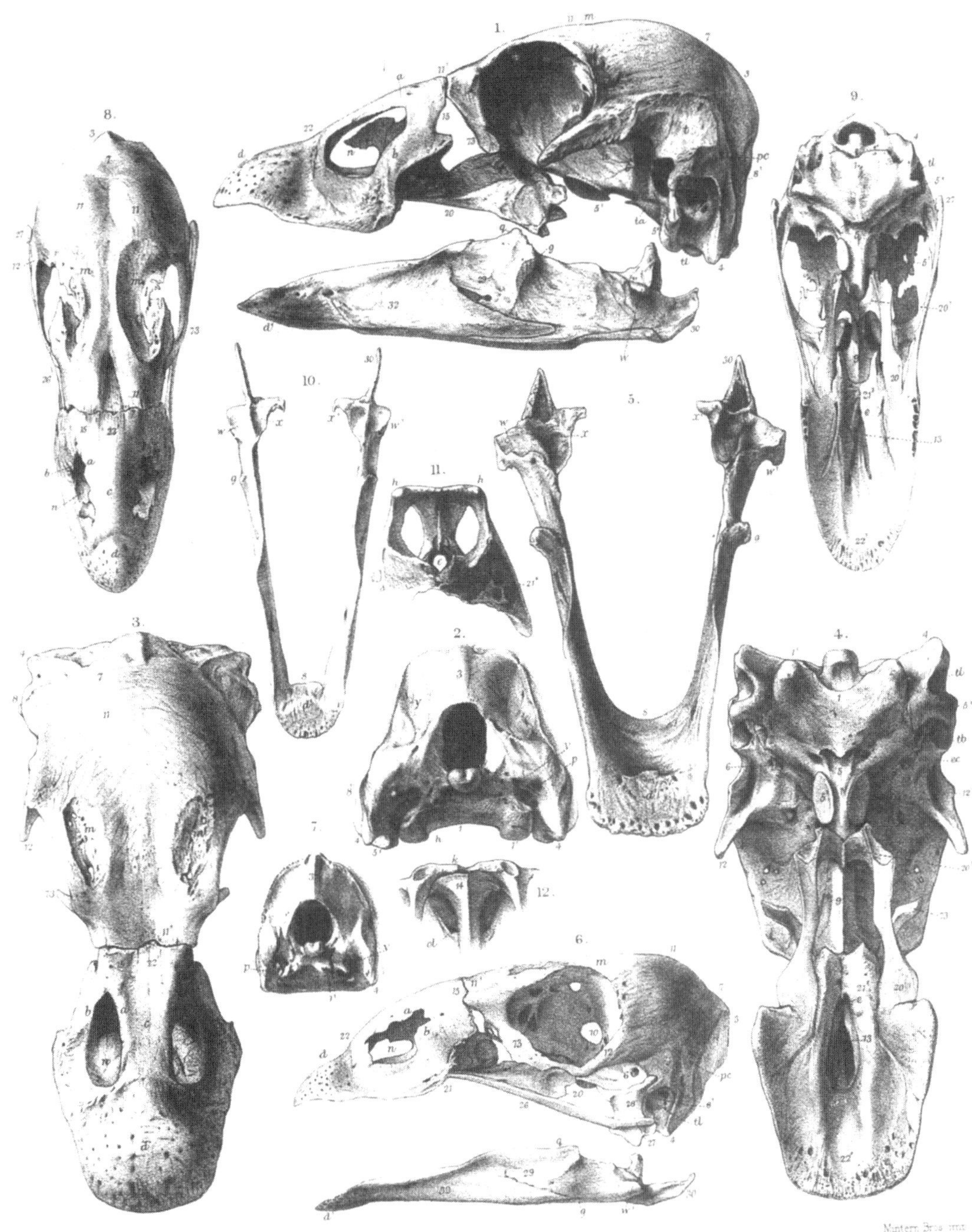

ed a very early branch of the Anseriformes, immediately following *Anseranas* (the Australian magpie goose). He therefore proposed that *Cnemiornis* be placed in its own family Cnemiornithidae, basal to the Anatidae. Worthy et al. (1997) challenged this conclusion on the basis of a further cladistic analysis of an enlarged data set and on a sequence of mitochondrial DNA from a fossil goose bone. They returned

Figure 7.5. Owen's plate portraying the skull of Cnemiornis *with that of* Cereopsis *(Owen 1879: pl. 101).*

Figure 7.6. Owen's plate portraying selected wing bones of Cnemiornis *and the tarsometatarsus of* Cnemiornis *with that of* Cereopsis *and* Tachyeres *(Owen 1879: pl. 104).*

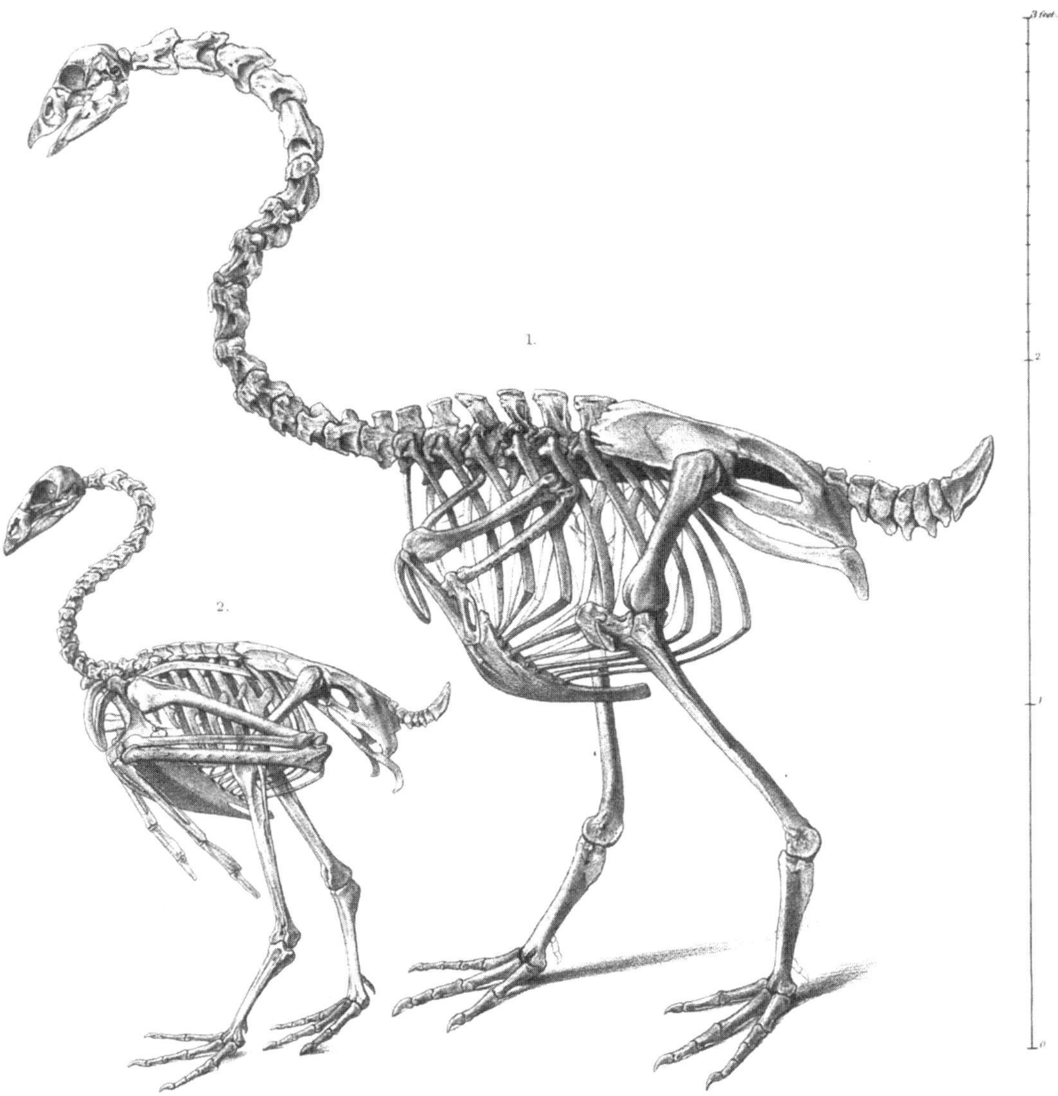

Figure 7.7. Owen's restoration of Cnemiornis *with a skeleton of* Cereopsis. *However, at the time Owen made this, he had mistaken the vertebrae and coracoid of* Aptornis *for those of* Cnemiornis *(Owen 1879: pl. 101). Nevertheless, it shows how very much larger the New Zealand bird was than its Australian ancestor.*

Cnemiornis to the anserines or geese and placed it as the sister taxon of *Cereopsis*.

In *Cnemiornis,* the ungual phalanx of pedal digit II is developed into a spur (see Fig. 7.8), perhaps indicating some aggressive behavior. *Cereopsis* has a spur on the wing, which they use when fighting (Marchant and Higgins 1990: 1198). In *Cnemiornis,* the wings were too reduced to provide an effective fighting tool. The New Zealand geese differ most obviously from their Australian counterpart in being bigger, having much smaller wings, and having the coincident reduced elements of the pectoral girdle (Fig. 7.9). *Cnemiornis* retained the unique salt gland impressions shared with its common ancestor with *Cereopsis,* albeit relatively reduced. The impressions dominate the top of its skull, and the glands allow *Cereopsis* to forage in salt-rich areas where no

TABLE 7.1.
Length data (mm) for long bones of *Cnemiornis gracilis* and *C. calcitrans*.[a]

Element	*C. gracilis*	*C. calcitrans*
Femur	126.2, 105–136.2, 6.88, 25	149.2, 126.1–162.5, 7.70, 20
Tibiotarsus[b]	225.1, 205–248, 11.01, 26	271.4, 249–291, 12.05, 26
Tarsometatarsus	118.0, 106.7–127.6, 5.81, 34	140.6, 130.1–151.3, 6.13, 19
Humerus	134.6, 123–146.2, 6.78, 23	160.6, 139.1–170.5, 7.78, 15
Ulna	104.0, 88.5–116.9, 7.28, 20	132.5, 124.3–143.2, 5.51, 9
Carpometacarpus	52.1, 47–58.3, 2.98, 19	—, 61.4–64.7, —, 5
Coracoid	60.1, 55.4–63.8, 2.20, 16	—, 63.6–75.9, —, 5

[a] Mean, range (minimum–maximum), standard deviation, *n*. Source: unpublished data of the authors.
[b] Length of the tibiotarsi were measured from the proximal articular surface to the distal extreme, rather than from the cnemial crest, which is broken in most specimens.

freshwater is available. They would be less useful in the salt-free grasslands of New Zealand. However, the bill of *Cnemiornis* is well adapted for grazing; its squared-off end could closely crop a sward (Figs. 7.2, 7.5). The periglacial landscapes of New Zealand would have had large areas of tussock grasslands and herb fields, where the New Zealand goose could graze. In Australia, *Cereopsis* grazing results in a very low, "mown" sward.

Body masses of the two species of *Cnemiornis* were calculated by allometric relationships with femur measurements given by Campbell and Tonni (1983) (Table 7.2). The New Zealand geese were big: mean estimated body masses were 15 and 18 kg for the North Island and South Island species, respectively. Both were therefore at least three times heavier than Canada geese (4.5–5.5 kg) and Cape Barren geese (5 kg). These extinct geese were as heavy as some of the smaller moa and the aberrant *Aptornis* (see Fig. 3.13 for bone comparison).

Both geese were flightless and lived only in areas where the vegetation included grassland and shrubland. Their bones are common in glacial-aged deposits from both islands, in sites where the only vegetation would have been grassland and shrubland. The preference or requirement for grassland probably meant that the North Island goose had a restricted distribution during the Holocene, when tall forests dominated by Podocarpaceae covered most lowland areas. Ecotones on dunes and forest margins in the eastern regions of low rainfall probably provided the only places they could live. They were therefore exterminated immediately after the arrival of Polynesians and have been found in only one archaeologic site. The South Island goose would have fared better during the Holocene because the rain-shadow zone of the Southern Alps contained large areas of grassland–shrubland–forest mosaic in Canterbury and Otago.

There is a single historic record from the North Island that may pertain to this goose. Cockburn-Hood (1875) related that in the previous seven years, a bird that resembled a goose in shape and that had

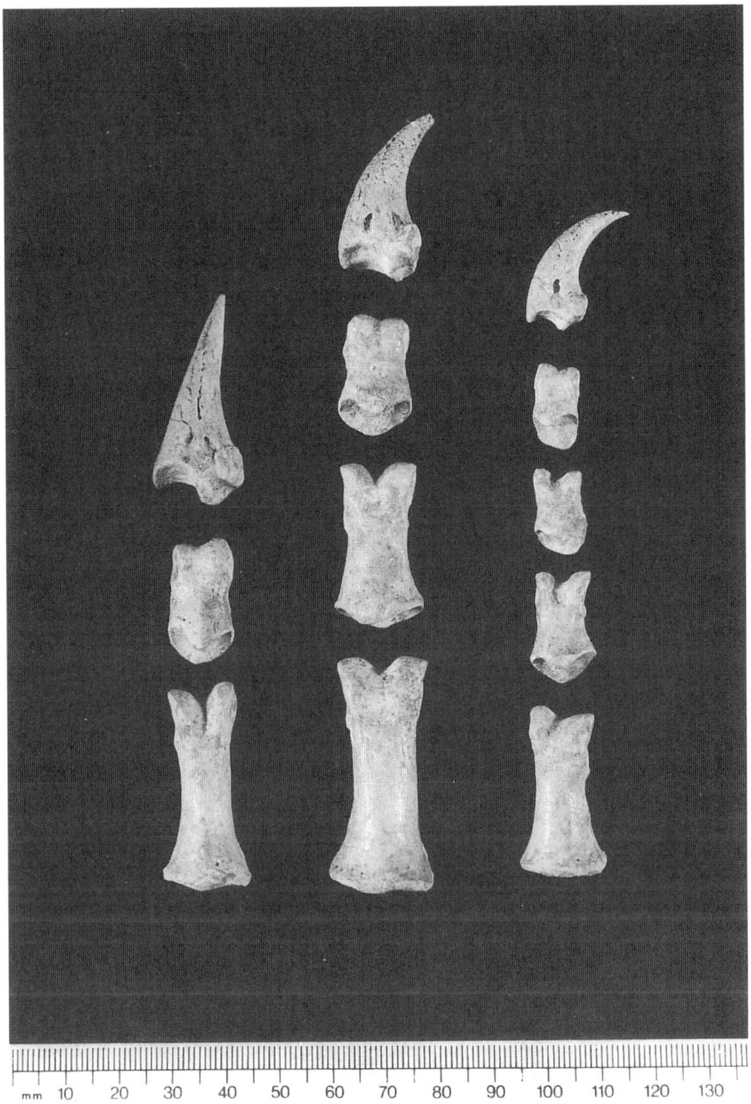

rufous plumage and was unable to fly was caught with dogs in the hills behind Opotiki. Apparently, it used to be common and esteemed for food. The reference to rufous plumage is interesting because *Cereopsis*, the living relative of *Cnemiornis*, has gray plumage. However, the small gooselike duck, *Chenonetta jubata*, has rufous plumage, and thus it might be expected that so did its relative *Euryanas*, which also was likely to have been much more common than the goose ever was. The goose was called "tarepo" by Māori (Martin 1886), but the source of this is not given.

Figure 7.8. Photograph of toe bones of Cnemiornis calcitrans *MNZ S35266, showing the development of the ungual of digit II into a spur. Photograph by J. Palmer. Photograph republished from Worthy et al. (1997) with permission of Cambridge University Press.*

It is tempting to think that the Cape Barren goose, which is recorded as an occasional vagrant from Australia, might reestablish a population in New Zealand. A few individuals were present in Canterbury during the 1990s.

Figure 7.9. Reconstruction of Cnemiornis. *This large, flightless goose stood about 1 m tall. Its small wings and tail attest to its lack of flying ability. Drawing by J. Winn.*

Malacorhynchus scarletti

Genus Malacorhynchus *Swainson*
Malacorhynchus scarletti *Olson, 1977 (New Zealand Pink-eared Duck)*

The New Zealand pink-eared duck (*Malacorhynchus scarletti*) was described from bill fragments (Fig. 7.10) from Pyramid Valley moa swamp, near Waikari in North Canterbury. The swamp has provided a large collection of fossil birds since excavations began there in 1939. It is most famous for moa remains, especially for many individual skeletons, but a diverse fauna of small birds was also present. Among the ducks was one of particular interest that Falla (1941) listed as *Malacorhynchus* sp. on the basis of terminal fragments of two premaxillae and a tibiotarsus. At that time, *Malacorhynchus* was known by a single Australian species, the nomadic pink-eared duck (*M. membranaceus*). Some 36 years later, Olson (1977a) reexamined these bones and erected a new species (*M. scarletti*) on an anterior premaxilla fragment and a

TABLE 7.2.
Mass of *Cnemiornis* species calculated from the minimum, maximum, and mean femur shaft measurements from the same samples for which length is given in Table 5.1 by use of the allometric relationships of Campbell and Tonni (1983).[a]

Taxon	Femur diameter	Mass on diameter (kg)	Range (kg)
C. gracilis	11.5 (minimum value)	8.03	3.04–14.4
	17.75 (maximum value)	16.55	7.99–25.62
	14.77 (mean)	12.18	5.31–20.05
C. calcitrans	16.4 (minimum value)	14.5	9.70–23.06
	20.1 (maximum value)	20.36	10.53–30.24
	18.33 (mean)	17.46	8.58–26.74

[a] The measured shaft widths were the minimum shaft width of the bone, with the maximum values of central shaft width being usually some 20% greater. Therefore, the calculated masses are likely to be minimal estimates.

subadult right mandible fragment. No other elements were referred to the species.

A reexamination of the Pyramid Valley fauna led to several other elements being referred to *M. scarletti* (Worthy 1995). However, it is still one of the rarest of New Zealand fossil species. A minimum of just four individuals was represented at Pyramid Valley, three of which were juvenile, indicating that the species bred there. Thirteen bones from at least two individuals are now known from dunes at Marfells Beach from beside a large, shallow coastal lagoon (Worthy 1998a). Some of these were from Polynesian middens and show that New Zealand's first inhabitants ate this duck. Only one other site is known in the South Island; two bones were found in a cave site in north Otago (Worthy 1998b). In the North Island, the species is known from a single bone from a swamp near Hastings and from several elements from Poukawa Swamp in Hawke's Bay. However, many anatid bones from Poukawa remain unidentified. A few bones have been found among the extensive collections from dunes on the Chatham Islands.

The morphology of the bones shows that *Malacorhynchus scarletti* was a filter feeder that dabbled at the surface, not a diver. The exceedingly broad, flattened bill of *M. scarletti* was superbly adapted to filter feeding of microcrustaceans, such as ostracods and cladocerans, and other aquatic animals. The leg bones exhibit none of the characteristics typical of divers. The femur was not anteroposteriorly bent, nor was the fibular condyle expanded as in *Aythya* and *Mergus*. The tibiotarsus did not have a high cnemial crest, nor was the shaft flattened anteriorly with a well-developed extension of the procnemial crest, as in the diving ducks such as *Aythya* and *Mergus*. Instead, these bones were similar to those of the dabbling ducks, such as *Anas superciliosa*. *M. scarletti* had a closed pneumatic foramen to the humerus, as in some diving genera—for example, *Aythya*—but this is probably of phylogenetic rather than functional significance. *M. scarletti* was thus a duck that lived in open-

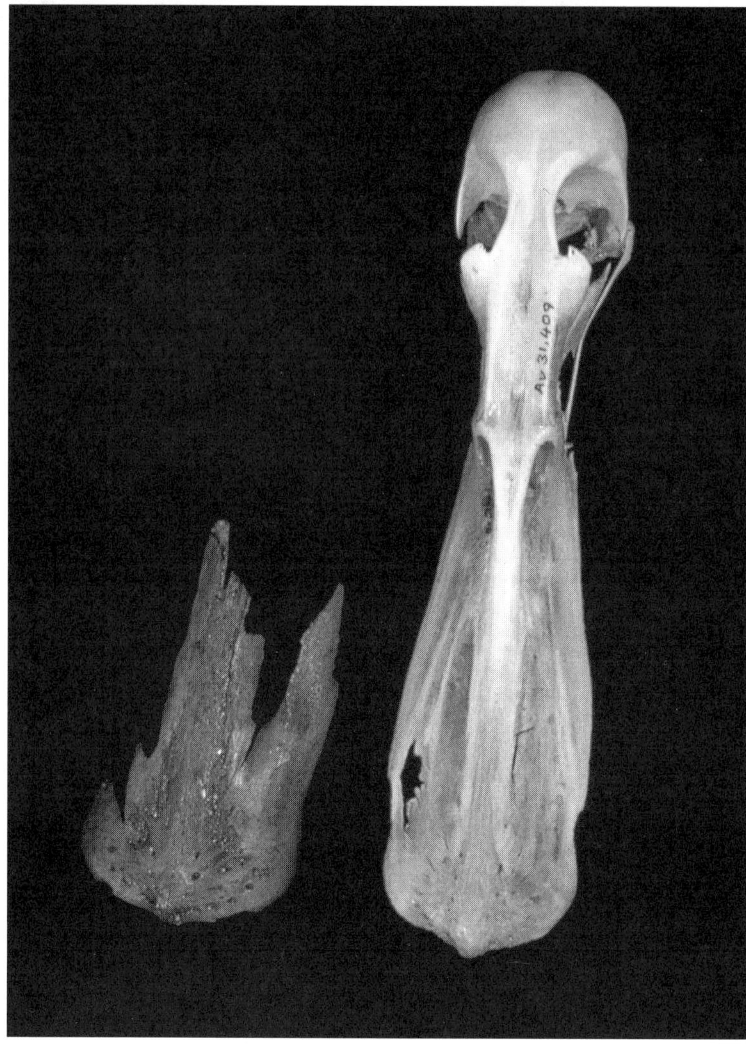

Figure 7.10. The holotype bill fragment of Malacorhynchus scarletti *(CM Av5855) compared to a skull of the Australian* M. membranaceus *(CM Av31409). Photograph by R. Morris.*

water habitats. No associated leg and wing elements have been found, but *M. scarletti* probably had leg bone–wing bone proportions similar to those of *Malacorhynchus membranaceus,* although it was considerably larger.

It could fly and so could travel between widely dispersed sites of suitable habitat, which makes the Chatham Islands records not too surprising. The many small lakes and the huge central lagoon also means that there were large areas of suitable habitat on the island for a filter-feeding duck. Chatham Island is easily within range for many birds; several species of birds have colonized it since 1900, taking advantage of newly available habitats and newly emptied ecologic niches. Individual ducks are known to have made the journey. A mallard *Anas platyrhynchos* (New Zealand bird banding scheme number 17-55566), banded at Lake Ohuia, northern Hawke's Bay, North Island, as a juvenile female on March 1, 1969, was recovered at Chatham Island on May 1, 1971. Two gray ducks, *A. superciliosa,* banded in the Waikato in North Island, were shot on Chatham Island in the mid-1950s (Williams 1981).

M. scarletti probably occupied the ecologic niche now taken by the shoveler *A. rhynchotis* in New Zealand wetlands. The shoveler was not recorded from Pyramid Valley (4,000–2,000 years before the present; Holdaway and Worthy 1997) and seems to have become established in New Zealand within the last millennium to become the 14th resident anatid.

Biziura delautouri

Genus Biziura *Stephens*
Biziura delautouri *Forbes, 1892*

Biziura was first recognized in the New Zealand fauna when Forbes (1892b: 417) designated a tarsometatarsus as the type of *Biziura delautouri.* The description was "a *Biziura* somewhat larger than *Biziura lobata,* the musk duck of Australia, an interesting species." Forbes clearly indicated that this right tarsometatarsus was associated with large numbers of moa bones in the Enfield site near Oamaru, South Island, New Zealand, and that he named it after Dr. H. de Lautour, of Oamaru, who helped acquire specimens from the site.

The species lapsed into obscurity for many years. It was not until 1969 that further elements were described. Then Scarlett (1969b) figured a right humerus from Marfells Beach and a right tibiotarsus and a right humerus from Poukawa Swamp in the North Island. However, Scarlett considered these bones to be indistinguishable from *Biziura lobata* of Australia and that the size differences were insignificant in this strongly sexually dimorphic species. This conclusion was supported by Harrison and Walker (1970). It was not until Olson (1977a) described two tarsometatarsi from Marfells Beach (Lake Grassmere) and noted differences in shape that distinguished them from *B. lobata* that the New Zealand form was once again recognized as distinct.

Fossil material of this species is rare. There is a single skeleton from dunes at Waikuku Beach by North Cape, a few bones from Poukawa Swamp in Hawke's Bay, some from both natural dunes and an archaeologic site at Marfells Beach in Marlborough, as well as the single specimen from the type locality at Enfield in Otago.

Fossil *Biziura* bones from New Zealand vary greatly in size, as expected from the marked sexual dimorphism of *B. lobata* (Olson 1977a). The bigger, presumably male bones are larger than those of modern males, and the smaller, presumably female bones are likewise larger than those of modern females. However, the proportions of the Waikuku specimen, the only one with associated bones of an individual, show that it had longer legs than the modern species. These differences and some differences in bone shapes (personal observation; Olson 1977a) support the classification of the New Zealand *Biziura* as a separate species.

Mergus australis

Genus Mergus *Linnaeus*
Mergus australis *Hombron and Jacquinot, 1841 (Auckland Island Merganser)*

The Auckland Island merganser is a typical merganser, a group whose center of distribution is in the Northern Hemisphere. The species isolated in New Zealand is a biogeographic enigma: the only other Southern Hemisphere species is the very rare *M. octosetaceus* in Brazil, Paraguay, and Argentina. By the time the New Zealand merganser came to the attention of science, it was already restricted to the Auckland Islands, a subantarctic group south of mainland New Zealand. About 30 birds were collected (Fig. 7.1), the last in 1902, and the species has not been seen alive since.

These birds lived in the quiet waters of the fjordlike inlets and harbors and at the heads of estuaries, venturing up into pools in rivers. They had the typical narrow bill with recurved serrations, and their diet comprised fish, crustacea, marine worms, and shellfish.

Mergus bones are rare in New Zealand fossil deposits, in part because they have rarely been recognized and in part because of the former distribution of the species in mainly coastal habitat. Bones have been recovered from dunes in Doubtless Bay (northern North Island), Delaware Bay and Marfells Beach (northern South Island), and Native Island (Paterson Inlet, Stewart Island). It has been recovered from middens on Ponui Island (Hauraki Gulf), Paremata (Wellington), Kakanui and Shag Mouth (north Otago), and Old Neck (Stewart Island). In the outlying islands, a *Mergus* was common in some deposits of Chatham Island (Fig. 7.2), indicating it was common in the large, saline Te Whanga Lagoon, where smelt and other small fish are abundant. Millener (1999) considered the Chatham birds to be an undescribed species a little smaller than mainland birds, with a shorter bill and more reduced wings. Although the New Zealand and Chatham fossils have not been compared yet, it is evident that there is considerable size variation in both populations (especially as seen in recently collected fossil samples from the Stewart Island area), so the few recent bones do not reflect the size range of individuals in the former population. The distinction or otherwise of the Chatham fossils from mainland specimens remains to be ascertained. Although it might seem unlikely that a sea duck would be distinct on different islands separated by only a few hundred kilometers, the pink-footed shags of the genus *Leucocarbo* have separate species on each of the subantarctic island groups.

Euryanas finschi

Genus Euryanas Oliver
Euryanas finschi *(van Beneden, 1875) (Finsch's Duck)*

The species *Euryanas finschi* (Fig. 7.2) was described as *Anas finschi* by van Beneden (1875) after he and Dr. Otto Finsch studied a series of bones from Earnscleugh Cave, Otago. Among living species, van Beneden found it was most similar to *Dendrocygna eytoni*, but—among other features—they were struck by its resemblance to the Miocene

Mionetta ("*Anas*") *blanchardi.* Van Beneden's article was later published in an abridged form (van Beneden 1877). A few years later, Lydekker was the first to note the similarity of *Euryanas* to the Australian wood duck (*Chenonetta jubata*) when he cataloged a skeleton (also from Earnscleugh Cave) and four crania as *Bernicla jubata* (Lydekker 1891: 106) without mention of *Anas finschi.* Oliver (1930) erected *Euryanas* to contain *Anas finschi,* and then and thereafter (Oliver 1955: 403), he maintained that the species was closely related to *Chenonetta jubata.* Falla (1953) regarded *Euryanas finschi* as the New Zealand equivalent of *Chenonetta jubata.* More recently, Livezey (1989b) made a cladistic study of the relationships of *Euryanas finschi* by use of skeletal characters and concluded that it was the sister group of Tadorninae + Anatinae. Unpublished data (Worthy and Olson, in preparation) challenges this conclusion and returns *Euryanas* to a sister-species relationship with *Chenonetta jubata.*

Finsch's duck *Euryanas finschi* is one of the most common species in late Quaternary fossil deposits of the eastern districts of both the South and North Islands of New Zealand. Hundreds of bones are found in some deposits, such as those at Castle Rocks (Hamilton 1893, 1894b), Earnscleugh Cave (Clark et al. 1996; Worthy 1998b), and Martinborough 1 (Yaldwyn 1956). *Euryanas finschi* has been demonstrated to be one of a key group of species that signify the presence of grassland–shrubland mosaics in New Zealand paleoenvironments (Worthy and Holdaway 1993, 1994b, 1995; Worthy and Mildenhall 1989).

In Australia, *Chenonetta* is common. It is primarily a terrestrial grazer in areas of grassland, pasture, and open or dense woodland near water, but not dense forest. On land, it can run about swiftly, but it swims awkwardly. It feeds only sparingly in shallow water on emergent vegetation. It winters in traditional camps and is usually found in flocks, which can be very large. It takes insects when available and nests in tree holes (Frith 1969; Marchant and Higgins 1990). It is thus the ecologic equivalent of *Euryanas finschi.* Finsch's duck was an anatid adapted for a terrestrial existence, with stout, gooselike legs, as is *Chenonetta* (Fig. 7.11). Similarly, its bill was short and stout, suitable for grazing, not broad, like that of a dabbling duck. Both the Australian and New Zealand birds were about the same size, similar to that of a mallard. *Chenonetta* has been recorded as a vagrant to New Zealand several times this century. Perhaps it too may reestablish a population.

Euryanas finschi has provided a rare example of a measurable rate of evolutionary change toward flightlessness. Large samples of bones from various deposits showed that wing length declined 10% between the Late Glacial 10–12 ky ago and the late Holocene, although leg length stayed constant (Worthy 1988b, 1997e). This example shows that wing reduction can occur not over the course of millions of years, but just in a few thousand years.

Pachyanas chathamica

Pachyanas chathamica *Oliver, 1955 (Chatham Island Duck)*

The extinct Chatham Island duck was somewhat larger than the mallard but substantially stouter. It was described in 1955 by W. R. B.

Figure 7.11. Reconstruction of Euryanas finschi *showing a presumed male, assuming it had similar plumage patterns to its close relative, the wood duck* Chenonetta jubata. *It was about the size of a mallard. Drawing by J. Winn.*

Oliver from isolated bones found in the sand dunes at Chatham Island. Its heavily constructed cranium (Fig. 7.2) is characterized by well-marked depressions for salt glands. This suggests that its main habitat on the Chathams were the saline waters of Te Whanga Lagoon and perhaps sheltered sections on the coast. Mallard ducks from estuarine localities have obvious salt-gland impressions, whereas individuals from inland, freshwater localities do not.

Material found more recently shows that the Chatham Island duck had a typical dabbling duck bill and relatively reduced wings. These were not reduced to the extent of obligate flightlessness. The relationships of this duck are unclear. Features of the skeleton such as a prominent capital shaft ridge and a raised dorsal tubercle on the humerus suggest a link with tadornids rather than with typical *Anas*.

Extant Native Species

The remaining members of the New Zealand waterfowl fauna are local equivalents of, or are populations of, more widespread species. For example, the anatids *Anas superciliosa, A. gracilis,* and *A. rhynchotis* are widespread in the southwest Pacific. The scaup *Aythya novaeseelandiae* is endemic to New Zealand but is closely related to other *Aythya* in the Northern Hemisphere. The shelduck *Tadorna variegata* differs little from the Australian counterpart.

The remaining species, the blue duck *Hymenolaimus malacorhynchos,* is usually considered to be a specialist torrent duck. It has a highly specialized niche feeding in riffles and rapids of streams and now is restricted to mountain areas of New Zealand. However, fossil deposits show that it also lived in the lowlands, outside its present range. In fact, Europeans caught them in the Waihopai Valley, Marlborough, in the 1850s. More interesting is the presence of their bones in pitfall deposits in potholes or caves sometimes up to 1 km from streams. Birds have also recently been found foraging long distances from streams. The species may have used, or still uses, forest-floor habitats away from flowing water, mostly at night. A terrestrial forest-floor feeding niche was also occupied by the brown teal (*Anas chlorotis*). Bones of *A. chlorotis* are often found in pitfall cave deposits at long distances from streams or other waterways. It is now known that these small teal retreat to the shelter of streams with overhanging vegetation during the day and at night move away from the stream to feed on the forest floor. Because they were always found on streams by diurnal human observers, it was naturally, but incorrectly, thought that that was where they foraged as well.

Order Pelecaniformes

Family Pelecanidae

Genus Pelecanus *Linnaeus*

Pelecanus conspicillatus *(Temminck, 1824) (Australian Pelican)*

Gilbert Archey was the first to record pelican fossils in New Zealand, after the discovery he and others made of a skeleton in a cave at Waikaremoana on November 30, 1930 (Archey 1931). Archey referred the fossil bones to the Australian pelican *Pelecanus conspicillatus*. Scarlett (1966) reported the discovery of further bones at Marfells Beach and at Lake Poukawa, Hawke's Bay, and gave subspecific status to the New Zealand bones as *Pelecanus c. novaezealandiae* because of their larger size and minor differences in shape of the pelvis. Rich and van Tets (1981) raised the New Zealand form to specific status mainly because of the broad pelvis, but perhaps they were more than a little swayed by the geographic separation of the New Zealand birds from the Australian counterpart. They showed that although the New Zealand form averaged larger, all specimens were within but near the upper limits of the size range of recent *P. conspicillatus* ($n = 20$) in all measurements except proximal width of the femur, which was greater in the New Zealand form. The purported wider pelvis was based on two specimens for which the $\log_{10}$ ratio of the pelvis width to synsacra length was within the range of recent specimens of *P. conspicillatus* ($n = 12$).

Worthy (1998a) considered that these differences were not specifically significant, noting that in total, the remains of only seven birds were represented in New Zealand deposits. No account was taken by Rich and van Tets (1981) of allometry, which would require that larger birds would likely have more robust pelves and hence a bigger proximal femur than smaller birds. Pelicans are gregarious species and usually congregate in flocks. They are also extremely good fliers and travel great distances to use ephemeral wetlands or to escape drought conditions (Marchant and Higgins 1990). In 1978–1979, drought conditions in central Australia saw a mass departure from that region, and coincidentally, birds were recorded far north of Australia in Fiji and in New Zealand (Marchant and Higgins 1990).

If such large birds in the past maintained breeding populations in New Zealand, then they should be more visible in the fossil record. Bones would be expected in the frequency of those of the similar-size *Cygnus*, but they are not. Only the following specimens are known: one bone from Puheke Beach, Northland; two bones of an individual from the Sunde archaeologic site on Motutapu Island, Hauraki Gulf; a skeleton from a cave at Lake Waikaremoana; and bones of two individuals from the extremely rich fossil site at Poukawa, Hawke's Bay (Rich and van Tets 1981; Scarlett 1966; Turbott 1990a). The many thousands of bones from Lake Grassmere have only yielded bones of two pelicans (Worthy 1998a). Such sparse records are more likely to be the result of occasional vagrants reaching New Zealand. For these reasons, Worthy (1998a) returned the New Zealand pelican fossils to *P. conspicillatus*. New Zealand did not have a resident population of pelicans.

Family Sulidae

New Zealand lies in temperate waters, so is outside the range of the majority of sulids. Around mainland New Zealand, only *Morus serrator*, the Australasian gannet, breeds now, and there is no evidence of any other species having done so. Boobies *Sula dactylatra* reach the New Zealand archipelago only at the tropical Norfolk Island and the Kermadec Islands. Van Tets et al. (1988) described *Sula tasmani* from fossil bones on Norfolk Island, but the bones are morphologically indistinguishable from those of *S. dactylatra* and represent the upper end of the size range of the population of that species formerly breeding there (Holdaway and Anderson, in preparation).

Family Phalacrocoracidae

The New Zealand archipelago has one of the most diverse shag faunas known, with 11 species (Appendix 1). There are three widely distributed species of *Phalacrocorax*, one *Stictocarbo*, and one *Leuco-*

carbo around mainland New Zealand. The Chatham Islands have an endemic *Stictocarbo* species, and each of the main subantarctic island groups and the Chatham Islands have an endemic *Leucocarbo* species.

A single fossil taxon was named by Forbes (1892d: 188–189). Like other taxa named in that unsatisfactory publication, the description is entirely wanting: "The species next described is a shag of greater dimensions than the largest New Zealand species, *Phalacrocorax novaezealandiae*; but until more material is available the author prefers to describe it under a variety of that species, i.e., *Phalacrocorax novaezealandiae*, var. *major*." No indication was given as to what element or elements Forbes had, nor was the site from which the specimen or specimens came noted. Because specific status was not advocated, the validity or otherwise is not of great importance. But a right humerus (NHM A4337) and left humerus (NHM A4338), labeled as from Te Aute Swamp, are present in the Palaeontology Department, Natural History Museum, London, and cataloged as syntypes of *Phalacrocorax major*. In the Museum of New Zealand, there is a right ulna (MNZ S23178) from the same Te Aute collection (Hamilton 1889), which thus might also be considered part of the syntype series for the form. Forbes's variety *major* would presently fall within *Phalacrocorax carbo*.

Leucocarbo carunculatus from the mainland of New Zealand is an example of a marine species whose range has markedly constricted in recent times. At present, the *Checklist of the Birds of New Zealand* (Turbott 1990a) lists king shags from mainland New Zealand in two species: *Leucocarbo carunculatus* and *Leucocarbo chalconotus*. This was largely influenced, it seems, by the fact that the two forms live at opposite ends of the South Island, *L. chalconotus* at Stewart Island is considerably smaller, and the population has a dimorphic plumage. However, birds that live on the Otago Peninsula have the same size range as those from northern South Island at the outer Marlborough Sounds and have other features that are intermediate between the other populations. Worthy (1996) described fossil bones of *Leucocarbo* from the northern extreme of the North Island, and bones have since been found in deposits from near North Cape. Other coastal deposits, such as at Delaware Bay near Nelson and Marfells Beach near Blenheim (Worthy 1998a), northern South Island, have common remains of this genus. The remaining populations are susceptible to human disturbance, and in view of the fossil record, they can now be seen as the relict parts of a former New Zealand–wide distribution.

Order Ciconiiformes

Family Ardeidae

New Zealand has few herons and bitterns (Appendix 1). The most common species seen now is *Ardea novaehollandiae*, the white-faced heron, but this only became a breeding species in the 20th century. The most popular heron is *Casmerodius (= Egretta) albus*, the white heron or kotuku, which, although of a widespread species elsewhere, has only a single breeding colony in New Zealand. Solitary birds disperse from this colony each year to all parts of New Zealand. Fossils are rare, but juvenile bones indicate a former breeding colony in the Far North area, at the top of the North Island. Only one other sustained breeding species of heron is present: the widespread *Egretta sacra* or reef heron, with a coastal distribution. However, most recently, *Nycticorax caledonicus,* or the Nankeen night heron, which has been a regular vagrant to New Zealand, began breeding in the Whanganui River area of North Island.

Bitterns are represented by one living species in New Zealand, the large *Botaurus poiciloptilus,* or Australasian bittern. This species seems to be absent from the fossil record, even from the huge wetland faunas of Lake Poukawa, which has many thousands of specimens (Horn 1983), so is assumed to be a recently established species. However, in the past, there was a diminutive bittern. *Ixobrychus novaezelandiae* (Potts 1871), the New Zealand little bittern, was recorded historically from a single skin from the North Island and fewer than 10 individuals from the South Island (Potts 1871); it is presumed extinct. Its leg bones are more robust than *Ixobrychus minutus,* which has been recorded as a vagrant to New Zealand. Fossil bones of *I. novaezelandiae* were first recorded from Lake Poukawa in Hawke's Bay as *Dupetor flavicollis* by Horn (1980), but they were reidentified as the New Zealand species by Millener (1991). Bones of this rare New Zealand species are known from Lake Poukawa, a midden near Wellington, and the Chatham Islands.

Podicipediformes Grebes

New Zealand has only two representatives of the cosmopolitan family of grebes: *Podiceps cristatus,* the crested grebe, and *Poliocephalus rufopectus,* the New Zealand dabchick. Grebe bones are relatively rare in the New Zealand fossil record, but only because few lacustrine sites have been sampled. They were present in the Lake Poukawa and Marfells Beach faunas (Horn 1983; Worthy 1998a).

The following quaint passage published by Buller (1878: 191–192) is an early record of dabchick (Oliver 1955: 92). The narrator was the botanist W. Colenso, whom Buller cited from the *Tasmanian Journal of Natural Science* for April 1845. It also is the foundation of the name *Fulica novaezealandiae* being in the synonymy of *Poliocephalus rufopectus.*

> A little below Ngaruawahie [modern-day Ngaruawahia] (in the Waikato district) we met a man in a canoe with a live and elegant specimen of the genus *Fulica.* I hailed the man and purchased the bird, which he had recently snared, for a little tobacco. It was a most graceful creature, and, as far as I am aware, an entirely new and undescribed species. Its general colour was dark, almost black; head grey and without a frontal shield; fore-neck and breast ferruginous red; wings barred with white; bill produced and sharp; feet and legs glossy olive; toes beautifully and largely festooned at the edges; eye light coloured and very animated. It was very fierce and never ceased attempting to bite at everything within its reach. I kept it until we landed, intending to preserve it, but as it was late, and neither material at hand nor time to spare, and the animal too, looking so very lovely that I could not make up my mind to put it to death, I let it go; it swam, dived, and

disappeared. From its not possessing a frontal shield on the forehead (which is one of the principal generic marks of the Linn. genus *Fulica*) it may possibly hereafter be considered as a type of a new genus, serving to connect the genera *Fulica* and *Rallus*. Not a doubt, however, in my opinion can exist, as to its being naturally allied in habit and affinity to the Fulicae; I have therefore named it *Fulica novae-zealandiae*. In size it was somewhat less than our European species, *F. arua*.

8 • The Grandest Eagle

Diurnal Predators

Diurnal birds of prey are often conspicuous members of fossil avifaunas from Quaternary deposits on continents (e.g., Howard 1932; Steadman and Martin 1984) and islands or ecologic islands such as the Gargano Peninsula in Italy (Arredondo 1976; Ballmann 1969, 1973). Island faunas from Cuba (Arredondo 1976) and the Bahamas (Olson and Hilgartner 1982) contained very large raptors that were apparently the largest predators in their particular ecosystems. Eagles and hawks were also prominent in the bird-rich terrestrial vertebrate faunas on prehuman Hawaii (Olson and James 1991).

Although there were no lions, leopards, bears, or wolves in New Zealand, there were still large predators that could attack and kill the largest herbivores. The top predator was Haast's eagle *Harpagornis moorei,* which was larger than the largest living eagle, the harpy (*Harpia harpyja*) of Central and South America. It seems also to have been larger than the extinct *Titanohierax borrasi* from Cuba (Arredondo 1976; Olson and Hilgartner 1982). Much smaller, but still the largest in its genus, was Eyles's harrier, which weighed more than a goshawk (*Accipiter gentilis*). The smallest of the diurnal raptors, the New Zealand falcon *Falco novaeseelandiae,* is the only survivor of the three. There is now evidence for two more unlikely predators, which were truly terrestrial and perhaps the most unusual component of an unusual avifauna. Stable isotope measurements (see Chapter 5) suggest that the two adzebills (*Aptornis otidiformis* and *A. defossor*) were flightless gruiform predators.

With avian rather than mammalian predators, the selection pressures on the herbivores were rather different than in most other places in the world. Hawks and eagles hunt mainly by sight, although some, such as the harriers, have keen hearing and can locate prey in dense vegetation or beneath ground cover. For potential prey, there were advantages to being nocturnal or crepuscular, to being well camouflaged, and to having behaviors that reduced visibility from above.

Several of the flightless birds that survive in New Zealand are at least partly nocturnal. The best known group (at least in the popular perspective), the kiwi, are almost wholly nocturnal on the main islands, and mostly so on Stewart Island. Even some of the species such as the kaka parrots that fly well are often active at night. It is difficult to determine the cause and effect of behaviors. One behavior of a surviving flightless bird seems to be a clear response to avian predators. A female South Island takahe *Porphyrio hochstetteri* incubating her clutch beneath a large grass clump will push her bright orange beak into the nest material, keeping both eyes clear, at the first sign of a shape in the sky (R. Morris, personal communication).

Accipitrid hawks have a poor sense of smell. Therefore, among the prey, there was no selection pressure against the evolution of odor for individual or site recognition, as there would have been in the presence of mammals, which hunt largely by scent. The rarest of the surviving flightless species, the kakapo parrot, has a characteristic smell that dogs find easy to trace. Kiwi are also easy prey for dogs, which can locate them at night and in their burrows. Petrels have a characteristic odor, as well as being more or less helpless in their nest burrows, so they were easy prey for introduced mammals, although their nocturnal behavior protected them against larger predatory birds. Small petrels were also vulnerable to predation by the two indigenous owls, but the larger species had no enemies on land, although perhaps adzebills dug some from their burrows.

Conversely, with no mammals to hunt, the New Zealand raptors were specialized bird eaters. There was no advantage in being a large terrestrial predator, such as the colossal flightless owls of prehuman Cuba, which could lie in wait for large rodents and insectivores and grab them with long talons (Arredondo 1976). The only mammals in New Zealand that visited the forest floor were two species of specialized bat, the greater and lesser short-tailed bats. They were nocturnal and too small to support a guild of giant terrestrial owls, but they were a favored prey of the laughing owl, which was neither flightless nor overly large.

In keeping with their position at the top of the vertebrate food chain in a bird-dominated environment, both large raptors had diverged markedly from other species in their lineages, and their specializations provide interesting case studies in evolution.

Systematics: The Accipitridae

Haast (1872c, 1879, 1881) and Owen (1879b) realized that the New Zealand eagle belonged with the other great eagles in the family now known as the Accipitridae but could offer no more precise indications of its relationships. This is not surprising because the systematics

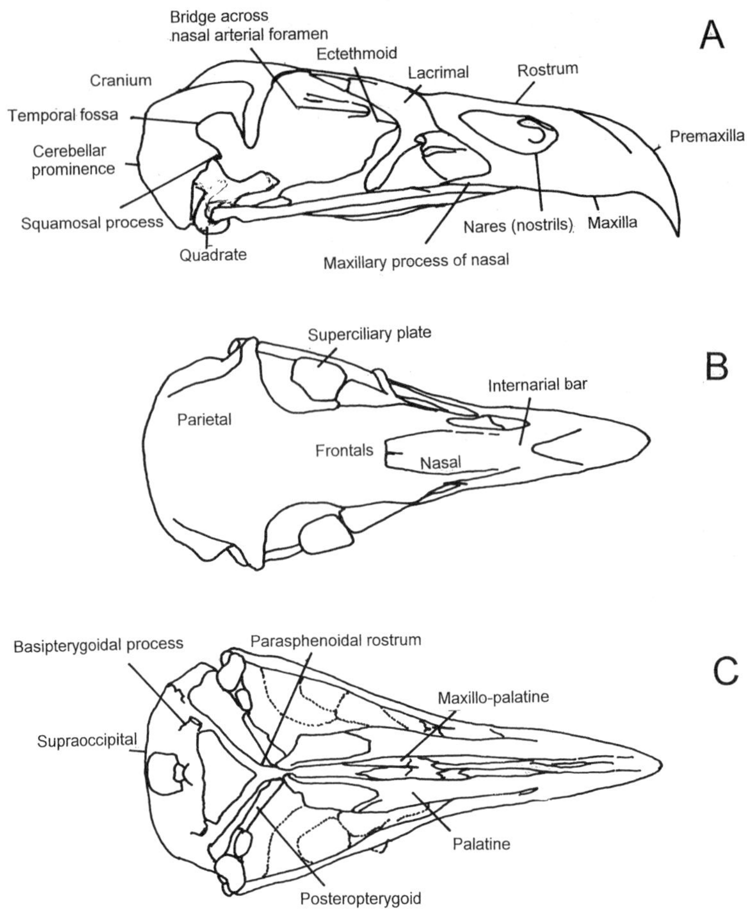

Figure 8.1. Skeletal features used in descriptions and diagnoses of Harpagornis moorei *and* Circus eylesi. Harpagornis moorei*: (A) Right lateral view of cranium and rostrum. (B) Cranium and rostrum, dorsal view. (C) Cranium and rostrum, ventral view. Not to scale.*

of the Accipitridae are still poorly known (Holdaway 1994). But what are accipitrids, and how do they differ from other birds with hooked beaks and sharp claws? Are all predatory birds the same?

Because much of the evidence for the relationships of the birds of prey comes from comparisons of parts of the skeleton (Holdaway 1994), and because all the endemic New Zealand accipitrids are extinct and are known only from fossil bones, names of parts of the skeleton referred to in this discussion are given in Figures 8.1–8.6.

What Defines an Accipitrid?

The family Accipitridae includes the diurnal birds of prey, except the falcons (Falconidae) and American vultures (Vulturidae). Pycraft (1902: 318–319) and Friedmann (1950: 64–65) have listed features of the skeleton that characterize the family. In the skull (Fig. 8.1), the palate is indirectly desmognathous: prevomers are fused together or are absent; maxillopalatines meet or are fused in the midline; and palatines and pterygoids are articulated with the parasphenoidal rostrum. Rarely—only in *Elanus*, or small kites—it is schizognathous; that is, it dif-

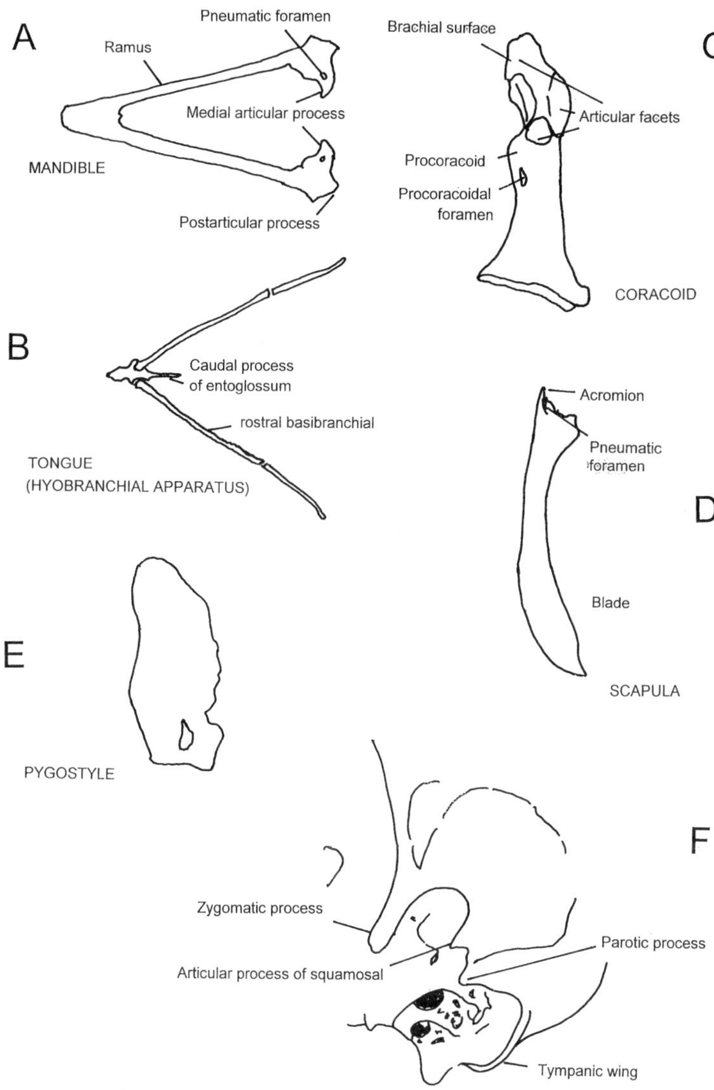

fers from the desmognathous state in that the prevomers are completely fused, the maxillopalatines do not meet in the center, and the vomer is not expanded anteriorly or closely applied to the palatines. The lachrymals (Baumel et al. 1993) have an accessory superciliary plate, which shields the eye from above; this is absent in the honey buzzards (*Pernis*) and in the vultures (e.g., *Gyps* and *Aegypius*). The so-called antorbital plate (ectethmoid), or anterior border of the orbit, is reduced and often tongue-shaped, articulating or even fused at its ventrolateral end with the ventral process of the lacrimal. The squamosal prominence (process) is weakly developed. In the beak, the ventral surface of the maxilla lacks a median bony ridge, and there is no "tooth" on the maxillary

Figure 8.2. Skeletal features used in descriptions and diagnoses of Harpagornis moorei *and* Circus eylesi. Harpagornis moorei: *(A) Mandible, dorsal view. (B) Hyobranchial apparatus, dorsal view. (C) Right coracoid, dorsal view. (D) Right scapula, lateral view. (E) Pygostyle, right lateral view.* Circus: *(F) Rostrolateral view of left tympanic region. Not to scale.*

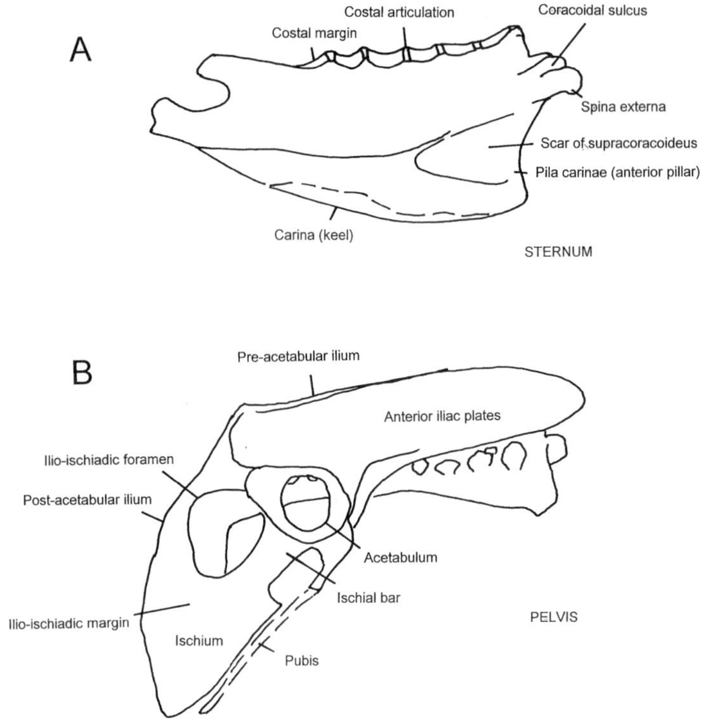

Figure 8.3. Skeletal features used in descriptions and diagnoses of Harpagornis moorei *and* Circus eylesi. Harpagornis moorei*: (A) Sternum, right lateral view. (B) Pelvis, right lateral view. Not to scale.*

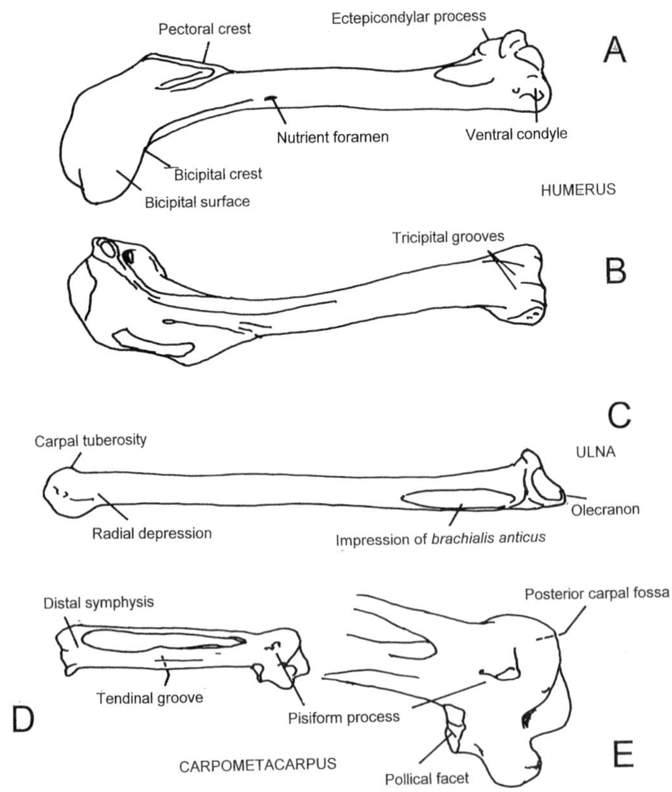

Figure 8.4. Skeletal features used in descriptions and diagnoses of Harpagornis moorei *and* Circus eylesi. Harpagornis moorei*: (A) Left humerus, cranial view. (B) Left humerus, dorsal view. (C) Right ulna, cranial view. (D) Left carpometacarpus, ventral view. (E) Left carpometacarpus, proximal detail, ventral view. Not to scale.*

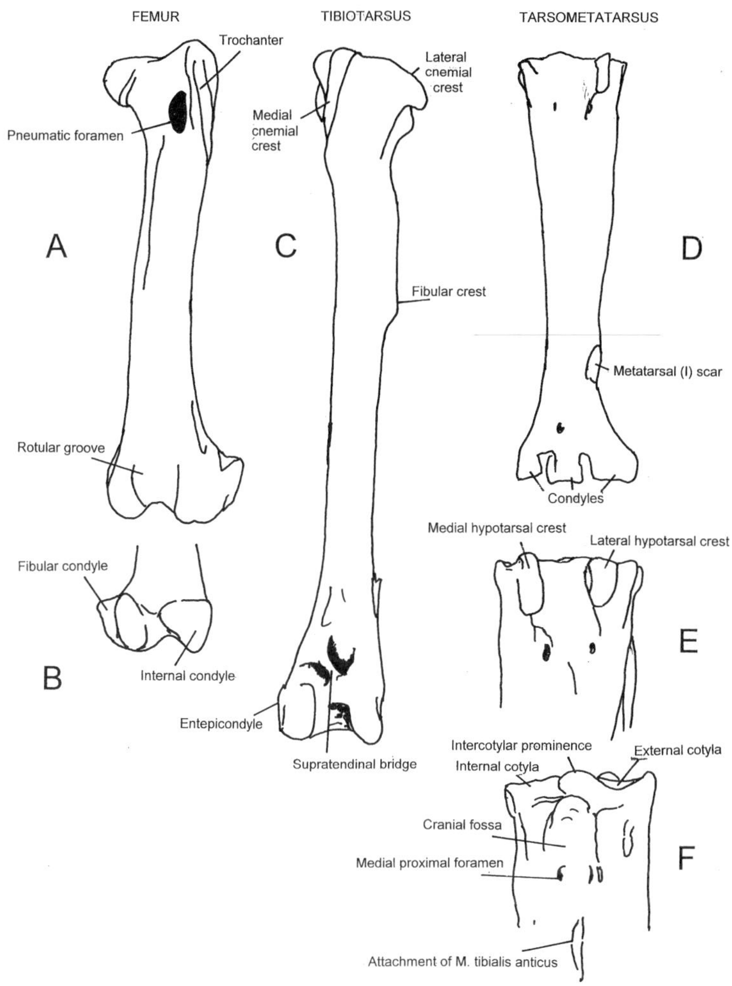

Figure 8.5. Skeletal features used in descriptions and diagnoses of Harpagornis moorei *and* Circus eylesi. Harpagornis moorei*: (A) Left femur, cranial view. (B) Left femur, distal, ventral view. (C) Left tibiotarsus, cranial view. (D) Left tarsometatarsus, caudal view. (E) Right tarsometatarsus, proximal caudal view. (F) Right tarsometatarsus, cranial proximal view. Not to scale.*

margin. The nasal canal is incompletely ossified, with a large pyriform internal opening that exposes the nasal septum. There is no central bony tubercle in the external nostril, but there is sometimes a cartilaginous tubercle or, if the nostril is slitlike, a tubercle at the lower posterior angle. If the external nostril is small and round, then there is no bony rim or tubercle. The ramus of the mandible has no fenestrae.

In the postcranial skeleton, the coracoids (Fig. 8.2) do not overlap at the sternum. The procoracoid is small and articulates with the scapula only; it is widely separated from the furcula. The thoracic vertebrae are not coalesced into a notarium. The sternum (Fig. 8.3) lacks a spina interna. In the pelvis (Fig. 8.3), the postacetabular ilium is shorter than the ischium and is much deflected from the dorsal line of the preacetabular ilium. The ischium is never produced backward as a spine, and the pubis is often vestigial. There is no bony loop on the proximocranial surface of the tarsometatarsus to restrain the flexor

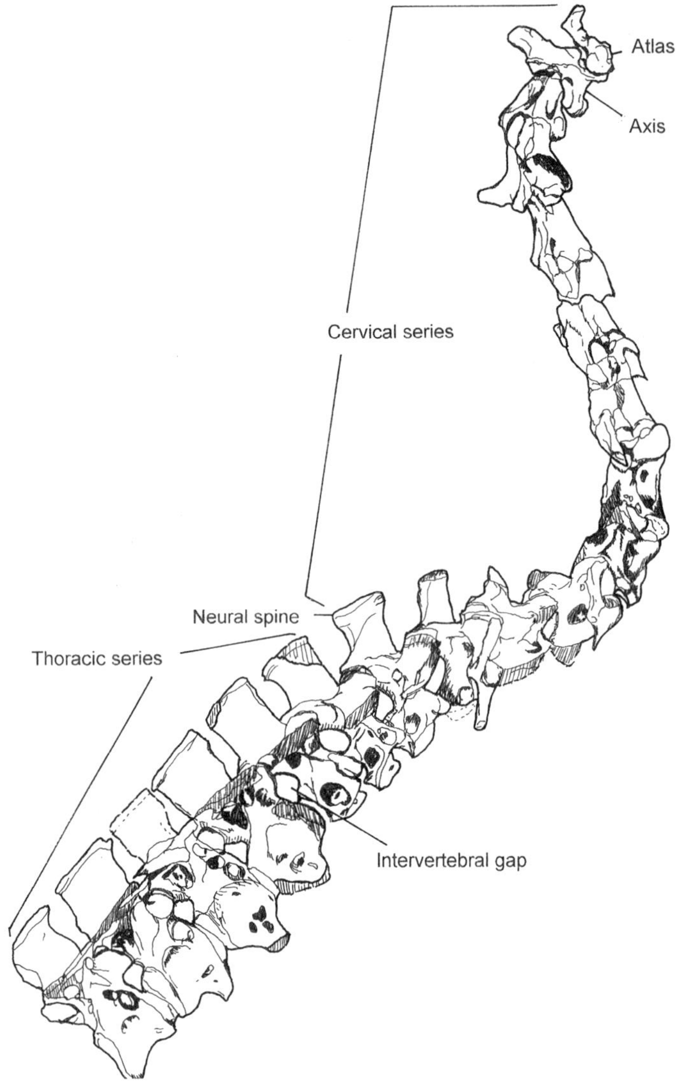

Figure 8.6. Skeletal features used in descriptions and diagnoses of Harpagornis moorei *and* Circus eylesi. *Right lateral view of rearticulated presacral vertebrae of* Harpagornis moorei *(MNZ S27773), showing cervical and thoracic series.*

tendon. The fourth toe is not reversible. The claws are graduated in size from the first (largest) to the fourth, and they are flat or rounded on the ventral surface.

Differences between Accipitrids and Falconids and Vulturids

In early descriptions, Haast's eagle was often referred to as a vulture, and several workers have suggested that it fed on carrion and was functionally a vulture. There are two separate lineages of birds for which the common name is vulture, and they are not closely related. The Old World (African–Asian–European) vultures are members of the Accipitridae. New World vultures belong to the family Vulturidae (Cathartidae of many authors) and are related to the storks rather than to the other birds of prey. Their possession of a hooked beak and predatory habits are convergent on the "true" birds of prey.

There are many anatomic, genetic, karyologic (differences in number and confirmation of chromosomes in the cells), and behavioral differences (Emslie 1988; Friedmann 1950; König 1982; Ligon 1967; Rea 1983) between the Accipitridae and the Vulturidae. In the skeleton, the main differences are that in the Vulturidae (but not the Accipitridae) the lachrymals (Fig. 8.1) are fused with the frontals, the nostrils are perforate, the maxillopalatines are thin and scroll-like, and the olfactory chamber is very large; there are basipterygoid processes; there is a deep notch in the ilioischiatic margin of the pelvis (Fig. 8.3); the hypotarsus (Fig. 8.5) has two shallow grooves; the tarsometatarsus has a large intercotylar prominence and two tubercles for the attachment of M. tibialis anticus (where the M. indicates that the name is of a muscle = Musculus); and the hallux or hind toe is elevated and nonfunctional.

In accipitrids, the olfactory chamber is small (the birds have a poor sense of smell), the basipterygoid processes (Fig. 8.1) are vestigial, and the nostril is imperforate—that is, there is no direct gap from one external nostril to the other. There is no notch in the ilioischiadic margin of the pelvis. Importantly for analysis of the relationships and habits of extinct species, there are clear differences in the tarsometatarsus: the articulation of the hallux (hind, or first, toe) is not raised above the level of the other toes, and it is fully functional and much more than half the length of the fourth toe; the tarsometatarsus is flattened and has strong lateral ridges (the medial one thin and produced), forming a trough for the large flexor tendons to the toes; and the hypotarsal ridges are large and separated by a deep groove.

The falcons (Falconidae) differ from the accipitrids in many points of the skeleton (Friedmann 1950; Jollie 1976, 1977a, 1977b, 1977c), as well as in behavior. For example, they kill by biting the head or neck of the prey, whereas accipitrids kill with the feet, and the excreta is dropped straight on to the perch, whereas accipitrids squirt the material away from the perch. There are also differences in karyology (the complement and structure of the chromosomes within the cell nucleii), the color and shape of their eggs (Tyler 1966), and the arrangement of their feathers, also known as pterylosis. The skeletal differences characterizing falconids that are visible in fossil material include the following: the bony palate is directly desmognathous (the maxillopalatine bones meet at the midline, and both the palatines and pterygoids articulate with the parasphenoid rostrum), the central vomer is expanded anteriorly, the lacrimal lacks a separate superciliary plate, the squamosal prominence is well developed, the ventral surface of the maxilla has a median bony ridge, the nasal bones are almost completely ossified, and the external nostril is small and slitlike with a central bony tubercle. In *Micrastur,* the forest falcons of South and Central America, the opening is large and has no tubercle. The premaxilla has a conspicuous "tooth" protruding ventrally from the lateral margin just posterior of the down-curved tip, which is obvious in the bill of the living falcon.

The main postcranial features that are characteristic of falcons are the large procoracoid process on the coracoid that articulates with the furcula, the "notarium," or fusion of all but one of the thoracic vertebrae into a single structural unit (although this is lacking in *Herpetotheres,* the laughing falcon), and the well-developed spina interna of the sternum. Perhaps most distinctive of all is the extra bar to the supratendinal bridge on the distal tibiotarsus, which produced a distinctive pattern of three perforations for the passage of the tendons.

By their skeletal characters, Haast's eagle and the harrier—or harriers, if the North and South Island populations are different—belong in the Accipitridae but to different lineages within that diverse family (Holdaway 1994). The New Zealand predators had evolved for an unknown length of time in an environment where they had no competition from mammalian carnivores. In doing so, they departed from the normal characteristics of their sister groups and were exploiting very large prey. Each entered an ecologic niche unlike that of the probable parent species. The relaxation of mammalian competition and the nature of the prey available to them led to extremes in size and power that were unmatched elsewhere. Studies of the raptors are continuing as more material about them and of their prey is found and analyzed. There may well be more surprises in store.

Haast's Eagle

Genus Harpagornis *Haast*

Neither Haast (1872c, 1874a), who described the huge New Zealand eagle and attributed it to a new genus, nor Owen (1879a), who fitted it into his scheme of the extinct species of New Zealand, was overly concerned about the relationships of that genus within the family, beyond the inescapable fact of its being a giant eagle. Indeed, until evidence for a larger accipitrid is found, it is the largest known eagle and one of the most spectacular components of the New Zealand fauna. It was the only eagle in the world to be the top predator in a complex ecosystem—in this instance, one consisting only of birds. It is unique that its prey were commonly 60–100 kg and sometimes up to 200 kg or more, the largest ratio of prey to predator mass of any terrestrial vertebrate predator. We therefore describe it in some detail.

In the 1890s, Professor T. Jeffery Parker, at Otago University, sought an opinion from someone who had worked on birds of prey and wrote to Dr. R. W. Shufeldt at the Smithsonian Institution, Washington, D.C. Shufeldt had studied skeletons of many of the large eagles and saw affiliations between *Harpagornis* and the two common Northern Hemisphere genera, *Aquila* and *Haliaeetus* (Shufeldt 1896). So Shufeldt agreed that it was an eagle, but because the two genera belong to separate lineages within the Accipitridae (Holdaway 1994), his conclusions were not very useful.

Matters rested there for all the birds of prey in New Zealand for many years. Until the 1980s, treatments of the relationships among the Accipitridae in general (e.g., Brown and Amadon 1968) relied mainly on external morphology such as plumage pattern, presence or absence of feathered tarsi, and details of wing proportions, tail proportions, and diet. Jollie (1976, 1977a, 1977b, 1977c) incorporated many aspects of the osteology of the group in his work, but as Olson (1985: 108) pointed out, the work is "prolix and idiosyncratic" and it is "a labor of love" to extract information from it.

Before work could begin on investigating the relationships of Haast's eagle within the Accipitridae, a preliminary analysis of the variation in the skeleton of as many genera in the family as possible was necessary (Holdaway 1994). The literature available included little else apart from Suschkin's work at the turn of the 20th century (Suschkin 1900, 1905) and papers on individual characters (including Ballmann 1969, 1973; Beddard 1889; Howard 1932; Miller 1937; Olson 1982, 1987b; Plótnick 1956). The preliminary study identified possible lineages within the family. It was based on a wider consideration of the skeleton than had been assessed before by a standard technique. A suggestion that Haast's eagle was related to *Spizaetus* (Brathwaite and Holdaway 1987) could not be refuted or supported by the analysis (Holdaway 1994) because little material of the southeast Asian species of *Spizaetus* was available for study. However, a South American *Spizaetus (S. coronatus)* was the next most derived genus on the phylogenetic tree, and the groups are clearly closely related. *Harpagornis* appeared as a sister group of *Aquila* (Holdaway 1991, 1994) (Fig. 8.7).

Although a relationship with *Aquila* conflicts with Oliver's (1945a, 1945b, 1955) conclusions, it agrees with more recent suggestions (e.g., Duff 1949a, 1949c) and with the biogeographic realities of the Accipitridae (Holdaway 1994). The only genera of large eagles found on the closest potential source areas (Australia and New Guinea) are *Aquila* and *Haliaeetus*. An aquiline ancestor is most likely, although it is the longest oceanic colonization by that genus. Despite persistent statements to the contrary (e.g., Millener 1999) and occasional records claimed of vagrants (Oliver 1955), no material of *Haliaeetus* is incontrovertibly known from New Zealand, although the genus had reached Hawaii. If the ancestor of *Harpagornis* was not present on the ancestral New Zealand land mass when it broke free from Antarctica and Australia, then it must have arrived across a large water gap.

Few other genera of accipitrids have colonized oceanic islands. Most appear to be deterred by even small water gaps. Of those that have reached out, *Accipiter* is found eastward as far as Tonga (Clunie 1981), and it made it as far as Norfolk Island (Meredith 1985) on the way to New Zealand. *Buteo* has reached the Galápagos (*Buteo galapagoensis*; Sibley and Monroe 1990) and Hawaii. The best overwater colonizers are harriers of the genus *Circus*. They have reached the Mascarene Islands in the Indian Ocean (*Circus maillardi*; Sibley and Monroe 1990) and many Pacific islands, at least as far as Hawaii, where *C. dossenus* is said to structurally mimic an *Accipiter* and was at first taken for one (Olson and James 1991), and New Zealand, where the resemblance of *C. eylesi* to a giant goshawk is very close and has also caused some doubt as to its correct generic placement (Holdaway 1989).

The relationships among the species included in *Aquila* have been

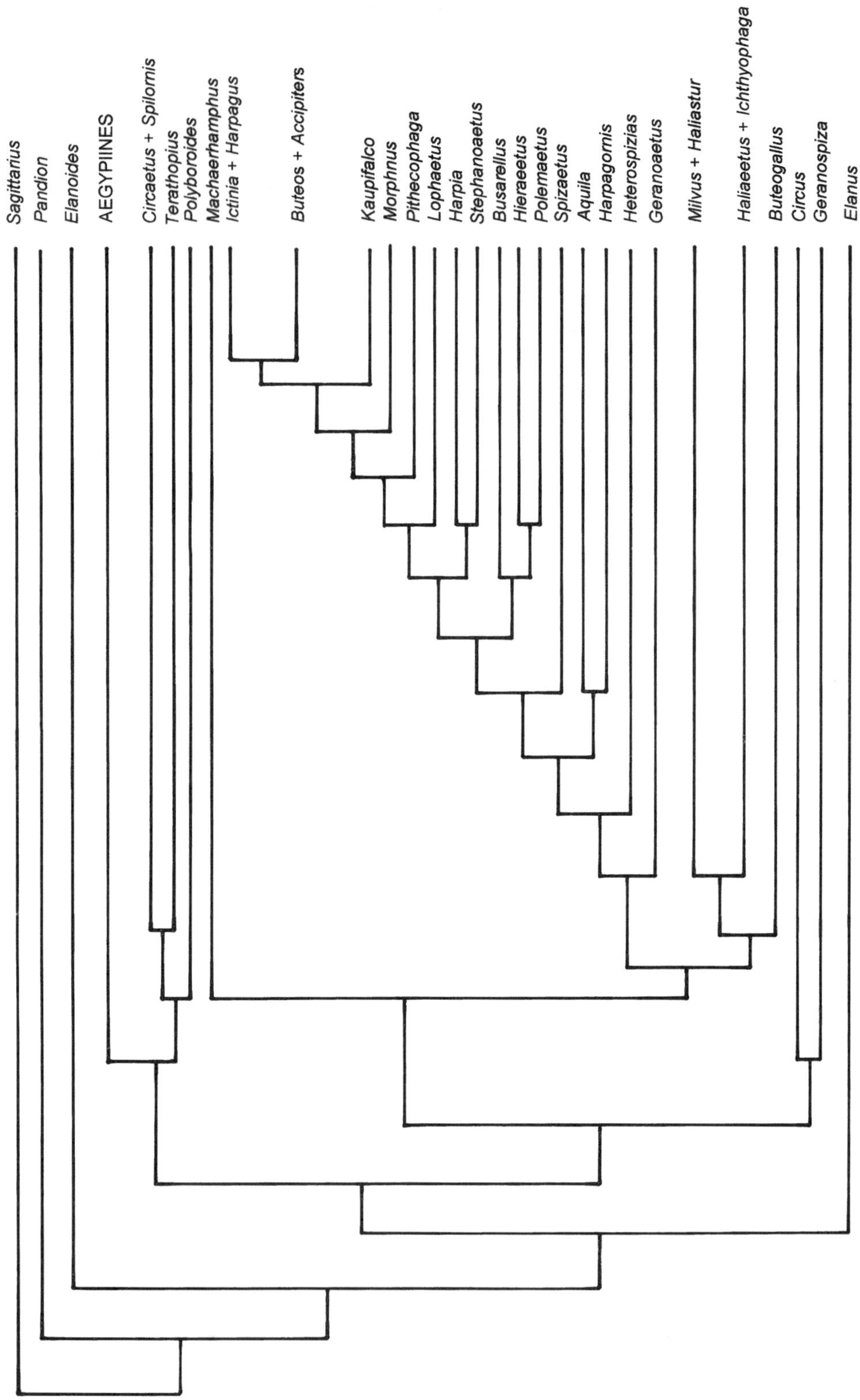
Sagittarius
Pandion
Elanoides
AEGYPIINES
Circaetus + Spilornis
Terathopius
Polyboroides
Machaerhamphus
Ictinia + Harpagus
Buteos + Accipiters
Kaupifalco
Morphnus
Pithecophaga
Lophaetus
Harpia
Stephanoaetus
Busarellus
Hieraeetus
Polemaetus
Spizaetus
Aquila
Harpagornis
Heterospizias
Geranoaetus
Milvus + Haliastur
Haliaeetus + Ichthyophaga
Buteogallus
Circus
Geranospiza
Elanus

based on plumage and proportions, not skeletal characters (Brown and Amadon 1968), so it is difficult to compare the variation on current knowledge of internal anatomy. Overall, aquilines are uniform in their skeletons, including the crania.

Previous diagnoses of the genus *Harpagornis,* apart from that by Haast himself, were published by Oliver. These differed. At first, his diagnosis was as follows: "Large extinct eagles with very strong legs but comparatively short wings. *Harpagornis* is closely related to the recent genus *Aquila.* Its distinctive features are the narrow skull, short ulna and stout metatarsus. As in *Aquila* the femur is longer than the tarso-metatarsus" (Oliver 1930: 392). Fifteen years later, his opinion had changed: "the giant eagle of New Zealand, was compared by Shufeldt with *Aquila.* By the form of the skull, however, it should be placed nearer to *Haliæëtus.* In fact, it is further from *Aquila* than to *Haliæëtus*" (Oliver 1945b: 137). Finally, he was sure: "Large extinct eagles with very strong legs but comparatively short wings. *Harpagornis* is closely allied to the recent genus *Haliaetus* [*sic*]. Its distinctive features are the narrow skull, short ulna and stout metatarsus" (Oliver 1955: 604).

In the cranium, the most obvious differences are in the length of the beak and how the elongation of the cranium was developed. Most of the change in proportions is in the preorbital section, especially in the narial region. The wing proportions also differ in that the ratio of the lengths of the humerus and ulna is greater than that in *Aquila,* which suggests that the mode of flight (and hence ecology) was not the same as that of the open-country *Aquila* eagles.

Formally, the genus *Harpagornis* is diagnosable as differing from *Aquila* by its much larger size; the narrow, low cranium (Fig. 8.1); elongated external nares; the presence of a well-developed scroll of bone around the external nostril (present in some individuals of *Aquila*); the posterior cranium sharply angular in lateral view; a broad maxillary process; caudal margin of Proc maxillare Os nasale entire, not notched to receive rostral angle of Os lacrimale as in *Aquila*; greater depth of the body of the thoracic vertebrae; ventral margin of the coracoidal sulcus more caudal than in *Aquila,* not obscuring dorsal margin in ventral view; pilum sterna visible inside sternal bowl as flat-topped ridge with lateral furrows, not visible in *Aquila;* spina externa deeper than broad, broader than deep in *Aquila;* relatively short ulna; steeper prepostacetabular angle than in *Aquila;* ilio-ischiadic foramen produced posterodorsally (a rounded triangle), oval in *Aquila*; ischial bar with distinct pit; anterior iliac plates produced dorsally above synsacrum, leaving deep dorsal fossa; and expanded iliac fossae.

The phylogenetic analysis based on skeletal characters showed that *Harpagornis* was most closely related to the widespread genus *Aquila* (Holdaway 1994). It is probably best interpreted as the sister taxon to *Aquila audax*—the extant Australian wedge-tailed eagle. However, the substantial modifications to the skeleton related to its size and large prey serve to separate it from all other members of the genus; hence, it is useful to distinguish the New Zealand bird at the generic level.

Figure 8.7. (opposite page) Main branches of a preliminary phylogeny of the Accipitridae based on analysis of skeletal characters. After Holdaway (1994), showing position of Harpagornis moorei *with respect to its sister group, the* Aquila *eagles.*

Harpagornis moorei *Haast, 1872 (Haast's Eagle)*

Type Material

The type material for *Harpagornis moorei* consists of a left femur, two pedal ungual phalanges, and a thoracic rib, all in CM Av5104, Canterbury Museum, Christchurch, New Zealand. Casts of all this material are in the Natural History Museum, London. The material was collected by F. R. Fuller at Glenmark, North Canterbury, in March 1871 while working for Julius Haast. The femur was designated as the lectotype in Holdaway (1990). The material preserved in Canterbury Museum under CM Av5102 was collected by the Haast party in July and August 1873 and is the type series of *Harpagornis assimilis,* a junior synonym of *H. moorei,* as it is a male of that species (Holdaway 1990).

Apart from the original description of the male as a different, smaller species, there is, in contrast to the situation in other New Zealand fossil birds (especially the harriers), little controversy over what to call Haast's eagle (Holdaway 1990, 1991), as is apparent in a brief list of names that have been applied in taxonomic and systematic publications.

1872 *Harpagornis Moorei* Haast, p. 193, pl. X, figs. 1, 4, 5; pl. XI, figs. 1, 1a, 2, 5

1874 *Harpagornis moorei* Haast; Haast, p. 62, pl. VII, fig. 1–6, pl. IX, fig. 1–3, 3a

Harpagornis assimilis Haast, p. 64, pl. VIII, figs. 1–7

1879 *Harpagornis moorei* Haast; Owen, 1, pp. 141, 142, pl. CV, figs. 1–4, pl. CVI, figs. 1–6, pl. CVII, figs. 1–7

1881 *Harpagornis moorei* Haast; Haast, p. 232, pl. XIII, figs. 1–4

1891 *Harpagornis moorei* Haast; Lydekker, p. 25

Harpagornis assimilis Haast; Lydekker, p. 25, as (queried) synonym of *Harpagornis moorei*

1893 *Harpagornis moorei* Haast; Hamilton, p. 92, pl. VII, C, D

Harpagornis assimilis Haast; Hamilton, p. 93, as male of *Harpagornis moorei*

1894 *Harpagornis assimilis* Haast; Hamilton, p. 227, pl. XXIII, figs. 1–3, as male of *Harpagornis moorei*

Harpagornis moorei Haast; Hamilton, p. 227, pl. XXIII, fig. 4

1898 *Harpagornis moorei* Haast; Beddard, p. 484

1907 *Harpagornis moorei* Haast; Rothschild, p. 85

1930 *Harpagornis moorei* Haast; Oliver, p. 392, unnumbered figure

Harpagornis assimilis Haast; Oliver, p. 394

1933 *Harpagornis Moorei* Haast; Lambrecht, p. 410, fig. 135, gives date of 1871

Harpagornis assimilis Haast; Lambrecht, p. 410, as synonym of *Harpagornis moorei,* gives date as 1873

1945 *Harpagornis moorei* Haast; Oliver, p. 137, fig. 46

Harpagornis haasti Haast; Oliver, caption to fig. 46 (lapsus)

1955 *Harpagornis moorei* Haast; Oliver, p. 604, unnumbered figure (specimen attributed to *H. assimilis* by Hamilton (1894b)

Harpagornis assimilis Haast; Oliver, p. 605

1964 *Harpagornis moorei* Haast; Brodkorb, p. 272

Harpagornis assimilis Haast; Brodkorb, p. 273 (as synonym of *Harpagornis moorei*)

1970 *Harpagornis moorei* Haast; Kinsky et al., p. 78

Harpagornis assimilis Haast; Kinsky et al., p. 78 (as synonym of *Harpagornis moorei*)

1972 *Harpagornis moorei* Haast; Scarlett, p. 11–12, pls. 3, 5, 7, 9, 10, 11, 20

1990 *Harpagornis moorei* Haast; Holdaway, pp. 39–47, includes *Harpagornis assimilis* Haast

Description

Apart from the original descriptions of the limited material available to Haast (1872a, 1874a, 1881), Owen (1875d, 1879a), and Hamilton (1893, 1894b), little has been published on the skeletal anatomy of Haast's eagle. Oliver (1930, 1955) gave abbreviated diagnoses of both nominal species (*H. moorei* and *H. assimilis*). He also described the structure of the palate as it was then known (Oliver 1945b). Scarlett (1972b) provided illustrations and measurements of the larger bones likely to be found in archaeologic sites, although, perversely, the only complete bones found so far in association with cultural deposits have been pedal phalanges, the occasional tarsometatarsus, the shaft of a tibiotarsus, and distal fragments of tibiotarsi and ulnae worked into tools.

A photograph of a mounted composite specimen from Castle Rocks in western Southland (Fig. 8.8) has been reprinted several times (e.g., Lambrecht 1933; Oliver 1930, 1955). The mount has deteriorated over the years, but Don Brathwaite used new photographs of it to make the first systematic calculations of the bird's body mass (Brathwaite 1992). Most writers who have dealt with the eagle (e.g., Duff 1949a, 1949c; Lambrecht 1933; McCulloch 1982; Rothschild 1907) have reprinted or reinterpreted previous descriptions. Errors crept in and accumulated in both the scientific and popular literature. Holdaway (1991) summarized previous descriptions and redescribed the species in comparison with the Australian wedge-tailed eagle *Aquila audax*.

The features of the skeleton, abstracted from that description, are given here to underpin both the systematic discussion and the ecomorphologic conclusions drawn later in this chapter. The main features of the principal bones are shown in the historically important lithographs published by Haast (1872c, 1874a) (Figs. 8.9, 8.10) and by Owen (1879a) (Figs. 8.11–8.13) and also in Figures 8.14–8.17. Owen's illustrations are lithographs and were prepared from photographs and

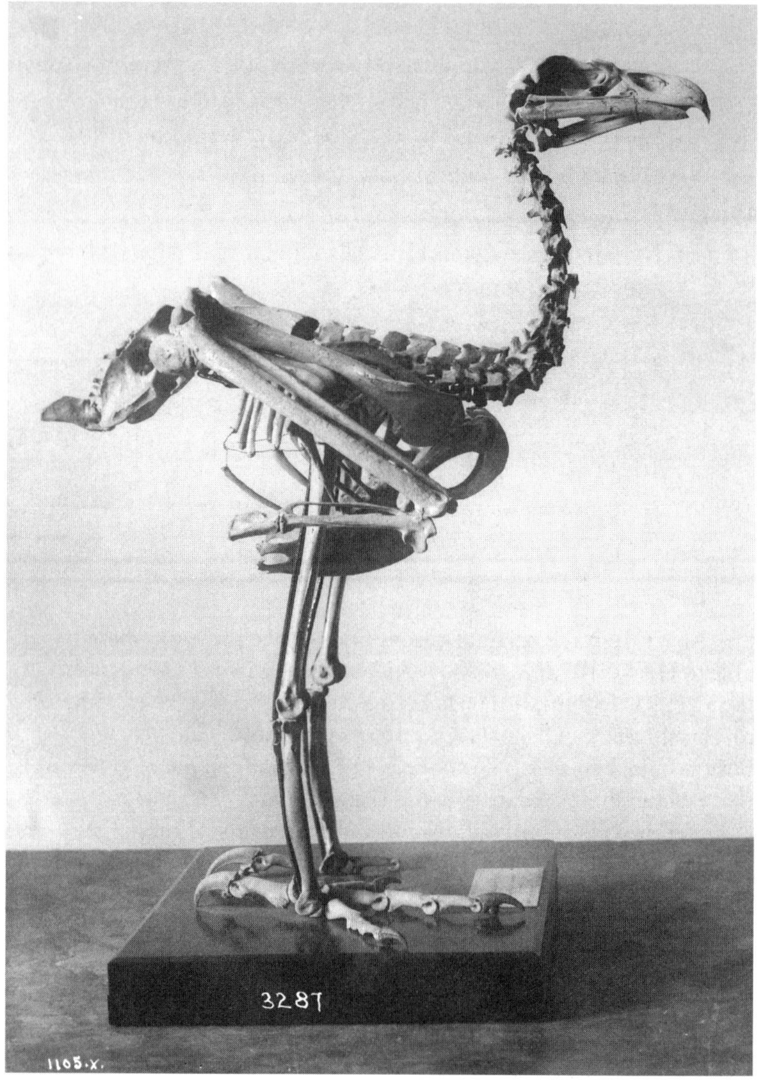

Figure 8.8. Photograph of mounted composite specimen from Castle Rocks, western Southland (OM C40.8, Av7473). Bones of darker tone (mainly pedal phalanges) are from a large individual represented by material in the Museum of New Zealand Te Papa Tongarewa (S2134). Collection of Museum of New Zealand Te Papa Tongarewa.

lithographs of the type series; the original material never left New Zealand. The dimensions of the bones are shown in Tables 8.1–8.9, with measurements made to bone landmarks that are illustrated in Figures 8.19 and 8.20.

Skull and Beak

The most characteristic feature of the skull of Haast's eagle (Figs. 8.1, 8.14) is the extension of the cranium and premaxilla in the region anterior to the orbit. The nasal region has in effect been stretched out, with the main bone struts lying at shallower angles to the axis of the skull than in other eagles. The main elongation relative to the proportions of the skull in *Aquila audax,* in contrast, is in the posterior cranium and nasal regions. There is no prefrontal process of the frontal. The cranium and rostrum (Figs. 8.1, 8.14) are elongated and com-

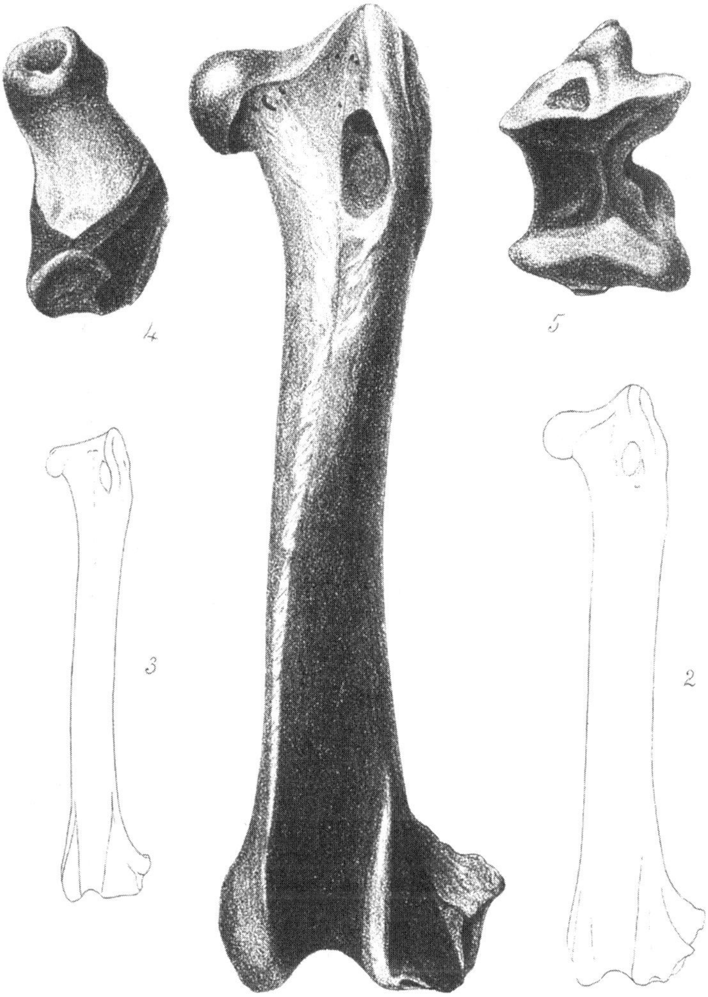

Figure 8.9. Lithograph of part of the type series of Harpagornis moorei. *Figured in Haast (1872c: plate X), in comparison with femora of* Haliaeetus leucogaster *(extant Australian white-bellied sea eagle) and* Circus approximans *(extant Australasian harrier).* Harpagornis moorei. *Left femur (CM Av5104) (1), cranial; (4) proximal; (5) distal.* Haliaeetus leucogaster. *Left femur: (2) cranial.* Circus approximans. *Left femur: (3) cranial. Drawn by Dr. Llewellyn Powell, pioneer physician and scientist of the European settlement at Christchurch, from the type series.*

pressed dorsoventrally. The line of the dorsal surface is more or less continuous across the cranium and premaxilla, with no break or angle at the base of the beak as in *A. audax*. Anteriorly, the maxillary process of the nasal in *Harpagornis* is broad, and its breadth is about equal to the depth of the internarial; in *A. audax*, it is much narrower than the depth of the internarial bar. In addition, the process has subparallel sides and lies at about 30° to the dorsal margin of the rostrum; in *A. audax*, the angle is 45–50°. In *Harpagornis*, the external nostrils are more than twice as long as they are high, but they are less than twice the height in *A. audax*.

The rostrum is elongated toward the tip. The curvature of the dorsal margin (as seen in lateral view) of the rostrum is centered about 0.3–0.5 times the depth of the rostrum below the ventral (tomial) margin of the premaxilla itself (as against on or dorsal to the margin in *A. audax*). Hence there is a much more gradual downward curvature

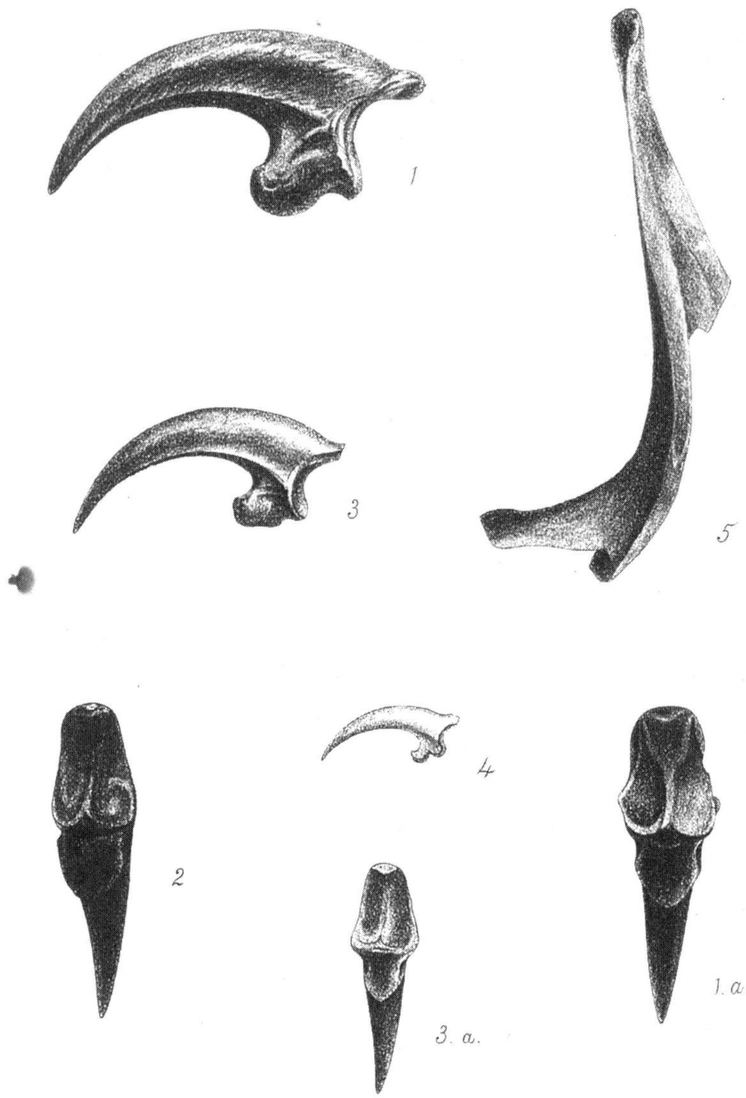

Figure 8.10. Lithograph of part of the type series of Harpagornis moorei. *Figured in Haast (1872c: plate XI), in comparison with ungual phalanx of* Aquila audax *(extant Australian wedge-tailed eagle) and of* Circus approximans *(extant Australasian harrier).* Harpagornis moorei. *Pedal digit I, ungual phalanx, left (CM Av5104): (1) lateral; (1a) proximal. Pedal digit II?, ungual phalanx, right (CM Av5104): (2) lateral. Rib III, right (CM Av5104): (5) cranial.* Aquila audax: *Pedal digit I, ungual phalanx, left: (3) lateral; (3a) proximal.* Circus approximans. *Pedal digit I, ungual phalanx, left: (4) lateral. From drawings by Dr. Llewellyn Powell.*

toward the tip. The ridge above the nostrils is swollen and strongly bowed. Inside the orbit, the bridge over the nasal artery and nerve is narrow and simple, with no folding. The ectethmoid plate (Fig. 8.1) is small and arises ventral to the lower margin of the nasal arterial foramen; its ventral margin is parallel to the skull axis. The nostril aperture is ossified anteriorly, dorsally, and ventrally, forming a partial tube, which contrasts with the condition in most species of *Aquila*. The remainder of the aperture is not covered with a bony plate, as it is in the large vultures such as *Aegypius* and *Gyps*. The posterior margin of the orbit is almost straight and is set at right angles to the skull axis to the tip of the postorbital process (Fig. 8.1), which tapers to a blunt point. In *Aquila,* there is a broad notch in the margin. There is no interorbital foramen.

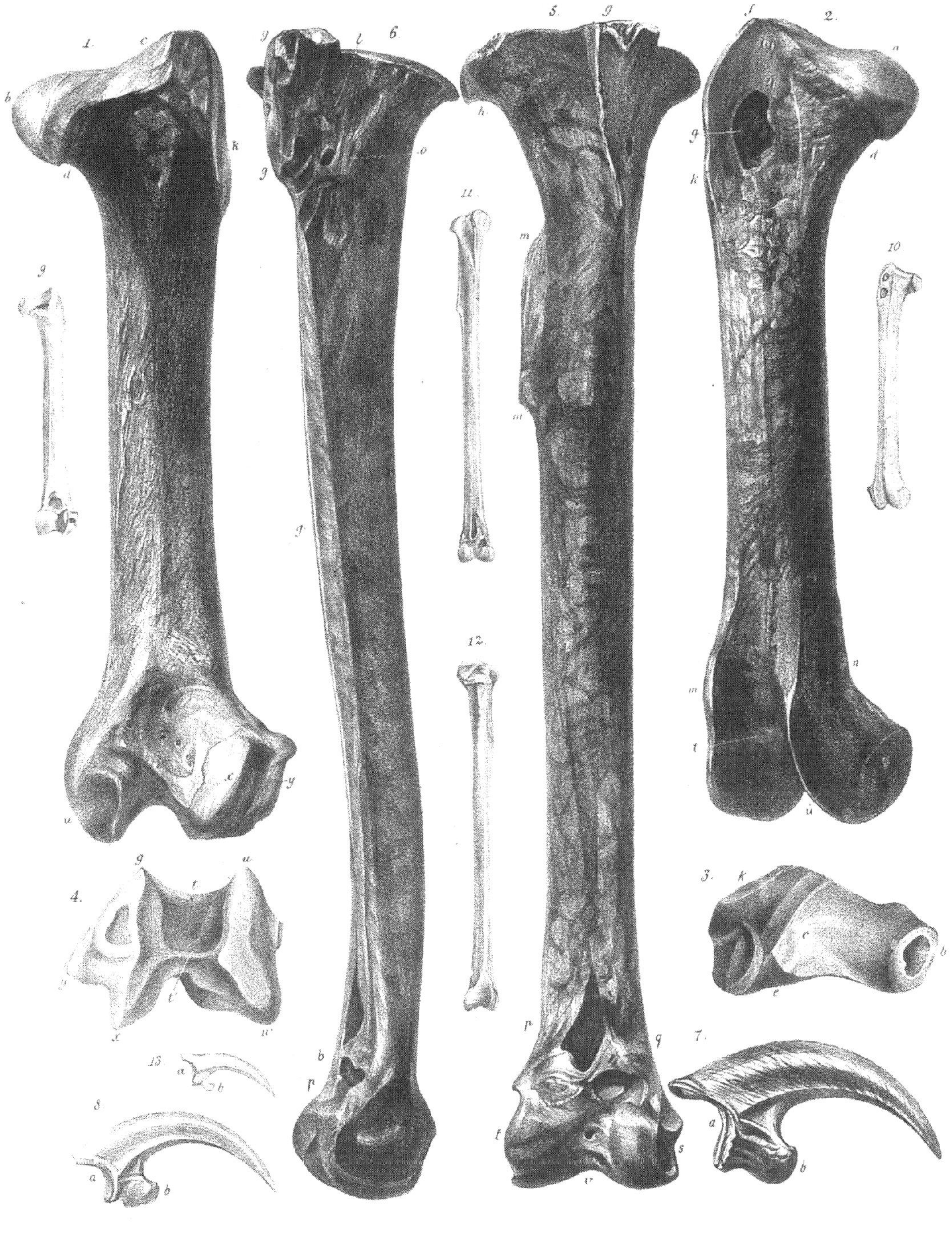

Figure 8.11. Owen (1879a: plate CVII). Harpagornis moorei. *Right femur: (1) caudal; (2) cranial; (3) proximal; (4) distal. Right tibiotarsus: (5) cranial; (6) medial. Pedal phalanx, digit I: (7) lateral.* Aquila audax *(extant Australian wedge-tailed eagle). Pedal digit I, ungual phalanx: (8) lateral.* Circus pygargus *(extant Montagu's harrier). Right femur: (9) caudal; (10) cranial. Right tibiotarsus: (11) cranial; (12) caudal.* Circus approximans *(extant Australasian harrier). Pedal digit I, ungual phalanx: (13) lateral. Lithographs prepared from photographs and lithographs of type series of* Harpagornis moorei *and* Harpagornis assimilis *(=* H. moorei *male); the original material never left New Zealand.*

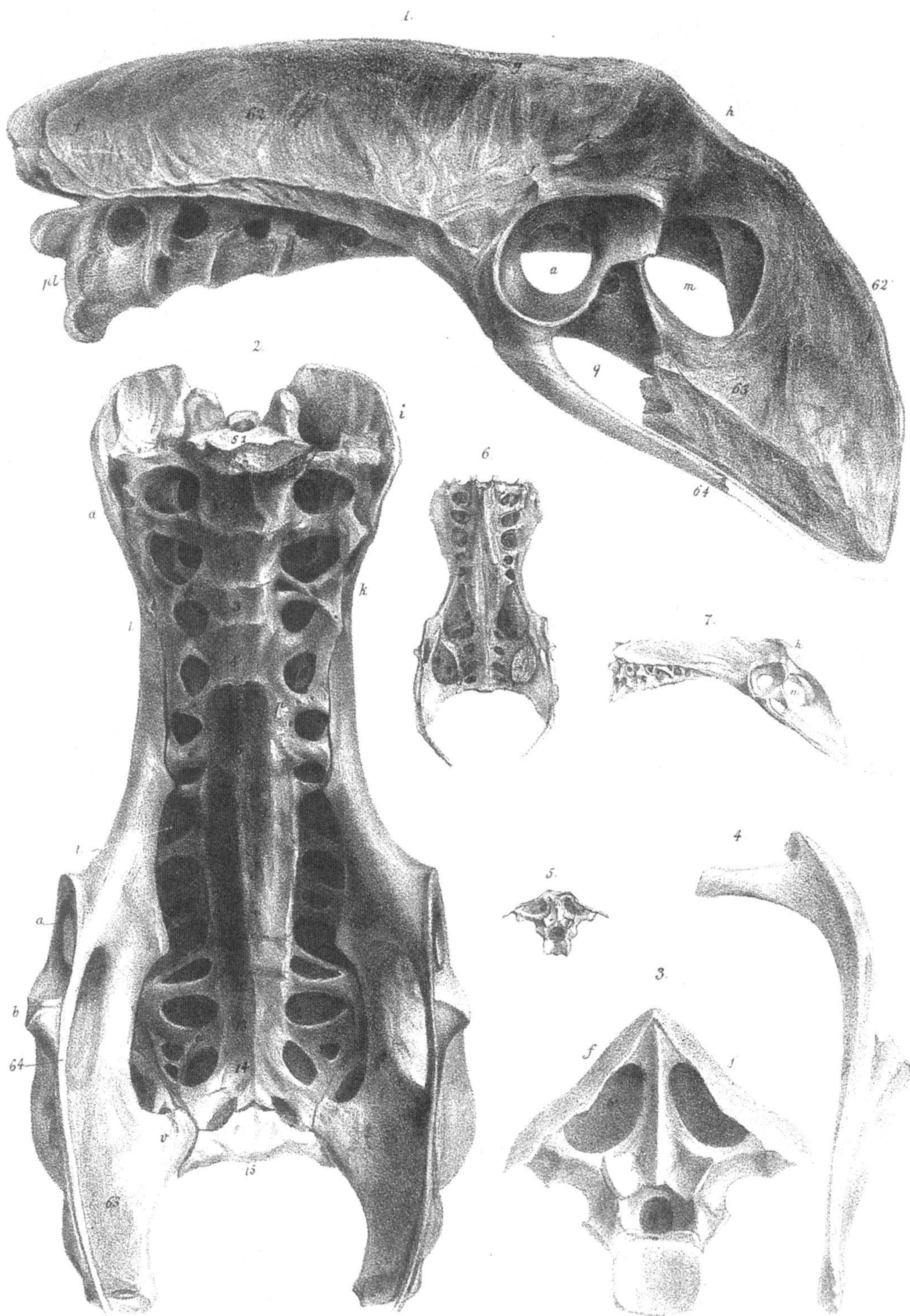
1.
62
h
a
m
62
pl
q
63
64
2
51
i
a
k
6
7.
4
5.
3.
f
f
a
b
64
v
14
15
63

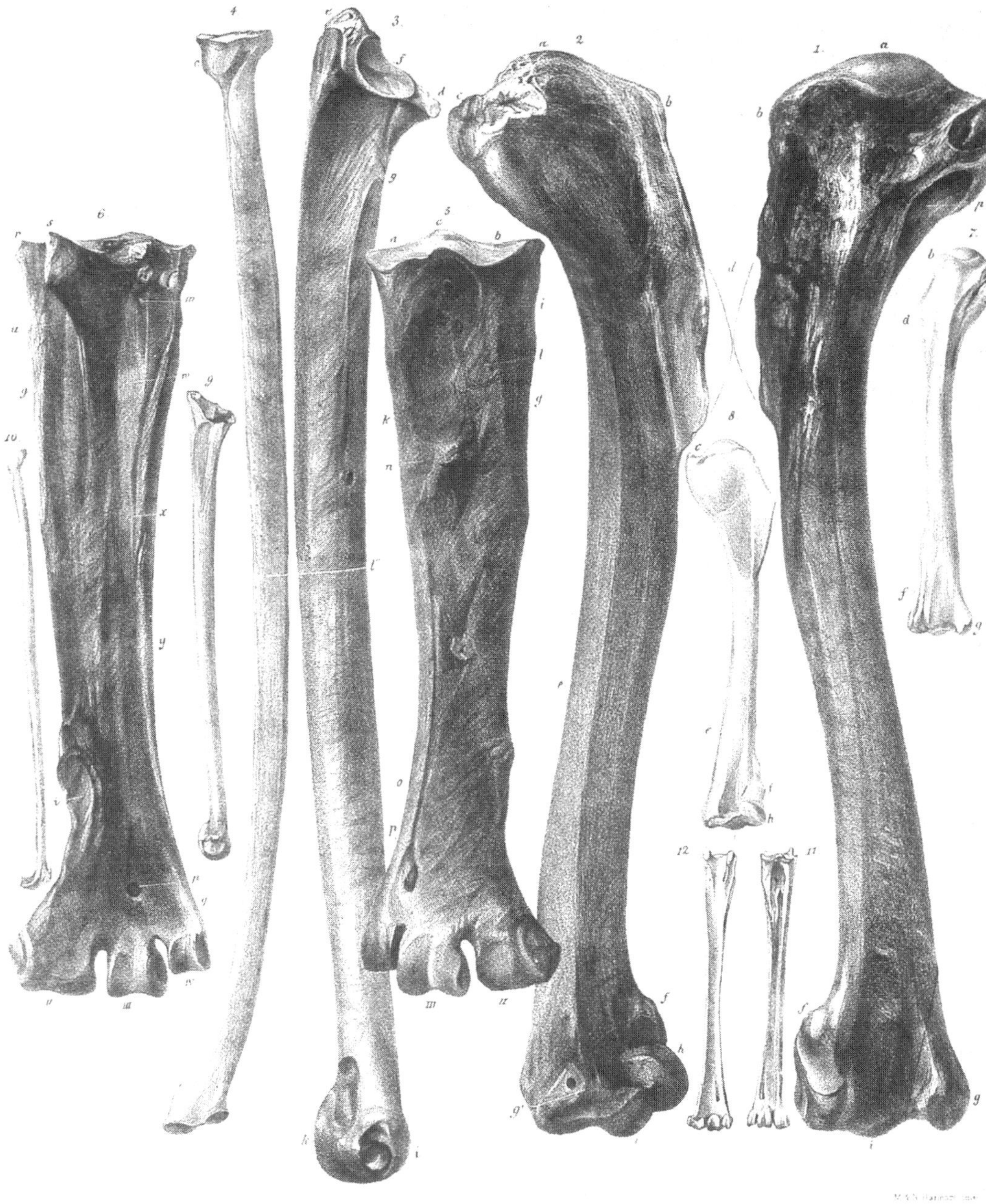

Figure 8.12. (opposite page) Owen (1879a: plate CV). Harpagornis moorei. *Pelvis: (1) left lateral; (2) ventral; (3) cranial. Left thoracic rib: (4) cranial.* Circus approximans *(extant Australasian harrier). Pelvis: (5) cranial; (6) ventral; (7) left lateral. Lithographs prepared from photographs and lithographs of type series of* Harpagornis moorei *and* Harpagornis assimilis *(=* H. moorei *male).*

Figure 8.13. Owen (1879a: plate CVI). Harpagornis moorei. *Left humerus: (1) cranial; (2) caudal. Left ulna: (3) cranial. Left radius: (4) caudal. Right tarsometatarsus: (5) cranial; (6) caudal.* Circus approximans *(extant Australasian harrier). Left humerus: (7) caudal; (8) cranial. Left ulna: cranial (9). Left radius: (10) caudal. Right tarsometatarsus: (11) cranial; (12) caudal. Lithographs prepared from photographs and lithographs of type series of* Harpagornis moorei *and* Harpagornis assimilis *(=* H. moorei *male).*

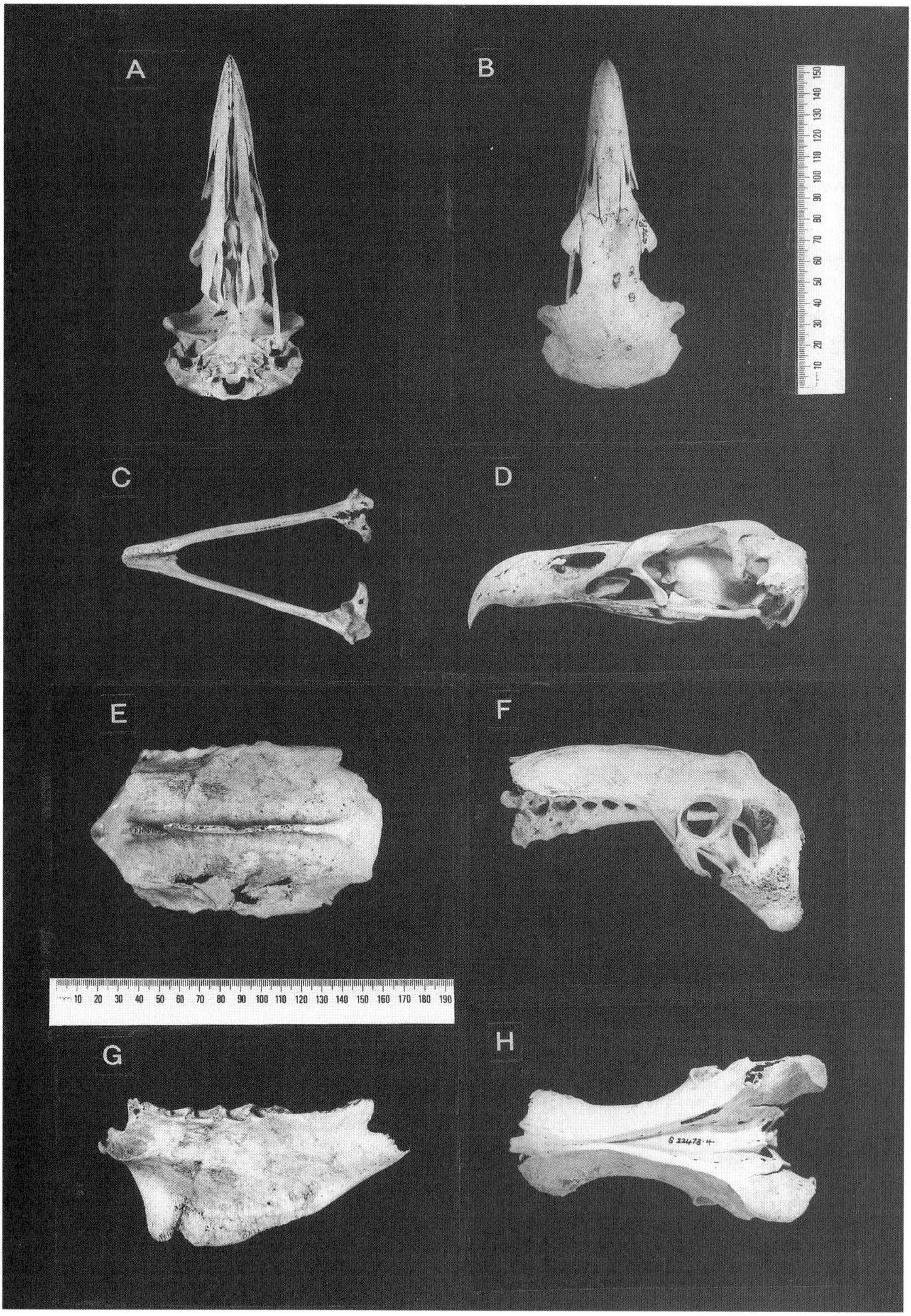
A
B
C
D
E
F
G
H
S 22473.4

At the rear of the skull, the temporal fossae (Fig. 8.1) are deep, providing space for large muscles, and the posteroventral angle of each fossa is acute. The rear of the cranium is produced and the cerebellar prominence is large, but the parietal and supraoccipital regions are straight in lateral outline, with the surfaces meeting at definite angles and not pronounced curves.

The superciliary plate (Fig. 8.1) is semielliptic and not fused to the lachrymal. The lachrymal consists of a dorsal posterior process above the orbit (supraorbital), a distally directed ventral process, and anteriorly articulates with the nasal. The lachrymal is modified from that typical in *Aquila* in that the dorsal surface of the supraorbital process is deeply troughed and not planar. The anterior angle of the lachrymal lies beside the maxillary process of the nasal, not within a notch as it is in other aquilines. The supraorbital process (Fig. 8.1) has a sharp process on the frontal margin and is notched at the rostral end of the flared distal blade. Both features are lacking in *Aquila,* which has instead a deep caudal incision on the process. The supraorbital process of the lachrymal, as measured from the anterior angle, is slightly shorter than the ventral process, and the greatest width of the lachrymals (across the supraorbitals) extends only half the width of the orbit as defined by the postorbitals.

In the palate, the posteropterygoid (Fig. 8.1) is proportionately much broader than in *Aquila.* The rostral side flange is also much longer, extending more than 50% of the bone's total length caudally. The body of the bone is deeply fluted ventrally and domed dorsally. The vomer is not known.

The quadrate (Fig. 8.1) is typical of that of *Aquila* eagles. Differences from the *A. audax* quadrate include the following: the orbital process is compressed; the tip is squared off; and the tip is equal in length to the otic process. In *A. audax,* the orbital process is shorter than the otic process, and the tip is rounded.

Tongue bones (Fig. 8.2) of the eagle are known from sites of Pleistocene and Holocene age. Their preservation gives some idea of the stability and quality of sites in New Zealand caves. The bones are unremarkable except for their size. The caudal process of the entoglossum is slender, and the rostral basibranchial is twice as long as the entoglossum. The total length of the hyobranchial apparatus is about 70% of the length of the mandibular ramus.

The mandible (Figs. 8.2, 8.14) is, according to Oliver (1930, 1955), rather narrow, and the symphysis is channeled. The medial articular process has a pneumatic foramen. The postarticular process is short and subrectangular and is at the base of the medial process. The medial process itself tapers gradually to a point, with the rostral edge of the process at about 90° to the axis of the ramus; its caudal margin curves gently toward the end of the mandible.

Figure 8.14. (opposite page) Harpagornis moorei, *MNZ S27773, cranium, mandible, sternum, pelvis: (A, B, D) Cranium in ventral (A), dorsal (B), and left lateral (D) views. (C) Mandible, dorsal view. (E, G) Sternum in ventral, anterior to left (E), and left lateral (G) views. (F, H) Pelvis in left lateral (F) and (H) dorsal views. Scale in millimeters. Photograph by J. Palmer.*

Axial Skeleton: Vertebrae and Pygostyle

The vertebrae (Fig. 8.6) are typical of accipitrids (Pycraft 1902) except for the great dorsoventral depth of the thoracic series and the greater size of the notches for nerve roots passing to the base of the wings and legs. The pygostyle (Fig. 8.2) is large, and although the sample size is small, there appear to be differences in shape between the sexes. In view of the possible differences in flight pattern suggested by the analysis of wing proportions and body masses, different areas for the attachment of tail feathers and different development of the muscles controlling the tail might be expected.

Pectoral Girdle and Wings

All the bones of the pectoral girdle and wings are robust and show no sign of being reduced or in any way deficient in strength or area for muscle attachment.

The sternum (Figs. 8.3, 8.14) is robust, with a large, thick carina or keel. The scar for the supracoracoideus muscle is well defined, as noted by Brathwaite (1992). The supracoracoideus muscle raises the wing but is critical only for vertical take-offs (Sy 1936). The deepest part of the carina is anterior, as in *Aquila,* not toward the center of the ventral margin, as in *Gyps* or *Aegypius*; the carina is also thicker than in the vultures. The pila carina (anterior pillar) is especially thick, as in *Aquila.*

In *Harpagornis,* the depth of the muscle attachment is maintained by the sternal "bowl" not extending ventrally, whereas in the vultures, which soar, the deeper bowl makes the carina shallower. The costal margin, to which the ribs attach, is half to two thirds the length of the anterior margin. There are six costal articulations; the coracoidal sulcus reaches second costal articulation from the front. The supracoracoidal sulcus reaches the fourth. The spina externa is deeper than broad in *Harpagornis* and narrower in *Aquila.*

In ventral view, the ventral margin of each coracoidal sulcus is more posterior than in *Aquila,* where it obscures the dorsal margin. In dorsal view, the anterior pillar is visible inside the sternal bowl as a flat-topped ridge with lateral furrows, but the pillar is not visible like this in *Aquila.*

Coracoid. The basic form of the coracoid (Figs. 8.2, 8.15) is typical for an accipitrid, but as with the other bones, it is much more massive. The articular facets are so placed that the lateral angle between the scapula and coracoid is more acute when articulated than in *Gyps* or other vultures, and it is at least as acute as in *Harpia.* There is a prominent triangular bulge on the dorsal surface of the coracoid for the major sternal ligament. This bulge in more sternally and laterally located in *Gyps,* and in *Aquila,* it is not raised. The procoracoid is prominent and is not deflected at the tip, as in vultures. All coracoids examined have a procoracoidal foramen.

Figure 8.15. (opposite page) Harpagornis moorei, *humerus, ulna, radius, carpometacarpus, scapula, coracoid (all MNZ S27773). (A, B) Left humerus in (A) caudal and (B) cranial views. (C, D) Right ulna in (C) cranial and (D) ventral views. (E, F) Left radius in (E) dorsal and (F) ventral views. (G, H) Right carpometacarpus in (G) ventral and (H) dorsal views. (I, J) Left scapula in (I) medial and (J) lateral views. (K, L) Right coracoid in (K) dorsal and (L) ventral views. Scale in millimeters. Photograph by J. Palmer.*

Scapula. The neck of the scapula is long (about 25% of the total length) and the dorsal margin of the blade is approximately in line with the neck. The angle in the blade is rather sharp (~30°), and the caudal margin is convex. There is a large pneumatic foramen beneath the acromion (Figs. 8.2, 8.15), which is broad and horizontal and has a foramen under the medial edge.

Humerus. As Haast (1874a) noted, the shaft of the humerus has a stronger sigmoid curve in the anteroposterior plane than in *Aquila,* much of the additional curvature being at the end of the pectoral crest

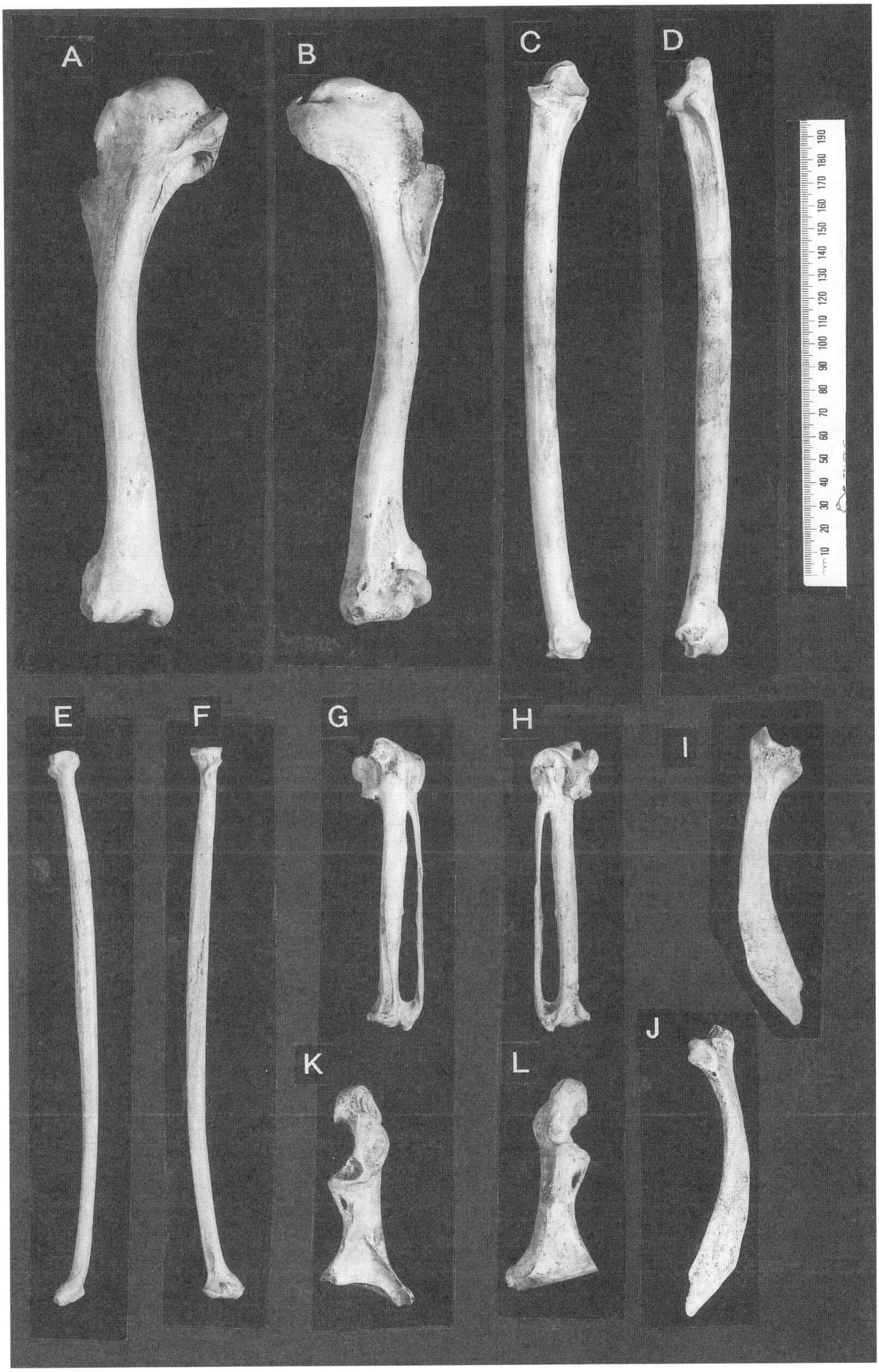
A
B
C
D
E
F
G
H
I
J
K
L

(Figs. 8.4, 8.15). All the muscle and ligament attachments are consistent with the attachment of strong flight muscles. All intermuscular scars are prominent, and the crests are thick and broad, not reduced in any way. In length, it is a greater proportion of the ulna length than in *Aquila.*

Ulna. As both Haast (1874a) and Owen noted, the ulna (Figs. 8.4, 8.15) is not especially long for the size of the bird, but it is thick and the articulations are large in proportion to the shaft width. Owen (1879a: 146) pointed out that what he called the "distal articular convexity indicates the extent of the evolutions of the manual part of the wing, with its great 'primary remiges,' in the action of flight." The bicipital attachment is level with the end of the prominence for the anterior articular ligament. The impression of the brachialis anticus extends more than one quarter the length of the shaft from the proximal end; the nutrient foramen (perforation for the nutritive artery) is about 33% along the shaft from the proximal end. There are 13 or 14 obvious papillae for secondary feathers on the caudal margin. At the distal end, there is a large carpal tuberosity, the tendinal pit is rectangular, and the groove is narrow.

Radius. Owen (1879a) noted that the radius (Fig. 8.15) is thicker in Haast's eagle in proportion to its length than in other eagles. Both Haast and Owen commented that the distal end was at a greater angle to the shaft than in other species. A thicker shaft and greater radius of curvature would provide a stouter wing, able to take larger loads.

Carpometacarpus. As with the other wing bones, the carpometacarpus (Figs. 8.4, 8.15) is a heavier version of the bone in other accipitrids, but it is proportionately shorter than in *Haliaeetus, Gyps,* and other soaring birds. The pisiform process is large, square-ended, and prominent. The pollical facet is notched proximally and distally, so that the surface is divided almost in two. The tendinal groove in the shaft is distinct and bifurcates about 25% of the length from the distal symphysis. The posterior carpal fossa is distinct and rectangular, and it reaches almost to the base of metacarpal III.

The ulnare and radiale are known from several specimens, as are the alar phalanges. They are typical for accipitrids but more robust.

Pelvic Girdle and Legs

In the pelvis (Figs. 8.3, 8.14), the prepostacetabular angle is steeper than in *Aquila.* The ilioischiadic foramen is produced posterodorsally to make a "rounded triangle," not oval, as in *Aquila.* There is a pit in the ischial bar immediately posterior to the acetabulum. The anterior iliac plates are produced dorsally above the synsacrum, leaving a deep dorsal depression in an area that is almost flat in *Aquila.*

The femur is massive, but apart from that, it is similar to that in other eagles (Figs. 8.5, 8.16). The size of the articulations increases with the size of the bone, so they are proportionately larger than in other accipitrids. The head is round and has a deep pit for the attachment ligament, which is offset to the proximal side. The shaft has the characteristic parasagittal bend (convex upward when the bone is horizontal), and the rotular groove for the extensor tendon across the knee is nar-

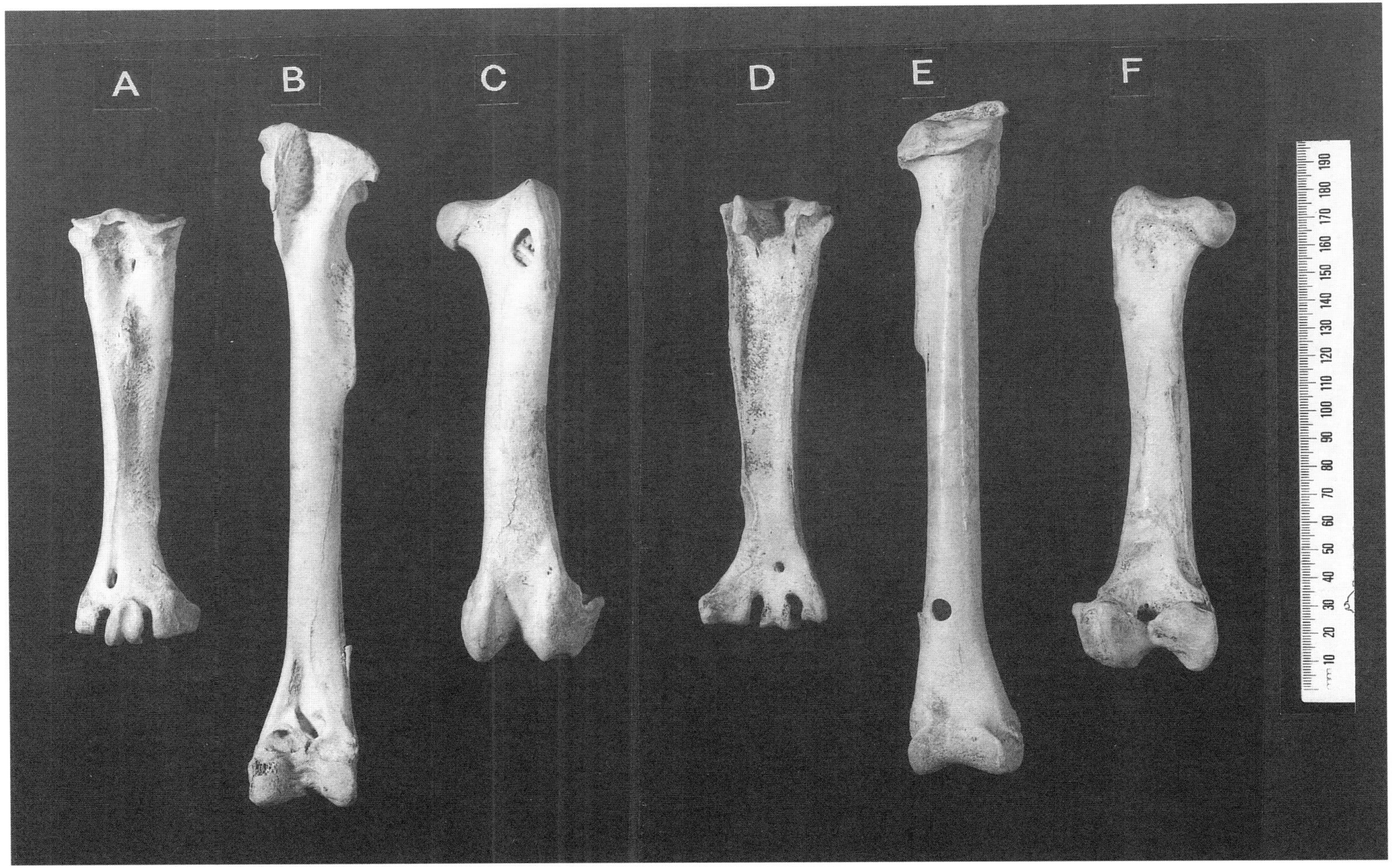

Figure 8.16. Harpagornis moorei, *femur, tibiotarsus, tarsometatarsus (all MNZ S27773). (A, D) Right tarsometatarsus in (A) cranial and (D) caudal views. (B, E) Left tibiotarsus in (B) cranial and (E) caudal views; the hole in caudal surface was the sampling site for mitochrondrial DNA extraction. (C, F) Left femur in (C) cranial and (F) caudal views. Scale in millimeters. Photograph by J. Palmer.*

row and deep. At the proximal end, the pneumatic foramen is large and "ear-shaped" (Haast 1872c: 193). Haast (1872c: 193) also noted that the lateral side of the trochanter is "very rough, showing that muscles of great strength and thickness must have been attached to it."

The tibiotarsus (Figs. 8.5, 8.16) has broad articular surfaces, and the medial cnemial crest is prominent. The shaft is straighter than in *Aquila,* although it is still slightly bent caudally at the proximal end. The fibular crest is relatively short but broad, and the fibula and tibiotarsus together form a broad junction with the distal femur. There are correspondingly conspicuous attachment surfaces for the crucial ligaments as the pits on the distal end of the femur. At the distal end, the supratendinal bridge narrows abruptly.

In the tarsometatarsus, the great strength of the leg and foot is most apparent. The bone (Figs. 8.5, 8.16) is much larger than in any *Aquila.* Particularly noticeable are the width and depth of the shaft, as well as the size of the condyle for the second toe. The cranial surface of the bone is more deeply excavated than in *Aquila,* leaving space for larger tendons and providing a stronger section to the bone shaft. The lateral hypotarsal crest is rectangular in caudal view, not triangular as in *Aquila;* similarly, the medial crest is about 25% of the width of the tarsometatarsus, which is more than in *Aquila.* Other differences are that the lateral flange for the condyle of the fourth toe flares laterally rather than being curved in, and the cranial edge of the internal cotyla overhangs the cranial fossa at the proximal end, which it does not do in *Aquila.* The attachment surface for the metatarsal of digit I is long and broad, implying a strong attachment to resist high loads imposed on the toe.

The metatarsal of the first toe is itself large: when articulated with the tarsometatarsus, the size of the metatarsal causes the base of the leg to be as deep as it is broad, with a large tunnel for the flexor tendons. Apart from their great size, the main feature of the pedal phalanges (Fig. 8.17) is the dominance of digits I and II. Although the ungual phalanges on the third and fourth toes are almost equal in size to the largest in *Aquila audax,* they are still much smaller than the first two. The first two phalanges of the second toe are not fused as in *Haliaeetus* (Olson 1982), but their articulation admits of almost no movement. When the toes grasp, the main load would be applied between the first and second toes. In Haast's eagle, the basal ventral prominence for the flexor tendon is the largest of any accipitrid, and the base of the claw blade itself is exceptionally deep.

Remarks

In most respects, *Harpagornis moorei* was a typical aquiline eagle. The main points of difference are those associated with a substantial increase in size. For example, the head length is appropriate for the body size of an accipitrid, but the added length is not a result of an overall increase in all parts of the head. Instead, it is obtained by the elongation of the proximal part of the beak, with greater angles between the bony struts; the width of the struts has increased to maintain

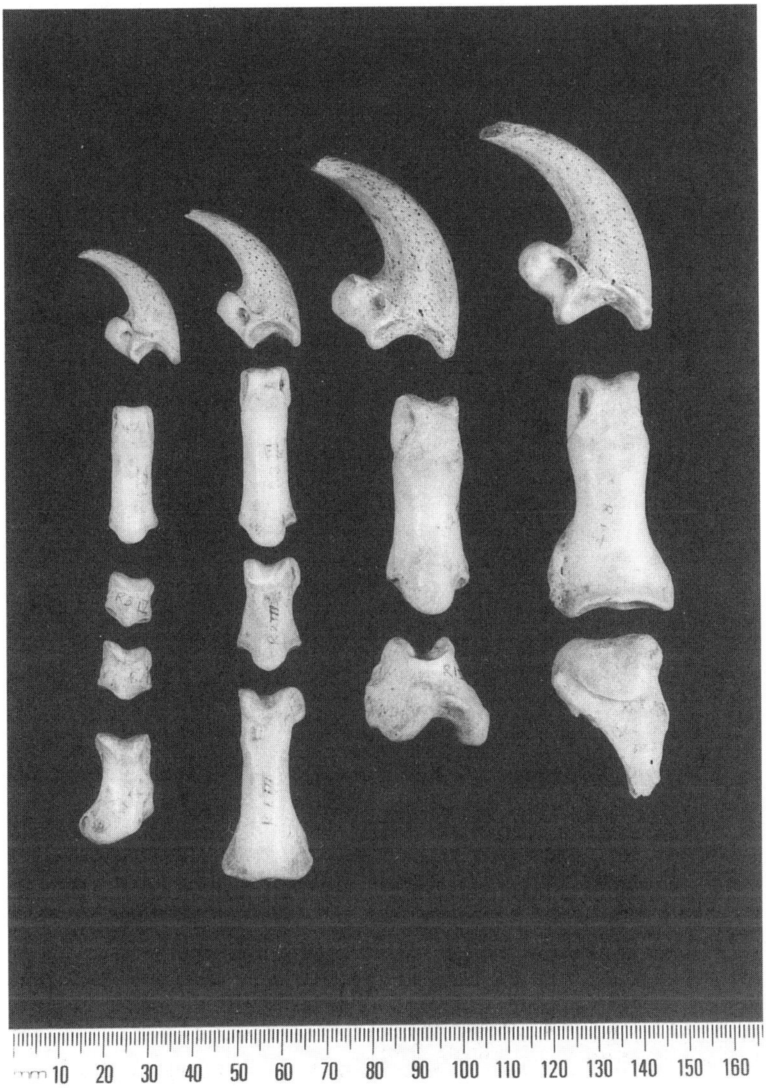

Figure 8.17. Harpagornis moorei, *right pedal phalanges and right first metatarsal (MNZ S27773). Left to right: toe IV, toe III, toe II, toe I (hind). Scale in millimeters. Photograph by J. Palmer.*

adequate strength. In the postcranial skeleton, the main differences from others in its lineage (and similarities with the harpy and other forest-dwelling eagles) is in the proportions of the wing bones. The ulna is relatively short, which shortens the wing and so reduces the oscillating mass and brings the center of the mass closer to the shoulder pivot. The pelvis is hypertrophied, and the ilioischiadic bend is the most acute in any accipitrid. That bend allows an arrangement of the upper leg muscle better suited to perching and gripping than to walking about on the ground, for which a straight or slightly bent pelvic axis is optimal.

The lower leg and toes are extremely robust but are similar in morphology and proportions to those in a typical *Aquila*. With these differences, Haast's eagle is separated morphologically from the aquilines by features associated with flapping rather than soaring flight, and

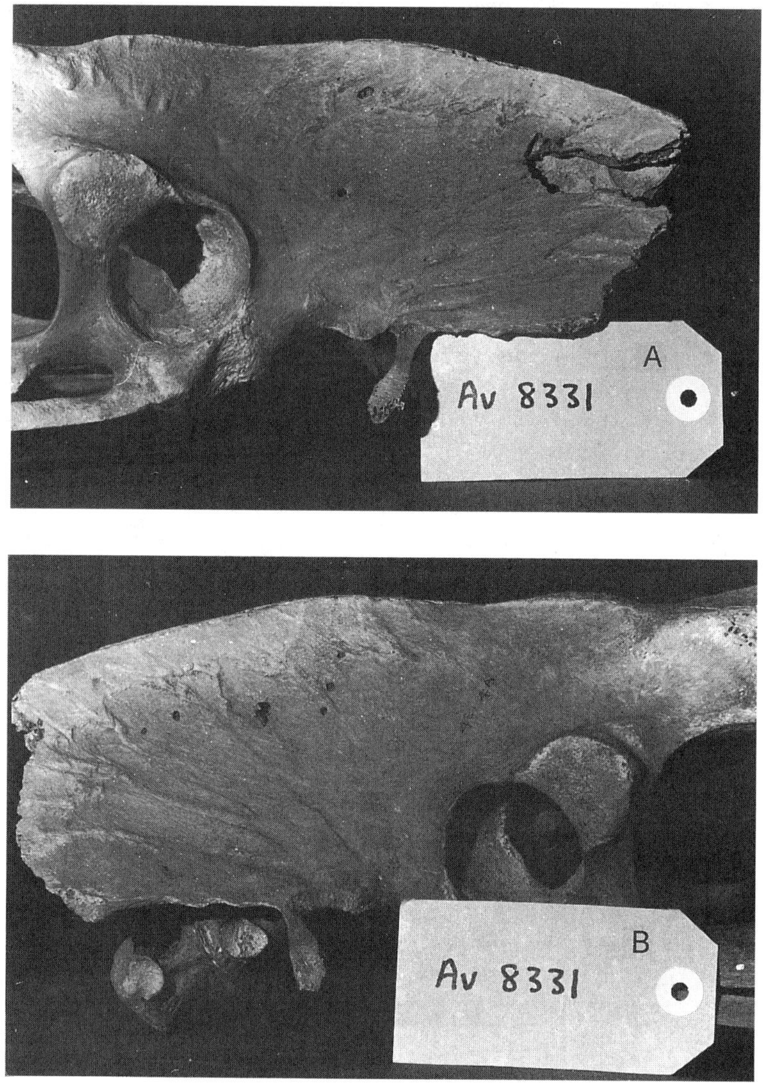

Figure 8.18. Anterior pelvis of Emeus crassus *(CM Av8331) from Pyramid Valley, North Canterbury, showing (A) mark of hind claw of Haast's eagle (*Harpagornis moorei*) and notch in anterior iliac plate caused by beak of eagle (or kea,* Nestor notabilis*) when excavating bone to expose tissue beneath, and (B) holes made by claws of three anterior toes (II–IV) of eagle while gripping pelvis. Photographs by R.N.H.*

with the catching and subduing of prey as large, or larger than, itself. Once that prey was caught, the elongated beak allowed it to reach into the carcass for the favored parts. Evidence from damage to moa pelves suggests that the eagle could open up the anterior intrapelvic cavity with its beak, to reach the kidneys and kidney fat of moa (Fig. 8.18).

Beyond Bones: History and Biology of Haast's Eagle

Discovery

The excavations at Glenmark Swamp in North Canterbury during 1868 gave Julius Haast the means to expand his new museum in Christchurch. Moa skeletons assembled from the huge collection of bones were sent around the world and in return for £20,000 worth of speci-

TABLE 8.1.
Dimensions (mm) of the cranium and premaxilla of *Harpagornis moorei*.

Statistic	Length	PML	PMW	POW	IOW	WTF	WSQ	CRD
Mean	159.6	83.6	20.8	70.2	26.0	50.4	66.4	47.1
Minimum	150.5	69.6	18.7	65.9	23.5	48.6	61.1	44.6
Maximum	166.9	91.0	23.8	76.0	28.3	53.7	70.2	50.0
SD	6.61	7.14	1.49	2.94	1.62	1.59	3.52	1.93
SE_{mean}	2.20	2.38	0.47	0.89	0.45	0.46	1.33	0.73
CV (%)	4.14	8.53	7.17	4.19	6.24	3.116	5.31	4.08
n	9	9	10	11	13	12	7	7

Abbreviations: CRD = cranial depth; CV = coefficient of variation; IOW = interorbital width; PML = premaxilla length to nasofrontal hinge; PMW = premaxilla width; POW = postorbital width; SD = standard deviation; SE_{mean} = standard error of the mean; WSQ = width at squamosal; WTF = width at temporal fossa.

TABLE 8.2.
Dimensions (mm) of mandible of *Harpagornis moorei*.

Statistic	Length	W	SL	WA	DA	DS	TC	TS	Dsur
Mean	122.7	73.0	25.5	27.5	9.8	14.7	103.7	99.7	9.0
Minimum	116.0	67.1	20.7	25.5	9.0	13.4	99.4	95.8	7.6
Maximum	130.8	77.6	27.6	29.2	10.3	15.5	111.7	107.1	10.2
SE_{mean}	1.96	1.26	0.66	0.42	0.18	0.27	1.90	1.69	0.36
CV (%)	4.52	4.47	8.56	4.87	5.28	5.23	4.48	4.15	11.17
n	8	7	11	10	8	8	6	6	8

Abbreviations: CV = coefficient of variation; DA = depth at anterior end of articular process; Dsur = depth at surangular process; SE_{mean} = standard error of the mean; SL = length of symphysis; TC = tip to coronoid process; TS = tip to surangular process; W = width; WA = width of articular process.

TABLE 8.3.
Dimensions (mm) of sternum of *Harpagornis moorei*.

Statistic	Length	Wstc	Wprox	Dtot	Dant	Lcar	Wpst
Mean	153.2	82.6	88.1	71.5	55.2	124.2	59.7
Minimum	140.3	63.1	83.5	57.5	47.4	110.3	55.2
Maximum	167.2	93.9	96.7	77.4	63.8	136.2	64.8
SD	12.48	11.680	5.29	4.78	2.25	13.05	4.83
SE_{mean}	5.58	5.22	2.37	2.39	2.14	7.53	2.79
CV (%)	8.15	14.15	6.0	6.68	9.51	10.51	8.09
n	5	5	5	4	6	3	3

Abbreviations: CV, coefficient of variation; Dant = depth over anterior margin; Dtot = total depth; Lcar = length of carina; SD, standard deviation; SE_{mean} = standard error of the mean; Wprox = anterior width; Wpst = posterior width; Wstc = width over sternocoracoidal processes.

TABLE 8.4.
Dimensions (mm) of pelvis of *Harpagornis moorei*.

Statistic	Length	Pw	Ww	Wint	Wpo	Wis
Mean	173.8	58.2	37.4	56.6	81.6	77.4
Minimum	157.0	52.4	33.1	50.7	75.4	59.4
Maximum	197.5	66.5	42.7	61.9	91.6	94.5
SD	16.51	6.05	3.71	4.28	6.82	17.57
SE_{mean}	7.38	3.03	1.40	1.75	2.78	10.14
CV (%)	9.5	10.40	9.92	7.55	8.35	22.7
n	6	3	6	6	4	3

Abbreviations: CV = coefficient of variation; Pw = proximal width; SD = standard deviation; SE_{mean} = standard error of the mean; Wint = width acetabulae; Wis = width across posterior ilioischiadic crest; Wpo = posterior width; Ww = width at waist.

TABLE 8.5.
Dimensions (mm) of coracoid of *Harpagornis moorei*.

Statistic	Length	Wf	Wn	Wst
Mean	92.2	27.9	15.8	38.1
Minimum	78.1	22.5	13.4	31.9
Maximum	106.6	31.4	17.8	45.2
SD	9.55	2.65	1.61	5.89
SE_{mean}	2.45	0.76	0.39	1.96
CV (%)	10.36	9.47	10.17	15.44
n	15	12	17	9

Abbreviations: CV = coefficient of variation; SD = standard deviation; SE_{mean} = standard error of the mean; Wf = width of furcular facet; Wn = width of neck; Wst = width of sternal facet.

TABLE 8.6.
Dimensions (mm) of scapula of *Harpagornis moorei*.

Statistic	Length	Wh	Wn	Wb	Lg
Mean	127.5	29.7	11.6	19.2	18.4
Minimum	116.7	25.6	10.7	16.8	14.0
Maximum	142.3	33.6	13.3	20.9	20.4
SD	9.39	2.62	0.82	1.72	2.35
SE_{mean}	3.83	0.83	0.26	0.70	0.96
CV (%)	7.36	8.83	7.07	8.96	12.75
n	6	10	10	6	6

Abbreviations: CV = coefficient of variation; Lg = length of glenoid; SD = standard deviation; SE_{mean} = standard error of the mean; Wb, maximum width of blade; Wh, width of head; Wn, width of neck.

TABLE 8.7.
Dimensions (mm) of furcula of *Harpagornis moorei*.

Statistic	W	Hdv	Hap	Wart
Mean	103.5	75.0	40.90	14.5
Minimum	95.3	63.0	33.7	13.9
Maximum	110.0	85.2	46.8	15.9
SD	5.96	7.62	6.64	0.82
SE_{mean}	2.67	3.41	3.83	0.37
CV (%)	5.76	10.16	16.22	5.62
n	5	5	3	5

Abbreviations: CV = coefficient of variation; Hap, "height" anteroposteriorly; Hdv, dorsoventral height; SD = standard deviation; SE_{mean} = standard error of the mean; W, overall width; Wart, width of articular facet.

TABLE 8.8.
Dimensions (mm) of long bones of the wing of *Harpagornis moorei*.

Bone	Statistic	Length	Proximal	Shaft	Distal
Humerus	Mean	232.4	52.7	41.3	17.9
	Minimum	208	42.3	37.1	15.6
	Maximum	259.9	58.3	47.3	21.0
	SD	16.25	4.53	3.01	1.69
	SE_{mean}	3.63	1.07	0.67	0.37
	CV (%)	6.99	8.59	7.27	9.46
	n	20	18	21/20	21
Ulna	Mean	254.4	29.0	—	19.1
	Minimum	228.1	23.8	—	15.7
	Maximum	281.5	32.6	—	23.2
	SD	15.79	2.72	—	1.78
	SE_{mean}	3.53	0.62	—	0.41
	CV (%)	6.21	9.38	—	9.32
	n	20	19	—	19
Carpometacarpus	Mean	117.3	31.0	—	21.1
	Minimum	105.5	25.5	—	18.1
	Maximum	131.8	34.8	—	24.9
	SD	8.36	2.95	—	1.88
	SE_{mean}	2.09	0.74	—	0.50
	CV (%)	7.13	9.51	—	8.93
	n	16	16	—	14

Abbreviations: CV = coefficient of variation; SD = standard deviation; SE_{mean} = standard error of the mean.

TABLE 8.9.
Dimensions (mm) of long bones of the leg of *Harpagornis moorei*.

Bone	Statistic	Length	Proximal	Shaft	Distal
Femur	Mean	162.8	42.2	18.3	45.4
	Minimum	140.3	36.0	15.3	36.2
	Maximum	176.0	47.5	52.5	19.9
	SD	10.55	3.78	5.12	1.55
	SE_{mean}	2.56	0.92	1.24	0.37
	CV (%)	6.48	8.95	11.29	8.50
	n	17	17	17	18
Tibiotarsus	Mean	235.6	44.6	34.1	14.2
	Minimum	213.1	32.2	27.3	11.8
	Maximum	255.1	51.7	39.8	16.6
	SD	13.07	5.10	4.02	1.58
	SE_{mean}	3.77	1.47	0.97	0.41
	CV (%)	5.55	11.4	11.79	11.07
	n	17	15	15	17
Tarsometatarsus	Mean	148.2	33.7	—	38.7
	Minimum	131.5	28.3	—	33.1
	Maximum	166.4	40.2	—	44.6
	SD	8.71	4.03	—	3.67
	SE_{mean}	1.90	0.90	—	0.78
	CV (%)	5.88	11.93	—	9.49
	n	21	20	—	22

Abbreviations: CV = coefficient of variation; SD = standard deviation; SE_{mean} = standard error of the mean.

mens (Buller 1888: xxv); grateful curators sent their surplus. But the booty was not won by the sweat of Haast's own brow. The deposit had been found by farm workers draining a swamp. At a depth of 2–3 m in the main drain, they had found well-preserved tree branches and large bones. When news of the discovery reached Haast in Christchurch 60 km away, an arrangement was struck with the manager of the sheep station. It was a simple matter to let the men continue digging the drains and to excavate the bones at the same time. So workmen from the Glenmark estate dug out and packed the bones, ready for the two-day wagon trip back to the museum. Under the estate manager, George Moore, men worked hard for their living (it was said that any man seen with a straight back in daylight hours was fired for slacking), and work progressed rapidly.

The only museum man to dirty his hands seems to have been Frederick Fuller, the taxidermist. It was he who first noted among the heaps of moa bones a few that belonged not to moa but to an enormous bird of prey—just a few bones at first: a femur, a rib, and a huge claw. Haast lost little time in writing a description (Haast 1872c), which was read first in May 1871 to the newly formed Philosophical Society of Canterbury. In the following year, it was published in the journal of its parent body, the New Zealand Institute. It was a bold departure for this immigrant geologist to challenge the tradition that had most of the new biologic treasures of New Zealand heading back to England to be described by Richard Owen or other researchers in the motherland. The first steps had been taken with some of the moa material, but for Haast to describe the eagle in New Zealand was a major advance.

The published description drew immediate reaction from the academic community. The Governor of New Zealand, Sir George Grey, always interested in natural history and a longtime confidante of Māori elders, pointed out that there was a Māori legend of an extinct predatory bird, the *Hokioi,* which Sir George now equated with the fossil eagle (Grey 1873). The description was as thorough as it could be with the limited range of material at hand: the skeletons of just two other birds of prey, the Australian wedge-tailed eagle *Aquila audax* and the Australasian harrier *Circus approximans.* Haast named the species *Harpagornis moorei* after the station manager and acknowledged Fuller as the original discoverer. Haast not only described the bones of the new eagle, he also speculated on its ecologic role, concluding that it had preyed on flocks of moa in the grasslands of Canterbury. The prevailing view until the mid-20th century was that the environment found by European explorers and settlers was the primitive, pristine environment of New Zealand. In this view, the tussock grasslands on the Canterbury Plains and on the basins and ranges of Otago had to have been the habitat of moa, which took the place of cattle and antelopes elsewhere. So Haast was fitting the eagle into the landscape of then-current opinion. It was significant, though, that he pictured the bird as a great predator and not a scavenger. The size of the claws seems to have impressed him; the claws were huge, as was the bird itself, judging from the femur. Haast obviously took the view that such a large claw would not belong to a scavenger.

In 1872, the workers returned to complete the drainage of Glenmark Swamp and promptly located more eagle bones in the original site as well as on the other side of the drain. From these more complete collections, Haast was able to describe the species in more detail. But the remains from the new site were smaller than the first specimen, so Haast referred them to a new species *H. assimilis.* In doing so, he recognized that the smaller bones were probably of a male of the first species. The apparent contradiction between the reasonable biologic view (male raptors are smaller than females) and publication of a new name may have resulted from a desire to name as many species as possible, as it certainly was for many others at that time, including Walter Buller, the leading New Zealand–born ornithologist of the day. In any event, Haast's smaller species was recognized in lists and texts of fossil birds for many years, almost always with the original caveat about its probably being only the male of *H. moorei.* Tradition is very powerful in taxonomy.

From the first, news of the giant eagle was noted by other observers, and each new discovery was reported in detail. The first records from sites other than Glenmark were made the same year as Haast's second paper. An American gold miner, B. S. Booth, drawn from the Californian and Australian gold fields to the new bonanza in southern New Zealand, excavated eagle remains from a neighboring claim (the "Cornishmen's" at the Hamilton diggings in central Otago. He prudently waited until the Cornish miners were not likely to disturb him. Also in central Otago, a prospector found a complete, apparently juvenile, eagle pelvis under an overhanging schist outcrop on the barren Obelisk Range. The bone seems to have been found on the surface—perhaps the remains marked the presence of an eyrie—but the exact location was not recorded and the site has yet to be rediscovered.

Haast missed a closer association of eagles with humans in his analysis of the artifacts from an excavation in a moa hunter site in Moa Bone Point Cave near Christchurch. Among the archaeologic material was an "awl" fashioned from the distal end of the tibiotarsus of a large bird. Haast identified the bird as a giant petrel, and he labeled the bone awl *Ossifraga gigantea,* which now would be one of the two species *Macronectes giganteus* or *M. halli,* both of which still occur off the eastern coast of the South Island. Some years later, the new director of the Canterbury Museum, Frederick W. Hutton, realized the error, crossed out Haast's identification, and penciled in *Harpagornis.* This fragment was the first indication that the eagle had been contemporary with the early Polynesian inhabitants of New Zealand. In view of Haast's interest in New Zealand archaeology and the major parts he played both in the earliest controlled excavations of an archaeologic site that was not merely a quest for booty and the scientific history of the eagle, it is ironic that he had in his hand a link between the two but did not recognize it.

Although Haast had described the eagle himself, he sent casts and photographs to Richard Owen in London, who redescribed the species from the casts and photographs as part of his series on New Zealand birds. His illustrated account (Owen 1875d) was published (Owen 1879a) in the compendium volume that included all of Owen's work on New Zealand birds; its inclusion rendered the title *Extinct Wingless Birds* rather inappropriate. In his description, Owen suggested that the eagle was a hunter and that it flew well, but he repeated Haast's observation (based, as we have seen, on limited material) that it had a proportionately short ulna.

As a footnote to history, Owen took pains to credit the discovery of the species to Fuller, referring to "Fuller's great extinct raptorial bird" and pointing out that it was Fuller who had, "amongst a quantity of moa bones" from Glenmark Swamp, found "a few small bones in an excellent state of preservation, which he at once correctly referred to a gigantic raptorial bird" (Owen 1879a: 142n). Obviously, the bones were "smaller" only by comparison with the moa bones. It is perhaps unfortunate that when a common name was coined for the species, Haast's eagle was chosen (Holdaway 1991). In retrospect, Haast had his credit and immortality in scientific nomenclature as the author of the species and George Moore is remembered in the name itself. Of the people directly involved in bringing the species to the world's attention, only Fuller, its real discoverer, missed out. Regrettably, it is probably too late to change the common name to Fuller's eagle.

Alexander McKay, a government field geologist and paleontologist, made the next discovery. He found eagle remains with moa bones and remains of forest trees in deeply buried and weathered gravels at Motunau, north of Christchurch. Haast had collected two bones from similar gravels inland near the original finds at Glenmark, so there was now further evidence that the eagle had been part of the local fauna for a considerable time.

None of the several reported occurrences of Haast's eagle in the North Island has withstood further scrutiny. One was based on a sheep bone, another on a pelican bone, and another on that of an individual of the extinct population of the black swan. A more persistent record was that by Augustus Hamilton in the late Pleistocene site of Te Aute in Hawke's Bay. The raptor bones from the site were shown to be of a huge species of *Circus*. Errors in labeling and cataloging perpetuated the mistake of attributing a bone in the Museum of New Zealand to Te Aute. Actually, a clue to the mistake had been available for more than a century. Hamilton (1893: 92) failed to mention the Te Aute specimens of *Harpagornis* in his detailed list of eagle bones accompanying the descriptions of the two eagles from Castle Rocks. In it, he stated that after the discovery of eagle bones at Glenmark, "Very few bones have since been found, except at Hamilton Swamp and Enfield." But in the same publication, a few pages later, he mentions the takahe *Notornis* bones from Te Aute. It was a tacit admission that he had misidentified the accipitrid material from Te Aute, never having seen eagle material at that time, and being unaware that there had been *two* huge raptors in New Zealand.

The purported eagle bones from near Hunterville, on the other side of the main ranges to Te Aute, were from the extinct original New Zealand population of the Australian black swan *Cygnus atratus,* which has become reestablished since the mid-19th century. That leaves only one other bone that could be of an eagle from the North Island. W. B. D. Mantell, son of the discoverer of the dinosaur *Iguanodon,* collected fossil bones from an early Polynesian site at Waingongoro on the southern coast of Taranaki, in the western North Island. The site is of Holocene age because the dunes were formed since the high sea stand about 5000–6000 years ago, and one of these bones is undoubtedly the bleached and leached first manus phalanx of *H. moorei.* However, Mantell also collected at other sites, including one at Waikouaiti, in the southeastern South Island, where there certainly were eagles. In the absence of other remains from the abundant beach, swamp, and lake deposits of Holocene age in the North Island, it seems likely that the Mantell bone was misplaced in the collections on its way to England. Therefore, there are no substantiated records of Haast's eagle from the North Island, despite the presence of numerous fossil sites analogous to those in the South Island that have yielded abundant remains of the bird. The only conclusion that can be drawn is that Haast's eagle was always confined to the South Island, despite the narrow water gaps between the three main islands of New Zealand.

The 1890s produced a new series of eagle finds in the South Island. A large swamp excavation at Enfield in North Otago produced large quantities of moa bones and those of smaller birds, including an unspecified number of eagle bones (Forbes 1892b, 1892c). As with most of Forbes's collections, there were no published reports of any substance, and most of the bones were taken to England. Several individual eagles seem to have been represented, but it is frustrating that more was not recorded at the time, so we do not know whether the collections include all the material that was found at the time.

Having moved south to take up the position as registrar at the University of Otago at Dunedin, Augustus Hamilton was responsible for recording and collecting eagle remains from several sites. One rich site was that at Castle Rocks in western Southland. Excavations by Hamilton (or, rather, the laborers that he and his farm-manager host employed) in 1891 and 1892 yielded the remains of two individuals, again a male and female. The female was one of the largest specimens ever found. Eagle bones were so prized that once word of the discovery of the bones reached the surface, Hamilton went down into the fissure himself to supervise their extraction. His reports on the large collection of fossils from the Castle Rocks stand as the most complete analysis of a Quaternary fossil fauna attempted in New Zealand until the mid-20th century. Hamilton collected eagle bones from two coastal sites, Warrington and Waipapa Point, where they are likely to have been from archaeologic contexts. Other finds in the 1890s were a single eagle bone among the large quantities of moa remains dug from Kapua swamp in North Otago in 1896, and a "nearly complete" (Parker 1897: 574) skeleton found in 1897 by sluicing at the gold workings of the Golden Point mine, in Deep Dell Creek in the schist tablelands of eastern Otago.

Hamilton was also responsible for capitalizing on the first find in the 20th century at Ngapara in North Otago. Through the interest of a local parson, Reverend James Standring, the deposit among large limestone slabs near a farm house was visited and reported on by Hamilton (1904a). Then there was a lull in reports of findings. Most of the large swamps had been drained, and cave exploration had not begun. No more eagles came to light for over 30 years, until the excavations at Pyramid Valley in North Canterbury were expanded in the late 1940s.

From its discovery, Haast's eagle had been the goal of fossickers and professional paleontologists alike. Most of the fossil sites discovered in the South Island after Glenmark had contained eagle remains. It is therefore something of a puzzle and an indication of the mystique of the great bird that they were still considered one of the rarest of finds in New Zealand Quaternary paleontology well into the 1980s. One of the most avid amateur collectors in the mid-20th century was William H. Hartree, Jr., who spent many years looking for their remains in the North Island. His search was doomed to failure—the birds were just not there. Or at least they were not there at the time the deposits he spent most time in were laid down.

Since 1985, finds of eagle bones have been much more frequent as cave systems in the South Island have become better known and the importance of old lake sediments and rockshelters was realized. The number of sites known has substantially increased, but almost all are within the known boundaries of the eastern and southern South Island. All the new finds in northwest Nelson except one, and a single record on the West Coast of the South Island, have been from sites of Otiran glaciation age. The exception was the spectacular discovery of an almost complete skeleton of a large female at 1500 m near the top of Mount Owen in northwest Nelson.

Early Views on Eagle Ecology: Anyone's Guess

The discovery of bones of a large eagle aroused considerable interest in the scientific communities in New Zealand and Britain. The size of the bird and its association with moa in swamp deposits led to speculations about its habits and role in the prehuman ecosystems. Forbes (1892d) considered that it was a carrion feeder that depended on the carcasses of dead moa—in other words, an eagle behaving like a vulture. Owen (1879a) had earlier taken the opposite view: that the eagle had actively preyed on the large, flightless birds that dominated the prehuman avifauna. Of the two viewpoints, by the 1980s, the first had gained some dominance in the popular literature, although little had been published on the species since Hamilton's papers at the end of the 19th century (Hamilton 1893, 1894b).

The perception of Haast's eagle as a scavenger—and thus one that depended on the population dynamics of prey species rather than exerting any population or evolutionary influences on the prey—has led to the importance of the eagle in the avifauna being underestimated. Studies of moa (Dinornithiformes) have dominated the literature of New Zealand fossil birds. The possibility that predator–prey relationships may have been important in shaping moa population biology or evolution, or that of the other large flightless species, has rarely been considered. Indeed, recent papers persist in stating that moa evolved in the absence of predators (e.g., Alexander 1989).

Two other features of Haast's eagle have occasioned comment: its relationships to other birds of prey and its preferred habitat. Haast (1872c, 1874a), Owen (1879a), Shufeldt (1896), and Oliver (1930, 1945a, 1945b, 1955) all commented on possible closest relatives in the Accipitridae, which presently is considered to contain about 220 species and 60 genera (Brown and Amadon 1968; Holdaway 1994; Stresemann and Amadon 1979). The taxonomic position of Haast's eagle, as presently known, was covered in a previous section of this chapter.

The second debate, on the eagle's preferred habitat, began when workers had little information on which to base paleoenvironmental reconstructions. The efforts were not helped when they sometimes overlooked evidence from their own excavations. In the late 19th century, and indeed until the middle of the 20th century, it was usually assumed that the vegetation encountered by European settlers in New Zealand was original (e.g., Cockayne 1928). Several hundred years of Polynesian settlement had apparently had little or no effect on the vegetation pattern. There had, at least since the most recent glaciation, been extensive grasslands on the plains and downlands east of the axial ranges of the South Island and wet forests on the West Coast, in which the "newly discovered" living and fossil faunas had lived. Haast's eagle was first perceived as a giant harrier that had followed flocks of moa across the eastern grasslands (Haast 1872c).

It is now well established that the natural late Holocene vegetation of New Zealand, even in the drier eastern areas, was forest or shrubland rather than grasslands (McGlone 1988, 1989). Closed or dense vegetation is a difficult environment for carrion eaters that seek prey by vision, but the view that Haast's eagle was a scavenger that depended on moa carcasses is still strong (e.g., McCulloch 1991). A further twist to the scavenger hypothesis was the perception that like other New Zealand birds, Haast's eagle had short wings and disproportionately large legs, and was evolving toward flightlessness (Duff 1949c). Just how a flightless carrion eater could find enough to eat was not explained. Any evolutionary changes that the species had undergone in the previous 20,000 years were a fine-tuning of the bird's flight ability and were not associated in any way with a trend toward walking, but that part of the story comes later.

The poor flight hypothesis originated with Haast's (1874a) observation, which Owen (1879a) repeated, that the ulna was relatively short by comparison with that of other accipitrids, such as the wedge-tailed eagle *Aquila audax* and Australasian harrier *Circus approximans*. Both species habitually glide or soar, but they were the only skeletons available to Haast when he described the species. With time, the "relatively" was forgotten, and the ulna was described simply as "short" by Oliver (1955), a condition that, if true, would have obvious implications for the bird's flying ability. In conjunction with the (apparently) long and (certainly) thick leg bones, the "short" ulna was interpreted by some as indicating a terrestrial mode of life (e.g., Millener 1984). However, none of the speculation on wing reduction and leg hypertrophy was supported by references to specimens or by comparisons with a wide range of other birds of prey (Fig. 8.21).

The ratio of humerus length to ulna length (which is a measure of the shape of the wing (rounded for flapping or elongated for gliding) in relation to body size (measured by femur length) shows that Haast's eagle had wings that were giant versions of those of a bird-eating hawk and entirely unlike those of a soaring vulture (Fig. 8.21A). Furthermore, comparison of the contribution of each of the three major bones to the total length of the bone of Haast's eagle wing with a range of very large raptors and other large flying birds (Fig. 8.21B) shows that Haast's eagle had wing proportions that were close to those of living species.

In 1975, a local amateur ornithologist, Don H. Brathwaite, presented new hypotheses on both the relationships and habits of Haast's eagle. His hypotheses were that the bird was a forest eagle related to the tropical hawk eagles, that it had the relatively short, broad wings and long tail characteristic of a forest raptor, that it could fly very well, and that it was an active predator. He published the results of his own research on the size and aspects of the biology of the eagle (Brathwaite

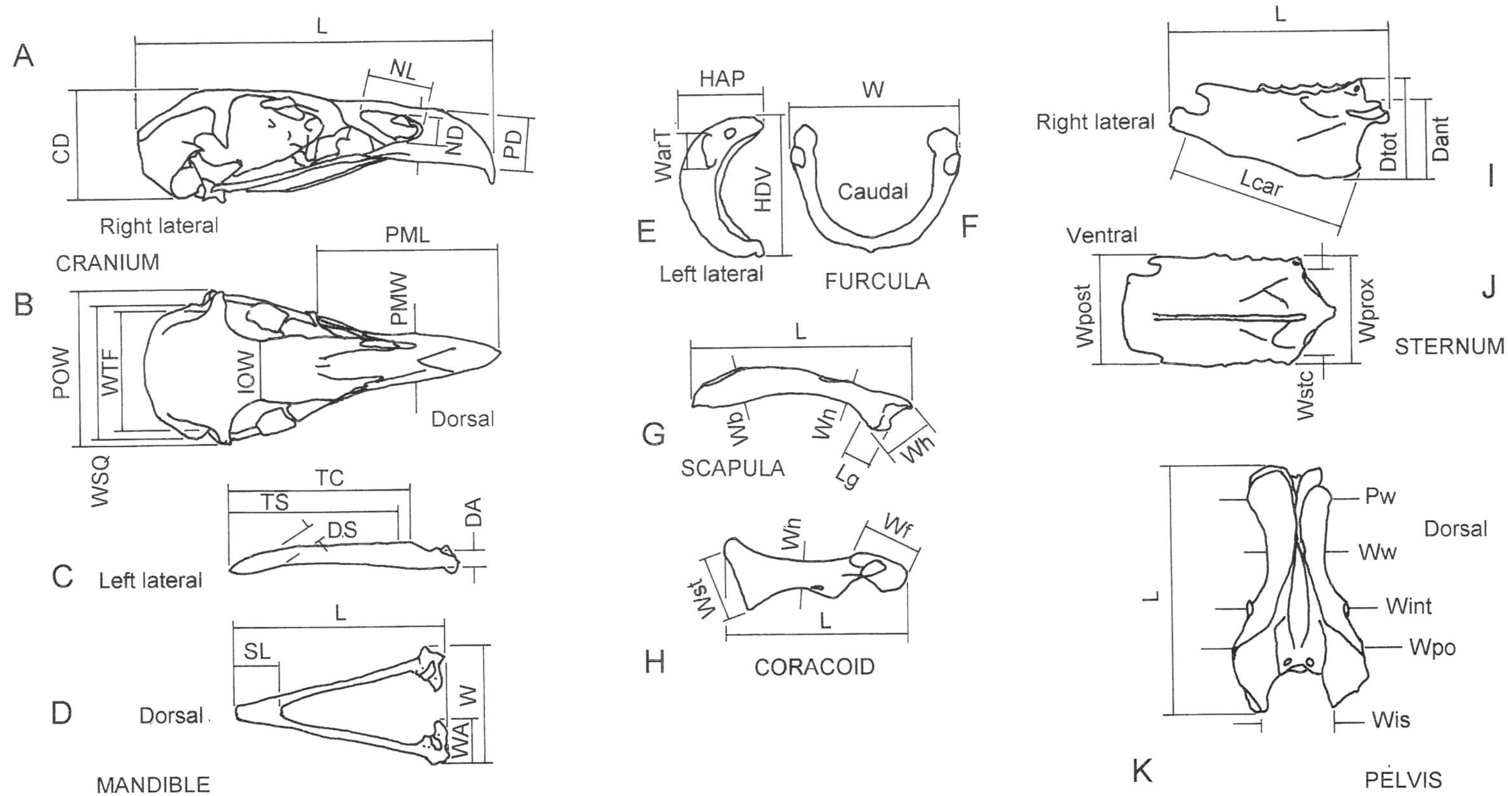

Figure 8.19. Landmarks and dimensions used for tabulated measurements of bones. Not to scale. (A, B) Cranium and premaxilla. CD = cranial depth; IOW = interorbital width; L = length; ND = nostril depth; NL = nostril length; PD = premaxilla depth; PML = premaxilla length; PMW = premaxilla width; POW = postorbital width; WSQ = width across squamosals; WTF = width across temporal fossae. (C, D) Mandible. DS = depth at proximal end of symphysis; L = length; SL = symphysis length; TC = tip to coronoid process; TS = tip to surangular process; W = width; WA = "width" of articular process. (E, F) Furcula. HAP = horizontal anterioproximal depth; HDV = dorsoventral depth; W = lateral width; Wart = width of articular surface. (G) Scapula. L = length; Lg = length of glenoid; Wb = width of blade; Wf = width of brachial facet; Wh = width of head; Wn = width of neck. (H) Coracoid. L = length; Wn, width of neck of shaft; Wst = width of sternocoracoidal facet. (I, J) Sternum. Dant = anterior depth; Dtot = total depth; L = length; Lcar = length of carina (keel); Wpost = posterior width; Wprox = proximal width; Wstc = width over lateral extremities of sternocoracoidal sulci. (K) Pelvis. L = length; P = proximal width; Wint = width across acetabulae; Wis = width across ischial extremities; Wpo = posterior width; Ww = width at "waist."

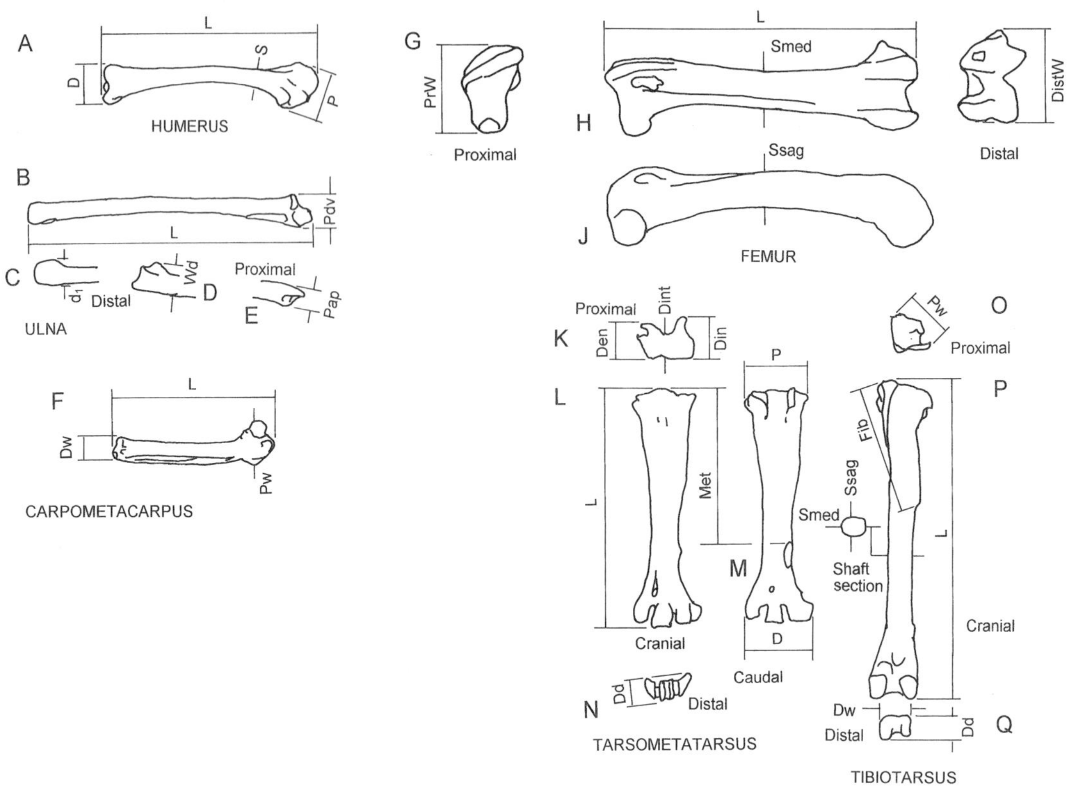

Figure 8.20. Landmarks and dimensions used for tabulated measurements of bones. Not to scale. (A) Humerus. D = distal width; L = length; P = proximal width; S = shaft width (dorsoventral). (B–E) Ulna. d_1, distal anterioposterior width; L = length; Pap = proximal anterioposterior width; Pdv = proximal dorsoventral width; Wd = distal dorsoventral width. (F) Carpometacarpus. L = length; Dw = distal width; Pw = proximal width. (G–J) Femur. DistW = distal width; L = length; PrW = proximal width; Smed = medial width of shaft; Ssag = shaft sagittal width. (K–N) Tarsometatarsus. D = distal width; Dd = distal depth; Den = depth of lateral hypotarsal ridge; Din = depth of medial hypotarsal ridge; Dint = proximal interhypotarsal depth; L = length; Met = distance from proximal end to proximal end of metatarsal articulation; P =proximal width. (O–Q) Tibiotarsus. Dd = distal depth; Dw = distal width; Fib = proximal end to distal end of the fibular crest; L = length; Pw = proximal width; Smed = shaft medial width; Ssag = shaft sagittal width.

1992). By then, the hypotheses had been the impetus for other research on the species (Holdaway 1991).

Although there were partial skeletons from Castle Rocks, and some of the smaller bones of the wing and skull had been found in Honeycomb Hill Cave in the 1980s (Holdaway 1991; Worthy 1993b), study of the eagle's overall structure and proportions and of the minor bones in the skeleton were revolutionized one (southern) summer's day in 1990. Dave Smith, a speleologist with an expedition to the marble massif of Mount Owen in northwest Nelson, was searching for new entrances to the extensive Bulmer Cave system. Working his way along the summit ridge at nearly 1500 m (Fig. 8.22), he decided to descend a narrow sinkhole. The entrance was only about a meter wide, but the shaft below went down 15 m before it ended in two small chambers.

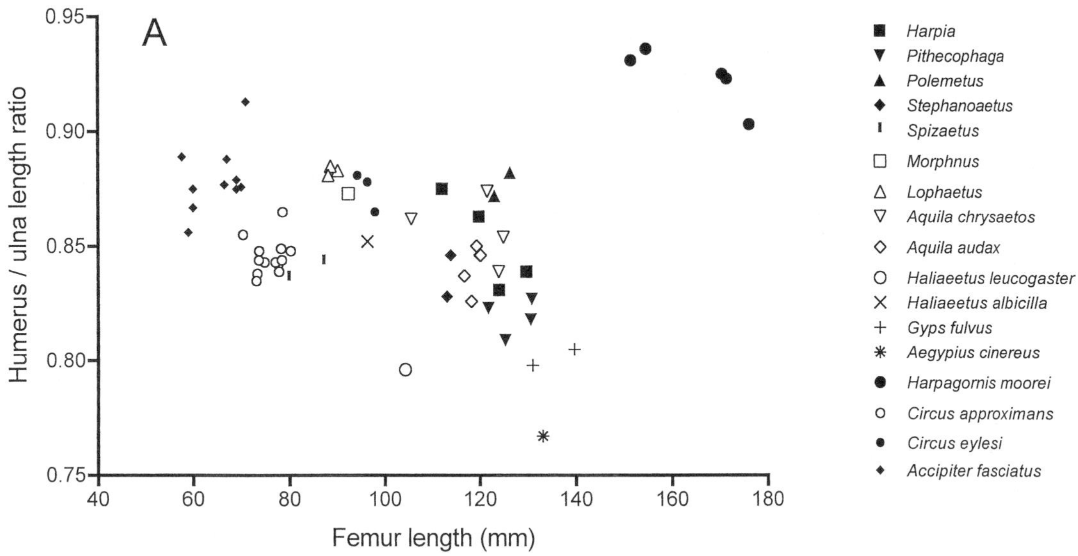

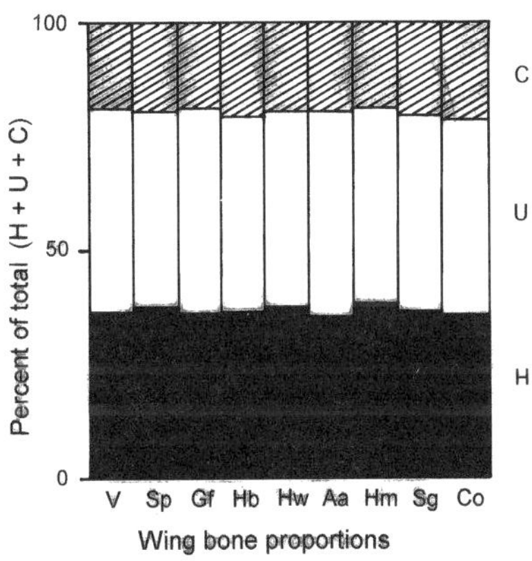

Figure 8.21. (A) (above) Ratios of wing bone lengths for Harpagornis moorei *against femur length (body size) in comparison to ratios for other large accipitrids. (B) (left) Proportions of total length of major wing bones occupied by each element lengths for* Harpagornis moorei *in comparison to those of other large accipitrids and other large flying birds. Aa =* Aquila audax *(Accipitridae); Co =* Cygnus olor *(Anatidae); G =* Gyps fulvus *(Accipitridae); Hl =* Haliaeetus leucocephalus *(Accipitridae); Hm =* Harpagornis moorei *(Accipitridae, extinct); Hw =* Haliaeetus leucogaster *(Accipitridae); Sg =* Spizaetus grinnelli *(Accipitridae, extinct); Sp =* Sarcoramphus papa *(Vulturidae); V =* Vultur gryphus *(Vulturidae).*

Spread across the rock fall in one chamber were some bones of a large bird. Mindful of the protocols for cave fossils, Smith left most of the remains in position, but he also realized that it was an unusual find. He collected some of the larger bones—the skull, sternum, a humerus, and a few others—and took them back to T.H.W. at the end of the expedition. They were obviously parts of the skeleton of a large female Haast's eagle.

The condition of the bones, the confined, isolated site, and Dave Smith's observations suggested that much of the skeleton might be there. The National Museum of New Zealand (now the Museum of New Zealand) funded a helicopter trip to the site to assess and collect the material. In spite of cloud and 50–60 k/h winds, the trip was successful and the checklist of eagle bones was almost completely filled (Fig. 8.23). The skeleton is complete apart from the scleral ossicles from around the eyes (a few were recovered), a caudal element of one side of

Figure 8.22. Mount Owen alpine basins, the last view seen by the Mount Owen eagle (MNZ S27773). Photograph by T.H.W.

the hyoid apparatus, the second phalanx of the major digit of the left wing, and the accessory bone at the carpal flexure usually present in accipitrids (Bock and McEvey 1969). One of the thoracic vertebrae was lacking the neural spine, but this too was found by Dave Smith on a later trip.

Most of the bones were in excellent condition, and almost all were in position where the eagle's body had decomposed, its limbs distorting as the muscles and tendons contracted. The pelvis was missing but was found in the next chamber, where it seems to have been taken by a weka, whose bones were also recovered from the cave. A radiocarbon date of 2160 ± 110 years (Rafter Radiocarbon Laboratory no. NZA 905) was obtained on the gelatin from one of the fibulae. From the location of the cave and the nature of its entrance, it is likely that the eagle was standing on the ridge, which overlooks two large mountain basins. At the time, these held populations of the upland moa *Megalapteryx didinus* and crested moa *Pachyornis australis*. The eagle probably fell into the hole accidentally, perhaps after breaking through a snow crust that obscured the entrance.

The Mount Owen eagle is the most complete of a very few skeletons found so far. One of these, from the Golden Point gold mine in Otago, was thought to be lost for a century before it was found to be in the Royal Museum of Scotland. The Castle Rocks birds (male and female, but from their different preservation, not a pair) have been mentioned. Another partial skeleton was found at Lake Grassmere, on

Clifford Bay, at the northeastern corner of the South Island, but immediately lost. J. R. Eyles, who had earlier found the moa hunter site at Wairau Bar, a few kilometers north of Clifford Bay, found the eagle when visiting the site in the mid-1940s with Dr. Roger Duff, later the director of Canterbury Museum. They put the bones in an old metal oven while they visited other parts of the site, and according to Eyles, they never went back to collect them.

*Figure 8.23. Bones of a large female Haast's eagle MNZ S27773 (*Harpagornis moorei*) in a cave (SO 209) on Mount Owen. Photograph by T.H.W.*

Body Mass

Many aspects of a species' biology and physiology can be predicted on the basis of its body mass (Peters 1983). Some of the parameters commonly estimated from body mass are metabolic rate (and hence an animal's food requirements), mobility, reproductive investment, and territory size to sustain itself and its young. For living species, the parameters may be useful in planning ecologic or physiologic studies, or in predicting the habitat requirements of rare or endangered animals (Kemp 1989).

Estimates of mass are especially useful in paleobiology, where the only other information on an extinct animal's physiology and ecology comes from comparative anatomy, and in paleoenvironmental reconstructions. Attempts to estimate body mass for extinct animals, from dinosaurs (e.g., Anderson et al. 1985) to moa (e.g., Alexander 1983a; Atkinson and Greenwood 1989; and this volume, Chapters 5, 8), and large flying birds (e.g., Campbell and Marcus 1992; Campbell and Tonni 1983; Hertel 1992; Kitchener 1993), have used mainly allometric scaling from limb bone dimensions. The regression equations in each instance were derived from empirical data on a variety of living taxa. Mass has also been calculated as a function of body volume estimated from trunk dimensions, measurement of body volume by use of models based on the best available restorations of the extinct animal (Alexander 1983a; Holdaway 1991; Kitchener 1993), and equations based on the relationship between skeletal mass and body mass (Anderson et al. 1979; Prange et al. 1979).

Previous estimates of the body mass of Haast's eagle had been made by Millener (1984) and Brathwaite (1992). Millener's suggestion was that a large individual may have had a mass of 18 kg, but no method of calculation was given. If true, the mass would make Haast's eagle by far the heaviest accipitrid known. Brathwaite (1992) compared the trunk volume of a mounted skeleton of the eagle with those of two other species and calculated body mass as a proportion of a typical published mass for the living species. As the mounted specimen is smaller than many—it was probably a male—he then scaled up the calculated mass, by use of wing bone lengths, to derive the mass of larger individuals. His estimates of live mass were about 11.5 kg for the male and over 14 kg for a large female. Although less than Millener's estimate, these values would still make Haast's eagle substantially heavier than the largest living eagle, the South American harpy (*Harpia harpyja*). The only raptorial birds of comparable or greater mass were the teratorns (Teratornithidae) of the Pleistocene of North and South America, which ranged from about 14 kg in *Teratornis merriami* to about 80 kg in the incredible *Argentavis magnificens* (Campbell and Tonni 1983).

More recently, three techniques were used to calculate the living body mass of Haast's eagle: by the water displacement of a full-size model (without feathers) (Fig. 8.24); from allometric relationships for the leg bones; and from a mass of the only complete skeleton of the eagle (allowing for the difference between dry and wet mass of the bones). The model was constructed to the proportions of the almost complete skeleton (MNZ S27773) from the late Holocene of Mount Owen in the northwestern South Island, using as a base dimension the ulna length of a large individual from Honeycomb Hill Cave in the same general area but of Pleistocene age. It was constructed by use of a

Figure 8.24. Full-size model of large female Harpagornis moorei *used to determine live body mass by displacement of water. The bird was modeled in plaster over a wire frame. Model and photograph by R.N.H.*

conservative body profile to provide a minimum value for the body mass. An extra 10–20 mm of muscle over the area of the sternum extended to cover the ventral third of ribs 3 and 4, as in *Vultur gryphus* (Fisher 1946), would have resulted in 0.75–1.25 kg greater muscle mass without distortion of the model outline. In addition, Olson and Feduccia (1979) point out that the main site of origin of the pectoral muscle is actually on the furcula and the coracofurcular membrane. The furcula is very large in *Harpagornis moorei,* which suggests that the pectoral muscle mass was probably substantially larger than is apparent from the size of the sternum and its keel alone.

The model displaced 0.0124 ± 0.0001 m^3, which corresponds to a carcass mass (without feathers) of 11.64 kg. Feathers would add about 850 g, giving a minimum body mass of at least 12.33 kg. A more realistic pectoral muscle profile would bring the body mass to at least 13.5 kg.

Various allometric relationships between body mass and leg bone dimensions were applied to Haast's eagle and a range of other large raptors. Results for femoral and tibiotarsal lengths, diameters, and circumferences according to relationships in Prange et al. (1979), Alexander (1983a), and Anderson et al. (1985) are shown in Figure 8.25 and Table 8.10. As an example, a bird with a femur length of 154 mm (at the low end of the range) would have a body mass of 12.81 kg (Prange et al. 1979), 9.66 kg (Alexander 1983a), or 9.47 kg (Anderson et al. 1985). These relationships were all developed from data on a wide range of taxa, mostly of relatively small body mass, but including such nonflying

TABLE 8.10.
Estimated live body mass (kg) of specimens of Haast's eagle for which major limb elements are securely associated.

	1	2		3					4	2
Specimen	FC	HL	FL	FL	FD	TBL	TBD	TML	TC	SW
S27773	13.1	4.1	17	12.7	13.7	7.2	13.3	6.8	17.1	12.6
C40.8	9.6	3.5	12.2	9.2	10.2	5.8	8.2	5.3	12.2	—
Av5102	9.1	3.5	12.4	9.7	9.5	5.6	9.1	6.0	12.4	—
DM2145	13.9	4.4	18.2	13.5	14.2	7.7	12.9	7.6	18.8	—

Abbreviations: 1 = Anderson et al. (1985); 2 = Prange et al. (1979); 3 = Alexander (1983a), model II; 4 = Campbell and Tonni (1983); FC = femur midshaft circumference; FD = femur midshaft sagittal diameter; FL = femur length; HL = humerus length; SW = skeleton mass; TBD = tibiotarsus midshaft sagittal diameter; TBL = tibiotarsus length; TC = tibiotarsus midshaft circumference; TML = tarsometatarsus length.

taxa as ratites. By use of the relationship derived by Hertel (1992) for living Vulturidae, which are related to storks but which have raptorial body form, a femur length of 154 mm corresponds to 12.22 kg. The data in Table 8.11 show that all of the relationships give body masses that are within the ranges for some large raptors. It is likely that a male Haast's eagle had a body mass of around 10 kg and a large female of 13–15 kg. If the relationship derived by Hertel (1992) holds for large accipitrids, a Haast's eagle with a femur length of 170 mm might have had a peak body mass of about 16.25 kg. The ranges are well above those for any living eagle, in the largest of which, the harpy (*Harpia harpyja*), a large female has a live mass of about 9 kg. Some scavenging accipitrids, such as the black vulture *Aegypius monachus,* may reach 12.5–14 kg (Weick and Brown 1980).

Campbell and Marcus (1992) presented new relationships between femur circumference and body mass for accipitrids. Use of the reduced major axis relationship yielded body mass estimates of 11–18 kg. The relationship predicts body masses for living large accipitrids within the known range for each species, so perhaps larger individuals were nearly twice the mass of a harpy eagle. It is possible, however, that the increased thickness of the limb bones in Haast's eagle may have been associated, at least in part, with the strength needed to subdue its prey, and the actual body mass may have been lower than predicted. Again, Campbell and Marcus (1992) point out that the allometric relationships have been developed from data on living birds, none of which approaches the mass of some of the extinct species, and so the mass of larger birds may be underestimated.

An allometric relationship for humerus length and body mass (Prange et al. 1979) gave results at variance with those from the pelvic limb. For a humerus length of 220 mm, that of a midsize Haast's eagle, the predicted body mass was about 4 kg, which is much lower than for all the other estimates (Table 8.11). In comparison, the predicted body mass from femur length of 154 mm (the lowest end of the range) would be about 13 kg, 9.66 kg, and 9.47 kg, according to

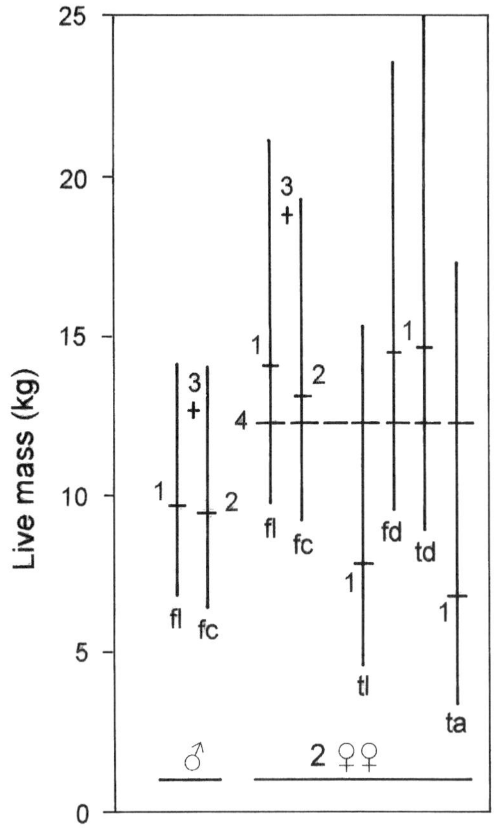

Figure 8.25. Live body mass (mean ± confidence interval), in kilograms, of Harpagornis moorei *estimated from allometric relationships between mass and measurements of the femur and tibiotarsus. Data are based on 1, the model II algorithm from Alexander (1983a); 2, Anderson et al. (1985); 3, Prange et al. (1979); 4, displacement model (estimate for bird with 174-mm femur; estimate probably 0.5–1 kg low). ♂, femur length 154 mm; first ♀ (third and fourth line from left) CM Av28366; second ♀, MNZ S22472. Abbreviations: fc = femur circumference; fl = femur length; ta = tarsometatarsus length; td = tibiotarsus shaft sagittal diameter; tl = tibiotarsus length.*

relationships given by Prange et al. (1979), Alexander (1983a), and Anderson et al. (1985), respectively. Predictions of body mass by use of humerus length do not seem to agree well with measurements of the mass of living birds (Tables 8.11–8.13).

When the humerus length to body mass relationship was applied to other raptors, it also produced consistently low predictions of mass. When applied to members of other families, the relationship produced widely varying results. For example, by use of data from Warham (1977), body mass was predicted for several petrels (Table 8.14). The results confirmed that humerus length is an inconsistent predictor of live body mass.

Ecomorphology

Did the Largest Eagle Fly or Walk?

Because Haast (1874a) and Owen (1879a) pointed out that the ulna in *Harpagornis moorei* was proportionately shorter than in *Aquila audax,* the Australian wedge-tailed eagle, and other accipitrids, several authors have doubted its ability to fly. Duff (1949c: 23) pointed to "the reduction of the wing-bones and the lengthened leg (the beginning of that fatal New Zealand tendency to change from a flying habit to a pedestrian one)," and Millener (1984) thought that it spent much of its

TABLE 8.11.
Predicted and measured live body mass (kg) for large accipitrids, based on the dimensions and algorithms used for Haast's eagle.

	Predicted live body mass (kg)									
	1	2		3					4	
Species	FC	HL	FL	FL	FD	TBL	TBD	TML	TC	Measured live mass (kg)
Harpia harpyja	4.71	1.98	5.29	4.09	4.97	2.28	4.61	2.10	5.99	4♂ 4.0–4.6 (A)
Harpia harpyja	7.54	3.13	7.94	6.07	7.92	3.98	6.82	3.19	8.60	2♀ 7.54, 7.6 (B)
Harpia harpyja	5.13	2.22	6.37	4.90	5.36	2.59	4.84	2.32	6.28	
Harpia harpyja	7.43	2.69	6.46	5.38	8.18	3.50	5.71	2.86	7.59	
Pithecophaga jefferyi	5.45	2.57	6.85	5.26	5.56	4.20	4.96	3.41	6.21	♀ 4.65 (C)
Pithecophaga jefferyi	7.21	2.79	7.38	5.65	7.20	4.36	5.98	3.20	7.83	
Polemaetus bellicosus	4.57	3.10	7.21	5.53	5.07	3.65	3.74	3.32	5.51	2 ♀ 5.92, 6.2; 3.01–5.66 Mean 3.97 (17, unsexed) (A)
Polemaetus bellicosus	6.07	3.26	8.13	6.21	6.41	3.42	4.16	3.55	5.92	♂ 3.8; ♀ 5.92–6.2 (C)
Polemaetus bellicosus	5.50	2.74	6.67	5.12	5.66	3.19	3.54	3.50	5.72	
Polemaetus bellicosus	6.79	3.34	8.10	6.18	7.44	3.66	4.61	3.75	6.95	
Stephanoaetus coronatus	5.31	2.05	5.42	4.19	5.77	2.47	4.38	2.05	5.72	2♂ 3.40, 4.12 (D); 2.8-4.1 (C)
Stephanoaetus coronatus	5.98	2.08	5.53	4.27	6.19	2.37	5.08	2.07	6.80	2♀ 3.18, 3.85 (D)
Spizaetus ornatus	2.04	0.89	2.63	2.08	2.28	1.33	1.57	1.51	2.14	2♂ 0.96, 1.0 (A)
Spizaetus ornatus	1.59	0.72	2.06	1.64	1.67	1.01	1.08	1.17	1.46	2♀ 1.39, 1.61 (A)
Morphnus guianensis	2.43	1.22	2.77	2.42	2.40	1.91	1.57	2.15	2.03	♀ 1.75 (A)
Lophaetus occipitalis	1.80	1.06	2.21	2.27	1.82	1.40	1.63	1.85	2.03	♂ 0.91-1.36 (C)
Lophaetus occipitalis	2.15	1.07	2.75	2.17	2.16	1.29	1.34	1.82	1.75	♀ 1.37-1.52 (C)
Lophaetus occipitalis	1.71	0.99	2.70	2.13	1.72			1.71		
Aquila chrysaetos	4.24	2.40	7.15	5.48	4.60	2.76	2.81	2.34	4.16	♂ 2.84-4.55 (C)
Aquila chrysaetos	4.08	2.34	6.99	5.36	4.25	2.86	2.64	2.27	4.10	♀ 3.84-6.66 (C)
Aquila chrysaetos	3.44	2.00	4.48	3.48	3.44	2.18	3.35	2.01	4.44	
Aquila chrysaetos	4.40	2.28	6.62	5.09	4.79	2.44	3.07	1.89	4.62	
Aquila audax	4.44	2.64	6.13	4.72	4.79	2.82	2.89	2.57	3.99	2.50–4.54 (up to 5.0) (C)

	Predicted live body mass (kg)									
	1	2		3					4	
Species	FC	HL	FL	FL	FD	TBL	TBD	TML	TC	Measured live mass (kg)
Aquila audax	4.96	2.77	5.92	4.56	5.07	2.83	3.74	2.66	5.12	
Aquila audax	4.61	2.70	6.40	4.92	5.07	2.95	3.44	2.45	4.56	
Aquila audax	4.57	2.77	6.41	4.93	5.26	2.96	4.27	2.77	4.99	
Gyps fulvus	8.23	5.55	9.75	7.41	9.22	4.54	6.97	2.81	8.34	♂ 8.18 (A) 6.2–8.5 (C)
Gyps fulvus	7.0	5.03	8.17	6.24	7.92	3.75	5.98	2.32	7.59	♀ 6.90 (A) 6.5–8.3 (C)
Aegypius monachus	9.13	5.10	8.54	6.51	9.914	5.46	6.53	4.60	9.14	♂ 7.0–11.5; ♀ 7.5–12.5 (C)
Haliaeetus albicilla	3.41	2.44	3.47	2.71	3.36	1.96	2.32	1.73	3.46	♂ 3.08–4.99; ♀ 3.65–6.56 (C)
Haliaeetus albicilla	3.85	2.69	4.33	3.37	3.91	2.21	2.64	2.13	3.72	♀ 2.48–2.80 (C)

Abbreviations: 1 = Anderson et al. (1985); 2 = Prange et al. (1979); 3 = Alexander (1983a), model II; 4 = Campbell and Tonni (1983); A = Brown and Amadon (1968); B = Fowler and Cope (1964); C = Weick and Brown (1980), model II; D = Brown et al. (1982); FC = femur midshaft circumference; FD = femur midshaft sagittal diameter; FL = femur length; HL = humerus length; TBD = tibiotarsus midshaft sagittal diameter; TBL = tibiotarsus length; TC = tibiotarsus midshaft circumference; TML = tarsometatarsus length.

TABLE 8.12.
Live body mass (kg) of a range of raptorial birds from three families, predicted from the algorithms listed in Table 6.11 for femoral and tibiotarsal measurements, compared to measured body mass for each species.

Family/species	Predicted live mass (kg) 1 CF	3 FL	FD	4 TC	Recorded mass (♂; ♀)
Pandionidae					
Pandion haliaetus	1.43	1.53	1.43	2.07	1.22–1.6; 1.25–1.9
Sagittariidae					
Sagittarius serpentarius	5.36	3.83	6.08	5.72	3.81; 3.41, 3.74–4.27
Accipitridae					
Milvus milvus	1.22	1.20	1.13	1.01	0.66–0.85; 0.75–1.08
Haliastur sphenurus	0.69	0.52	0.62	0.51	0.65; 0.76–0.92
Haliaeetus albicilla	4.04	4.65	3.91	3.77	3.08–4.99; 3.65–6.56
Ichthyophaga ichthyaetus	3.06	3.37	3.14	3.56	1.59; 2.27–2.7
Gypohierax angolensis	1.43	1.38	1.34	1.46	1.36, 1.45, 1.71; 1.71
Neophron percnopterus	1.71	1.42	1.57	2.14	1.58–2.18
Gypaetus barbatus	5.00	5.01	5.56	5.12	4.57–5.72
Necrosyrtes monachus	2.12	1.75	1.99	2.41	1.52–2.10
Aegypius monachus	9.26	7.16	10.20	8.87	7–11.5; 7.5–12.5
Circaetus gallicus	1.93	1.72	1.88	1.99	1.82, 1.98; 1.30, 1.82, 2.11
Terathopius ecaudatus	2.34	2.20	2.34	2.58	1.93–2.95
Spilornis cheela	2.20	1.53	2.22	2.26	0.76; ? 1.60

Abbreviations: 1 = Anderson et al. (1985); 3 = Alexander (1983a), model II; 4 = Campbell and Tonni (1983); CF = femur circumference; FD = femur midshaft sagittal diameter; FL = femur length; TC = tibiotarsus midshaft circumference. Recorded mass is from Brown and Amadon (1968).

TABLE 8.13.
Live body mass (kg) estimated from dry skeleton mass, using a transposed version of the algorithm in Prange et al. (1979). Live mass and skeleton mass for taxa other than Haast's eagle are from Prange et al. (1979).

Species	Dry skeleton mass	Live mass	Predicted mass LCL (95%)	Mean	UCL (95%)	Mass range
Heterospizias meridionalis	0.060	0.763	0.703	0.928	1.225	0.83–1.04
Necrosyrtes monachus	0.143	1.882	1.582	2.088	2.756	1.52–2.10
Neophron percnopterus	0.138	2.134	1.530	2.020	2.666	1.58–2.18
Haliaeetus leucocephalus	0.272	4.082	2.883	3.806	5.024	4.00–6.30
Harpagornis moorei	0.981	—	9.55	12.61	16.64	—

Abbreviations: LCL, lower confidence limit; UCL, upper confidence limit, as 0.32 ± mean for confidence limits for skeleton mass from live mass. Mass range (live masses) from Brown and Amadon (1968).

TABLE 8.14.

Comparison between live mass (g) predicted from humerus length (mm) (Prange et al. 1979) and live mass for a range of species of petrels (Procellariiformes). Humerus length and live mass are from Warham (1977).

Family	Species	Predicted mass	Live mass	Range	Ratio, predicted/live
Diomedeidae	*Diomedea exulans*	14,477	8677	6636 ± 1184	1.67
	Thalassarche melanophris	5106	3515	2911 ± 284	1.45
	Thalassarche chrysostoma	4904	3300	3300 ± 427	1.49
Procellariidae	*Macronectes giganteus*	4474	4549	—	0.98
	Procellaria aequinoctialis	1690	1134	1134 ± 102	1.49
	Procellaria cinerea	1301	1018	1018 ± 115	1.28
	Puffinus griseus	754	849	819 ± 76	0.89
	Puffinus puffinus	435	426	426 ± 46	1.02
	Puffinus assimilis	263	238	238 ± 11	1.11
	Fulmarus glacialis	999	763	—	1.31
	Thalassoica antarctica	680	641	—	1.06
	Pterodroma lessoni	832	586	586 ± 40	1.42
	Pterodroma inexpectata	446	329	329 ± 31	1.36
	Pterodroma mollis	458	276	276 ± 27	1.66
	Pachyptila vittata	281	196	196 ± 17	1.43
	Pachyptila desolata	221	159	159 ± 11	1.39
	Pelecanoides urinatrix	123	124	119 ± 10	0.99
Hydrobatidae	*Pelagodroma marina*	47	47	—	1.00
	Oceanodroma leucorhoa	80	39	39 ± 1	2.05
	Garrodia nereis	28	32	32 ± 5	0.88
	Hydrobates pelagicus	43	28	—	1.5

time on the ground. In addition, a carrion-eating rather than predatory lifestyle has been mooted by, for example, McCulloch (1991) and Millener (1984). Although Haast's eagle undoubtedly ate carrion when that was available, as do almost all eagles, there is strong direct and indirect evidence that it was an extremely powerful predator.

Indirect evidence for the eagle's flying and predatory abilities comes from its body conformation and proportions, in comparison with living raptors, whose flight and food habits are known. The predatory abilities of raptors were termed predation potential by Voous (1969), who measured the differences among species of South American raptor in terms of their body size, claw size, and other attributes. These factors can be determined for the New Zealand raptors. In addition, obligate carrion feeders have different morphometrics than active predators because the functional requirements of the two ways of life are different. A multivariate analysis of the dimensions of all the major bones in the skeleton for a range of species with known habits revealed interesting patterns within the large raptors. If Haast's eagle was an obligate

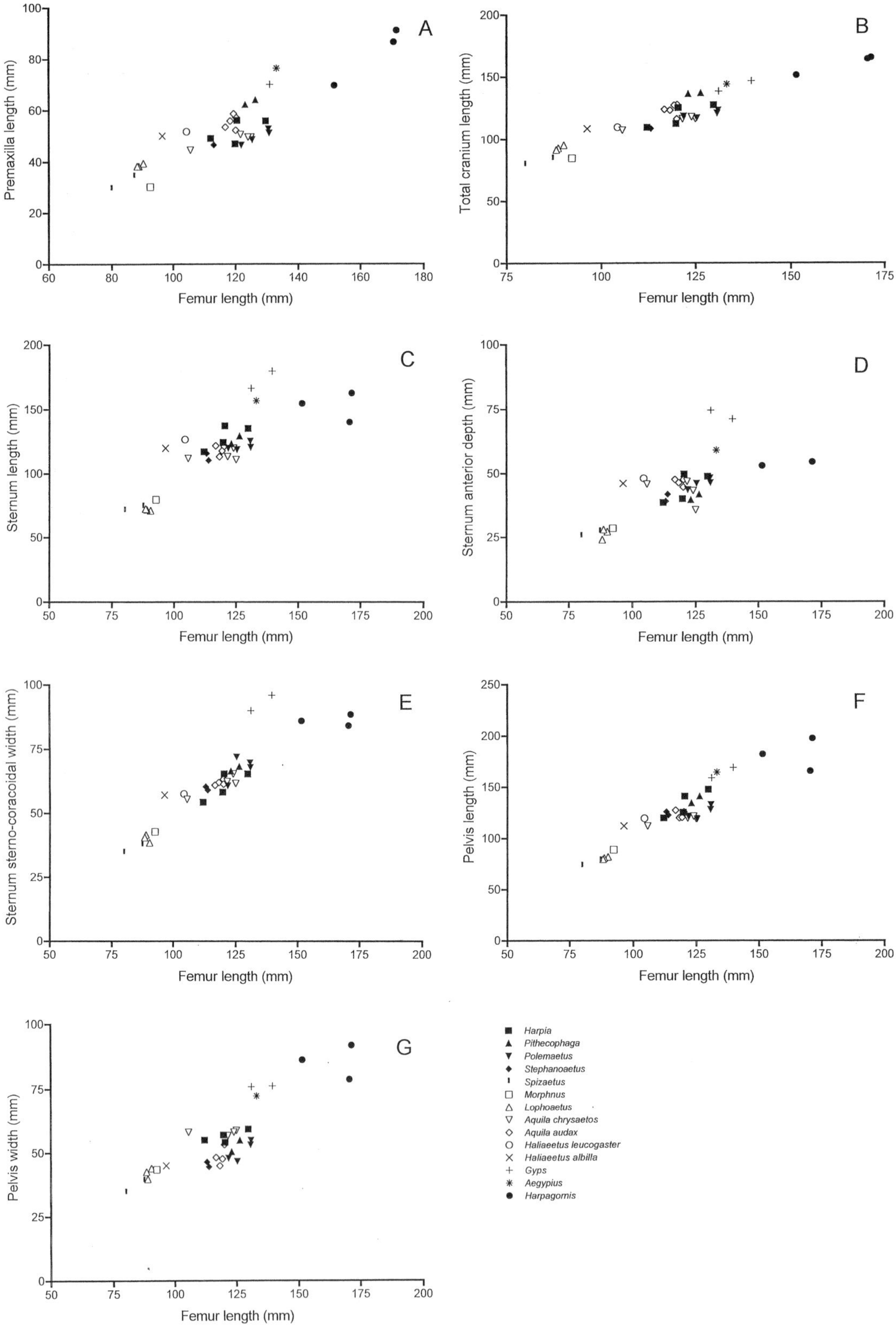
A
Premaxilla length (mm)
Femur length (mm)
B
Total cranium length (mm)
Femur length (mm)
C
Sternum length (mm)
Femur length (mm)
D
Sternum anterior depth (mm)
Femur length (mm)
E
Sternum sterno-coracoidal width (mm)
Femur length (mm)
F
Pelvis length (mm)
Femur length (mm)
G
Pelvis width (mm)
Femur length (mm)
Harpia
Pithecophaga
Polemaetus
Stephanoaetus
Spizaetus
Morphnus
Lophoaetus
Aquila chrysaetos
Aquila audax
Haliaeetus leucogaster
Haliaeetus albilla
Gyps
Aegypius
Harpagornis

scavenger, its body conformation was likely to have been similar to that of living scavengers.

If indeed the wings were smaller and legs larger than expected in a raptor of its size, then the hypothesis that it was losing the power of flight would be supported. Smaller "weapons" and reduction in those parts of the skeleton associated with strong grip and killing ability would support carrion feeding.

Body Form

It is as difficult to reconstruct the proportions of a fossil species as it is to estimate its living body mass. The body conformation can be determined by comparing the parts of the skeleton with corresponding parts of other large birds of prey. Comparison is made easier when differences associated with body size are taken into account. By use of femur length as a measure of body size, graphs of cranial and limb bone lengths and widths for large eagles and vultures (Figs. 8.26–8.29) show that Haast's eagle had cranial, wing, and leg dimensions that could be expected for an eagle of its size.

Haast's eagle was obviously strongly sexually dimorphic in size, more so than even *Harpia.* In a size series of animals, to maintain a 10% difference in linear size, the greater absolute difference in size between the sexes arises simply because of the sheer size of the bird. For example, by use of femur length as the measure of body size, as before, if a male bird has a femur 100 mm long, the female would be expected to have a femur 110 mm long. But a female in a species in which the male's femur measures 150 mm would have a femur 165 mm long to maintain the 10% difference. In fact, analysis of the few data on the large eagles suggests that differences in the linear measurements are of that order. In four *Harpia,* the femur length was 112, 129.7, 119.8, and 120.4 mm, respectively. If the individual with the 112-mm femur was a small male and the two that had a femur 120 mm long were female, then the difference is a little over 7%.

Unfortunately, the sex of the skeletons used in the analysis was not known. If the assumed male harpy was a reasonably large individual of its sex, then the larger female's femur (129.7 mm) was 16% longer. In Haast's eagle, femur length is 150–173 mm, which fits with the hypothesis of similar difference between the sexes. But in doing so, it results in a larger absolute size gap between sexes. The underlying reason for the difference presumably depends on the ecologic or behavioral factors that have resulted in female accipitrids being larger than males (Brown and Amadon 1968). Perhaps a smaller proportional difference would not be biologically significant. But for the largest birds of prey, such as Haast's eagle, maintaining the difference begins to impose other stresses on the way the bird can function and behave. *Harpagornis* was approaching the upper limit of body mass for powered flight (Greenewalt 1962) because of biophysical constraints. Supporting area (wing area) increases only as the square of bone lengths, but body mass increases with the cube of those lengths. There comes a point where the amount of muscle available is not sufficient to power flight and the wing area is insufficient to support the mass. Above this mass, birds (or indeed any

Figure 8.26. (opposite page) Bivariate plots of bone lengths against femur length for 14 taxa of large raptor. (A) Premaxilla. (B) Total skull. (C) Sternum length. (D) Sternum anterior depth. (E) Sternum sternocoracoidal width. (F) Pelvis length. (G) Pelvis maximum width. Taxon symbols at bottom right-hand corner of figure.

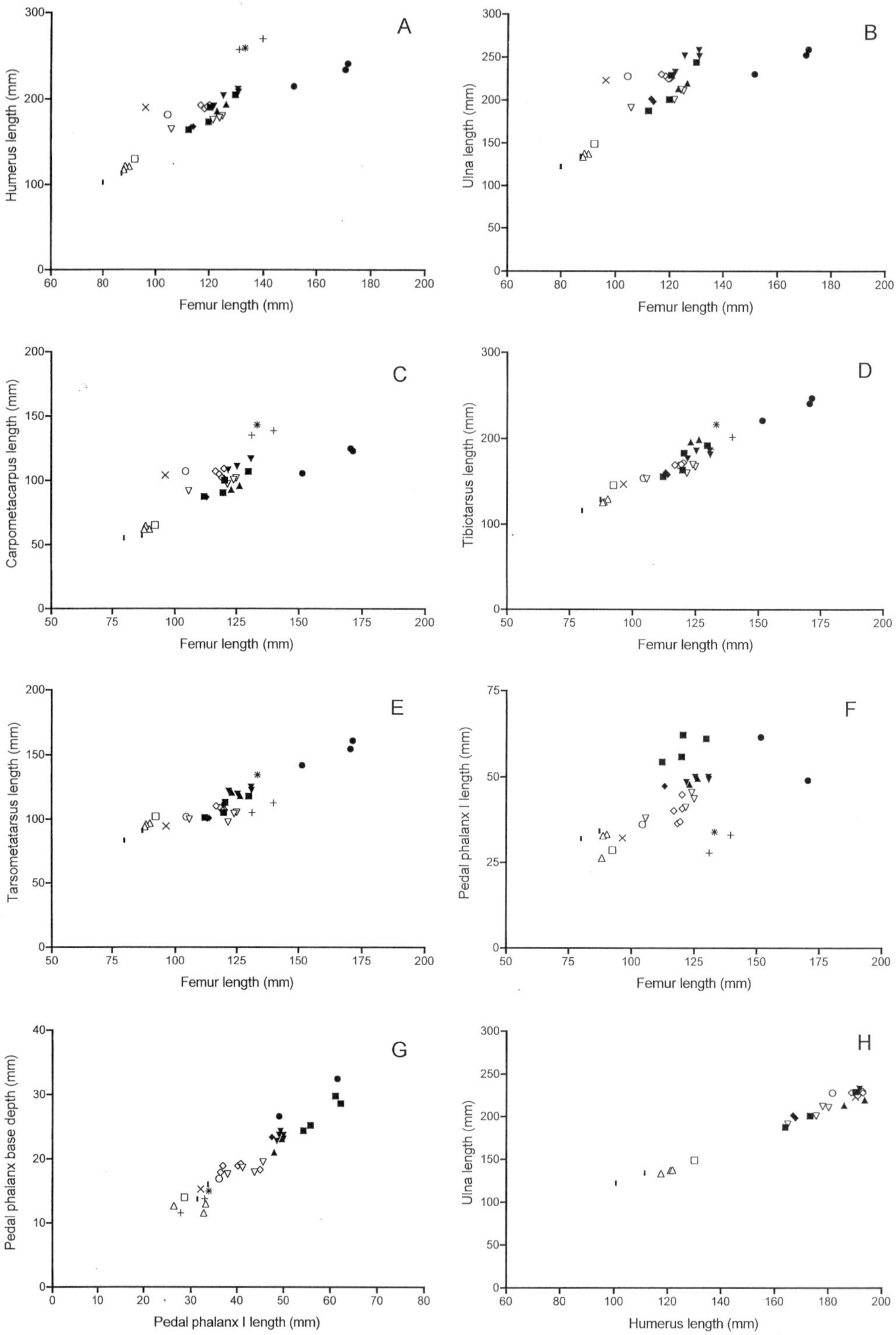

Figure 8.27. Bivariate plots of bone lengths against (A–G) femur length and (H) humerus length. (A) Humerus. (B) Ulna. (C) Carpometacarpus. (D) Tibiotarsus. (E) Tarsometatarsus. (F) Pedal phalanges. (G) Flexor tendon against phalanx length. (H) Ulna. Symbols as in Figure 8.26.

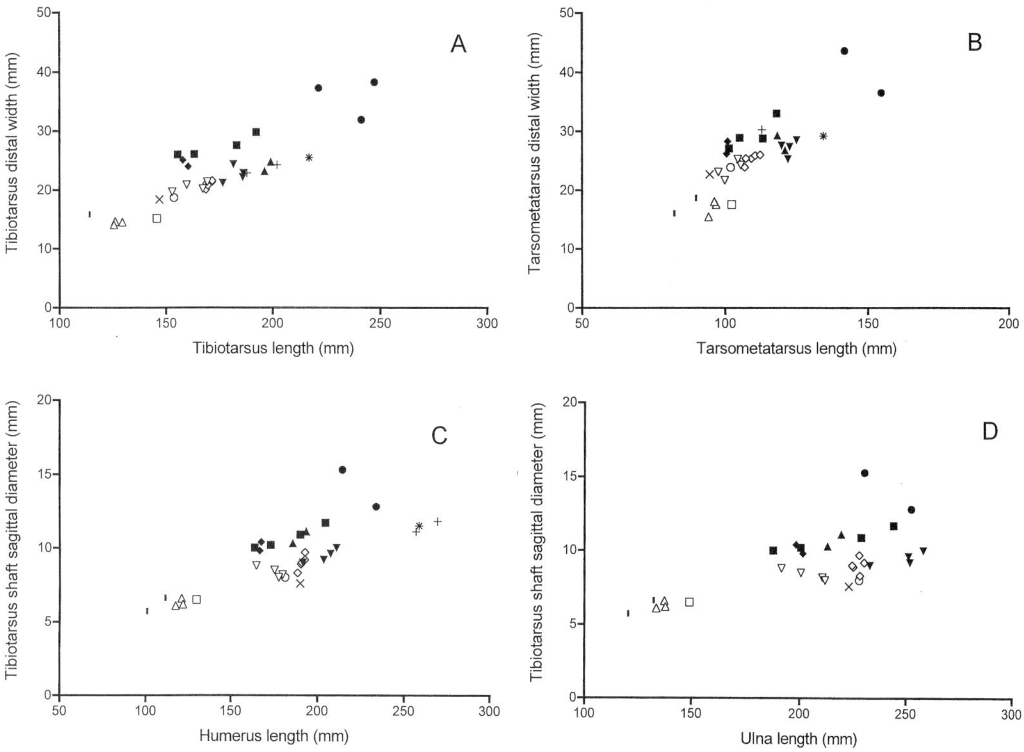

Figure 8.28. Bivariate plots of limb bone proportions. (A) Tibiotarsus width against length. (B) Tarsometatarsus distal width against length. (C) Tibiotarsus shaft diameter against humerus length. (D) Tibiotarsus shaft diameter against ulna length. Symbols as in Figure 8.26.

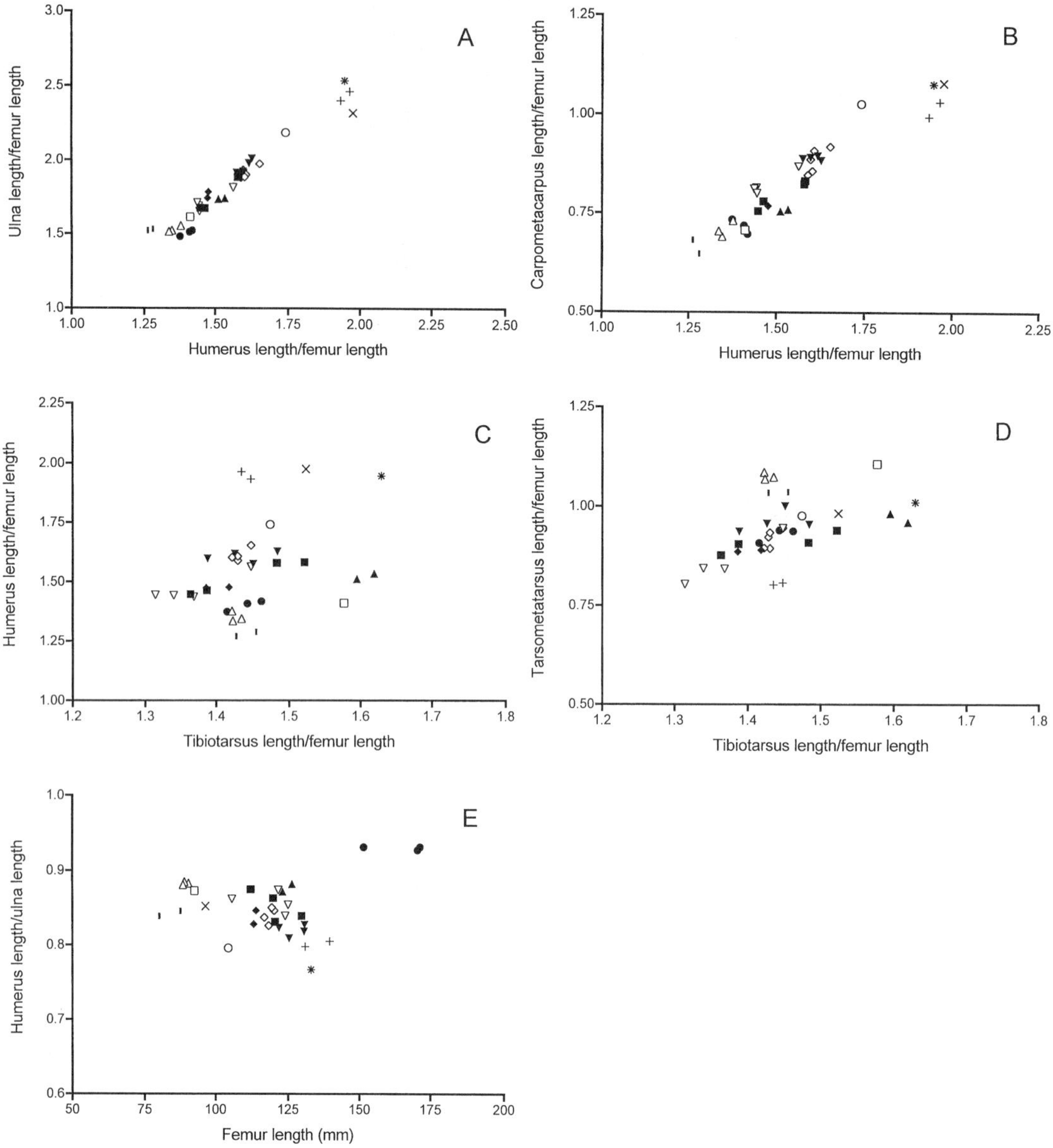

Figure 8.29. Normalised limb bone lengths (lengths divided by femur length). (A) Ulna against humerus. (B) Carpometacarpus against humerus. (C) Humerus against tibiotarsus. (D) Tarsometatarsus against tibiotarsus. (E) Humerus/ulna lengths against femur length.

flying animal, including pterosaurs) are restricted to gliding flight and taking off in strong winds or from cliffs (Campbell and Tonni 1983).

Even though Haast's eagle was about as large as it could possibly be and still function as an eagle, all its body parts were in proportion. As expected from its body size, the head and premaxilla (beak) were substantially bigger than even the larger of the two vultures, the black vulture *Aegypius cinereus* (Fig. 8.26A, B). They were not out of proportion, but such a large beak could have exerted a powerful pull.

Two other parts of the axial skeleton, the sternum and pelvis, are also useful indicators of the body form of a bird and the power of its flight and legs. Again, the dimensions for the sternum in Haast's eagle lie along the same line as the other great eagles (Fig. 8.26). Three measures of sternum size are shown: total length (Fig. 8.26C), anterior depth (Fig. 8.26D), and anterior width at the lateral edge of the sternocoracoidal impressions (Fig. 8.26E). In each, the dimension in Haast's eagle is greater than for the eagles in the sample but less than that for the two vultures. A vulture sternum is very long, and the deepest point of the keel is not at the anterior end but somewhat posterior to that. Vultures habitually soar in low-energy flight to seek prey by use of thermals in the air, rather than flapping to maintain lift. The deepest part of their breast muscles is set back, to one third to one half from the anterior edge of the keel, whereas in Haast's eagle, it is well forward beneath the coracoidal sulci. Although large and robust, the sternum was light and well pneumatized, and the "bowl" of the body of the bone is deep.

Forest eagles such as *Harpia harpyja* and *Stephanoaetus coronatus* flap with powerful beats of the wings, so the deepest part of the sternum and greatest bulk of pectoral flight muscle is well forward, to maximize the downward pull on the humerus. The anterior position of the maximum depth of the keel allows a direct line of application of the downward force of M. pectoralis on the base of the humerus and also a strong pronation of the wing leading edge during the downstroke. In vultures, the more posterior position of the greatest depth of the keel would tend to reduce the efficiency of the downstroke by increasing the angle through which that muscle must act (Fig. 8.30). Eagles in the genus *Aquila* are intermediate in that they have powerful flapping flight and can also glide for long periods. The Haast's eagle sternum is very like the other eagles in proportions, which suggests that it had the power for flapping flight. However, birds with that configuration can still glide well; for example *Pithecophaga jefferyi* is a forest eagle that glides above the canopy, and *Polemaetus bellicosus* hunts while soaring; both have forest eagle–type sterna. The muscle scars and attachment eminences on the wing bones were large, suggesting the insertion of large, powerful muscles. These areas are especially obvious on the manus, as Owen (1879a) noted long ago.

In pelvis length, Haast's eagle also follows the trend of the other eagles but again is an extreme example. The pelvis in the female is longer than in either of the vultures, and in the male it is the same size, which fits with its body mass being comparable to the vultures' (Fig. 8.26F). The vultures are not sexually dimorphic. At its maximum, just posterior to the acetabulum, the pelvis in Haast's eagle is rather wider than might be expected from the other eagles, which are all similar, regardless of body size (Fig. 8.26G). Where the Haast eagle pelvis differs from that of the large vultures is in the angle between the cranial (preacetabular) and caudal (postacetabular) sections. The greater angle in the pelvis axis is associated in birds of prey with the strength of grip of the toes, because the muscles have better and more direct lines of action. Conversely, it inhibits freedom of movement in walking (most terrestrial birds have almost flat pelves). The angle of the pelvis in Haast's eagle is consistent with the size and robustness of the leg bones and of the claws.

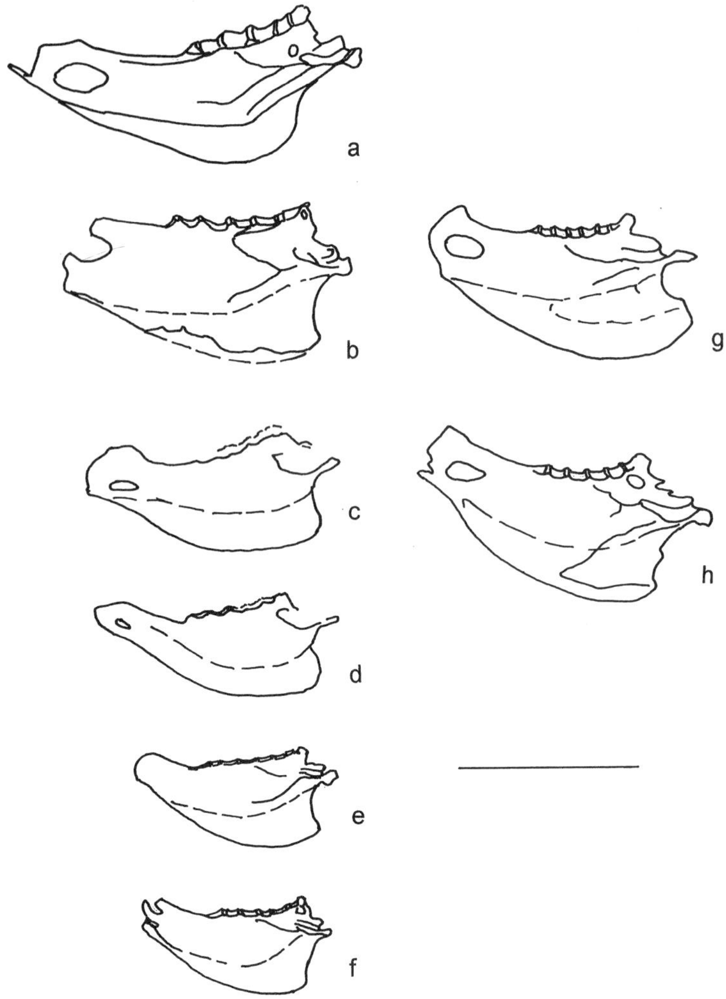

Figure 8.30. Sterna of a range of large raptors from two families, Vulturidae (a) and Accipitridae (b–h), with a range of flight styles, in right lateral view, showing depth of keel and orientation of anterior edge of keel. (a) Andean condor Vultur gryphus. *(b) Haast's eagle* Harpagornis moorei. *(c) Harpy eagle* Harpia harpyja. *(d) Philippine eagle* Pithecophaga jefferyi. *(e) Australian wedge-tailed eagle* Aquila audax. *(f) Australian white-bellied sea eagle* Haliaeetus leucogaster. *(g) Cooper's hawk* Accipiter cooperii. *(h) Egyptian vulture* Necrosyrtes monachus. *Scale bar = 100 mm (g and h not to same scale, to assist comparison of keel shape with those of the larger birds).*

In the length of its ungual phalanges (claws), Haast's eagle is simply a larger version of the living eagles (Fig. 8.27). Although its claws were proportionately shorter than those of *Harpia* (whose claws are long for its body size), they were significantly longer than those in the two vultures. What Haast's eagle claws lacked in extra length, they made up for in robustness and the power that could be applied to the point. That power is proportional to the depth of the attachment boss for the flexor tendon at the base of the claw. The greater the distance between the center of rotation of the base of the claw and the attachment of the tendon, the greater the force that could be applied. In this dimension, the vultures and Haast's eagle are at the opposite ends of the distribution (Fig. 8.27G), and *Harpia* is more in proportion to its body size. The vultures have very little gripping power and do not use the feet, except in fighting and in pushing against a carcass to gain a better pull with the beak. *Harpia* catches mammals of up to its own body mass. Haast's eagle could apply even greater forces to the ends of its thicker claws—forces large enough to drive the points through several centimeters of moa flesh and into, and often through, the bone.

Large forces in the claws require robust leg bones to transmit and to resist reactions to both the claw forces and struggling prey. Both the tibiotarsus and tarsometatarsus in Haast's eagle were as long as expected from the body size (Fig. 8.27), but they were somewhat thicker, as measured by the width of the distal end of each bone; the distributions of both are set above the general trend in the living species. In comparison to the two vultures, Haast's eagle had a much thicker tibiotarsus (Fig. 8.28). When measured against wing bone lengths, the difference in structure between the large vultures and Haast's eagle is most obvious (Fig. 8.28C, D). The soaring and gliding species (vultures and *Aquila* eagles) have perceptibly thinner tibiotarsi than eagles of equal size, in proportion to the length of both the humerus and ulna (Fig. 8.28C, D). In this respect, Haast's eagle was an extreme version of the living forest eagles. Its leg bone length proportions were normal for large eagles (Fig. 8.28), but its tibiotarsi were stouter. The outliers are the Philippine eagle *Pithecophaga jefferyi* and the black vulture *Aegypius cinereus,* which have long tibiotarsi, and the griffon vulture *Gyps fulvus,* which has short tarsometatarsi, both relative to body mass. *P. jefferyi* uses its long legs to catch prey in tree holes, and the vultures spend much time on the ground, around carrion.

The wing has been the most controversial part of the anatomy of Haast's eagle. That is because flight style and strength is reflected in the lengths and relative proportions of the major wing bones. From the time of Haast's original statement that the ulna of his new species was relatively short, the proportions of its wing bones have been used to downplay its flying ability. The three wing bones that constitute most of the length of the wing skeleton are the humerus (which transmits the heavy loads from the flight muscles to the wing and the reactive loads of the air pressure and drag back to the body), the ulna (which carries the secondary feathers producing most of the wing lift), and the carpometacarpus (to which are attached the primary feathers that develop the thrust to fly). In humerus length, Haast's eagle was much as expected from the dimensions of other large eagles (Figs. 8.27, 8.29). The large vultures, however, have much longer humeri than their body mass might suggest. That extra length allows more area for a soaring wing. The gain in area is at the cost of a long lever that is more difficult to flap efficiently as the center of aerodynamic pressure is farther from the body axis, and the resistance to rotation (angular momentum) is greater in a longer arm.

Soaring birds typically have long ulnae that give a long, narrow (high aspect ratio) wing with low drag that economizes on energy for sustained flight. Vultures have disproportionately long ulnae, both in terms of body size (Fig. 8.27B) and humerus length (Fig. 8.27H). Indeed, ulnae of *Aquila* and *Polemaetus* are also slightly longer than those of the forest eagles for body size (Fig. 8.27), which indicates their preference for soaring. Haast's eagle had ulnae that were in keeping with the main trend of the forest eagles (Fig. 8.27B), especially in terms of humerus length (Fig. 8.27H).

Haast's eagle also had carpometacarpi that were about as long as expected for its body mass (Fig. 8.27). Eagles that can soar, such as *Polemaetus bellicosus* and *Aquila audax,* have carpometacarpi that are slightly longer than expected for their body mass, thus enabling a large attachment surface for larger primary feathers, giving the gliding wing more length and area. Extreme development of the hand is seen in specialist soarers, the vultures *Gyps fulvus* and *Aegypius funereus.*

In overall body form, Haast's eagle had the same proportions as living eagles but was much larger. The greater size was not accompanied by the adaptations to soaring flight that are typical of large vultures; rather, Haast's eagle shared the body—and especially wing—plan of the largest living forest eagles. The wings were powerfully built, and the mass rotating about the shoulder joint was kept close to the body axis, where the forces could be applied more directly; the bending moments on the humerus were kept to a minimum.

The pelvis and legs were adapted to providing a strong grip and especially great force in the flexion of the claws. The legs were powerfully built and the bones were thick. The large circumference gave added resistance to the bending and torsional loadings required for a strong grip and imposed by large, struggling prey. For these reasons, the legs, feet, and claws were far more robust than in the largest living vultures and even stronger than those in the fearsome harpy and crowned eagles.

Flight

As noted above, the massive size of Haast's eagle has prompted suggestions that the bird could not fly well; its large size has even prompted the extreme view that it was becoming flightless. Only one flightless raptorial bird is known: the now-extinct immense owl *Ornimegalonyx oteroi* of Cuba (Arredondo 1976), whose femur was slightly shorter than that of the smallest Haast's eagle. The owl had a flat sternum with a vestigial keel, characteristic of flightless species in all groups in which they occur. Nothing about the individual bones in the skeleton of Haast's eagle described above suggests anything other than that it was a powerful flapping flier. To recapitulate, the sternum in Haast's eagle has a fully developed and robust keel, which is deeper and thicker than that in *Harpia.* And the structure of the pectoral girdle is that of a fully flighted species, with robust coracoids and scapulae, which meet at an acute angle rather than the obtuse angle with downward-angled sternum typical of flightless birds. Another clue supporting the notion that Haast's eagle was a strong flier is the expansion of the neural canal in the thoracic vertebrae at the shoulder, which is fully as marked as in *Harpia* and other eagles and is not reduced as in flightless or poorly flighted species (Giffin 1990). The larger neural canal at the shoulder provides space for the thicker spinal cord at that point, which results from the large number of nerves needed to control the wing.

Brathwaite (1992) conjectured that the relatively short ulna in Haast's eagle was consistent with its having a short, broad wing. He combined this with the observation that the scar for the wing-raising supracoracoideus muscle was large compared with that in two other eagles of open country (*Aquila audax* and *Haliaeetus leucogaster*), to suggest that the bird had a powered upstroke to its wing that allowed it to take off from confined spaces (Brathwaite 1992). He rejected the view that Haast's eagle flew poorly, although he conceded that its great mass would have stretched its physical and aerodynamic abilities. He concluded that the strong legs could have allowed it to spring into the air from the ground, in the manner of *Stephanoaetus coronatus,* which can take off almost vertically from the forest floor (Brown 1982).

Olson and Feduccia (1979) pointed out that the large pectoral muscle in birds arises from the furcula and the coracoclavicular membrane, the strong fascia or sheet of connective tissue between the anterior sternum, the coracoids, and only from those parts of the carina and body of the sternum that are not occupied by the supracoracoideus. Hence, although the size of the sternal carina is important in determining the size of the pectoral muscle and thus the flight ability of a bird, it is not the only, or even perhaps the major, factor. When the extent of the insertion area available on the furcula, fascia, and the junction between the left and right pectoral muscles is taken into account, the volume of pectoral muscle in *Harpagornis* can be seen to be more than adequate to allow sustained, strong flight.

As noted above, much of the sternal carina is occupied by the insertion of the supracoracoideus muscle. However, although part of the function of the supracoracoideus is to raise the wing—indeed, that is usually the primary function attributed to that muscle (e.g., George and Berger 1966: 22)—severing the tendons attaching the muscle to the humerus does not materially affect the bird's flight (Sy 1936). In the absence of the supracoracoideus, the wing is raised by the deltoid complex (mainly the deltoideus major) and by the effects of weight around the shoulder joint. What the bird could not do was take off from flat ground. The observation agrees with the conclusion of Hartman (1961), who pointed out that there was a relationship between the mass of the supracoracoideus and the ability of the bird to make a quick takeoff.

Multivariate Analysis

Multivariate analyses allow an examination of the overall body plan in relation to those of other species. The technique of principal component analysis reduces the variation in all characters—in this case bone measurements—to three variables that allow the differences in proportions among species to be examined as a whole. By this method, species of known feeding technique or flight type are seen to group together; its position in the three-dimensional space defined by the new variables indicates the likely characteristics of an extinct species for which these cannot be directly observed or measured. In short, there would be expected to be distinct groups for predators and scavengers and for gliders and flappers. In an analysis including *Harpagornis,* plots of the first three components (Fig. 8.31) for the lengths and other

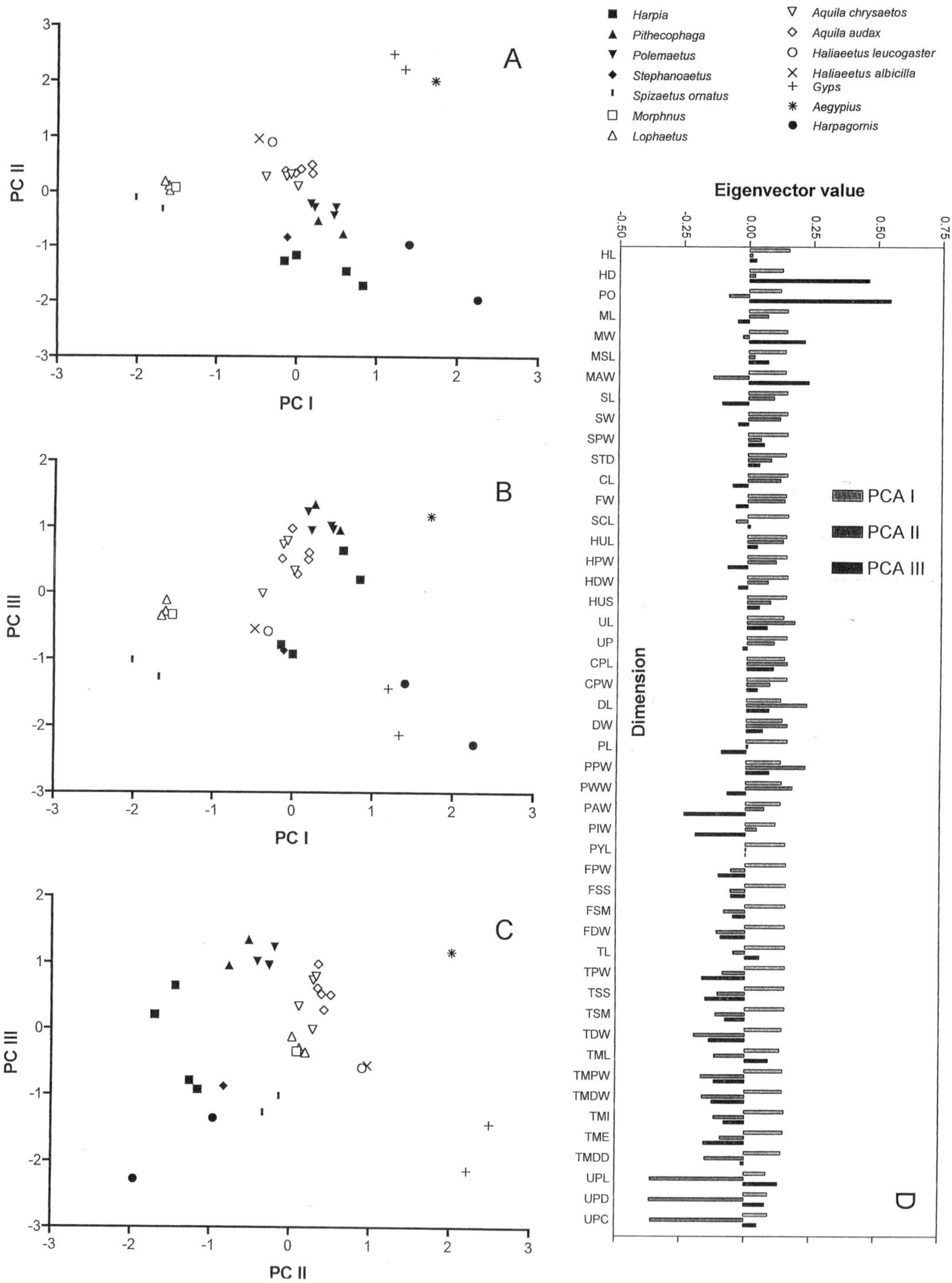

Figure 8.31. Plots of scores from principal component (PC) analysis of skeletal dimensions of large raptorial birds. (A) Scores for PC II plotted against scores for PC I. (B) Scores for PC III plotted against scores for PC I. (C) Scores for PC III plotted against scores for PC II.

(caption continued on next page)

dimensions of 49 bones in 15 species in 11 genera of large accipitrid whose favored foods and flight mode are known. The first component accounted for 81.6% of the variance of the independent variables, the second component another 9.7%, and the third 2.2%.

The results show that *Harpagornis moorei* was most similar to *Harpia* in its body form, with some important differences. The two individuals that were complete enough for the analysis were part of the group of large forest eagles; indeed, they were the most extreme members of that group. The species was not close to either of the two *Aquila* eagles, despite the genus *Aquila* being the sister group (in phylogenetic relationship) of *Harpagornis* (Holdaway 1994). The *Aquila* formed a group regardless of which pair of components was plotted. *H. moorei* differed from the two species that most closely approached it in size, the large accipitrid vultures *Aegypius* and *Gyps,* which had high scores on component II, for which *Harpia* and Haast's eagle had the lowest scores (Fig. 8.31A).

The first principal component (PC I) is fundamentally a size axis, with the presumed female *Harpagornis moorei* from Mount Owen being the largest of the sample. PC II had higher positive loadings for the length of the ulna and the first alar digit (Fig. 8.31D), which are measures of wing length (and hence gliding ability); the species with the highest scores were the two *Gyps* vultures and *Aegypius monachus.* Negative scores were related to the dimensions of the leg bones, particularly the distal width of the tibiotarsus and tarsometatarsus and the proximal width of the tarsometatarsus. These are measures of the robustness of the lower leg and foot, so scores on the PC II axis contrasts the importance of the longer wings of the vultures with their insignificant claws to the shorter, broader wings and massive legs and claws of the eagles such as *Pithecophaga jefferyi, Stephanoaetus coronatus,* and *Harpia harpyja* that catch and kill large prey. On these axes, *H. moorei* appears as a very large member of the latter group. One additional factor contributing to the negative scores for the largest eagles was the width of the mandibular articulation (Fig. 8.31D), which suggests that the larger prey are also linked with a stronger jaw apparatus to deal with the prey.

The separation of the four *Harpia* specimens into two groups of two suggests that they represent the two sexes. There is a similar separation of the two *Harpagornis* individuals that supports the a priori attribution to sex that has been adopted on the basis of dimensions. *Polemaetus bellicosus,* a powerful African eagle that combines a soaring flight mode with handling of relatively large prey items (Brown and Amadon 1968), forms a link group between the largest eagles and the typical *Aquila* eagles. The two species of *Haliaeetus* lie very close to each other, with as little difference as between different individuals of the two species of *Aquila.* All three of the smaller accipitrids apparently have similar wings and claws and hence form a tight grouping.

When PC III was plotted against PC I (Fig. 8.31B), the pattern was rather similar, except that the two *Gyps* vultures were much nearer *H. moorei* in loading on PC III than on PC II. Positive loadings on PC III were related to the size of the head, with species with higher scores

(D) Values for eigenvectors for dimensions for first three PCs, indicating contribution to total PC score for each component. CL = coracoid length; CPL = carpometacarpus length; CPW = carpometacarpus proximal width; DL = manus major digit proximal phalanx length; DW = alar major digit proximal phalanx width; FDW = femur distal width; FPW = femur proximal width; FSM = femur shaft medial diameter; FSS = femur shaft sagittal diameter; FW = furcula width; HD = cranial depth; HDW = humerus distal width; HL = total head length, including premaxilla; HPW = humerus proximal width; HUL = humerus length; HUS = humerus shaft width (dorsoventral); MAW = length of mandible articular process; ML = mandible length; MSL = mandible tip to surangular process; MW = mandible width; PAW = pelvis acetabular width; PIW = pelvis ischial width; PL = pelvis length; PO = post-orbital width; PPW = pelvis proximal width; PWW = pelvis "waist" width; PYL = pygostyle length; SCL = scapula length; SL = sternum length; SPW = sternum proximal width; STD = sternum total depth; SW = sternum width; TDW = tibiotarsus distal width; TL = tibiotarsus length; TMDD = tarsometatarsus distal depth; TMDW = tarsometatarsus distal width; TME = tarsometatarsus lateral hypotarsal ridge depth; TMI = tarsometatarsus medial hypotarsal ridge depth; TML = tarsometatarsus length; TMPW = tarsometatarsus proximal width; TPW = tibiotarsus proximal width; TSM = tibiotarsus shaft medial diameter; TSS = tibiotarsus shaft sagittal diameter; UL = ulna length; UP = ulna proximal width; UPC = ungual phalanx pedal digit I chord length; UPD = ungual phalanx pedal digit I depth at base; UPL = ungual phalanx pedal digit I length.

having much deeper heads and beaks than those with negative scores: *H. moorei* has one of the "lowest" crania of any accipitrid for its size. High negative scores were related to the dimensions of the pelvis and leg bones, but not the claws, and hence the proximity of *Gyps* and *Harpagornis* when wing size and claw size were not included.

Plotting PC III against PC II (Fig. 8.31C) provided the best discrimination of the functional groups of eagles and vultures because the general confounding element of size was eliminated and the loadings on both axes reflect the relative size of head, wings, legs, and claws. The three great eagles that kill (or killed) large prey (*Harpia, Stephanoaetus,* and *Harpagornis*) form a marginal cluster. Both taxa for which both sexes are represented also show large separations between the sexes in their physical attributes. With size out of the equation, the forest eagle *Spizaetus ornatus* is closest to *Harpagornis* and *Stephanoaetus* in conformation, which is a partial vindication, at least, of Brathwaite's hypothesis of the relationships of *Harpagornis.* Structurally, at least, *Harpagornis* was a giant version of the spizaetine body form. That this form has been recognized tacitly in the past is shown by the fact that *Stephanoaetus* has sometimes been placed in or near *Spizaetus.* For example, Brown and Amadon (1968: 707) thought that it was very closely related to *Spizaetus,* "of which it is a specialized offshoot."

The other large eagles separate out into three groups (Fig. 8.31C): *Pithecophaga* was close to *Polemaetus,* the two *Aquila* were indistinguishable and close in conformation to *Lophaetus* and *Morphnus,* and the two *Haliaeetus* species were again very close. The large vultures that occupy the "long-winged" end of the PC II axis are obligate soarers with weak claws, and hence the birds at the other end of the distribution are more likely to have flapping flight and to wield large claws. So *Harpagornis* appeared from this analysis to be a supereagle whose body form is most similar to that of *Harpia* and *Stephanoaetus,* the species that take the largest live prey in comparison to their own body mass. Apart from the shallow cranium, it had none of the characteristics of an obligate carrion eater.

Application of another multivariate analysis technique (discriminant function analysis) shows that when its dimensions are contrasted with those of eagles that habitually either flap or glide, Haast's eagle appears as a "super flapper" (Fig. 8.32A). When canonical variables for wing dimensions are plotted against those for leg bones (Fig. 8.32B), there are two clusters corresponding to the flappers and gliders. Paradoxically, the two Haast's eagles are separated in the plot. They were part of neither group; rather, they occupied points between the clusters but offset to either side of the axis joining the clusters. The leg dimensions of one individual (the smaller male) were closest to those of the flappers; the opposite applied to the wing dimensions. Because the individuals differed markedly in size, probably in relation to the sex of the individuals, one interpretation of the data is that the two sexes had different proportions and could even have had different flight modes. These results reinforced the conclusions from the PC analysis that the larger eagles, including all raptors except the vultures, exhibit a marked

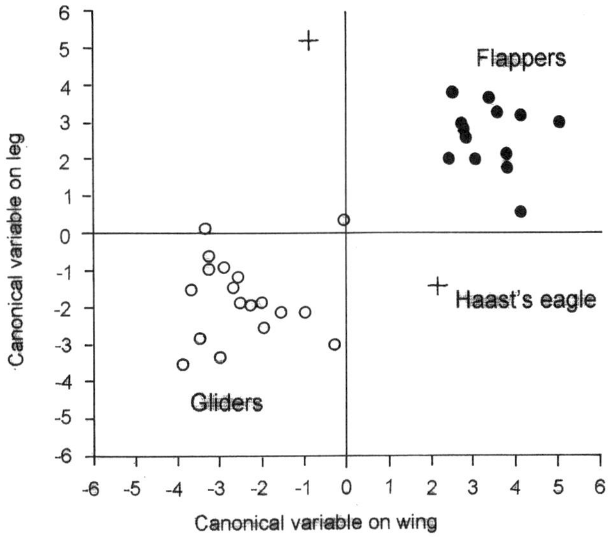

Figure 8.32. (A) (below) Plot of scores on canonical variables of dimensions contrasting flapping and gliding eagles in a discriminant function analysis of dimensions of several large raptors and vultures compared to Harpagornis moorei.
(B) (left) Plot of canonical variables of dimensions contrasting wing and leg bones in a discriminant function analysis of various large raptors and vultures compared to Harpagornis moorei.

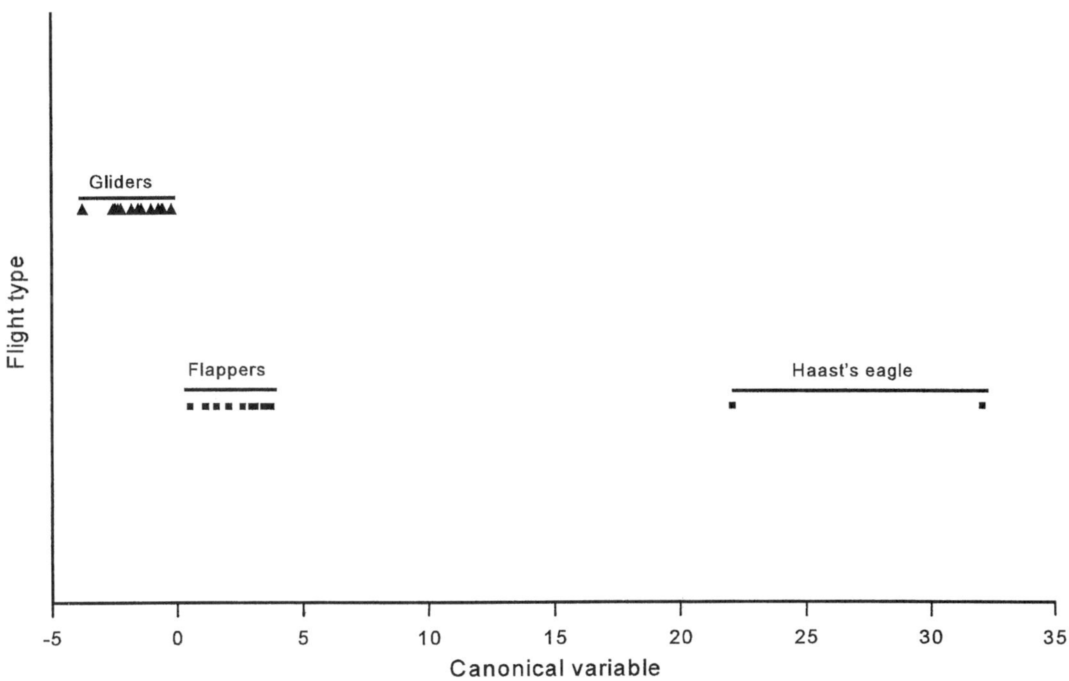

sexual dimorphism in body size, but the sexes of Haast's eagle apparently also had different feeding and flight modes.

The only conclusion can be that Haast's eagle has normal body form and proportions for an eagle of its size. The comparisons with other species also show that it had a body form most similar to the large predatory eagles and that it was better adapted for flapping flight than for soaring. But it definitely was not becoming flightless.

Flight Ability: Physiology and Energetics

Inferences of the feeding ecology of Haast's eagle depend on the bird's ability to fly. Although the structure of the bones and the pectoral girdle indicate strongly that the species flew well, the species was so large that predictions of the physiologic requirements for flight are of great interest. Flying ability can be assessed by calculating the power required for flight from the size of the flight muscles and the power available from them, and from the ability of the bird to sustain this power output.

Three independent methods for estimating the power required were used, those of Tucker (1973, 1974) and Greenewalt (1975), which rely on conventional steady-state aerodynamics, and that of Rayner (1979a, 1979b), which is based on vortex theory. Surplus power was calculated by the method of Caccamise and Hedin (1985). Parameters, such as wingspan and equivalent flat plate area, which are required for applying the power equations to Haast's eagle, were calculated from relationships given by the respective authors and from empirical estimates obtained from reconstructions. The mass of the major flight muscles was estimated from relationships given in Greenewalt (1962) and compared to the amount suggested by the size and shape of the sternum and the pectoral girdle, particularly the furcula.

Figure 8.33. (opposite page) Curves of power required (in watts) for a range of large birds, for flight velocities up to 25 m/s, according to two theories of flapping flight: broken lines = Rayner (1979b, 1979c); solid lines = Tucker (1974), both sets assuming 20% efficiency. Solid circles at lowest point of curve, velocity of minimum power; solid circles to right of lowest point (tangential to origin), velocity of minimum transport (cost per unit distance traveled). (a, f) Harpagornis moorei, *mass 13 kg, wingspan 2.43 m. (b)* Harpagornis moorei; *mass 13 kg, wingspan 2.6 m (c, h)* Harpagornis moorei, *mass 10 kg, wingspan 2.14 m (equivalent to a large female* Harpia harpyja*). (d, g)* Vultur gryphus, *mass 12 kg, wingspan 3.0 m. (e, i)* Diomedea epomophora, *mass 8 kg, wingspan 2.82 m. (j)* Branta canadensis, *mass 5.6 kg. (A–C) Curves of maximum sustainable power available (in watts) for birds with pectoral muscle mass scaling to body mass according to the relationship in Greenewalt (1962), at optimum velocity and flap rate. (A) Mass, 13 kg. (B) Mass, 10 kg. (C) Mass, 5.5 kg.*

Application of the general relationship for pectoral muscle mass for different body masses ($M_m = 0.17\ M^{1.0}$, Greenewalt 1962) suggested that a 13-kg bird (hence a female Haast's eagle) would have had about 2.2 kg of pectoral muscle, or 2.01 kg when the alternative value of 15.5% of body mass is used. Similarly, the supracoracoideus mass would have been about 200 g (1.6% of body mass). If, as the size of the scar for the supracoracoideus muscle on the sternum suggests, that muscle was better developed than in other species, its mass would have been higher, but the total pectoral mass is unlikely to have exceeded 2.25 kg. The profile of the full-size displacement model suggests that at a muscle density of 1060 kg/m^3 (Weis-Fogh and Alexander 1977), there would have been sufficient attachment area and volume for this mass of muscle.

The curves of power required for flight versus flight velocity for birds of 10 and 13 kg with wingspans of 2.14, 2.43, and 2.6 m (calculated from dimensions of wing bones) are given in Figure 8.33, along with curves for two other large flying birds, the condor (*Vultur gryphus*), which soars, and the southern royal albatross (*Diomedea epomophora*), which uses the winds over the Southern Ocean in so-called dynamic soaring. Curves were plotted for power required according to equation 2 in Tucker (1974) and Rayner (1979a, 1979b). Both sets of curves were based on a muscle efficiency of 20%. The maximum sustainable power output for a bird with 2.2 kg of pectoral muscle and for muscle masses equivalent to the two lower body masses are also plotted on Figure 8.33.

The area above a power-required curve cutoff by a curve of power available represents the power envelope for which flight is possible (Fig. 8.33). The velocities at the limits of the cutoff area represent the maxi-

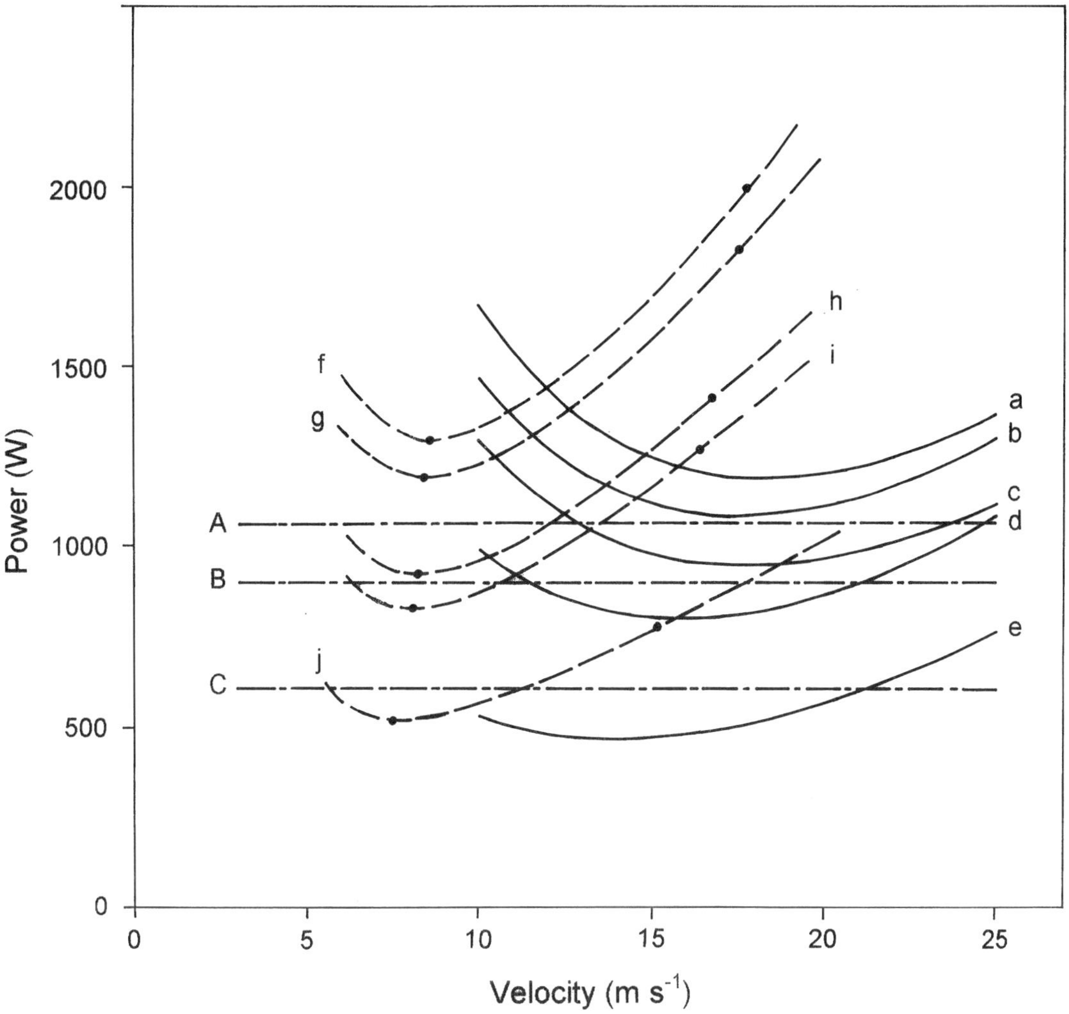

mum and minimum velocities of steady-state flight. The curves do not coincide for the 13-kg bird, but Tucker (1974) noted that many of the predictions (of the momentum jet theory) are accurate to only about 10%–20%. The power-required curves used here could well be too high for birds of this mass. If so, this would bring the power-required curve for a large female below the power-available curve. The lighter male had the same mass as that of a female *Harpia* that obviously flies well.

The most economical flight speed (lowest point of the power-required curve) increases as body mass becomes greater (Fig. 8.33). The vortex theory (Rayner 1979a, 1979b, 1979c) suggests a lower velocity for minimum power (all the Rayner curves are to the left of the Tucker curves). A female Haast's eagle with a 2.6-m wingspan would, according to the equation given by Tucker (1974), fly most economically at about 17 m/s (nearly 60 km/h), but at 8.5 m/s (30 km/h), according to that of Rayner (1979c).

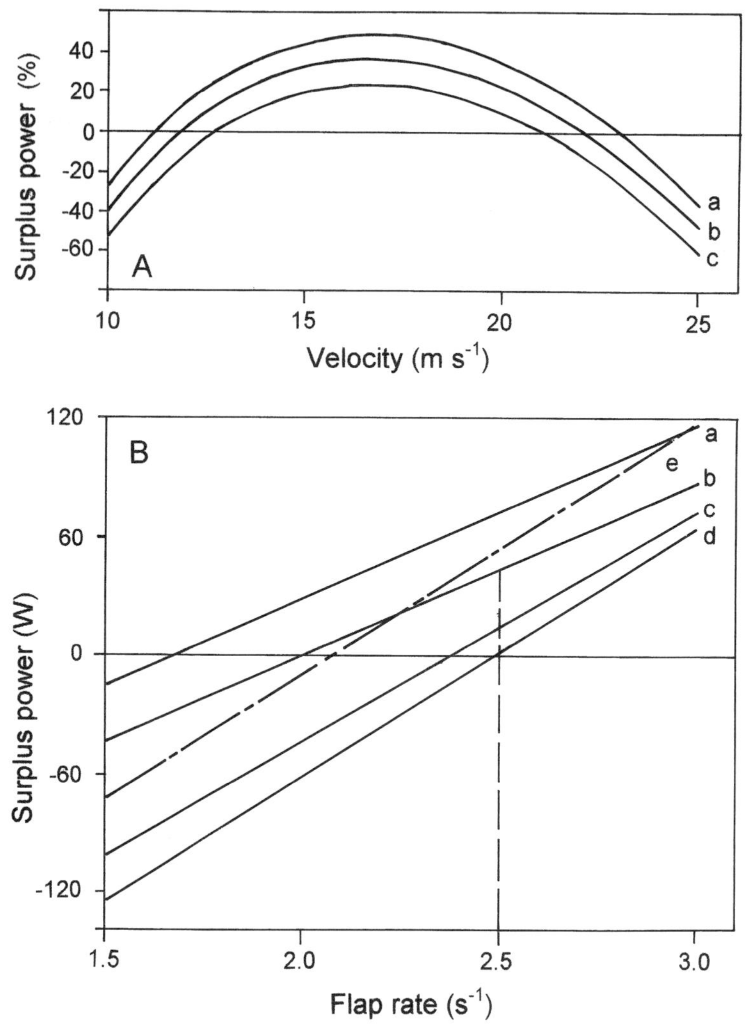

Figure 8.34. (A) Surplus power (in watts) for a 13-kg bird with 2.2 kg of pectoral muscle and a wingspan of 2.43 m (mass estimate for average female Harpagornis moorei, *with shorter-than-average wingspan), for flight velocities of 1–25 m/s, at three flapping rates within the range recorded for Andean condors (*Vultur gryphus*) (McGahan 1973): (a) 2.0 flaps/s, (b) 2.1 flaps/s, and (c) 2.2 flaps/s. At low (≤11–13 m/s) and high (≥21–23 m/s) flight velocities, there is predicted to be insufficient power for sustained flight. (B) Variation in surplus power available for birds of 9–13 kg body mass, for flap rates of 1.5–3.0 flaps/s, from the method of Caccamise and Hedin (1985). Vertical broken line = rate sustained by* Vultur gryphus *for 53 flaps (McGahan 1973). (a) Mass 9 kg, wingspan 2.43 m. (b) Mass 9 kg, wingspan 2.14 m. (c) Mass 11.5 kg, wingspan 2.14 m. (d) Mass 13 kg, wingspan 2.14 m. (e) Mass 13 kg, wingspan 2.43 m (estimated for female* Harpagornis moorei*). Horizontal line at 0 W surplus power indicates the break-even point, at which each bird can just sustain flight.*

Power required for flight increases rapidly at low airspeeds (Fig. 8.33, both sets of curves), and takeoff would be the most difficult phase of flight in very large birds. Figure 8.34A shows the surplus power available from the pectoral muscles (in a 13-kg bird) at different flight velocities for three flap rates (2, 2.1, and 2.2 flaps/s) that are within the known range for *Vultur* (McGahan 1973). Figure 8.34B shows the variation in surplus power with flap rate for birds of different body masses and wingspans. The curves show that for realistic flap rates, the pectoral muscles could have had surplus power available in steady flight, for ascent, for maneuvering, or for carrying food or nesting materials for individuals of *Harpagornis moorei* of all reasonable combinations of body mass and wingspan (Fig. 8.35).

That the wingspans and flap rates used for *Harpagornis* in the analyses are within the bounds of possibility is shown by a comparison of the predictions of general equations developed by Greenewalt (1962,

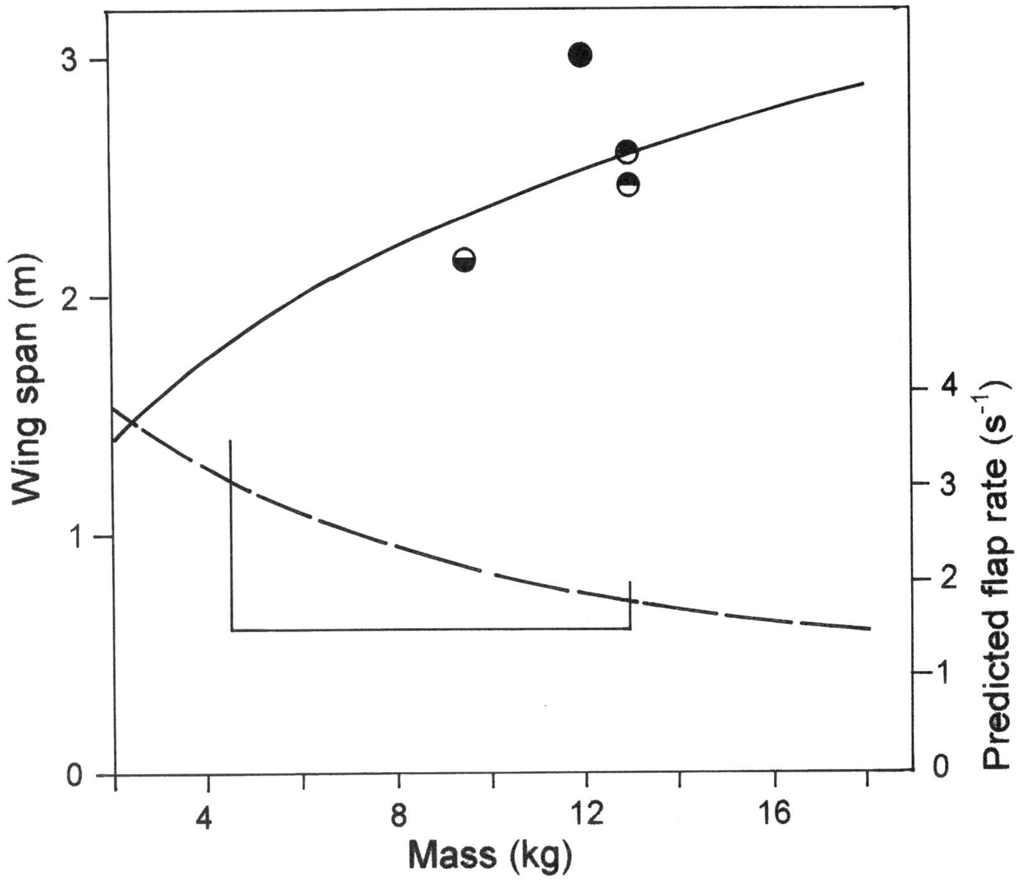

Figure 8.35. Predicted wingspan (solid line, measured in meters) and flap rate (broken line, measured in flaps per second) for birds of more than 2 kg of body mass, from relationships in Greenewalt (1962). Solid circle = Vultur gryphus; *top-filled circles =* Harpagornis moorei; *bottom-filled circle =* Harpia harpyja, *female. The section cut off the curve for the predicted flap rate indicates the range of body masses predicted for* Harpagornis moorei *from various allometric relationships between leg bone dimensions and body mass (see Fig. 8.25).*

wingspan) and Pennycuick (1975, flap rates). The 3-m wingspan of a 12-kg *Vultur* lies well above Greenewalt's curve (span = 1.1 $M^{0.33}$), whereas spans calculated from wing bone lengths for the mass range of Haast's eagle lie close to the curve (Fig. 8.35). Maximum flap rates predicted from Pennycuick (1975, $f_{max} = aM^{-0.33}$) for the body mass range of Haast's eagle and *Vultur* lie well below those (up to 3 beats/s, with a rate of 2.5 beats/s sustained for 53 beats) measured in flying *Vultur* (McGahan 1973).

Other Parameters: Predictions and Possibilities

Most biologic characteristics are related in some way to body size. Within the limits of the data on which allometric relationships can be developed, predictions can be made (Peters 1983) that allow a more complete picture of the living eagle to be compiled. Parameters such as wing beat frequency, flight speed, territory size, clutch size, and food requirements can all be surmised on the basis of the relationships and by analogy with living species in the same phylogenetic group. Table 8.15 summarizes various predicted physiologic and ecologic parameters for raptorial birds the size of Haast's eagle, assuming a body mass

TABLE 8.15.

Physiological and ecological parameters estimated for a large female *Harpagornis moorei* from allometric relationships in Berger and Hart (1974), Calder (1974), and Peters (1983, various sources cited within).

Parameter	Berger and Hart (1974) Value	Calder (1974) Value	Peters (1983) Source	Peters (1983) Value
Ventilation (V)	42.31 L min^{-1}			
Frequency				
Maximum	209.6 min^{-1}			
Minimum	54.14 min^{-1}			
$\dot{V}O_2$	925 mL O_2 min^{-1}	5.5 L kg h^{-1}		
	(=4.27 L kg h^{-1})			
$Power_{cont,max}$	308.2 W (=265 kcal h^{-1})			
Heart rate				
During flight	6.05 s^{-1}			
After flight	5.96 s^{-1}			
Maximum	5.67 s^{-1} (nonflight)	86.4 min^{-1} (rest?)		
From heart mass	6.06 s^{-1} (H_m = 64.24 g)			
	5.63 s^{-1} (H_m = 97.83 g)			
Heart mass	97.88 g (42.13, 227.42)	95.41		
	100.83 g			
	64.24 (54.17, 76.18)			
O_2 $pulse_{flight}$	0.70 µL O_2 $beat^{-1}$			
Cardiac output (Q_{min})	4.58 L min^{-1}			
Metabolism				
Existence, H_{ex}		32.86 W		
Flight, H_{flight}		343.9 W		
		1102.1 W		
Thermoneutrality (T_{lc})		–38.4°C		
Food requirements		79 W/d (=0.207 kg)	Schoener	76 W/d
Prey mass (raptor)			Schoener	1.94 (1.13, 1.33) kg

Parameter	Berger and Hart (1974) Value	Calder (1974) Value	Peters (1983) Source	Value
Large-prey raptor			Vzing	166.2 (119, 231.9) kg
Egg mass			Cabana et al.	0.429 (0.403, 0.456) kg
			Blueweiss et al.	0.383 kg
			Ricklefs	0.221 (0.203, 0.242) kg
Clutch mass			Cabana et al.	1.44 (1.29, 1.61) kg
			Blueweiss et al.	1.375 kg
Hatchling mass			Blueweiss et al.	0.193 kg
Incubation time			Blueweiss et al.	42.2 d
			Rahn and Ar	94.9 (75, 120) d
Age at first flight			Ricklefs	85 d
Life span			Linsted and Calder	26.52 (25.19, 27.91) y
Home range		15.39 km^2	Schoener	152.6 (23.5, 992.5) km^2

of 10 kg for the male and 13 kg for the female. Other life parameters such as heart rate, heart size, oxygen consumption ($\dot{V}O_2$), flight metabolic rate (H_{flight}), existence metabolic rate (H_{ex}), clutch mass, incubation time, food requirements, home range, lifespan, and other physiologic and ecologic parameters were calculated from relationships given in Berger and Hart (1974), Calder (1974), and Peters (1983). These predicted values can be compared with those for living birds, given in the summaries by Brown and Amadon (1968) and Brown et al. (1982).

Wing Loading

Haast's eagle was one of the heaviest flying birds. How did such a large bird become airborne and stay there? In a comparison of the estimated wing loadings for a range of very large flying birds (Table 8.16), the wing loading of Haast's eagle can be seen to have exceeded those for other raptors, but values for both male and female were substantially less than those for the other three birds. Data for large forest eagles are rare, so only one species is used as representative of those. The span loading of Haast's eagle was approached only by *Diomedea exulans,* a high-aspect ratio, dynamic soarer. Both the high wing and span loading of Haast's eagle imply a high stalling speed and, possibly, low maneuverability.

During the Late Glacial transition, from 14,000 to 10,000 years ago, fundamental changes took place in the structure of the vegetation of the South Island. Forest and forest–shrubland mosaics dominated where a few thousand years before there was grassland and shrubland with scattered patches of forest. The physical environment for large flying birds changed, and there is some evidence that Haast's eagle changed in response to the new conditions.

Many island birds are, as we have seen, flightless. Others, such as the saddlebacks (*Philesturnus* spp.), can fly for only short distances. Wing reduction and flightlessness are inextricably linked, and Duff (1949c) thought that the relatively short ulna of Haast's eagle indicated that the bird was gradually becoming flightless. Although the variation in lengths of all the major elements is in the range expected for a sexually dimorphic species, there is considerable overlap when samples from Holocene and Pleistocene sites are pooled. When geologic age is taken into account, it is apparent that Pleistocene eagles had significantly longer ulnae than more recent individuals (Table 8.17). The humerus and carpometacarpus may have been longer, but the differences on the small sample size were not statistically significant.

Total length of wing bones declined by 6.2% between 20,000 and 10,000 years before the present, but neither that nor the 2.3% reduction in leg length observed in available samples is statistically significant. The lack of difference between femur lengths through time indicates that body size and mass did not change. The Holocene wing, with its shorter ulna, had to carry the same mass as the longer wing had some thousands of years earlier. But it is normal for birds that must fly through dense habitats to have shorter wings than their relatives that

TABLE 8.16.
Various flight parameters for living and fossil taxa of large flying birds.

Taxon	Sex	Mass (kg)	Span (m)	Area (m^2)	Wing loading ($g\ cm^{-2}$)	Span loading ($g\ cm^{-1}$)	Aspect ratio, ($span^2\ area^{-1}$)	Source
Aquila audax		4.5	2.25	0.573	0.785	20.0	8.84	R.N.H., unpublished data
Haliaeetus leucogaster		2.8	2.20	0.643	0.435	12.7	7.54	R.N.H., unpublished data
Circus approximans		0.7	1.23	0.194	0.360	5.7	7.8	R.N.H., unpublished data
Harpagornis moorei	♂	9.5	2.14	0.821	1.158	44.4	5.6	R.N.H., unpublished data
	♀	12.5	2.43	0.959	1.306	51.5	5.7	R.N.H., unpublished data
Stephanoaetus coronatus	♂		2.09		0.687		6.0	Brown et al. (1982)
Gymnogyps californianus		9.5		1.316	0.72			
		10.2		1.392	0.73			Campbell and Tonni (1983)
Teratornis merriami		13.7		1.75	0.78			Campbell and Tonni (1983)
Argentavis magnificens		79.8		6.96	1.15			Campbell and Tonni (1983)
Cygnus olor		11.6		0.68	1.70			Poole (1938)
Otis tarda		8.95		0.57	1.56			Campbell and Tonni (1983)
Diomedea epomophora		8.70	(2.82)	0.496	1.75	30.8	(14.7)	Warham (1977)

live in open country. It did not mean that Haast's eagle was headed for flightlessness.

The effect of the smaller area for support resulting from the shorter ulna was offset by the shorter moment arms of the wing, which increased the effect of the breast muscles in overcoming the loads, and by the lower oscillating mass of the wing itself, which allowed faster wing beats and lowered the loads on joints and ligaments. The shortening of the ulna changed the geometry of the load-bearing surface and the relative position of the downward force applied by the pectoral muscle. The dimensional differences are small, but with the loads involved in lifting the extreme body mass of Haast's eagle, small changes in distance between points of application of loads would be important. Because the humerus stayed substantially the same length through time, the pectoral muscle insertion would have effectively moved further out toward the spanwise center of lift as the ulna shortened. The ulna bears the secondaries, which provide the greatest lift-producing area. With a shorter ulna, the center of lift would also have been brought closer in to the body. Thus a greater effective bending moment would have been applied for the same muscle loading to compensate for the increased spanwise loading occasioned by the shorter ulna. If all the major wing bones had shortened in concert, the pectoral muscle loading would have increased substantially. The differential shortening of the ulna appears to have been a compensation allowing an overall shortening of the wing when muscle strength was near its limits.

But why develop a shorter wing? Because there is no evidence that Haast's eagle was evolving toward flightlessness, there must have been other advantages. The most obvious is the change in vegetation structure. Eagles that live in open environments have long wings, optimized for soaring and gliding. Conversely, even the largest forest-dwelling eagles have relatively short but very broad wings, and their flight is fast and highly maneuverable. With the increasing complexity of the airspace in which it flew, selection pressures on Haast's eagle seem to have favored individuals that could flap faster (another factor that depends on wing length), turn more quickly, and fit through gaps.

Claws and Beak: The Potential to Be a Predator

Haast's eagle had huge claws, even in proportion to its own great size. In a large female, the bony claw is over 60 mm long, about the same length as in *Harpia,* but the base of the claw is deeper (14.4 mm cf. 12–13 mm). On the bases of an outline of *Harpia* claws in Voous (1969) and the length of the ungual phalanx, the horny claw in a female *Harpia* extends nearly 30 mm past the bone, and it probably did the same in Haast's eagle. The massive claws clearly indicate Haast's eagle's potential as a predator. The claws clearly distinguish the eagle from the largest accipitrid vultures, whose short, slender, blunt claws are not used for gripping or killing. The force applied at the point of each claw is related directly to the depth of the attachment boss for the flexor tendon, and as the grip is made, the moment arm on the claw increases, increasing the force at the end of the claw (Goslow 1972). In Haast's eagle, the attachment for the claw flexor tendon is further from the

TABLE 8.17.
Mean length (mm), *n*, coefficient of variation (%), and significance of the differences between means of major wing and leg bones of Haast's eagle in the period from the Otiran Glacial maximum to the late Holocene.

Element	Otiran	Holocene	Significance
Wing			
Humerus	236.1 (8, 7.0%)	224.5 (10, 5.0%)	NS
Ulna	267.2 (8, 4.6%)	243.4 (8, 5.1%)	0.01 (*t* test)
Carpometacarpus	120.1 (7, 7.2%)	114.7 (8, 7.3%)	NS
Leg			
Femur	162.5 (7, 6.2%)	161.8 (8, 7.5%)	NS
Tibiotarsus	240.6 (6, 6.4%)	234 (7, 5.3%)	NS
Tarsometatarsus	152.7 (7, 6.6%)	147 (14, 6.2%)	NS

Abbreviation: NS = not significant at 5% level (one-tailed *t* test).

claw's articulation than in any other raptor, living or extinct (Table 8.18, Fig. 8.36), so a great force could be transmitted to the point of the claw. To take that load, the base of the claw had to be deeper than in other eagles, even in the smaller toes. Other large forest eagles that take large prey have claws that had a lower ratio of flexor depth to claw length and hence a weaker grip than in Haast's eagle (Table 8.18).

The larger claws of the first (hind) and second toe were those that inflicted the most damage. Total claw length and body mass are correlated with prey size in raptors (Voous 1969), and the long, deep-based claws of Haast's eagle are evidence that it could catch and kill large, struggling prey (Figs. 8.36, 8.37).

Aspects of behavior such as how the strike was made on a bird the size of an adult moa can usually only be inferred from observations of living species more or less closely related to the extinct species that is being studied. Sometimes those inferences can be tested and even shown to be wrong. For the attack of large eagles on large ratites, there is one observation: a wedge-tailed eagle (*Aquila audax*) attacked a young emu (*Dromaius novaehollandiae*) by grasping and holding the emu's neck (Nevinson 1991). From that, it might be assumed that Haast's eagle was likely to adopt a similar tactic and strike at the vulnerable head and neck. Instead, there is grim evidence that Haast's eagle aimed for a bigger target. Actually, a comparison of the relative size of the eagle's foot and moa pelves (Fig. 8.38) turned out to be far closer to the truth.

The spectacular fossil evidence shows that the eagle used its great claws to grasp the hindquarters of moa and to kill by crushing the bone and causing massive bleeding. Over a dozen moa pelves are known that bear severe damage from eagle claws (Figs. 8.18, 8.39). Moa species involved include the largest, *Dinornis giganteus*. The marks range from pinpricks to crushing indentations, punctures, and rents up to 75 mm long and 10 mm wide. In all instances where it is possible to determine which foot was involved, assuming the greatest damage was inflicted by the huge hind toe, which is also set toward the middle of the eagle's

TABLE 8.18.

Total length of claw (mm) and depth of flexor attachment (mm), and ratio of depth to length in individuals of large raptors, including large forest eagles, an extinct giant raptor from Cuba (*Titanohierax*; data from Arredondo 1976), Haast's eagle, and two accipitrid vultures.

Species	Common name	Length (mm)	Flexor depth (mm)	Ratio
Harpagornis moorei	Haast's eagle	61.5	32.4	0.53
		49.0	26.6	0.54
		59.3	31.3	0.53
Harpia harpyja	Harpy eagle	54.4	24.4	0.41
		61.1	29.8	0.49
		55.9	25.2	0.46
Pithecophaga jefferyi	Philippine eagle	47.9	21.0	0.44
		49.6	23.1	0.47
Stephanoaetus coronatus	Crowned eagle	47.4	23.4	0.49
Titanohierax borrasi	Extinct Cuban eagle	~57	~26	0.46
Gyps fulvus	Griffon vulture	33.1	13.8	0.42
		27.8	11.6	0.41
Aegypius cinereus	Black vulture	34.0	15.0	0.44

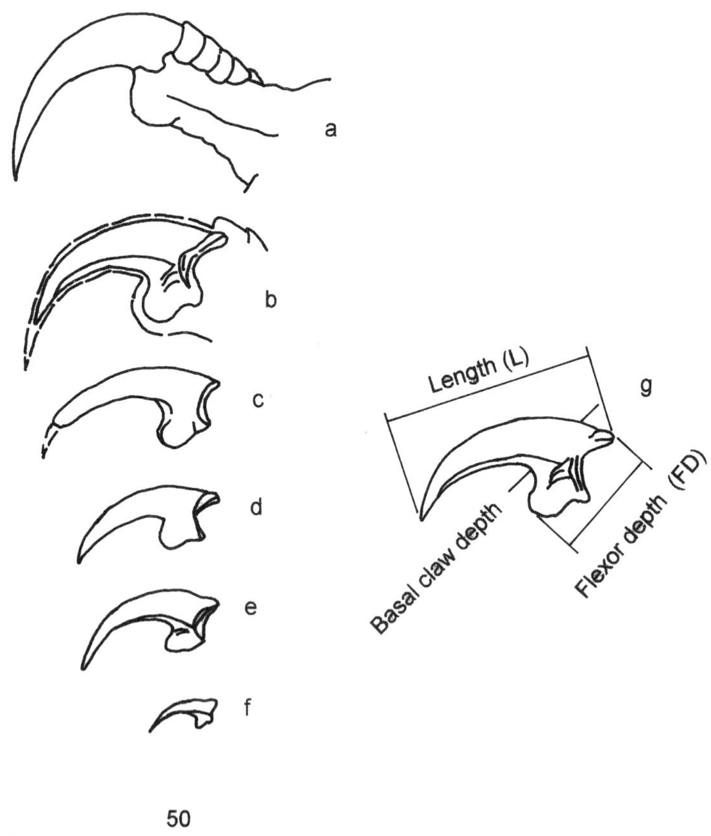

Figure 8.36. Right lateral view of ungual phalanx of pedal digit I of large extinct and extant raptors: (a) Harpia harpyja *(intact toe). (b)* Harpagornis moorei, *with outline of keratinous sheath and toe suggested by broken lines. (c)* Titanohierax borrasi. *(d)* Pithecophaga jefferyi. *(e)* Aquila audax. *(f)* Circus approximans. *(g)* Harpagornis moorei, *showing dimensions used in analysis. Scale bar = 50 mm. Outline of* Titanohierax borrasi *from Arredondo (1976), where it is included under* Aquila.

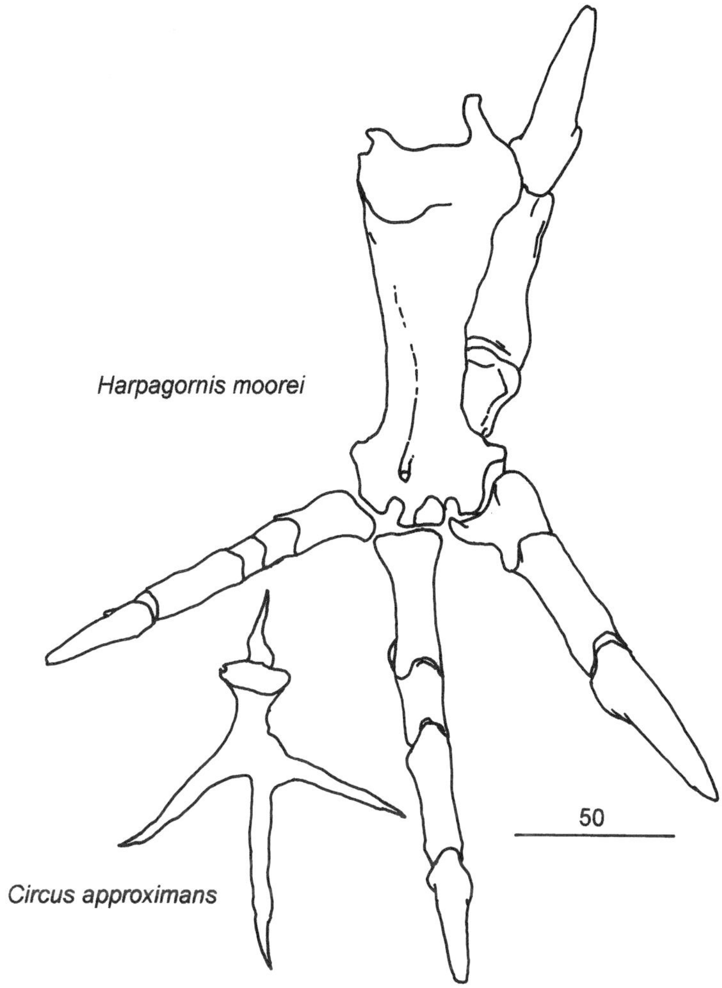

Figure 8.37. Articulated foot of Harpagornis moorei *beside intact foot of the Australasian harrier* Circus approximans. *Scale bar = 50 mm.*

body, marks are consistent with the eagle having struck from the side. One foot gripped the anterior pelvis, and the other was then free to grip the upper thorax or the base of the neck. For example, if the marks on the pelvis show that it was struck by the right foot (the hind toe mark opposite the left of the three opposing claw marks), the eagle had approached from the left, and the left foot was then able to strike further forward on the body. Several pelves bear evidence of multiple strikes, as if the eagle were altering its grip or making repeated grabs at the moa to inflict more damage. Goslow (1971) reported that of seven species of raptor for which he recorded the strike in slow motion, four species made one or more rapid strikes with one foot. Evidence that Haast's eagle did the same is etched in the bones.

The claws invariably penetrated the thickest bone of the iliac plate, and the edges of the holes and rents were crushed inward. That the damage to the bone was inflicted after penetration of the plumage, the skin, and up to 50 mm of flesh only makes it more dramatic. Most of the major damage is concentrated where the bone overlies the kidneys, as well as branches of the air sacs. Penetration of the subiliac space almost certainly resulted in severe hemorrhage in the renal cavity, with damage to the vital abdominal air sacs, with consequences to both blood loss and respiration. The other foot, placed farther up the body, probably near the base of the neck, would be in a position to damage anterior air sacs, nerves, and major blood vessels. The moa would have succumbed to loss of blood and to respiratory problems, because birds take air directly into the abdominal air sacs from which it is passed forward through the lungs. Many would have doubtless drowned in their own blood.

Having caught and killed its prey, Haast's eagle was then faced with feeding on a carcass much larger in proportion to its own size than any other active predator. It is therefore not surprising that the head and beak had the proportions of a scavenger as defined by Hertel (1995). The apparent conflict between having the claws of a predator and the head of a vulture is resolved by considering the size of the carcass dealt with by the species used by Hertel in his analysis. Avivores elsewhere typically eat birds of about the same or less than their own body mass, and even the mammalivores such as *Harpia* do not regularly tackle species larger than themselves (Brown and Amadon 1968). In contrast, most of the potential prey for Haast's eagle was up to 10 or even 20 times its own body mass, and the species would have regularly dealt with carcasses of a size only encountered by raptors elsewhere as carrion.

Of the largest living eagles, the African crowned eagle *Stephanoaetus coronatus* takes the largest prey, both absolutely and relative to its own body mass. It is known to catch and kill half-grown antelopes with body mass up to 18–20 kg, about five times its own mass of 3.2–4.1 kg (Brown 1982; Weick and Brown 1980). Brown (1982) also reports that it is a struggle for the eagle to subdue the largest prey. The crowned eagle has, for its size, very large and strong legs and feet to cope with the loads imposed by its hunting methods and prey. The huge claws of Haast's eagle were wielded by massively strong legs and feet. Although the tibiotarsus was only as long as expected for an eagle of its body mass, the bone was relatively much thicker. Most workers have commented on the robustness of the legs. Rather than being evidence for weak flight and incipient flightlessness, their thickness and the deep muscle scars indicate the strength of grip and ability to resist the stresses applied by a struggling prey much larger than the predator itself.

The antelopes eaten by crowned eagles may be up to 20 kg in body mass, but they have small bodies. The four legs contribute much to the mass, so no part of the carcass is very far from its surface. Moa carcasses, like those of the larger ungulates, had much food value within a deep body cavity, and the muscle blocks and connective tissue were substantially larger and therefore more resistant to damage. An eagle regularly killing and eating ostriches would have to have the same head proportions of the vultures gathered around a wildebeest carcass nearby. The long, narrow beak was similar to that in *Gyps* vultures, which feed within the body cavity of large ungulates. Although the total skull length (cranium plus beak) in Haast's eagle was in proportion to its body size, much of the additional length was achieved by an elongation in the nasal region. Haast's eagle is unique among the large eagles in having a scroll of bone partially surrounding and protecting the nasal opening. It is analogous to the plate covering much of the external nares in the accipitrid vultures, and it may have been an adaptation that reduced the chance of damage to the nasal area or entry of foreign material to the nasal passages while feeding within a carcass. Eagles normally pull at tissue on the outside of a carcass and do not immerse the head and bill in the meal.

A longer, more slender beak with a protected nostril simply meant that Haast's eagle could deal efficiently with large prey and does not of itself indicate that the bird was a scavenger. Indeed, it is entirely expected, given the apparent ability of the eagle to kill prey many times its own mass. Although it was phylogenetically one of a group of open-country eagles, Haast's eagle was closer in its functional morphology to the largest forest eagles. Its flight pattern and hunting techniques were probably similar to forest eagles rather than to its nearest relatives, judging from the body proportions and morphology as well as the habitats where its remains and those of its prey have been found. In bill form, Haast's eagle was convergent with the accipitrid vultures. Taken together, the mosaic of features shows how the morphologic plasticity characteristic of accipitrids obscures the phylogenetic relationships within the group. There seems to have been sufficient variability in the genetic bases for the structures of wings, legs, and bills within the ancestral population that gave rise to Haast's eagle, and in its behavior, to allow it to converge on two different trophic and locomotory modes at the same time. By doing so, it was able to more efficiently exploit a supply of large prey in a mammal-free environment.

Moa may have been only occasional prey for Haast's eagle, but each kill represented much more meat than even a pair of eagles could have eaten before it rotted away. It is unlikely that other eagles would have been welcome at the kill; the aquiline eagles tend to be territorial and do not form feeding aggregations as do, for example, the sea eagles of the genus *Haliaeetus*. However, in the areas where eagles lived, several other species might have helped themselves to the largesse. Kea (*Nestor notabilis*) are known to eat meat and fat from carcasses and even live sheep, and the mainland form of the New Zealand raven *Corvus* sp. could well have included carrion. Even the South Island adzebill (*Aptornis defossor*) may have scavenged dead moa when they were available. The indigenous meat flies would have made short work of the remnants.

Distribution and Habitat

Remains of Haast's eagle have been found in more than 50 sites in the South Island (Table 8.19; Fig. 8.40). Most of the material has come from natural deposits (those without a cultural context), but artifacts made of eagle bone have been found in two archaeologic deposits, Wairau Bar and Moa Bone Point Cave, and worked and unworked bone has been found in others. The Marfells Beach specimens still in

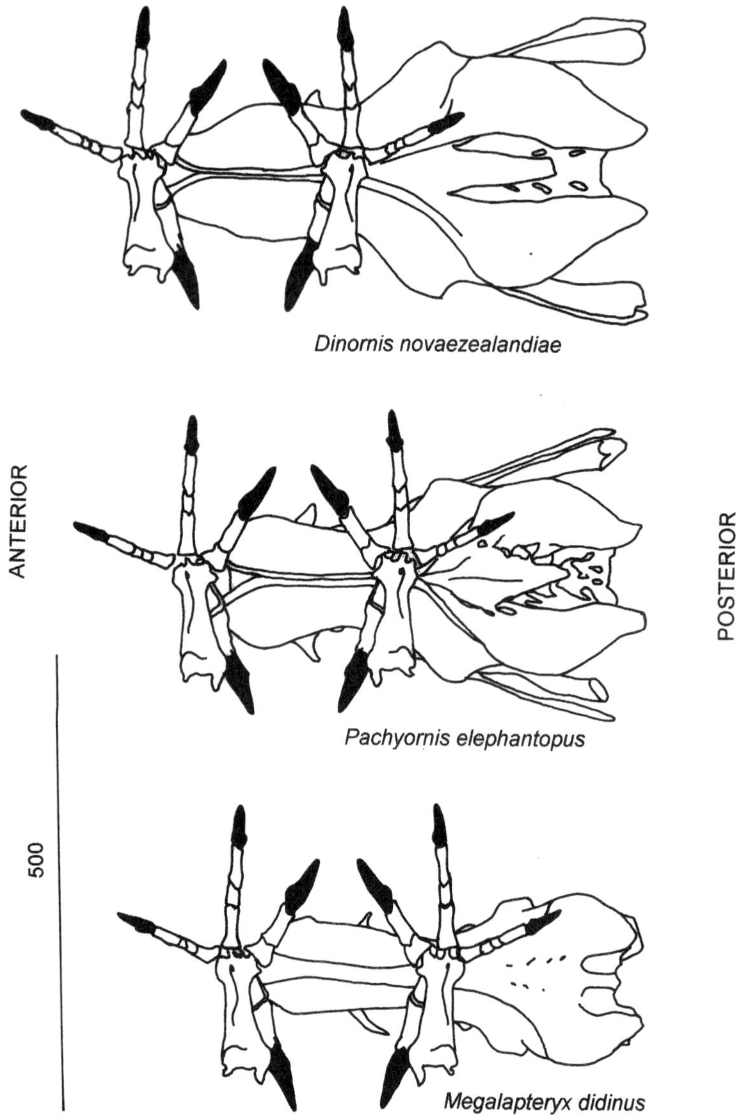

Figure 8.38. Dorsal view of articulated bones of both feet of Harpagornis moorei *in comparison with dorsal views of pelves of a range of moa species. Scale bar = 500 mm.*

collections are not from an archaeologic context, but those from Orepuki, Waipapa Point, Wakapatu, and Warrington could well be. The bones from Papatowai and Old Neck had clear cut marks but had not been further processed. Old Neck is on Stewart Island, and the shaft of a tibiotarsus found there was almost certainly taken from the South Island as industrial bone (Worthy 1998e).

The natural sites provide the best indication of the former distribution and habitat of Haast's eagle, because there is usually no doubt that the individuals represented lived in the immediate area of the deposit and formed part of a population residing there. That assurance is heightened by knowledge of the usual density of large raptors in relatively undisturbed environments. Usual home ranges vary but often

Figure 8.39. Pelvis of Dinornis giganteus *from Glencrieff Swamp excavated by the authors, showing punctures in the iliac plate made by the talons of Haast's eagle. Photograph by T.H.W.*

include 15–20 km^2 of habitat (Brown and Amadon 1968). Consequently, the total population in any given area must be limited, and the chances of a vagrant being preserved outside its normal range is even smaller.

Most eagle sites are in areas of relatively low relief and at low altitude. More than 80% of sites are below 500 m (Holdaway 1991). However, most of the swamp sites in which fossils have been found are at these low altitudes. Although there are large areas of the South Island above the tree line, only a very small proportion of it has karst—and hence the possibility of fossil deposits—in steep, high terrain. Most inland sites are in areas of subdued relief, even if they are at a relatively high altitude. The highest, on a steep-sided ridge at 1600 m on Mount Owen, overlooks broad montane basins above the present tree line in a marble massif. Other sites above 1000 m on the Obelisk and Dunstan ranges in central Otago are in mature mountains whose bedrock is Mesozoic schist and whose slopes are relatively gentle. The sites with sharpest relief are at Oaro, in southern Marlborough, and at the Golden Point gold mine at Deep Dell Creek; the first is coastal and the second is at the bottom of a steep-walled valley cut through the low relief of an elevated peneplain. That eagle bones have been found in such terrain almost everywhere that suitable deposits exist, but not in other areas, including the North Island and the West Coast and northwest of the South Island in the Holocene, despite the presence of large and diverse fossil deposits, suggests that the distribution of eagle fossils closely

Figure 8.40. Distribution of sites in which eagle have been found. Note their absence from the North Island. Sites of both Pleistocene (triangles) and Holocene (circles) age are shown.

represents the former distribution of the bird and not just of suitable deposits.

It is difficult to be certain about the terrain surrounding deposits that are older than the beginning of the Holocene. New Zealand was not only physically much larger because of the 120-m lower sea level at the coldest time during the Otiran Glacial, but many parts of the South Island were heavily glaciated. Outwash gravels from the glaciers covered large areas of the eastern lowlands and form much of the lowland on the western side of the axial ranges. The montane basins inhabited by eagles just over 2000 years ago on Mount Owen were filled with valley glaciers and ice fields 20,000 years ago, probably until 12,000 years ago. Pleistocene deposits in the east include estuarine deposits, peat deposits, weathered stream gravels, lacustrine sediments, and cave fill sediments. In most instances, the age of the stream gravels has not been determined, so the altitude and setting of the deposit when it was laid down are unknown. The deposits at Cape Wanbrow were deposited behind the beach at sea level in the Pleistocene, between 40,000 and 120,000 years ago. The associated faunas indicate that although the vegetation was quite open, at least some shrubland was present at the sites. The presumed Pleistocene Motunau peat beds contained the remains of trees, including stumps in situ and hence the former presence of tree-size vegetation, and the Pleistocene peat deposits within the gravel terraces in Glenmark Stream are likely to have been formed under similar conditions.

Preferred Habitat: Vegetation

Over most of the area occupied by Haast's eagle in the Holocene, the vegetation was a mosaic of forest, shrublands, and riparian grasslands along the great braided rivers of the eastern South Island. Forest types ranged from fairly dense podocarp associations dominated by totara and kahikatea, and matai. Lower rainfall east of the main axial ranges meant that the forest lacked the multistory structure of the lowland forests on the West Coast and in most of the North Island. Instead, there were many small-leaved shrubs and lianas. Edge habitats (ecotones) were common because the rivers frequently changed course and eliminated areas of forest. Grasses, shrubs, and several seral stages of forest then developed on the abandoned gravel riverbeds and flood scars, creating a greater structural diversity than in the wet forests elsewhere. In addition, the soils were younger and more fertile than the leached and podzolized soils under wet forest. The higher nutrient levels seem to have supported a richer diversity of large herbivores, and probably in larger numbers, than did forests in western and northern regions of both main islands.

Mosaic vegetation on the downlands, away from the rivers, resulted from the different degrees of exposure of ridge and valley. Hot, dry northwest winds in summer and cold southerlies in winter restricted taller matai forest to the valleys. It was replaced on the slopes by shrublands, and tussock grasslands occupied the exposed ridges. The foothill ranges were covered in southern beech forest. Above the tree line were extensive subalpine shrublands, a band of tussock grassland, and open herb fields on the upper slopes. In Central Otago, there were larger areas of grassland and shrubland because of the almost continental climate and the lowest rainfall in New Zealand (about 250–300 mm per year). At the southern limit of the range, the plains bordering Foveaux Strait and the dissected hill country of the Catlins had denser, damper forest than in the east, but there were extensive areas of swamp and duneland, and the hills carried a drier forest.

Eagles seem to have been relatively common throughout these areas; at least their remains have been found in deposits of all kinds: from coastal dunes to lowland swamps, to rock overhangs in the dry center of the island. In some areas, eagles also lived where there were extensive mountain basins above the tree line.

TABLE 8.19.
Sites from which bones of Haast's eagle have been recovered.

Site no.	Site name	MNI	Type	Age
1	Albury Park	1	Swamp	H
2	AR 144	1	Cave	P
3	Cannibal Bay	1	Dunes	H
	Cape Wanbrow			
4	"Old Rifle Butts" J41/f8212, 8213, 8230, 8710	4	Back beach	P
5	"Shirley Creek," J41/f213	1	Lagoon	P
6	Castle Rocks	2	Cave	H
7	Chatto Creek, Otago	1	Lacustrine	P
8	Dunstan Range	1	Overhang	H
9	Enfield	3	Swamp	H
10	Glencrieff Swamp	2	Swamp	P-H
11	Glendhu Station site 6	1	Cave	H
12	Glenmark Creek	1	Peat	P?
13	Glenmark Creek 2	1	Peat	P?
14	Glenmark/East	1	Swamp	H
15	Glenmark/West	1	Swamp	H
16	Golden Point mine, Deep Dell Creek	1	Peat?	H?
17	Hamilton Swamp	2	Swamp	H
18	Hamilton gully	1	?	
	Hawke's Cave		Cave	
19	Site 5	1		
20	Site 3	2		
21	His Cave, Oparara	5	Cave	P
22	Hodges Creek Cave System	2	Cave	
23	Hollywood Cave	1	Cave	P
	Honeycomb Hill Cave, Oparara			P
24	E entrance	1	Cave	P
25	Eagle Roost	2	Cave	P
26	Graveyard	4	Cave	P
27	Hives extension	1	Cave	P
28	Hospital Flat site 4	1	Cave	H
29	Kakanui Beach	1	Swamp	H
30	Kapua	1	Swamp	H
31	Kings Cave	1	Cave	H
32	Marfells Beach	3	Dune	H

Site no.	Site name	MNI	Type	Age
33	Marfells Beach midden	1	Dune, A	H
34	Moa Creek, Otago	1	Lacustrine	P
35	Motunau	2	Peat	P
36	Mount Harris	1	Swamp	H
37	Mount Owen	1	Cave	H
38	Ngapara	2	Overhang	H
39	Oaro	1	Swamp	H
40	Obelisk Range	1	Overhang	H
41	Old Neck, Stewart Island	1	Dune, A	H
42	Orepuki	1	Dune	H
43	Papatowai	1	A	H
44	Pyramid Valley	4	Swamp	H
45	Sim's Cave	1	Cave	P
46	Spring Gully, North Otago	1		H
47	Sumner	1	A	H
48	Tuarangi Cave	1	Cave	P
49	Waipapa Point	1	Dune	H
50	Wairau Bar	1	A	H
51	Wakapatu	1	Dune, A?	H
52	Warrington	1	Dune, A?	H
	Total	77		

Abbreviations: A = archaeological; H = Holocene; P = Pliocene.

The South Island Eagle? Absence from the North and Stewart Islands

All but one of the South Island deposits of Holocene age that have yielded eagle bone are in the east and south of the island. Although there are now many records of Haast's eagle from northwest Nelson and one from the West Coast, only the almost complete skeleton from near the summit of Mount Owen is of Holocene age. The Holocene distribution of Haast's eagle seems to have been confined to the drier vegetation zones to the east of the Main Divide in the South Island, with some birds in the high-altitude shrublands and herb fields of the mountains. There is no evidence that the species lived in the wet, multistoried Holocene forests of the North Island or the West Coast of the South Island. Several substantial fossil deposits are known from lowland North Island, from the dunes near North Cape to the spring hole swamp at Makirikiri, and the lake edge deposit at Poukawa in Hawke's Bay. None of these has yielded eagle bones, although similar, or far smaller, sites almost always contain some eagle material in the South Island.

An important implication of the relative fixity of the composition of the eagle fauna is that the range of herbivores in the North Island did not include some of the apparent key species present in the South Island. Those that were present, such as *Euryanas finschi,* were often less abundant than in the South Island over large areas. In the North Island during the Holocene, the nearest analog to the vegetation type characteristic of the eastern South Island was the coastal shrubland, which was most extensive on the dunelands in the Far North. Elsewhere, the strip was narrow and the total biomass of moa and other species occupying it was much lower than on the plains and downlands of the South Island.

An "Eagle Fauna": The Assemblage and Probable Prey

Regardless of the age or setting of the site, fossils of Haast's eagle are normally found in association with a particular range of other extinct species, which has been termed informally the eagle fauna. Members of this assemblage include the moa species *Pachyornis elephantopus, Euryapteryx geranoides, Emeus crassus,* and *Dinornis giganteus,* the extinct goose *Cnemiornis calcitrans,* the terrestrial Finsch's duck *Euryanas finschi,* and the South Island adzebill *Aptornis defossor.* Takahe *Porphyrio hochstetteri* are often included as well. The implication is that these species preferred a similar habitat and that the eagle preyed on some of the herbivores. At higher altitudes, such as on Mount Owen, the range of taxa was less, and the lowland moa were replaced by upland moa *Megalapteryx didinus* and crested moa *Pachyornis australis.*

Haast's eagle was specialized for killing large vertebrates and dealing with their carcasses. There was probably a lower limit on the size of their prey, set by their ability to catch and manipulate it with the huge talons. In the absence thus far of evidence from food remains around a former eyrie or feeding station, the prey is assumed to have been the terrestrial birds of about 1 kg body mass and larger, up to the giant moa *Dinornis giganteus* (200 kg or more). One of the most abundant species throughout the eagle's range was *Euryanas finschi,* and it is likely to have been a staple in the eagle diet. Eagles would also have eaten other species such as large rails *Porphyrio hochstetteri,* weka *Gallirallus australis,* New Zealand coot *Fulica prisca,* and Hodgens' gallinule *Gallinula hodgenorum*), parrots (kea *Nestor notabilis* and kaka *N. meridionalis*), and New Zealand pigeon *Hemiphaga novaeseelandiae.* The nocturnal habits of kakapo *Strigops habroptilus* and the kiwi *Apteryx* spp. may have reduced their exposure to the eagle, but it probably also hunted at dawn and dusk.

Most eagle sites contain a single bird, but a significant number (about 25%, Table 8.19) contain the remains of more than one, and up to five, eagles. However, any attempt to compare the abundance of eagles to the abundance of the probable prey species represented in the same deposits must await much more detailed analysis of the population densities of prey species in the different habitats occupied by the eagle. That analysis, by its very nature, is highly speculative. The best results may be obtained from predictions of territory size from allometric relationships that are based on body mass (Peters 1983), but empirical assessments of the carrying capacity of some vegetation types for large herbivores are, theoretically at least, possible.

Population Size

Large raptors have large territories, typically 10–20 km^2 or more per pair, although denser concentrations have been known in areas of particularly dense prey populations (Brown and Amadon 1968). Over the area of suitable habitat, which might have included up to 50% of the area of just over 150,000 km^2, the population of eagles could have been 3800–7600 breeding pairs. With an allowance of 20% for young birds yet to claim a territory, the eagle population at human arrival may have been about 15,000 individuals at most. By the same method of density of breeding pairs and area of suitable habitat, Fox (1978b) estimated that the total population of New Zealand falcons was 3000–4500 breeding pairs in the early 1970s.

If each pair killed one moa per month, the total predation pressure on the moa populations each year would have been substantial. It is unlikely that the moa population, even if all species were included, could have withstood that level of predation (Holdaway and Jacomb 2000; McGlone, Holdaway, and Wilmshurst, in press), which suggests that most of the eagle diet was smaller, more abundant birds. Moa were taken when the opportunity arose, probably mainly juveniles.

Moa and Haast's eagle had coexisted for at least 120,000 years if the interglacial age of some of the Hillgrove faunas at Cape Wanbrow is correct; obviously, they had achieved some measure of balance. Moa undoubtedly were killed and eaten by the eagle, but any bird down to a Finsch's duck would have been taken, just as *Aquila audax* takes parrots but has been seen to catch a juvenile emu (Nevinson 1991).

Relationship with People—And Time of Extinction

Artifacts made from fresh eagle bone have been found in two sites, industrial bone (with cut marks) in another two, and bones in probable association with cultural remains in another four or five sites. There is no question that the species survived into Polynesian time in New Zealand, around A.D. 1300, but when it became extinct is unknown.

Eagles not unexpectedly made an impression on the Polynesian settlers. Although it can never be known for certain, it is possible that eagles killed people, just as people killed eagles. The birds were certainly powerful enough to do so, and their claws were much larger than those of the Mongolian golden eagles that can—and do—kill the falconers who fly young birds against wolves. Accustomed as they were to preying on large bipeds, the eagles could have mistaken the newcomers for a bizarre new form of the old quarry, a mistake that would have been all the easier if the human was wearing a sealskin cloak or one made of moa feathers. Whatever the reason, eagles seem to be among the birds depicted in rock paintings from the earliest period of Polynesian settlement (Fig. 8.41). The most recent find, which is difficult to photograph satisfactorily because it appears only when the rock surface

Figure 8.41. An early Polynesian (A.D. 13th–14th century) depiction of an eagle in a rockshelter in South Canterbury, South Island, New Zealand. Photograph by T.H.W.

is damp from dew or mist, is reported to be an undoubted representation of an eagle, and a very fine one.

There are a few reports from the 19th century that just might be very late records of Haast's eagle. Haast himself thought he saw an eagle in the Canterbury mountains in the late 1860s. A very large bird blundered into his tent in the dark, perhaps attracted to the fire (Haast 1948). But it is difficult to see what even a small population of eagles could have been eating in that area where all the ground birds were already extinct or extremely rare, by Haast's own observations (Haast 1879).

The most believable account is that recorded in the reminiscences of Charlie Douglas, an explorer, surveyor, and one-time gold prospector in the remote southern end of the West Coast of the South Island. He reported having seen and shot two large hawks, whose wingspans were up to 9 feet (nearly 3 m), in the Landsborough Valley of South Westland in the late 1860s (Pascoe 1957). Douglas's surveys show that he was a meticulous observer, he did not seek publicity, and he certainly never claimed that he had seen the "extinct" eagle. The report is the more credible because it refers to an area and a time when there were still populations of kakapo and kiwi and perhaps even takahe in the high-altitude tussock grasslands. A few years later, the kakapo and takahe had gone. The wingspans reported by Douglas are much greater than those of either the living Australasian harrier or the reconstructed span of the extinct giant Eyles's harrier *Circus eylesi.* The identity of Douglas's birds remains a mystery.

Who Killed Haast's Eagle?

Just as the saber cats, lions, and cheetahs of North America died out when most of their prey became extinct at the end of the Pleistocene, Haast's eagle was vulnerable to loss of its prey and to habitat destruction. Removal of most of the habitat by fire before A.D. 1400 (McGlone, Holdaway, and Wilmshurst, in press) and the extinction of most of the suitable prey at about the same time (Holdaway 1999a) would have severely limited the eagle population by the middle of the 14th century. Almost all of the lowland areas of eagle habitat had been burned or hunted out by A.D. 1350 (Holdaway 1999a; Holdaway and Jacomb 2000; McGlone, Holdaway, and Wilmshurst, in press). There is no secret to the extinction of Haast's eagle. It required a specific habitat and was specialized to take very large prey. Just as the saber cats faded away when the mammoths and mastodonts were eliminated from North America, so Haast's eagle was unable to transfer its attentions to smaller prey when the large birds were killed off by people. The eagles would have come into conflict with people by competing for the same food species. Certainly, people used eagle bone for tools when they were available. As we have noted before, no doubt when an unwary human ventured into the territory of a pair of eagles, the person was as likely to be the prey of a species long adapted to take large bipeds as they were to be the predator themselves. In the final analysis, however, people proved to be too much for the eagle and the world it had inhabited.

9 • Other Birds of Prey

The Largest Harrier

Genus Circus *Lacépède*
Circus eylesi *Scarlett, 1953 (Eyles's Harrier)*

The second of the great predatory birds of the New Zealand late Quaternary, after Haast's eagle, was a giant harrier (*Circus*). It is more mysterious even than Haast's eagle because until recently, its remains were rare in collections. Although more material has been found in the past decade, including three partial individuals, there is still much less material than for the supposedly rare eagle. Two sites, Poukawa and Pyramid Valley, each had the remains of several individuals. At Poukawa, the birds were represented by small fragments, whereas the Pyramid Valley bones were as well preserved as the other material. There is still not enough material present to be able to tell whether the North and South Island populations were different races or species.

Discovery

Augustus Hamilton's collection from Te Aute included the remains of a large harrier hawk. Henry Ogg Forbes referred to these bones as *Circus hamiltoni* and *C. teauteensis.* No further material was found until the late 1940s, during extensive excavations at Pyramid Valley. Scarlett (1953) based his *Circus eylesi* on a partial cranium and other bones taken from there. In the early 1960s, T. Price, an amateur archaeologist, conducted an intensive series of excavations beside Lake Poukawa in southern Hawke's Bay. The deposit was formed beside a lake surrounded by forest and shrubland. Almost all the harrier remains from Lake Poukawa are fragments, but they represent a large number of individuals. In fact, the number of individuals at Lake Poukawa and Pyramid Valley suggests that the harrier was not uncommon in suitable habitat.

Harrier remains have been found in at least 10 other sites in the North Island. These include apparent associations with archaeologic sites (at Hotwater Beach and Kaupokonui), but the other eight have been in natural deposits in caves or under overhangs. Of the 14 sites known so far in the South Island, half are in caves, three are in swamps, and the remaining four are in dune sites, two of which may be associated with archaeologic sites. Another site has been identified by the presence of prey remains. As cave exploration developed, the amount of material available for study has increased, mainly because most of the relatively complete individuals have come from cave deposits. Examples of this are the two individuals from Hobson's Tomo on Takaka Hill and the most recent (March 1998) specimen from just beneath the Taupo ignimbrite in Hukanui 7a in Hawke's Bay.

Taxonomic Issues

Compared with the minor issue of the male of Haast's eagle being described as a second species (Haast 1874a; Holdaway 1990), the taxonomic history of the New Zealand harriers is long, tedious, and still unresolved. More than for any other species, Henry O. Forbes's penchant for publishing minimal descriptions had a major effect on the taxonomy of the New Zealand harrier (or harriers, because it is still uncertain whether there were one or two species). Augustus Hamilton's collection from Te Aute (Hamilton 1889) included the remains of a large harrier hawk. Forbes had these bones when he named two species *Circus hamiltoni* and *C. teauteensis*. But just which bones he used is not known, because he did not refer to any specimens.

Forbes's (1892d: 186) descriptions were inadequate even by the standards of the day. He described both as being "much larger than living harrier," but then did not say how they differed from each other. Among the other inadequacies in his "description," Forbes did not give an explicit locality for the material or nominate type material. The specific epithet he chose for the second species has been taken to indicate the provenance of the material.

In his catalog published at the same time, Lydekker (1891a) listed only *Circus gouldi* (= *C. approximans*) from New Zealand deposits. The three bones he listed (with collection numbers), all purchased with the Walter Mantell collection in 1855, include the distal two thirds of a left tibiotarsus (32245a), a complete tarsometatarsus (32232a), and an "imperfect" left ulna (32241). The tarsometatarsus was 103 mm long, and, if Lydekker's illustration (1891: fig. 6) is to scale, the distal width is about 15 mm. These measurements suggest that the bone should be referred to the New Zealand species and not to *C. approximans*. All three were reportedly from the surface deposits at the mouth of the Waingongoro River in southern Taranaki. The provenance is not completely certain because the eagle bone in the same collection probably came from Waikouaiti or another site visited by Mantell in the South Island. But the material shows that Forbes and Hamilton were not the first to see bones of the New Zealand harrier; Mantell had preempted them, as he had, again unknowingly, preempted Haast with his eagle. The whereabouts of these bones of *Circus eylesi* is presently unknown; the collection of the Natural History Museum in London contains only those brought there by Forbes.

Lambrecht (1933) considered *C. hamiltoni* to be a *nomen nudum,* a species name without an adequate description or reference specimen. He gave the locality for that species as the "Mittelinsel," or Middle Island, the old name for the South Island, and gave Hamilton as the collector. However, of *C. teauteensis,* he wrote,

> Im British Museum fand ich einen Tibiotarsus aus dem Plistozän von Neuseeland, Mittelinsel, der aber noch beschrieben werden müsste. Artname nach dem Fundort: Te Aute swamp.
>
> Lydekker bemerkt im C.F.B., S. 20, dass die einzige Form von Neuseeland Circus Gouldii Bonaparte ist. Mit den von Forbes als nomina nuda erwähnten Formen, von denen wenigstens das letztere valid sein kann, liegen nun drei Circus-Arten von hier vor.

This may be translated as follows:

> In the British Museum I found a tibiotarsus from the Pleistocene of the Middle Island [= South Island] of New Zealand, that should be described. The species name is derived from the locality where the specimen was found: Te Aute.
>
> Lydekker noted in the Catalogue of fossil birds p. 20 that the only form of *Circus* from New Zealand is *Circus gouldi.* The forms described as *nomina nuda* by Forbes, of which at least the last might be valid, brings to three the number of species of *Circus* described from there.

Hence Lambrecht treated both of Forbes's species as *nomina nuda* although he explicitly cited only *C. hamiltoni* as such. Lambrecht (1933) was mistaken in locating Te Aute in the South Island. The provenance of the tibiotarsus is known because it is still in the Natural History Museum's paleontology department and clearly labeled. The bone was labeled in 1922 when it was transferred from the zoology department; the label identifies the bone as one of the cotypes of *Circus teauteensis* and that it was from Te Aute. Casts of the bone are also present; one was cataloged in 1895 and the other in 1921, and both also carry this information. So undoubtedly Lambrecht had this information. The taxonomic standing of the tibiotarsus, if any, depends on whether Lambrecht's text can be interpreted to mean that he definitely identified the tibiotarsus with the name *C. teauteensis.*

Oliver (1930) ignored the fossil hawks, apart from a sentence saying that fossil material may eventually be shown to be from an extinct species. Later, he reduced both of Forbes's names to *nomina nuda* (Oliver 1955) and accepted Scarlett's name on the basis of South Island material. Dawson (1958b) rediscovered the Forbes material in the British Museum (Natural History), as it then was known, and reported that he had located the type material, meaning the material that Forbes had brought to the museum. Dawson maintained that he had established the validity of the Forbes names for the harriers, but he presented no information in support of the assertion. After Dawson's visit, the Forbes collection fell into obscurity again.

Brodkorb (1964) accepted *C. teauteensis* Forbes, listed *C. hamiltoni* (from the South Island) as a *nomen nudum*, and synonymized *Circus eylesi* Scarlett, 1953, with *C. teauteensis*. Brodkorb (1964) also noted that the tibiotarsus and an unspecified type for *C. hamiltoni* were both in the British Museum. Because there is no information in Forbes's scientific publications regarding the origin of the material he named *C. hamiltoni*, the reason for accepting *teauteensis* over *hamiltoni* may have been Lambrecht's interpretation of 30 years before: that the tibiotarsus in the British Museum was a type. Although Forbes (1892a, 1892b, 1893c) did not say which specimens he was looking at, it is possible to identify the bones in the collections as the putative types because they are labeled as such, and have been since at least 1894 (Worthy 2000). Their existence does not, however, validate Forbes's names because the bones had to have been described in conjunction with the name. This was not, and has not yet, been done. The confusion has spilled over into the working literature: Kinsky (1970) accepted both *C. teauteensis* Forbes, 1891, and *C. eylesi* Scarlett, 1953, as North and South Island species of harrier, respectively. The most recent New Zealand checklist (Turbott 1990a) retains only *C. eylesi*, but it introduces the red herring of a possible change in the genus for that. The Forbes material in the British Museum collections has recently been relocated (Worthy 2000), but the taxonomic tangle has yet to be completely unraveled.

Systematics

Whether there was one or two species of harrier in New Zealand before human contact is still unresolved. Although most specimens from the North Island allow the measurement of only the widths of elements, it is apparent that the North and South Island forms of harrier differed significantly in size. The difference was sufficient for the males of the North Island form (judging from the size distributions and the assumption that the smaller individuals in both populations were male) to be as large as the South Island females. Such a difference probably warrants taxonomic recognition, but too little material is available as yet from the North Island for the forms to be adequately diagnosed and described.

Names have, however, been proposed for the North and South Island forms, but only the most recent for the South Island has been accompanied by a description or adequate indication. An apparent change in size of individuals within the South Island during the interval between the last glacial maximum and the late Holocene has compounded the problem. In this chapter, the New Zealand harriers are treated as one species for convenience, although it is probable that the two forms will eventually be recognized at least at the subspecific level.

Although Scarlett (1953) referred the species to the genus *Circus* in his original description, he later was impressed by the similarity of its wing proportions to those of the bird hawks in *Accipiter*. He did not publish this opinion but relabeled the fossil material in the Canterbury Museum as *Accipiter eylesi*. Holdaway (1989) followed this interpretation, and Diamond (1984) and others relied on that source in turn.

Not an Accipiter, *but Why?*

The structure of the cranium shows conclusively that the species was correctly assigned to *Circus*. The accipitrine wing proportions were a response to the forest and shrubland habitat in prehuman New Zealand. Harriers (*Circus*) normally hunt by gliding over open country or marshlands, dropping on prey located by sight or especially by sound. They have a characteristically expanded tympanic wing in the cranium (Figs. 8.2, 9.1), which presumably allows a refined sense of sound direction, much as in most fully nocturnal owls (Martin 1990).

The holotype cranium of *Circus eylesi* is damaged, but enough material is present to show that it differs from *Accipiter* in having a lateral ala tympanica (Fig. 8.26). The term *ala tympanica* is, according to Baumel et al. (1993: Osteologia, n. 84), used for the "flared part of os parasphenoidale that bounds tympanic cavity ventrally." Here it is called the tympanic wing, and it may be analogous to the tympanic wing in owls, where it is associated with highly directional hearing. In harriers, the os exoccipitale is flared laterally and forms the posterior wall of the tympanic cavity (i.e., it is Jollie's [1977] posterior tympanic process). The lateral ala tympanica is especially prominent in *C. approximans,* where it partially overlaps the articulation of the quadrate in lateral view. Harriers can locate prey by sound, even over a closed canopy.

Actually, there is a range of differences in the skeleton by which *Circus* and *Accipiter* can be distinguished. Some of these characters were noted by Olson and James (1991); in the list that follows, these characters are indicated by the notation OJ. In the cranium (Figs. 8.2, 9.1), the rostral ala tympanica is prominent in *Circus,* but it does not arise as steeply from the edge of the basiparasphenoidal plate or overtop the parotic process as it does in *Accipiter.* The zygomatic process and articular process of the squamosal are separated from the tympanic wing in *Circus*; in *Accipiter,* they are joined by a bony bridge. The tympanic margin is notched in anterolateral view in *Circus* but is smoothly convex in *Accipiter.*

In *Circus,* the lateral margin of the bridge over the nasal arterial foramen rounds out into the dorsal margin of the ectethmoid plate at the ventral extremity of the bridge, whereas in *Accipiter* the lateral edge of the ectethmoid arises about halfway along the bridge. The ectethmoid plate itself in *Circus* extends ventrolaterally as a gradually narrowing blade, the outline of which varies between species. However, in *Accipiter,* the lateral margin of the ectethmoid descends vertically from the nasal foramen bridge and the extension is ventral, so its basal length is greater than the lateral extension of the blade. The angle between the line from the parotic process to the lateral extremity of the tympanic wing and the sagittal plane of the cranium is obtuse in *Circus* and acute in *Accipiter.*

The procoracoid foramen (Fig. 8.2) is invariably lacking in *Accipiter.* In *Circus,* some species lack the foramen; in those that have it, some individuals lack it (Olson 1978b). Wide, open foramina are present in the coracoids in the type series of *Circus eylesi* from Pyramid Valley in

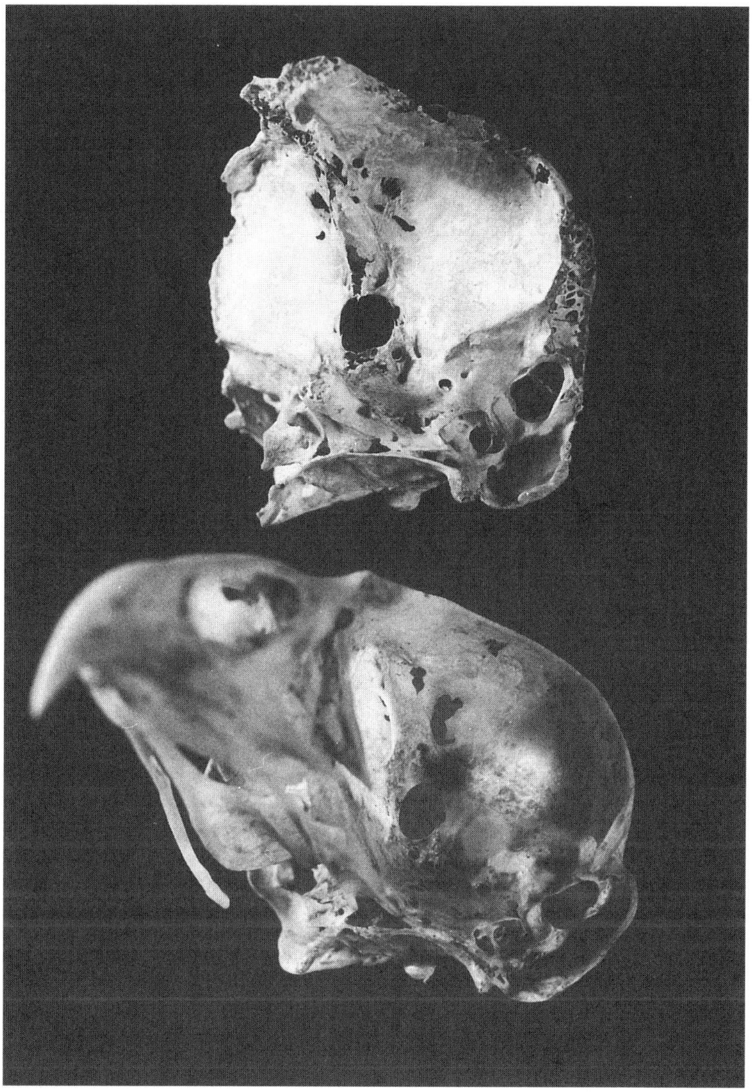

Figure 9.1. Three-quarter views of the cranium of Circus eylesi *(top) and* Circus approximans *(bottom) showing the presence of the diagnostic tympanic wing (see Fig. 8.2) in both and the differences between the two. Photograph by J. Palmer.*

the Canterbury Museum (Av5690, left side; Av5716, right side). The North Island populations also had the foramen. In the sternum (Fig. 8.3), the sternocoracoidal sulcus extends caudally to the fifth costal eminence (counting from the cranial end) in *Circus eylesi* but to between the sixth and seventh in *Accipiter fasciatus*.

In the humerus (Figs. 8.4, 9.2), the bicipital surface is more elongate and not rounded in *Circus* (OJ), and the bicipital crest is indistinct in palmar view, tapering gradually into the shaft (OJ). The ventral condyle of *Circus* is deeper dorsoventrally in distal view, and in anconal view, it is oriented more nearly parallel to the shaft (OJ). Also at the distal end of the humerus, the ectepicondylar process is situated more proximally in *Circus*, and the area of tricipital grooves is more excavated and flattened (OJ). On the ulna, the olecranon (Figs. 8.4, 9.2) is

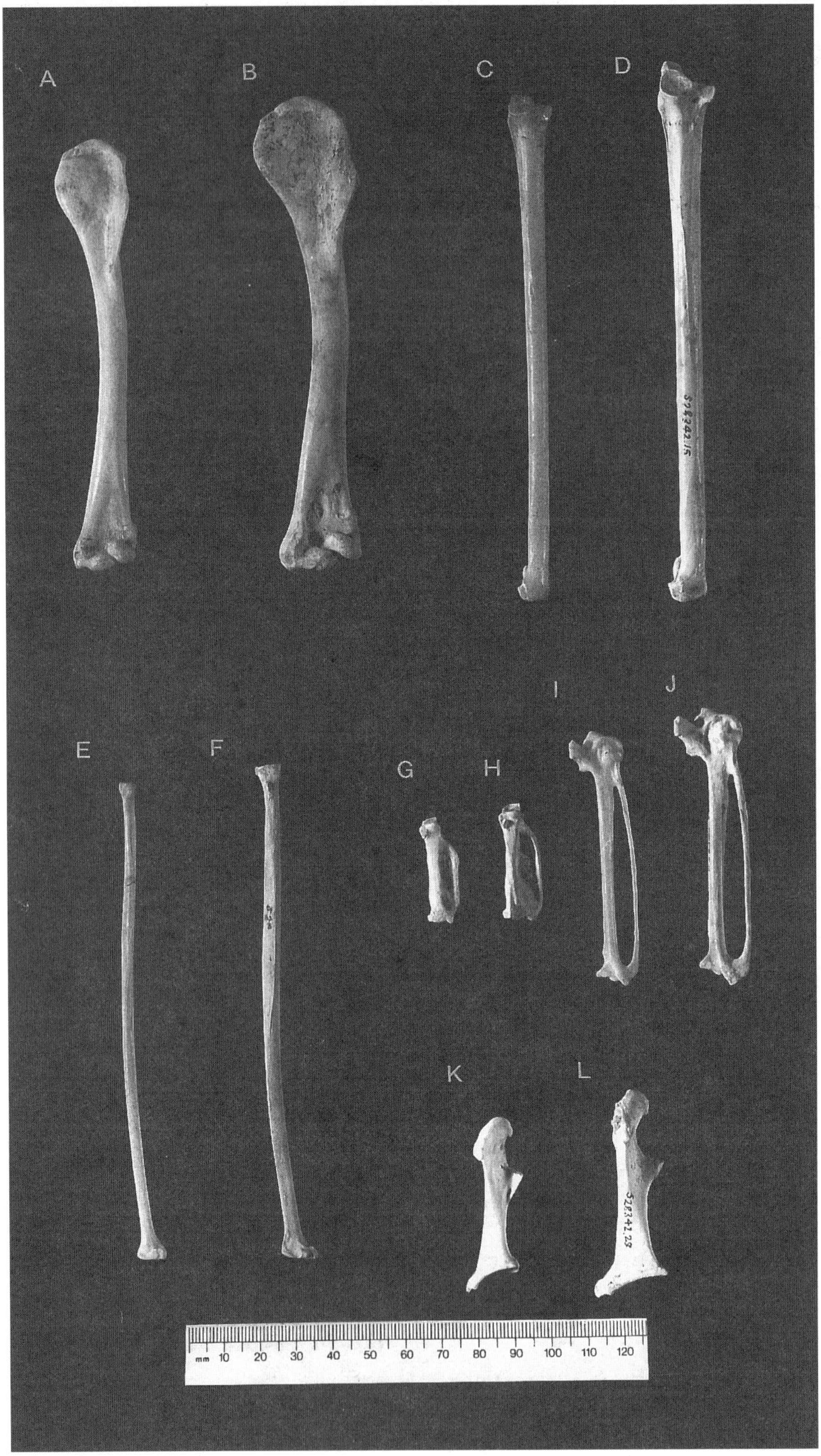
A
B
C
D
E
F
G
H
I
J
K
L
mm 10 20 30 40 50 60 70 80 90 100 110 120

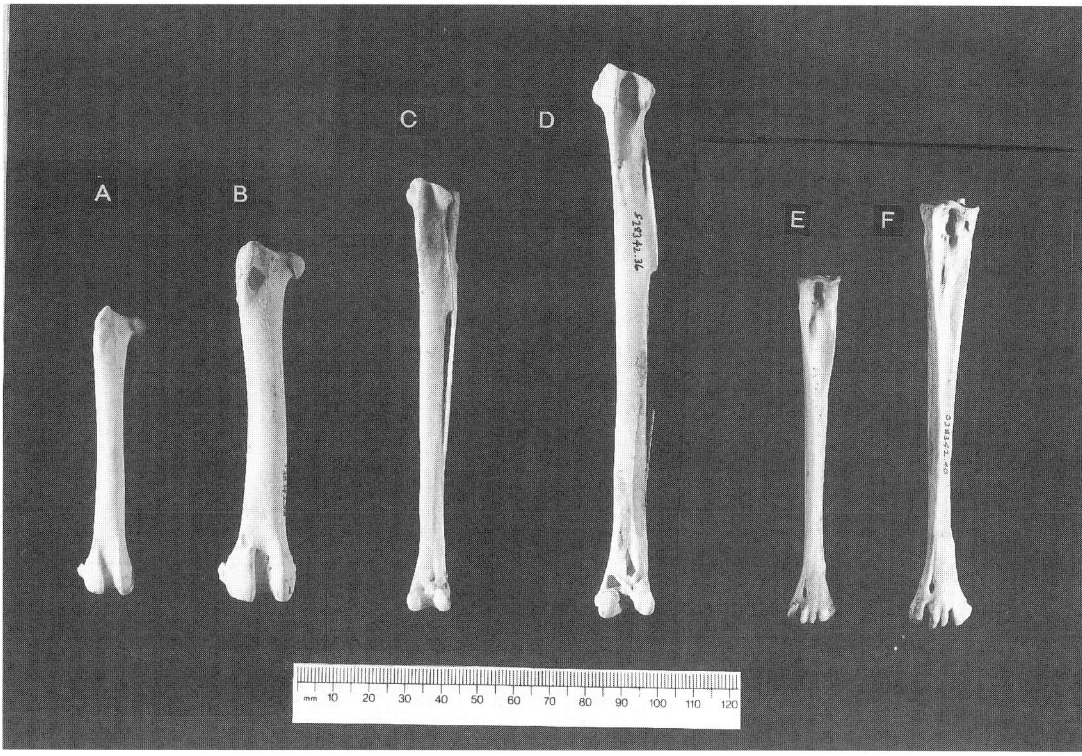

Figure 9.2. (opposite page) Wing bones of Circus eylesi *(right of each pair) compared with those of a female* C. approximans *(left of each pair). (A, B) Left humerus. (C, D) Left ulna. (E, F) Right radius. (G, H) Right manus phalanx 2.1. (I, J) Right carpometacarpus. (K, L) Right coracoid. Photograph by J. Palmer.*

Figure 9.3. (above) Leg bones of Circus eylesi *(right of each pair) compared with those of a female* C. approximans *(left of each pair). (A, B) Right femur. (C, D) Left tibiotarsus. (E, F) Right tarsometatarsus. Photograph by J. Palmer.*

not as prominent in *Circus,* in which genus it is also angled ventrally rather than projecting straight proximally (OJ).

Of the hindlimb bones, the femur (Figs. 8.5, 9.3) is usually more slender and gracile in *Circus* than in *Accipiter* (thick in *Circus eylesi*), the proximal end is markedly narrower in lateral view, and the trochanter is less expanded laterally than in *Accipiter* (OJ). In *Circus,* the rotular groove is narrower; at the distal end, the external condyle does not project so far beyond the fibular condyle in lateral view (OJ). At the distal end of the tibiotarsus (Figs. 8.5, 9.3), the edge of the internal condyle angles strongly laterally, exposing more the internal surface of the shaft. The proximal end of the tarsometatarsus (Figs. 8.5, 9.3) is relatively narrower in most *Circus* (OJ).

When Was a Harrier an Eyles's Harrier? A Problem of the Sexes

Although Eyles's harrier was larger than the Australasian harrier, in the South Island, bones of the male Eyles's harrier are about the same length as those of the female Australasian harrier. The Canterbury Museum collections contain several harrier specimens from Pyramid Valley that Ron Scarlett had identified as *C. approximans.* In the North Island, many of the smaller harrier bones from Lake Poukawa were also attributed to that species by Millener (1991) and Horn (1983). Harriers are like other accipitrids in that they are markedly sexually dimorphic.

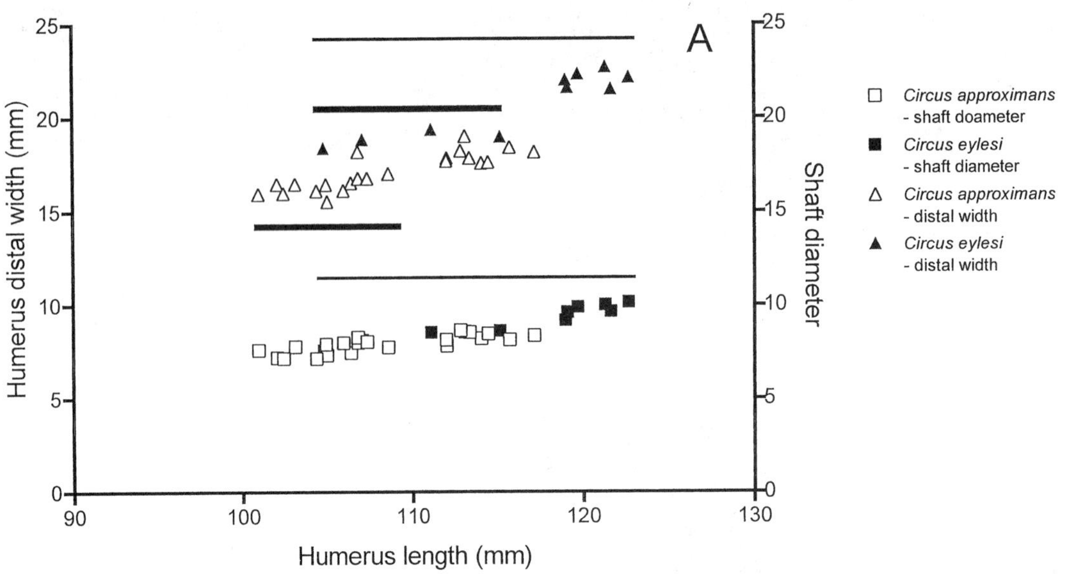

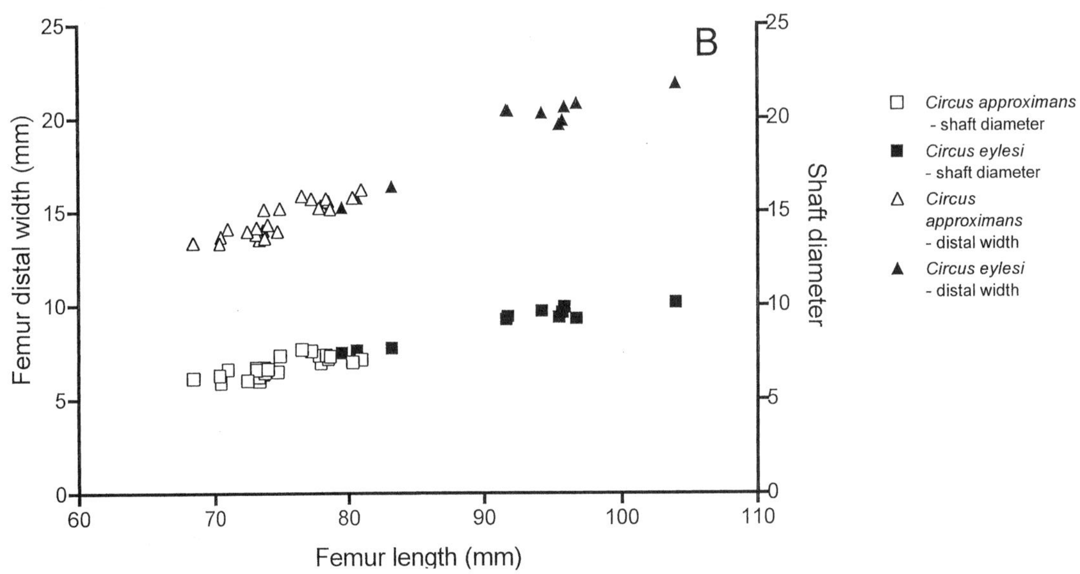

Figure 9.4. Dimensions of the humerus and femur in species of Circus *in New Zealand, showing overlap between female* C. approximans *and ?male* C. eylesi. *(A) Humerus proximal width against humerus length. (B) Femur proximal width against length.*

Examination of the Pyramid Valley material showed that those attributed to *C. eylesi* exhibited only the degree of size variation normal for a sexually monomorphic species. Further examination of the supposed *C. approximans* material showed that although the bones were similar in length to those of *C. approximans,* they were much more robust. It was apparent that the "*C. approximans*" material belonged to the missing male Eyles's harrier (Fig. 9.4). When that knowledge was applied to the collection of fragments from Lake Poukawa, it was found that the same situation applied there. Some material attributable to *C. approximans* was present, but only in the layer laid down in the past 1000 years.

The measurements useful for discriminating the two taxa include the thickness (shaft and distal) of the main bones. The lengths of the wing elements of the two species broadly overlap.

The two taxa did not differ only in size. Other morphologic differences, unrelated to allometry, exist. First of these is the difference in proportions of the wing bones, mentioned before. The expanded tympanic wing in the cranium is present but less well developed in *C. eylesi* than in *C. approximans,* which may mean that the larger species did not rely on hearing as much as typical harriers do.

The anterior rim of the sternum (Fig. 8.3) is thicker and the manubrial spine more upturned in *Circus eylesi.* The other bones in the pectoral girdle are also enlarged versions of those in *C. approximans*—the scapula in particular. However, there is a morphologic difference at the cranial end of the coracoid. In *C. eylesi,* the broader brachial surfaces are rotated dorsally, so most of the face of the triosseal canal is visible in dorsal view. That would result in a change in angle for the tendons from the pectoral muscles to the wing, which may be associated with the change in flight pattern signaled by the difference in humerus–ulna ratio.

The major wing bones (Fig. 9.2) are of similar length in the two species, despite the substantial difference in body mass. It is not surprising, therefore, that the wing bones of *Circus eylesi* are 20%–30% thicker than the corresponding element in *C. approximans.*

Because the humerus is the wing bone closest to the body, it has to transmit both the power of the pectoral muscles to the wing and the oscillating flight loads and lift back to the body. Flapping flight and high body mass require the humerus to be robust, to be anchored solidly to the shoulder, and to be strongly attached to the ulna and radius, which carry the main lifting feathers. As in Haast's eagle, the humerus in *C. eylesi* is exceptionally stoutly built. The proximal end of its humerus is appreciably larger than the distal end, whereas the proximal end of the *C. approximans* humerus is only marginally broader than the distal. Correspondingly, the muscle and ligament attachments are larger in proportion to the bone length in the larger species. At the distal end, the depression for attachment of the brachialis anticus muscle is deeply excavated and clearly outlined in the palmar (anterior) face of the bone; in the lighter, soaring species, it is a shallow depression, grading smoothly into the surrounding surface of the shaft.

Other points of difference include the nutrient foramen being distal to the pectoral crest in *C. eylesi* and the pectoral crest itself being broader and having a much sharper angle than in *C. approximans.* Differences in breadth of shaft and articulations are obvious, with the humeral head an elongated oval in *C. eylesi.* At the distal end, the ectepicondylar process in *C. approximans* diverges from the dorsal margin of the shaft, so it is visible in anconal (caudal) view. The process in *C. eylesi* is in line with the shaft margin. On the internal side of the bone, the entepicondylar process is more rounded and expanded in the larger bird.

For identification of fragments of bone, a useful feature of the ulna is the depth and width of the pit of the radial depression below the

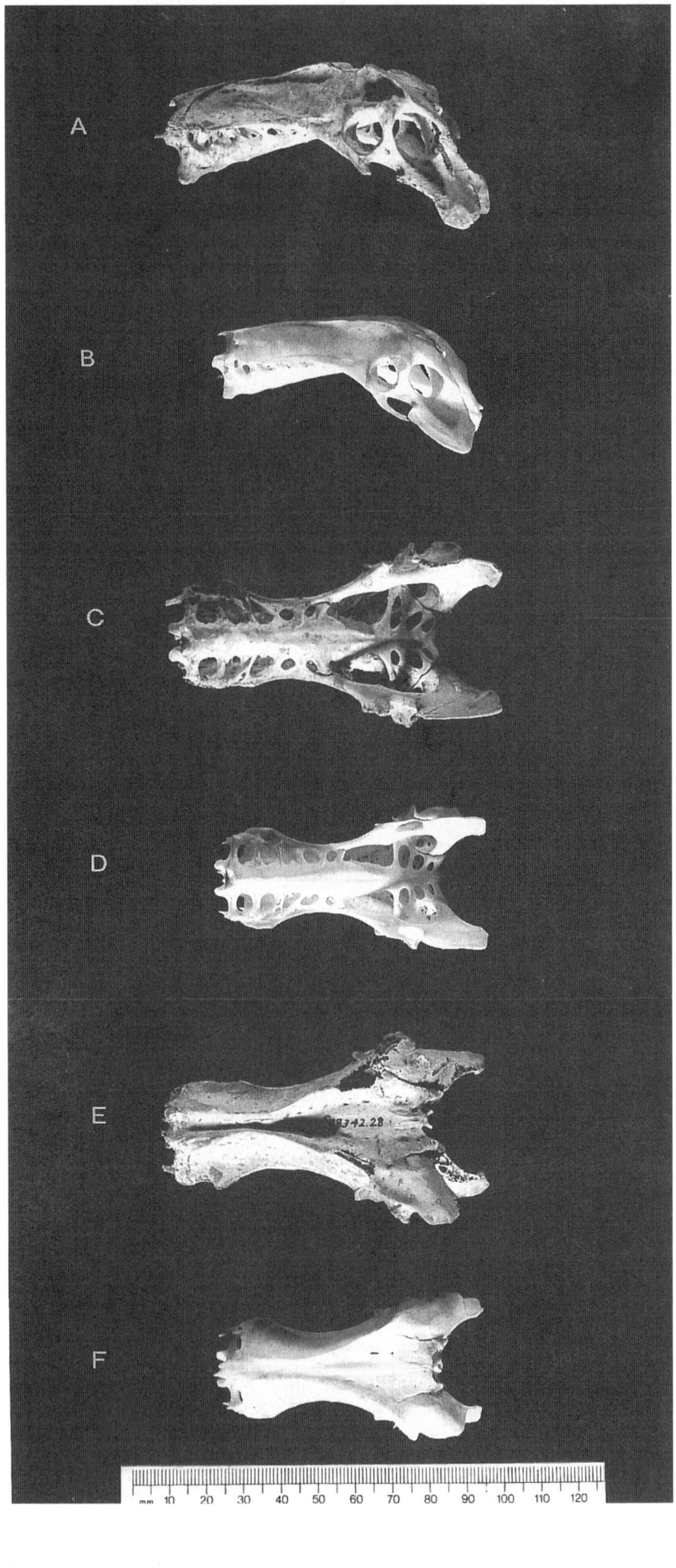

Figure 9.5. Pelves of Circus eylesi *(A, C, E) compared with a female* C. approximans *(B, D, F) in (A, B) lateral, (C, D) ventral, and (E, F) dorsal views. Photograph by J. Palmer.*

articular surface. It is deep in *C. eylesi* and shallow and small in *C. approximans*. At the proximal end, *C. eylesi* has a deep pit ventral to the radial cotyla that *C. approximans* lacks, and *C. eylesi* has an olecranon that extends farther proximally from the ulnar cotyla. The nutrient foramen is further down the ulnar shaft in *Circus eylesi*.

The major differences between the radius of *Circus eylesi* and *C. approximans* are obvious in Figure 9.2E, F: the shaft is much thicker, although the bones are the same length, the proximal and distal articulations are proportionately larger, and the intermuscular line is strongly expressed. In the hand, the carpometacarpus is not markedly longer than in *C. approximans*, but the proximal extensor process is much larger. Phalanx I of the major digit has a wider "blade" with a deep trough, providing a larger surface area for attachment of primary feathers.

The pelvis of *C. eylesi* is not markedly longer than that in *C. approximans*, but it is much more robustly built (Fig. 9.5). It is much wider across the posterior ilia (Fig. 8.3), and there is a deep trough between the anterior iliac plates caused by the greater width of the iliac plates themselves. That width, and an expansion of the posterior ilio-ischiadic blade, provided more surface area for the attachment of the leg muscles. A most conspicuous difference between the two pelves in lateral view is the size of the acetabulum, which received the massive head of the femur in *C. eylesi*.

It is in the femur that the best indication of the size difference and much greater power of *C. eylesi* can be seen. The bone is much thicker than that in *C. approximans*. The rim of the femoral head is rotated toward the axis of the bone instead of being at an angle to it in *C. approximans*, and its proximal end is much more at right angles to the bone axis (Fig. 9.3).

Another measure of the greater strength of the leg is that the fibular crest occupies a greater proportion of the lateral margin of the tibiotarsus in *C. eylesi*, where it is a third of the length of the bone; in *C. approximans*, it is about 28%. The shaft is proportionately broader in *Circus eylesi*, as are the distal condyles. The entepicondyle is also proportionately larger in the larger species.

In *Circus approximans*, the medial margin of the tarsometatarsus has a double angle that links two more or less parallel sections, a wider section from the articular surface, and the narrow shaft. The margin is a smooth curve in *Circus eylesi* so that the proximal two thirds of the tarsometatarsal shaft forms an elongated triangle with the apex down. As a result of the greater size of the medial hypotarsal process in *Circus eylesi*, the medial proximal foramen is distal to the lateral foramen, whereas in *Circus approximans*, they are at the same level. Both the distal width and the width of the condyle for the first toe are proportionately greater in *Circus eylesi* than in *C. approximans* as a consequence of the much greater development of the toes and claws in the larger species.

Size and Flight Pattern

The New Zealand harrier was much larger than the Australasian harrier in most dimensions (Tables 9.1–9.2). The same allometric rela-

TABLE 9.1.
Summary statistics (mm) of axial bone length measurements of *Circus eylesi* from the North and South Islands in comparison with those of *C. approximans*. Data given: mean, *n*, standard deviation, range, standard error, and coefficient of variation.

	Element				
Taxon	Cranium	Coracoid	Scapula	Sternum	Pelvis
C. eylesi	85.4	49.7, 4, 1.76	76.5	86.8	88.1, 2, 1.75
North Island	1	47.3–51.2	1	1	86.8–89.3
		0.88, 3.54			1.24, 1.99
C. eylesi					
South Island	79.4	47.4, 5, 1.23	66.6, 3, 0.32	83.6, 2, 1.05	82.8, 3, 2.3
	1	46.1–49.1	66.3–66.8	82.8–84.3	81.3–85.4
		0.55, 2.60	0.18, 0.48	0.75, 1.26	1.33, 2.78
C. approximans	74.4, 20, 2.34	39.6, 21, 7.52	54.3, 16, 3.84	61.7, 19, 3.96	62.4, 18, 3.98
	70.3–77.8	35.5–40.8	47.7–60.1	55.1–69.4	56.7–71.0
	0.52, 3.1	1.64, 19	0.96, 7.1	0.91, 6.4	0.94, 6.4

tionships as for the eagle and moa suggest that the body mass of a female Eyles's harrier was 3–3.5 kg. Of 13 species of harrier, most have body masses of about 0.5 kg; female *C. aeruginosus* may reach 1.1 kg (Brown and Amadon 1968). A large female *Circus approximans*, the Australasian harrier, has a mass of about 0.75 kg, so the New Zealand bird—almost certainly its sister species—was four times heavier. The homogeneity of body size of all the other harriers suggests that all other harriers depend on a low wing loading to be able to soar slowly at low altitudes. The greater mass of Eyles's harrier would have precluded it flying in the same manner. Therefore, the alteration of the wing proportions to those more like a fast-flapping species was a response both to the more closed environment and to the larger body mass resulting from selection for the ability to take large prey.

In the ratio of ulna length to humerus length, Eyles's harrier converged on goshawks in the genus *Accipiter* (Table 9.3). Accipiters are the classic woodland, bird-eating hawks, and they cannot maintain the soaring, gliding flight typical of harriers. Olson and James (1997) have suggested that the extinct *Circus dossenus* of Hawaii also had the same wing proportions as goshawks. Indeed, those researchers had as much trouble in assigning the bones to the correct genus as Ron Scarlett had with *Circus eylesi*. But whereas Olson and James (1992) had to work with isolated bones and some elements from different islands, the associated skeletons of the New Zealand bird clearly show its *Accipiter*-like proportions.

Distribution and Habitat

Bones of the North and South Island forms of the harrier are much less common than those of Haast's eagle, but unlike the eagle, there were harriers in both main islands in the Holocene. Contrary to the claim by Dawson (1958b), there is no material of *Circus eylesi* from the Chatham Islands. The only diurnal raptor on that group was the New Zealand falcon (*Falco novaeseelandiae*). The South Island form of *Circus eylesi* has not been found in deposits on the West Coast in Holocene deposits or in Southland. In the east and on Takaka Hill, it was sympatric with Haast's eagle. On Takaka Hill, it was present in the late Pleistocene, and there are no records that are unequivocally Holocene there. At higher altitudes on Mount Arthur, a few kilometers to the south of Takaka Hill, the harrier has been found in a site of Holocene age, where it was living in subalpine shrubland. On present evidence, which includes a sample of less than 25% of the number of sites on which the eagle distribution is based, Eyles's harrier was largely confined to the drier forests and shrublands of the east and south of the South Island during the Holocene. There were some birds in subalpine habitats as well. The distribution is close to coinciding with that of Haast's eagle, except that there are as yet no records from central Otago and Southland, despite there being many apparently suitable sites in those areas. At present, the southernmost record is at Mount Harris Swamp, north of the mouth of the Waitaki River, at nearly 45°S.

In the North Island, the harrier has been found in deposits from dunes on the Coromandel Peninsula, in caves near Waitomo, dunes in south Taranaki, rockshelters in inland Hawke's Bay, Lake Poukawa, and the drained Lake Te Aute in the northern Wairarapa. Most, if not all, of the Waitomo sites are also likely to be Pleistocene in age. At Lake Te Aute, the bones came from Locality 17 (Hamilton 1889), which Worthy (2000) argues is of Holocene age, and not the main moa deposit that is definitely of Late Glacial age. They are therefore equivalent in age to the Lake Poukawa specimens. All others, the coastal sites, those at nearly 900 m in Hawke's Bay and the lowland Lake Poukawa deposit are of Holocene age in the North Island. On the Coromandel Peninsula, the forest is one of the most floristically rich in New Zealand and is a typical wet, multistory forest. However, there was a population of *Euryapteryx curtus* in the shrubland on the coastal dunes. It may be that the harrier lived in this coastal vegetation and was rare or absent in the wetter inland forests of the west and north of the North Island. The record from Kaupokonui on the South Taranaki coast is also from an area where there was a band of coastal low forest and shrubland, again with *Euryapteryx curtus* and other shrubland species, including takahe. The Poukawa site is in the driest part of the North Island; from what is known of the flora and fauna in the area during the Holocene, shrubland was scattered through the forest.

Biology

In stark contrast to the cloudy issues surrounding the taxonomic status of the harriers in the North and South Islands, there is clear and graphic information on their prey and how they processed it. The rockshelters and shallow caves investigated by William Hartree on the slopes of Hukanui ridge in inland Hawke's Bay contained many bones

TABLE 9.2.
Summary statistics for element length (mm), in geographical and temporal populations of *Circus eylesi* (as used here), compared with *Circus approximans*. Data given: mean ± standard deviation, *n*, standard error, coefficient of variation, and range.

Taxon	Cranium	Humerus	Ulna	Carpometacarpus	Femur	Tibiotarsus	Tarsometatarsus
Circus approximans	74.36 ± 2.34, 20	108.53 ± 4.72, 24, 100.9–117.2	128.85 ± 5.66, 22, 117.4–138	64.51 ± 2.80, 24, 60.5–69.7	74.85 ± 3.28, 24, 68.4–80.9	112.43 ± 4.32, 21, 103.9–119.7	91.63 ± 3.68, 23, 86.2–98.1
Circus eylesi North Island	85.4 1	130.89 ± 3.10, 3, 1.79, 2.369, 127.6–133.8	143.19 ± 4.98, 6, 2.03, 3.48, 133.8–147.5	72.43 ± 4.31, 6, 1.76, 5.95, 64.5–77.0	97.21 ± 4.94, 4, 2.47, 5.08, 96.9–97.8	153.63 ± 2.08, 4, 1.04, 1.35, 151.8–156.4	118.52 ± 2.21, 6, 0.90, 1.86, 114.5–118.9
Circus eylesi South Island Pleistocene		136.95 1	150 2	77.93 1	104.0 1	153.8 1	119.7 1
Circus eylesi South Island Holocene	79.35 1	120.85 ± 1.51, 5, 0.675, 1.25, 119.1–122.7	136.34 ± 3.29, 7, 1.24, 2.41, 131.2–140.4	69.52 ± 2.3, 7, 0.87, 3.31, 65.8–67.9	92.67 ± 5.256, 7, 1.99, 5.67, 80.6–95.7	137.91 ± 5.8, 8, 2.05, 4.21, 138.3–143	110.41 ± 3.27, 6, 1.33, 2.96, 107.2–114.4

TABLE 9.3.
Intrawing ratios (ratio of ulna length to humerus length) for individuals of *Circus eylesi*, *C. approximans*, and *Accipiter fasciatus*.

C. eylesi	*A. fasciatus* (brown goshawk)	*C. approximans*
1.134	1.143	1.192
1.139	1.137	1.185
1.120	1.117	1.184

of medium-size birds not found in such abundance elsewhere. Hartree and his helpers made substantial collections of these bones, most of which were deposited in Canterbury Museum.

When the material was reexamined, it became apparent that damage to the bones, which had initially been taken to be normal wear and abrasion, told a different story. In a large collection of New Zealand pigeon (*Hemiphaga novaeseelandiae*) bones from Hukanui 7a, the element defining the minimum number of individuals was not a large limb bone such as the ulna or femur, but rather the sternum. Sterna are relatively fragile and are normally not well represented in collections. Another unusual feature was that most of the sterna themselves consisted only of the anterior pila carina, with ragged edges where the bowl and keel were once attached. The explanation was made clear by one sternum that retained part of the main body of the bone, in which was impressed a triangular hole that matched the plan view outline of the end of a harrier's beak.

Other elements were then examined more carefully and the representation of body parts in the collection compared with that in an intact skeleton. Instead of long bones dominating, there were abnormally large numbers of distal wing elements and few scapulae or intact coracoids, sterna, pelves, or femora. Preservation conditions that allowed the retention of the digits of the wing should have also meant that pelves, for example, would stay intact. If a bird skeleton can be imagined laid out with wings and legs extended, it was clear that the collection consisted mostly of the outer wings and legs, where the bones were also undamaged, and that elements from the body and proximal wings and legs were underrepresented and usually damaged (Fig. 9.6, Table 9.4). Of the 238 pigeon bones in the collection, 55.9% were damaged and 44.1% were whole. In comparison to the large collection of New Zealand pigeon bones from Pyramid Valley, which is a swamp deposit unbiased by predator choice, the proportions of the different elements preserved differ markedly (Fig. 9.6).

In summary, the unusual characteristics of the Hukanui 7a pigeon collection are as follows: distal wing elements are largely intact and well represented. Humeri are less abundant than would be expected from their size; several were broken, with the proximal and distal ends clipped off or only fragments of the ends remaining. None of the scapulae has a blade, with only the humeral end remaining. Nine of the 22 coracoids are intact, and the others lack the sternal end; only the anterior end of most of the sterna remains. The six synsacra are broken, "bitten off" near the caudal end, and synsacra are otherwise represented only by about six isolated acetabulae. The first phalanx of the major digit of the hand (MII/1) is uncharacteristically abundant, and most have puncture marks. Of the vertebrae, the cervical series are the most common, and only one is caudal; none of the fused vertebrae were recovered, but these are common in other deposits such as Pyramid Valley. The thoracic vertebrae in the collection have been "nibbled." There are many pedal phalanges and two pygostyles. There are only two complete femora and a single shaft; the rest are represented by

TABLE 9.4.
Damage to New Zealand pigeon bones from Hukanui 7a, inland Hawke's Bay.

			p		whole		d		-p -d		-p		-d	
Bone	%MNI	%Intact	L	R	L	R	L	R	L	R	L	R	L	R
Coracoid	67	27.7	6	5	6	7					1		2	1
Scapula	61.1	18.0	11	11		2			2	1			3	3
Humerus	100	22.2	5	5	5	5	2	6	1	1		4		
Ulna	67	38.9	1		8	8	2					3	1	
Radius	72.2	44.4	3	1	9	8	1		1				1	1
Carpometacarpus	83.3	77.8			16	11	1			1	2			1
Femur	61.1	11.1	3	5		2	5	7		1			1	1
Tibiotarsus	22.2	0					1	3	1	2				
Tarsometatarsus	33.3	22.2			2	5		2	1		1	1		
MII/1	50	33.3	1		4	7		1					4	2
Total (238)			30	27	50	55	12	19	6	6	4	8	12	
Percentage of total				24.0		44.1		13		5.0		5.0		8.8

Abbreviations: %Intact = percentage of total MNI represented by intact examples of that element; %MNI = percentage of the total MNI represented by the minimum number of that element; d = distal end only; -d = lacking distal end; L = left; -p -d = shaft lacking both proximal and distal ends; -p = lacks proximal end; p = proximal end only; MNI = minimum number of individuals; MII/1 = first phalanx of the major digit of the hand; R = right; whole = whole bone.

isolated proximal and distal ends. There is only one (basi)cranium, in contrast to the several in the Pyramid Valley collection. Tibiotarsi (three shafts and three distal ends) are drastically underrepresented. There are few tarsometatarsi, and the damaged ones lack the proximal end. Most of the abundant carpometacarpi lack the minor metacarpal.

Obviously, collection bias can be rejected as a reason for the underrepresentation of long bones because the technique was efficient enough to retain much smaller elements. The simplest explanation is that the meatier parts of the body were subject to greater damage as the harrier sought to remove muscle and organs from the carcass. Any bone ingested by the harrier was destroyed by the highly acidic stomach fluids; the pH of stomach fluids is lower in accipitrids than in owls, so much less ingested bone is preserved. As a consequence, an Eyles's harrier site mainly contains the material not eaten by the bird, whereas an owl site contains the material the bird did eat.

Details of damage to the humerus provided the final clue to the identity of the predator. Most humeri had part of the bicipital crest removed, or there were punctures in it. Sometimes there were two punctures, and some of the punctures were triangular. The evidence, starting with the abundant but severely damaged sterna through the overrepresentation of distal (nonmeaty) bones of the limbs and the details of damage to the proximal limb bones, showed that the pigeon bones were the remains of kills of large raptors. Pigeons were not

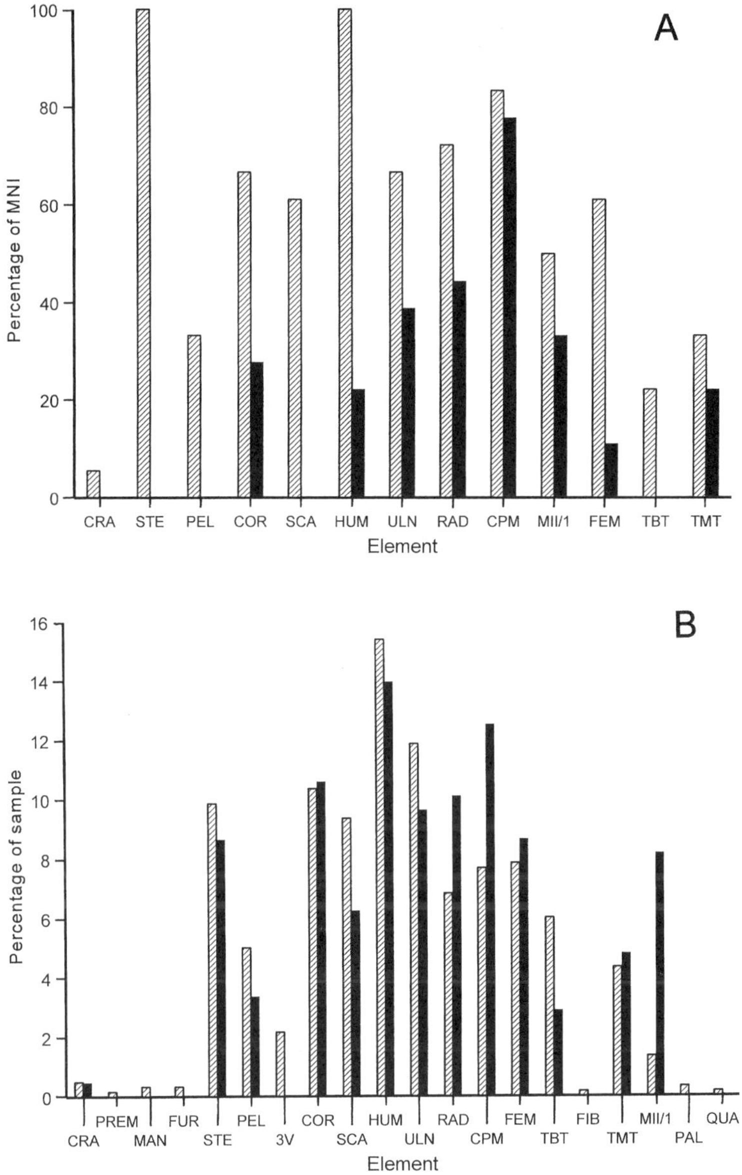

Figure 9.6. Survival of elements in a deposit of harrier prey remains compared with survival in a nonpredator site in a lake. (A) Predator site. The proportion of elements is expressed as a percentage of the minimum number of individuals (hatched bars) determined from the most abundant elements (sterna and humeri). Proportions of intact elements (solid bars) are contrasted against elements represented by fragments. More of the peripheral elements were intact. Minimum number of individuals for the site was 18 from sterna and humeri. None of the sterna was intact, whereas most of the carpometacarpi were intact. (B) Proportions of elements surviving in assemblages from the nonpredator lacustrine site at Pyramid Valley (hatched bars; n = 599*) compared with the representation in the predator site at Hukanui 7a (solid bars;* n = 208*). The predator site displays enrichment in peripheral elements. Vertebrae, pedal phalanges, and fragments were omitted from sample sizes. 3V = notarium; COR = coracoid; CPM = carpometacarpus; CRA = cranium; FEM = femur; FUR = furcula; HUM = humerus; MAN = mandible; MII/1 = first phalanx of second digit of wing; PAL = palatine; PEL = pelvis; PREM = premaxilla; QUA = quadrate; RAD = radius; SCA = scapula; STE = sternum; TBT = tibiotarsus; TMT = tarsometatarsus; ULN = ulna.*

normal prey for the laughing owl, whose prey waste deposits are now well known; nor were kokako, or Finsch's duck, or kaka, which made up most of the rest of the collection. A larger, diurnal predatory species was indicated. The discovery of bones of Eyles's harrier in the site confirmed that the harrier was present in the region, as did, eventually, the collection there of an almost complete skeleton of the harrier. Because the taphonomy clearly indicates that a majority of the fossils in these sites were predated, because the species present and their relative abundance show that the predator preferred larger diurnal prey, and because damage patterns are dissimilar to those known for *Sceloglaux* or *Falco*, the fauna in site Hukanui A is attributed to *Circus eylesi*.

The larger form of the harrier lived with smaller moa than the South Island birds, 20–40 kg as opposed to 40–120 kg. Eyles's harriers in the North Island were as large as some subspecies of golden eagles, so it is entirely possible that they killed and ate small moa. Some of the moa pelves found in rockshelters at Puketitiri in Hawke's Bay have the characteristic damage of a predator, exposing the kidneys. Kea, which damaged bones in the same way, were not found in the North Island at that time, so the damage—and perhaps the moa's death—may be attributed to the harrier.

Extinction

There are no historical records of giant harriers, so it can be assumed that the species became extinct sometime during the Polynesian period. Its extinction poses some problems, not the least being the fact that some of its prey species remained abundant until the 19th century and beyond. So unlike Haast's eagle, a shortage of favored prey may not have been a factor. Habitat loss, however, may well have been a major problem. In the South Island, most of the lowland forest and shrubland in the eastern areas where harrier remains have been found was destroyed early in the Polynesian period. Similarly, the coastal vegetation around the Coromandel Peninsula and southern Taranaki was also damaged by fire or occupied for horticulture from an early date. However, substantial areas of forest similar to that inhabited by the harriers in inland Hawke's Bay existed in the southeastern North Island until the 1870s.

Another factor not usually taken into account with birds of prey is their palatability to humans. *C. eylesi* has been found in Polynesian middens. It was obviously palatable. If it also had the characteristically New Zealand tendency to K selection, loss of adults to human predation could have significantly affected the population.

In undisturbed environments that lack predatory mammals, many birds that normally nest in trees instead nest on the ground. Harriers normally nest on the ground anyway. The presence of prey remains in rockshelters and cave entrances suggests that the New Zealand harriers may have chosen nest sites typical of their genus, even in forest and shrubland. The records of Australasian harriers nesting in trees, or at least on stumps, in New Zealand indicates that the New Zealand species may have become more like the spotted harrier of Australia, which habitually nests in trees. In either case, their eggs and chicks could also have been vulnerable to rat predation, especially if the adults were accustomed to leave the chicks unattended for even short periods.

When grassland became the common landscape in New Zealand, the harriers, which had adapted well to forest and shrubland, would have been at a disadvantage: they had gone too far down the path of flapping flight supporting a heavy body to easily revert to the typical harrier tactic that now serves the Australasian species well.

The Other Harrier: Reinvasion by an Ancestor?

Despite its present abundance, there is no evidence that the Australasian harrier (*Circus approximans*), which presently breeds in New Zealand, was part of the prehuman fauna. None of the bones of *C. approximans* identified in the collection from Pyramid Valley in the South Island proved to be of that species. All that material has been reidentified as *Circus eylesi*. Fossils of *Circus approximans* have been found at Marfells Beach in Marlborough, but there is no reason to suppose that they predate Polynesian settlement. Fossils of *C. approximans* found in the Lake Poukawa deposit in the North Island, where there was a substantial number of *C. eylesi* bones from the lower layers, were all from the superficial layer, which Horn (1983) thought was laid down after the arrival of the Polynesians.

With the resident giant harrier and a lack (or rarity) of suitable habitat, Australasian harriers that survived the flight across the Tasman Sea before human settlement would not have found a vacant niche. They would not therefore have been able to colonize until lasting environmental change had begun. Although Australasian harriers have been videotaped taking kokako (*Callaeas wilsoni*) chicks beneath the canopy in rainforest in the Urewera National Park in the northeastern North Island, its typical habitat is open grassland, low shrubland and forest, and coastal swamps. These habitats were rare in New Zealand before anthropogenic firing removed much of the forest cover.

As one of the two species of harrier that breed closest to New Zealand (the other is the Australian spotted harrier) and the only one to have reached other islands in the Pacific as a vagrant, *C. approximans* is likely to share a common ancestor with the founding population that became the giant New Zealand species. In a situation that parallels that of the takahe and pukeko, it seems that when the specialized first immigrant succumbed to the consequences of human settlement and when new habitats replaced old, the way was clear for a new colonization by the relatively undifferentiated descendants of the original source population.

An Almost Colonist

Genus Accipiter *Brisson*
Accipiter *cf.* fasciatus *(Brown Goshawk)*

Bird hawks of the genus *Accipiter* were not as successful as the harriers in crossing sea barriers, but they did reach east as far as Tonga in the tropical Pacific. Their closest approach to New Zealand was on Norfolk Island, 600 km to the northwest of North Cape. They may actually have reached the main islands, only to find them occupied by the harrier masquerading as an *Accipiter*. Similar attempted invasions by other species that regularly cross the Tasman Sea were apparently repelled by a full ecological house, as we shall see later among the songbirds.

Hawks were reported on Norfolk Island by Europeans in the first convict settlement established there in 1789. Two kinds (gray and blue) were mentioned by King (1786–1790). A bird very similar to the Australian brown goshawk (*Accipiter fasciatus*), which is actually slate gray (Pizzey 1980), has been identified in both natural and archaeologic deposits at Kingston (Holdaway and Anderson 2001; Meredith 1985, 1991). At present, its closest population is in New Caledonia. Bones of a smaller species are known (P. Gaff, personal communication), which may be the other bird seen by King. By 1800, hawks were apparently extinct on the main island, but a population of one species seems to have survived on Philip Island for a few more decades. It was recorded as being a predator of the Norfolk kaka (*Nestor productus*) that survived there until the late 1840s (Gurney 1854).

The goshawk fed on a range of birds, including Pycroft's petrel (*Pterodroma pycrofti*), which was abundant on the island, the Norfolk pigeon (*Hemiphaga spadicea*), the Norfolk Island rail (*Gallirallus* sp.), and other terrestrial species and petrels. Remains of all these have been found in deposits where some of the bones show signs of damage during dismemberment that are very similar to those described for *Circus eylesi* in New Zealand. Surprisingly, no bones of any of the several species of tern and noddy, which are common on and around the island today, have been found in the predator deposits on Norfolk Island.

Pycroft's petrels could have been taken as they prepared to leave for the sea in the morning. Pigeons, rails, and parakeets would have been taken in the forest of Norfolk Island pines (*Araucaria heterophylla*). Terns and noddies are extremely capable fliers and may have been too nimble for the goshawk to catch. It is surprising that the white terns, which nest on open branches of the pine trees, were not vulnerable to goshawks.

Phantom Chatham Island Sea Eagle

Forbes Strikes Again

In early 1892, Henry Ogg Forbes collected many fossils from the dunes on Chatham Island during a Canterbury Museum expedition; most of these Forbes took to England when he returned there later in 1892. When he examined the Forbes collection in the 1950s, Elliot Dawson found some eagle bones, apparently part of the Chatham Island collection, that he identified as a species of *Haliaeetus* related to the Northern Hemisphere *H. leucocephalus* and *H. albicilla* (Dawson 1961). Harrison and Walker (1973) described the material as being a new species of eagle, but placed them in the genus *Ichthyophaga*, which is otherwise represented by two species in the Old World tropics. Olson (1984) examined the material again and agreed with Dawson, transferring the species to *Haliaeetus*.

The existence of a Chatham Island sea eagle, related to Northern Hemisphere species and not to the Australian *H. leucogaster*, has been accepted by Turbott (1990a) and by Holdaway (1989). Paradoxically, it was also accepted by Millener (1991, 1996, 1999), who studied the available material and pointed out (Millener 1996: 119) that the bones "may not be distinguishable from the Alaskan race of the bald eagle [*H. leucocephalus*]." Indeed, it is doubtful that the bones came from the Chatham Islands in the first place, because Forbes is known to have visited and collected from middens in British Columbia, and many of the labels in the Forbes collection were added in the 20th century on the assumption that all the material came from the Chatham Islands. Many tens of thousands of fossil bones have been collected from the dunes of Chatham Island since Forbes's visit in 1892, but no other bones of an eagle have been found. Despite Millener's (1996, 1999) cautious retention of the species in the Chathams fauna, there can be no doubt that the so-called Chatham Island sea eagle *Haliaeetus australis* never existed. It was based on material of other species collected elsewhere than on the Chatham Islands.

The Lone Survivor among the Raptors
Order Falconiformes
Family Falconidae

Genus Falco *Linnaeus*
Falco novaeseelandiae *Gmelin, 1788 (New Zealand Falcon)*

The New Zealand falcon is the only one of four diurnal raptors in the late Holocene avifauna of New Zealand, including Norfolk Island, to survive today. It has been described as having the form of a Eurasian hobby (*Falco subbuteo*) but with behaviors like those of peregrines (Brown and Amadon 1968). It is a fierce and fearless predator, and both sexes do not hesitate to attack people who go near the nest.

Fox (1978a, 1978b) recognized three different forms of the New Zealand falcon, a large, pale "eastern" falcon in the South Island, a "bush" falcon in the wet forest of the North Island and the western South Island, and a "southern" falcon, on the Auckland Islands. However, in his redescription of the species, Fox (1988) declined to formally separate them as subspecies. Bones of the New Zealand falcon are known from fossil collections from Chatham Island (in the Canterbury Museum), but the species was extinct on the island when Europeans settled in the early 19th century. Most is known about the eastern form, which lives in habitats that allow of easy observation.

Falcons today eat mostly small birds and insects. They can and do kill birds up to the size of New Zealand pigeons (~600 g). Fossil deposits accumulated by falcons have a different species composition and pattern of breakages than those that are attributed to laughing owls and Eyles's harrier. One feature of some sites is the presence of rangle stones (Worthy and Holdaway 1995), which are kept in the crop to assist the breakdown of food and which are cast up with pellets of waste. Other features are the presence of diurnal lizards, mainly skinks, rather than the nocturnal geckos typical of laughing owl deposits; the general rarity of other nocturnal species among their prey; and the different butchering techniques. Prey assemblages attributed to falcons rather than the laughing owl *Sceloglaux* were in open overhangs rather than crevices or cavities and included mainly small, diurnal prey whose bones exhibit a distinctive pattern of damage.

All avian predators digest the bones of their prey to some extent, but as Andrews (1990) showed, falcons characteristically cause far greater digestion damage than owls. Moreover, in owl sites, some examples of all elements are whole, whereas in falcon deposits, few bones are whole. The most common elements, which form the basis for estimates of minimum numbers of prey individuals, are a few of the most robust elements only; the other elements are usually severely underrepresented.

Prehistoric prey assemblages of the falcon are known from eastern South Island and include a wide range of species. However, falcons favored parakeets and quail, taking birds up to kokako and pigeon in size more rarely. Most sites usually contain a few bats, which shows that falcon hunted in the near dark, then as now. Where preservation conditions are favorable, numerous beetle remains indicate that falcons had also included invertebrates in their diet.

Its small size, ability to live in open environments on small prey, and fierce defense of the nest have allowed the falcon to survive in the face of environmental change that removed the other raptors. Its survival may signal the importance of specialization to habitat and particularly to prey among the factors that led to the extinction of the eagle and harrier. Being a small generalist is a successful strategy when the rules of the ecologic game change radically.

Nocturnal Birds of Prey

Order Strigiformes

A report that a species of *Tyto,* probably *Tyto alba,* was present in New Zealand (Scarlett 1967b) has been shown to have been a misidentification of the endemic strigid *Sceloglaux albifacies* (Millener 1983). Therefore, there were only two species of owl in New Zealand. One of them has been vital to progress in understanding the composition and distribution of most of the small vertebrates in prehuman New Zealand and the changes wrought in that fauna by the Pacific rat.

Family Strigidae
Genus Sceloglaux *Kaup*
Sceloglaux albifacies *Gray, 1845 (Laughing Owl)*

The laughing owl (Fig. 9.7) is placed in a monotypic, endemic genus, but there has been no recent analysis of the relationships of the hawk owls of the southeast Asian region. Two races have been described, but the plumage differences are minor (if they exist), and size differences seem to be a clinal increase from north to south (Gill 1996b). The relationships of the bird are not well known, and the skeleton has yet to be compared with those of other genera in the Australasian region. The bird is of medium size for an owl, standing about 400 mm high, with feathered tarsi, large claws, and conspicuous feathered disks around the large, dark eyes (Fig. 9.7).

The importance of this extinct owl to vertebrate paleobiology in New Zealand cannot be overstated. It was the major accumulator of fossils of small vertebrates, and thus it is the most important source of

information on the small birds, lizards, frogs, and large invertebrates in the late Holocene fauna. It was a generalist predator, taking any species that was on or near the forest floor at night, dawn, or dusk. Diet varied according to region; small petrels were prominent in some areas but absent and replaced with other species such as the parakeets in others.

If not actually extinct, the owl must be very rare and endangered. No specimens have been obtained since 1914, and reports of its presence in several places in the South Island remain unconfirmed. The laughing owl was relatively common in the South Island when Europeans arrived (Oliver 1955; Turbott 1990a; Williams and Harrison 1972). Its status in the North Island in the 19th century is not clear, but specimens and sightings indicate its presence over much of the southern half of the island before 1890.

*Figure 9.7. Photograph of a laughing owl (*Sceloglaux albifacies*) taken sometime between 1909 and 1912 by Cuthbert Parr at Raincliff in South Canterbury, South Island, New Zealand. The fledgling is holding a mouse in its beak, put there to provide a point of interest in the photograph. Photograph reproduced courtesy of the Parr family and Museum of New Zealand Te Papa Tongarewa.*

Old Bones and Owl Biology

After predator deposits were first recognized (Worthy and Holdaway 1994a), the first step with each new site was to determine which of several predators was responsible. Species likely to produce deposits of small vertebrates included the extant New Zealand falcon and morepork owl, the laughing owl, and the extinct Eyles's harrier. The prey size range for the harrier was likely to include significantly larger species than for the other three, whose prey might be expected to be about the same size. "Signatures" were developed on the bases of location and kind of site, the composition of the fauna, the representation of body parts in the deposit, and the kind of damage to individual bones (Fig. 9.8). Although falcon sites are now known in several areas, the species composition of the prey fauna and the damage to the bones is significantly different from those characteristic of owls (Worthy and Holdaway 1996a).

The high representation of nocturnal species (Fig. 9.9) such as the endemic short-tailed bats (*Mystacina* spp.) and geckos in the genus *Hoplodactylus* is a feature of owl deposits in New Zealand, whereas falcon sites contain numbers of the diurnal *Oligosoma* skinks and day-flying birds. The first laughing owl deposit to be identified, at Hermit's Cave on the West Coast of the South Island (Worthy and Holdaway 1994a), was dominated by the remains of fairy prions (*Pachyptila turtur*), small petrels that visit their nest burrows at night and whose fledglings move about on the forest floor before leaving for sea. There were also many bones of the nocturnal terrestrial frogs (*Leiopelma*) in Hermit's Cave. However, in areas where there were no frogs (as in the lowlands of the eastern South Island), petrels, or other characteristic nocturnal animals, details of the bone representation and damage are required before the predator responsible can be firmly identified. Although the smaller morepork is now known to take birds in the size range preferred by laughing owls when they are available (Anderson 1992), most of the owl deposits on the main islands have been attributed to laughing owls.

Once the identity of the predator had been confirmed as far as possible, the composition of the fauna and representation of species with different ecology may be used to start to piece together the biology and behavior of the owl in that area at the time the deposit was laid down. For example, once it was established that laughing owls produced the prey remains found in Predator Cave on Takaka Hill in northwest Nelson (Worthy and Holdaway 1996a), it was possible to make some inferences about the environment around the cave, as well as the hunting time and behavior of the owls (Holdaway and Worthy 1996).

The owls on Takaka Hill lived in a closed-canopy forest dominated by southern beech (*Nothofagus*) over a relatively open understory. There were no clearings or even shrublands near the sites, so the birds had to live and feed in the forest. Reports in the 19th century of owls in more open habitats in the eastern South Island have been used to suggest that laughing owls preferred open country, perhaps at forest edges (Williams and Harrison 1972). The fact that most of the deposits attributed to them were in areas with tall vegetation suggests that the owls lived wherever there was food. They survived for several centuries in the eastern South Island after the lowland forest was almost completely destroyed and the indigenous fauna replaced by Pacific rats because the owls were not specialist hunters: they changed to a diet of rats. Deposits

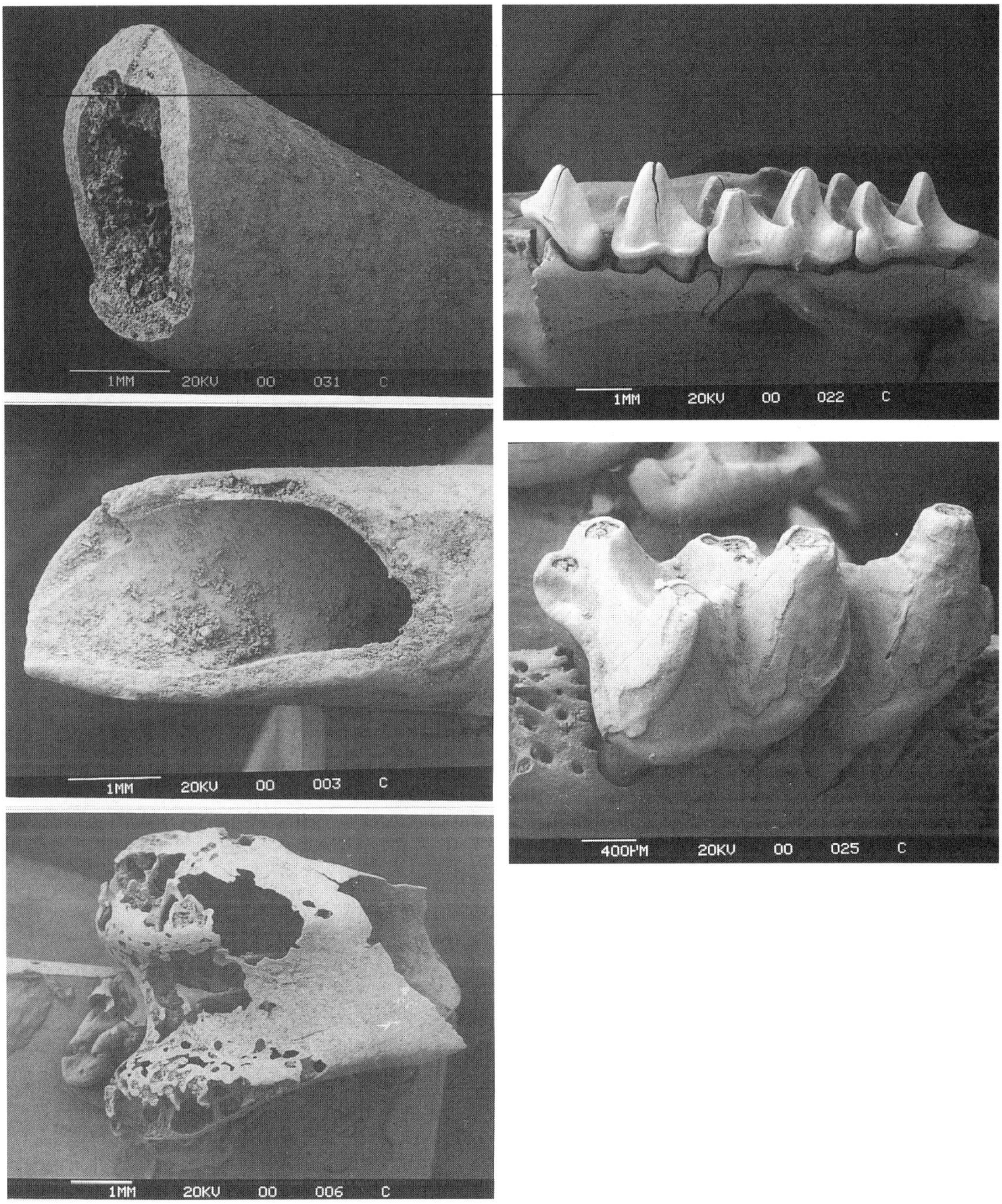

*Figure 9.8. Digestion damage to bird and bat bones from Hermit's Cave laughing owl (*Sceloglaux albifacies*) deposit. (A) (left) Fairy prion (*Pachyptila turtur*). (Top) Natural break in weathered undigested bone, with fracture plane at right angles to the bone axis and the fracture surface with a "floury" texture. (Middle) Greenstick fracture partly rounded by digestion; the fracture plane is at a shallow angle to the bone axis and is curved along the fracture edge. (Bottom) Heavily digested distal femur with bone thinned to extreme extent before disintegration, exposing trabecular bone distally. (B) (right) Teeth of the endemic bat* Mystacina robusta. *(Top) Right maxillary tooth row in labial view with P^1, P^2, M^1, and M^2, showing erosion of bone around roots. Cracking probably resulted from a combination of weathering and acid corrosion during digestion. (Bottom) Detail of right mandible, with M_1–P_2 in labial view showing extensive corrosion of the enamel. Electron micrograph courtesy of University of Canterbury, Christ church.*

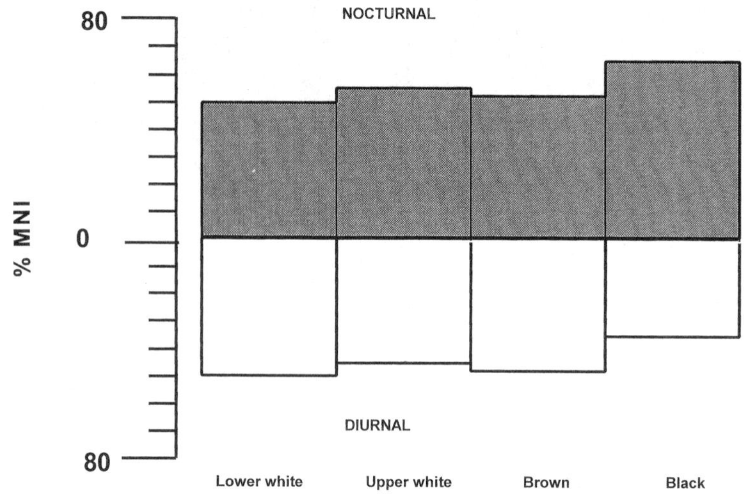

Figure 9.9. Proportional representation of nocturnal and diurnal prey in the layers in Predator Cave. After Holdaway and Worthy (1996: fig. 6).

known so far on Takaka Hill show that the owls roosted and nested in cave entrances, although they probably also used hollow trees. Elsewhere, where there were cliffs, narrow, dry rock crevices were favored (Fig. 9.10).

Diet: From Frogs, Lizards, Birds, and Bats—To Rats

Although not as large as some of the Northern Hemisphere owls—and it did eat many kinds of large insects (Table 9.5)—the laughing owl could catch and kill birds up to the size of weka and occasionally ate the other New Zealand owl, the morepork. Regardless of where they were living, laughing owls seem to have taken the largest prey they could handle. The total diet was, however, diverse and included large insects, frogs, lizards, (mainly young) tuatara, birds, and bats (Table 9.6; Fig. 9.11). Historical records of large earthworms in the diet (Smith 1884b) have not been confirmed by the identification of fossil setae in the sediments in the cavities, but this may be because little of that material has been subjected to microscopic examination. Insect cuticle preserves well in the drier sites, so there is no reason why setae might not also be preserved. The New Zealand worm fauna includes species up to 1 m long—a source of protein that was probably not eschewed. Some of the sight records of the owl refer to their flying down from cliffs to catch large worms in a small swamp as food for the chicks (E. Parr, personal communication).

The composition of the diet was different in different areas at different times and depended on the forest-floor fauna available around a site at the time. In the older deposit at Predator Cave on Takaka Hill, birds contributed about 50% of the individuals in the owl diet. Short-tailed bats contributed another 15%–20%, and 20% or more were

Figure 9.10. Photographs of the Gowan Hill and Ardenest owl sites. (A) The nest crevice at Gowan Hill, near Waikari, North Canterbury, after excavation of about 300 mm depth of dry sediment with many triturated bones of the owl's prey. (B) Ardenest owl site, near Waikari, North Canterbury, with T.H.W. sieving sediment from the nest cavity between the large boulders. The nest contained portions of an introduced plant and therefore was used at least until the mid- to late 19th century. Photographs by R.N.H.

geckos; the remainder were skinks and frogs (Fig. 9.11). Of the birds, parakeets comprised 35%–45% of the bones in the deposit. Numbers of individuals must be counted against the biomass they represent. For example, wrens were more abundant than snipe in the Predator Cave deposit, but the biomass of snipe was nearly as high as that of wrens (Fig. 9.12). The contribution of both to the owl's diet in that forest was lower than the parakeets, which provided more than 50% of the bird biomass.

In other areas, parakeets were not as important as in the higher altitude beech forest, where parakeets were particularly abundant. On the West Coast, much earlier, petrels were important in the owl diet, as they probably were in the east before the populations of small species crashed in response to predation by the Pacific rat (Holdaway 1999b). On Takaka Hill, petrels seem never to have been an important part of the fauna. In sites in the eastern South Island, the parakeets were still important prey, but other taxa, such as the New Zealand quail, were important, at least in sites that were formed after the arrival of Polynesians and the destruction of forest. Snipe and the South Island piopio were also common there, the piopio more so than on the West Coast or on Takaka Hill. The presence of quail and the pipit in reasonable num-

TABLE 9.5.
The invertebrate species found in laughing owl deposits in Eastern South Island.

Species	Status	North Canterbury	South Canterbury
Carabidae (carabid beetles)			
Metaglymma monilifer		Y	
Metaglymma sp.			Y
Megadromus rectangulus		Y	
Megadromus antarcticus		Y	?
Megadromus sp. ?*temukensis*			Y
Megadromus (very large species)	Extinct		Y
Holcaspis subaenea		Y	
Holcaspis delator			Y
Holcaspis augustula			Y
?Small *H. subaenea*		Y	
Oregus sp. indet.		Y	
Oregus aereus			Y
Mecodema sulcatum		Y	
Mecodema costellum lewisi	Locally extinct	Y	
Mecodema costellum ssp.	Locally extinct		Y
Mecodema fulgidum			Y
Mecodema brittoni			Y
Lucanidae			
Lissotes reticulatus		Y	Y
Elateridae (click beetles)			
Amychus sp. cf. *granulatus*	Extinct on SI	Y	
Thoramus sp.			Y
Agrypnus variabilis			Y
Trogosittidae			
Lepidopterix nigrosparsa			Y
Colydiidae			
Enarsus sp.			Y
Tenebrionidae			
Mimopeus ?lateralis		Y	
Mimopeus costellus		Y	
?*Mimopeus opaculus*		Y	Y
Mimopeus sp. nov.		Y	
Artystona sp.			Y
Scarabaeidae			
Odontria sp. cf. *striata*		Y	Y
Pericoptes sp.			Y

Species	Status	North Canterbury	South Canterbury
Histeridae (live in bird nests)			
?*Saprinus* sp.			Y
Sp. indet.		Y	
Cerambycidae			
Somatidia sp.		Y	
Dermaptera: Forficulidae			
Forficula auricularia (earwig)		Y	
Curculionidae (weevils)			
Hydramphus tuberculatus	Extinct		Y
Ectopsis ferrugalis	Extinct on SI	Y	Y
Anagotus stephenensis	Extinct on SI	Y	Y
Anagotus rugosus	Locally extinct	Y	Y
Megacolabus garviensis			Y
Tychanus verrucosus		Y	Y
Psepholax sulcatus		Y	Y
Strongylopterus hylobioides		Y	Y
Tychanopais fougeri		Y	Y
Crisius longulus		Y	Y
Crisius ornatus		Y	Y
Crisius posticalis			Y
Didymus impexus		Y	
Clypeolus cineraceus		Y	
Phrynixius terreus		Y	Y
Echinopeplus insolitus			Y
Epitimetes grisealis		Y	
Epitimetes lutosus		Y	
Epitimetes cuprealis		Y	
Sargon suturalis		Y	
Sympedius sp.			Y
Irenimus aequalis		Y	
Species introduced to New Zealand			
Otiorhynchus ovatus		Y	
Listroderes difficilis		Y	

Abbreviations: SI = South Island; Y = present. North Canterbury data are from the Gowan Hills, Timpendean, and Ardenest sites near Waikari (Worthy and Holdaway 1996b); South Canterbury data from Worthy (1997b: appendix 5). Most are probable prey items. Invertebrate remains survive only in very dry conditions.

TABLE 9.6.
List of vertebrate species found as prey in laughing owl sites in four South Island regions.

Species	West Coast Hermit's Cave	Takaka Hill Predator Cave	North Canterbury	South Canterbury
BIRDS				
Apteryx "Eastern"				Y
Apteryx sp. juveniles	Y		Y	Y
Procellariids				
Pachyptila turtur	Y			Y
Pterodroma inexpectata			Y	Y
Pterodroma cookii			Y	Y
Puffinus sp.			Y	Y
Puffinus spelaeus	Y			
Pelecanoides urinatrix	Y		Y	Y
Garrodia nereis			Y	Y
Oceanites sp.	Y			
Ducks				
Anas chlorotis				Y
Euryanas finschi	Y		Y	Y
Parrots, owls, and pigeons				
Strigops habroptilus			Y	Y
Nestor meridionalis			Y	
Nestor notabilis				Y
Cyanoramphus spp.	Y	Y	Y	Y
Ninox novaeseelandiae		Y	Y	Y
Aegotheles novaezealandiae	Y	Y	Y	Y
Hemiphaga novaeseelandiae			Y	Y
Rails, waders, and gamebirds				
Gallinula hodgenorum			Y	Y
Gallirallus australis	Y		Y	Y
Coenocorypha iredalei	Y	Y	Y	Y
Charadrius bicinctus			Y	Y
Charadrius novaeseelandiae			Y	
Sterna albostriata			Y	
Coturnix novaezelandiae			Y	Y
Passerines				
Acanthisitta chloris	Y	Y	Y	Y
Xenicus sp.	Y	Y	Y	Y
Traversia lyalli		Y		Y
Pachyplichas yaldwyni	Y	Y	Y	Y
Bowdleria punctata				Y
Prosthemadera novaeseelandiae		Y	Y	Y
Anthornis melanura		Y	Y	Y
Petroica australis	Y	Y	Y	Y
Petroica macrocephala	Y	Y	Y	Y
Mohoua ochrocephala	Y	Y	Y	Y
Mohoua novaeseelandiae		Y		Y

Species	West Coast Hermit's Cave	Takaka Hill Predator Cave	North Canterbury	South Canterbury
Gerygone igata		Y	Y	Y
Rhipidura fuliginosa		Y	Y	Y
Anthus novaeseelandiae	Y	Y	Y	Y
Callaeas cinerea	Y		Y	Y
Philesturnus carunculatus		Y	Y	Y
Turnagra capensis	Y	Y	Y	Y
Introduced passerines				
Alauda arvensis		Y		Y
Carduelis carduelis			Y	Y
Carduelis chloris				Y
Emberiza citrinella				Y
Turdus merula			Y	Y
Turdus philomelos				Y
Sturnus vulgaris			Y	Y
Prunella modularis				Y
Passer domesticus				Y
Fringilla coelebs				Y
OTHER VERTEBRATES				
Fish				
(*Anguilla* and *Gobiomorphus*)	Y		Y	Y
Frogs				
Leiopelma cf. *hamiltoni*	Y	Y		
Leiopelma markhami		Y		
Tuatara, geckos, and skinks				
Sphenodon sp.			Y	Y
Hoplodactylus cf. *maculatus*	Y	Y	Y	Y
Hoplodactylus cf. *granulatus*		Y		Y
Hoplodactylus cf. *duvaucelii*		Y	Y	Y
Naultinus sp.				Y
Skink sp.	Y	Y	Y	Y
Bats				
Mystacina tuberculata		Y	Y	
Mystacina robusta	Y	Y	Y	Y
Chalinolobus tuberculatus		Y		
Rodents				
Rattus exulans		Y	Y	Y
Rattus rattus/norvegicus			Y	Y
Mus musculus		Y	Y	Y
Rabbit				
Oryctolagus cuniculus			Y	Y

Abbreviation: Y = species present. Data for West Coast (Worthy and Holdaway 1993), Takaka Hill (Worthy and Holdaway 1994b), North Canterbury (Worthy and Holdaway 1996b), and South Canterbury (Worthy 1997b). More introduced species are represented in South Canterbury deposits because laughing owls survived in that area longer than elsewhere. Fish are rare in the deposits, but their presence indicates that the owl was capable of snatching fish from shallow water.

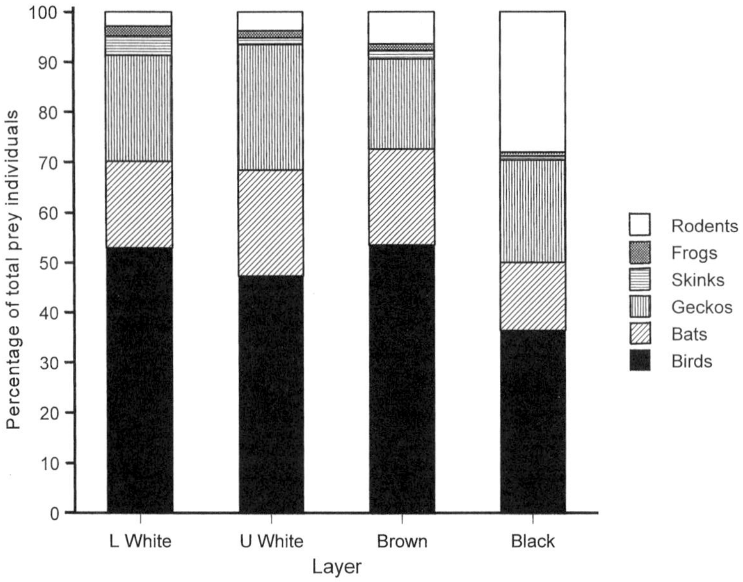

*Figure 9.11. (top) Relative proportions of numbers of total prey in samples from Predator Cave, Takaka Hill, South Island (after Holdaway and Worthy 1996: fig. 1). Black layer youngest (to ca. A.D. 1900); lower white oldest (to early Holocene). Relative age of four layers is affected by the downflow of material. Although most of the material in the lower white layer is older than that in the upper white, and so on, younger material has been shown to have spilled into each of the lower layers. Hence, discrimination between faunas before and after the arrival of the Pacific rat (*Rattus exulans*) is less than perfect.*

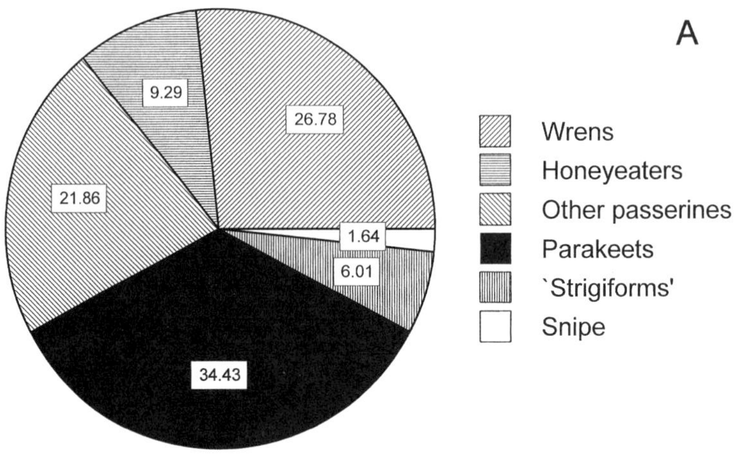

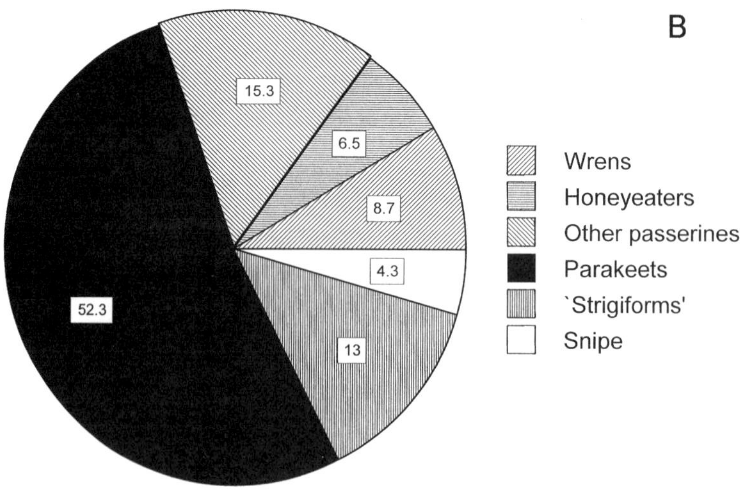

Figure 9.12. Proportions of total minimum number of individuals (MNI) and biomass for birds in the Predator Cave deposit to show the relative importance of different taxa in the diet of laughing owls on Takaka Hill, northwest Nelson, in the Holocene (after Holdaway and Worthy 1996: 552, fig. 2). (A) (center) Percentage of total MNI for each taxon. (B) (bottom) Percentage of total biomass for each taxon.

bers in eastern sites, in company with large numbers of Pacific rat bones, shows the great changes that followed human contact with New Zealand. After the arrival of the Pacific rat, the rodent formed the bulk of the diet as it replaced the indigenous fauna.

The diet of the laughing owl changed as the environment and fauna around them changed. Shifts in the abundance of the main food species—and the eventual extinction of some—are tracked in a few sites (notably Predator Cave) where the deposits cover a period of several thousand years. Other sites record the diet for shorter periods, at different times, and provide more information on the availability of species before and after the first human contact with the New Zealand archipelago. For example, the Hermit's Cave site provides a record that the species lived along a river on the West Coast 17,000 years ago, when the coastline was several kilometers further away than now. On the basis of the preferred habitats of birds in the fauna, the vegetation was then a mosaic of grassland, shrubland, and isolated patches of forest on sheltered, north-facing slopes.

The Hermit's Cave owls ate one species of leiopelmatid frog (*Leiopelma hamiltoni*), lizards, the greater short-tailed bat (*Mystacina robusta*), and at least 14 species of land bird, from tiny flightless wrens to the New Zealand owlet-nightjar and snipe. However, the bulk of its food there consisted of seabirds. Three species—the fairy prion (*Pachyptila turtur*), common diving petrel (*Pelecanoides urinatrix*), and Scarlett's shearwater (*Puffinus spelaeus*)—together accounted for 1265 (92%) of 1375 bird bones identified in the deposit and 37 (60.7%) of 61 individuals of all bird species. Four of the other bones (from one individual) were of a storm petrel (Worthy and Holdaway 1994a). Many of the prion bones were juvenile, so the owls may have been preferentially taking the fledglings as they prepared for their first flights, in February or March (Heather and Robertson 1996).

Owl Deposits as Sources of Information on the Paleobiology of Other Species

If a generalist feeder such as the laughing owl leaves identifiable food residues, the contents of the deposits can be used to reconstruct the fauna in the area as well as the biology of the owl itself. Until the laughing owl deposits were identified, little was known of the distribution and relative abundance of the smaller vertebrates in New Zealand. Now there is sufficient information for some regional and habitat differences to be discerned. That in turn has enabled a better understanding of the particular habitat requirements of extinct species and the rarer extant species for which there is little contemporary information.

In the Predator Cave deposit, three acanthisittid wrens (*Acanthisitta chloris, Xenicus* spp., and *Pachyplichas yaldwyni*) together contributed about 25% of the birds; the fourth (*Traversia lyalli*) was less than 1%, and a fifth (*Dendroscansor decurvirostris*) was not represented. All these tiny birds, which were primarily insectivores, ate similar foods. Flight pattern—and indeed flightlessness—were not important in the prey choice because only one species common in the owl deposit was flightless, as were the two that were not well represented.

The differences in abundance in the owl deposit probably reflect the abundance of the wren species in the environment or perhaps the fact that different species were in such different habitats that they were not, or only rarely, encountered by owls.

Traversia survived into European times on Stephens Island, off the northern end of the Marlborough Sounds, about 100 km across Tasman Bay from Takaka Hill. There it lived in a coastal forest that was probably physically more similar to the eastern South Island dry forest–shrubland mosaic than to the wet forests in the Marlborough Sounds and elsewhere in northwest Nelson. The predator-free Maud Island within the confines of the Sounds might also have been expected to have supported a remnant population; the absence of *Traversia* there may have been determined by the island's wet forest association rather than the drier low forest on exposed Stephens Island. *Traversia* is also abundant in the owl deposits in the eastern South Island and was found in a cave above the tree line on Mount Owen south of Takaka Hill. Its rarity in the Predator Cave deposit, in closed wet forest, indicates the potential usefulness of the predator deposits—used carefully—in elucidating the ecology of extinct species.

A lack of time depth in one site may be countered by sites of different age nearby. Because the taphonomy is the same in all deposits formed by one species, their contents should be comparable even if they were formed thousands of years apart, allowing for possible changes in the local fauna during that time. Even when one of the deposits was formed by another process, comparisons of deposits formed at different times may be useful. The rarest of the New Zealand fossil birds, the acanthisittid wren *Dendroscansor decurvirostris,* has so far been found in four sites. One of these, Hobson's Tomo, is on Takaka Hill. The wren was flightless, and its long bill suggests that it was a bole gleaner (Millener and Worthy 1991). With these attributes, it should have figured in the owl deposits, if it had been present when they were formed. Its absence from the owl deposits suggests that the habitat was not suitable for it when they were accumulating.

A radiocarbon date from the basal layer of the Predator Cave deposit showed that the site became a home for laughing owls at the end of the Otiran Glacial. Owls in the area before then either did not have access to the cave, or their prey deposits have not been preserved. The material of *D. decurvirostris* from Hobson's Tomo consists of 24 bones of three individuals, which is a large representation for a tiny passerine in a pitfall trap. The bird must have been relatively common when that deposit was forming, only 200 m from Predator Cave. The material in Hobson's Tomo dates from over 14,000 years before the present, when the top of the marble massif was clothed by a cold-climate shrubland, and it is in that shrubland that *D. decurvirostris* seems to have lived. During the Holocene, it was probably restricted to high-altitude shrublands, where fossil deposits are rare.

The end came for the laughing owl much later than for most of its original prey. Relatively large and aggressive, it could cope with rats, but the mustelids introduced to combat the rabbit plague in the late 19th century were too much. The species declined rapidly after 1890

(Williams and Harrison 1972). Its legacy is the rich store of information it left in the clefts and crevices where it lived and raised its young. These are being lost through damage by rabbits and Australian brush-tailed possums (*Trichosurus vulpecula*), which scratch out the contents of crevices that are wide enough for them to enter. Wider platforms on cliff faces are eroded by sheep trampling, and the rock faces themselves erode away. Deposits in cave entrances can be disturbed or destroyed by human visitors who never even see the tiny shards of bone in the dust or clay. In a few decades, the only relic of the laughing owl's existence will be the skins and skeletons in museums and the catalogs of its prey.

Genus *Ninox* Hodgson
Ninox novaeseelandiae (Gmelin, 1788) (Morepork)

The morepork is one of the group of small to medium-size "hawk owls" in the genus *Ninox* that ranges from Southeast Asia as far as the main islands of New Zealand. Although it does not reach the Chatham Islands or any of the other outlying islands in the archipelago, the New Zealand species is well established in native forests on the main islands and on the closer offshore islands. In some areas, it enters introduced vegetation. The smaller of the two owls, it has been considered to be mainly insectivorous (Heather and Robertson 1996), but on islands where there are no mammalian predators and where there are thriving populations of petrels and small forest birds, the morepork catches birds up to the size of saddlebacks (*Philesturnus rufusater*) and little shearwaters (*Puffinus haurakiensis*) (Anderson 1992). On these islands, the owl sometimes nests on the ground, whereas on the main islands, it nests in tree holes or clumps of epiphytes (Heather and Robertson 1996).

In the prehuman avifauna, it would have taken prey at the lower end of the range used by the laughing owl. Its smaller size and greater aerial ability also allow it to include a wider range of invertebrates than the larger species. It is often seen hawking moths in the early evening or around lights, and it includes more arboreal birds in its diet. It was not immune to predation by the laughing owl (e.g., Worthy and Holdaway 1996a).

Order Caprimulgiformes
Family Aegothelidae

Genus Aegotheles *Vigors and Horsfield*
Aegotheles novaezealandiae (Scarlett, 1968)
(New Zealand Owlet-nightjar)

The large, distinctive New Zealand owlet-nightjar was described by Ron Scarlett in a new genus, *Megaegotheles,* on the basis of its larger size and apparently reduced wings. When more material became available, Rich and Scarlett (1977) retained the generic distinction and suggested that there was a cline in size. Olson et al. (1987) transferred it to *Aegotheles*, pointing out that the reduced wings and robust legs were associated with a diminished power of flight and were not sufficient to warrant a separate genus. The differences between *A. novaezealandiae* and *A. cristatus* are no different in degree from those separating the flightless takahe *Porphyrio hochstetteri* and *P. mantelli* from the flighted pukeko *P. porphyrio*.

Ron Scarlett knew of the existence of an unknown bird in the fossil collections from caves (Fig. 9.13) for some time before he was able to get comparative material on which to diagnose the new species (Scarlett 1968a). The presence of an owlet-nightjar in the New Zealand fauna was unexpected (Rich and Scarlett 1977). The original material had been collected from 17 sites, 12 in the South Island and 6 in the North Island, 5 of which were in a small area in inland Hawke's Bay. The type locality is the spectacular Harwood's Hole pothole cut in Ordovician marble on Takaka Hill, but most of the other sites from which the species is known are small caverns and rockshelters. By the time Rich and Scarlett (1977: 6) provided a more complete description of the skeleton, another eight sites had been found in the North Island and three more in the South Island, bringing the totals to 22 and 8, respectively.

Rich and Scarlett (1977) showed that individuals of the South Island population were larger than those north of Cook Strait, which they attributed to the difference in mean temperature between the islands. They also pointed out that in addition to its being larger than the Australian species (the term "gigantism" hardly seems appropriate for a bird of less than 200 g), the New Zealand owlet-nightjar may have been flightless, or nearly so.

The authors pointed out that although the New Zealand bird appears to be larger than the Australian *A. cristatus*, the differences are in only part of the skeleton. However, as the femur was one of the bones identified as being larger—34% longer than in *A. cristatus*—the New Zealand species obviously had a considerably higher body mass. Rich and Scarlett (1977) estimated that body mass by assuming that the mass scaled as the cube of linear measurements. The 34% longer femur in *A. novaezealandiae* meant a mass of 96.2–131.4 g. Estimates of the body mass of *A. cristatus* by use of allometric relationships from femur length, femur diameter, and humerus length closely matched the mass of living birds (Table 9.7). If the estimate based on femur length is applicable to the New Zealand species, then it would have had a mass of about 150 g, three times the mass of *A. cristatus* and somewhat higher than the estimate of Rich and Scarlett (1977).

Femur circumference apparently overestimates the mass of *A. cristatus* (Table 9.7), but it gives a similar mass to that derived from femur length in *A. novaezealandiae*. Rich and Scarlett (1977) concluded that the robust legs of *A. novaezealandiae* meant that it was "adapted for a terrestrial existence." An indication of the relative robustness of the femur is shown by the body mass of 740 g corresponding to the femur diameter. The extra strength implied by the thick femur suggests that the bone was taking loadings over and above those normally imposed by simply supporting the body mass while perching. The Australian owlet-nightjar catches insects on the ground (Pizzey 1980). The New

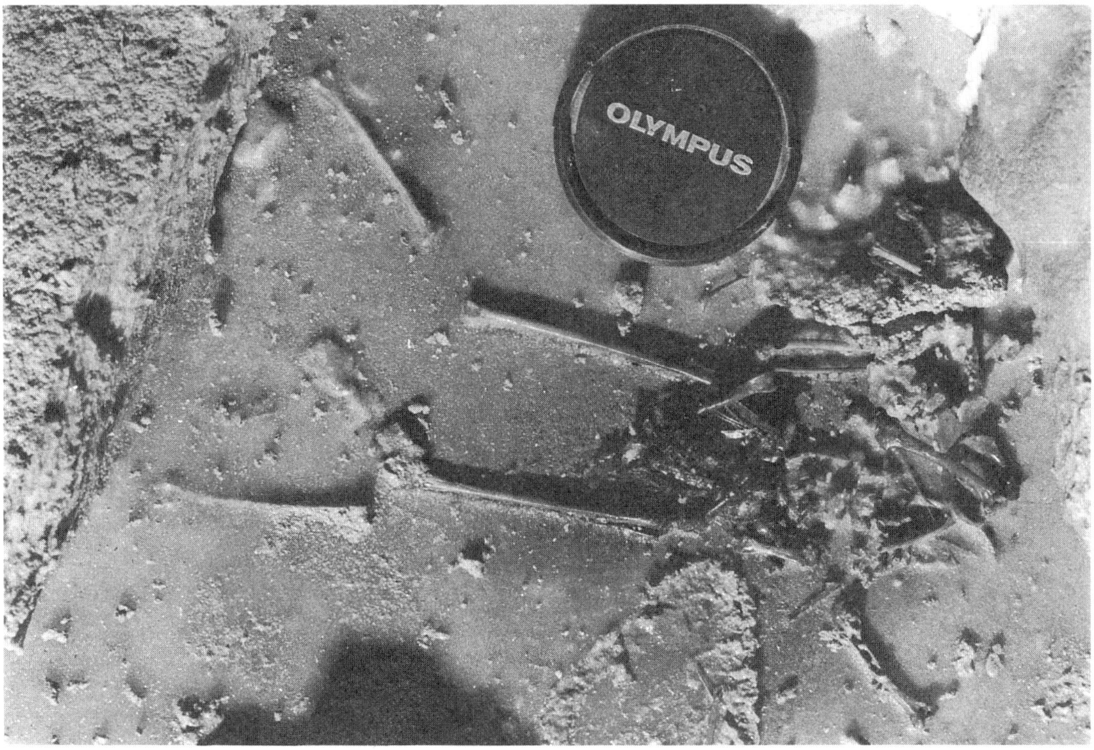

Zealand species probably did the same, and it could cope with larger species and spend more time there. Its presence in accumulations of prey remains of the laughing owl supports its having spent a lot of time on the forest floor at night, where the owl did much of its hunting (Holdaway and Worthy 1996).

*Figure 9.13. Skeleton of a New Zealand owlet-nightjar (*Aegotheles novaezealandiae*) in Pukeroa Cave, Waitomo, North Island. Photograph by I. Millar.*

Rich and Scarlett (1977) also pointed out that the pectoral girdle was reduced in *A. novaezealandiae,* but they noted that their analysis of wing loading could not conclusively indicate that the bird was flightless. The reduced wing and breast musculature might have meant that the body mass was actually below that estimated from the femur dimensions. Body mass estimates from humerus length for the two species (Table 9.7) show that the wing of *A. cristatus* scaled with body mass (the estimate matches that from the femur dimensions), but that the wing was shorter than expected in *A. novaezealandiae.* The humerus length for the New Zealand species would have been appropriate for a bird with a mass of a third to a half of its likely mass, but that does not automatically indicate flightlessness. As Rich and Scarlett (1977) pointed out, some birds have high wing loadings but can still fly well. The presence of a strong keel on the sternum is evidence for some flying ability. For the New Zealand owlet-nightjar, in an environment where the main predators were aerial, a terrestrial life was possible and it may have flown rarely, with a labored wing beat. When a predator such as the Pacific rat arrived, the owlet-nightjar would have been exposed to predation pressures that it had never previously experienced.

TABLE 9.7.
Estimated body mass (g) of the New Zealand and Australian owlet-nightjars, using different dimensions of supportive long bones.[a]

Long bone	*Aegotheles cristatus*	*Aegotheles novaezealandiae*
Femur length	46.6–56.8	128–161
Femur diameter	52.7 (44.4–62.4)	740 (127–2130)
Femur circumference	71.2 (59.3–85.6)	147 (118–183)
Humerus length	46–48.7	51.7–82.7

[a] Live masses for *A. cristatus* are as follows: ♂, mean 44 g, standard deviation 4.81; range 35–54, $n = 22$; and 44 g, 10.6, 21–65, $n = 17$. Femur length estimate, Prange et al. (1979); femur diameter, Alexander (1983a); femur circumference, Anderson et al. (1985); humerus length, Prange et al. (1979).

The New Zealand owlet-nightjar probably also ate small vertebrates such as frogs and lizards as well as being a nocturnal insectivore. It has been found in deposits where forest or at least dense shrubland was the dominant vegetation, and it was widespread in both main islands and on at least one of the larger nearshore islands. One of the features of the New Zealand environment was the numbers of large invertebrates on the forest floor. These are still prominent on predator-free islands today. A reduction in the number of invertebrates might have played some part in the demise of the owlet-nightjar, but predation is the more likely cause.

Although a possible record in a cultural context has been reported (McCulloch 1994), this is the result of later occupation of a laughing owl site by people (Worthy and Holdaway 1996b). The owlet-nightjar probably became extinct before or very shortly after Polynesian settlement. There are tantalizing mid-19th-century records of a tame owl ("*Athene parvissima*") smaller than *Ninox novaeseelandiae* in forests on both sides of the axial ranges in the South Island (Potts 1882). This has been thought at times to refer to remnant populations of the owlet-nightjar (D. H. Brathwaite, personal communication), but the records have never been confirmed. The bones of no owls other than the morepork and laughing owls have been found in New Zealand. A wing, supposedly of a caprimulgid, was found on the West Coast in the 1960s but has been lost.

Summary: Avian Predators in New Zealand

As a suite of large herbivorous birds came to dominate the New Zealand vertebrate fauna, the way was open for any bird of prey that could reach the main islands to specialize in their capture. Because raptors usually take a range of prey determined at the larger end by their ability to subdue the prey, and at the smaller end by whether the return is worthwhile or the prey is too small to handle (Jaksić and

Braker 1983), there would always be an advantage to taking larger prey. Up to the limit at which flight is possible, body mass and body strength would enable larger, more economical prey to be killed and processed. The energetic advantages of large size in lowering the amount of food required per unit mass and in maintaining body temperature in a cool climate would also encourage increasing size in the raptors, as it seems to have done in the moa, whose body size fluctuated with changing climate. As sample sizes of the raptors from Pleistocene-age deposits increase, hints of changes in the body size through time may prove to be real.

Both Haast's eagle and Eyles's harrier were much larger than their closest relatives. Both probably depended to some extent on moa for food, the harrier taking the place of Haast's eagle in the more closed vegetation of the North Island, where the smaller average size of the moa species was within the capacity of its attainable body size to subdue. When the supply of large prey dwindled and then vanished, these highly specialized raptors, like the vultures that preyed on the elephants and ground sloths of North America, became victims of their own specialization.

10 • A Plethora of Rails

Order Gruiformes

Family Rallidae

The ground bird fauna of New Zealand is dominated by members of the family Gruiformes, especially various species of rails. New Zealand lacks cranes and storks, but many of the rails were large and several were flightless. The extant species, which are mostly small and not endemic to New Zealand, consist of the small spotless and marsh crakes (*Porzana tabuensis* and *P. pusilla,* respectively), the banded rail (*Gallirallus philippensis*), and the purple swamphen or pukeko (*Porphyrio melanotus*). All three species are found at least in Australia and elsewhere in the southwest Pacific region. Only two species of New Zealand's endemic rail fauna survive: the South Island takahe and the weka.

Many of the endemic species were derived from *Gallirallus*-like ancestors (Andrews 1896a; Olson 1975), but they came to differ greatly in size and particularly in bill form. As noted by Olson (1975: 77), "There is as much diversity in skull morphology in this group as may be observed throughout the rest of the family Rallidae. A comparative study of the skulls of these forms would probably provide as instructive an example of morphological adaptive radiation from a common stock as can be found in any group of non-passerine birds."

Brief introductions to each genus and species of rail is given below to illustrate their importance in New Zealand's fossil record.

Genus Porphyrio *Brisson*
Porphyrio mantelli *(Owen, 1848) (North Island Takahe)*
Porphyrio hochstetteri *(Meyer, 1883) (South Island Takahe)*

The takahe was originally described by Richard Owen in the new genus *Notornis* from fossils collected at Waingongoro (South Taranaki), North Island, New Zealand (Owen 1848a). He named it *Notornis mantelli* after Walter Mantell, who had collected the material (Owen 1848b). Because it was known from only a few bones and because it was of large size, the takahe was presumed to be extinct. Owen believed it to be closely related to members of the genus *Porphyrio*. Shortly after the description was published, Mantell associated the Māori name of "takahe" with the fossils, because in 1849 he had bought from among a sealer's wares the skin of a bird, which he recognized as belonging to the takahe, or *Notornis*. The skin was shipped from New Zealand in May 1850 and arrived in London in October 1850 (Andrews 1986). The discovery caused considerable excitement in the European scientific community because the takahe had been named from fossils, yet obviously, some birds still survived. A second bird was shot in 1851, a third in 1879, and a fourth in 1898. But after that, no more were seen for 50 years.

The third specimen was purchased by the Dresden Museum for £105, where Meyer (1883) studied it and bestowed the name *Notornis hochstetteri* because the skeleton differed from that of *N. mantelli*. Perhaps if the advice of Alfred Newton, professor of zoology at Cambridge University, given a few years earlier when he settled the question on the distinctiveness of *Gallirallus modestus*, had been heeded earlier, the distinctiveness of the South Island takahe would have been detected sooner. Newton advised that naturalists should obtain the skeleton of any supposed new bird so as to clarify its relationships (Buller 1875: 511), but the second species of takahe was by then considered extinct because no birds had been seen for several decades.

The North Island takahe probably survived into the late 19th century. In 1894, a *Notornis* was caught in the Ruahine Mountains by the surveyor Morgan Carkeek and taken to the home of Roderick McDonald at Horowhenua, just north of Wellington (Phillipps 1959). The local Māori identified it as the rarely seen mohoau. Phillipps obtained his information from Roderick's son, Hector, who was present when the bird was brought to the house. He said that the bird had blue feathers and resembled the pukeko. Martin (1886) lists the Māori name for this species as moho.

In 1948, the South Island takahe was rediscovered as a beleaguered population persisting in a subalpine area in Fiordland. Through years of ecologic research, the species became well known. The significance of the obvious skeletal differences between it and the North Island form were forgotten or disregarded. A series of informal taxonomic rationalizations from the mid-20th century onward (e.g., Kinsky 1970) saw the relegation of the original two species to subspecies and their removal to the genus *Porphyrio*, as *Porphyrio mantelli mantelli* (extinct North Island race) and *Porphyrio mantelli hochstetteri* (South Island race) (Ripley 1977) (Fig. 10.1). The population of *P. hochstetteri* had declined steadily until active management stemmed its decline in the 1990s. By 1999, it was still rare, with about 200 birds divided between the wild but managed Fiordland population, captive breeding pairs, and two transferred populations on predator-free islands.

Most recently, morphologic and DNA analyses of the bones of the North Island form have shown that Meyer had been correct. The two

Figure 10.1. Reconstruction of Porphyrio mantelli *showing its taller, leaner stature compared with* Porphyrio hochstetteri. *Drawing by J. Winn.*

forms are different and are not even sister species. The North Island bird is larger than the South Island one and has a body shape closer to that of *Porphyrio melanotus* than the other takahe, indicating an independent derivation from a proto-*Porphyrio* (Trewick 1996b). Also, mitochondrial DNA (mtDNA) comparisons show that the two takahe are paraphyletic with respect to *P. melanotus*—that is, the two takahe species do not form a monophyletic pair and are each inserted at separate points within the clade of other *Porphyrio* species (Trewick 1997b). The conclusion must be made that each arose from an independent colonization of New Zealand by a proto-*Porphyrio* (Trewick 1996b; Trewick and Worthy 2001). Trewick's work has been challenged by Livezey (1998), who used a morphologically based cladistic analysis and found that the two takahe were sister species. However, Livezey (1998: fig. 1) appears to have merged as one the very taxa of *Porphyrio* (*madagascariensis, melanotus,* and *seistanicus*) that separated the takahe species in Trewick's phylogenetic tree. When he added *P. mantelli* to the analysis, it was to add it into a tree with a constrained typology that only included two taxa—*P. hochstetteri* and other undifferentiated *Porphyrio*—so he did not in fact test Trewick's hypothesis.

The South Island takahe is a large (3 kg), flightless gallinule that feeds mostly on grass tillers. The surviving natural population eats a restricted range of tussock (clump grass) species, augmented by selected dicotyledons, grass seeds (when available), and fern rhizomes in winter. Its choice of food reflects the most energy-rich food sources available in its present harsh subalpine habitat, undoubtedly a behavioral adaptation to survive there. The degree to which takahe are a specialist inhabitant of cool grasslands and the extent to which this contributed to the species' prehistoric decline have been hotly debated (e.g., Beauchamp and Worthy 1988; Mills et al. 1984, 1988; Trewick and Worthy 2001). Fossils from Holocene sites throughout the South Island show that takahe were formerly widespread in lowlands. Translocated birds on islands now feed on a variety of grasses and dicotyledons and so belie the specialized feeding habits of the subalpine population. That population may be specialized solely to satisfy the energy requirements its colder environment demands. Takahe became restricted to Fiordland after Polynesian settlement, and the species range has continued to contract through European time, even after its rediscovery. The range has contracted until the species survives only in a restricted part of the Murchison Mountains, and even there, the population is only stable because of the addition of captive-bred individuals each year.

The North Island takahe was also widespread in the lowlands. In the Holocene, both species appear to have been most common in mosaics of grassland, shrubland, and forest. In the last Glacial, their expanded ranges matched the greater availability of grasslands. Like other New Zealand large flightless birds, takahe are K-selected species with a low annual productivity. They would have been unable to survive even low levels of predation, so it was undoubtedly human predation that exterminated the North Island species and restricted the South Island takahe to the most inhospitable extremity of its former range by the time Europeans arrived. Since then, additional predators such as stoats (*Mustela erminea*), other mammalian predators, and competing mammalian herbivores have added to the pressures on the population.

Genus Gallirallus *Lafresnaye*
Gallirallus australis *(Sparrman, 1786) (Weka)*

The second surviving endemic rail is the weka (*Gallirallus australis*), a large (male, 1 kg; females, 0.7 kg) derivative of the banded rail (*G. philippensis*), which is widespread in the southwest Pacific. The weka formerly was found only on the North, South, and Stewart Islands of New Zealand and some nearshore islets. Four weakly differentiated subspecies of weka have been recognized: *G. greyi* for the North Island, *G. australis* for the western and northern South Island birds, *G. hectori* for those in the eastern South Island (now only surviving as an introduced population on the Chatham Islands), and *G. scotti* on Stewart Island. The North Island weka is seriously endangered if not extinct on the mainland, and the populations of the western weka are also declining rapidly. *G. scotti* is nearly extinct on Stewart Island itself and is only "safe" on a few offshore islets. Only *G. hectori* on the Chatham Islands is not endangered. It is only safe on islands that lack stoats and ferrets, which are major predators of weka.

Bones of weka are common in fossil deposits throughout New Zealand at all altitudes, from coastal to subalpine sites. *Gallirallus australis* is extremely variable in size. In the past, this led to the naming of two further fossil species. *Ocydromus* (= *Gallirallus*) *minor* was erected by Hamilton (1893) for bones smaller and more slender than those of the living *Gallirallus*. Southern birds, either the fossils found in South Canterbury, Otago, and Southland, or those living on Stewart

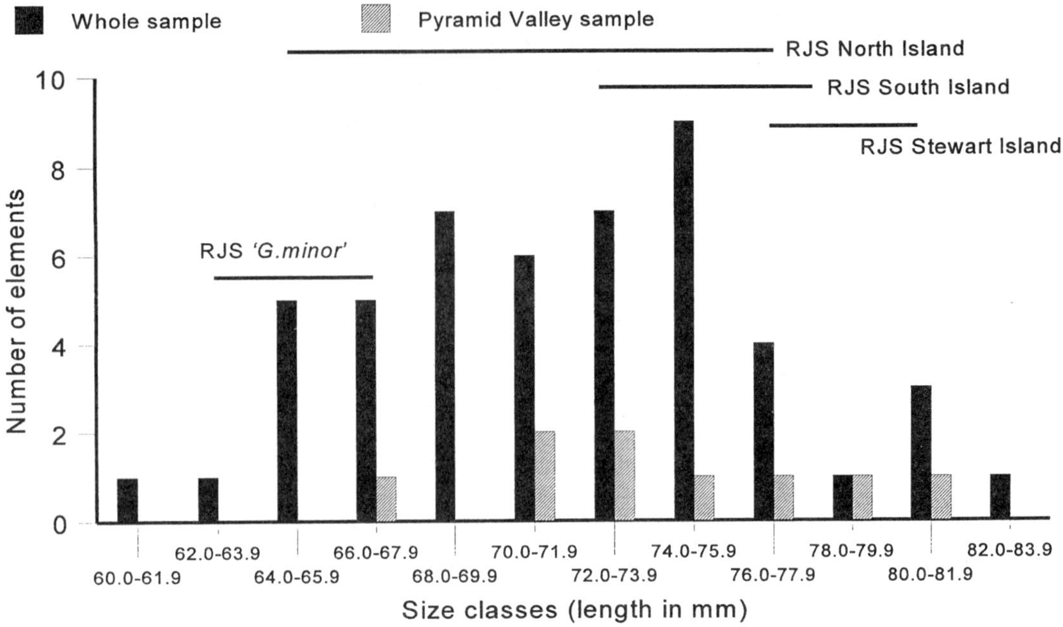

Figure 10.2. Histogram of lengths of weka femora. The horizontal bars denote the size range given by Scarlett (1972b) for samples from North, South, and Stewart Islands and for Gallirallus minor. *After Holdaway and Worthy (1997: fig. 10b), with the permission of the Royal Society of New Zealand.*

Island, are smaller than western South Island and North Island birds. Although many bones were allocated to *O. minor* subsequently, there is no justification for a separate name because the size range of *O. minor* falls within that of living weka (Olson 1975). To define *O. minor* on the basis of its size would result in an aberrant size distribution for *G. australis* (Fig. 10.2). For these reasons, *Ocydromus minor* was synonymized with *Gallirallus australis* by Holdaway and Worthy (1997) (Fig. 10.3).

An almost forgotten species was erected for a bone said to be larger than those of the common weka. Forbes (1892a, 1892b) named *Ocydromus insignis* on the basis of a tibiotarsus from Enfield (a swamp site in Otago) that was said to far exceed in size any existing New Zealand form. Forbes stated specifically that only one rail bone was found in Enfield. When this was found in the Forbes collection in the British Museum (BM A6542), it was designated as the holotype of *Ocydromus insignis* Forbes, 1892 (Worthy 1998b). However, it is indistinguishable in morphology from tibiotarsi of *Gallirallus australis,* and at 104.3 mm long, it is within the upper bound of the size range for that species. Therefore, *Ocydromus insignis* Forbes was taken to be a junior synonym of *Gallirallus australis* Sparrman.

Gallirallus dieffenbachii *(G. R. Gray, 1843) (Dieffenbach's Rail)*

Dieffenbach's rail was obviously closely related to the banded rail, but it differed by its substantially stouter pelvic girdle, relatively smaller wings, and decurved bill (Buller 1874: figs. 1, 2). Unlike the banded rail, it was flightless. Millener (1996) estimated that *G. dieffenbachii* weighed

Figure 10.3. *A weka* Gallirallus australis *skeleton found in Honeycomb Hill Cave. Photograph by T.H.W.*

about twice as much as the banded rail's 170 g. Ernst Dieffenbach collected the single known living specimen in 1840, and the species became extinct soon after the introduction of cats. The fossil record shows that its distribution included Chatham, Pitt, and Mangere Islands (Fig. 10.4).

The genus *Nesolimnas* was erected for this bird by Andrews (1896b), who cited substantial differences in the shape of the bill and tarsometatarsus from that in *G. philippensis*. The culmen is convex along the whole length, and the sharply decurved tip is unlike that in other species. The tarsometatarsus is shorter and relatively stouter than in *G. philippensis*. Although such differences respectively represent specializations for feeding and are associated with flightlessness and do not justify distinction at genus level, they are adequate to demonstrate that it is a good species.

Trewick (1997a, 1997b) studied this species by use of both morphology and mtDNA. He concluded that it is most closely related to *Cabalus modestus* and that they arose from the same volant ancestor. Trewick's work clearly supports *G. dieffenbachii* being a distinct species, against the *Checklist of the Birds of New Zealand* (Turbott 1990a), which lists it as a form of *G. philippensis*. The checklist treatment was made despite the descriptions and views promoted by Andrews (1896b), Olson (1975), and Scarlett (1975a), in each of which specific distinction was advocated. One can only conclude that the significant osteologic differences were ignored and that the single study skin was not consid-

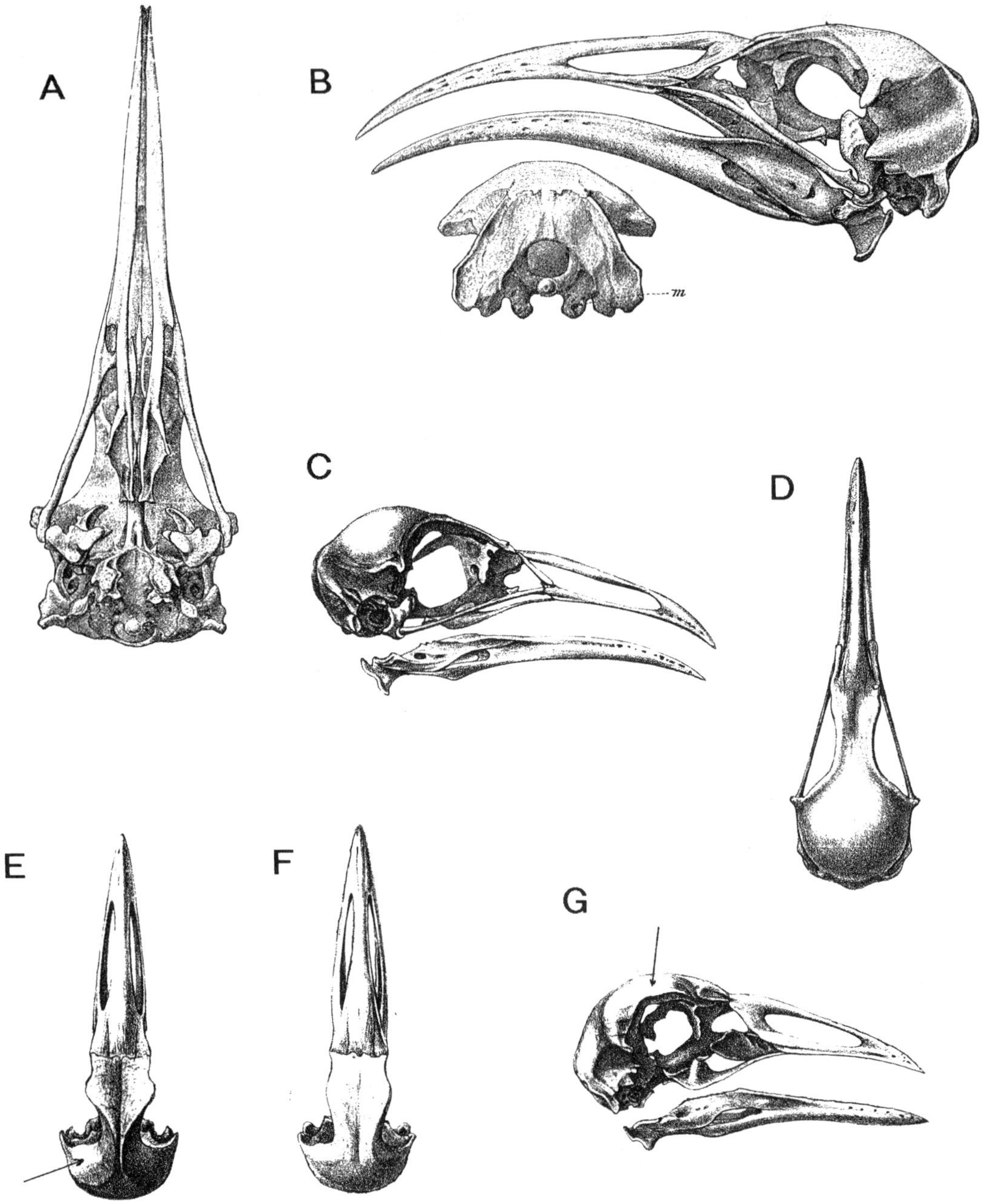

Figure 10.4. Skulls of some Chatham rails. (A, B) Diaphorapteryx hawkinsi. *(C, D)* Gallirallus dieffenbachii. *(E, F, G)*, Fulica chathamensis. *After Andrews (1896a, 1896b). Note variation in the size of the salt gland impressions (arrow) in* Fulica*—big in E, small in F.*

ered sufficient evidence that there had been a distinct species, or that since it was extinct it did not matter to biologists primarily concerned with the living fauna.

Cabalus *Hutton 1874*
Cabalus modestus *(Hutton, 1872) (Chatham Island Rail)*

Often listed in *Rallus* or *Gallirallus,* this small rail has been extinct on the Chatham Islands since about 1900. It was formerly sympatric with both *G. dieffenbachii* and *Diaphorapteryx hawkinsi* on Chatham and Pitt Islands and with *G. dieffenbachii* on Mangere Island. It survived into European time only on Mangere Island. H. H. Travers collected the first three specimens, an adult female and a chick, and a further adult (preserved in spirits) in 1871 (Hutton 1872a; Travers 1873). Hutton (1872a) named the form and described the plumage of first the adult in detail and then the chick briefly. The descriptions were slightly enlarged in Hutton (1873a). From this, we can presume the adult female and parent to the chick was the type. Later, Hutton said,

> last year I described a new species of Rail, from the Chatham Islands, under the name *Rallus? modestus* (Trans. N.Z. Inst., V., p. 233), at the same time expressing an opinion that it would form the type of a new genus. Since then the skeleton of the other specimen obtained by Mr H. Travers has been prepared by the late Dr Knox, and an inspection of this skeleton has so confirmed my previous opinion that I no longer hesitate to place it in a new genus, of which the following is a diagnosis. (Hutton 1874: 108)

This second specimen was probably the specimen preserved in spirits, and from the following description, it was apparent that the sternum, 12 cervical and 9 dorsal vertebrae, femur, humerus, coracoid, and scapula at least were in this "skeleton." Hutton (1874: 108) then proceeded to say, "Both of the birds obtained by Mr H. Travers were full-grown, one accompanied by her young one, and the other containing well-developed ova; they were both exactly alike in colour and dimensions." Therefore, Hutton twice clearly reported that two adult birds and a young had been obtained by Travers in 1871. Buller (1879: 362) reported that "several specimens of this now well-established form have been received from the Chatham Islands," and several years later (Buller 1893), he reported the obtaining of a further pair from Mangere Island. The 1871 chick survives as a mount in the Museum of New Zealand, and two mounted skins of adult birds are in Canterbury Museum. The whereabouts of the type adult female and the skeleton are unknown.

The early history of nomenclature of this species illustrates the professional jealousies that prevailed between various authorities of the day. Despite Travers's statement that the chick had been associated with a pair of adults of which he had only been able to catch the female, Buller, to whom a skin (presumably that of the chick) was sent after Hutton named it, tried to discredit the species by calling it the young of *G. dieffenbachii* (e.g., Buller 1874). Hutton responded by describing the skeleton of the second adult specimen in detail and erecting a new genus *Cabalus* for the species (Hutton 1874). Hutton was most impressed by the long, down-curved bill, the reduced sternum, and the unusual loose plumage that almost completely lacks barbules. Hutton then forwarded a skin and the skeleton to Professor Alfred Newton at Cambridge. His support of Hutton's view forced Buller to retract (Buller 1875: 511). As noted before, Newton's letter stressed the importance of securing the skeleton of any bird supposed to be new or unknown and stated that the species was perfectly good and clearly was within the *Ocydromus* (= *Gallirallus*) group of rails. Within 25 years of its discovery, collectors and cats had exterminated this, the smallest of the Chatham rails.

Cabalus modestus was only the size of an American robin *Turdus migratorius* and was like the other Chatham rails in having a long, down-curved bill. Travers and others noted that it was nocturnal, and the collector W. Hawkins stated that the birds nested in holes in the ground and fed mainly on sandhoppers. Its long bill would have been suited to probing for these terrestrial amphipods and for insects. Bones of this species were rare in collections until recently, when a spectacular deposit containing thousands of bones was found in the small Te Ana a Moe, a cave on Chatham Island itself. This amazing resource was used by Trewick (1997a) in a morphologic study of *Cabalus modestus*, which demonstrated, among other things, that the bill length of this small rail was sexually dimorphic. Two distinct peaks in the bill-length distribution were not matched by differences in body size as indicated by long bone lengths, so it seems likely that males and females had different length bills.

The relationships of *C. modestus* were recently studied by Trewick (1997b) by use of mtDNA sequencing. He found it to be most closely related to *Gallirallus dieffenbachii*, the next largest rail on the Chathams. These species were sister taxa to *Rallus owstoni*, and together they formed a clade that was the sister group of *Rallus sylvivestris* and *Gallirallus philippensis*. However, *Capellirallus* and *Diaphorapteryx* were not included in the study, and the insertion of these taxa into the tree could markedly alter the conclusions. However, the present data suggest that, as Andrews (1896b) had intimated, all the Chatham rails probably evolved from a proto–*Gallirallus philippensis*.

Genus Capellirallus *Falla*

Capellirallus karamu *(Falla, 1954) (New Zealand Snipe-rail)*

Capellirallus was a small rail whose remains were first discovered by speleologists in Karamu Cave, southwest of Hamilton, North Island. Its final resting place was immortalized by the name "The Bird Cage" on the map of Karamu Cave. While still in situ, the type specimen was photographed with only candles for its illumination before being collected by Henry Lambert (Fig. 10.5). *C. karamu* was about the size of a banded rail but had a spectacularly long bill. At nearly 70 mm, the bill relative to the bird's size was the longest of any rail. Also, unlike the banded rail, this small species was entirely flightless, as its small wings show (Fig. 10.6). So far, remains of this species have been

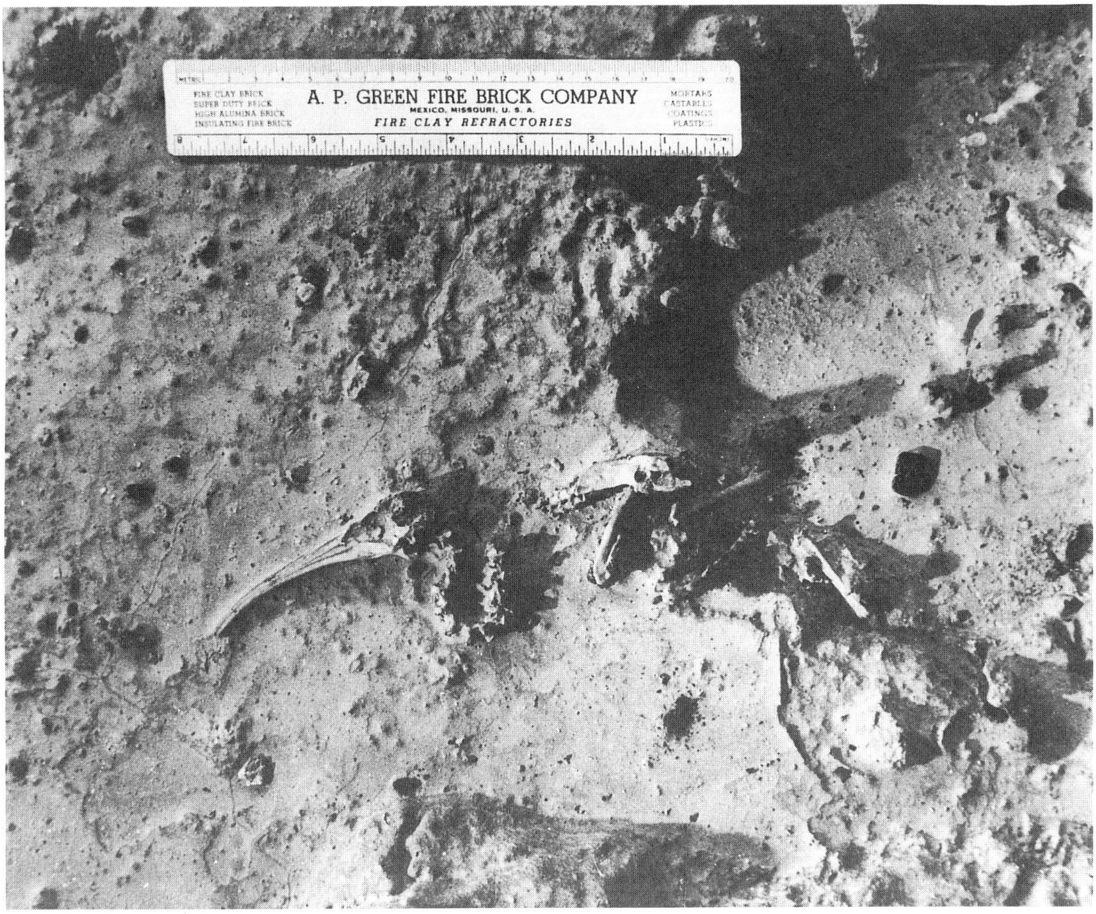

Figure 10.5. Photograph of the type specimen of Capellirallus karamu *in Karamu Cave before collection. Photograph reproduced courtesy of Waitomo Caves Museum.*

found only in the North Island and are most common in sites in western areas where wetter closed canopy forests prevailed. The long bill converges on the form of those in snipe and kiwi. It seems likely that the bird foraged by probing the soft soils, deep litter, and mosses of the wet podocarp forests. Eastern forested areas, from which it was absent, were seasonally dry, so soils are likely to have been dry and difficult to probe for at least parts of the year, and the litter would have been much shallower.

The relationships of *Capellirallus* remain unstudied, but it is usually considered to be close to *Gallirallus*. It bears a striking similarity to *Cabalus modestus* of the Chatham Islands, which also has a long, down-curved bill. This similarity was supported by a recent morphologic study (Livezey 1998), but uncertainty as to how much the similarity was a result of convergence resulting from their shared loss of flight ability caused Livezey to be conservative and maintain them in distinct genera. *C. karamu* was extinct before Europeans arrived and so was probably an early victim of predation by the Pacific rat *Rattus exulans*, which had been introduced by Polynesians possibly as early as about 2000 years ago (Holdaway 1996a).

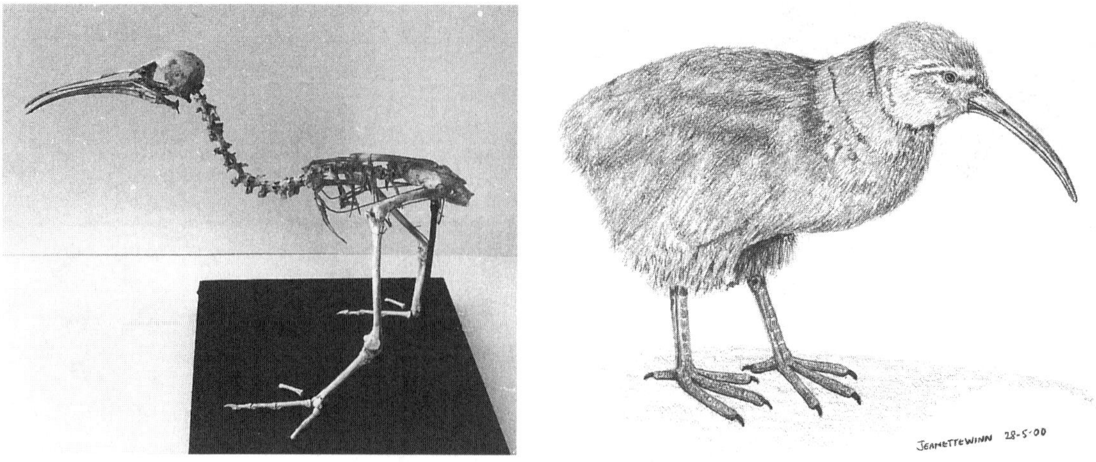

Figure 10.6. (a) (left) Mounted skeleton of Capellirallus karamu *in Waitomo Caves Museum from Tapuae Weka Cave. Photograph by T.H.W. (b) (right) Reconstruction of* Capellirallus karamu. *Drawing by J. Winn.*

Genus Diaphorapteryx *Forbes*
Diaphorapteryx hawkinsi *(Forbes, 1892) (Giant Chatham Island Rail)*

The giant Chatham Island rail was first described in the genus *Aphanapteryx,* otherwise known only from the Mascarene Islands in the Indian Ocean as *A. hawkinsi,* in a one-sentence telegram to the journal *Nature* (Forbes 1892a) and further reported in Forbes (1892e, 1892f). However, Forbes erected *Diaphorapteryx* for the species in the same year (Forbes 1892g), then just as quickly—but after another short note on the subject (Forbes 1892h)—reverted to *Aphanapteryx* (Forbes 1893a). The first substantive description of *Diaphorapteryx* was of its skull (Forbes 1892g), and this was later reprinted in the journal *Ibis* (Forbes 1893e).

It was, however, left to Andrews (1896a: 83) to describe the skeleton of *Diaphorapteryx* in detail and to compare it to other rails. He concluded that *D. hawkinsi* was "merely an Ocydromine rail, in which the reduction of the wings has advanced somewhat farther than in the living forms, and has been accompanied by a lengthening of the beak and a considerable increase in bulk." He considered that *Diaphorapteryx, Gallirallus australis,* and *Cabalus modestus* were probably all derivatives of *Gallirallus philippensis,* as it is now known. He found that although *Diaphorapteryx* and *Aphanapteryx* were superficially similar (both had large, pointed, down-curved bills and were of similar size), the two genera were quite different in details of the skull, pelvis, wing bones, and particularly the tarsometatarsus (Andrews 1896a). The similarity represented to Andrews a remarkable instance of convergent evolution.

Andrews noted that both the length and the curve of the beak varied considerably and that the size of the leg bones was also highly variable (Tables 10.1–10.3). Such variation, Andrews observed, was normal in flightless birds, but it had been underestimated in the Dinornithidae because of the practice of naming forms that varied slightly in size. This was an interesting insight, not revisited, it seems, for 80 years.

TABLE 10.1.
Skull measurements (mm) of *Diaphorapteryx* from Andrews (1896a).

Variable	Length bill tip to occiptal condyle	Zygomatic width	Frontal width	Premaxillae length	Mandible length
Mean	133.5	45.8	18.0	92.9	117.9
Standard deviation	7.81	2.49	1.30	4.83	7.26
Minimum	120	41	16	85	108
Maximum	144	49	20	100	127
n	14	13	14	14	11

TABLE 10.2.
Long bone measurements from Andrews (1896a).

Bone	Mean	Range	*n*
Femora	97.4	90–106	33
Tibiotarsi	130	122–140	28
Tarsometatarsi	69	63–78	55
Humeri		57–68	

TABLE 10.3.
The wing length of rails related to leg length (= 100). After Andrews (1896a).

Andrews's names	Names in modern use	Wing length as proportion (%) of hindleg length
Diaphorapteryx hawkinsi	*D. hawkinsi*	~44
Ocydromus fuscus	*Gallirallus australis*	46.5
Tribonyx mortieri	*Tribonyx mortieri*	48
Erythromachus leguati	*Aphanapteryx leguati*	~52
Cabalus dieffenbachii	*Gallirallus dieffenbachi*	52
Cabalus sylvestris	*Rallus (Tricholimnas) sylvestris*	58
Hypotaenidia celebensis	*Gallirallus celebensis?*	59.5
Rallus aquaticus	*Rallus aquaticus*	64
Fulica atra	*Fulica atra*	82

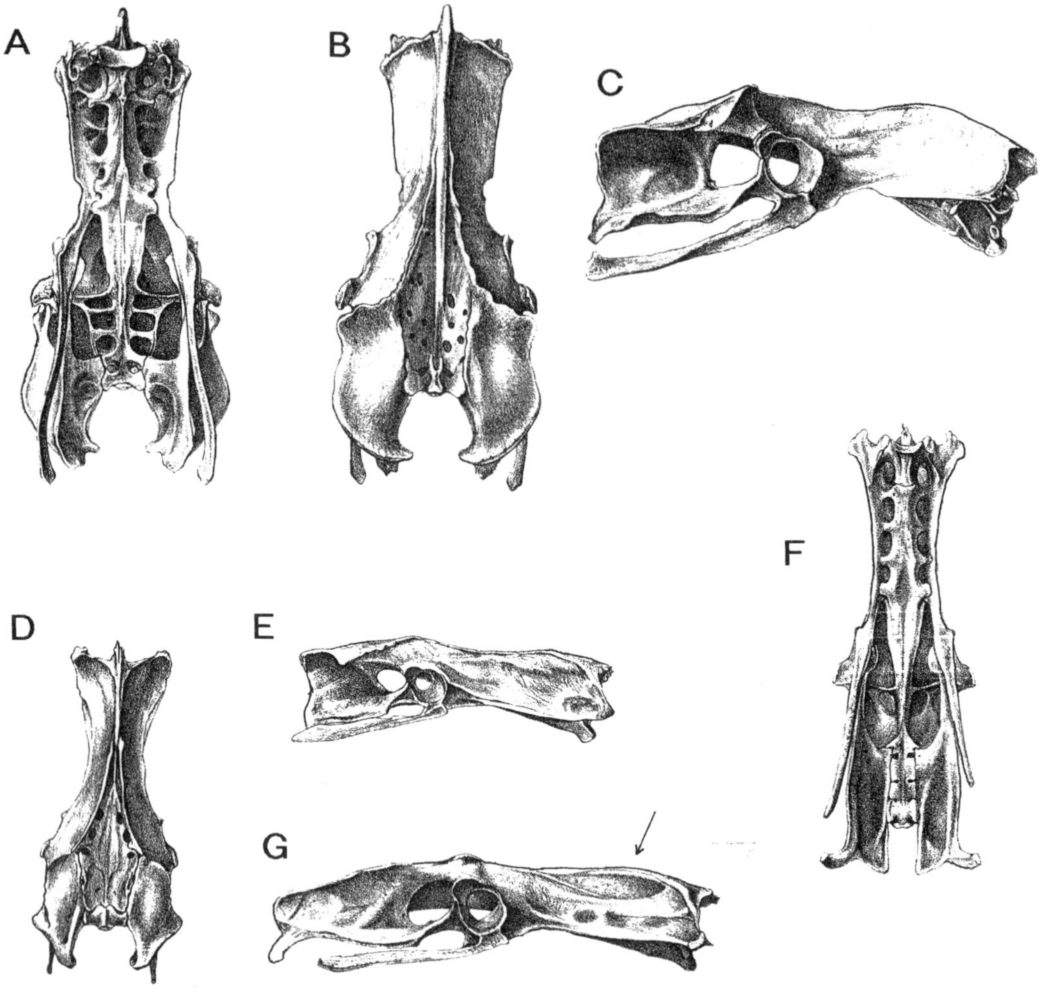

Figure 10.7. Pelves of some Chatham rails.
(A–C) Diaphorapteryx hawkinsi. *(D, E)* Gallirallus dieffenbachii. *(F, G)* Fulica chathamensis. *After Andrews (1896a, 1896b). Note unfused ilium (arrow) in* Fulica.

Perfectly preserved fossil bones of *Diaphorapteryx hawkinsi* have been collected from dune deposits on Chatham Island and are, indeed, still quite common there (Figs. 10.7, 10.8). The species is also present in fossil deposits on Pitt Island. It was extinct by the time of European colonization in the early 1800s, so it probably fell victim to human predation in the previous 450 years, during the period of Polynesian occupation. Its remains are common in the archaeologic sites investigated so far.

Genus Gallinula *Brisson*
Gallinula hodgenorum *(Scarlett, 1955) (Hodgens' Waterhen)*

For a species described as recently as 1955, Hodgens' waterhen holds something of a record in New Zealand for the number of name changes it has undergone, having been listed in no fewer than seven

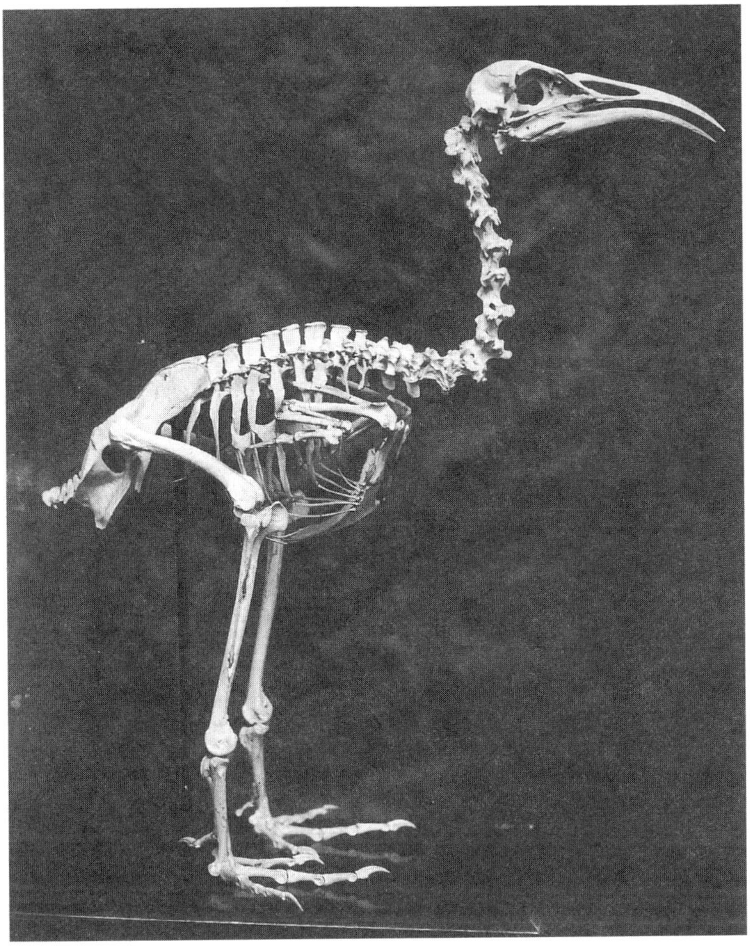

Figure 10.8. Mounted skeleton of Diaphorapteryx hawkinsi. *Photograph reproduced courtesy of Museum of New Zealand Te Papa Tongarewa, B4905.*

genera and under three specific names (Olson 1975, 1986, 1987a; Turbott 1990a). Ron Scarlett first described the species from bones collected in the famous Pyramid Valley swamp and named it after the Hodgen brothers (Scarlett 1955b). Oliver (1955) examined the type series and demonstrated that the affinities of the fossils were with gallinules. Because he lacked specimens of the skull or beak at the time, Oliver relied on the structure of the pelvis. In *Gallinula, Tribonyx,* and *Fulica,* the dorsal edge of the anterior ilium is not fused to the medial dorsal ridge of the sacrum over the greater part of the length, whereas it is in *Rallus, Gallirallus,* and their near relatives (Fig. 10.9).

Oliver noted that the fossils were closely similar to *Tribonyx* but chose to erect the new genus, *Pyramida.* Later, Scarlett, no doubt swayed by the similarity of the leg structure, transferred *hodgeni* to *Capellirallus,* but he failed to address the significant differences pointed out by Oliver (Scarlett 1970b). Meanwhile, Scarlett gave leg bones of the same species found in the North Island the name *Gallirallus hartreei* (Scarlett 1970a). When Olson (1975) examined the New Zealand rails, he located crania and premaxillae of a small gallinule closely related to

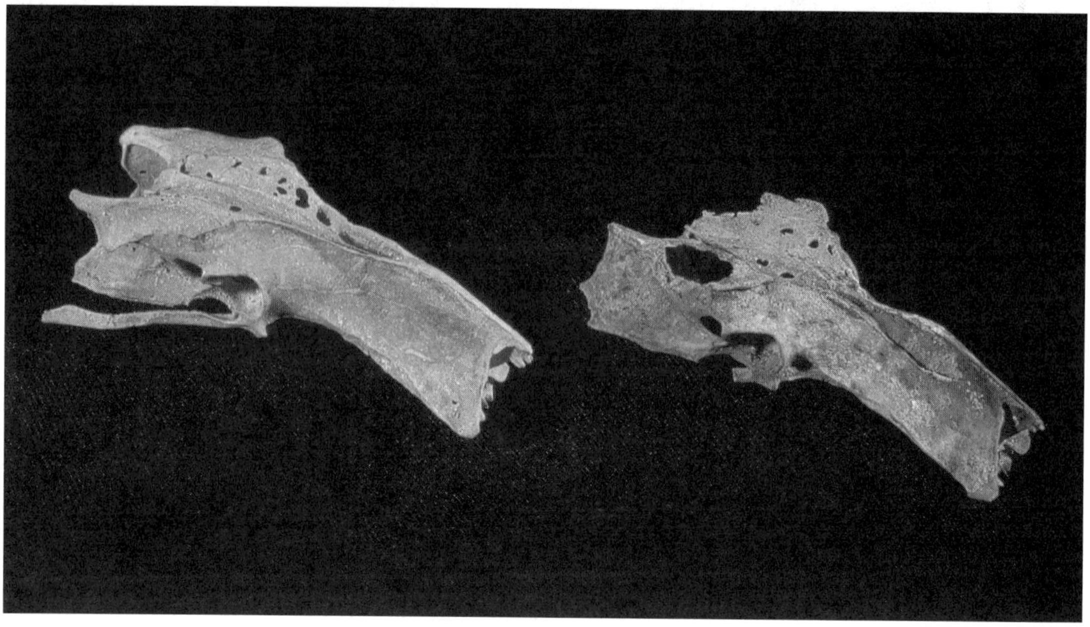

Figure 10.9. Lateral view of pelves of Gallirallus dieffenbachii *(MNZ S?) (left) and* Gallinula hodgenorum *(MNZ S35511) (right). Note the unfused anterior ilium in* Gallinula. *Photograph by R. Morris.*

the Australian subgenus *Tribonyx* of *Gallinula* in the same collections as those containing the leg bones referred to the species *hartreei.* Olson found *hartreei* to be indistinguishable from *hodgeni,* and because of the cranial elements, he concurred with Oliver that the species was a gallinule and transferred it to *Gallinula* as *G.* (*Tribonyx*) *hodgeni.* A spat involving the desirability of nomenclatural exactitude versus stability of nomenclature then ensued (Mayr 1986; Olson 1976, 1986, 1987a; Scarlett 1975a, 1976a), which culminated in the current usage *Gallinula hodgenorum.*

Genus Fulica *Linnaeus*
Fulica chathamensis *Forbes, 1892 (Chatham Island Coot)*
Fulica prisca *Hamilton, 1893 (New Zealand Coot)*

Coots, whose skeleton differs most from other rails in their elongate, narrow pelves, were first noticed in the New Zealand fauna by Forbes (1892f), when he described a large, extinct form from the dune deposits of the Chatham Island as *Fulica chathamensis.* Shortly thereafter, Forbes established the genus *Palaeolimnas* for the species, supposedly on the basis of distinctive cranial characters. Milne-Edwards (1896) briefly examined *F. chathamensis* and figured many bones, and he verified that it was distinct from the Mascarene *Fulica newtoni.* However, he used the similarity of Mascarene and Chatham forms of this and other rails to advocate a former land connection. In 1896, the species was described in detail by Andrews (1896b), who used the abundant, well-preserved material by then available.

One of the remarkable attributes of the Chatham coot skeleton is that the interorbital area preserves large depressions for the salt glands. This feature varies in its development but is probably related to the presence of the large, saline Te Whanga Lagoon in the center of Chatham Island. Other birds inhabiting this lagoon, including the swan and extinct duck, also had enlarged salt gland impressions, whereas populations of the same or similar species living in nonsaline environments elsewhere do not. Skulls of the New Zealand coot have only small depressions (see below). With regard to its ability to fly, Andrews observed that the depth of the sternal keel was reduced (15% of length) relative to those of the volant *F. atra* (25%) and *F. cristata* (23%). However, he judged the coracoscapular angle to be the same as in the volant species, not opened as is normal for flightless species, and he found the furcula to be relatively stouter than in the volant species. He also noted that although the wing bones were relatively shorter than in *F. atra* or *F. cristata*, the humerus was stouter and had a longer pectoral crest (crista deltopectoralis) and a more prominent inner trochanter (tuberculum dorsale). Similarly, the ulna, radius, and carpometacarpus, although shorter, were stouter than in the two volant species. None of these characteristics is usual for a bird that has lost the ability to fly. Andrews (1896b: 266) concluded, "it seems probable that *Palaeolimnas* may still have been capable of heavy flight for short distances."

Coot bones discovered about the same time on mainland New Zealand at Castle Rocks in the South Island were named *Fulica prisca* by Hamilton (1893). According to Rothschild (1907), this species differed from *F. chathamensis* by not being flightless and in being slightly smaller. As Lambrecht (1933) pointed out, however, there are no differences in size. Oliver (1955) stated that the New Zealand birds differed from *Fulica* in the more curved form of the skull, the presence of deeply marked depressions for the nasal glands over the orbit, and the pronounced pneumaticity of the frontals, and so concurred with their generic distinction as advocated by Forbes and Milne-Edwards. However, Oliver (1955: 599) found that the differences "are not more than would be expected in a single species," so he accepted only one New Zealand taxon.

In a study of the fossil rails of New Zealand, Olson (1975) returned these coots to *Fulica* because the differences between them and other members of *Fulica* were shown to be related to large size and flightlessness and thus were insufficient to support generic distinction. He found that *F. chathamensis* and *F. prisca* were virtually identical in size, and he considered that the low depth of the carina on the sternum meant that they were incapable of flight. Olson did note that the crania of *F. chathamensis* were highly variable in the degree of development of the salt gland impressions and that crania of *F. prisca* were peculiarly domed posteriorly, but he had only two crania of the latter form to examine. The only consistent difference that Olson found to separate these forms was that the brachial depression of the humerus in *F. chathamensis* was always deeper and more extensive than in *F. prisca*, and he advocated their subspecific distinction.

Millener (1980), in contrast, could not distinguish between humeri of the mainland and Chatham Island birds but did, however, detail minor differences in the shape of the sternum and tibiotarsus. He noted that the manubrial spine on the sternum of *F. chathamensis* was distinctly rounded anteriorly and separated the coracoidal sulci, whereas in *F. prisca,* the manubrial spine did not lie between the coracoidal sulci, was broader and notched anteriorly, and had a ventral spine. For the proximal tibiotarsus, he found that a ridge connected the fibular crest to the ectocnemial crest in *F. prisca,* whereas in *F. chathamensis,* the homologous ridge went from the fibular crest to the protuberance below the external articular surface. Also, the cnemial crest was more elongate proximally in *F. chathamensis,* and concordantly, the groove between the inner and outer cnemial crests was deeper than in *F. prisca.* On the basis of these distinctions, Millener advocated subspecific separation, stating explicitly that any elevation to specific rank should await further study. It was therefore surprising that the forms were treated as species in the most recent *Checklist of the Birds of New Zealand* (Turbott 1990a), citing only Millener's publications.

In view of the insubstantial differences reported by Millener, we reexamined the differences between these coots. As noted by Olson (1985), the postcranial bones overlap widely in length, even in the small samples used here (Tables 10.4, 10.5). The differences Millener described in the tibiotarsus were found to relate to size of the bone. Larger specimens from both mainland and Chatham sites had a more elevated ridge, which tended to be more anteriorly situated relative to the protuberance below the external articular surface. In small specimens, the ridge went below the protuberance. The syntypic series (syntypes are a series of specimens used to describe a species but from which no holotype was selected) of *F. prisca* from Castle Rocks included the entire range of variation.

Examination of many sterna revealed that the shape of the manubrial spine also varied greatly: in *F. chathamensis,* the spine was pointed, rounded, or square, with the square ones either notched or not, just as in *F. prisca.* Similarly, some manubrial spines had a ventral spike, whereas others did not. The syntype series from Castle Rocks includes two sterna with squared and notched manubrial spines, one merely square and one rounded. Clearly, the form of the spine varied between individuals. A difference that Millener did not describe but that was evident in the bone he figured was that sterna from mainland New Zealand specimens had a prominent pneumatic foramen that was lacking in sterna from the Chatham Islands. However, the feature is also variably present; in some it is prominent (CM Av5589 and CM Av9758), and in others it is lacking altogether (CM Av11557 and CM Av12289). However, the foramen was not observed in any Chatham Island sterna.

We reexamined the cranial difference noted by Olson (1975) and found that all crania of *F. prisca* have a more elevated vault posterior to the orbit than do crania of *F. chathamensis.* In lateral view, the vault of *F. chathamensis* rapidly declines in depth from the posteroventral region toward the interorbital area, whereas that of *F. prisca* is more rectangular (Fig. 10.10). Although no substantial differences have so far been noticed in the shape of individual skeletal elements, we found proportional differences as detailed in Table 10.6. The humerus–femur

TABLE 10.4.
Average measurements (mm) for *F. prisca* derived from bones of mixed individuals from the syntype series (MNZ 379 and MNZ 990) and Waikari Cave (MNZ S33715 and MNZ S33721), augmented by bones of known individuals (CM Av5589, CM Av19148, CM Av23129, MNZ S32448, and MNZ S32647).

Variable	Humerus	Ulna	Carpometacarpus	Femur	Tibiotarsus[a]	Tarsometatarsus
Mean	91.47	72.80	47.54	83.21	142.76	89.12
Standard deviation	4.19	3.81	1.90	3.88	5.49	3.94
Minimum	85.5	68.3	45.5	78.0	133.4	82.0
Maximum	98.4	79.7	50.7	89.6	154.0	95.8
n	16	15	12	17	23	18

[a] Tibiotarsus length was measured from the proximal articular surface, not total length.

TABLE 10.5.
Summary statistics for lengths (mm) of postcranial elements of *F. chathamensis*.[a]

Variable	Humerus	Ulna	Carpometa-carpus	Femur	Tibiotarsus[b]	Tarsometa-tarsus	Sternum length	Sternum depth
Mean	95.60	78.70	49.88	83.98	154.19	92.43	76.38	20.09
Standard deviation	3.33	2.41	1.84	3.25	4.07	2.07	1.34	1.40
Minimum	90.4	74.6	47.0	79.7	148.1	88.1	74.0	17.0
Maximum	100.4	82.0	52.3	89.5	161.5	94.7	78.2	21.1
n	13	11	11	12	13	12	9	7

[a] Specimens used were as follows: CM Av5280, CM Av6746, CM Av29355, CM Av3311, CM Av5267, CM Av5269, CM Av5270, CM Av5271, CM Av5275, CM Av5277, CM Av5290, MNZ 384, and MNZ 7937. All are individual skeletons.
[b] Tibiotarsus length was measured from the proximal articular surface, not total length.

ratio for three individual *F. prisca* skeletons and the ratio of the means for the samples in Table 10.4 were in the range 108.2–109.9. This was outside the range for a sample of 12 individual *F. chathamensis* (111.2–115.9), so *F. prisca* had a shorter humerus relative to the femur. Similarly, *F. prisca* had a relatively shorter wing than *F. chathamensis*. A detailed comparison of the skeletal morphology of *F. prisca* and *F. chathamensis* is overdue, but in the meantime, we consider that the different cranial shapes and the proportional differences in the postcranial elements support specific status for Chatham coots as distinct from mainland New Zealand coots, as they were originally described (Fig. 10.11).

Whether or not Hamilton (1893) was correct in believing that *F. prisca* could fly is important because if mainland birds could fly, then

TABLE 10.6.
Average proportions of various skeletal elements derived from individuals for *Fulica chathamensis*, *F. prisca*, and *Porphyrio melanotus*.[a]

Species and value	Humerus/femur	Wing[b]/femur length	Sternum diameter/length
Fulica chathamensis			
Mean	114.07	266.85	26.37
Standard deviation	1.38	3.74	1.55
Minimum	111.28	260.00	22.97
Maximum	115.95	270.77	27.55
n	12	10	7
Fulica prisca			
Individual 1			29.97
Individual 2	108.21		
Individual 3	108.72		
Individual 4	109.62		
From mean values of *prisca* bones in Table 10.4	109.93	254.56	
Porphyrio melanotus			
Mean	106.97	274.49	28.74
Standard deviation	3.06	8.98	1.28
Minimum	101.00	256.70	27.29
Maximum	111.71	286.86	31.36
n	11	11	10

[a] Individual specimens used were as follows: *chathamensis*: CM Av5280, CM Av6746, CM Av29355, CM Av3311, CM Av5267, CM Av5269, CM Av5270, CM Av5271, CM Av5275, CM Av5277, CM Av5290, MNZ 384, and MNZ 7937; *prisca*: CM Av5589, CM Av19148, CM Av23129, and MNZ S32448; *P. melanotus*: THW#1 (private collection), CM Av25202, CM Av37030, CM Av37016, CM Av15755, CM Av12636, CM Av10185, CM Av12634, MNZ 25124, MNZ 23535, and MNZ 23979.

[b] Wing is humerus + ulna + carpometacarpus length.

the population on Chatham Island may not have been genetically isolated from New Zealand birds. *Fulica chathamensis* has often been tacitly assumed to be flightless (Gill and Martinson 1991; Olson 1985), but Andrews (1896b) concluded otherwise. The relative proportions of its pectoral skeleton and stoutness of its wing bones suggests that it probably did fly.

Both *F. cristata* and *F. atra* are highly aquatic forms and may not be the best analogs for the terrestrial New Zealand coots. Fossil bones of New Zealand coots are often found in sites not associated with aquatic habitats, and the general proportions of *Porphyrio melanotus* are much more similar to those of the New Zealand coot, and so it may be that the New Zealand coot had similarly terrestrial habits. Although *P. melanotus* is usually thought of as a poor flier, this is not true. Although ungainly, it is an accomplished flier and can travel long distances. It colonized the Chatham group in the 20th century, New Zealand in only

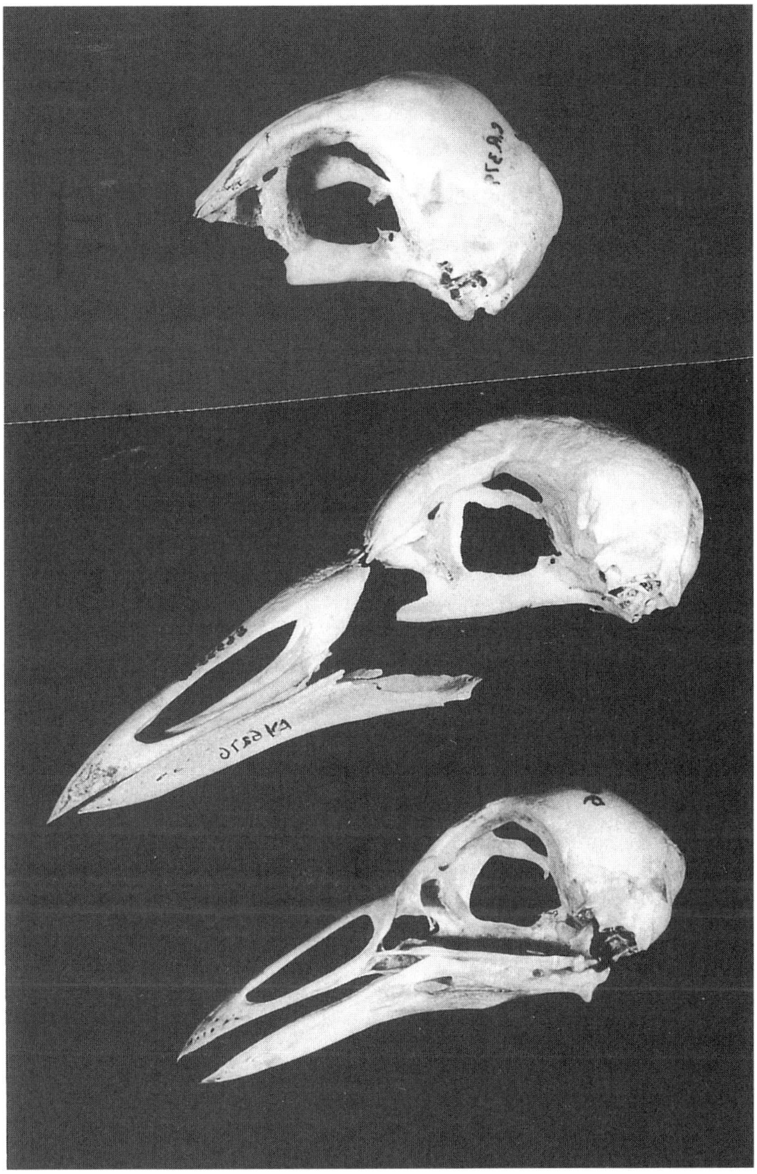

the last few centuries, and many Pacific islands. One specimen measured for this study was obtained from L'Esperance Rock in the Kermadec group, to which it is assumed to have recently flown at least as far as from New Zealand—a distance of about 600 km.

We compared the proportions of humerus length to femur length, wing length to femur length, and the keel depth to length ratio of the sternum in the New Zealand coots with those of *P. melanotus* (Table 10.6). Both *F. prisca* and *F. chathamensis* had relatively longer humeri compared with the femur than *P. melanotus,* but total wing length (humerus + ulna + carpometacarpus) as a proportion with femur length

Figure 10.10. Comparison of the crania of Fulica *species in lateral view. (Top)* Fulica prisca *(MNZ S379). (Middle)* F. chathamensis *(CM Av5276). (Bottom)* F. atra *(CM Av5269). Note the much smaller salt gland impressions in* F. prisca *and the way the cranial vault is more elevated in its posterior part. Photographs by R. Morris (*F. chathamensis *and* F. atra*) and J. Palmer (*F. prisca*).*

Figure 10.11. Reconstruction of Fulica prisca. *Drawing by J. Winn.*

was less. The outer wing bones of *P. melanotus* are relatively longer. Nevertheless, the wing–femur length average ratio for *F. prisca* was only just below the minimum for the sample of individual *P. melanotus,* and the entire range for this ratio in *F. chathamensis* fell within that of *P. melanotus*. The ratio of sterna depth–length ranged from about 23%–28% for *F. chathamensis*, which is only just below that for *P. melanotus* (28%–31%). Analysis of these data shows that both *F. chathamensis* and *F. prisca* had only relatively slightly smaller wings and sternal carina than *P. melanotus*. As *P. melanotus* flies well, it seems probable that both *F. prisca* and *F. chathamensis* could fly, so Andrews and Hamilton were probably right.

An early record of a living "coot" in New Zealand was republished by Buller (1878). William Colenso had described the bird in the *Tasmanian Journal of Natural Science* for April 1845 as a new species of coot *Fulica novae-zealandiae*. However, it is a good description of a dabchick, as Oliver (1955: 92) has noted, and so this name belongs in the synonymy of *Poliocephalus rufopectus* (Gray 1843).

Family Aptornithidae

Genus Aptornis *Mantell*[1]
Aptornis otidiformis *Owen, 1844 (North Island Adzebill)*
Aptornis defossor *Owen, 1871 (South Island Adzebill)*

The massive gruiforms known as *Aptornis* have had confused nomenclatural and taxonomic histories. *Aptornis otidiformis* was first listed as a moa in the Dinornithidae by Owen, and for a long time, elements of its skeleton were confused with those of the similar-size extinct goose *Cnemiornis*. *Aptornis* lived only on the two main islands of New Zealand. It was extinct by the time Europeans arrived, but its presence in the earliest middens indicates Māori hunted them.

One other species name has been used in conjunction with *Aptornis*. A skull was excavated by W. W. Smith at Albury in South Canterbury (Buller 1888; Smith 1891), which Buller forwarded to Owen. Buller (1888: xxiii) quoted a letter from Owen, which indicated that it was of a larger form than hitherto seen and that if further bones offering corroboratory characters were found he "should, however, rejoice . . . in introducing to our Zoological Society an *Aptornis bulleri*." Hutton (1896a: 642) refers several bones to this taxon with the note that it is "equal in size to *A. defossor*." However, the name was never promoted with a description and is therefore a *nomen nudum*.

As early as 1866, Parker considered *Aptornis* to be a gruiform most similar to the trumpeter *Psophia* (Parker 1866). Fürbringer (1888) and Beddard (1898) placed *Aptornis* among the gruiforms, just after the cranelike birds *Cariama, Psophia,* and *Eurypyga,* near the kagu *Rhynochetos jubatus*. They reached this conclusion mainly because *Aptornis* has inwardly curved schizorhinal nostrils (which means the nostrils extend cranially past the nasal process of the premaxilla), and as in *Rhynochetos* and cranelike birds, it has a partially ossified nasal septum. The solid interorbital septum of *Aptornis* is seen in *Psophia* among gruine birds, and the form of the palatines is between that of *Rhynochetus* and *Psophia*. Beddard (1898) concluded that while the outline of the skull of *Aptornis* is very like that of some of the large rails its affinities had been more correctly diagnosed by Fürbringer. On the basis of the structure of the quadrate and the pterygoid, Lowe (1926) concurred with Fürbringer (1888) and Beddard (1898) and regarded *Aptornis* "as so primitive that it is doubtful if it could be included in either the porphyriine or ocydromine series, but is ancestral to both and nearer the Gruimorph complex" (Lowe 1926: 178).

More recently, Cracraft (1982) and Olson (1985) also suggested that *Aptornis* was related to the kagu (*Rhynochetos jubatus*). Specifically, Cracraft (1982) suggested that the trumpeters (*Psophia*), seriemas (*Cariama*), sunbitterns (*Eurypyga*), kagu, and *Aptornis* formed a monophyletic clade, or natural group, exclusive of other gruids and therefore rails. A further viewpoint was presented by Weber and Hesse

1. On February 2, 1848, Mantell (1848a) preceded Owen (1848a) in his use of *Aptornis* on April 22.

(1995), who found particular features of the mandible and quadrate that suggested a relationship with galliform and anseriform birds exclusive of gruids. The degree to which such features could be related to function was not addressed. Weber and Hesse noted that the eminentia articularis, which is a tubercle on the outer side of the quadrate, uniquely articulates with the zygomatic process of the squamosal in *Aptornis* and that a homologous tubercle was otherwise only seen in galloanseres. But we note that a large similar tubercle is present on quadrates of *Porphyrio*. The proposed galloansere relationship may relate entirely to convergences related to function.

The relationships of *Aptornis* have recently been reexamined by Houde et al. (1997) with mtDNA analyses. They found that *Aptornis* was a gruiform that branched off the main evolutionary lineage immediately before the rails. Moreover, they found that *Psophia* was closely associated with *Aptornis* on the tree. They found no support for the suggestion that *Aptornis* was related to the kagu, which was always found to be close to the sunbittern *Eurypyga* as well as usually basal on the tree and well removed from rails, *Aptornis,* and *Psophia.* There was also no support for the group defined (see above) by Cracraft (1982), so it seems probable that these species have been variously associated in the past because of convergences in their form that do not reflect phylogeny. For example, schizorhinal nostrils have developed convergently several times in neognath birds (Weber and Hesse 1995), and thus it is a weak indicator of relationships.

Most recently, Livezey (1998) presented a phylogenetic analysis of gruiforms on the basis of morphologic characters that found *Aptornis* and *Rhynochetos* to be most closely related. However, the degree to which noncomparable characters affected this placement was untested and is likely to be significant, because all of the 189 characters of the integument and at least 106 of the 361 skeletal characters were essentially not comparable either because of their complete absence (sclerical ossicles, furcula, and manus) or because the bones were so vestigial in nature that character placement is likely to be spurious (humerus, ulna, radius, carpometacarpus, coracoid, and scapula). Livezey's placement of *Aptornis* with *Rhynochetos* may well be due to morphologic convergences, making the comment of Houde et al. (1997: 138) pertinent: "The phylogenetic informativeness of postcranial morphological characters is all but obliterated by gigantism and the shift of locomotory dependence from the wings to the legs in this flightless bird." Furthermore, the specialization in the structure of the skull related to feeding in this predator further obscures any phylogenetic signal there might be; neither would the fact that, to our knowledge, a complete skull has never been studied—one with the large lachrymals and pterygoids present.

At present, the phylogenetic relationships of *Aptornis* must still be considered to be unresolved. It is clear that it is a gruiform, it has no close relatives, and it is highly specialized toward a particular feeding niche and further modified by degeneration of structures associated with flight. Its role in the functioning New Zealand ecosystem also remains a mystery. Owen (1879a: 301) ventured that

> the peculiarities of the skull in *Aptornis* may be considered in relation to the food and the work to which its long adze-like beak was put. I infer this work to have involved frequent strong and deep thrusts into the ground, and that the quest was for animal, not vegetable matters. I have heard casually and vaguely of the great number, size and unusual colour of the earthworms of New Zealand; and it is probable that a rich field here remains to be explored by the helminthologists.—The strange appearance of the parasitic cryptogram, *Sphaeria* sp., when it has achieved its growth at the cost of the caterpillar infested, has made us familiar with the burrowing habits of at least two species of New Zealand nocturnal Lepidoptera (*Cheiragria virescens* and *Ch. rubroviridens*) at the larval stage of their existence. Such larvae and earthworms were probably part of the food of *Aptornis*.

Other evidence supports Owen's suggestions concerning the diet of *Aptornis*. In two remarkably complete skeletons in the Otago Museum, in which such small elements as the carpometacarpus and glossohyals are preserved, a small set of gizzard stones was present (35 stones, ~16 g; 23 stones, ~8 g). Species that eat fibrous plants have relatively large gizzard stone sets, whereas predators have few or no stones. Most significantly, analyses of bone gelatin reveal stable isotope values for nitrogen (^{15}N) consistent with *Aptornis* having been a predator (see text box, Chapter 5).

Description of the Skeleton

The two species of *Aptornis* differ mainly in size: the North Island species was smaller than its South Island congener (Table 10.7). Within species, there was little size variation, as shown by relatively small coefficients of variation.

Skull. The skull of *Aptornis* is characterized by its large size—it is longer than the femur or about the same length as the tibia—and its remarkable bill (Figs. 10.12, 10.13). The highly vascularized bill is large, pointed, down-curved, and robust and is supported by a robust maxillary process of the nasal and robust quadratojugals. The skull has large lachrymals, which are not ankylosed to the skull and so are found detached and have rarely been recognized (cf. Fig. 10.14 and Fig. 10.15 of a complete skull). The lachrymals enclose the front half of the orbit. The orbit is relatively small and probably housed an eye only about 15 mm wide in a skull about 90 mm wide. The olfactory capsule lies between the lachrymals and is large, medially divided, and serviced by a single olfactory nerve. The quadrates are locked into position in the cranium by downgrowths of the squamosal, so they are nearly always found with the cranium. Weber and Hesse (1995) noted that *Aptornis* had, uniquely among birds, a distinct articular facet between the zygomatic process of the squamosal and the quadrate. But, we note *Porphyrio* has a similar facet. The structure contributes to the extraordinary robustness of the bill. The ventral development of the basisphenoid plate was greater than that seen in any other bird and provides the attachments for huge muscles to wield the head. The anterior cervical vertebrae were correspondingly large and broad with a high neural

TABLE 10.7.
Measurements for leg bones of *Aptornis.*

Species and value	Femur length	Femur shaft diameter	Tibiotarsus length	Tarsometatarsus length	Humerus length	Ulna length
Aptornis otidiformis[a]						
Mean	161.45	17.66	217.79	95.79	111.01	44.7
Standard deviation	5.79	0.70	8.32	4.14	5.23	—
Minimum	142.6	16.2	203.2	86.9	102.6	44.6
Maximum	175.1	19.2	232.5	104.4	129.4	44.8
n	68	80	45	75	25	2
Coefficient of variation	3.59	3.99	3.82	4.33	4.71	—
Aptornis defossor[b]						
Mean	181.21	19.47	246.63	108.68	128.35	49.95
Standard deviation	7.68	1.38	11.25	5.19	5.84	0.35
Minimum	167.4	16.7	219.9	96.5	119.3	49.7
Maximum	193.2	22.6	265	116.3	137.1	50.2
n	25	27	39	32	8	2
Coefficient of variation	4.24	7.11	4.56	4.77	4.55	0.71

[a] Livezey (1994) gives the length of a carpometacarpus as 18.1 mm.
[b] Livezey (1994) gives lengths of 22.4 and 20.0 mm for two carpometacarpi, and an Otago Museum specimen is 20.9 mm long.

spine, to provide the attachment areas for some of these muscles. There were 22 vertebrae, of which vertebrae 18–22 articulate with ribs and have tall neural spines, thus constituting the thoracic series. However, vertebra 17 had a small fused rib. Vertebra 15 had a robust ventral process coincident with its position at the base of the neck loop, and by vertebra 18, there were three well-developed ventral processes. Vertebrae 19–22 each had a single anteriorly curved ventral spine.

The pelvis was large and robust, with a significant posterior decurvature, and had tall neural spines in the anterior synsacrum (Fig. 10.16). The anterior ilium was thin in its central dorsal region and usually did not form a continuous plate fusing to the dorsal margin of the anterior sacral spines. The posterior ilium and the fused ischium enclosed a pocket that opened anteriorly as in rails. Two pairs of ribs articulated with the pelvis (Fig. 10.17).

Leg Bones. The leg bones were robust, and the tarsometatarsus was, uncharacteristically for a gruiform, very short (Table 10.7, Fig. 10.18). The femur shaft was straight, had a low trochanter, and had a characteristic ventral ridge from the medial condyle to the shaft. The tibiotarsus was stout and straight, and it had a low cnemial crest (Fig. 10.17). The tarsometatarsus had two hypotarsal ridges that were fused

Figure 10.12. Articulated skeleton of Aptornis defossor *(CM Av6016) in lateral view. Photograph by R.N.H.*

posteriorly to enclose a single canal, within which a medial constriction suggests that two tendons were enclosed within it. Rails usually have three hypotarsal ridges, none of which encloses a canal. Short toes with relatively flat unguals indicate little dexterity of the foot. No unguals were modified into spurs.

Wing Bones. The wing bones were very reduced, which meant that the typical avian shape of most of the constituent elements had been altered considerably. The wing bones were relatively smaller even than those of the dodo. The femur–humerus ratio was very similar in each species: 1.45 versus 1.43. The ulnae were tiny, with the humerus, itself much reduced, some 2.5 times longer than the ulna. The carpometacarpus had degenerated to the extent that the major and minor metacarpals were fused so that there was no intermetacarpal space, as in the elephantbirds of Madagascar, but otherwise known from carinate birds only for the moa-nalos (*Thambetochen, Ptaiochen,* and *Chelychelynechen*) of Hawaii (Livezey 1994). Unlike any bird, however, the phalanx for digit 1 was fused to the carpometacarpus, creating a short, spurlike process, but it is doubtful that this was reflected in any external feature of the wing. The primary feathers insert in the wing along the carpometacarpus and secondaries in the ulna, and because these elements were so reduced, it is certain that no wing would have been visible in the external plumage of *Aptornis,* as in kiwi.

Pectoral Girdle. The sternum of *Aptornis* was greatly reduced and had a triangular shape with no lateral processes. There was almost no keel, it being represented by a low central ridge. Deep sulci remain for the articulations with the coracoids, but there was a deep, broad invagi-

Figure 10.13. Reconstruction of Aptornis. *Note the absence of a visible wing. Drawing by J. Winn.*

nation between them. There were four articulations for sternal ribs. Often the coracoids are found ankylosed to the sternum. The furcula has not been described and may have been lost altogether, and the scapula was reduced.

The distribution of the sites in which *Aptornis* has been found indicates that it preferred the drier eastern regions of both islands during the Holocene. During the last Glacial, its range included western areas. It therefore seems to have preferred shrublands and grasslands (dry or wet), although it seems to have avoided the subalpine shrubland habitats. Its remains have never been found above 1000 m. This distribution has a bearing on determining what *Aptornis* may have eaten. It probably did not feed on frogs because frogs lived in areas largely outside of this distribution. But as Owen surmised, many large inverte-

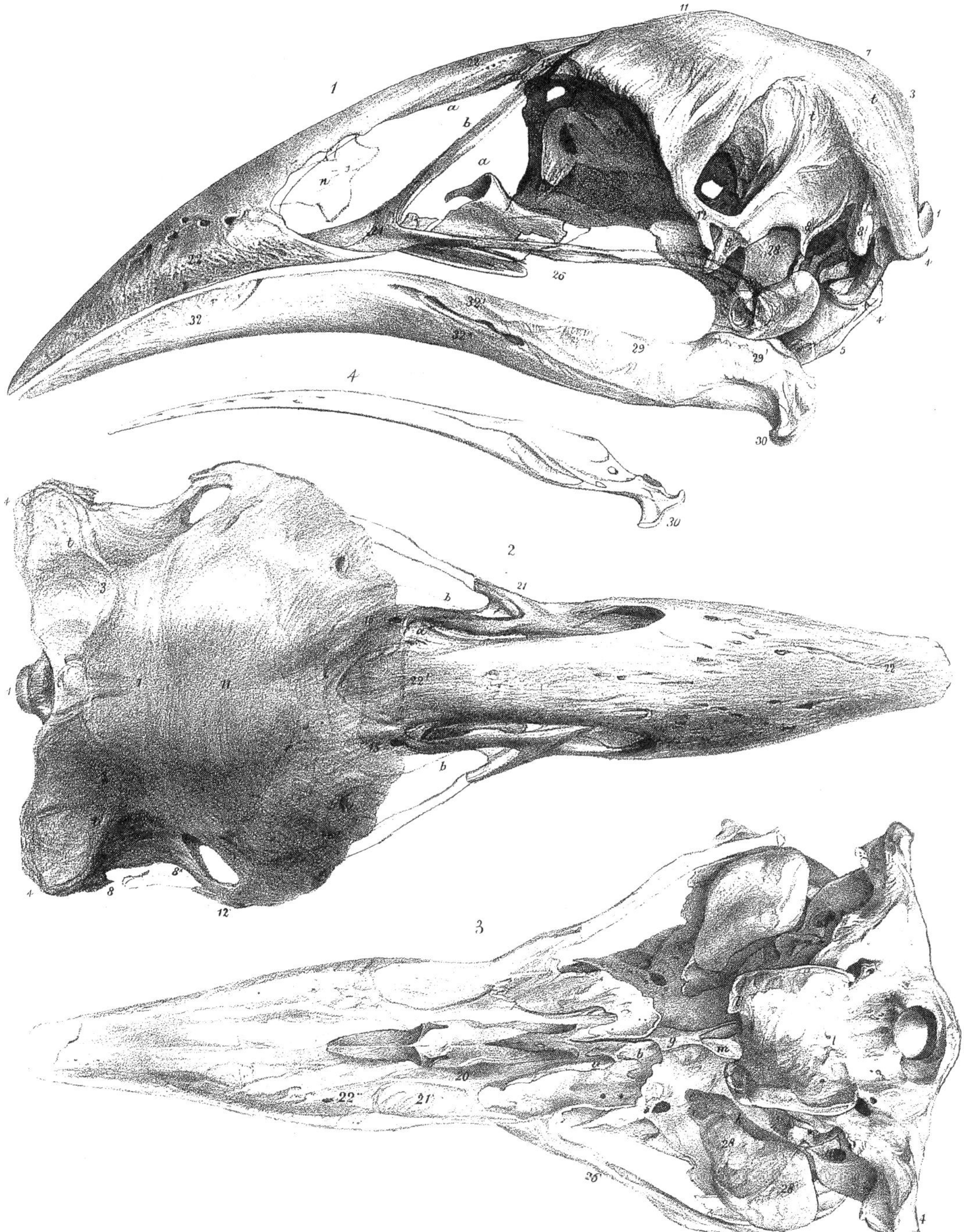

Figure 10.14. Views of the skull of Aptornis defossor *(from Owen 1879: pl. 83). This skull lacks the large detachable lachrymals.*

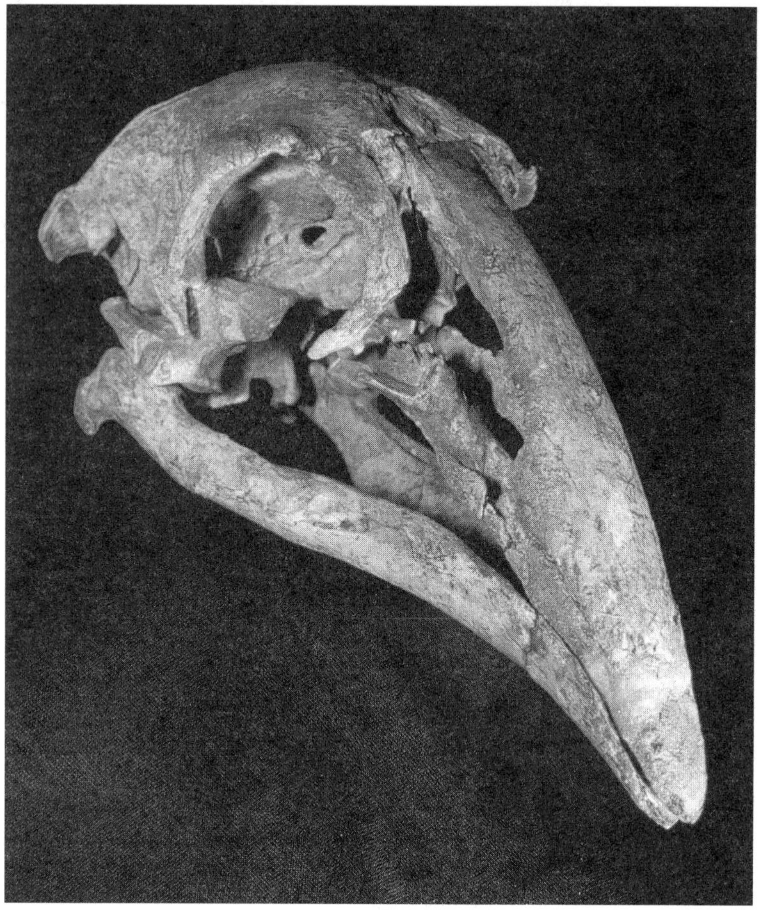

Figure 10.15. View of a complete skull of Aptornis *showing large lachrymals. Photograph by R. Morris.*

Figure 10.16. (opposite page) Views of the pelvis and sternum of Aptornis defossor *(from Owen 1879: pl. 89).*

brates, though not worms, would have been available, as well as lizards and tuatara. Possibly even Finsch's duck or seabirds could have been taken during their nesting seasons.

Aptornis defossor had an ossified glossohyal (Worthy 1989f) that supported a presumably large tongue. One glossohyal measured 56 mm long by 10 mm wide, and another (OM Av6727) was 8.4 mm wide and more than 45 mm long, and was associated with a bird with a femur length of 191 mm. The thyroid, reported by Worthy (1989f) and associated with the complete glossohyal, was 22.6 mm long and 17.9 mm wide. A thyroid with OM Av6727 was 21.1 mm long by 17.2 mm wide. OM Av6727 also has a small basihyal (12.1 mm long, 6.3 mm deep, and 3.8 mm wide) and two ossified hyoids.

The measurements for the femur shaft enable predictions of the body mass of *Aptornis* to be made (Table 10.8). The North Island birds averaged about 16 kg and probably ranged 8–20 kg, whereas the South Island birds were only slightly larger, at 18 kg (range, 10–25 kg). As predicted from similar-size tibiotarsi, these masses are similar to those for *Harpagornis, Cnemiornis,* and some smaller individual moa (see Fig. 3.13).

PL. LXXIX.

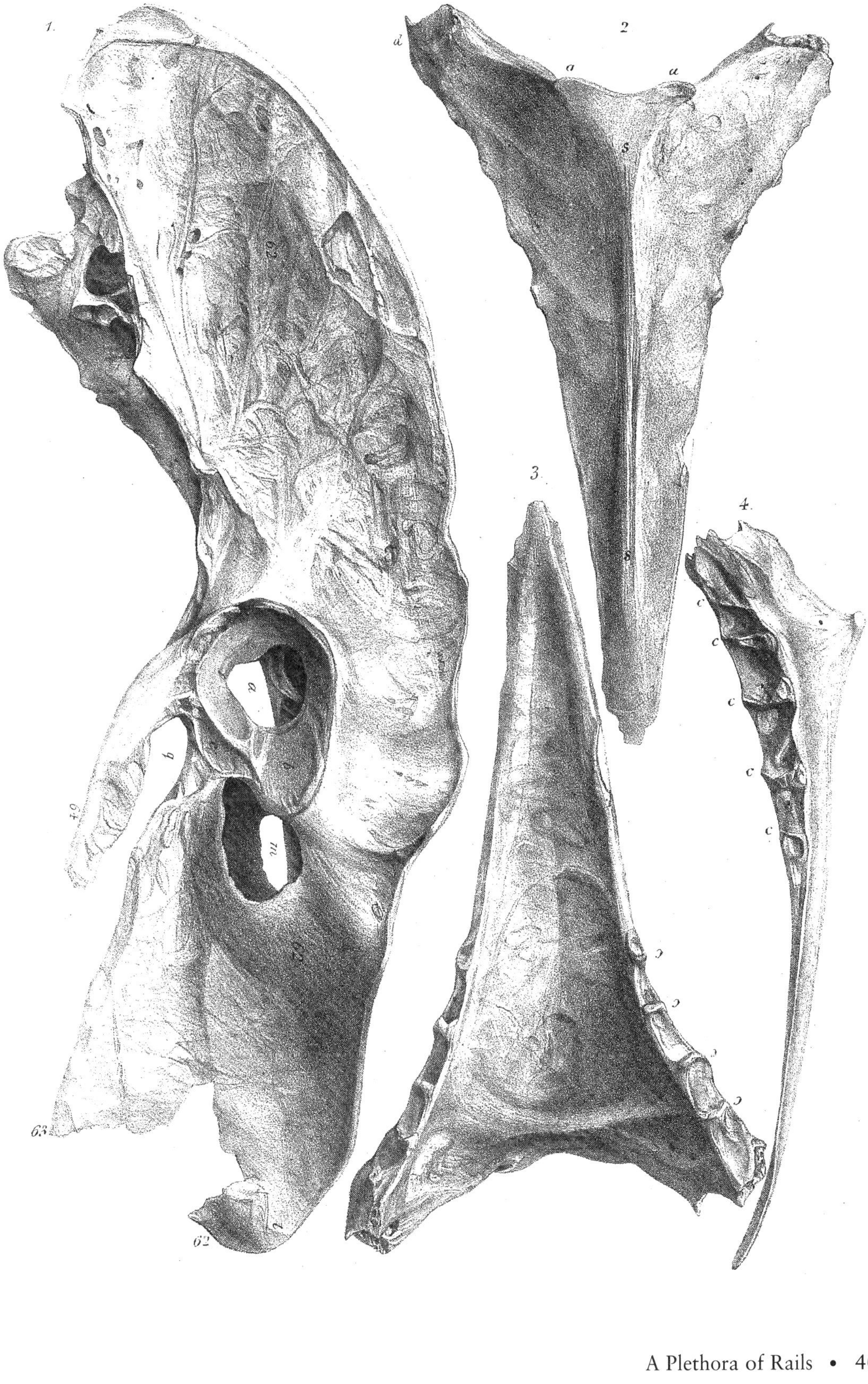

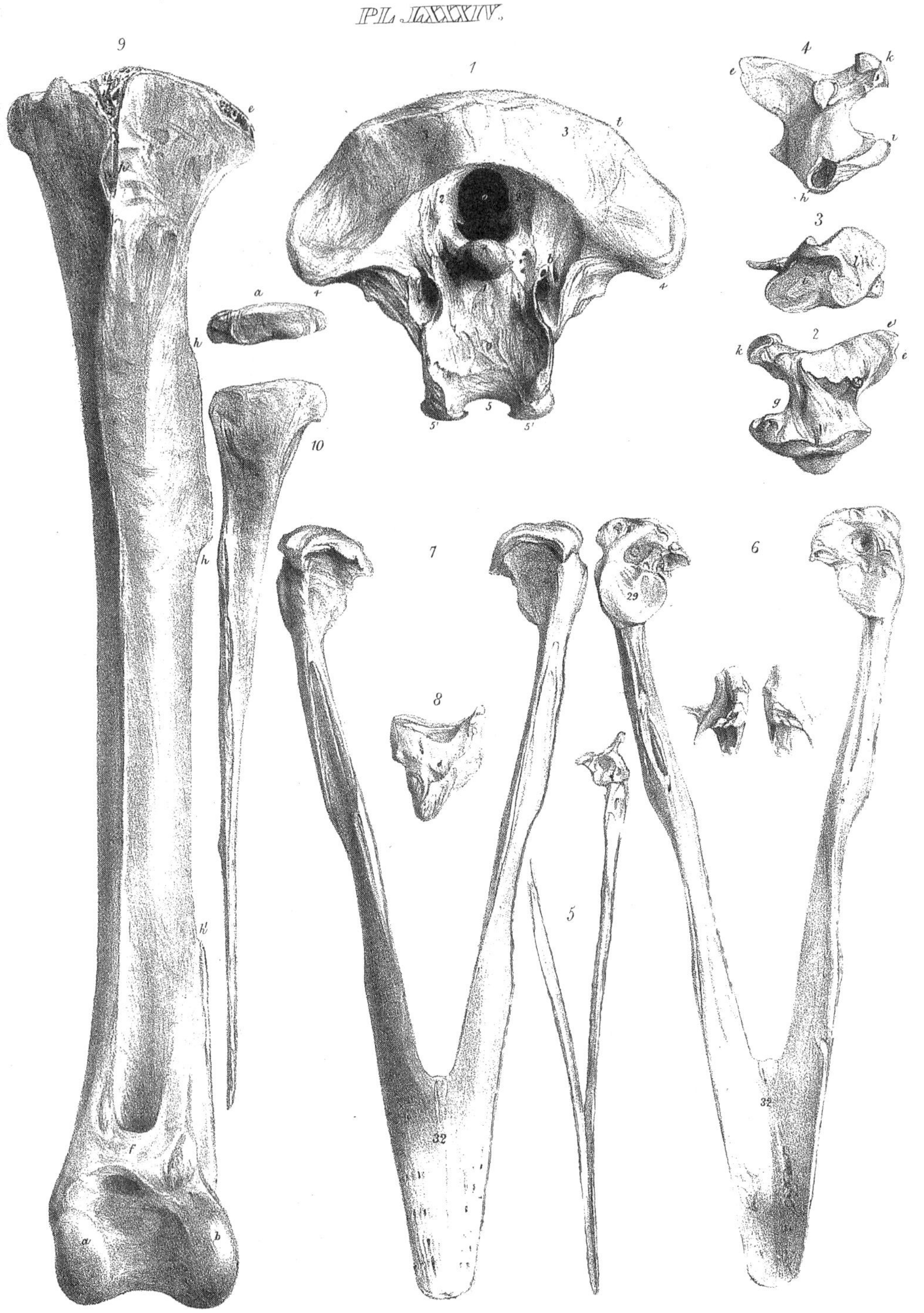

Figure 10.17. Views of various bones of Aptornis defossor *(from Owen 1879: pl. 84). (1) Cranium in posterior view. (2–4) quadrates in medial, ventral, and lateral views, respectively. (5) Mandible of kiwi (*Apteryx*) for comparison. (6–8) Mandible in dorsal, ventral, and posterior view of articular end, respectively. (9) Left tibiotarsus in anterior view. (10) Left fibula in lateral view.*

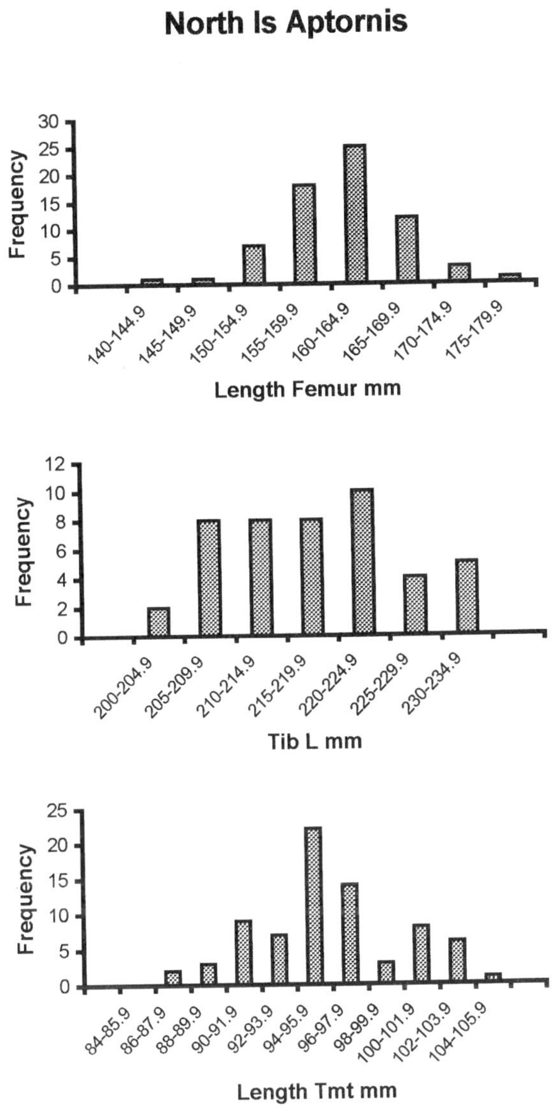

Figure 10.18. Histogram of lengths of femora, tibiotarsi (Tib), and tarsometatarsi (Tmt) for Aptornis otidiformis *based on data in Table 10.7. Distributions are approximately normal; there is no bimodal nature, suggesting there was no overt sexual dimorphism.*

Order Galliformes

New Zealand has few representatives of this family, which is prominent in faunas of the Northern Hemisphere.

Family Phasianidae

Genus Coturnix *Bonnaterre*
Coturnix novaezelandiae *Quoy and Gaimard, 1830*
(New Zealand Quail)

The small New Zealand quail was closely related to the Australian *Coturnix pectoralis*. Historically, it was recorded throughout the North and South Islands, as well as on Great Barrier Island and some other

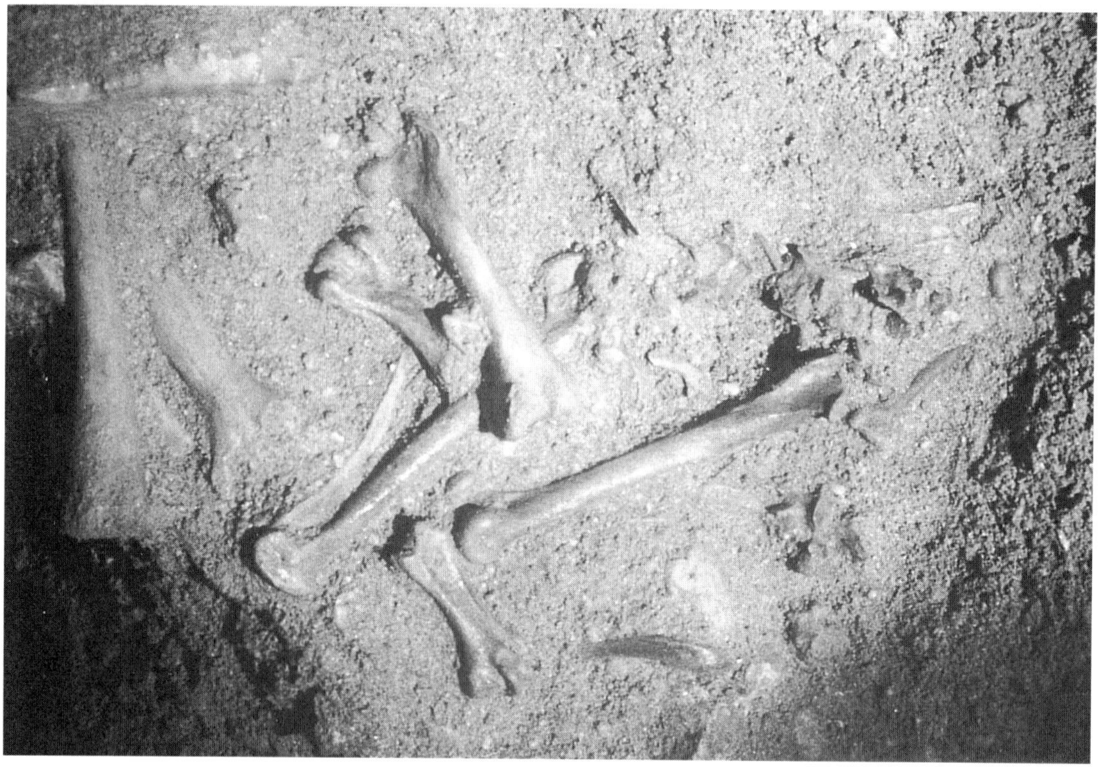

Figure 10.19. Aptornis *bones exposed in layer 3 in Excavation 2, The Graveyard, Honeycomb Hill Cave. Photograph by T.H.W.*

inshore islands. It was abundant in the grasslands and fern lands of the drier parts of the South Island in the early half of the 19th century. However, its numbers rapidly diminished in the 1850s and 1860s, and it was extinct sometime in the 1870s. It was shot in large numbers, but it is more likely that its extinction was caused by the spread of *Rattus rattus*, which spread through the country at this time.

The fossil record indicates that it was formerly distributed throughout the mainland of New Zealand, and its bones are particularly common in falcon deposits of the eastern South Island (Worthy 1997b; Worthy and Holdaway 1995). It is known from the last Glacial deposits in several areas. This is one of the key species for inferring the presence of grassland or short shrubland in the vegetation of an area.

Family Megapodiidae

Genus and Species Unknown

The megapodes, especially the genus *Megapodius*, were widely distributed in the southwest Pacific, from Australia and Papua New Guinea through the Solomon Islands and Tonga to as far east as Samoa (Jones et al. 1995). Many islands that now lack these birds formerly had populations, often of extinct species (e.g., Steadman 1989a, 1993a, 1993b).

The closest that megapodes came to New Zealand appears to have been to the Kermadec Islands. Cheeseman (1891) recorded a descrip-

TABLE 10.8.
Predicted body mass for *Aptornis* species using femur diameter (D) measurements given in Table 10.7 (algorithms used are as for moa and *Cnemiornis*).

Species and value	Estimated mass (kg)	Calculated range
Aptornis otidiformis		
Femur D (mean)	16.4	7.9–25.5
Femur D (maximum)	18.9	9.5–28.5
Femur D (minimum)	14.4	6.5–22.7
Aptornis defossor		
Femur D (mean)	18.9	9.51–28.5
Femur D (maximum)	22.4	12.0–32.7
Femur D (minimum)	14.9	7.0–23.6

tion by Mr. Johnson, an early settler on Raoul (Sunday) Island, of a bird that made mounds of sand and decayed leaves 0.6–0.9 m (2–3 feet) high in which it laid its eggs. The species became extinct after the 1870 volcanic eruption. Although no specimen is known to survive, this record most likely pertains to a megapode. The veracity of Johnson's observation has been recently validated by the discovery of a *Hemiphaga* bone on Raoul Island in an archaeologic site. Pigeons exactly like the New Zealand species were described by Johnson, as reported in Cheeseman (1891), but no specimen was ever collected and the population is now extinct (Worthy and Brassey 2000).

11 • Shorebirds, Parrots, and Passerines

Order Charadriiformes: Waders, Gulls, Terns, and Skuas

With its long coastlines, numerous islands, and extensive braided riverbeds, New Zealand has a diversity of charadriiform birds. Although there were few breeding species of waders per se, they are all endemic; several are rare and endangered by the loss of wetland habitat and the prevalence of predatory mammals. About 20 species of Arctic wader, of which 9 are common, regularly migrate to New Zealand. For example, at present, the wintering population of bar-tailed godwits (about 100,000 birds) is a large proportion of the world population, and indeed most of the eastern race *Limosa lapponica baueri* winters in New Zealand (Sagar et al. 1999). Terns are more diverse and include both cosmopolitan and endemic species. Three gull and one skua species breed in the area, although three Northern Hemisphere jaegers regularly reach New Zealand. Unlike most other groups, the waders are conspicuous birds in accessible habitats, and their population sizes are better known than for many but the rarest of the other New Zealand birds (Sagar et al. 1999).

Family Scolopacidae

Genus Coenocorypha *G. R. Gray*
Coenocorypha aucklandica *(Gray, 1845)*
(Auckland Island Snipe)
Coenocorypha meinertzhageni, *Rothschild, 1927*
(Antipodes Island Snipe)
Coenocorypha huegeli *(Tristram, 1893) (Snares Island Snipe)*
Coenocorypha iredalei, *Rothschild, 1921*
(South Island Snipe)
Coenocorypha barrierensis, *Oliver, 1955*
(North Island Snipe)

Coenocorypha pusilla *(Buller, 1869) (Chatham Island Snipe)*
Coenocorypha chathamica *(Forbes, 1893) (Forbes's Snipe)*

Snipe were widely distributed throughout the New Zealand archipelago. The New Zealand snipe were well differentiated into several allopatric forms, though at present, the subantarctic forms and mainland New Zealand ones are only recognized as subspecies of *C. aucklandica* in the *Checklist of the Birds of New Zealand* (Turbott 1990a). However, we follow Higgins and Davies (1996) and accept a species-level distinction for several of the forms. Although formerly widespread on the North and South Islands of New Zealand, snipe were extinct by the time Europeans arrived there. The Northern form survived until 1870, when a single specimen was collected live on Little Barrier Island near Auckland. The southern form survived until recently on islets around Stewart Island, where the last population succumbed to black rats in 1964. A recent undescribed addition to the well-differentiated forms found on the Auckland Islands, the Snares Islands, and the Antipodes Island is a population found surviving on an islet off Campbell Island. Campbell Island itself has had rodents on it for many years, and so any snipe there are long extinct.

New Zealand snipe are most similar to Strickland's snipe *Gallinago stricklandi* of the Straits of Magellan and Chile and occupy similar forested habitats. They are, in contrast, little like their Northern Hemisphere relatives. They are not waders in the conventional sense; rather, they live within forest or shrublands, where they probe soils for a variety of invertebrates. They are small (~100 g) and have a long (5 cm) bill. The surviving forms can fly but are essentially terrestrial in habit, and they are mainly nocturnal. The surviving populations are only where there are no introduced mammalian predators of any kind.

The surviving Chatham Island snipe is a well-differentiated species about 20% smaller than other forms, and it has a 3-cm bill. It was formerly on all islands in the group but now survives only on Mangere and South East Islands. Deposits throughout the Chatham archipelago contain bones of another, larger snipe with a bill 5 cm long. Forbes (1893f) named it *Gallinago chathamica,* and the types (two crania) are in the British Museum in London. The bill length is variable because the premaxilla length (tip to a line drawn across the proximal ends of the outer nasal processes) ranges 77–63 mm (Lowe 1915). Lowe (1915) compared the osteology of *C. pusilla* and *C. chathamica* to that of other snipe and concluded that they are generalized or primitive scolopacines, and as such are living fossils. There are numerous differences in detailed osteology from other snipe, but one notable feature is that, unlike in all other snipe, the descending process of the lachrymal is not connected to the squamosal, so the orbit lacks a continuous bony surround.

Extralimitally, the peculiar New Zealand snipe were possibly also represented in New Caledonia, because Balouet and Olson (1989) referred three bones from there to this genus. An undescribed species has also been reported from Norfolk Island (Meredith 1991), who considered it to be close to "*C. aucklandica,*" but he considered the New Zealand snipe to be monotypic. Thus the *Coenocorypha* snipe constituted a spectacular insular radiation with probably as many as nine species in the New Zealand region. Among terrestrial birds, this is one of the most species-rich radiations in the New Zealand region, exceeding or equaling that of the *Gallirallus* rails.

Family Recurvirostridae

Genus Himantopus *Brisson*
Himantopus novaezelandiae *Gould, 1841 (Black Stilt)*

The black stilt was the only stilt in New Zealand before Polynesian settlement (Holdaway 1995). In the past few hundred years, the Australian pied stilt (*Himantopus leucocephalus*) has colonized and is now widespread, with a population of about 30,000 birds (Sagar et al. 1999). In contrast, black stilt numbers have declined dramatically, and 72 adults were known in the wild in 1994 (Sagar et al. 1999). Of the several causes for the decline that have been put forward, predation at the nest by introduced mammals has been identified as the major factor. Loss of habitat is a lesser problem, and the black stilts remain genetically, morphologically, and behaviorally distinct from pied stilts (Maloney and Murray 1999).

The black stilt has shorter, stouter legs and a shorter bill than the pied stilt. The all-black plumage of the adults would absorb heat better in the cold, windswept habitat of glacial riverbeds and lakeshores. Black stilts dominate pied stilts in interspecific interactions, but they breed in solitary pairs and hence do not have the protection of a colony. Although the species is now confined to the Mackenzie Basin, where pairs breed on open gravel riverbeds or near ponds, it was once widespread. Fossils of adults and juveniles have been found at Pyramid Valley, indicating breeding where the small lake was surrounded by tall matai forest and shrubland. It was also present, and presumably breeding, along narrow streams in forest in inland Hawke's Bay, where its bones have been found in a deposit formed by one of the predatory birds (Worthy and Holdaway 2000).

Family Charadriidae

Genus Charadrius *Linnaeus*
Charadrius obscurus *Gmelin, 1789 (New Zealand Dotterel)*
Charadrius bicinctus *Jardine and Selby, 1827 (Banded Dotterel)*
Charadrius novaeseelandiae *Gmelin, 1789 (Shore Plover)*
Charadrius frontalis *(Quoy and Gaimard, 1830) (Wrybill)*

We replace the generic name *Anarhynchus* with *Charadrius* for the wrybill following Holdaway et al. 2001. The wrybill is a typical *Charadrius* plover, differing only by having a bill that is always bent to the right (Burton 1972) (Fig. 11.1). The deflected bill appears to be a specialization that allows the bird to feed more effectively around cobbles in the gravel riverbeds on which it breeds (Pierce 1979). It feeds on insect larvae that attach to the undersides of stones. In the southern winter, the population migrates to the mudflats of the great harbors in Auckland and Northland. Here individuals feed and roost with migrant waders from the Arctic and with other New Zealand waders that migrate between the North and South Islands, especially Finsch's oyster-

catcher (*Haematopus finschi*). A few wrybill fossils have been found in coastal dunes in the eastern South Island.

*Figure 11.1. Wrybill (*Charadrius frontalis*). Photograph by R. Morris.*

The New Zealand shore plover (*Charadrius novaeseelandiae*) vanished from the main islands in the 19th century (Davis 1994). Its range reduction had begun before that, because fossils have been found in South Island laughing owl sites beside rivers many kilometers from the coast (Worthy and Holdaway 1996b). Historically, the species was confined to coastal habitats. Natural populations of the shore plover are now confined to small islands in the Chatham group, where they live on rocky wave platforms. The present population is about 100 birds (Davis 1994). There are a few managed populations on islands elsewhere. As a wading bird, the shore plover is unusual in building its nest under cover—beneath thick vegetation or under boulders. In its present range, hiding the nest would provide protection from skuas (*Catharacta skua*). Most of its former natural range was outside the known limits of skua breeding; if the habit of constructing a covered nest was also characteristic of the mainland populations, it must be seen as an adaptation to the presence of avian predators in general. When Pacific and Norway rats reached New Zealand, the shore plover was extremely vulnerable to predation at the nest. By the time Europeans started recording birds in Canterbury (the mid-1840s), the former inland populations indicated by fossils in laughing owl sites were gone. The species was confined to the coast, apparently over much of the country, as Cook's second expedition found them in both Fiordland and the Marlborough Sounds, and there were records into the early 19th century in several parts of the North Island.

Both the other species are more "normal" species of plover. The New Zealand dotterel is characteristic of open sand beaches in northern North Island, where the present population is about 1400 birds (Dowding and Murphy 1993). The South Island population apparently bred also in the alpine zone, around mountain tarns, but is now extinct.

On Stewart Island, the remnant southern population (about 65 birds) is endangered and nests in the higher altitude herb fields; it feeds and winters on the beaches and mudflats (Dowding and Murphy 1993). Fossil of the species are rare, found mainly in a few dune deposits, but fossils were recently found in a laughing owl site in the high-altitude (600–700 m) Gouland Downs of northwest Nelson.

The most abundant of the endemic waders, the banded dotterel, has a population of about 50,000 birds, 30,000 of which migrate to Australia in the southern fall and winter (Sagar et al. 1999). The remaining birds in the South Island mostly move north from their breeding grounds on beaches, riverbeds, salt flats, and lake margins. Its fossil remains are rare; all have been from dunes and laughing owl deposits.

Family Stercorariidae

Genus Catharacta *Brünnich*
Catharacta skua *Brünnich, 1764 (Subantarctic skua)*

Family Laridae

Genus Larus *Linnaeus*
Larus dominicanus *Lichtenstein, 1823 (Kelp Gull)*
Larus novaehollandiae *Stephens, 1826 (Red-billed Gull)*
Larus bulleri *Hutton, 1871 (Black-billed Gull)*

Genus Sterna *Linnaeus*
Sterna albostriata *(G. R. Gray, 1845) (Black-fronted Tern)*
Sterna caspia *Pallas, 1770 (Caspian Tern)*
Sterna striata *Gmelin, 1789 (White-fronted Tern)*
Sterna fuscata *Linnaeus, 1766 (Sooty Tern)*
Sterna vittata *Gmelin, 1789 (Antarctic Tern)*
Sterna nereis *(Gould, 1843) (Fairy Tern)*

Genus Anous *Stephens*
Anous stolidus *(Linnaeus, 1758) (Common Noddy)*
Anous tenuirostris *(Temminck, 1823) (Lesser Noddy)*

Genus Procelsterna *Lafresnaye*
Procelsterna cerulea *(Bennett, 1840) (Blue-grey Noddy)*

Genus Gygis *Wagler*
Gygis alba *(Sparrman, 1786) (White Tern)*

The subantarctic skua is presently treated as a race of the great skua. It breeds on the islands to the south and east of New Zealand, including Stewart Island. Pairs have nested on the South Island. Although fossils are rare and are all found in dune deposits or middens, it is likely that skuas were more abundant on the main islands when there were more extensive seal, sea lion, and penguin colonies at which to scavenge.

Of the three species of gull, one, the black-billed gull, is endemic and adapted to nesting on the braided rivers of the South Island. The others have circumpolar, or nearly so, distributions in the Southern Hemisphere. Early reports suggest that the kelp gull was rather uncommon when Europeans arrived. Its numbers have built up substantially in response to the greater food supply from garbage and fishing wastes. Kelp gulls in New Zealand behave rather like corvids elsewhere, scavenging road kills and, frequently, on farmland as well as coastal sites. Most colonies are on coastal dunes or salt flats, but others are on braided riverbeds; some pairs breed beside high country tarns. A few isolated pairs nest on rocky coasts. Fossils of kelp gulls are comparatively rare for such a large bird and almost confined to dune deposits, but they have been found twice in swamp deposits associated with moa and other birds, indicating that the scavenging habit is of long standing.

Major colonies of the New Zealand race of the red-billed gull are located near intense marine upwellings where large quantities of shrimplike euphausiids are concentrated (Mills 1969). Elsewhere, small numbers nest at river mouths and on rocky islands. It is difficult or impossible to separate the bones of this species from those of the endemic black-billed gull. Bones in dunes are likely to be of the coastal red-billed species rather than the black-billed species, which breeds inland and visits the coast in autumn and winter. The black-billed gull is one of the species (with the black-fronted tern, black stilt, and wrybill) that is characteristic of the braided gravel riverbeds of the South Island. It was probably more abundant during the Otiran glaciation, when there were much larger areas of riverbed for nesting and feeding. The spread of introduced weeds such as lupins on the gravel banks has reduced the amount of habitat available for all the specialist riverbed species.

The terns can be divided into a group of tropical species and those that live farther south in temperate New Zealand. The sooty tern, white tern, and three species of noddy nest on the subtropical islands of the Kermadec and Norfolk groups. There are few or no fossils of these species in the large collections made from both natural (probably predator) and archaeologic sites on Norfolk Island. Their abundance and general distribution in the southwest Pacific indicates that the lack of fossils results from the selective nature of the fossil deposits rather than a former absence of any or all of the species.

Further south, there are five species. Two are endemic—the black-fronted and white-fronted terns—and the rest have more widespread distributions. The black-fronted tern, whose total population is about 5000 birds (Heather and Robertson 1996), nests only on a few major rivers in the eastern South Island and is primarily an insectivore. After breeding, some birds move to the North Island, but others stay on the South Island coast. This small tern is threatened by predators such as ferrets and European hedgehogs, which eat the eggs. Fossils of black-fronted terns have been found in laughing owl sites.

The Caspian tern reaches the southern limit of its huge breeding range in Southland. In the past, it may have been confined to the northern harbors and long beaches because its large bones have been found only in Northland and not in extensive collections from dunes farther south. Isolated pairs nest at many points along the coast, but there are colonies of a few dozen pairs at some favored places. The fairy tern is now extinct in the South Island and reduced to fewer than 30 birds on the beaches of the northern North Island. Disturbance and predation have been identified as the main causes of its decline (Heather and Robertson 1996). In the South Island, it was reported in the 19th century as breeding on the riverbeds in Canterbury.

From Northland to the Auckland Islands, the white-fronted tern is the most common tern. It breeds at many points along the coast, in pairs and in large colonies, with some of those sites associated with red-billed gull colonies. Some birds fish up the larger braided rivers of the South Island to as far as the gorges into the foothills of the main ranges. This is one of several New Zealand birds (others being Hutton's shearwater, the Australasian gannet, and the banded dotterel) that migrate west in winter, rather than north. Most juvenile white-fronted terns and a few adults migrate to Australia in autumn and winter. The white-fronted tern is almost entirely restricted to New Zealand as a breeding species, but a few birds have bred in Australia, on islands in Bass Strait, since 1979 (Heather and Robertson 1996). As with the other terns, white-fronted terns are rare in fossil deposits, found mainly in a few dune sites.

Order Psittaciformes: Parrots on Ice

Family Psittacidae

Parrots reach their southernmost distribution on the New Zealand subantarctic islands. On the main islands, they formed an important component of the forest avifauna. One species, the kakapo *Strigops habroptilus,* is one of the largest and most unusual parrots in the world. It is now among the rarest and most endangered of all birds. The large *Nestor* parrots, as well as parakeets in the genus *Cyanoramphus,* are represented by different species on the main islands and on the Chathams and Norfolk Island. *Cyanoramphus* was also widespread in the South Pacific, from New Caledonia to Tahiti, until the effects of human colonization caused the loss of the populations in eastern Polynesia. Parrots and parakeets were, as historical records attest and judging from the abundance of their remains, extremely abundant in habitats from the coastline to the alpine herb fields. Only one species, the yellow-crowned parakeet *C. auriceps,* is still widespread on the mainland islands. The others are either rare and endangered or confined to remnant populations on offshore islands. Species of *Nestor* have become extinct on Chatham and Norfolk Islands in the past 1000 years. The Norfolk bird survived into the mid-19th century, but the as-yet-unnamed Chatham species was extinct before European contact with the island group.

Subfamily Strigopinae

Genus Strigops *G. R Gray*
Strigops habroptilus *G. R. Gray, 1845 (Kakapo)*

The kakapo is, at 2–2.5 kg, the heaviest parrot in the world, and it is the only flightless one (Figs. 11.2, 11.3). It is a folivore and frugivore and has a lek breeding system (Merton et al. 1984), apparently the only

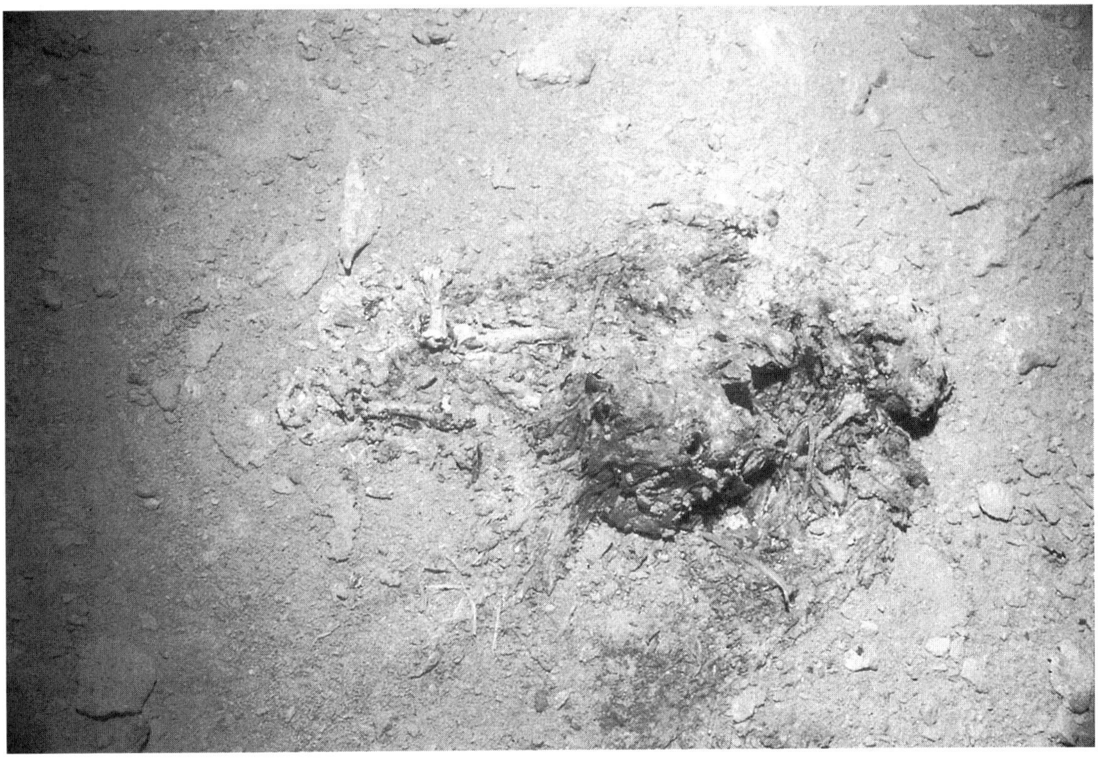

*Figure 11.2. Uniquely preserved kakapo (*Strigops habroptilus*) photographed in situ in Megamania Cave, northwest Nelson. The dry environment has preserved the shape of the wings and tail as organic deposits on the cave floor. Photograph by T.H.W.*

parrot to do so. Males form track systems along ridges and excavate shallow bowls in which they call at night to attract females. The deep, booming call is amplified by inflated air sacs and can carry several kilometers. Females incubate and rear the young alone. Birds walk for kilometers through the forest each night to visit known food trees.

A moss-green, mottled plumage and nocturnal habits would have protected them from the harrier and Haast's eagle. Most birds of prey have a poor sense of smell, but predatory mammals use scent in hunting. In their absence, kakapo developed a strong scent, which is convenient for scientists, who use muzzled dogs to find them. For most kakapo, their scent and extreme placidity have spelled disaster. Despite their large and powerful beak, kakapo are helpless in the face of stoats and cats, which can find them by scent as well as sound. The incubating female is most vulnerable, and the last population on the South Island consisted exclusively of males. Rats, even the small Pacific rat, are also implicated in the decline of the kakapo, especially in the North Island. Video monitors at nests have shown that Pacific rats take eggs and chicks, even when the female is on the nest. Parrot eggs are small for the size of the birds, and kakapo eggs are within the size range that Pacific rats can open (Holdaway 1999b).

Kakapo were abundant in many areas of the main islands before human settlement. They appear to have been most common in areas with higher rainfall (>2000 mm), where the podocarp forests were dominated by rimu trees, which may be important in bringing the

*Figure 11.3. Skeleton of kakapo (*Strigops habroptilus*) from Honeycomb Hill Cave. Photograph by T.H.W.*

females to breeding condition. However, they are also common in fossil sites in wet montane *Nothofagus* forests. Kakapo were not common in areas where matai dominated in the forest. Early records suggesting that the species bred only in alternate years were disbelieved until intensive studies in Fiordland and with the remnant population on Stewart Island showed that females came into breeding condition only in years when the podocarps fruited heavily. Management of the highly endangered species (there were fewer than 65 individuals in 1999) has concentrated on providing supplemental food to the females to maximize their egg production and ensuring they are in predator-free environments.

Their fossils are abundant in deposits west of the axial ranges. Kakapo are the most common species in sites in the extensive karst around Waitomo Caves, as well as in caves in the montane forests of northwest Nelson. Conversely, they are very rare in dune deposits in both main islands. No kakapo fossils have been found on Stewart Island. Their abundance in the karst is also a function of the abundance of pitfall traps and the propensity of the flightless birds to fall in and then find themselves unable to climb out. Unlike other species, which tend to avoid caves and hence being trapped in them, kakapo nested and roosted in cave entrances. Though the birds are now gone, piles of their coprolites can be found in many caves, which are an untapped resource for paleodiet studies.

Subfamily Nestorinae

Genus Nestor *Lesson*
Nestor productus *(Gould, 1836) (Norfolk Island, or Long-billed, Kaka)*
Nestor meridionalis *(Gmelin, 1788) (New Zealand Kaka)*
Nestor *sp. (Chatham Kaka)*
Nestor notabilis *Gould, 1856 (Kea)*

The Norfolk Island kaka became extinct in the 19th century. Few observations were made of its biology before that, except that it nested in holes, often among rocks, and spent much time on the ground. New Zealand kaka feed on fruit, leaves, and insects. They also tap tree sap from selected totara (*Podocarpus totara*) and other trees (Heather and Robertson 1996). To get beetle larvae from branches, kaka will spend over 60 minutes excavating the wood with their strongly hooked beaks, with the reward a single grub.

Kaka are one of New Zealand's most beautiful birds. Their head is pale gray, ear coverts are washed with yellow, cheeks are reddish, upper parts are mainly olive brown, and under parts are olive brown with crimson fringes. The underwings are a bright red. The plumage of North Island birds tends to be darker, especially on the head and back, than that of South Island birds (Higgins 1999). This plumage makes kaka almost invisible in the evergreen forests of New Zealand. But when kaka lift their wings to fly, the bright red wing linings flash a warning through the dark forest. With cryptic plumage and partly nocturnal habits, they were well adapted to avoid predation from the giant harrier. Predatory mammals are another story. The kaka is endangered by stoat predation on the North and South Islands. On Stewart Island, they are, like the brown kiwi, still abundant: there are no stoats or other mustelids there.

The Chatham Island species flew poorly (Millener 1999). In the absence of predators other than the falcon, the species seems to have spent much of its time on the ground.

Fossils of kaka are found regularly throughout New Zealand, although they are most common in dune deposits, especially on main Chatham Island. In cave deposits—usually pitfall traps—bones of the volant kaka are always far less common than those of kakapo. In the North Island, kaka are present in the harrier prey deposits.

Although it is popularly thought of as the "mountain parrot," there are still a few populations of kea in lowland forest, particularly in South Westland (O'Donnell and Dilks 1986). Before the eastern lowland forests and shrublands were destroyed, kea were as abundant there as kaka (Holdaway and Worthy 1997). The kea is now confined to the South Island, and there are no Holocene fossil records of them in the North Island. One Pleistocene record from the Waitomo area shows that the species was more widespread when there were extensive areas of subalpine vegetation during the Otiran glaciation. Kea is the only *Nestor* species in the mid-Otiran to last interglacial deposits on the coast at Cape Wanbrow near Oamaru. Kea bones are common in fossil deposits of Pleistocene age in The Graveyard (Honeycomb Hill Cave), which were deposited when a shrubland and forest mosaic prevailed about the cave. In Holocene sites, kea are widespread in the South Island, from the lowlands to the subalpine zone.

Kea have a generally lighter plumage than kaka. Their plumage is olive to dark olive green with brilliant orange wing linings, but it lacks the silvery white crown. Kea eat fruit and insects, although they also eat flesh. In the past, they have been persecuted as sheep killers. Kea were attracted to the sheep stations in the 19th century and learned to scavenge flesh and fat from drying skins. That was supposed to have led to some individuals perching on the backs of live sheep—especially those trapped in snow drifts—and pecking through the skin on the back to get at the kidney fat. Sheep either died of loss of blood or infection from the wound. It may have been an older habit. Sheep trapped in snow, or resting on the ground, have a similar silhouette to that of a large bird. Kea may have simply transferred from a prey that was no longer around to another that was kindly provided by European settlers.

Supporting this transference hypothesis is the presence of kea in swamp deposits where there were trapped moa. Some of the moa, as well as the goose skeleton from Chatto Creek, exhibit damage to the cranial part of the pelvis, where deep notches have been cut back into the iliac plates from the anterior margin. In some instances, the damage may have been caused by eagles, but it is as likely that kea also sought kidney fat from carcasses of the large herbivorous birds. Further evidence for the carnivorous tendencies of kea is video footage of individuals that have learned to take chicks of Hutton's shearwater from their burrows in the subalpine grassland of the Seaward Kaikoura Range. One was seen to repeatedly look into the burrow, gauge the distance back from the entrance where the shearwater chick was standing, and then dig into the burrow from directly above the chick, killing it with a bite to the head.

Subfamily Platycercinae

Genus Cyanoramphus *Bonaparte*
Cyanoramphus novaezelandiae *(Sparrman, 1787) (Red-crowned Parakeet)*
Cyanoramphus auriceps *(Kuhl, 1820) (Yellow-crowned Parakeet)*
Cyanoramphus unicolor *(Lear, 1831) (Antipodes Island Parakeet)*
Cyanoramphus forbesi *Rothschild, 1893 (Forbes's Parakeet)*
Cyanoramphus cookii *(G. R. Gray, 1859) (Norfolk Island Green Parrot)*
Cyanoramphus malherbi *Souancé, 1857 (Orange-fronted Parakeet)*
Cyanoramphus erythrotis *(Wagler, 1832) (Hochstetter's Parakeet)*

The New Zealand parakeets are a closely related group that seems to have been still actively speciating when humans arrived and began altering many habitats. Extralimital species on New Caledonia (C.

saisseti) and in the Tuamotus (*C. zealandicus, C. ulietanus*) were distinct in their plumage, but all the New Zealand forms resemble each other and are genetically close (Triggs and Daugherty 1996). The most widespread species, the red-crowned parakeet, once had populations from subtropical Lord Howe Island to the Auckland Islands in the subantarctic. It now appears to be almost extinct on the two main islands and is extinct on Lord Howe Island. Long regarded as a race of *C. novaezelandiae,* the Norfolk Island *C. cookii* has now been recognized as a full species (Boon et al. 2001; McAllan and Bruce 1989), part of a subgroup including the New Caledonian *C. saisseti. C. cookii* is now intensively managed. *C. novaezelandiae* is still abundant on nearshore island groups about North Island and South Island. The Lord Howe Island population has also been recognized as a species, *C. subflavescens.* The *Cyanoramphus* parakeets seem to have radiated relatively recently, particularly in New Zealand proper, and reproductive barriers have not been sufficient to prevent hybridization when habitats have been damaged (e.g., Nixon 1994).

Yellow-crowned parakeets are still present in larger tracts of native forest on the main islands. They feed in the canopy and on tree trunks, as well as on insects, seeds (particularly beech seeds in the South Island), and fruit.

Cyanoramphus malherbi was accepted as a distinct species by Triggs and Daugherty (1996), contrary to earlier suggestions that it is a color morph of *Cyanoramphus auriceps* (Nixon 1981; Taylor et al. 1986). Boon et al. (2000) present new biochemical evidence and further biological evidence (assortative mating) supporting the recognition of *C. malherbi* as a distinct species most closely related to *C. novaezelandiae.* However, it survives precariously with a very restricted distribution only in the South Island.

The red-crowned parakeet seems to spend more time on the ground than the yellow-crowned and favors scrub or shrubland (Nixon 1994; Taylor 1975), which might explain its decline to extinction on the main islands, where there are now many predators. On mammal-free offshore islands, red-crowned parakeets are common. All the parakeets are hole nesters and the young are noisy in the nest, which makes it relatively easy for stoats to find them. Breeding success is low during mast years in the *Nothofagus* forests, which are their main remaining habitat, because house mouse numbers increase rapidly, which in turn allows the stoat population to increase. When the mouse population crashes, there is a higher population of stoats, which then switch their attention to birds. Most of the hole-nesting species suffer extremely high rates of predation in beech mast years (Elliott 1996a; O'Donnell 1996). The difference between the extents of the declines of the two species may be related to the greater exposure of the red-crowned parakeet to predation by cats as well as stoats in nonmast years. That effect would be exacerbated by the red-crowned parakeet's being larger and requiring a larger nest hole entrance.

Parakeets were important in the prehuman fauna as insectivores and seed eaters. Most of the seeds would have been crushed by the strong bills, but some at least may have survived to be transported away from the parent plant. They were important prey for laughing owls and falcons. Their bones are common in dune deposits throughout New Zealand. In caves, they are relatively rare in pitfall deposits but can be numerous in predator deposits.

We accept *C. forbesi* as a separate species because of its larger size and distinctive plumage (Higgins 1999; Oliver 1955) and biochemical differences (Boon et al. 2000). The level of hybridization between *C. forbesi* and *C. novaezelandiae* (Taylor 1975) has obscured its status (Boon et al. 2000; Triggs and Daugherty 1996). Some island forms traditionally regarded as races of *C. novaezelandiae* may also be specifically separable. For example, Boon et al. 2001 present biochemical evidence that the Macquarie Island and Antipodes Island populations belong to a single species (*C. erythrotis*) distinct from *C. novaezelandiae* that is found on the Auckland Islands.

The Antipodes parakeet (*C. unicolor*) is endemic to that island, which has an area of 2025 ha. The Antipodes Island parakeet shares the island with the smaller *C. erythrotis*. Both species eat insects as well as grass seeds, and they frequent penguin colonies, where they scavenge from carcasses and broken eggs. They also take storm petrels, which they apparently capture from in or near their burrows. There is no woody vegetation on the island, and both parakeets nest in petrel burrows, in the bases of the *Poa* tussock grasses, which are up to 2 m high, and in the bases of *Polystichum* ferns.

Order Columbiformes: Fruit Pigeons

Family Columbidae

Genus Hemiphaga *Bonaparte*
Hemiphaga novaeseelandiae *(Gmelin, 1789) (Kereru)*
Hemiphaga spadicea *(Latham, 1802) (Norfolk Island Pigeon)*
Hemiphaga chathamensis *(Rothschild, 1891) (Parea, Chatham Island Pigeon)*

Genus Gallicolumba *Heck*
Gallicolumba norfolciensis (Latham, 1801) *(Norfolk Island Ground Dove)*

Pigeons are represented by only two genera in the New Zealand fauna, with three species in one and a single species in the other. A small ground dove has been found as fossils on Norfolk Island and is probably the same species seen and painted by the first European settlers in the late 18th century and given the name *Gallicolumba norfolciensis*. The three species of *Hemiphaga* are large, closely related fruit pigeons. They are widespread in the northern parts of the archipelago, as far south as Stewart Island. Apart from the Chatham Islands, pigeons are not found on any of the southern islands. They are important seed dispersers in forests in their respective ranges. On the main islands, podocarp forests advanced after the Otiran glaciation. It is still unclear whether the spread was from local refugia or from the northern North Island, but the rates, given the limitations of radiocarbon dating, were within the capacity for spread by pigeons, given their normal flight distance and gut passage time for pigeons (M. McGlone, personal communication). Pigeons, which feed heavily on podocarp fruits, could have been instrumental in the rapid extension of the podocarp forests with the ameliorating climates after 14,000 years before the present. With the kokako and piopio, they were also a major seed disperser in the forest and shrublands.

The New Zealand pigeon is abundant in fossil deposits east of the main ranges in both islands. This abundance is linked to the presence of faunas accumulated by predators because pigeons were not only favored prey of the extinct harrier but were also taken by the laughing owls and New Zealand falcon. Many of the sites where the species is most common were under matai (*Prumnopitys spicata*) forest. Although the bird is common today (where not illegally hunted) in podocarp forests dominated by rimu (*Dacrydium cupressinum*), the matai forest and associated areas with abundant seasonal food resources provided by the foliage and flowers of *Sophora* spp., *Houheria* spp., and *Aristotelia* spp. seem to have been the preferred habitat. The matai is the richest fruiting podocarp. Numbers were probably much higher in that forest than in rimu or other forests today.

Pigeons were also abundant on the Chatham Islands. Rich fossil deposits in the Chatham Island dunes are often dominated by pigeon bones, unlike deposits on the mainland. The Chatham Islands species is conspicuously larger than the mainland or Norfolk Island species, and the bill is larger and more brightly colored. After a prolonged decline, the population is recovering after measures were taken to reduce the number of feral cats and rats in the species' stronghold at the southern end of main Chatham Island. The only natural predator of the pigeon on the Chathams was the local population of the New Zealand falcon (*Falco novaeseelandiae*), which is now itself extinct.

Farther north, the New Zealand pigeon reached Raoul Island in the Kermadec group. It was reported in the 19th century by European settlers, but confirmation was forthcoming only recently with the discovery of a humerus indistinguishable from the mainland form (Worthy and Brassey 2000). Volcanic activity is regular and violent on the islands. The eruptions are of rhyolite, with accompanying pyroclastic flows that devastate and bury much of the landscape (Latter et al. 1992; Wilson 1985; Wilson and Walker 1985). It is likely that most, if not all, the terrestrial avifauna is subjected to cycles of extinction and recolonization. This is seen in the lack of endemicity in the avifauna in contrast to the older, basaltic islands of the Norfolk group 600 km to the west.

New Zealand pigeons were a favored food of Māori from the earliest settlement. The dramatic reduction in the area of favored eastern forests before A.D. 1500 would have diminished the population from the precontact level. Mainland populations rapidly declined further with the introduction of firearms in the 19th century. Restrictions on hunting pigeons were made within 20 years of the establishment of a European government in 1840. However, they are still illegally hunted, especially in widespread parts of the North Island.

Order Cuculiformes

Both the New Zealand cuckoos are migratory. They breed in New Zealand and spend the southern winter in the Pacific Islands. Both species are parasitic on small passerines. They are rare in fossil deposits. The larger long-tailed cuckoo was probably immune from predation by laughing owls, even though it is partially nocturnal, because it lives predominantly in the canopy, where the owl did not regularly hunt, and because the cuckoo is a large and powerful predator in its own right. It has not yet been found among the food remains of Eyles's harrier. Shining cuckoos are diurnal, cryptic, and have ventriloquial calls, and hence they probably escaped predation by both the major accumulators of food residue deposits.

Family Cuculidae

Genus Eudynamys *Vigors and Horsfield*
Eudynamys taitensis *(Sparrman, 1787) (Long-tailed Cuckoo)*

Genus Chrysococcyx *Boie*
Chrysococcyx lucidus *(Gmelin, 1788) (Shining Cuckoo)*

The long-tailed cuckoo parasitizes the species of *Mohoua, M. albicilla* in the North Island, and *M. ochrocephala* and *M. novaeseelandiae* in the South Island. Other hosts are apparently rarely used. Little is known of its behavior or breeding, however. After spending the southern spring and summer in New Zealand, it winters throughout much of the South Pacific, from the Bismarck Archipelago east to the Marquesas. Long-tailed cuckoos are insectivores, but they also take lizards and eggs and young of both the host species and other birds.

The shining cuckoo is the smaller of the two New Zealand cuckoos and is insectivorous. Its usual breeding host species is the grey warbler (*Gerygone igata*). It winters in the western South Pacific from the Bismarcks to the Solomon Islands. Shining cuckoos reach the Chatham Islands, where they parasitize the Chatham Island warbler (*Gerygone albofrontata*).

Order Passeriformes: Southern Songbirds
Suborder Oligomyodi

Family Acanthisittidae (New Zealand Wrens)

Genus Acanthisitta *Lafresnaye*
Acanthisitta chloris *(Sparrman, 1787) (Rifleman)*

Genus Xenicus *G. R. Gray*
Xenicus longipes *(Gmelin, 1789) (New Zealand Bush Wren)*
Xenicus gilviventris *Pelzeln, 1867 (New Zealand Rock Wren)*

Genus Traversia *Rothschild*
Traversia lyalli *Rothschild, 1894 (Lyall's Wren)*

Genus Pachyplichas *Millener*
Pachyplichas yaldwyni *Millener, 1981 (South Island Stout-legged Wren)*
Pachyplichas jagmi *Millener, 1981 (North Island Stout-legged Wren)*

Genus Dendroscansor *Millener and Worthy*
Dendroscansor decurvirostris *Millener and Worthy 1991 (Long-billed Wren)*

The acanthisittid wrens are small, apparently almost tailless birds, which form a suborder of their own within the songbirds that is endemic to New Zealand. Their relationships have been debated at length; at present, they are seen as a separate, early passerine branch that was isolated in New Zealand. Only *Acanthisitta chloris* and *Xenicus gilviventris* survive today, although *X. longipes* and *Traversia lyalli* both survived into the European era. *X. longipes* became extinct in the early 1960s, when the last populations on offshore islands were eliminated by rats, and *T. lyalli* did not survive the removal of forest from its last refuge island in 1894.

The acanthisittids range from less than 10 g to an estimated 40 g in body mass. Plumage patterns vary from smooth browns, grays, and greens to dappled, but apart from small patches of yellow, the living species are not brightly colored. Both extant species are almost entirely insectivorous. In the surviving species of *Xenicus*, the legs are proportionately large, as they were in the extinct genera. *Acanthisitta* is almost entirely arboreal, but *X. gilviventris* is largely terrestrial.

The wrens are, in their own way, as characteristic of the New Zealand fauna as are the moa. Like those large herbivores, the wrens are of great antiquity, and they display a radiation of the same order as the moa. According to biochemical evidence, both groups began their present radiation about the same time, in the late Oligocene, when New Zealand was reduced to a series of low islands (Cooper and Cooper 1995). To cap it all, four of the six wren species were flightless. Only one other flightless songbird is known, a bunting (*Emberiza*) known from fossil remains on the Canary Islands (Rando et al. 1999).

The rifleman (*Acanthisitta chloris*) is common in the South Island and even lives in areas of light forest and introduced vegetation in farmland and the outskirts of cities in some areas (Fig. 11.4). It is the smallest of the seven species and builds its domed nest in tree crevices (Heather and Robertson 1996). In the North Island, it has a local distribution on the mainland and is absent from much of the island (Heather and Robertson 1996).

The rock wren is now almost confined to subalpine and alpine shrubland and herb fields. It reaches the coast to the west of the divide, where boulder-strewn rivers provide suitable habitat. It appears to have formerly lived in lowland forest and in both main islands, if the material from the North Island is correctly identified. It is now restricted to habitats with few or no rodents and where there are piles of rocks and dense vegetation in which to shelter from other predators.

*Figure 11.4. Rifleman (*Acanthisitta chloris*), New Zealand's smallest bird. Photograph by R. Morris.*

The other species of *Xenicus,* the bush wren (*X. longipes*), is known from fossil deposits on both main islands and was present historically on Stewart Island. It was last known from one population on Big South Cape Island in the early 1960s, but the population was endangered by the introduction of black rats to the island and a translocation attempt was not successful (Bell 1978).

A remnant population of a fourth species, Lyall's wren (*Traversia lyalli*), was discovered on Stephens Island, at the northern entrance to Cook Strait, by the keepers of a lighthouse established in 1894. Only a few (16–18; Millener 1989) specimens were collected—by a keeper's cat, by the keepers themselves, and by professional collectors who then rushed to the island. As with the rock wren, the Stephens Island population of Lyall's wren was a relict. Its peculiarly flattened cranium, flatter, wider bill, and distinctive leg bones are not uncommon in laughing owl deposits and other sites on both main islands (e.g., Worthy and Holdaway 1994b, 1996a).

Although the cat has received all the blame for the extinction that so rapidly followed the wren's discovery, at the same time, the island was cleared of much of its forest to establish a farm for the lighthouse station. The human collectors and almost total deforestation are rarely mentioned as causes of the wren's extinction, but they are, in this instance, as likely as the direct predation to have caused it. Three other species that are now either rare or extinct were found on the island when surveyors for the new station visited Stephens Island in 1893. South Island piopio, kokako, and saddleback were all present, as were, in the absence of Pacific rats, large weevils, giant weta, Hamilton's frog (*Leiopelma hamiltoni*), tuatara, geckos, and seabirds such as fairy prions (*Pachyptila turtur*). The island was preserved in the following year and cats were removed, but not before all the forest birds had become extinct and the frogs and lizards reduced to remnant populations. Thus the last chance of survival of a remnant of an undamaged New Zealand forest ecosystem was lost forever—almost in sight of the country's capital, where debate on the establishment of reserves was gathering steam (Galbreath 1989; Holdaway 1996).

The stout-legged wrens (*Pachyplichas yaldwyni* and *P. jagmi*) were described from material from the South and North Islands, respectively. The differences between them were simply matters of size, and the larger samples now available suggest that only a single, flightless species was involved. The systematics of the genus needs reexamination. Bones of the stout-legged wren had been found in various sites, such as Castle Rocks, as long ago as the end of the 19th century. In the absence of a thorough analysis and sufficient skeletal material of other songbirds in the local museums, most of these bones were identified as other, living, species familiar to the workers dealing with the bones, such as fernbirds *Bowdleria*. New material obtained from Honeycomb Hill Cave in the 1980s allowed a full description and the recognition of a new genus of wren (Millener 1988). Since then, the species has been found in many laughing owl sites, and often in pitfall deposits as well, in both islands.

Dendroscansor decurvirostris, the extinct, flightless, long-billed wren, is the rarest fossil species in the New Zealand avifauna. It is known from only four sites and from fewer than six individuals (Millener and Worthy 1991; Worthy 1998d; Worthy and Holdaway 1994b). It was apparently absent from the lowlands of the eastern South Island, where rich laughing owl sites would surely have contained it if it had been present. It is not known from the few North Island sites with rich faunas of microvertebrates. Paleoecologic reconstructions of the areas around the sites where it has been found suggest that the long-billed wren lived in high-altitude shrubland and perhaps in the montane beech forest (Millener and Worthy 1991; Worthy and Holdaway 1994b).

Suborder Polymyodi

Family Motacillidae

Genus Anthus *Bechstein*
Anthus novaeseelandiae *(Gmelin, 1789) (New Zealand Pipit)*

Ostensibly, the New Zealand pipit has one of the widest breeding distributions of any songbird. It was named from specimens collected in New Zealand, but, as Richard's pipit, it breeds in Australia, Asia, and western Europe and is even on the British bird list. Most populations over this vast area are probably distinct local species masquerading under the name of the New Zealand bird. Indeed, in New Zealand, there are several island populations of pipits, which have usually been treated as races of the mainland species, or as having no differences at all, which warrant recognition as full species. Those on the subantarctic southern islands have quite distinct plumage. They are isolated, live in harsh environments, and exploit food resources such as the insects associated with seal and penguin colonies.

The New Zealand pipits are, along with the fernbirds, the only songbirds in New Zealand that have the double tricipital fossa in the humerus that is a characteristic of the Northern Hemisphere radiation of passerines (Sibley and Ahlquist 1990: 571). All the other groups have a single fossa, which is shared with the southern corvid radiation.

In the fossil record, pipits are indicators of the presence nearby of shrubland or tussock grassland. They do not live on close sward but prefer some woody vegetation in their territory. Today, they live in subalpine grassland and shrubland, rough pasture on cleared land, and dunes, and they can penetrate dense forests along road verges. Undoubtedly they exploited such line habitats along streams and land slips in the past.

Family Sylviidae

Genus Bowdleria *Rothschild*
Bowdleria punctata *(Quoy and Gaimard, 1830) (New Zealand Fernbird)*
Bowdleria rufescens *(Buller, 1869) (Chatham Island Fernbird)*
Bowdleria caudata *(Buller, 1894) (Snares Island Fernbird)*

Fernbirds are small songbirds with frayed tails; the birds resemble the Australian grassbirds and the Eurasian and African *Cisticolas*. They are typical at present of swamps and low shrubland within their range —anywhere there is dense, tangled vegetation that provides protection from predators. Occasionally, they have been found in tall forest (O'Donnell and Dilks 1986), and it is probable that they had broader habitats in the past, before mammalian predators were introduced.

The fernbird radiation is cryptic; the races and species are separated by subtle differences in plumage. There is still doubt about the taxonomic status of the most distinctive population of the mainland species, the Codfish Island fernbird. The Chatham Island and Snares

Island species are quite distinct. Unfortunately, the Chatham fernbird became extinct around the turn of the 20th century, when the forest and shrubland on its last refuge island (Mangere) were removed for stock grazing and cats were introduced.

All fernbirds share a highly modified pelvis and strong hindlimbs that appear to be specialized for a semiterrestrial lifestyle (Olson 1990a); the large surface area and down-curved posterior portion of the pelvis provide attachment points for strong muscles of the legs and toes. Fernbirds need that strength for grasping tillers while clambering through dense thickets of rushes and sedges.

Family Pachycephalidae

Genus Mohoua *Lesson*
Mohoua albicilla *(Lesson, 1830) (Whitehead)*
Mohoua ochrocephala *(Gmelin, 1789) (Yellowhead)*
Mohoua novaeseelandiae *(Gmelin, 1789) (Brown Creeper)*

The three species in the genus *Mohoua* are small, arboreal insectivores. Two species—the yellowhead and brown creeper—live in the South Island, whereas the whitehead is confined to the North Island and its outliers.

The yellowhead, often known as mohua, is now absent from most of its former range on the South Island, and it has been extinct on Stewart Island since 1930 (Heather and Robertson 1996). Yellowheads and whiteheads prefer high forest (red beech *Nothofagus fusca* for yellowheads), and brown creepers can also be found in lower scrub, even isolated patches in rough grassland.

Mohoua live in family parties. All three species are cooperative breeders: the young of the previous year help rear the new brood. They are also foster parents for the endemic long-tailed cuckoo, but few cuckoos are now able to find nests of the yellowhead and most depend on the other two.

The two smaller species feed by gleaning in the branches and in the leaves at the end of branches. Yellowheads do too, but they are specialized to feed in the debris that accumulates in the crotches of branches. A bird grips the bark with one foot, braces its tail against the bark, and scratches with the other foot, stopping to look periodically and pick up its invertebrate prey. The skeleton of the yellowhead, and to a lesser extent that of the other two species, is modified for strong grasping, with the pelvis bent dorsoventrally and the posterior part broadened to hold large muscles. The pelvis is modified in much the same way in the other two species, which is similar to that in the fernbirds, but *Mohoua* belongs to the corvid radiation, and the similarity is convergent. The feet are large in proportion to the body, and the shafts of the tail feathers project beyond the barbs and are stiffened, which provides a prop for the body. The shower of dead leaves, bark, and lichen falling to the forest floor from a feeding flock of yellowheads alerts both human observers and predators to the presence of the feeding birds (Soper 1972a: 49). All the species need strong legs and powerful toes, because whiteheads and brown creeper hang upside down while gleaning the undersides of leaves.

Members of the *Mohoua* species all call constantly while moving around. Brown creeper males sing antiphonally. Two males sit a few centimeters apart and sing the same song loudly at each other, one bird following the other with a delay of a fraction of a second.

On the mainland, the *Mohoua*s used to form mixed-species flocks with saddlebacks, fantails, and parakeets. Mixed-species flocks still form in the few places where all these species coexist, such as Little Barrier Island (Heather and Robertson 1996). Feeding flocks on the North and South Island now include silvereyes (*Zosterops lateralis*), which colonized from Australia in the mid-19th century.

Bones of all the *Mohoua* species were rare in collections until laughing owl deposits were exploited; the bones were found to be common in these deposits.

Family Acanthizidae

Genus Gerygone *Gould*
Gerygone igata *(Quoy and Gaimard, 1830) (New Zealand Grey Warbler)*
Gerygone albofrontata *Gray, 1844 (Chatham Island Warbler)*

The two small warblers are part of a radiation centered on Australia and New Guinea; up to 19 species have been described. They are insectivorous, gleaning from leaves and often hovering off the end of branches to reach the outermost leaves. The New Zealand warblers build pendulous domed nests with a side entrance, and both species are the main hosts of the shining cuckoo *Chrysococcyx lucidus*.

Gerygone has some of the smallest bones of any New Zealand bird, and its bones, like those of other small passerines, are rare in deposits. Most have come from recent discoveries in laughing owl sites.

Family Monarchidae

Genus Rhipidura *Vigors and Horsfield*
Rhipidura fuliginosa *(Sparrman, 1787) (Grey Fantail)*

Fantails are small sallying and aerial flycatchers that also search for food among litter on the forest floor. The New Zealand populations of the grey fantail are still abundant, especially in altered forests. Fantails are multibrooded and short-lived, which are continental (and Australian) strategies that allow the species to coexist with a suite of mammalian predators. Their fossils are extremely rare except in predator deposits. Some individual owls seem to have concentrated on a few prey species; one near Karamea on the West Coast of the South Island had eaten many individuals of the fantail.

Family Petroicidae

Genus Petroica *Swainson*
Petroica australis *(Sparrman, 1788) (South Island Robin)*
Petroica longipes *(Garnot, 1827) (North Island Robin)*

Petroica traversi *(Buller, 1872) (Black Robin)*
Petroica multicolor *(Gmelin, 1789) (Scarlet Robin)*
Petroica toitoi *(Lesson, 1828) (North Island Tomtit)*
Petroica macrocephala *(Gmelin, 1789) (South Island Tomtit)*
Petroica dannefaerdi *Rothschild, 1894 (Snares Island Tomtit)*

The Australian robins in the genus *Petroica* have reached most islands in New Zealand, from the subtropics to the subantarctic. Petroicids radiated twice in the New Zealand archipelago (Fleming 1950). Most different from the Australian forms are the two robins, *Petroica longipes* and *P. australis* on the North and South Islands and the black robin *P. traversi* on the Chathams. These species are larger than the so-called tomtits, which have usually been considered to be races of *P. macrocephala*. Most are less abundant than when Europeans arrived but are still widespread. The exception is the black robin, which in the 1980s was reduced to a single fertile female ("Old Blue," named after her color band combination) and a few males. Habitat deterioration had threatened the remaining natural population on a tiny precipitous island. Relocation to another predator-free island did not stop the decline, but when Old Blue was the sole reproductive female, intensive, innovative management techniques, including cross-fostering and double-clutching, brought the species back from the dead. The total population is now more than 250 birds on two islands (D. Merton, personal communication).

The mainland robin populations have declined markedly since the 19th century. At present, numbers vary from year to year in the South Island east of the Southern Alps, probably in response to changes in the abundance of stoats and rats. On predator-free offshore islands, robins are still abundant. The susceptibility of robins to rat predation was demonstrated when the population of robins on Big South Cape Island off Stewart Island vanished in the early 1960s after black rats were accidentally introduced. Codfish Island, a few kilometers north of Big South Cape Island, is larger and has a more diverse vegetation, and snipe, saddlebacks, bush wrens, and robins had never been reported from there. All those species were present on Big South Cape before the rat plague (Bell 1978). However, there were Pacific rats on Codfish Island and not on the smaller island. Although Pacific rats probably reached Codfish Island only at the beginning of the 19th century, with the advent of modern muttonbirding (harvesting the young of some species of petrel for food; Anderson 1997a; Holdaway 1999b), they were almost certainly responsible for the extinction of those and perhaps other species and for the diminution of others, such as Cook's petrel and the South Georgian diving petrel (Holdaway 1999b).

Tomtits are found on all the larger island groups. The range of variation suggests that most of the island forms warrant species status (Holdaway et al. 2001) although all are presently regarded as subspecies of *P. macrocephala*. The black tomtit, *P. dannefaerdi*, is confined to The Snares, an island group of less than 3 km^2. The others enjoy larger ranges, all of which are on islands with at least low forest of southern rata (*Metrosideros umbellata*) and tree daisies (*Olearia* spp.). Tomtits live in forest or tall shrubland. They are insectivorous and spend some time on the ground, but not as much as the robins (McLean et al. 1994). They have up to three broods each year; with a normal clutch of four eggs, they can withstand higher rates of predation than other New Zealand passerines.

Fossils of *Petroica australis* and *P. longipes* are common in pitfall deposits, which is a reflection of the species' preference for feeding on the ground, among litter and fallen branches. Bones of *P. macrocephala* are regularly found in laughing owl sites but are rare in other kinds of deposit. There is no fossil record yet of the genus on other island groups within the New Zealand archipelago.

Family Zosteropidae

Genus Zosterops *Vigors and Horsfield*
Zosterops albogularis *Gould, 1837 (White-breasted White-eye)*
Zosterops tenuirostris *Gould, 1837 (Slender-billed White-eye)*

There are no honeyeaters on the Norfolk Island group, but there are two white-eyes which may have occupied the small honeyeater niche to the exclusion of later immigrants. The Australian silvereye *Zosterops lateralis* colonized New Zealand in the 19th century but has not managed to enter natural forests where there is a relatively complete avifauna, including all three endemic honeyeaters, as on some island reserves (Diamond and Veitch 1981; McCallum 1982). It is excluded from Aorangi Island in the Poor Knights group, which has a high population of bellbirds (*Anthornis melanura*) (Bartle and Sagar 1987).

Family Meliphagidae (Honeyeaters)

Genus Prosthemadera *G. R. Gray*
Prosthemadera novaeseelandiae *(Gmelin, 1788) (Tui)*

Genus Anthornis *G. R. Gray*
Anthornis melanura *(Sparrman, 1786) (New Zealand Bellbird)*
Anthornis melanocephala *G. R. Gray, 1843 (Chatham Island Bellbird)*

Genus Notiomystis *Richmond*
Notiomystis cincta *(Du Bus, 1839) (Hihi)*

Four species of honeyeater are known in the present and fossil faunas of New Zealand. One, the hihi or stitchbird, was found only in the North Island. The Chatham Island bellbird was confined to that group. The other two were more widespread: the tui still breeds as far north as the subtropical Kermadec Islands and as far south as the subantarctic Auckland Islands. Unusually, there is no historical or fossil record of a honeyeater on Norfolk Island, where their place seems to have been occupied by two endemic species of white-eye (*Zosterops*). Honeyeaters have reached most of the other island groups in the Southwest Pacific and even as far north as Hawaii.

The tui is the largest of the New Zealand species. It is regularly found in fossil deposits, especially predator sites, even in the eastern South Island, where the species is rare today. Today, they are most abundant in the North Island. Tuis dominate the other two honeyeaters at flowering trees and are aggressive toward other species as well. The bellbird is complementary in distribution and abundance to the tui. It is extinct in Northland and uncommon in most places farther south in the North Island. It is abundant in the South Island, even east of the main ranges. Bellbirds are less common than tuis in most fossil deposits, which is probably because their bones are much smaller.

Hihi were found throughout most of the North Island at European contact. Like the bellbirds, they vanished first from the northern North Island and were extinct on the North Island altogether by about 1885 (Heather and Robertson 1996). A small population survived on Little Barrier Island off the northeastern coast of the North Island and has recovered well after the feral cats were removed in the early 1980s (Veitch 1982, 1983).

Family Turnagridae

Genus Turnagra *Lesson*
Turnagra tanagra *(Schlegel, 1865) (North Island Piopio)*
Turnagra capensis *(Sparrman, 1787) (South Island Piopio)*

These two distinct species have usually been regarded as races of a single species in recent literature. The plumage differences alone are sufficient reason to regard them as separate species (Holdaway et al. 2001). Both the specific names were founded in error. Anders Sparrman, naturalist and artist on James Cook's first voyage to the South Seas and New Zealand, mixed his specimens and mistakenly thought that the South Island piopio had been collected in South Africa. The North Island species was named in the belief that it was related to the tanagers of North and South America. Instead, the piopios are a uniquely New Zealand group whose affiliations within the songbirds are still controversial. Their almost certain extinction is therefore one of the major losses of the New Zealand late Holocene extinction event.

The piopios have often been considered to be part of the Australasian pachycephaline (whistler) lineage (e.g., Mayr 1967; Sibley and Monroe 1990). Alternatively, a relationship with the bowerbirds and cnemophiline birds of paradise has been favored (Olson et al. 1983). Recently, Christidis et al. (1996) presented biochemical evidence (sequences of the mitochondrial cytochrome *b* gene) that they interpreted as placing *Turnagra* closer to the bowerbirds (Ptilonorhynchidae) than to the golden whistler (*Pachycephala pectoralis*, a pachycephaline) or the birds of paradise (Paradisaeidae).

Features of the skeleton, pterylosis, and plumage pattern resemble those of other passerine groups, particularly the whistlers. The plumage pattern of the North Island piopio is very close to that of the brown whistler (*Pachycephala simplex*) of Australia—a plumage pattern far from typical of bowerbirds; nor is it as colorful as a bird of paradise. A feature of the postcranial skeleton that may indicate the relationships of the piopios is the prominent, proximal caudal foramen in the femur. This foramen is restricted to a few families of passerines, the pachycephalines (Pachycephalidae: *Colluricincla, Pachycephala, Oreoica,* and *Pitohui*), bulbuls (Pycnonotidae: *Hypsipetes, Criniger, Nicatur,* and *Phyllastrephus*), orioles (Oriolidae: *Sphecotheres* and *Oriolus*), some Madagascan vangas (Vangidae: *Falculea, Leptopterus,* and *Tylas*), *Newtonia* (Sylviidae: also from Madagascar), *Chloropsis* (Irenidae: a leafbird), and a few honeyeaters (Meliphagidae: *Melidectes* and *Meliphaga*). Of these, the pachycephalines, orioles, and honeyeaters are geographically closest to New Zealand.

One important feature of the biology of the whistlers is that they seem to be good at crossing ocean barriers. There are species of whistler—and apparently many races of a few species—on many islands and island groups, as far east as Fiji and Samoa. A population of whistlers on Norfolk Island, 600 km to the northwest of the North Island, is usually treated as a race of the golden whistler, which is almost ubiquitous in the Southwest Pacific.

Piopio eggs have a pattern of spots at the larger end, whereas the eggs of bowerbirds are polished and immaculate, and those of birds of paradise are heavily streaked. Piopio built firm, well-constructed cup nests within a few meters of the ground, but bowerbirds build skimpy nests of twigs high up in trees. Like pigeon nests, those of bowerbirds often are so light in construction that the eggs can be seen from below. Whereas piopio laid an egg clutch of two to four eggs, birds of paradise and bowerbirds usually lay one or two eggs.

The songs of both species of piopio were described as sustained and varied, and among the most beautiful in the New Zealand forest (Buller 1888). Buller's description of the song of a captive bird and that of the North Island species as being composed of at least six bars (Andersen 1926) suggests a song of some complexity—indeed, of great power and beauty. In strong contrast, bowerbird and bird of paradise calls and songs are uniformly loud and harsh or peculiar. None of them sings a song attractive to the human ear.

Although few European observers were lucky enough to be familiar with the piopios, their reports do not contain any hint of the building of bowers or use of display areas. The lack of these alone is not sufficient evidence to deny a relationship with bowerbirds, but in concert with the other differences, we doubt a close phylogenetic link. Perhaps a wider survey of the pachycephaline lineage will find further genetic heterogeneity than is presently apparent. The apparent closeness in the mitochondrial DNA of *Turnagra* and the bowerbirds may be an artifact of the groups chosen by Christidis et al. (1996) for their comparisons.

It would help if there were live birds to study and to sample for biochemical and behavioral evidence of their relationships. Both species are now regarded as extinct, even though rumors of sightings persist. Both were widespread in their respective islands until the 1890s, but their populations were reported to be declining before that (Buller 1888). In the North Island, the piopio vanished first from the northern peninsula and was last reported in the dense forests of the central hill country. Historically, the South Island piopio was found mainly to the west of the axial ranges, where it was reported to be abundant until the 1880s (Buller 1888), but that was probably because most of the forest was still there. In the fossil record, the species is more characteristic of the drier eastern forests (Worthy and Holdaway 1996b). Potts (1882) studied it in the Canterbury mountains and on the West Coast, although he noted that it had largely left the forests of Banks Peninsula by 1870. He saw them hawking insects over a large river, flying out from the riparian shrubs in the heat of a summer afternoon. Others saw piopio take insects and fruit, and an early conservationist recorded that they ate spiders and food scraps and could shell oats (Henry 1903). The birds were apparently omnivorous. The bill was stout but only superficially like that of the catbirds: there is no "tooth" on the cutting edge.

Callaeatidae

Genus Callaeas *J. R. Forster*
Callaeas wilsoni *(Bonaparte, 1851) (North Island Kokako)*
Callaeas cinerea *(Gmelin, 1788) (South Island Kokako)*

Genus Philesturnus *Geoffroy St.-Hilaire*
Philesturnus rufusater *(Lesson, 1828)*
(North Island Saddleback)
Philesturnus carunculatus *(Gmelin, 1789)*
(South Island Saddleback)

Genus Heteralocha *Cabanis*
Heteralocha acutirostris *(Gould, 1837) (Huia)*

The family Callaeatidae includes five species of crowlike or starlinglike songbirds that all have fleshy wattles at the angle of the jaw, strong legs, and poor flight. Fleming (1962a) considered them to be part of the original avifauna of New Zealand. Their relationships seem to be with the southern corvid radiation, but their systematic position is unresolved (Sibley and Ahlquist 1990).

All the callaeatids were vulnerable to introduced predatory mammals. The North Island kokako is the only callaeatid that is still present on the main islands of New Zealand (Fig. 11.5). Its populations are endangered by predation of adults, chicks, and eggs by rats, stoats, and brush-tailed possums (Heather and Robertson 1996; Innes and Hay 1995). Intensive conservation and management have stabilized a few populations. The only "safe" populations are those on offshore islands. The South Island kokako is extremely rare, if not extinct. Reports from the 19th century, when it was still present in many forested areas, suggest that it preferred a different habitat than did the North Island species.

Kokako were one of the largest prey species of the laughing owl (Holdaway and Worthy 1996), and because of their size, they were taken relatively infrequently. In the North Island harrier prey deposits such as Hukanui 7a, the North Island kokako was one of the common species, after the New Zealand pigeon and Finsch's duck.

North and South Island saddlebacks are similar enough to have

*Figure 11.5. A male North Island kokako (*Callaeas wilsoni*) showing overlapping wattles that are characteristic of the family. Photograph by R. Morris.*

been considered to be races of the same species in most recent lists. Their plumage differences are significant, however, including the presence of a juvenile plumage in the South Island *P. carunculatus*. There are differences, too, in their breeding biology and behavior that seem to make it impossible for South Island saddlebacks to coexist with Pacific rats on small islands (Lovegrove 1996a). Attempts to translocate South Island saddlebacks to islands with populations of Pacific rats have all failed. North Island saddlebacks, although markedly smaller than its southern relative, have thriving populations on islands with and without the rat (Lovegrove 1996a). Even on the North Island itself, the North Island saddleback could not contend with predation by more than one rodent species, and Lovegrove (1996b) suggests that its population there had started to decline soon after Norway rats were established, early in the 19th century. Like starlings, the saddlebacks are specialized for probing, and the stout, chisel-pointed beak can be opened after insertion to pry apart moss or loosen a piece of bark to expose an insect. A long retroarticular process on the mandible gives the jaw opening muscles better leverage.

Saddlebacks are one of the most common passerines in fossil deposits, particularly in predator sites in areas presently or formerly under dense forest.

Largest of the callaeatids was the huia (*Heteralocha acutirostris*), which was confined to the North Island (Fig. 11.6). It was uncommon when Europeans arrived, although in some areas such as the so-called 70-Mile Bush in the Wairarapa, it was familiar to the settlers felling the forest to make dairy farms (Phillipps 1959). By European times, its range had already shrunk away from the northern peninsula, where it was part of the fauna preserved in the Tokerau sand dunes. Farther south, there are few historic records from areas west of the North Island axial ranges (Medway 1968; Phillipps 1959). Despite reports of living huia in the north of the South Island (Phillipps 1959), the species has never been identified in the rich fossil deposits south of Cook Strait (Worthy and Holdaway 1993, 1994b, 1995, 1996b). Only a few individuals are known from the extensive pitfall deposits in the karst surrounding Waitomo Caves. Huia are also unknown or rare in fossil deposits in the central North Island in Hawke's Bay, where Eyles's harrier was taking pigeons, kokako, and kaka, all birds about the same size as a huia. It seems to have preferred habitats that are not well sampled by the deposits known at present.

The huia is one of the more widely known of New Zealand birds because of the extreme sexual dimorphism of the bill, which is the most divergent between sexes known in any bird species, and because the white-tipped black tail feathers were highly prized by the Māori. Males had the shorter bill, which was very like that of the saddlebacks, but in the female, it was much longer and decurved. The difference was not only in the bone; the rhamphotheca grew way past the end of the bony maxilla and mandible to produce a pliable implement able to penetrate holes made by wood-boring beetle larvae. The significance of the difference to the feeding strategies of the sexes has been debated for over a century (Jamieson and Spencer 1996; Moorhouse 1996), but as the huia became extinct early in the 20th century, the problem is unlikely to be resolved. Even though the huia is mentioned in many textbooks of ornithology and biology because of its beak, little is known about the bird itself.

Family Corvidae

Genus Corvus *Linnaeus*
Corvus moriorum *Forbes, 1892 (Chatham Island Raven)*
Corvus *sp. (New Zealand Raven)*

The New Zealand ravens are another group whose taxonomy still suffers from H. O. Forbes's inability to finish a scientific project or

*Figure 11.6. The magnificent elongate bill of a female huia (*Heteralocha acutirostris*). Photograph by R. Morris.*

paper. The name for the Chatham Island bird is well enough established, although his later removal of the species to a new genus—*Palaeocorax*—was unwarranted on present evidence. The mainland birds were significantly smaller than those in the Chathams, and the South Island birds were rather larger than those in the North Island. The species name *antipodum* was proposed by Forbes for the North Island form, but, as with the harriers, it is not certain which bones he was examining at the time (Forbes 1893e). The distinction of the Chatham raven at the species level fits with the general pattern that birds there had been isolated from the parent species on the main islands for

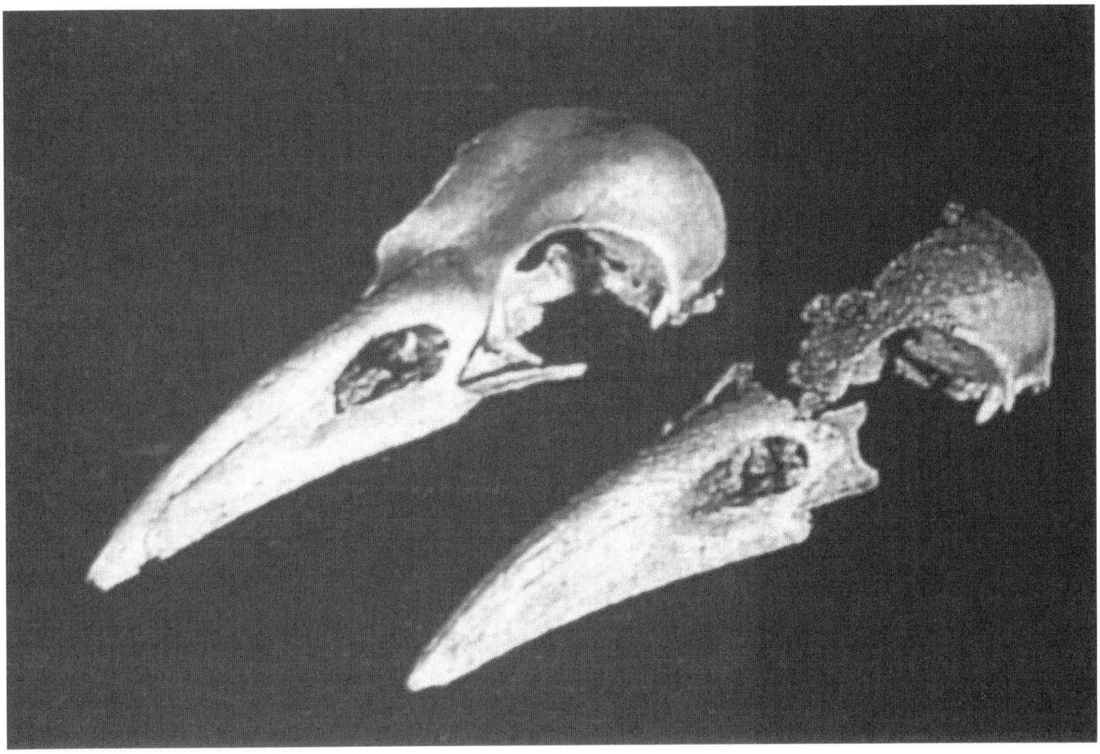

substantial periods of time. That in itself suggests that suitable habitat for terrestrial birds has existed in the Chatham Islands through glacial–interglacial cycles.

New Zealand ravens were large crows with long, broad bills (Fig. 11.7) that were not as arched as those of some of the Hawaiian crows (Olson and James 1992). All forms are now extinct. Their remains are most common in coastal sites, both on the main islands and on the Chatham Islands. The South Island form was a typical part of the eastern fauna during the Holocene. It is found in deposits of Pleistocene age on the West Coast and in deposits of both Pleistocene and Holocene age in the east and south. A single bone is known from the subantarctic Auckland Islands, so it is not clear whether it represents a resident population or a vagrant from the mainland. Only a single individual was found in Pyramid Valley, and the species is uncommon at Poukawa in the North Island, where tens of thousands of bones were preserved. On the coast, it may have frequented the seal and penguin colonies or fed in the intertidal zone, as does the Tasmanian forest raven *Corvus tasmanicus*. Alternatively, it may have depended on fruit, like the New Caledonian crow *Corvus moneduloides,* but it is difficult to understand why a fruit eater would have been most common in coastal forest and shrubland when fruit was distributed throughout the forest.

Figure 11.7. Skulls of Corvus moriorum *(MNZ S28679) (left) from Chatham Island, and* Corvus *"mainland" (MNZ S27925) (right) from Metro Cave. Photograph by R. Morris.*

12 • Oceanic Birds in the Terrestrial Environment

Order Procellariiformes

This discussion of oceanic birds is limited because of the patchy knowledge of the fossil faunas of most of the offshore islands. Information on those is incomplete, rudimentary, or absent. Obviously, there have been some changes, even on remote islands in the subantarctic, where predatory mammals have been introduced. Other, more subtle influences have been the effects of fishing for wet fish and squid in the surrounding oceans on the food chain, removal of whales as significant parts of the food chain, bycatch (the accidental capture of nontarget species) of migratory seabirds by North Pacific fisheries, and lately changes in oceanic water patterns resulting from global climate change. We concentrate on the main New Zealand islands and the immediate offshore islands. Comments on the birds of the Chatham Islands, Norfolk, and the Kermadecs are included where there is information from fossil sources.

The conservation of petrels and other migratory seabirds is one of the greatest environmental challenges facing the world because, as with waterfowl, their ranges transcend political boundaries. In addition, the main threats involve already controversial multinational fishing industries.

Petrels are oceanic birds that feed on plankton, fish, or squid. They are usually nocturnal on land, where they arrive in the evening and depart at or before first light and so avoid most diurnal predators. Worldwide, most breeding colonies are on islands that lack mammalian predators. When predators such as cats and rats are introduced, petrel populations usually decline dramatically, often to extinction, if no management measures are taken. A classic example of a species that was decimated by humans and predators is the cahow (*Pterodroma cahow*) of Bermuda, which was rediscovered on small islands after many years when it was thought to be extinct. In the New Zealand area, the most dramatic example of human impact in European times is the extirpation of the providence petrel (*Pterodroma solandri*) on Norfolk Island in the 1790s by the newly established European penal colony, after HMS *Sirius,* the supply ship, was wrecked on the reef (Clarke 1790) and other sources of food also were erratic. Other extinctions occurred there before European contact, including the total loss of an unsuspected—and very large—colony of Pycroft's petrel (*Pterodroma pycrofti*).

Petrels have a strong, musky odor, all but the largest species nest in burrows, all lay one egg, and most leave the egg and chick for long periods. Although there is no evidence yet that albatrosses have bred on the North and South Islands, representatives of the other petrel families certainly did: a diving petrel, a prion, two or three storm petrels, at least two gadfly petrels (*Pterodroma*), two *Procellaria* petrels, and more than five species of shearwater (*Puffinus*). They ranged in size from the tiny storm petrels (≥20 g) to large shearwaters (*Puffinus* spp.) and *Procellaria* petrels of over 1 kg. Almost all of them are now extinct on the main islands. A few hold on as tiny remnants in places that mammalian predators cannot reach. Only two of the largest species are in more than remnant colonies; one of those seems to be declining in numbers.

At least three petrel species were confined to the mainland—Hutton's and Scarlett's shearwaters (*Puffinus huttoni* and *P. spelaeus*), and the Westland petrel (*Procellaria westlandica*). The first still breeds above an altitude of 1500 m in the Seaward Kaikoura Mountains, where the returning birds often locate their previous burrow beneath up to 2 m of snow. Scarlett's shearwater bred in wet forest on the West Coast but is now extinct. The Westland petrel still breeds there, and its population may have increased because an offshore fishery has provided an extra source of food for the breeding birds. For a possible fourth petrel species, a storm petrel, the picture is less clear because a link has yet to be firmly established between bones in predator and other deposits and three skins taken from the New Zealand coast in the 19th century

Only those species that nested on the main islands, including the Chatham group, and those close offshore are listed here; these and species on the subantarctic and subtropical islands are listed in Appendix 1. The importance of petrels in the New Zealand fauna cannot be overestimated. In sheer numbers alone, they were most significant parts of the bird fauna. Those numbers and the petrel habit of nesting in burrows in huge colonies meant that they also influenced the terrestrial environment more than any other group of vertebrates. Indeed, as related in Chapter 14 (Guilds and Regional Faunas), a substantial fraction of the nutrient flow through the terrestrial food chain came directly from the oceanic food chain. The link has been detected only recently, as the number and variety of petrels breeding on the main islands have become apparent.

Family Diomedeidae

Genus Diomedea *Linnaeus*
Diomedea sanfordi *Murphy, 1917 (Northern Royal Albatross)*

Genus Thalassarche *Reichenbach*
Thalassarche bulleri *(Rothschild, 1893) (Southern Buller's Albatross)*
Thalassarche eremita *Murphy, 1930 (Chatham Island Albatross)*
Thalassarche *sp. (Northern Buller's Albatross)*

Although northern royal albatrosses (*Diomedea sanfordi*) breed on the mainland of the South Island at Taiaroa Head, near the city of Dunedin, there is as yet no fossil evidence for this or any other species breeding on the main islands before European contact. Bones of large *Thalassarche* albatrosses are found commonly enough in dunes and archaeologic sites in the eastern South Island to suggest that there were breeding colonies nearby, but until remains of prefledging juveniles are found, the record probably indicates larger populations elsewhere that were visiting a coast where there were more whales and seals and probably more offal and fish. There is, however, evidence for a range reduction in the northern royals breeding on the Chatham Islands. Fossils of adults and juveniles are common in deposits on Pitt Island, where the species no longer breeds, the second largest in the group. Polynesians visited the island before European contact and continued to visit the other breeding colonies on small islets to collect young birds and eggs into the 20th century (Millener 1999).

Family Procellariidae
Subfamily Procellariinae

Genus Puffinus *Brisson*
Puffinus carneipes *Gould, 1844 (Flesh-footed Shearwater)*
Puffinus pacificus *(Gmelin, 1789) (Wedge-tailed Shearwater)*
Puffinus bulleri *Salvin, 1888 (Buller's Shearwater)*
Puffinus griseus *(Gmelin, 1789) (Sooty Shearwater)*
Puffinus gavia *(J. R. Forster, 1844) (Fluttering Shearwater)*
Puffinus huttoni *Mathews, 1912 (Hutton's Shearwater)*
Puffinus spelaeus *Holdaway and Worthy, 1994 (Scarlett's Shearwater)*
Puffinus haurakiensis *Fleming and Serventy, 1943 (New Zealand Little Shearwater)*

Genus Pelecanoides *Lacépède*
Pelecanoides urinatrix *(Gmelin, 1789) (Northern Diving Petrel)*
Pelecanoides georgicus *Murphy and Harper, 1916 (South Georgian Diving Petrel)*

Genus Procellaria *Linnaeus*
Procellaria parkinsoni *G. R. Gray, 1862 (Black Petrel)*
Procellaria westlandica *Falla, 1946 (Westland Petrel)*

The shearwaters, two large petrels, and diving petrels were major elements of the late Holocene avifauna of New Zealand. Four species of shearwater (*Puffinus gavia, P. huttoni, P. spelaeus,* and *P. griseus*), two *Procellaria*s, and the common diving petrels bred on the North and South Islands. Three more shearwaters (*P. carneipes, P. haurakiensis,* and *P. bulleri*) and the other diving petrel nested on inshore islands, and another shearwater (*P. pacificus*) is still one of the most common petrels on the subtropical islands.

Shearwaters: Puffinus *spp.*

This genus has the largest range of body size of any group of petrels, which together have the widest range of body mass for any of the orders of birds, from 20–30 g in small storm petrels to some of the largest flying birds at over 10 kg. In New Zealand, *Puffinus* range in size from closely related species in the little shearwater (*P. assimilis*) superspecies group to the large and aggressive sooty shearwater (*P. griseus*).

On Norfolk Island and in the Kermadecs, the wedge-tailed shearwater (*Puffinus pacificus*) still tries to nest on the main islands. It is large enough to avoid predation by rats, but cats on Raoul Island exterminated the population there and kill most of the attempted recolonists. They also kill many of the birds attempting to nest on Norfolk. On offshore islets such as the Meyer Islands, this species is still abundant. It was part of the diet of Polynesian people on both Raoul and Norfolk Islands. Although the flesh-footed shearwater (*Puffinus carneipes*) nests as far south as islands in Cook Strait, its fossils are not known from likely colony sites in the North or South Islands. Beach cast specimens are common in Northland dune deposits.

Buller's shearwater (*Puffinus bulleri*) breeds on only one island group, the Poor Knights, off Northland, where the colony was not discovered until 1915 (Harper 1983). The bird itself had been known from beached specimens since 1884, and birds wintering off California had been shot in the early 20th century (Loomis 1918). The only fossil record, from Castle Rocks in western Southland, is far outside its present range. It is an unusual place for a vagrant, but other petrels nested nearby, so it is not impossible that it was attracted to those. On Aorangi (68 ha), the bird population was small when it was discovered in 1924 because the island had been occupied by Māori until late 1823 (Harper 1983), and the last of the pigs that they had introduced were not shot until 1936 (Bartle 1968). No introduced mammals were ever present on the larger island, Tawhiti Rahi (132 ha), where the shearwater remained abundant. The bird colony grew to about 100,000 pairs by 1968 (Bartle 1968), and the total population (breeding on less than 200 ha) is now estimated to be about 2.5 million birds, including prebreeders (Heather and Robertson 1996).

By the 19th century, sooty shearwaters (*P. griseus*) still bred in millions on the islands around Stewart Island. The formerly large colonies on the main islands had by then been reduced to remnants. On the North and South Islands, only tiny groups remain on promontories and coastal cliffs, where they are under threat from cats and ferrets. On islands around Stewart Island, sooty shearwaters are still abundant. Farther south, the breeding population on the 328 ha of The Snares was estimated at 2.75 million pairs (Warham and Wilson 1982), and there are many more on most of the other subantarctic islands. In the southern fall, sooty shearwaters migrate to the North Pacific and the Bering Sea, where thousands are killed as bycatch in the salmon fisheries. The return migration takes many birds south along the California coast and thence to New Zealand in late September.

On the islands around Stewart Island, sooty shearwater young have been harvested each summer since at least the early 19th century, and perhaps longer (Anderson 1996, 1997a). Polynesians visiting Stewart Island in the 14th century to crop the sooty shearwaters nesting there took along moa haunches for food on the trip from their homes on the southern South Island (Worthy 1998e).

Three of the smaller shearwaters were important in the terrestrial environment, but evidence of their former presence and abundance has only recently been obtained. Judging from the abundance of fossils in caves (Fig. 12.1), there were large breeding colonies of *P. spelaeus* on the West Coast of the South Island (Worthy and Holdaway 1993). The youngest radiocarbon date for the species is about 600 years before the present (B.P.). Because it is not known from the few archaeologic sites that have been excavated in its nesting range, the species seems to have become extinct early in the Polynesian period. It was the smallest of the *gavia* superspecies, and it is morphologically and dimensionally distinct from both fluttering and Hutton's shearwaters (Holdaway and Worthy 1994). Its body mass (~150 g) was well within the range accessible to Pacific rats.

Fluttering and Hutton's shearwaters bred mainly east of the main ranges in the South Island. Radiocarbon dates on shearwater bones show that there were active colonies there within the past 1000 years. A few remnant colonies may have survived until European times—there are stories of birds having been shot on misty nights—but no skins were ever preserved. Both these species, too, are vulnerable to predation by any of the introduced predators. In contrast to Scarlett's shearwater, both the eastern species survive—Hutton's shearwater as a relict population nesting in subalpine grassland above 1500 m in the Seaward Kaikoura Mountains, and the fluttering shearwater is still breeding on many island refuges in sheltered waters around the northern North Island and Cook Strait.

Hutton's shearwater is one petrel that does not migrate to the North Pacific. Most migrate to Australia in winter and return to the colonies in spring. In some years, the colony area in the subalpine zone is still covered with up to 2 m of snow. The birds somehow find their burrow under the snow, then sit and wait until the snow has melted enough before they clear it out and prepare to lay eggs. The colonies have declined in extent and number in the 20th century (Sherley 1992). This, a continuation of the long-term decline, includes a contraction from the Inland Kaikoura Ranges, which were still being used as a source of muttonbirds in the 19th century. Until the discovery of the

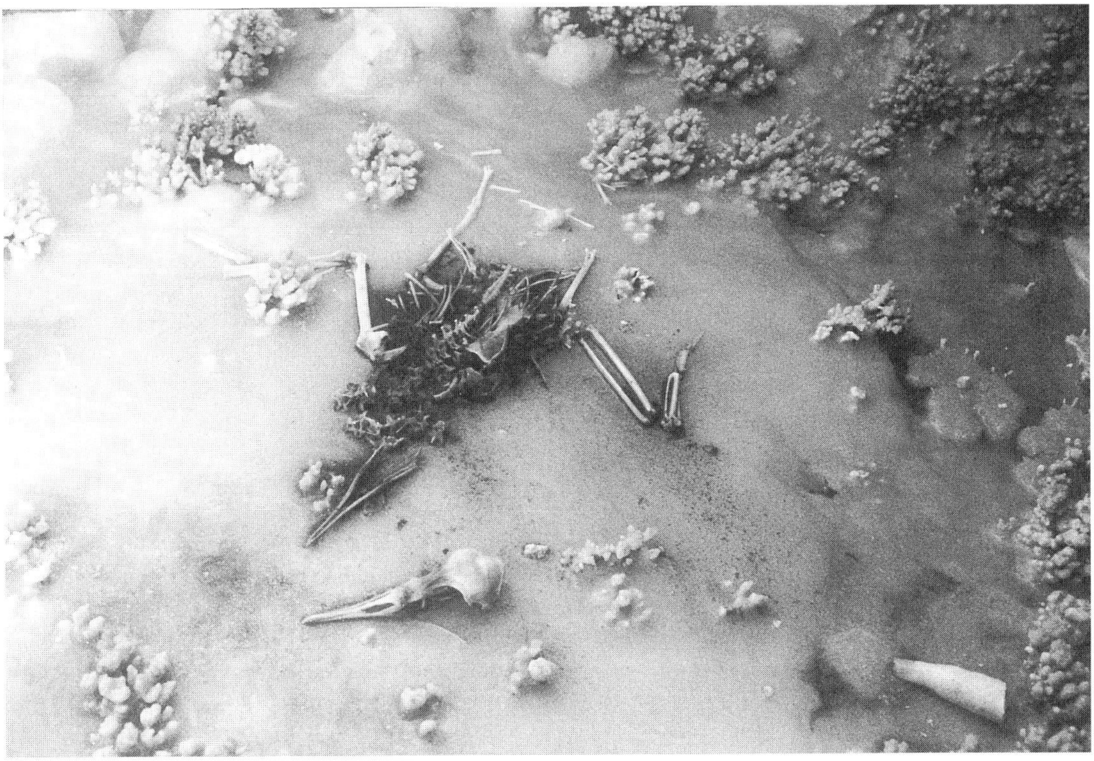

Figure 12.1. A skeleton of Puffinus spelaeus *in Te Ana Titi [cave] radiocarbon dated at 1960 ± 38 years before the present. Photograph by T.H.W.*

colonies in the mid-1960s, the breeding site was thought to be farther south, in the subantarctic islands. As explained in Chapter 11, muttonbirding is the harvesting for food of the young of some species of petrel; there is a traditional and now semicommercial take of the sooty shearwater chicks on islands in and around Foveaux Strait.

The eggs of the smallest local shearwater, the New Zealand little shearwater (*Puffinus haurakiensis*), are vulnerable to rats (Booth et al. 1996). There are a few fossil records from coastal parts of the North Island. Populations on northern offshore islands are starting to recover since the removal of Pacific rats in 1996. Islands where remnant populations survived to the present are likely to have been colonized by rats relatively recently; rats would be likely to decimate colonies in 150–250 years (Holdaway 1999b).

Diving Petrels: Pelecanoides *spp.*

Whereas most petrels fly above the sea and dive into it or alight on it to feed, diving petrels seem to almost ignore its presence. They have short, strong wings and fly with rapidly beating wings just above the surface of the water. A shearwater may dip its wingtip into a swell, but diving petrels often just keep on flying straight into the rising wave, through it, and out the other side. Thoresen (1967) was one of the first biologists to notice that Pacific rats could affect a population of diving petrels. He noticed that a colony on a sea stack off Stanley Island in the Mercury group was surviving better than one on Stanley Island itself, and he related that to observations of broken eggs in burrows that rats had just left.

Common diving petrels bred in several places on the main islands, including the wet western and dry eastern coasts of the South Island. Their bones are regular components of laughing owl deposits throughout the South Island (Worthy 1997b, 1998b; Worthy and Holdaway 1994a, 1995, 1996b). The common and South Georgian diving petrels both bred on Stewart Island, but the latter was confined to the extensive dunes in the west of that island (Worthy 1998c). Diving petrels are sedentary and feed over the continental shelf and in inlets, harbors, and the straits separating the three largest islands. *P. georgicus* burrows in sand dunes and so has a more limited choice of sites than *P. urinatrix*, which nests in burrows in consolidated soil beneath forest or coastal shrubland. There is a remnant population of South Georgian diving petrels on Codfish Island, off the west coast of Stewart Island, where Pacific rats may have arrived less than 200 years ago (Holdaway 1999b). However, the formerly large population on Stewart Island survived only until shortly after Pacific rats reached the island about 500 years B.P. Trampling by cattle and rabbit burrowing have been implicated in the extinction of a colony recorded on Enderby Island in the Auckland group in the 1940s. Fossils indicate South Georgian diving petrels also formerly nested in the sand dunes of Chatham Island.

Procellaria *Petrels: The Largest Coped Best*

Four species in the genus *Procellaria* nest in the New Zealand archipelago, two on the main islands. So far as is known, the breeding range of the Westland petrel (*P. westlandica*) has never been much larger than it is now (Worthy and Holdaway 1993): a few colonies along a 10-km strip of the West Coast of the South Island. The range of the black petrel (*P. parkinsoni*) is relict, active burrows being confined now to two islands off the northern North Island. Into the 20th century, the species still attempted to breed in forested hills in several widespread places in both main islands (Oliver 1955), and their fossils have been found near the colonies of Westland petrels. The remaining colonies are on islands in the Hauraki Gulf; in the nonbreeding season, the black petrel migrates across the Pacific and feeds near Central America.

These large species were, like the sooty shearwater, able to withstand the activities of all but the largest and fiercest of the predatory mammals for some time. But eventually those predators, as well as the pigs brought by explorers and settlers, have removed most colonies. Only the larger of the two species and that which nests in winter in the roughest, wettest terrain on the West Coast have survived. In recent decades, its population has even increased with the new food supply of waste from offshore fishing fleets; its numbers have tripled since the late 1950s (Heather and Robertson 1996). In the summer, it disperses widely in and out of New Zealand waters, and some birds reach the eastern Pacific. The Westland petrel was discovered only in the 1940s, when students at a school at Barrytown on the West Coast wrote to a scientist, Robert Falla, who had talked about black petrels on a radio program for schools. The students pointed out that "their" black petrels nested in burrows in the forest behind their school in winter, not summer, as he had described. A visit showed that the birds also differed in size and beak coloration, so he described the population as a new subspecies of *P. parkinsoni*. It is now recognized as being a full species, with different habits.

Subfamily Fulmarinae

Genus Macronectes *Richmond*
Macronectes halli *(Gmelin, 1789) (Northern Giant Petrel)*

Genus Pachyptila *Illiger*
Pachyptila turtur *(Kuhl, 1820) (Fairy Prion)*
Pachyptila vittata *(G. Forster, 1777) (Broad-billed Prion)*
Pachyptila crassirostris *(Mathews, 1912) (Fulmar Prion)*

Genus Pterodroma *Bonaparte*
Pterodroma pycrofti *Falla, 1933 (Pycroft's Petrel)*
Pterodroma cookii *(G. R. Gray, 1843) (Cook's Petrel)*
Pterodroma nigripennis *(Rothschild, 1893) (Black-winged Petrel)*
Pterodroma axillaris *(Salvin, 1893) (Chatham Petrel)*
Pterodroma inexpectata *(J. R. Forster, 1844) (Mottled Petrel)*
Pterodroma cervicalis *(Salvin, 1891) (White-necked Petrel)*
Pterodroma neglecta *(Schlegel, 1863) (Kermadec Petrel)*
Pterodroma solandri *(Gould, 1844) (Providence Petrel)*
Pterodroma macroptera *(Smith, 1840) (Grey-faced Petrel)*
Pterodroma magentae *(Giglioli and Salvadori, 1869) (Chatham Taiko/Magenta Petrel)*
Pterodroma *sp. (Unnamed Petrel, Chatham Islands)*

Many of the fulmarine petrels, including the snow petrel, southern fulmar, Antarctic petrel, and cape petrel live and nest at high southern latitudes and reach the coasts of the main islands of New Zealand as vagrants or wrecked on beaches. Petrels, even though they live in the open ocean, can still be affected by bad weather; storms prevent birds from feeding, and the weaker ones die. Those that are washed up on beached are said to be "wrecked," and when large numbers of dead and dying birds come ashore, it is said to be a "wreck." Such occurrences can potentially bias the fossil record in sand dune areas, and the remains of birds cast up have been taken as evidence for a species breeding in an area where it is only a migrant. Indeed, most fossil records from dunes are of these migrant individuals.

The southern giant petrel is another species that breeds too far south to be of interest here, but the northern, or Hall's, giant petrel breeds as far north as Stewart Island and was probably common off the coasts in the past. Giant petrels are fearless and powerful. They follow ships, take offal from fishing boats, and feed from whale and seal carcasses. The prions and *Pterodroma* petrels are extremely placid and almost helpless when on the ground. In short, they are supremely vulnerable to predation by rats. On the North and South Islands, large breeding populations that undoubtedly existed before human contact were almost gone by the time of Polynesian settlement 800 years B.P., as their bones are rare in "mainland" middens. Their remains are common in predator deposits such as Hermit's Cave (South Island) and Te Waka (Hawke's Bay, North Island) that date from before the arrival of Pacific rats. In more recent sites, their bones are rare. Fossils in inland deposits indicate that Cook's and mottled petrels and fairy prions were breeding in widespread localities in the North and South Islands before the first indirect effects of people.

Prions: Genus Pachyptila

One species, the fairy prion, bred in many inland areas of the South Island but not, apparently, the North Island, before the arrival of the Pacific rat. Its remains appear in Pleistocene laughing owl sites, such as that in Hermit's Cave near Charleston on the West Coast. Fairy prions breed in summer, and the young birds taken by owls on the West Coast would have been caught on the forest floor in late February or early March, as they prepared to leave the colony. Fairy prions are still abundant on offshore islands, but only on those that lack all predators. One of the largest colonies is on Stephens Island, at the western end of the narrows of Cook Strait. Other species became extinct on the island when it was taken over for a lighthouse station in the 1890s. However, the prions can burrow in grassland, are as long-lived as most petrels, do not reach maturity until six or seven years of age, and were in huge numbers to start with, so the removal of forest and the presence of a few cats were not the problems they were for Lyall's wren, the South Island piopio, the South Island saddleback, or the South Island kokako, which all vanished before the cats were removed. The mutually exclusive distributions of fairy prions, diving petrels, and storm petrels on the one hand and Pacific rats on the other are telling evidence for the predatory abilities of the small rodent.

Many other species breed in the subantarctic and reach the shores of the main islands as vagrants and beach wrecks. The broad-billed prion nests on islands around Stewart Island and on the Chathams. Two prions on the Chathams, the broad-billed (*P. vittata*) and fulmar (*P. crassirostris*) prion, were subject to predation by people and rats, and populations on the larger islands were wiped out. Broad-billed prions are now common again on the smaller islands of the Chatham group, where they pose problems for the much rarer Chatham petrel (*Pterodroma axillaris*), whose burrows they appropriate.

Gadfly Petrels: Pterodroma *spp.*

Some of the gadfly petrels are migratory and visit the North Pacific and Central America in their nonbreeding seasons. They are strong fliers and most have a wheeling, bounding flight that takes them high above the waves, unlike the shearwaters, which normally remain close to the surface. They feed on cephalopods, euphausiids, and fish and are almost totally defenseless on land.

No fossils of Pycroft's petrel (*Pterodroma pycrofti*) have been confirmed on the North Island itself. The species survives in small colonies on islands in the Hauraki Gulf and Bay of Plenty. However, this historical distribution and abundance is misleading. Archaeologic and natural deposits on Norfolk Island contain large numbers of bones of this species, which seems to have been one of the most abundant petrels on that group (Meredith 1991). The New Zealand populations must be relicts, diminished by the Pacific rats that reached their islands after Polynesian settlement (Holdaway 1999b). Removal of the rats from some islands has resulted in greatly enhanced breeding success for these birds. But their recovery to former levels will take decades, if not centuries.

Cook's petrel (*Pterodroma cookii*) was one of the most abundant birds on the main islands before human contact. Their fossils occur in predator and other sites throughout the South Island and in the North Island in inland locations from Waitomo to Hawke's Bay, at altitudes from near sea level to 900 m. The few applicable radiocarbon dates show that the petrel was present on the mainland within the past millennium. With the loss of its mainland colonies, Cook's petrel has become globally threatened. The remaining colonies on Codfish (near Stewart Island) and Little and Great Barrier Islands (in the Hauraki Gulf) declined either in the presence of Pacific rats or cats, or both, on these islands. On Little Barrier Island, from which cats were eradicated in the early 1980s, the colony is now relatively safe and is the largest, with about 50,000 pairs. The 200 pairs on Codfish Island, down from an estimated 20,000–40,000 in the 1930s, now do not have to contend with Pacific rats, which were eradicated in 1998, but the 20 or so pairs on Great Barrier Island are still unprotected. Cook's petrels spend the southern winter off the California coast.

As with the larger white-necked petrel, the black-winged petrel (*Pterodroma nigripennis*) recovered after the removal of goats and the recovery of the vegetation, and this species probably now numbers several million. Birds are actively prospecting for new breeding sites in the Chatham Islands, northern New Zealand, and Norfolk Island. On Norfolk Island, they may be exploiting the niche left vacant by the extinction of Pycroft's petrel in early European times.

The Chatham Islands have the dubious distinction of being home to two of the rarest petrels in the world, and the former home of a third, which is known only from fossil bones in natural dune deposits and in archaeologic sites. *Pterodroma axillaris* (Chatham petrel) is now confined to Southeast Island, where the population is variously estimated at about 200 (West 1994) and 500–1000 individuals (Heather and Robertson 1996), which contest for nest burrow space with the far more numerous broad-billed prion.

In the forest at the southern end of main Chatham Island, the few remaining Chatham taiko (*Pterodroma magentae*) are now protected from rats and cats (mustelids are not present on the Chathams). The numbers of these birds had continued to decline from the low numbers left after centuries of muttonbirding by the Moriori (descendants of the earliest Polynesians in New Zealand who colonized the Chathams in the 14th century), Māori, and European settlers in the 19th century. The rediscovery of the taiko, also known as the Magenta petrel (named after the ship carrying the Italian research expedition that obtained specimens of an unknown petrel in the southwest Pacific in the early 1860s), was the result of a quest by New Zealand naturalist David Crockett begun in the late 1940s. While assisting the curator of Canterbury Museum after school, David found that petrel skulls from the Chatham Islands did not match those of any of the known species from the islands. He was aware that there were stories of a different, black-and-white petrel that had been seen by older residents of Chatham Island, and he thought that there might be a connection between those stories, the skulls, and the mysterious Magenta petrel. For many years, he organized progressively more sophisticated expeditions to the remote area where residents recalled seeing the birds. After a few sightings of the birds in the powerful floodlights directed into countless misty nights, finally, on New Year's Day 1978, the first taiko was caught, photographed, banded, and released. About 30 years after he started the quest, David Crockett held a live bird in his hands (Crockett 1994). Now the species is the subject of an intensive conservation effort that has spilled over into assisting some of the other remaining Chatham species, such as the Chatham Island pigeon, that had reached very low numbers.

One species of petrel on the Chatham Islands has not been as lucky as either the Chatham petrel or the Chatham taiko. Bones of an undescribed species of *Pterodroma* have for years been confused with those of the mottled petrel, which bred on the North and South Islands and some subantarctic islands but has never been seen on the Chathams (Holdaway et al. 2001; Tennyson and Millener 1994). The species seems to be related to others from the subtropical Pacific, which is not impossible because the Chatham Islands lie on the subtropical convergence. Petrels from both cold- and warm-water groups nest there and feed in the water masses on each side of the front. Final determination of the affinities of this enigmatic petrel may need to await the fortuitous discovery of associated cranial and postcranial elements of the skeleton.

Although the mottled petrel (*Pterodroma inexpectata*) has not shared the fate of the Chatham Island bird, it has declined from its former status as one of the most abundant birds in the North and South Islands. On the main islands, it is now reduced to a single colony on a small island in Lake Hauroko, near the shores of Foveaux Strait. Mottled petrels still breed on islands to the south of New Zealand, but their population, as with those of many other species, is probably orders of magnitude lower now than before humans arrived in the archipelago. The loss in terms of nutrients not brought ashore to the young soils of New Zealand is incalculable, as is the effect on the oceanic food chain, both in New Zealand waters and in the northern Pacific, where the mottled petrel spends the nonbreeding season. Modern scientists studying nutrient flows, plankton, and fish stocks are working with systems that differ greatly from those of only 2000 years ago, before Polynesians and the Pacific rats removed vast petrel populations (and those of other seabirds) from the Pacific equation.

At the northern edge of the New Zealand fauna (Holdaway et al. 2001), four of the largest *Pterodroma* petrels were also affected by

Polynesian and European influences. And once-rare petrels have had another chance. In the mid-1960s, the white-necked petrel (*Pterodroma cervicalis*) was discovered breeding on Macauley Island, the second largest of the active volcanoes that form the Kermadec chain. The island then had the appearance of a bowling green, brought about by a thriving population of feral goats left there as an emergency food supply by 19th-century sailors. Once Macauley Island in the Kermadecs was cleared of goats, the white-necked petrel population recovered rapidly. It is now abundant and increasing in numbers.

On Raoul Island, the northernmost and largest of the Kermadec chain, the Kermadec petrel (*Pterodroma neglecta*) was abundant until the early 20th century. A huge colony on Raoul Island in the Kermadecs was exterminated by people, cats, and Norway and Pacific rats by the mid-20th century. There are still populations on the Meyer Islands off Raoul Island's northwest coast, and on Balls Pyramid near Lord Howe Island, but the petrel is now one of the rarest in the southwest Pacific. The species was also common in Norfolk Island before humans arrived. It appears in archaeologic sites and may have been seen by the first European settlers in the late 18th century, but no specimens were obtained. Kermadec petrels are now recolonizing, with a few pairs breeding on Philip Island. The species nests on the surface and not in burrows and so was even more vulnerable than was usual for members of the genus.

The providence petrel (*Pterodroma solandri*) is famous for taking the brunt of the settlers' hunger in 1790 and 1791 on Norfolk Island. Many thousands of adults and young were taken, and the population was exterminated by the start of the 19th century. The loss was so complete that it was over 100 years before the identity of the birds nesting on mountains on Lord Howe Island with the extinct population on Norfolk was established. Only the Lord Howe Island population was thought to survive until a few pairs were discovered breeding on the badly eroded Philip Island off the south coast of Norfolk Island. Providence petrels are less common than Pycroft's petrel in archaeologic sites on Norfolk, which are primarily near the south coast, so it appears that the providence petrels were always confined to the higher areas around Mount Pitt at the northern end of the island.

Grey-faced petrels (*Pterodroma macroptera*) are almost cosmopolitan in midsouthern latitudes, although this name certainly refers to a superspecies group of at least two different-size allospecies. There are breeding colonies in the south Atlantic and Indian Oceans, around Australia, and in northern New Zealand. They still breed in small colonies on coastal cliffs and stacks in the northern half of the North Island and were probably once common in the coastal forest and shrubland. Since before European contact, their main colonies have been on islands in the Hauraki Gulf and Bay of Plenty, where their young were harvested as northern muttonbirds into the early part of the 20th century. These are among the largest *Pterodroma*s and are less susceptible to rat predation than the smaller birds. However, rats are still causing declines on island and mainland populations (Imber 1978).

Family Oceanitidae

Genus Oceanites *Keyserling and Blasius*
Oceanites maorianus *(Mathews, 1932)*
(New Zealand Storm Petrel)

Genus Garrodia *Forbes*
Garrodia nereis *(Gould, 1840)*
(Grey-backed Storm Petrel)

Genus Pelagodroma *Reichenbach*
Pelagodroma marina *(Latham, 1790)*
(White-faced Storm Petrel)
Pelagodroma albiclunis *Murphy and Irving, 1951*
(Kermadec Storm Petrel)

Three skins of a storm petrel with a plumage pattern like the so-called *pealea* form of Wilson's storm petrel (*Oceanites oceanicus*) were collected in the 19th century, reportedly off the east coast of the South Island near Banks Peninsula. On the basis of these specimens, Mathews erected a species (*Pealeornis maoriana*). Oliver (1955) removed it to *Oceanites* as *O. maorianus. Oceanites oceanicus* does not breed in the New Zealand region, and the skins may be of the *Oceanites* identified in fossil sites in both the North and South Islands (Worthy 2000). Although the skins were obtained in the 19th century, no colonies have been found. This may not be significant, because neither were colonies of any other petrel in inland Canterbury, despite fossil evidence of their Late Holocene presence. Perhaps more revealing is the observation that no nests of another New Zealand storm petrel (the Kermadec storm petrel *Pelagodroma albiclunis*) have been found either, even though it is seen regularly near suitable breeding islands in the group. The connection between the fossils and the three skins will be difficult to prove because there are few bones left in the skins to compare with the fossils, but what appears to be certain is that a species of *Oceanites* storm petrel whose bones differ consistently from those of that genus elsewhere in the Southern Hemisphere once bred on both North and South Islands of New Zealand (Worthy 2000).

White-faced storm petrels (*Pelagodroma marina*) still breed abundantly on small, rat-free islands around the coast of the main islands and on some outlying islands, with a separate species in the Kermadecs. The white-faced storm petrel formerly bred on the South Island and perhaps the North Island, but the small number of bones in collections suggest that it was either in small numbers or not a major food of laughing owls. The grey-backed storm petrel *Garrodia nereis* still nests on subantarctic islands, but there are records on the main islands that indicate former breeding colonies.

Order Sphenisciformes: Classic Winged Swimmers

Family Spheniscidae

Genus Megadyptes *Milne-Edwards*
Megadyptes antipodes *(Hombron and Jacquinot, 1841) (Yellow-eyed Penguin)*

Genus Eudyptula *Bonaparte*
Eudyptula minor *(J. R. Forster, 1781) (Little Penguin)*

Genus Eudyptes *Vieillot*
Eudyptes pachyrhynchus *J. E. Gray, 1845*
(New Zealand Crested Penguin)
Eudyptes sp. *(Undescribed Chatham Island Crested Penguin)*

Although other species of penguin reach the main islands as vagrants, only three breed on the main islands of New Zealand and another, extinct, species was on the Chathams. The ranges of the two larger species, the yellow-eyed (*Megadyptes antipodes*) and New Zealand crested penguin (*Eudyptes pachyrhynchus*), have contracted greatly since Polynesian colonization. Both penguins are now rare and endangered. The yellow-eyed penguin is confined as a breeding species to a few localities in the eastern and southern South Island as far north as Banks Peninsula, but not in Fiordland. New Zealand crested penguins now nest only in Fiordland and the southern West Coast. A few pairs breed on Stewart and Codfish Islands and other small islands around Stewart Island, but its range does not include subantarctic islands, whereas the yellow-eyed has populations on the Auckland and Campbell Islands. Yellow-eyed penguins prefer to nest in shrubland and forest inland from sandy beaches; New Zealand crested penguins nest in caves and among rocks and tree roots on forested, rocky coastlines.

Fossils of both species have been found as far north as the northern end of the South Island (Worthy 1997d). It is likely that they once nested in the North Island as well. Identifications of fossil material of large penguins from archaeologic sites have not been checked. At Lake Grassmere, and in other archaeologic and natural sites in the eastern South Island, reports of erect-crested and rockhopper penguins have been shown to refer to material of the two present species. Both species were subject to human predation, and their reduction in range and numbers parallels that of the seals and the *Leucocarbo* shags, with relict populations in the area with the harshest climate and lowest human population in pre-European New Zealand.

Tennyson and Millener (1994) have recognized that crested penguin bones from dune deposits in the Chathams belong to a new, extinct species. It appears in archaeologic sites and was probably one of the victims of human occupation of the group.

The little penguin (*Eudyptula minor*) still breeds at many points along the coast of the main islands and on many, if not most, offshore islands (Heather and Robertson 1996). There are even breeding populations in harbors within the boundaries of major cities. However, some populations have declined rapidly in the past two decades as a result of predation by ferrets, dogs, and cats. Only colonies inaccessible to these predators at the base of high cliffs seem to be stable. Their vulnerability was forcibly underlined in January 2001, when two dogs killed 55 birds in a colony at Oamaru in a single night. The dogs returned two nights later and killed a further 15 penguins before they themselves were killed. The little blue penguin is abundant in dune deposits throughout New Zealand.

Summary

The role of the seabirds in the prehuman—natural—avifauna of New Zealand and in the ecology of the main islands in general has become clear only with the recent discovery of a diverse range of breeding species on the main islands, often apparently in huge numbers. The remnant population of Hutton's shearwater in the Seaward Kaikoura Range contains about 150,000 pairs, occupying a tiny fraction of the species' former range. It is probably no exaggeration to say that there would have been hundreds of millions, if not billions, of petrels nesting in the New Zealand region before humans made contact. The effects of these in terms of soil nutrients, and hence vegetation patterns, higher numbers of invertebrates, and prey for predatory birds, can only be guessed. A full understanding of the paleoecology of New Zealand will rest on our ability to estimate the impact of the loss of the petrels.

13 • Bats to Frogs

Bats

The only land mammals known from New Zealand are three species of bat in two genera. All are or were small microchiropterans. The larger fruit bats (Pteropodidae) are absent, apart from rare stragglers from Australia.

Family Mystacinidae

Genus Mystacina *Gray*
Mystacina tuberculata *Gray, 1843 (New Zealand Lesser Short-tailed Bat)*
Mystacina robusta *Dwyer, 1962 (New Zealand Greater Short-tailed Bat)*

Family Vespertilionidae

Genus Chalinolobus *Peters*
Chalinolobus tuberculatus *(Foster, 1844) (New Zealand Wattled Bat, Long-tailed Bat)*

Of these bats, *Chalinolobus tuberculatus* is the New Zealand representative of a genus that is widely distributed in Australia, New Guinea, New Caledonia, and Norfolk (Flannery 1995). It is an aerial insectivore and is still widespread, although endangered, in New Zealand. Fossils of it are rare and all are from caves (Figs. 13.1–13.3).

In contrast, *Mystacina* is in a family, Mystacinidae, that is presently endemic to New Zealand. It has a unique assemblage of morphologic features that have puzzled systematists for more than a century. In that time, it has been placed in 7 of the 18 extant families of bats (Kennedy et al. 1999; King 1990). Analysis of results obtained by use of albumen immunology (Pierson et al. 1986) suggested that *Mystacina* was an

Figure 13.1. Photograph of a skeleton of a Mystacina tuberculata *that died while hanging on the wall. Gardners Gut Cave, Waitomo, North Island. Photograph by T.H.W.*

early branch off the superfamily Phyllostomoidea (the Noctilionidae of Kennedy et al. 1999) and was most closely related to the South American Noctilionidae, Mormoopidae, and Phyllostomidae. The observed divergence of *Mystacina* from the South American forms suggested that *Mystacina* had dispersed to New Zealand before 35 million years ago. More recently, Kennedy et al. (1999) analyzed mitochondrial DNA and found strong support for its classification in the Noctilionoidea, currently restricted to the neotropic Americas. Analysis of the data suggests the divergence of modern chiropteran families radiated about the Cretaceous–Tertiary boundary and that *Mystacina* was closest to the Mormoopidae with a divergence time of 45 million years ago.

Figure 13.2. Photograph of a skeleton of a *Mystacina robusta* collected as shown from Millars Waterfall Cave, Waitomo, North Island. Photograph by T.H.W.

However, the origin of *Mystacina* in South America has been overturned by the recent discovery of fossils in Australia. Although many bat fossils are known from Australia, the identity of some remained obscure until Hand et al. (1998) described three new species in a new genus *Icarops,* in the family Mystacinidae, from the early–middle Miocene of northern Australia. It was suggested that *Icarops* and *Mystacina* are sister taxa, which implies their common ancestor must be of at least late Oligocene age. Because of the prevailing wind direction, and because of the historical fact that most dispersal events have been from Australia to New Zealand, it seems probable that *Mystacina* resulted from a dispersal event out of Australia in the Late Oligocene or

Figure 13.3. Photograph of Mystacina tuberculata *foraging on the forest floor, Codfish Island. Photograph by R. Morris.*

early Miocene, when New Zealand was closer to Australia. Also, at this time, New Zealand was emerging from a prolonged period of widespread inundation by shallow seas. This does not negate a relationship to the Noctilionoidea but does require that the members of this group all became extinct in Australia and only survived in South America.

In New Zealand, only one species of mystacinid (*M. tuberculata*) that reached the Quaternary survives, and it is endangered. Less than a decade ago, it was thought to be much rarer, but its cryptic habits have obscured a healthier distribution. It still survives widely on both North and South Islands and some offshore islets. Fossil and historical records show the former existence of a much larger species. *Mystacina robusta* was described, and then as a subspecies, in only 1962 (Dwyer 1962) from specimens from Big South Cape Island off Stewart Island. By 1965, ship rats had invaded that island, and the bat (along with several birds) was extinct. Much later, it was accepted as a good species (Hill and Daniel 1985), mainly because of its demonstrated sympatry with the smaller *M. tuberculata*. Examination of fossils of large mystacinids from throughout New Zealand led Worthy et al. (1996) to conclude that they all were referable to *M. robusta*. Of considerable interest was the demonstration that the specimens exhibited a North–South size cline, with the larger specimens in the warmer north. The cline is counter to the suggestion that the larger species had evolved as a southern, more cold-adapted, form. Many smaller bats, including *Mystacina,* are not true or full homeotherms, and their body temperature cools during roosting (they do not hibernate). Before becoming active again, the bat must warm up, and to heat a larger body in a cold climate is energetically more costly than it is to heat a small one. Conversely, as one goes north and the ambient temperature increases, a larger body size is allowable for a given amount of stored energy.

Mystacinids are odd. They feed on the ground, hopping and flitting from one spot to another, often burrowing deeply into the litter in search of prey (Daniel 1979, 1990). The behavior is rodentlike or shrewlike. It seems these bats have taken the niche usually occupied by these small insectivores of continental regions. Mystacinids even excavate tunnels in rotten wood in search of food and for roost sites (Daniel 1990). They are known to pollinate the flowers of the saprophytic angiosperm *Dactylanthus taylori* (Webb and Kelly 1993). This rare plant has no photosynthetic tissues and only becomes visible when its flowers rise out of the forest floor litter. They are highly nectariferous and are apparently the only bat-pollinated flower to be found on the ground. The larger *M. robusta* was observed chewing on the meat of muttonbirds (*Puffinus griseus*) hanging up to dry, and it is believed to have preyed on nestling birds (Daniel 1990). It was even found roosting in petrel burrows (Daniel 1990). Usually mystacinids roost in hollow trees or caves, and the propensity for the latter has resulted in their fossil record being quite rich. The surviving lesser short-tailed bat is the only microchiropteran known to have a lek behavior. Males occupy strategic trees and attempt to attract females with all-night singing stints (Daniel 1990); the songs are partly audible to human ears.

In the prehuman ecosystem, mystacinids were much more common than they are now. They were significant prey for the laughing owl (*Sceloglaux albifacies*), and mystacinid bones are abundant in its prey accumulations (Worthy 1997b; Worthy and Holdaway 1994b, 1995, 1996b). Falcons also preyed on mystacinids, although far less frequently. It seems that the aerial feeding mode of *Chalinolobus* made this bat far less obtainable to owls and falcons, as its bones are rare in their prey accumulations.

Family Sphenodontidae

Genus Sphenodon *Gray*
Sphenodon punctatus *(Gray, 1842) (Tuatara)*
Sphenodon guntheri *Buller, 1877 (Tuatara)*

Among the most distinctive vertebrates of New Zealand are the sphenodontids *Sphenodon* spp., or tuatara (Fig. 13.4). These reptiles are the last living descendent of a worldwide group that appeared in fossil time back in the Jurassic, early in the dinosaur era. The rest of its group died out some 60 million years ago. Although New Zealand has no known pre-Pleistocene fossil record of sphenodontids, it must be assumed that tuatara have lived here since the rift with the rest of Gondwana, some 80 million years ago. In all that time, they have remained little changed. There are currently two species accepted, which are morphologically cryptic and can only be distinguished electrophoretically by allozyme analysis. DNA studies have failed to support the

Figure 13.4. Tuatara Sphenodon punctatus, *New Zealand's archaic reptile that survives only in the absence of predators on offshore islands, grows to about 60 cm maximum length. Photograph by R. Morris.*

distinction, and the differences may yet be considered as due to variation in the allopatric population concerned. After all, what is normal in such an ancient and unique lineage?

Tuatara are New Zealand's largest reptiles. They were predators preying on any large invertebrate (there were many large beetles and orthopterans in rodent-free New Zealand) and smaller vertebrates they could catch. Tuatara are, however, nocturnal. In prehuman New Zealand, the laughing owl was its chief predator. In most of the prey accumulations of this owl, the bones of hatchling to midsize tuatara are common. Only the adults, at over 400 mm total length, were safe.

The remaining populations of tuatara are now threatened. They have been extirpated from the mainland islands of New Zealand, and survive only on predator-free (i.e., rat-free) nearshore islands.

In the latest Holocene, they were present throughout the North and the South Islands, reaching nearly to 47°S. In central Otago, they were found up to 500 m above sea level, where deep snow covers the ground for long periods during winter. That they could cope with such harsh conditions is partly attributable to their being able to fast for considerable periods of time. They also have a prolonged egg development; incubation exceeds 12 months. They appear sluggish but can move rapidly when the need arises, particularly to grab prey. They are large, reaching slightly over 600 mm total length, and bigger males tip the scales at 1 kg. Like other New Zealand animals, they are long-lived.

Thompson et al. (1992) found that tuatara could live to be more than 60 years old in wild populations.

Tuatara live in burrows, which they excavate or enlarge from preexisting features. All surviving populations are on nearshore islands; the fact that petrels also occur on most of these islands has led some researchers to see a symbiotic arrangement. It is true that on occasion, a tuatara has been found in the same burrow as a seabird such as the fairy prion (*Pachyptila turtur*), but how long each cohabitation lasts is a matter of conjecture. The tuatara concerned may have fed in the last week and may not be hungry. Tuatara are often observed catching and eating these and the slightly smaller diving petrels (*Pelecanoides urinatrix*). Space on the islands is at a premium, and burrowing species need soil, which brings different species into proximity.

The Holocene fossil record of tuatara on the North and South Islands suggests that the present association with seabirds is only coincidental. Tuatara were widespread, but they seem to have been abundant only where well-drained soils were frequent. Their remains are common in sand dunes. One dune area where tuatara were most common was Tokerau Beach. The 10 km of dunes backing this beach in Doubtless Bay, Northland, are rich in fossils, but there is no evidence for petrel colonies. There is no petrel eggshell. Bones of various petrels are present but in frequencies no greater than would result from beach-wrecked specimens, which are still frequent. The absence of petrels is perhaps hardly surprising, because the dunes were a favorite nesting ground of the moa *Euryapteryx curtus*. Trampling by these large birds would have not been conducive to the survival of burrows in soft sand. Tuatara were also abundant in forests on the Pleistocene gravels of Takaka Valley in northwest Nelson, where there were no procellariids (Worthy and Holdaway 1994b). In the areas of former huge petrel colonies on the South Island's West Coast, such as on the Fox River, tuatara were exceedingly rare.

Lizards

Family Scincidae

Genus Cyclodina *Girard*
Cyclodina aenea *Girard, 1857*
Cyclodina alani *(Robb, 1970)*
Cyclodina macgregori *(Robb, 1975)*
Cyclodina oliveri *(McCann, 1955)*
Cyclodina ornata *(Gray, 1843)*
Cyclodina whitakeri *Hardy, 1977*
Cyclodina northlandi *Worthy, 1991*
(Plus Two Undescribed Species)

Genus Oligosoma *Girard,*
Oligosoma chloronoton *(Hardy, 1977)*
Oligosoma fallai *(McCann, 1955)*
Oligosoma gracilicorpus *(Hardy, 1977)*

Oligosoma grande *(Gray, 1845)*
Oligosoma homalonotum *(Boulenger, 1906)*
Oligosoma inconspicuum *(Patterson and Daugherty, 1990)*
Oligosoma infrapunctatum *(Boulenger, 1887)*
Oligosoma lineoocellatum *(Dumeril and Dumeril, 1851)*
Oligosoma longipes *Patterson, 1997*
Oligosoma maccanni *(Patterson and Daugherty, 1990)*
Oligosoma microlepis *(Patterson and Daugherty, 1990)*
Oligosoma moco *(Dumeril and Bibron, 1839)*
Oligosoma nigriplantare *(Hardy, 1977)*
Oligosoma n. polychroma *(Patterson and Daugherty, 1990)*
Oligosoma notosaurus *(Patterson and Daugherty, 1990)*
Oligosoma otagense *(McCann, 1955)*
Oligosoma smithi *Gray, 1845*
Oligosoma stenotis *Patterson and Daugherty, 1994*
Oligosoma striatum *Buller, 1871*
Oligosoma suteri *(Boulenger, 1906)*
Oligosoma waimatense *(McCann, 1955)*
Oligosoma zelandicum *(Gray, 1843)*
(Plus about Five Undescribed Species)

Family Gekkonidae

Genus Hoplodactylus *Fitzinger*
Hoplodactylus chrysosireticus *Robb, 1980*
Hoplodactylus delcourti *Bauer and Russell 1986 (?New Zealand)*
Hoplodactylus duvaucelii *(Dumeril and Bibron, 1836)*
Hoplodactylus granulatus *(Gray, 1845)*
Hoplodactylus kahutarae *Whitaker, 1985*
Hoplodactylus maculatus *(Gray, 1843)*
Hoplodactylus nebulosus *(McCann, 1955)*
Hoplodactylus pacificus *(Gray, 1842)*
Hoplodactylus rakiurae *Thomas, 1981*
Hoplodactylus stephensi *Robb, 1980*
(Plus about 13 Undescribed Species)

Genus Naultinus *Gray*
Naultinus elegans *Gray, 1842*
Naultinus gemmeus *(McCann, 1955)*
Naultinus grayii *Bell 1843*
Naultinus manakauensis *(McCann, 1955)*
Naultinus rudis *(Fischer, 1882)*
Naultinus stellatus *Hutton, 1872*
Naultinus tuberculatus *(McCann, 1955)*

New Zealand has only four genera of lizards, but these have a recently discovered high diversity within them (Daugherty et al. 1994). There are only two gecko genera (*Naultinus,* 7 species, which includes *Heteropholis,* and *Hoplodactylus,* 9 plus 13 undescribed species) and two skink genera (*Cyclodina,* 9 species, and *Oligosoma* [= *Leiolopisma*], 21 plus at least 5 undescribed species; Fig. 13.5). This diversity

represents a 64% increase in the size of the known fauna since 1980. Many of the known species are still undescribed.

Many lizards now have restricted distributions. This is especially true of the larger species. The largest extant gecko is *Hoplodactylus duvaucelii,* which is now restricted to predator-free (i.e., rat-free) offshore islands. However, the Late Holocene fossil record indicates it was widespread through the North and South Islands. Similarly, the skinks of the genus *Cyclodina* contain four extant, named, large taxa. All are nocturnal forest-dwelling species, and all are now restricted to rat-free offshore islands, except for a single vulnerable population of one species that survives in a boulder pile in the southern North Island. The fossil record of these larger *Cyclodina*s shows all were formerly widespread in the North Island (Worthy 1987d). Like the tuatara, many New Zealand lizards are likely to be long-lived. Thompson et al. (1992) reported a gecko (*Hoplodactylus duvaucelii*) that was at least 36 years old.

Figure 13.5. Photograph of an unidentified skink skeleton in Ruakuri Cave, Waitomo Caves, North Island. Photograph by T.H.W.

A single fossil species of skink has been described from New Zealand. *Cyclodina northlandi* (Fig. 13.6) was possibly 50% larger than the largest extant species, *C. alani,* and was apparently confined to the area north of Auckland in North Island (Worthy 1991a). Although no fossil geckos have been described, *Hoplodactylus delcourti* was based on a single unprovenanced stuffed specimen in the Musée d'Histoire Naturelle de Marseille in France (Bauer and Russell 1986). This is the world's largest gecko, at about 370 mm snout–vent length. However, it

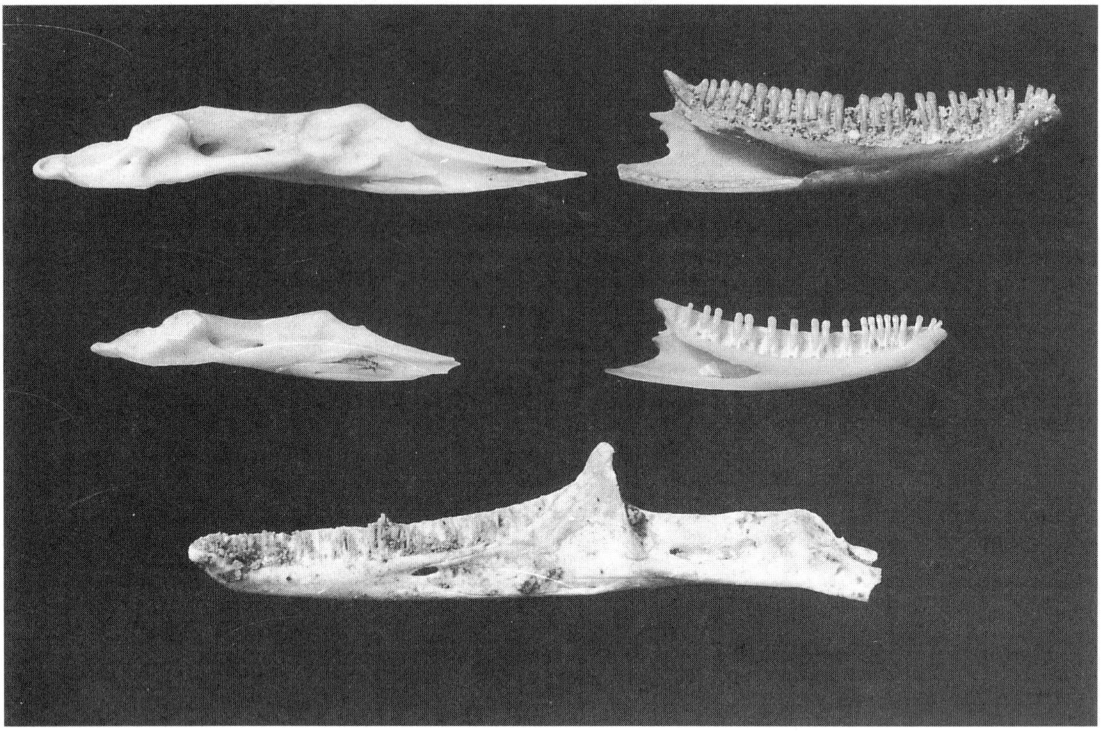

Figure 13.6. Dentaries from Cyclodina northlandi *(upper and lower), compared to* Cyclodina alani *(center). Published with permission from the Royal Society of New Zealand from Worthy (1991a: fig. 3).*

is only provenanced to New Zealand by the coincidence that *Hoplodactylus* is only known from New Zealand. No bones in the huge collections available could be of this species, with the single exception of one specimen that Bauer and Russell (1986) interpreted as a cloacal bone without actually examining it. The bone is not referable to any class of vertebrates with certainty; it could be one of the several vestigial bones in a rat, duck, or tuatara, and considering that thousands of bones have been obtained from the site (Earnscleugh Cave) from these animals, but not a single other gecko bone, its referral to *H. delcourti* is most unlikely to be correct (Worthy 1998b). With the recent analysis of laughing owl prey accumulation sites from the South Island, particularly from the previously fossil-poor areas of Canterbury, vast numbers of small vertebrate bones have been examined without a single bone attributable to *H. delcourti.* If it ever existed in New Zealand, it was not in the South Island. Similarly, large fossil faunas have now been obtained from Northland, Hawke's Bay, Waitomo region, and southeastern regions of the North Island, with no sign of this species.

Frogs

Family Leiopelmatidae

Genus Leiopelma *Fitzinger*
Leiopelma archeyi *Turbott, 1942*
Leiopelma hamiltoni *McCulloch, 1919*

Leiopelma hochstetteri *Fitzinger, 1861*
Leiopelma waitomoensis *Worthy, 1987*
Leiopelma markhami *Worthy, 1987*
Leiopelma auroraensis *Worthy, 1987*

New Zealand, superficially a group of oceanic islands, has frogs. As a rule, such islands lack amphibians. They are some of the biological evidence of the continental origin of the island country and of its ancient origins. Placed in their own family (Leiopelmatidae), they are, along with the Ascaphidae of the American Rockies, the most primitive living frogs. Characters of the skeleton, including the presence of 10 and not fewer vertebrae and the fact that the centra are hollow at either end (amphicoelous), ally these frogs with Jurassic fossils.

But although it is of ancient lineage, one group of leiopelmatids has a specialized reproductive strategy. Frogs of the *Leiopelma hamiltoni/archeyi* lineage have no tadpoles in the normal sense (Fig. 13.7). The young develop in the eggs right through the tadpole stage and hatch as small, tailed froglets. The adult male broods the eggs until they hatch, and after hatching, the froglets ride on the male's back. These frogs are terrestrial. Although they require a damp environment to live in, they do not need water in the form of pools or streams. The surviving member of the other lineage, *L. hochstetteri,* is facultatively aquatic and needs water in either seepages or tiny streams for breeding. It is usually found within a few meters of such habitat. All leiopelmatids are nocturnal. The three surviving species are all small—not more than 45 mm snout–vent length.

However, as seen in the lizards, some of the larger members of the group are recent losses from the New Zealand fauna. There were formerly three other species, all considerably larger than the extant forms. One reached 100 mm snout–vent length (Worthy 1987b). The largest, *L. waitomoensis* (Fig. 13.8), was restricted to the North Island and was allied to the *hamiltoni/archeyi* group. Somewhat incongruously, it is usually found in fossil sites where streams were accessible. The other two species were squat, toadlike forms indeed. Their front legs were nearly as robust as the hindlimbs. They were allied to *L. hochstetteri,* and presumably they walked and did not jump. From their fossil distribution, it would seem that they had highly terrestrial habits. One (*L. markhami*) was distributed throughout the North and South Islands, except in the driest eastern regions, whereas the other (*L. auroraensis*) is known only from Fiordland. No fossil frogs are known from any of the offshore or nearshore islands of New Zealand. However, two nearshore islands provide the last refuge for *L. hamiltoni,* the largest surviving terrestrial form. The loss of these frogs and reduction in the distribution of *L. hamiltoni* occurred in the latest Holocene.

As diverse as these groups of vertebrates discussed in this chapter are, the bats, lizards and frogs all share one thing. The nocturnally active species of terrestrial habits have suffered the greatest declines, and the larger members of each group have suffered most, some to the point of extinction. The surviving members of each group with these factors (biggest, nocturnal, and terrestrial: the largest gecko *H. duvau-*

Figure 13.7. Leiopelma hamiltoni *on Maud Island, Marlborough Sounds, South Island. Photograph by T.H.W.*

Figure 13.8. Arranged skeleton of Leiopelma waitomoensis. *Scale bar = 30 mm.*

celii, the four largest extant *Cyclodina* skinks, and the frog *Leiopelma hamiltoni*) are now restricted to islands without rodents of any sort. The extinction of these forms and the range restriction of the others is undoubtedly a result of predation by *Rattus exulans,* which was introduced about 2000 years ago. The ravages of this rodent extended well beyond its well-documented penchant for birds.

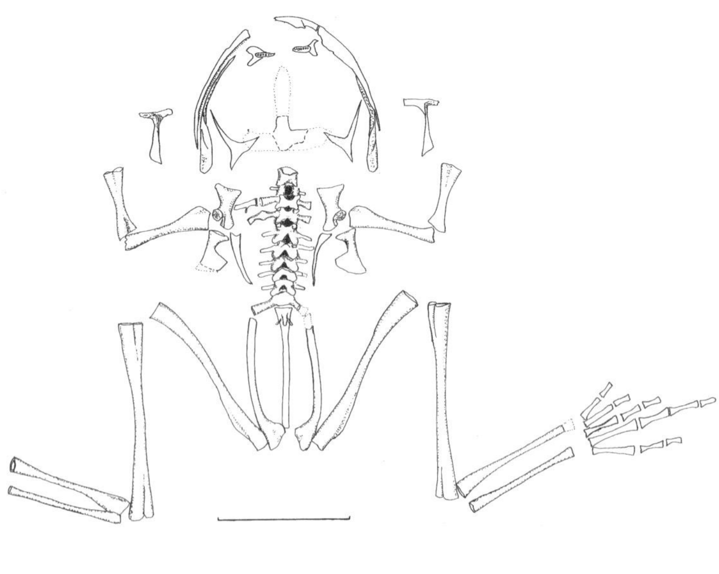

14 • Guilds and Regional Faunas

Although most studies of New Zealand Quaternary paleobiology have concentrated on individual species or groups of related species, ultimately these species, and all the other components of the former environments, had to live together. The relative lack of studies on local and regional groupings of species probably results from the fact that it is even more difficult to study the extinct assemblages than it is to work on individual extinct species. The difficulties in turn result from the fossil deposits—the raw materials for research—being several steps removed from the living systems that they represent, and they include only a few of the species of animal, and usually none of the plants, at that site. Each species is sampled in a different way by the different fossil deposits, such as caves, swamps, and predator. To put the species in their environment, proxy information on climate and vegetation and analogy with surviving systems are necessary.

At the present level of understanding, it is misleading to refer to the groups of species that are found together in New Zealand Quaternary fossil deposits as "communities." A community is an assemblage of species of animals, plants, and other organisms that occupy a particular environment and that interact with each other and their environment, thus forming a definable entity with its own structure and development. Few of these attributes can be detected at the distance in time that separates us from the living communities that contributed to the fossil deposits. What we can see—and study—are the assemblages found in the fossil deposits themselves. We can observe the patterns of their composition through space and time by dating the time of deposition of the fossil remains. Patterns of association can then be compared with results of studies of recent assemblages. By use of this information and information from a wide range of other disciplines, we can then form a picture or reconstruct the past environments and communities at various levels of detail, then infer regional patterns of change with climatic fluctuations.

One way of integrating the information on the distributions of animals, plants, climate, and terrain into a picture of the past environment is to look at the groups of species that feed in the same way or at the same places. Ultimately, all organisms interact with their environment in their own way, but similar organisms often interact with their environment in similar ways. Such a group of species that have closely similar niches, and especially diets and the way they obtain their food, are termed "guilds." The structure and relationships within the fauna can best be presented as an hierarchy, from guilds and other groupings of individual species through regional faunas (which can also be seen as temporal faunas) and faunas of constituent islands to the total fauna of the archipelago. Rather than there being a single assemblage or "New Zealand fauna" found throughout the main islands, as has often been assumed or implied (Atkinson and Millener 1991), distinct assemblages were associated with each island and island group (Holdaway, Worthy, and Tennyson 2001) and in major geographic, vegetation, and altitude zones within the main islands. The distribution of assemblages changed through time as the vegetation pattern responded to the glacial–interglacial climate cycles. Although their distributions changed, the composition of each assemblage largely did not. For example, the same species inhabited cool, dry forest-shrubland mosaics on the West Coast during the Otiran Glacial as lived in vegetation with similar physical structure on the downlands and plains east of the Southern Alps during the Holocene.

The farther one goes back in time, the more difficult it is to tease out the details of changes in environments and faunas. New Zealand is a unique laboratory in which to study such changes because the major extinctions occurred so recently. Events equivalent to the Pleistocene–Holocene extinctions in North America are still occurring in New Zealand, as remnants of the megafauna struggle to survive the latest changes in their environment. Without extinctions as such to worry about, it is easier to trace changes in the distribution of faunal assemblages further back in time, including those accompanying the end of the most recent glaciation, between 14,000 and 10,000 years ago.

Much has been made of the fact that in North America 10,000 years ago the faunas included species' associations unknown at present (Graham 1993; Graham and Grimm 1990; Graham and Lundelius 1984; Guthrie 1984). These "disharmonious associations" have been used to argue that the latest glacial–interglacial cycle was unique. However, Alroy (1999) has convincingly shown that the incidence of disharmony has always been low and that not only was it present in the last Glacial period, but it was also present in similar frequencies in all faunas before and after then. One problem with testing these proposals is that events that are apparently contemporaneous within the limits of the available dating technologies could have occurred decades, even centuries, apart. Even worse, as is obvious from almost any Quaternary fossil deposit in New Zealand where more than one specimen has been dated, even bones that are resting on each other may be from animals that died hundreds or even thousands of years apart in time. When time-transgressive deposits such as these are buried by some later event, without dating each and every bone it would be quite possible to conclude that the species present were contemporaries at the site, even if they still exist elsewhere and are known to require markedly different habitats. It is possible that such effects are the main basis for apparent differences between the interpretations of faunal change in New Zealand and North America.

In New Zealand, it seems that the physical structure of the vegetation rather than factors in the physical environment governed faunal distributions. For example, the moa *Pachyornis australis* is a montane species in the northern South Island. In Southland, it is present in the lowland fauna in the dunelands. The cooler dunelands along the Foveaux Strait coast may have supported vegetation with a similar structure or composition to that in the subalpine zone. The tree line is lower in the far south, and there were probably corridors of suitable habitat linking subalpine areas with the lowlands. Such corridors were less likely to exist in the north of the island, where extensive montane forests covered the slopes of the main ranges and the tree line was higher.

Another example is that of the upland moa (*Megalapteryx didinus*) population in the Punakaiki karst on the West Coast. Elsewhere, it was also an almost exclusively montane species; it is certainly absent from deposits formed under dense lowland rainforest elsewhere in the western South Island. Steepness of the terrain, open patches in the forest, and the richer vegetation along the rockfalls below the river cliffs in the karst landscape, with access to the Paparoa Range, may have allowed it to penetrate the lowland forests that excluded it elsewhere on the coast. The karst area is one of the few places where *Megalapteryx didinus* is sympatric with *Anomalopteryx didiformis*; elsewhere, they replace each other altitudinally. Populations of *Megalapteryx didinus* in Otago may also have been able to reach suitable areas at lower altitude via habitat corridors.

The North Island had a rather different mix of major habitat types. A more subdued relief, and a less dramatic climatic division than between the eastern and western areas in the south, meant less contrast in the vegetation between areas. During the Holocene, a dense, multistoried forest like that in the north and west of the South Island covered much of the North Island, and the bird assemblage appears to have been much the same. One major difference between the islands was the area of shrublands and grasslands induced on and around the central plateau by ashfalls that sometimes covered tens of thousands of square kilometers. The scale of the glaciation was much less than in the South Island and was confined mainly to the southern ranges and the two highest volcanoes.

Until recently, it has been difficult to make sense of the assemblages and ecologic relationships within the prehuman avifauna because little was known about the smaller species. For the South Island at least, many of those gaps are rapidly being filled with the discovery and analysis of predator sites (Holdaway and Worthy 1996; Worthy and Holdaway 1994a, 1996a). A reasonably satisfactory list of the breeding species in the New Zealand avifauna at the time of first human contact was not available until sufficient sites with different taphonomies had been analyzed to be able to say with some certainty which species were present. Just as important, several species, such as the Australasian harrier and bittern, which had been regarded as being part of the original Holocene fauna, have been shown to have colonized after Polynesian settlement began in the late 13th century (Anderson 1991; Holdaway and Worthy 1997; Holdaway et al. 2001).

The original Holocene avifauna must be the basis for ecologic and evolutionary interpretations of the fauna because it was those species and assemblages that had coevolved in New Zealand for the past several tens of thousands of years. The fauna had managed to resist further immigration until major vegetation changes followed Polynesian fires. Even today, relatively intact forest bird communities continue to exclude recent immigrants, such as the grey-backed silvereye (*Zosterops lateralis*), which may be present and breeding on the outskirts of the forest stand (Diamond and Veitch 1981; McCallum 1982). Species have been found in discrete associations that must reflect assemblages in the living fauna because they almost always occur in sites with similar taphonomy and in areas with the same vegetation when the deposits were being formed. This interpretation contrasts with the last major review of fossil faunas, made on the basis of information gathered before intensive surveys were undertaken from the early 1990s (Atkinson and Millener 1991). In treating all species recorded from a region as having been of equal abundance and importance in the local fauna, Atkinson and Millener (1991) followed the traditional view of New Zealand ecology.

It is now evident that pooling species lists without regard to site age, site taphonomy, or habitat (as well as, significantly, relative abundance) obscures real patterns in the distribution and association of species (Holdaway and Worthy 1997; Worthy and Holdaway 1996b). However, Atkinson and Millener's (1991) approach of grouping the extinct and living species in broad guilds has proven to be useful (Holdaway and Worthy 1996, 1997). True, the allocation of some species, such as the laughing owl (Worthy and Holdaway 1996b) and adzebills, to guilds has had to be radically altered with the discovery that both were predators on other vertebrates rather than insectivores and herbivores, respectively, but the changes reflect information that was not available in 1990.

Formalizing the guilds is one way of making explicit, testable hypotheses about the role of extinct species. In addition to facilitating the investigation of assemblage structures by guild composition, the more detailed information available now has allowed the use of numeric techniques, such as clustering and canonic analysis (Holdaway 1991; Holdaway et al. 2001). One problem with the approach is that extinction of whole guilds makes it difficult to reconstruct, even in broad perspective, the synecology of the undisturbed fauna because there are no living analogs to draw on for comparative information.

The association of species and faunas discerned subjectively on the basis of co-occurrence in faunas of different ages (Worthy and Holdaway 1993, 1994b, 1995, 1996b; Worthy 1997b, 1998b) is now supported by objective evidence of co-occurrences in sites with similar taphonomy. When the different taphonomies of different sites are taken into account, a picture of the different assemblages can be built up. If the two main site types—predator and nonpredator—are used, distributions and associations of species at both ends of the size range can be examined. If during cluster analysis faunas from all deposits (predator and nonpredator) are pooled, the first major branches separate two assemblages of species representing the different kinds of deposit. Hence, because the different taphonomies involved directly influence the results, it is reasonable to analyze the sites separately. Put simply, small predators would not take large vertebrates, so moa are seldom represented in owl sites, and the smallest species are greatly underrepresented in swamp, dune, and normal cave deposits.

The Main Islands Were Similar but Different

One limitation of present information is that substantially more is known about the South Island faunas than about those of the North Island or the outlying islands. Enough is known, however, of the distribution of species and former habitats for a basic picture of the North Island faunas to be assembled. Future research will undoubtedly result in changes of detail and emphasis, but the overall pattern should not change drastically. Where the species were the same as or closely related to South Island species, the guilds were probably structured in similar ways. For example, the wet forest terrestrial herbivore guild in the North Island included *Dinornis novaezealandiae* and *Anomalopteryx didiformis*, as it did in the same habitat on the South Island west coast. In the wet forests, according to isotopic evidence, *Dinornis novaezealandiae* fed on plants that grew beneath the closed canopy, whereas *Anomalopteryx didiformis* probably relied on plants growing where the canopy was broken, such as around tree falls and slips and along streams.

Although most guilds were substantially the same in the North and South Islands, the different species in some groups meant that the details differed slightly between the islands in some. Obvious examples are the absence of yellowheads (*Mohoua ochrocephala*) and brown creeper (*M. novaeseelandiae*) from the North Island and absence of whitehead (*M. albicilla*) from the South Island. Huia (*Heteralocha acutirostris*) and kea (*Nestor notabilis*) were confined to the North and South Islands, respectively. Other differences were the presence of the snipe rail (*Capellirallus karamu*) in the North Island only and Haast's eagle (*Harpagornis moorei*) in the South Island only. Hence, the composition of the arboreal insectivore, terrestrial omnivore, and predator guilds differed between the two islands, the predator guild most significantly.

Guilds

Several of the basic trophic/behavioral guilds were common to both islands, especially those associated with freshwaters or the sea. The members varied somewhat, particularly the petrels. Marine species depend on different water masses, and their breeding colonies are usually close to the feeding grounds that they use during nesting. The South Island faunas and guilds are better known than those in the North Island, so the descriptions and discussions that follow concentrate on those. Differences in composition and other information from the North Island fauna are added where available.

Seabirds

Of the truly marine birds, New Zealand has penguins, petrels, gulls and terns, marine shags (*Stictocarbo* and *Leucocarbo*), and gannets (including the subtropical boobies). Three species of penguin bred on the main islands, the little penguin (*Eudyptula minor*) and two larger birds, the New Zealand crested penguin (*Eudyptes pachyrhynchus*) and the yellow-eyed penguin (*Megadyptes antipodes*). All three species nest under cover of dense vegetation or in burrows or caves. None of the three breeds in the dense colonies on open ground that are characteristic of other species in the subantarctic and Antarctic. The little penguin also breeds on the Chatham Islands, and the yellow-eyed penguin has populations on the Auckland and Campbell Island groups. The New Zealand crested penguin breeds (or bred) on the main islands and those around Stewart Island and is one of five species of crested penguin that are confined, or nearly so, to the archipelago—Chatham Island penguin (*Eudyptes* sp.), Snares crested (*E. sclateri*), erect-crested (*E. robustus*), eastern or Filhol's rockhopper (*E. filholi*), and the royal penguin (*E. schlegeli*). Of these, the yet-to-be-described Chatham Island bird is now extinct.

Little penguins feed in coastal waters, often close to shore. Yellow-eyed penguins seem to travel farther out from the coast, but most of them still spend each night on shore. The crested penguins leave their breeding sites after nesting and do not return until the next season. Although it is obvious that they spend that time at sea, no one knows where any of the crested penguins live for most of the year. Penguins can feed much deeper in the water column than flying birds such as petrels. Their ability to move to different feeding areas is, however, more limited, so when food supplies fail because of climatic factors such as the El Niño–Southern Oscillation, many starve. Their role in the terrestrial ecosystem was similar to that of the petrels in bringing oceanic nutrients to shore. Many fewer individuals and the primarily coastal breeding sites meant that their contribution was probably much less, however. Individual penguins would have been one source of food for Eyles's harriers and even Haast's eagles hunting along the forested edge of sandy beaches.

The petrel guilds (marine feeders) seem to have been important on all coasts and in all parts of the extended archipelago. While at sea, they subdivide their apparently homogeneous environment by exploiting prey found at various depths in separate water masses. For example, common diving petrel rarely venture far from land, taking small crustaceans in the surface waters. South Georgian diving petrels feed in the open sea at the edge of the continental shelf. Among the small shearwaters, fluttering shearwaters (*P. gavia*) dive for small fish in coastal waters, usually in harbors and inlets, whereas the slightly larger Hutton's shearwater (*P. huttoni*) feeds farther out on the continental shelf and on open coasts; it even makes long-distance migrations to southern Australian waters and beyond. What part the small Scarlett's shearwater played is unknown. Its wing conformation and the location of the breeding colonies on the West Coast of the South Island suggest that it fed over oceanic waters and may also have been migratory (Holdaway and Worthy 1994).

On land, petrels bred in both single-species and mixed colonies, from lowlands to the subalpine zone. For the terrestrial environment, their underlying importance may have been in the supply of nutrients to the typically nutrient-poor New Zealand soils (Hawke et al. 1999; Okazaki et al. 1993) and the mechanical alteration and aeration of soils (Okazaki et al. 1993). More directly, however, their sheer numbers were also obviously important in the terrestrial food chain. There are still remnants of this lost link in the ability or habit of kea to hunt, kill, and eat Hutton's shearwater chicks in the colonies above the winter snow line in the Seaward Kaikoura Range. Laughing owls certainly depended largely on small petrels on the West Coast (e.g., Hermit's Cave and Road Cave), and probably elsewhere, before all the small species became extinct under predation by the Pacific rat (Holdaway 1999b). Aside from the living birds, the guano, broken eggs, and dead chicks provided nutrients for members of the invertebrate food chain. All members of the guild can be regarded as being at the base of the terrestrial food chain, not participating in the chain above its lowest level. A parallel is the breeding migration of salmon in North America (Kline et al. 1993), but there, the nutrients are largely dispersed in freshwaters, except for the salmon taken by eagles, bears, or scavengers.

The marine shags are mainly inshore feeders, flying up to 20 km from their colonies to dive for school fish. Each species of the genus *Leucocarbo* occupied a discrete geographic range, so there was overlap—more than one species in the guild—only with the spotted shag (*Stictocarbo punctatus*) on the main islands and the Pitt Island shag (*S. featherstoni*) in the Chathams.

Around the fish shoals, the marine fishing guild included the Australasian gannet (*Morus serrator*) near the main islands and the masked booby (*S. dactylatra*) in the subtropical Norfolk and Kermadec groups.

Piscivores

If the oceanic species—petrels, gannets, and marine shags—are excluded, the New Zealand avifauna included few piscivores. The guild included a sea duck (*Mergus australis*), the coastal shags and freshwater cormorants, two herons, the little bittern, and, in part, the grebes and kingfisher. To that list may be added the black-fronted tern and Caspian terns (*Sterna albofrontata* and *S. caspia*), which fished in braided streams, estuaries, and coastal lakes. A combination of the lack of phylogenetic diversity and the character of the New Zealand freshwater fish fauna probably left only a limited range of options and opportunities. Whereas it is relatively easy in evolutionary terms for a species to become larger or smaller—or even flightless—a radical change of diet may be less easy to achieve. The phylogenetic diversity of the avifauna was controlled by the ability of different groups to cross the water barriers or to survive in situ through changing geographies.

Although depauperate in species, the freshwater fish fauna was rich in individuals (McDowall 1978). The genera *Galaxias, Retropinna,* and *Gobiomorphus* contribute most species, but one of the most abundant species was *Prototroctes oxyrhynchus,* the New Zealand grayling. Although some species in the other genera are now rare, the grayling is the only species that is now almost certainly extinct. Most freshwater fish in New Zealand are catadromous—that is, they live in freshwater but the hatchlings go out to sea and return as larvae. Two species of eel, *Anguilla dieffenbachii* and *A. australis,* were also abundant. Both eels migrate thousands of kilometers to spawning grounds off Samoa. Species of *Galaxias* spawn in the tidal stretches of rivers and vast runs of larval fish ("whitebait") once migrated upstream in spring. The numbers of all the whitebait species have declined dramatically with the removal or damage of wetlands, and large runs occur now only on the West Coast of the South Island. *Gobiomorphus* spp. breed both inland and near the coast, and their young join the influx of galaxiids in spring. Schools of mullet and the New Zealand sea trout or kahawai (*Arripis trutta*) schooled in the estuaries, whereas black flounder (*Rhombosolea retiaria*) still live not only in estuaries and river mouths but also follow rivers up to 250 km inland (McDowall 1990), which is as far as it is possible to reach from the sea in the narrow main islands.

In New Zealand, only the smelts of the genus *Retropinna* habitually swim high in the water column in lakes; the others are mostly benthic or cruise in shallow streams. Some New Zealand freshwater fish can move between river systems and up considerable waterfalls, reaching the higher stretches of streams. Eels move readily across country through wet vegetation and appear in many landlocked ponds and lakes. Eels may have provided much of the diet of the freshwater cormorants *Phalacrocorax carbo* and *P. melanoleucos.* They also appear occasionally in the diet of laughing owls. Both cormorants have survived persecution by sport fishermen for nearly 150 years.

The grayling (*Prototroctes oxyrhynchus*) was reported in the 19th century to have formed dense shoals in some rivers, but it declined rapidly after the introduction of brown and rainbow trout and its existence has not been recorded since the 1920s (McDowall 1978). However, it was sufficiently abundant to provide a food resource for the first European settlers and may be assumed to have been important for species such as the cormorants. As an algal feeder, it may have filled a niche now vacant in New Zealand freshwaters. Young fish may also have been part of the diet of terns and gulls.

One indication that total abundance of fish may have affected the range of large piscivorous birds is the absence of permanent populations of pelicans. Although they were—and are—regular vagrants to New Zealand, the absence of remains of juveniles in sites that should have been nesting sites, such as Lake Grassmere in Marlborough, indicates that breeding was sporadic, if it occurred at all. A large population of pelicans may have required a greater standing crop of shoaling surface-feeding fish than was available in the New Zealand fauna.

Into Freshwater but Not into Fish

The New Zealand waterfowl fauna is unusual because many of the waterfowl lived mainly on land. These species form part of terrestrial guilds. Of the aquatic waterfowl—those that gain most, if not all, of their food from standing or running water—one, the merganser (*Mergus australis*) was a piscivore. The others fed on insects, molluscs, and aquatic vegetation. Scarlett's duck (*Malacorhynchus scarletti*) was the only specialized filter feeder, sieving the surface waters for small organisms. Grey teal (*Anas gracilis*) and grey duck (*Anas superciliosa*) dabbled for water plants and small animals in the shallow water, and the scaup (*Aythya novaeseelandiae*) dove to take bottom-living organisms. The large musk duck (*Biziura delautouri*) lived in larger, deeper swamps and lagoons, and the black swan was abundant on lakes, lagoons, and estuaries. Perhaps best known of all is the blue duck *Hymenolaimus malacorhynchos,* a specialist feeder of invertebrates in rapids and riffles in rivers. It is, however, now known to make nocturnal excursions into forest hundreds of meters from streams to forage in litter, which explains its presence in fossil sites at such distances from streams.

Two grebes were widespread. The New Zealand dabchick (*Poliocephalus rufopectus*) bred even on small ponds surrounded by forest, such as Pyramid Valley in North Canterbury. The reason for its extinction in the South Island is not known. The small ponds and lakes it inhabits in the North Island are superficially the same as many in the South Island and one of the centers of population, in the Wairarapa, is nearly at the same latitude as the northern South Island in similar, highly modified countryside. The southern crested grebe (*Podiceps cristatus*) was widespread on lakes and lagoons. From the scarcity of its fossils, it is unlikely to have been as numerous as any of the ducks.

Around the edges of pools and lakes, and along riverbeds, black stilts (*Himantopus novaezelandiae*) and plovers (*Charadrius novaeseelandiae, C. bicinctus,* and *C. frontalis*) fed on insect larvae on the water surface, on the mud, and around stones. Black stilts bred at Pyramid Valley, set among forest and shrubland, as well as on open riverbeds and at estuaries. The white egret (*Casmerodius albus*) was more widespread. Its diet of invertebrates and small fish is varied with small terrestrial vertebrates, such as lizards and small birds.

Terrestrial

Herbivore

Allowing for their being composed mainly of extinct species, the guilds containing moa and the other large herbivores are reasonably well known. The eastern (Holocene) assemblage included the giant moa *Dinornis giganteus,* the eastern moa *Emeus crassus,* the heavy-footed moa *Pachyornis elephantopus,* and the stout-legged moa *Euryapteryx geranoides.* To the west of the main ranges there were the large bush moa *Dinornis novaezealandiae,* the slender bush moa *Dinornis struthoides,* and the little bush moa *Anomalopteryx didiformis.* The little bush moa also occupied the higher hillsides of the inland ranges and places such as Banks Peninsula, where there were areas of taller, closed-canopy forest. At higher altitudes still, there were crested moa *Pachyornis australis* and the upland moa *Megalapteryx didinus,* which replaced the little bush moa in the submontane forest, shrublands, and herb fields. One species, the slender bush moa *Dinornis struthoides,* seems to have been such a generalist herbivore that it was found in small numbers almost throughout New Zealand. Analysis of sites in northwest Nelson reveals that *Dinornis novaezealandiae* also used the subalpine shrubland, a versatility of habitat use that allowed it to survive on the West Coast during the Otiran glaciation, when wet forest had restricted distributions.

In contrast, the habitat preferences of the other moa appear to have been more specific. For example, *Euryapteryx geranoides* seems to have been more common on low hill country within its range, whereas *Pachyornis elephantopus* preferred shrubland or grassland on plains or gentle slopes. Its small congener in the North Island, *P. mappini,* was seemingly most abundant in the vegetation around swamps and lakes, where there were many creepers and low shrubs. It appeared in the Hukanui deposits (inland Hawke's Bay) during the period after the major volcanic ashfalls, when shrubland replaced much of the standing forest for 100–200 years. *Euryapteryx curtus* has been found in situations that suggest that it lived in lower forest, or shrubland, but not necessarily near swamps. Within its range in the North Island, extensive areas of shrubland were mainly found along the coast, particularly around Northland, along the southern Taranaki coast, and along the east coast of Wairarapa. *Dinornis giganteus* was found in similar sites in both islands. It may have been the large herbivore cropping the taller shrubs and creepers along the forest–shrubland ecotone. Both *Dinornis giganteus* and *Euryapteryx curtus* used the seral shrublands that existed after each eruption in the central North Island.

For many scientists, moa were the only terrestrial herbivores in New Zealand. However, other species, although smaller, were equally or even more specialized for a diet of vegetation. Two of the smaller herbivores survive today, and they illustrate how closely some species depend on details of the vegetation and how others were more versatile. The kakapo (*Strigops habroptilus*) appears to feed the fruit and leaves of some of the podocarps, especially rimu, and other gymnosperms to chicks, but they can substitute other taxa such as *Nothofagus* when podocarps are not available (Merton 1999; Merton et al. 1999; Powlesland et al. 1992; Trewick 1996a). In this lek species, the male contributes nothing apart from its genes; the female has to accumulate enough fat and nutrients to produce the eggs and raise the young on a vegetarian diet. Its difficulty in doing this in other than exceptional years for fruiting suggests that the apparent abundance of plants in the cool temperate zone may be deceptive. If some of the moa required especially high levels of food quality for breeding, those requirements could well have limited their abundance by requiring large feeding ranges in all but the most productive habitats.

It has been proposed that the other surviving large herbivore, the South Island takahe (*Porphyrio hochstetteri*), also had a restricted diet (Mills et al. 1984). In summer, the relict population in the Fiordland mountains feeds mostly on the tillers and seeds of a tussock grass *Chionochloa pallens,* and in winter, it feeds on rhizomes of the fern *Hypolepis* (Mills et al. 1984). Birds transferred to small islands and other lowland areas feed readily on other plants, including introduced clovers (*Trifolium* spp.), and even on small vertebrates such as geckos (A. H. Whitaker, personal communication). Many—even most—of the fossil occurrences of *P. hochstetteri* in the Holocene are in lowland areas where the vegetation was either dense shrubland or tall forest (Holdaway and Worthy 1997; Worthy 1997b, 1998b, 1998d; Worthy and Holdaway 1996b).

The behavior of the kakapo blurs the distinction between terrestrial and arboreal guilds. Although kakapo cannot fly (their flight has been described by the well-known naturalist photographer R. Morris as simply breaking their fall), they spend much of their time feeding above the ground in trees and tree ferns, reaching considerable heights and out onto slender branches. They can also move from tree to tree without descending by grasping the twigs of a neighboring tree with the beak, transferring the grip to one foot, then moving across (R. Morris, personal communication). Other species bridge the ground–tree barrier from the other direction. The kokako (*Callaeas wilsoni* and *Callaeas cinerea*) are members of the species-poor arboreal herbivore guild. They can fly, albeit poorly, but they move through the forest by running and bounding on long, strong legs. Starting low on a tree, kokako run and hop upward, feeding along the way, until they reach the top and glide down to the lower branches of the next tree. Kokako also spend a portion of their time on or near the ground, in the lower shrubs, or grubbing in moss on fallen logs or the forest floor. Mainland populations of the North Island kokako (*Callaeas wilsoni*) may have used the ground much more in the past, if anecdotal observations point to real differences in behavior (St. Paul and McKenzie 1974).

The closest relative of the extinct New Zealand geese (*Cnemiornis calcitrans* and *C. gracilis*) is the Australian Cape Barren goose (*Cereopsis novaehollandiae*), which grazes on grasses and herbs. Both New Zealand species are likely to have had a similar diet because their beaks are, if anything, better adapted to cropping low sward vegetation.

Neither species is common in fossil deposits (Worthy et al. 1997). All the occurrences are in areas where there would have been some grassland: around coastal lagoons, along rivers or around ponds, or in the drier parts of central Otago. These geese were perhaps more specialized in their habitat requirements than most New Zealand birds and were not a major part of the herbivore guild over most of the country in the Holocene.

Two further species may be included here to fill out the list of terrestrial herbivores, but both may have incorporated some animal matter in their diets. The New Zealand coot (*Fulica prisca*) and Hodgens' waterhen (*Gallinula hodgenorum*) were typical members of their genera, if larger. Their diets might be expected to have been similar to those of their relatives in Australia, always remembering that the New Zealand birds were much more terrestrial. Australian wetland birds tend to be nomadic to cope with the temporary and unpredictable nature of wetlands in Australia. The Australian coot (*Fulica atra*) has colonized New Zealand since the 19th century; its diet mainly consists of plants, but it includes some aquatic invertebrates (Heather and Robertson 1996). Ridpath (1969) lists aquatic vegetation, seeds, invertebrates, and frogs as the diet for the dusky moorhen (*Gallinula tenebrosa*). The New Zealand *G. hodgenorum* probably included all of these but may have also taken lizards, as does the takahe. Whether the mix of foods places the two rails among the herbivores or omnivores (discussed in the next section) may be resolved by analysis of the stable isotopes in their bone proteins, but this work has yet to be done.

In contrast, although Holdaway and Worthy (1996) included Finsch's duck (*Euryanas finschi*) among the terrestrial omnivores, information from feces attributed to the species and from the stable carbon signature of its bone gelatin ($\delta^{13}C$ values of –24 to –26) show that it was a terrestrial herbivore. It was found in many areas of the country, although it was mostly absent from faunas deposited under closed-canopy podocarp forests, found in regions receiving more than ~1500 mm of rainfall. Its very ubiquity suggests that it was a major element in the prehuman fauna, both as a consumer of low vegetation and (probably) fallen fruit and as prey to the two largest predators. Its close relative in Australia, *Chenonetta jubata,* eats leaves and seeds of grasses, sedges, and legumes (Frith 1969), and Finsch's duck probably consumed much the same mix in New Zealand. Worthy (1988b) showed that it was approaching flightlessness, a condition shared with all the other terrestrial herbivores. Digesting vegetation requires a large gut to be efficient, which in turn means a large body (with the benefit of a lower mass-specific metabolic rate to lower the energy requirements), all of which lead to flightlessness if there is no selection pressure from predators to retain flight.

Another species misassigned in Holdaway and Worthy (1996) is the paradise shelduck. This species is thought to be primarily vegetarian (Heather and Robertson 1996) and should be regarded as part of the terrestrial herbivore guild. Throughout most of the Holocene, the shelduck was mainly present in coastal areas with Finsch's duck inland. With the extinction of the latter, the shelduck, one of the two surviving species of the guild—and by far the most abundant—actually extended its range in response to the increase in suitable open habitat provided by farmland (Heather and Robertson 1996; Williams 1971). Paradise shelducks are also cavity nesters: one is known to have taken this to extremes. It nested in an old, abandoned house, gaining access via a broken window, then jumping down onto the floor, scuttling across the room so entered, down the hall, into another room, and flying up onto a wardrobe, where it had its nest (M. Williams, per R. Morris).

Omnivore

Of the other terrestrial ducks, the brown teal (*Anas chlorotis*) is extant. Remnant populations on the main islands and offshore islands, and the closely related *A. aucklandica* and *A. nesiotis* in the subantarctic, eat insects and marine invertebrates (Heather and Robertson 1996). The blue duck (*Hymenolaimus malacorhynchos*) also spends part of its time well away from streams or ponds. In the past, before mammalian predators arrived, the species may have been even more terrestrial. Its normal diet is presently aquatic insects (Wakelin 1993; Williams 1979), but recent observations support fossil evidence that the bird forages far from running water (Worthy and Holdaway 1994b). Its diet may be more diverse than is apparent from analyses of feces from birds living on streams.

The small teal are nocturnal (Weller 1974; Williams 1995). On the southern islands, the impetus for nocturnal behavior may come from the abundant skuas (*Catharacta skua*) and perhaps, on the Auckland Islands, the local population of the New Zealand falcon (*Falco novaeseelandiae*). On the main islands, the nocturnal behavior of these small ducks, as with many other species, may have been in response to predation pressure from the diurnal raptors.

Oliver (1955) lists grass and seeds as the food of the New Zealand quail (*Coturnix novaezelandiae*). However, the diets of the three Australian quail include insects (Frith 1969; Pizzey 1980), so it is likely that the New Zealand quail was equally catholic in its choice. It seems to have been confined to the riparian grasslands and shrublands. Its range within the main islands probably increased greatly with the environmental changes induced by Polynesian settlement (Worthy 1997b).

The last of the major members of the omnivore guild is the weka (*Gallirallus australis*), whose diet incorporates everything from invertebrates to fruit, intertidal invertebrates, eggs, reptiles, and ground-nesting birds (Carroll 1963; Heather and Robertson 1996). On some offshore islands, weka populations have been blamed for the declines of other indigenous birds (e.g., Blackburn 1968). Most of these weka populations have been introduced within the past 200 years, and the interactions may result from novel contacts between weka and other species that would not have been sympatric in the prehuman system. It is more likely that weka by themselves are not to blame for these declines. We question how populations on the main islands could coexist with many of the same small species there yet cause their decline on islands. Some of the effects on small islands attributed to weka predation are, however, the result of predation by the Pacific rat (*Rattus exulans*) (Holdaway 1999b). One of the species that Blackburn (1968) suggested had been eliminated by weka on the islands around Stewart Island, the banded rail (*Gallirallus philippensis*), is still found on the North and South Islands in places rarely sampled by the fossil record, and there is a fossil from the Pleistocene of Honeycomb Hill Cave (Worthy 1993b), so the broad sympatry of banded rails and weka on large islands has prevailed for at least 20,000 years. On the main islands, there was a much greater range of potential foods, and there were predators of the weka themselves. In addition, weka in the presence of other rails may have occupied a narrower niche than in the present much-altered environment, thus separating weka from some of those species now apparently vulnerable to weka predation.

As noted for the terrestrial herbivores, the potential to expand a feeding niche in an island environment with a limited number of competitors has resulted in some species being difficult to assign to a guild. The New Zealand and Chatham Island coots (*Fulica prisca* and *F. chathamensis*) and Hodgens' waterhen (*Gallinula hodgenorum*) must also be recognized as possible members of the terrestrial omnivore guilds. Species in both genera take plant and animal matter, and individuals of both have been incorporated in fossil deposits far from water. In New Zealand, it was possible that any one of these was a specialist herbivore or insectivore. Given the propensity of even the apparently specialized herbivore, the takahe, to take small vertebrates and large invertebrates as well as a wide range of plant material, as well as the propensity of the kakapo to be almost entirely vegetarian when others of its order in New Zealand are partly carnivorous, much more information from isotopic analysis and gut contents will be needed to establish the role of these rails.

Arboreal Insectivores

Gleaners

Apart from a number of small passerines that feed primarily by gleaning (searching the surfaces of leaves and branches for insects), several others also exploited arboreal insects as well as their normal diet of nectar, fruit, or leaves. The constitution of the guild differed between the islands, but North and South Island representatives of the genus *Mohoua* played a prominent role in both: the whitehead (*Mohoua albicilla*) in the North Island, and the yellowhead (*M. ochrocephala*) and brown creeper (*M. novaeseelandiae*) in the South Island. Whereas the whitehead searches for insects in the outer branches of canopy and subcanopy trees, the two South Island birds divide the niche. The brown creeper feeds near the outer parts of trees, and the yellowhead specializes in tree-crotch faunas while opportunistically taking insects from other places.

Also feeding among the leaves at the end of branches, and often hovering to reach the outermost leaves in the canopy, the grey warbler (*Gerygone igata*) is common in shrubland and low forest. On the Chatham Islands, it is replaced by a slightly larger species (*Gerygone albofrontata*) with similar habits.

At present, the New Zealand fernbird (*Bowdleria punctata*) is almost totally confined to swamps, coastal marshes, and dense shrubland. Records from tall forest in south Westland (O'Donnell and Dilks 1986) suggest that fernbirds may have occupied other habitats more extensively in the past. They may well have been part of the forest arboreal gleaner guild before rats and other predators eliminated populations in all but the densest vegetation.

Of the endemic New Zealand wrens, the rifleman (*Acanthisitta chloris*) gleans small insects from bark and foliage. There were few observations of the feeding habits of bush wrens before the last populations became extinct in the 1960s. Records suggest that they spent more time in the outer foliage than did the rifleman (Heather and Robertson 1996) and exploited invertebrates in mosses and ferns on the forest floor (Oliver 1955). On the islands around Stewart Island, the last surviving race of bush wren even hunted insects down petrel burrows (Guthrie-Smith 1925). This observation may indicate the former range of resources used by wrens on the main islands before the petrel populations were eliminated and highlights the difficulty of reconstructing the guild structure in the extinct fauna.

The rifleman and the other surviving species (the rock wren *Xenicus gilviventris*) are allopatric in most parts of their ranges. One is arboreal, and the other is terrestrial and lives in treeless environments, either with densely branched shrubs or in boulder piles entirely lacking in woody vegetation.

Of the other groups of passerines, the three honeyeaters, stitchbird (*Notiomystis cincta*), tui (*Prosthemadera novaeseelandiae*), and bellbirds (*Anthornis melanura*), also glean insects from leaves, branches, and trunks when nectar supplies are scarce. As another example of the seemingly general reluctance for many New Zealand birds to confine themselves to a single way of obtaining food, four of the callaeatids (both saddlebacks and both kokako) also glean insects from bark (saddleback) or leaves (kokako) (Powlesland 1987) and saddlebacks will eat nectar.

Bark and Wood Probers

The classic probers in the New Zealand avifauna were the huia (*Heteralocha acutirostris*) and North and South Island species of saddleback (*Philesturnus rufusater* and *P. carunculatus*). The long-billed wren (*D. decurvirostris*) (Fig. 14.1) also probed, but because it was flightless, it was probably as much part of the terrestrial fauna as of tree trunks (Millener and Worthy 1991). The crania and mandibles of the huia and saddlebacks are similar; the saddlebacks are miniatures of the huia. All three species have long retroarticular processes on which are (or were) inserted powerful muscles with which they could open the bill against the resistance of the bark, rotten wood, or moss into which it had been thrust (Burton 1974). Starlings have the same conical beak and structure of the mandible. They use their beak to extract food from the surface soil, forcing the bill into the ground and opening it underground to grasp the prey.

Figure 14.1. Dendroscansor decurvirostris, *the rarest of the extinct endemic New Zealand wrens (Acanthisittidae), as it may have appeared in life. Reproduced from Millener and Worthy (1991: fig. 3) with permission from the Royal Society of New Zealand.*

There are few descriptions of the feeding techniques of huia because the bird became extinct some time after 1907 (Heather and Robertson 1996). The apparent cooperation between male and female in obtaining wood-boring beetle larvae—the male tearing at the wood and opening the larval tunnels and the female probing deeply into the tunnels with her long, pliant bill—has become part of ecologic folklore. Unfortunately for tradition, the passage in Buller (1888) that is usually quoted in support of cooperative feeding specifically states that they did not cooperate while feeding. Males and females seem to have exploited different microhabitats, and no contemporary observations provide evidence of "cooperation" (Jamieson and Spencer 1996; Moorhouse 1996).

Woodpeckers do not cross the famous Wallace's Line (or, indeed, any of the other biogeographic dividing lines that have been suggested in the eastern Indonesian islands). Wallace's Line was named after Alfred Russel Wallace, the codiscoverer of evolution by natural selection and a pioneer biogeographer, who was the first to recognize the major biological boundary between Oriental and Australasian faunas that exists in what is now eastern Indonesia. Other boundaries, based on different groups of animals and plants, have been suggested since, but the recognition of Wallace's Line remains one of the pivotal points in the science of biogeography. To the east of the line, the role of woodpeckers is taken by other groups of birds in the oak forests of New Guinea, the eucalypt woodlands of Australia, and the podocarp and *Nothofagus* forests of New Zealand, that feed on wood-boring beetle larvae, albeit in rotten rather than sound wood. In New Zealand, two species in different orders masqueraded as woodpeckers. One was the male huia, already discussed as a prober. The male could also tear at bark and wood to expose the tunnels of beetle larvae. Wood-boring insect larvae are so important as sources of protein in the present diet of the kaka (*Nestor meridionalis*) that a bird has been known to spend up to 81 minutes to extract a single longhorn beetle larva (Beggs and Wilson 1987a). Kaka also attack the bark and wood of trees for sap. They score particular southern rata (*Metrosideros umbellata*) trees to make the sap run, leaving other, apparently identical, trees, alone (O'Donnell and Dilks 1989).

Kingfishers also dig into rotting or soft wood, but they do so only to construct a nest hole. Kingfishers seem to have been rare in New Zealand before Polynesian settlement and probably were restricted to a coastal distribution.

Insectivore–Nectarivore

The nectar-feeding guild includes the kaka parrot (*Nestor meridionalis*) as well as the three honeyeaters. Most of them are part-time nectarivores. The stronghold of the nectarivores in that niche is the north of the North Island, where the diversity of trees and shrubs (mostly angiosperms) is much greater than in the southern North Island or anywhere in the South Island, which has a higher percentage of endemic herbs (McGlone 1985). The stitchbird's dependence on fruit (Gaze and Fitzgerald 1982) implies that it was probably most abundant in the Northland forests, where the diversity of shrubs and trees is highest (McGlone 1985), but both it and the bellbird were extirpated there early in the European period. Few fossil deposits are known in Northland in which honeyeaters are likely to have been preserved (no laughing owl sites have been found so far), so its former abundance is unknown.

Major sources of nectar in southern forests were the mistletoes (Loranthaceae). O'Donnell and Dilks (1994) reported that 70% of feeding observations on kaka in southern Westland in January were on mistletoes, which provided about 46% of the nectar taken by kaka.

Most of the remainder came from one of the canopy trees, the southern rata (*Metrosideros umbellata*), and from an epiphytic orchid (*Earina autumnalis*).

A major source of plant sugars in the *Nothofagus* forest is honeydew—the exudation of the sap-feeding sooty beech scale insects (*Ultracoelostoma assimile* and *U. brittani*)—which appears as droplets on the end of the caudal filaments of the insects, which protrude 10–30 mm from the bark, and the black fungus covering the tree trunks. Both South Island honeyeaters use honeydew extensively, and it is also important to kaka (Beggs and Wilson 1987b). At present, tui are rare in the beech forests east of the main divide (Bull et al. 1985). Before the forests and shrublands of the plains and downlands were destroyed by Polynesian fires, tui were as abundant there as the bellbird (Holdaway and Worthy 1997).

Frugivore–Folivore

The arboreal herbivore guild included pigeons, kokako, and kaka. Pigeon will fly some distance (up to 18 km) to fruiting or budding trees (Clout et al. 1986, 1991), and the birds are important for the dispersal of some podocarps. Although kokako take invertebrates when feeding chicks, they feed mainly on leaves throughout the year, and they eat fruit when it is available (Heather and Robertson 1996). Kaka take fruit and seeds (O'Donnell and Dilks 1994); beech seed has been identified as critical to their breeding in *Nothofagus* forest (Wilson et al. 1998).

Nothofagus seeds are dry; succulent fruits are uncommon in *Nothofagus* forests. One of the principal sources of fruit for birds in these forests are the mistletoes (Loranthaceae). Ladley and Kelly (1996) showed that birds are important to the dispersal and germination of mistletoes.

Arboreal–Terrestrial

Insectivore

Although the food and habitat requirements of the rifleman and rock wren have been studied, the extinct wrens are more problematic. However, they were probably both arboreal as well as terrestrial. Only the bush wren (*Xenicus longipes*) and Lyall's wren (*Traversia lyalli*) survived into European time, Lyall's wren only on a small island. The bush wren was mostly arboreal; the others probably fed both on trees and the ground. Few observations were made of the habits or food of Lyall's wren. It was never seen to fly, which is not surprising given the rudimentary keel on its sternum (Millener 1989). Lighthouse keepers and collectors saw it on the ground among rocks. By the time they made the observations, most of the forest on Stephens Island had been cleared, so the bird's habits in a forest environment were not recorded. It probably was quite versatile; it occupied subalpine habitats far above the tree line as well as lowland forest.

Despite Millener's (1989) assertion that Lyall's wren was the only flightless songbird, it was already known that another, the stout-legged wren (*Pachyplichas yaldwyni*), had such an extremely reduced keel to its sternum and strong legs (Millener 1988) that it too was obviously flightless. Indeed, the next species to be described, the long-billed wren (*Dendroscansor decurvirostris*) (Millener and Worthy 1991), also had an extremely reduced sternal keel (i.e., small breast muscles) and stout legs. All three flightless species would have belonged to both the terrestrial and arboreal guilds. With its longer bill, *D. decurvirostris* would also have probed in bark and among tightly set leaves, food sources that put it in the bark-probing guild as well. Although there is also a record of long-billed wren from Takaka Hill, it is doubtful that it was present during the Holocene. If it was, its habits or habitat prevented it from being eaten by the laughing owls, which produced copious deposits of prey remains (Worthy and Holdaway 1994a, 1994b). During the Pleistocene, there would have been a low shrubland in the area, which is high and exposed, so the wren is likely to have preferred rocks, dense shrubland, or elfin forest. Many subalpine shrubs have thick, stringy bark.

The tits on the two main islands, *Petroica toitoi* and *P. macrocephala,* sortie from perches to pick up insect larvae from the ground or low branches. Each is sympatric with a species of robin (*P. australis* or *P. longipes*), which spend much of their time actually on the ground. Even the apparently obligate aerial insectivore, the fantail, will feed on the ground, even beneath the litter (R.N.H., personal observation) and is known to eat fruit occasionally as well (R. Morris, personal communication).

Frugivore–Folivore

Parakeets (more so the red-crowned parakeet) eat grass seeds, flowers, and fruit of low shrubs and herbs. They are also insectivorous and will eat meat, fat, and the contents of eggs when available. Because kakapo depend on fruit at some times of the year, they can be included here.

Omnivore

Of the omnivores, one of the most surprising is the kea (*Nestor notabilis*). Kaka (*Nestor meridionalis*) will take insects, but kea are known to actively hunt and kill other birds. In the past, it is likely that kea also killed moa already trapped in swamps or snow drifts, or at least shared in the spoils of an eagle kill. They may well be responsible for some of the characteristic damage to the anterior pelvis of moa in several deposits, which has been interpreted as being left by hook-beaked birds damage that exposed the kidneys and other organs of dead moa. Although scavenging of bird carcasses has been known for several years, the Antipodes parakeet was observed digging storm petrels from their burrows and eating them (Greene 1999). The added protein and fat component gained by scavenging and predation facilitates the survival of these parrots in cold habitats.

Both piopio were insectivorous as well as frugivorous. They were notably omnivorous around the camps of miners and explorers in the 19th century. Buller (1888) recorded insects, fruit, and leaves, and Henry (1903) noted a predilection for spiders. As noted below, piopio were flexible enough in their habits to have included hawking for insects in their feeding repertoire (Potts 1882).

Most crows are omnivores, although some are strict herbivores. In keeping with the trend for generalist behavior in New Zealand birds, it is likely that the New Zealand raven fed on fruit and insects when these were available. Its close relationship with coastal sites suggests that it also fed in and around seal and penguin colonies and on beach debris. The few birds that lived inland may have scavenged around moa carcasses.

Aerial Insectivore

There were no swallows or swifts in the original New Zealand avifauna, so the aerial insectivore guild was open to others. The only specialist species was the grey fantail (*Rhipidura fuliginosa*), one of a group of monarch flycatchers in Australia and the South Pacific. Potts (1882) observed the South Island piopio (*Turnagra capensis*) hawking insects. Tui (*Prosthemadera novaeseelandiae*) also occasionally hawk for flying insects, as do tomtits (*Petroica* spp.).

Predators

Apart from the terrestrial herbivores, the predatory birds were among the most interesting of the guilds. They included the largest known eagle, and harrier hawk, and surely the weirdest predatory birds of recent times, the adzebills.

Flying

Haast's eagle was found only in the South Island, which must be a reflection of the availability of suitable prey and the habitat it could hunt in. Dense, multitiered forests would have held too few prey for it and contained too few openings from which it could launch an attack. The African crowned eagle (*Stephanoaetus coronatus*) also takes comparatively large prey from the ground, and it typically hunts at dawn or dusk from a perch near a water hole. That scenario would fit the association of eagles and moa at Pyramid Valley, which was surrounded by tall matai (*Prumnopitys taxifolia*)5 forest.

In the North Island, the areas of richer mosaic habitat were smaller, which would have limited the population of an already thinly spread species. The eagle's place seems to have been taken by a larger form of Eyles's harrier, which was large enough to have preyed on the smaller North Island moa as well as a harrier's normal diet of pigeons, kokako, Finsch's duck, and kaka. Almost all of the South Island moa were large enough to be out of contention as prey for the Eyles's harriers there, but they were all within the ability of Haast's eagle to catch and kill.

The New Zealand falcon took the smaller species of bird, skinks, and insects, as it still does. Falcons also take prey up to and including the pigeon in size, so the whole range of vertebrates in New Zealand had one or more predators. Although they are active mainly during the day, falcons have been observed hunting after dark. Fossil prey accumulations include the strictly nocturnal mystacinid bats and petrels.

Flightless

The adzebills were identified as consumers of herbivores via analysis of stable isotopes of carbon and nitrogen in their bone gelatin (Chapter 5, text box). From their size and the remarkable adaptations of the jaw articulation, they are unlikely to have been solely insectivores, although even then the diet could have been varied with giant weta and weevils. Larger prey would have included lizards, tuatara, frogs, and small birds, perhaps petrels. Although there is no direct evidence for the time of day they were active, the adzebills are included here among the diurnal predators because of their small eyes and lack of evidence for an acute sense of smell (their nasal regions are relatively small, as in most birds). Some of their probable prey, such as tuatara and geckos, were active at night. However, the huge bill and powerful neck of the adzebills equipped them well to dig out burrows or wrench open rotten logs, and the large lachrymals provided protection to the eyes during such activities. As yet, we have no direct evidence of diet, such as preserved gizzards or prey remains from a nest site, that would tell us exactly which form of animal food adzebills ate, but their role as predators appears to have been firmly established (R.N.H., unpublished data).

Nocturnal Guilds

Aside from the typically nocturnal groups, such as owls, owlet-nightjars, kiwi, and petrels, many other New Zealand birds seem to have been active at night, or at least at dusk or dawn. Mention has already been made of the kaka, kakapo, and kea, but other surviving species are also largely nocturnal. For example, brown teal and the subantarctic teal feed mostly at night. On the southern islands, teal might be nocturnal because of predation by skuas in the daytime, and on the mainland islands of prehuman New Zealand, brown teal were avoiding predators.

Terrestrial Predator and Insectivore

This guild included the unique New Zealand tuatara (*Sphenodon* spp.). Young tuatara actively feed both day and night, but adults, although they occasionally bask during the day, only feed at night, when they join the small rails and kiwi in taking insects including weta, carabid beetles, and giant weevils. If, as is suspected (Rich and Scarlett 1977), the New Zealand owlet-nightjar flew poorly, then it too may have foraged on the forest floor for insects and even frogs and small lizards.

Terrestrial Insectivores

In New Zealand, a common evolutionary theme in the rails—and indeed among other ground birds—has been the development of long, probing bills. New Zealand is unusual in having high populations of terrestrial amphipods (Talitridae), which may process much of the plant detritus in forest soils (Duncan 1994). In places that lack rats and hedgehogs, there are still abundant faunas of invertebrates. Potential food for insectivores (or more accurately, microcarnivores) included one of the largest faunas of terrestrial snails (Climo 1975), large earthworms (Annelida: Lumbricidae, Megascolecidae) (Stout 1973), and insects, and no specialist mammalian insectivores, although short-tailed bats (*Mystacina* spp.) did their best to fill that niche as well as several others (Daniel 1979). Kiwi (*Apteryx* spp.) and rails (*Capellirallus karamu* and *Cabalus modestus*) evolved long, probing bills (the snipe, *Coenocorypha* spp., presumably already had them) and can exploit the litter fauna. The living species of litter and soil probers are nocturnal or crepuscular. With only avian predators, it is likely that the extinct taxa also fed in thick cover or emerged only at night or on overcast days.

Kiwi almost constitute a guild of their own. Worldwide, they are the most recognized of New Zealand birds, but little was known of their biology until the early 1970s. Their taxonomy is still controversial (Baker et al. 1995; Holdaway et al. 2001), but at least five species seem to have been present in the late Holocene. North and South Island brown kiwi (*Apteryx mantelli* and *A. australis*) and great spotted (*A. haastii*) and little spotted (*A. owenii*) kiwi survive (the last only as managed populations), but formerly, there was at least one species of kiwi in almost all parts of the main islands, from dense lowland rainforests to montane grasslands. They are still versatile, occupying habitats from beaches to subalpine scrub and from dense indigenous forests to plantations of introduced pines. They eat litter insects, but they also take berries and freshwater crayfish from streams, and (where they have access to it) they probe along the tide line for amphipods (Talitridae). All are in danger from introduced predators (Marchant and Higgins 1990), especially stoats, cats, and sometimes dogs. Attacks by dogs, such as the much-publicized one that killed several hundred North Island brown kiwi in a month (Taborsky 1988), damage local populations, but the overall decline is a result of long-term lack of recruitment caused by stoats (*Mustela erminea*) eating chicks and juveniles. Until some way can be found to remove stoats from the 260,000 km^2 of the main islands of New Zealand, the only viable populations of kiwi will be those on mammal-free islands and in mainland islands where the numbers of stoats and other predators are kept low by intensive trapping and poisoning.

In addition to the biting, blood-sucking sandflies (blackflies, Austrosimulidae), one other group of invertebrates fought back. Terrestrial leeches are now almost confined to predator-free islands, where they feed on petrels and penguins, among other birds, attaching to the web of skin between the toes (R. Morris, personal communication). When the mainland populations of petrels were in the billions, the leech population would have been similarly vast—a resource not available today. Some of the insectivores may well have included leeches in their diet, and vice versa.

Arboreal–Terrestrial Insectivore–Nectarivore

This multitalented guild was represented at night mainly by lizards, both geckos (Gekkonidae: *Hoplodactylus* spp.) and skinks (Scincidae: *Oligosoma, Cyclodina*); geckoes of the genus *Naultinus* and other *Oligosoma* skinks were diurnal. The species diversity within all these genera was grossly underestimated until biochemical techniques became available (Chapter 13). Even now, the precise habitat and feeding ecology of many species are little known. Some of the skinks have proven to be far more arboreal than previously thought. Mostly, however, it is the geckos that live in trees. All the geckos eat insects and will take nectar when it is available. Some species may be important in pollinating trees and shrubs. The population density of geckos on northern offshore islands without mammalian predators reaches astonishing levels, with a gecko per square meter of tree surface on flowering pohutukawa (*Metrosideros excelsa*) (Whitaker 1987). Similar densities would have been present on the two main islands before Pacific rats arrived. Also, populations of skinks on rodent-free islands reach densities never seen on the mainland now. The guild may have been a highly significant resource for predators such as tuatara and adzebills, and even perhaps the owlet-nightjar, and was important for the laughing owl. It is now obvious that the lizards were a larger part of the New Zealand equation than has been appreciated. On Takaka Hill, geckos constituted about 25% of vertebrate prey items in the diet of laughing owls (Holdaway and Worthy 1996; Worthy and Holdaway 1994b).

Most skinks in *Oligosoma* are diurnal, at least now, in the presence of nocturnal mammalian predators. Skinks in the genus *Cyclodina* were, or are, mainly nocturnal and confined to the North Island; on the main North Island, at least six species are extinct. The largest of these, *Cyclodina northlandi,* is globally extinct (Worthy 1987d, 1991a), as is another smaller species, *Oligosoma gracilicorpus. Cyclodina whitakeri* survives in a precarious population in a single boulder bank in a small area near Wellington, but it is otherwise absent from the North Island. It and others (*C. alani, C. oliveri, C. macgregori,* and *O. homalonotum*) survive on northern offshore islands.

Aerial–Arboreal–Terrestrial Insectivore–Nectarivore–Frugivores

The most unusual terrestrial vertebrate guild in New Zealand does not involve birds but two species of bat. The guild of aerial–arboreal–terrestrial insectivore–nectarivore–frugivore animals was formed by the mystacinid bats, *Mystacina robusta* and *M. tuberculata.* Only *M. tuberculata* survived on the mainland into European times (but see below), and the breadth of its niche is astonishing. It takes nectar and it catches insects in flight and on the forest floor and on tree limbs (Daniel 1979). There are reports of short-tailed bats chewing flesh and fat from petrel carcasses. In the absence of insectivorous mammals and nectivorous bats, the short-tailed bats have become the ultimate bat generalists. Feeding on and above the forest floor, on the limbs of trees, and even down petrel burrows, only insects flying above the canopy are not included in their prey. The extinction of the greater short-tailed bat leaves the populations of the smaller species as the only members of this unique guild. O'Donnell et al. (1999) reported the discovery of a population of short-tailed bats, only the second population known to remain on the South Island, in Fiordland, where they are sympatric with long-tailed bats (*Chalinolobus tuberculatus*). These short-tailed bats differ from individuals in populations of lesser short-tailed bats elsewhere and from specimens of the greater short-tailed bat, and their taxonomic status has yet to be clarified (O'Donnell et al. 1999). There may be unrecognized levels of diversity in this guild, as there are in the lizard guilds.

Aerial

The long-tailed bat (*Chalinolobus tuberculatus*) is the only nocturnal aerial insectivore in New Zealand. Individuals will fly up to 20 km from a solitary or small communal roost in a hole in a branch high in a forest tree or cave to feeding grounds in open areas along rivers, over the forest canopy, or over shrubland, and then return to a different roost in the same home area (Sedgeley and O'Donnell 1999).

Predators

At night, the forests and shrublands were the realm of the laughing owl (*Sceloglaux albifacies*), morepork (*Ninox novaeseelandiae*), and the owlet-nightjar (*Aegotheles novaezealandiae*). Their prey ranged from insects and earthworms to birds up to nearly a kilogram in mass. Only the larger birds, and those that lived in the canopy, were safe at night. Larger birds (over 1 kg) in New Zealand tend to be at least partially nocturnal; at night, they were protected from the attentions of Haast's eagle and from the laughing owl by their size.

Much of the laughing owl's prey, including the two species of short-tailed bat (*Mystacina robusta* and *Mystacina tuberculata*), lived or foraged on or near the forest floor. In that respect, the laughing owl was part of the terrestrial insectivore–predator guild as well. Few New Zealand birds seemed able to confine themselves to one way of earning a living—or needed to.

Despite repeated suggestions that most of the New Zealand avifauna was somehow inferior to species that evolved on continents, a myth started by Charles Darwin himself (Darwin 1884), species such as the laughing owl showed that they were able to cope with changes in their diet and habitat. They survived for centuries after substantial habitat change. There is now good evidence that the laughing owl switched prey after the arrival of the Pacific rat. Analysis of the owl deposits on Takaka Hill in the northern South Island show that the representation of indigenous fauna declined over the past 2000 years, at the same time as the number of rats in the sample increased dramatically. By the time Europeans arrived, the bulk of the laughing owl's diet consisted of the Pacific rat and the small birds that could withstand rat predation.

Extinction of the Pacific rat over most of the South Island (there is a remnant population in Fiordland) happened at the same time as black rats and mustelids spread through the island. Although the events were simultaneous, and although the relative effects of a reduction in food supply and predation by mustelids can never be separated in field data, predation was the major factor in the owl's decline. The new rodents and other birds, such as starlings (*Sturnus vulgaris*), chaffinches (*Fringilla coelebs*), song thrushes (*Turdus philomelos*), and blackbirds (*T. merula*), were spreading through the forests and replaced the Pacific rat in the owl's diet. Prey assemblages from sites in the region where owls survived longest, South Canterbury, do in fact contain a high proportion of introduced passerines as well as mice and European rats.

Owlet-nightjars elsewhere are typically insectivores. The large New Zealand species could have been a minor predator of small birds, lizards, and frogs. In the wetter parts of the main islands, the extinct frogs may have formed a significant component of the forest floor fauna, eating insects and being eaten by owlet-nightjars (probably), laughing owls, and moreporks.

Structure and Composition of Two Regional Faunas

Possible guilds have been worked out for two different geographic and ecologic areas. For Takaka Hill in northwest Nelson (Worthy and Holdaway 1994b) and in North Canterbury (Holdaway and Worthy 1997), adequate information has been collected on which to base some interpretation of the composition of faunas in two areas differing in their climate, vegetation, and topography. Knowledge of the geologic ages of the deposits and the composition of the microvertebrate faunas as well as the megafauna has allowed a step forward from previous attempts, which were hindered by lack of data on the time scale of deposition and by the scarcity of specimens for species smaller than ducks (Atkinson and Millener 1991). Similarity of terrain and climate, and, judging from pollen data, vegetation, indicates that the fauna of North Canterbury may also have been present in South Canterbury. Further south, in Otago, more differences might be expected because of the greater relief and more extreme climate.

The Canterbury sites were in shrubland and mixed podocarp–hardwood forest on plains or low hills, with beech on the steeper slopes and shrubland and tussock grassland on the exposed ridge tops. In summer, the area is subjected to strong, warm, dry northwest winds; the temperature frequently exceeds 30°C. In winter, cold, southerly winds bring rain and occasional snowfalls. The Takaka Hill fauna, as it was in submontane beech–hardwood, was not entirely representative of the wet western forests. Both regions showed changes with time. Takaka Hill is at a greater altitude than Pyramid Valley, and the rainfall is both greater and more evenly spread through the year. The time depth for the faunas was greater on Takaka Hill because of the dominance of old cave deposits, but only sites of Holocene age were used to build a picture of the prehuman fauna.

A representative sample of the Holocene fauna in the eastern South Island is available in the Pyramid Valley (Holdaway and Worthy 1997), augmented by predator sites in the surrounding hills (Worthy and Holdaway 1996b). These provide evidence for the presence of smaller birds and petrels that were rarely or never preserved in the lake sediments. Most of the predator deposits date from the past 2 millennia, but the lake deposit represents a relatively brief period about 2000–4000 years before the present (B.P.) Macrofossil and microfossil evidence from the deposit shows that the environment near Pyramid Valley was relatively constant from about 5000 years B.P., until Polynesian fires removed the surrounding forest about 650–700 years ago.

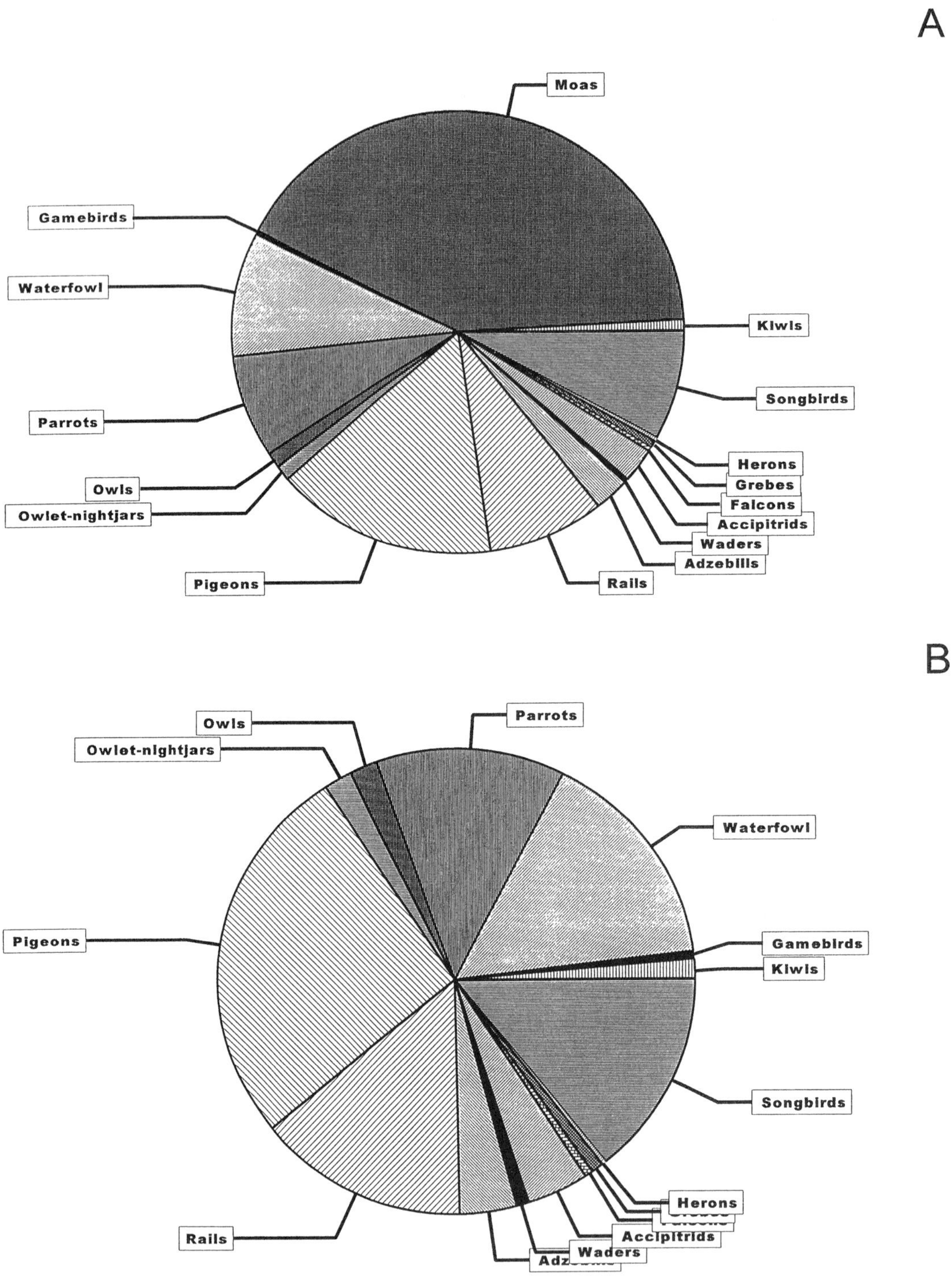

Figure 14.2. Representation (percentage of the minimum number of individuals) of major taxa in the Pyramid Valley, North Canterbury, lake deposit. (A) Including moa. (B) Excluding moa.

The swamp deposit fauna is dominated by the large ground birds (moa, kiwi, waterfowl, and rails), which together make up over 50% in terms of numbers of individuals (Fig. 14.2A). The other major components were the several parrots, the pigeon, and the songbirds, as can be seen in Figure 14.2B. Other groups, including the owls (and owlet-nightjar), raptors, and waders, were minor elements numerically but were important in rounding out the fauna. Despite the nature of the deposit, the fauna was overwhelmingly a terrestrial assemblage. Even most of the waterfowl were terrestrial in that they spent much time away from standing or running water.

Even aside from the waterbirds, the vertebrate fauna was as rich in species as anywhere in New Zealand (Table 14.1). There was a complete, functional fauna, with all trophic levels represented and the whole range of habitats occupied. In the forest and shrublands in the valleys and shrubland on the adjacent slopes of North Canterbury, four species of moa (*Emeus crassus, Euryapteryx geranoides, Pachyornis elephantopus,* and *Dinornis giganteus*) were common. The single specimen of *Dinornis struthoides* shows that this generalist and widespread species had a small population in the eastern downlands. A single goose (*Cnemiornis calcitrans*) was probably a stray from the grassland and shrubland on the higher slopes or on nearby riverbeds. It was apparently not a common part of the local fauna: one specimen was recovered, compared with over 180 moa.

On Takaka Hill during the Holocene, the vertebrate fauna was rather different to that in the drier eastern lowlands. Several of the guilds had different constituent suites of species (Table 14.2). For example, the terrestrial herbivores were limited to three species (all moa) instead of seven (five moa), which probably reflects the lower productivity of the forest at a substantially higher altitude (800 m compared with 330 m at Pyramid Valley), and the annual rainfall is about 2100 mm, whereas at Pyramid Valley it is about 650 mm. The forest on Takaka Hill was primarily composed of *Nothofagus* beeches, with mountain totara (*Podocarpus hallii*) and with a relatively even, closed canopy. The moa taxa (Table 14.2) were typical of wet western forests. The takahe required forest edges and shrublands, and the goose was restricted to grasslands and open habitats, so their absence is on Takaka Hill is understandable. The Holocene forests on the Hill also lacked two terrestrial omnivores, the waterhen and coot that were abundant at Pyramid Valley.

Two guilds or ecologic groups that were important in Canterbury were almost completely absent from Takaka Hill. The absence of standing water meant that the aquatic guild had no habitat. Perhaps surprisingly in view of their abundance elsewhere in northwest Nelson (Worthy 1993b) and on the West Coast (Worthy and Holdaway 1993), petrels were almost totally absent from the Takaka Hill fauna (Worthy and Holdaway 1994b). Two species only, *Pterodroma inexpectata* and *Puffinus spelaeus,* were represented in the deposits. The exact locality for the first species is not known and may have been collected in Takaka Valley, but several individuals of the small *Puffinus* were recovered

TABLE 14.1.
Broad guilds of vertebrates at Pyramid Valley during the late Holocene, before human intervention (after Holdaway and Worthy 1997: table 5).
DIURNAL

AQUATIC	TERRESTRIAL	ARBOREAL	ARBOREAL–TERRESTRIAL	AERIAL	PREDATOR
	Herbivore	Insectivore	Insectivore	Insectivore	
Malacorhynchus scarletti[a]	*Emeus crassus*[a]	*Acanthisitta chloris*[b,c]	*Xenicus* sp.[d]	*Rhipidura fuliginosa*[b]	*Circus eylesi*[a]
Anas gracilis[c]	*Pachyornis elephantopus*[a]	*Gerygone igata*[b,c]	*Traversia lyalli*[a]		*Harpagornis moorei*[a]
Anas superciliosa[c]	*Euryapteryx geranoides*[a]	*Mohoua novaeseelandiae*[c]	*Petroica macrocephala*[c]		*Falco novaeseelandiae*[c]
Aythya novaeseelandiae[c]	*Dinornis struthoides*[a]	*Mohoua ochrocephala*[d]	*Petroica australis*[c]		*Aptornis defossor*[a]
Himantopus novaezelandiae[d]	*Dinornis giganteus*[a]	*Bowdleria punctata*[d]	*Philesturnus carunculatus*[d]		
	Cnemiornis calcitrans[a]				
Poliocephalus rufopectus[d]	*Tadorna variegata*				
Casmerodius albus[d]	*Euryanas finschi*[a]				
	Porphyrio hochstetteri[d]				
	Omnivore	Insectivore–nectarivore	Frugivore–folivore		
	Anas chlorotis[a,d]	*Anthornis melanura*	*Cyanoramphus novaezelandiae*[d]		
	Coturnix novaezelandiae[a]	*Prosthemadera novaeseelandiae*	*Cyanoramphus auriceps*[c]		
	Gallirallus australis[d]		*Strigops habroptilus*[d]		
	Fulica prisca[a]				
	Gallinula hodgenorum[a]				
	Coenocorypha cf. *iredalei*[a,b]				
		Frugivore–folivore	Omnivore		
		Hemiphaga novaeseelandiae	*Nestor notabilis*[d]		
		Callaeas cinerea[a]	*Turnagra capensis*[a]		
		Nestor meridionalis[c]	*Corvus* Mainland[a]		

NOCTURNAL OR NOCTURNAL–CREPUSCULAR

TERRESTRIAL	ARBOREAL–TERRESTRIAL	AERIAL–ARBOREAL–TERRESTRIAL	OCEANIC	AERIAL	PREDATOR
Insectivore	Insectivore–nectarivore	Insectivore–nectarivore–frugivore	Insectivore		
Apteryx owenii[d]	*Hoplodactylus* sp.[d]?	*Mystacina tuberculata*[d]	*Oceanites maorianus*[a,b]	*Chalinolobus tuberculatus*[b,c]	*Aegotheles novaezealandiae*[a]
Apteryx eastern[a]			*Pelecanoides urinatrix*[b,d]		
Sphenodon sp.[d]			*Pterodroma inexpectata*[b,d]		*Sceloglaux albifacies*[a]
			Pterodroma cookii[b,d]		*Ninox novaeseelandiae*
			Puffinus gavia/huttoni[b,d]		

[a] Globally extinct species.
[b] Species recorded in predator sites nearby but not in the Pyramid Valley deposit itself.
[c] Reduced numbers locally.
[d] Locally extinct.

TABLE 14.2.

Broad guilds of vertebrates on Takaka Hill during the Holocene, before human intervention (modified from Holdaway and Worthy 1996: table IV).

DIURNAL

TERRESTRIAL	ARBOREAL	ARBOREAL–TERRESTRIAL	AERIAL	PREDATOR
Herbivore	Insectivore	Insectivore	Insectivore	
Anomalopteryx didiformis[a]	*Acanthisitta chloris*	*Xenicus longipes/gilviventris*[b]	*Rhipidura fuliginosa*	*Circus eylesi*[a]
Dinornis novaezealandiae[a]	*Gerygone igata*	*Pachyplichas yaldwyni*[a]		*Falco novaeseelandiae*[c]
Dinornis struthoides[a]	*Mohoua novaeseelandiae*	*Traversia lyalli*[a]		
	Mohoua ochrocephala[c]	*Petroica macrocephala*		
		Petroica australis[c]		
		Philesturnus carunculatus[b]		
Omnivore	Insectivore–nectarivore	Frugivore–folivore		
Hymenolaimus malacorhynchos[c]	*Anthornis melanura*	*Cyanoramphus novaezelandiae*[b]		
Anas chlorotis[b]	*Prosthemadera novaeseelandiae*	*Cyanoramphus auriceps*[c]		
Euryanas finschi[a]		*Strigops habroptilus*[b]		
Gallirallus australis[b]				
Coenocorypha cf. *iredalei*[a]				
Insectivore	Frugivore-folivore	Omnivore		
Oligosoma sp.	*Hemiphaga novaeseelandiae*	*Nestor notabilis*		
Oligosoma sp.	*Callaeas cinerea*[b]	*Turnagra capensis*[a]		
	Nestor meridionalis[c]			

NOCTURNAL OR NOCTURNAL–CREPUSCULAR

TERRESTRIAL	ARBOREAL–TERRESTRIAL	AERIAL–ARBOREAL–TERRESTRIAL	AERIAL	PREDATOR
Insectivore	Insectivore–nectarivore	Insectivore–nectarivore–frugivore	Insectivore	
Leiopelma hamiltoni[b]	*Hoplodactylus maculatus* gp	*Mystacina tuberculata*[b]	*Chalinolobus tuberculatus*[c]	*Aegotheles novaezealandiae*[a]
Leiopelma markhami[a]	*Hoplodactylus granulatus* gp	*Mystacina robusta*[a]		*Sceloglaux albifacies*[a]
Apteryx owenii[b]	*Hoplodactylus duvaucelii*[b] gp			*Ninox novaeseelandiae*
Apteryx spp. (large)[c]				

Abbreviation: gp = species group. [a] Globally extinct species. [b] Locally extinct species. [c] Reduced numbers locally.

from one site on the slopes of the steep-sided Gorge Creek Valley on the Hill. The absence of large petrel colonies on the top of the plateaulike hill itself would have meant that there was no fertilization of the soils by guano, and laughing owls depended on species such as parakeets as prey.

Although the nocturnal predator guild on Takaka Hill was the same as that in the eastern lowlands, it lacked the largest diurnal predator, Haast's eagle. If the South Island adzebill was present on Takaka Hill during the Holocene, as a recent find suggests, then it was rare there. One difference between the nocturnal faunas was the presence of *Mystacina robusta* on Takaka Hill, where it was sympatric with its smaller congener. Among the smaller, arboreal species, the similarities were much greater. One difference is that although the raven was rare at Pyramid Valley, it was absent from the Hill. The nocturnal insectivores, apart from the kiwi, where there was one large and one small species in each area, were completely different. At Pyramid Valley, there were tuatara and no frogs, and on Takaka Hill, it was quite the reverse: one of the frogs (*Leiopelma markhami*) was large and probably terrestrial in habits (Worthy 1987c).

The main conclusion that can be drawn from this comparison is that most of the larger species of bird were limited in their distribution by habitat preferences that seem to have been much stricter than those of the smaller birds. The same conclusion applies, on a much smaller scale, to the bats, amphibians, and lizards. The overall greater diversity, particularly of large species, in the drier, lower, eastern area agrees with the observations of Olson and James (1982) in Hawaii: that dry lowland forests had higher levels of diversity and of numbers of individuals than did montane forests. In most places, the dry lowland forests and shrublands were eliminated before ecologists were able to study them. The only window to that lost world, and with it understanding of some basic ecologic patterns, is the study of fossil faunas (and floras) that reflect the natural distribution of life on the planet.

Major Outlying Islands: Chatham Islands, Norfolk Island, and Auckland Islands

Among the outlying islands, the Chatham group, Norfolk, and the Auckland Islands had sufficiently diverse avifaunas to have multispecies terrestrial guilds. On the Chatham and Auckland Islands, there were no bats, and the only reptile on those groups, so far as is known, is a single species of skink on the Chathams. Others may have been present; no sites suitable for the preservation of lizard bones have so far been analyzed.

Norfolk Island

The Norfolk Island avifauna had a strong representation of seabirds, with several terns (but no gulls) and various burrowing and surface-breeding petrels (Holdaway and Anderson 2001; Meredith 1985, 1991). The only pelecaniform was a widespread species of tropical *Sula*; there were no cormorants. The ground-feeding guild included a flightless *Gallirallus* rail, perhaps other rails, and a local form of *Coenocorypha* snipe. Arboreal insectivore guilds were dominated by small passerines. There were two arboreal herbivores—the pigeon and kaka—and both may have spent much time on the ground before humans arrived with the Pacific rat. The kaka was reported by early European settlers to live among rocks on Philip Island. Perhaps the most unusual of the guilds in its composition was the nectarivores, represented by two *Zosterops* and no members of the Meliphagidae at all. Given the near ubiquity of honeyeaters on western Pacific islands, their absence from the relatively old land mass of Norfolk Island is remarkable, given that honeyeaters occur on islands all around Norfolk, even Raoul in the Kermadecs 600 km to the east of Norfolk, which periodically experiences violent volcanic eruptions. Norfolk is downwind of the Australian mainland, the center of honeyeater distribution, and much closer than the main islands of New Zealand.

A bat (*Chalinolobus* sp.) and a gecko (*Phyllodactylus guentheri*) had reached the island, but the bat is almost certainly extinct and the lizard is confined to offshore islets.

Chatham Islands

On the Chatham Islands, where the forests consisted mainly of four plant species that rarely exceeded 10 m in height, the arboreal gleaners were restricted strictly to the local species of bellbird (*Anthornis melanocephala*), the tui (*Prosthemadera novaeseelandiae*), and the Chatham Island warbler (*Gerygone albofrontata*). Both species of parakeet also take insects, the red-crowned parakeet more often on or near the ground, in addition to fruit, seeds, and leaves (Nixon 1994). Fantails (*Rhipidura fuliginosa*) were the aerial insectivores.

The three rails on the Chatham Islands were separated in their feeding habits by being different sizes and having different length bills. The smallest species even subdivided its own feeding niche: it had a sexually dimorphic bill (Trewick 1997a). The evolutionary history of the group must antedate the formation of Pitt Strait and the channels between the islands, with apparent stasis after the barriers were formed, because all three flightless species were present either side. The bathymetry suggests that the islands of the group have been isolated from each other for at least 10,000 years. The Chatham Island ground prober guild also included two sympatric species of snipe (*Coenocorypha pusilla* and *C. chathamica*), again separated by size, so the invertebrate fauna in the peaty soils and leaf litter must have been substantial.

Divisions between arboreal and terrestrial guilds were blurred, as they often were on the mainland islands. The rails were flightless and hence totally terrestrial. The Chatham Island kaka (*Nestor* sp.), on the other hand, could fly but was apparently well adapted to feeding on the ground. Its pelvis has some similarities to that of the flightless kakapo (Millener 1999), and the Chatham Island pigeon (*Hemiphaga chathamensis*) flies well but spends much time on the ground (Millener and Powlesland 2001).

The seabird guild on the Chathams was overwhelmingly dominated by procellariiforms (petrels and albatrosses), of which there were 17 species as contrasted with only two species of sea shag (*Leucocarbo onslowi* and *Stictocarbo featherstoni*) and no gannets. Feeding on the wealth of seabirds was the falcon, the only diurnal raptor on the group. The southern skua (*Catharacta skua*), a relative of the gulls, was, however, a surrogate raptor, preying on small and medium-size petrels and occasionally shorebirds. Skuas accumulate large middens of prey remains at feeding stations and around their nests (Tennyson and Millener 1994). Analysis of these middens is useful in determining the former presence of small petrels on an island—and indeed of the possible presence of cryptic nocturnal species today. Owls did not reach any of the islands to the east and south of the main New Zealand islands.

Auckland Islands

Both the tui and bellbird (the mainland species) reached the subantarctic Auckland Islands, where they vary their diet of nectar from rata (*Metrosideros*) blossoms with insects. These are the southernmost honeyeaters (Meliphagidae) in the world. They live in a stunted southern rata (*Metrosideros umbellata*) forest with the red-crowned parakeet (*Cyanoramphus novaezelandiae*) and the Auckland Island tomtit (*Petroica marrineri*), which among them share all the arboreal insectivore and herbivore (parakeet) niches on the islands.

One prober is present, the Auckland Island snipe (*Coenocorypha aucklandica*), which is now confined to predator-free islands. Also on the Aucklands is an endemic species of flightless duck (*Anas aucklandica*), which formerly fed in the grasslands as well as the tide line, as it still does on islands in the group that lack introduced cats. A population of the southern merganser was, until its extinction in the early 20th century, part of the marine piscivore guild, whose other member is the local species of marine shag, the Auckland Island shag (*Leucocarbo colensoi*). The only other terrestrial species are a pipit (*Anthus* sp.), a rail (*Dryolimnas muelleri*), and a wading bird, the Auckland Island dotterel (*Charadrius exilis*). One bone of a raven has been reported from the group (Holdaway et al. 2001). Whether it was even a raven or a vagrant of either the mainland or Chatham Island species or a resident population of either is not known. In its poverty of terrestrial species, the Auckland Island avifauna is a depleted extension of the mainland fauna and not a separate entity, as is that of the Chatham Islands.

Other Outlying Islands

The outlying islands are dominated by the seabirds. Restricted species lists of other groups provided space for some to become generalists in their feeding niche. An example is that the two parakeets (Hochstetter's and Antipodes parakeets) on Antipodes Island eat seal meat and fat, the spilled contents of penguin eggs, and grass seeds and insects. The Antipodes parakeet (*Cyanoramphus unicolor*) has even been recorded catching and eating storm petrels (*Garrodia nereis*) in exactly the same way that kea kill and eat Hutton's shearwaters. With the abundance of potential prey and the dearth of available predators,

TABLE 14.3.
Losses of guilds and members of guilds from the Holocene avifauna at Pyramid Valley, North Canterbury (after Holdaway and Worthy 1997: table 6).

		Species richness					
		Diurnal			Nocturnal		
Location	Food[a]	Prehuman	Present	Loss (%)	Prehuman	Present	Loss (%)
Aquatic	SD	1	0	100			
	D/S	2	1	50			
	W	2	0[d]	100			
	D	2	0	100			
Terrestrial	H	7	0	100			
	O	8	1	87.5			
	I				3	0	100
Arboreal	I	5	1	80			
	I-N	2	1	50			
	F-F	3	1[b]	67			
Arboreal–terrestrial	I	5	0	100			
	F-F	3	0	100			
	O	3	0	100			
	I-N				1	0	100
Aerial[c]	I	1	1	0			
Aerial–arboreal–terrestrial	I-N-F				1	0	100
Predator		4	1[b,d]	75	2	1[b]	67
Total		47	7	85	7	1	86

Abbreviations: D = diver; D/S = dabbler/shore feeder; F-F = frugivore–folivore; H = herbivore; I = insectivore; I-N = insectivore–nectarivore; I-N-F = insectivore–nectarivore–frugivore; O = omnivore; SD = specialist dabbler; W = wader. Nocturnal includes crepuscular–nocturnal. Taxa introduced by humans are not included.

[a] Feeding method and site for aquatic taxa.

[b] Occasional to very rare resident near Pyramid Valley.

[c] Other taxa known from other fossil sites near Pyramid Valley (Worthy and Holdaway 1996b).

[d] Taxon replaced by self-introduced taxon (W, *Himantopus leucocephalus*; predator, *Circus approximans*); replacements not included in calculation of extinctions.

groups such as the parrots may have begun to occupy part of the small predator niche. Of the New Zealand parrots, only the kakapo seems to have been entirely herbivorous.

On each of the islands except Macquarie Island, a species of *Coenocorypha* snipe was the main (or only) terrestrial insectivore. On The Snares, the black tomtit (*Petroica dannefaerdi*) fed in the litter (McLean et al. 1994). There was another flightless duck (*Anas nesiotis*) on Campbell Island. It and an endemic snipe there have survived Norway rats and cats only on offshore islets, the duck on Dent Island off the northwestern coast and the snipe on Jacquemart Island off the southern coast, where it is protected by vertical basalt cliffs. Both lack predators.

Extinctions and Guilds

The late Quaternary extinctions affected the major guilds to different degrees. Almost all the terrestrial herbivores became extinct. Only the paradise shelduck (*Tadorna variegata*) and South Island takahe (*Porphyrio hochstetteri*) remain, the latter only with intensive management, whereas Finsch's duck, the geese, North Island takahe, and 11 kinds of moa did not survive the early impact of Polynesian settlers. With them went most of the terrestrial omnivores—if not totally (as with the coot), then the mainland populations, as in the brown teal (*Anas chlorotis*). Only the weka (*Gallirallus australis*) survives, but it is endangered everywhere on the main islands. Of the predators, the four diurnal raptors and the adzebills, only the falcon remains.

Losses and Percentage Losses—Major to Catastrophic

Examples from the faunas that are best studied show the extent of the changes in faunal composition over the past 2000 years. At Pyramid Valley, the losses in most of the original guilds have been catastrophic (Table 14.3). Half (7 of 14) of the diurnal guilds identified in the prehuman avifauna are extinct in the area, and the minimum loss in 6 of the other 7 is 50% (overall loss, 87%). The only guild to escape is the aerial insectivores, represented by only the flycatcher *Rhipidura fuliginosa*, a species that has large clutches and is multibrooded. Only 7 of the original minimum of 47 diurnal species survive in the area; most of the rest are globally extinct. For nocturnal guilds, the situation is even worse, with three of four guilds now missing entirely, and the fourth reduced by two thirds (overall loss, 86%; only one of seven species remaining).

On Takaka Hill, the extinctions were similar in pattern but less severe in terms of guilds surrendered (Table 14.4). The differences result mainly from the retention of areas of seminatural vegetation on Takaka Hill, which has allowed some of the smaller forest species to remain in small populations, with one or two species left in guilds that had no chance of survival in the totally altered environment near Pyramid Valley. Only one diurnal guild (the terrestrial herbivores) has ceased to exist in the area, and all but one of its constituent species are globally extinct. It is likely that a second guild, the terrestrial omnivores, became

TABLE 14.4.
Losses of guilds and members of guilds from the Holocene avifauna on Takaka Hill, northwest Nelson (after Holdaway and Worthy 1996: table V).

		Species richness					
		Diurnal			Nocturnal		
Location	Food	Prehuman	Present	Loss (%)	Prehuman	Present	Loss (%)
Terrestrial	H	3	0	100			
	O	5	1	80			
	I	2	2	0	4	0	100
Arboreal	I	4	3	25			
	I-N	2	2	0			
	F-F	3	1	67			
Arboreal–terrestrial	I	6	2	67			
	F-F	3	1	67			
	O	2	1	50			
	I-N				3+	2+	~30
Aerial	I	1	1	0	1	1	0
Aerial–arboreal–terrestrial	I-N-F				2	0	100
Predator		2	1	50	3	1	67
Total		33	15	55	13+	4+	69

Abbreviations: F-F = frugivore–folivore; H = herbivore; I = insectivore; I-N = insectivore–nectarivore; I-N-F = insectivore–nectarivore–frugivore; O = omnivore. Nocturnal includes crepuscular–nocturnal.

extinct on Takaka Hill in the late 1990s. The weka *Gallirallus australis* disappeared then and now survives only in more remote areas to the west.

Two other guilds on Takaka Hill (arboreal folivores–frugivores and arboreal terrestrial omnivores) have lost two thirds of their members. In total, 15 of the minimum of 33 diurnal species in the Holocene forest on Takaka Hill survive in the area today. Two nocturnal–crepuscular guilds have been lost, and another has declined by two thirds (Table 14.4). Only 4 of 13 species survive there (with the minimum known level of diversity in the lizard fauna).

Causes

The losses reflect the transformation in two stages (Polynesian firing and then European settlement) of the original forest–shrubland vegetation via induced grasslands into a European-style agricultural landscape. In concert with the results of the serial introduction of mammalian predators (Holdaway 1999b), the vegetation changes resulted in the elimination of the rich vertebrate fauna of the eastern South Island. Three species that self-introduced from Australia (pukeko, harrier, and pied stilt) replace congeners in the extinct fauna, but their presence does not mitigate the losses, contrary to Wilson (1997). Although there now might be as many species of birds in New Zealand as before, the additions were not replacements and are not in any way compensations for the interrupted evolutionary lineages that were lost. In addition, the losses in other groups of animals although unknown and largely unlamented would outweigh any apparent gain in diversity in a limited group such as birds.

The causes of the extinctions are covered in some detail in Chapter 15. Briefly, the timing of the losses was determined by the time of arrival of humans and the predatory mammals they brought with them, starting with the Pacific rat, introduced by transient visitors about 2000 years ago (Holdaway 1996a, 1999a, 1999b; Holdaway and Beavan 1999; Holdaway and Jacomb 2000). Other factors were the size of the bird (which determined its vulnerability to different predators or choice as food by people) and habitat (Holdaway 1999b). It was a cascade of disaster for the indigenous fauna, few species of which have survived in anything approaching the numbers they enjoyed before people first saw the islands.

Faunas: Putting It All Together

In contrast to the orthodox view of the vertebrate fauna of New Zealand being homogeneous in time and space, analysis of well-dated deposits has shown that there were distinct and distinctive regional faunas. The composition and distribution of these faunas can be related to the vegetation and topographic patterns in the undisturbed environment. The boundaries between habitat types were in most instances not sharp, as is denoted by the way the boundaries are depicted in Figure 14.3. However, each major habitat seems to have been associated with a characteristic group of species, as would be expected elsewhere.

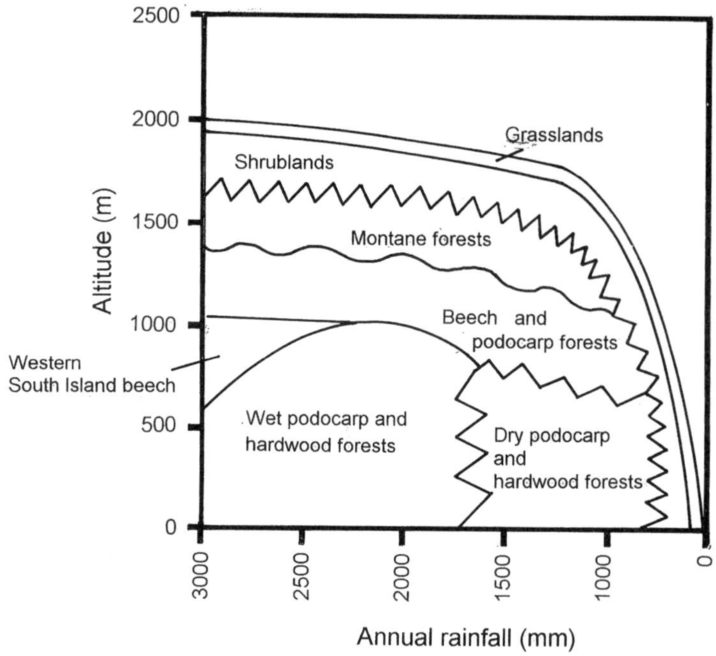

Figure 14.3. Distribution of generalized major vegetation types in New Zealand, with respect to altitude and rainfall. The x-axis uses an unequal scale. Zigzag and wavy lines indicate interdigitation of habitats with altitude and climatic variation through the main islands.

South Island

General

The Holocene fauna of the whole archipelago consisted of at least 245 breeding species (Holdaway et al. 2001). Of these, 112 bred in the South Island. It did not take long after the start of a concerted program of radiocarbon dating the sites and assemblages to determine that, as might be expected, some species were always found together, and some were seldom or never in the same place at the same time. With samples now available from throughout the South Island, some preliminary analysis of the species groupings—faunas—is now possible.

In the simplest case, the degree of association of species large enough to be preserved in the least favorable environments of preservation (all nine kinds of South Island moa, a big kiwi [great spotted or brown kiwi], eastern kiwi, South Island goose, South Island adzebill, Haast's eagle, South Island takahe, and coot, 16 species in all) was assessed with a multivariate clustering technique (TWINSPAN [two-way indicator species analysis]; Hill 1979) that classifies species and sample sites by the similarity of the combinations of species that each site contains. Of the two major groups, one contains 6 species and the other 10. *Megalapteryx didinus, Anomalopteryx didiformis, Dinornis novaezealandiae, Dinornis struthoides*, and *Pachyornis australis* were in one group (with large *Apteryx*), and *Euryapteryx geranoides, Pachyornis elephantopus, Emeus crassus*, and *Dinornis giganteus* were in the other group, along with the remainder of the smaller species (Fig. 14.4A). The two moa faunas were separated in time and space. The group containing *Megalapteryx didinus* was found in sites that were mostly west of the Southern Alps, with a few in the high country to the east and in western Southland. The group containing *Euryapteryx geranoides* was confined to the eastern South Island during the Holocene; the only sites with that fauna to the west of the main ranges were all of Otiran Glacial age (Fig. 14.4B).

The end branches in Figure 14.4A show further that three of the moa in the eastern group were more often associated than was any with the remaining species (*Euryapteryx geranoides*). The three species are more common in lowland sites (<200 m), or at least those with a low relief. *Euryapteryx geranoides* is more often found in sites on hill country above 200 m. Of the birds other than moa, *Aptornis defossor* was more commonly found with *Porphyrio hochstetteri* than with other species included in the sample. The goose *Cnemiornis calcitrans* was more often associated with the three lowland moa than with *Porphyrio hochstetteri* and *Aptornis defossor*, or with *Euryapteryx geranoides*. Haast's eagle (*Harpagornis moorei*) was found most often in sites that contained *Euryapteryx geranoides*. That may reflect the wider distribution of eagle remains in central Otago and Southland, away from the center of distribution of the other moa. *Apteryx* "eastern," the undescribed large kiwi found in deposits in Canterbury and Otago, was more often associated with the *Pachyornis elephantopus* group of moa. In this analysis, *Fulica prisca* was part of the group that included *Aptornis defossor, Porphyrio hochstetteri, Harpagornis moorei*, and *Euryapteryx geranoides*.

In the western group, there are also patterns that probably represent actual habitat preferences. *Dinornis struthoides* was at the base of that branch, which reflects the wider distribution of that species, which has been recorded—in low numbers—in many places east of the mountains. The little bush moa (*Anomalopteryx didiformis*) is often associated with the large kiwi (*Apteryx australis* and *A. haastii*) and seems to have shared their habitat in the wetter forests, including the eastern slopes of the main ranges. The grouping of *Megalapteryx didinus* and *Dinornis novaezealandiae* reflects their joint presence in some of the wet West Coast forests as well as in high-altitude shrublands. Both reached eastern areas in small numbers in the higher hill country. In contrast, *Pachyornis australis* seems to have been restricted to high-altitude shrublands and the coastal shrublands of Southland.

These, then, are the bases for the eagle and western faunas, the two major groupings of birds (and other vertebrates) in the South Island throughout the past 100,000 years. The distributions of the groupings changed as the major vegetation types responded to climatic fluctuations, but the composition of each fauna has remained remarkably constant. Many details of the habitat requirements of species within each fauna still need to be worked out. But some progress has been made. Examples are the discovery that *Anomalopteryx didiformis* may have preferred openings in the dense western forest and that *Dinornis novaezealandiae* was at home both in the interior of the dense forest or in more open upland shrublands.

Finsch's duck (*Euryanas finschi*), a species that was apparently

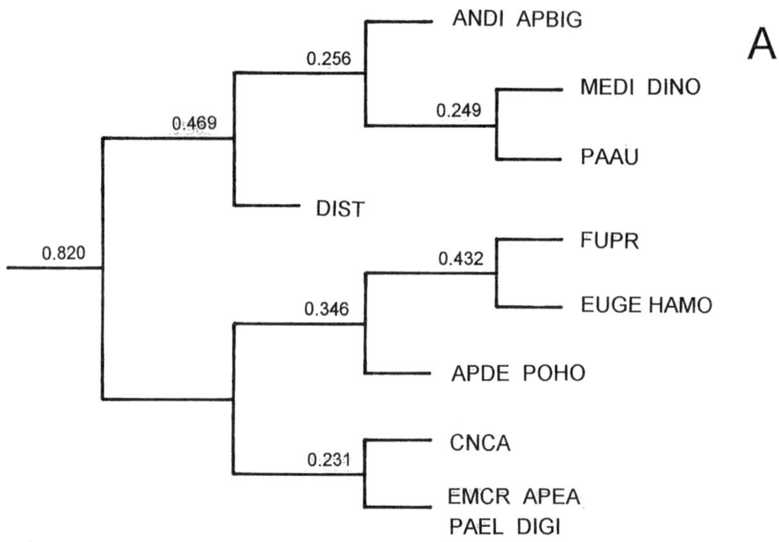

*Figure 14.4. (A) (above) TWINSPAN dendrogram showing classification of species and hierarchic relationships between the species groupings for presence or absence of all the South Island moa species plus large kiwi (*Apteryx haastii *or* A. australis*),* Apteryx *eastern South Island,* Cnemiornis calcitrans, Aptornis defossor, Porphyrio hochstetteri, Fulica prisca, *and* Harpagornis moorei. *Taxa are as follows: ANDI* = Anomalopteryx didiformis*; APBIG* = Apteryx australis *or* A. haastii*; APDE* = Aptornis defossor; *APEA* = Apteryx eastern; *APOW* = Apteryx owenii; *CNCA* = Cnemiornis calcitrans; *DIGI* = Dinornis giganteus; *DINO* = Dinornis novaezelandiae; *DIST* = Dinornis struthoides; *EMCR* = Emeus crassus; *EUGE* = Euryapteryx geranoides; *FUPR* = Fulica prisca; *HAMO* = Harpagornis moorei; *MEDI* = Megalapteryx didinus; *PAAU* = Pachyornis australis; *PAEL* = Pachyornis elephantopus; *POHO* = Porphyrio hochstetteri.

(B) TWINSPAN (opposite page) dendrogram showing classification of sites containing the same species as in A and the hierarchic relationships between the site groupings. Values at each branch are eigenvalues for the division point. Site abbreviations in order from top to bottom of diagram: CAV4WC = Cave 4 (W); DOUCK = Doubtful Creek Cave (W); JAWS = Jaws Cave (W); PUNTOM = Punakaiki Tomo (W); EARLGC = Earl Grey Cave (TH); RATHOL = Ratite Hole (TH); HOLHOL = Holocene Hole (MtC); HONGR2 = Honeycomb Hill Cave Graveyard Site layer 2 (Op); FIREWO = Firewood Creek (Oc); CAV1WC = Cave 1 (W); MADCAV = Madonna-Equinox Cave System (W); NETTRE = Nettletrench Cave (W); UNIWON = Unique Wonder Cave (W); VENCAV = Venturer Cave (W); KAIRUR = Kairuru Cave (TH); NGARUA = Ngarua Cave (TH); STREOH = Sterndale Road Overhang (SC); HOSPF3 = Hospital Flat Site 3 (Oc); HOBTOM = Hobson's Tomo (TH); HODCKT = Hodge Creek Tomo (MtA); BABLAG = Babylon Cave, Passage of Thoth, lag deposit (W); BABOLD = Babylon Cave, Passage of Thoth, old sediments (Pleistocene) (W); HERMIT = Hermit's Cave (W); KAIEXT = Kairuru Cave Extension (TH); COBDEN = Cobden Fissure Site (W); COMCAV = Commentary Cave (TV); KIWIHO = Kiwi Hole (TH); LTWIN = Little Winter Cave (TH); CAITOM = Cairn's Tomo (W); CAV6WC = Cave 6 (W); TORCAN = Tory Canyon Cave (W); IRVTOM = Irvine's Tomo (TV); POT = Pothole 3 (W); STRUTH = Struthoides Cave (TH); WINCAV = Winter Cave (TH); METALL = Metro Cave (all sites) (W); IRVTBR = Irvine's Tomo beyond rockfall (TV); HONEAR = Honeycomb Hill Cave Eagle's Roost Site (Op); TEANAT = Te Ana Titi [cave] (W); SIMTOT = Sim's Cave total fauna (TV); HONH&H = Honeycomb Hill Cave, His & Her Cave (Op); TOMBWC = Tomo B (W); BONE = Bone Cave (TH); GLEND6 = Glendhu Station Site 6 (Oc); HAWEN = Hawke's Cave Entrance (TH); HODCKT = Hodge Creek Tomo (MtA); MERCAV = Merino Cave (MtC); EUMURS = Euan Murchison's Rockshelter (Wai); TIMTOM = Timaru Tomo (SC); CRBDS = Craigmore Bird Rockshelter (SC); FREGUL = Frenchman's Gully Rockshelter (SC); SMIPRE = Three Mile Road Predator Site (SC); STNDE = Station Deposit (Oc); HODSTB = Hodge Creek Cave Streambed (MtA); HONG34 = Honeycomb Hill Cave Graveyard Site layers 3 and 4 (Op); AR1442 = AR144 Cave layers 1 and 2 (Op); HOSPF4 = Hospital Flat Site 4 (Oc); GBCCAV = Golden Bay Cement Cave (TV); IDAVAS = Ida Valley Swamp (Oc); TORPOT = Pothole beside Torture Track (MtC); GLENMS = Glenmark Swamp (Wai); FORHIT = Forest Hills Tomos (So); LIMCAV = Limehills Cave (So); HAMOTA = Hamilton Swamp (Oc); RIVNAR = Riverton Narrows Site (So); KAUNA = Kauana Swamp (So); WAKADU = Wakapatu Dunes (So); SCALAG = Scaife's Lagoon (Oc); EARNSC = Earnscleugh Cave (Oc); HAMWIN = Hamilton's Swamp, Winton (So); PAERAU = Paerau Swamp (Oc); ARDGOW = Ardgowan Swamp (Oco); 5FORKS = Five Forks Swamp (Oco); ENFIED = Enfield Swamp (Oco); TIMPRS = Timpendean

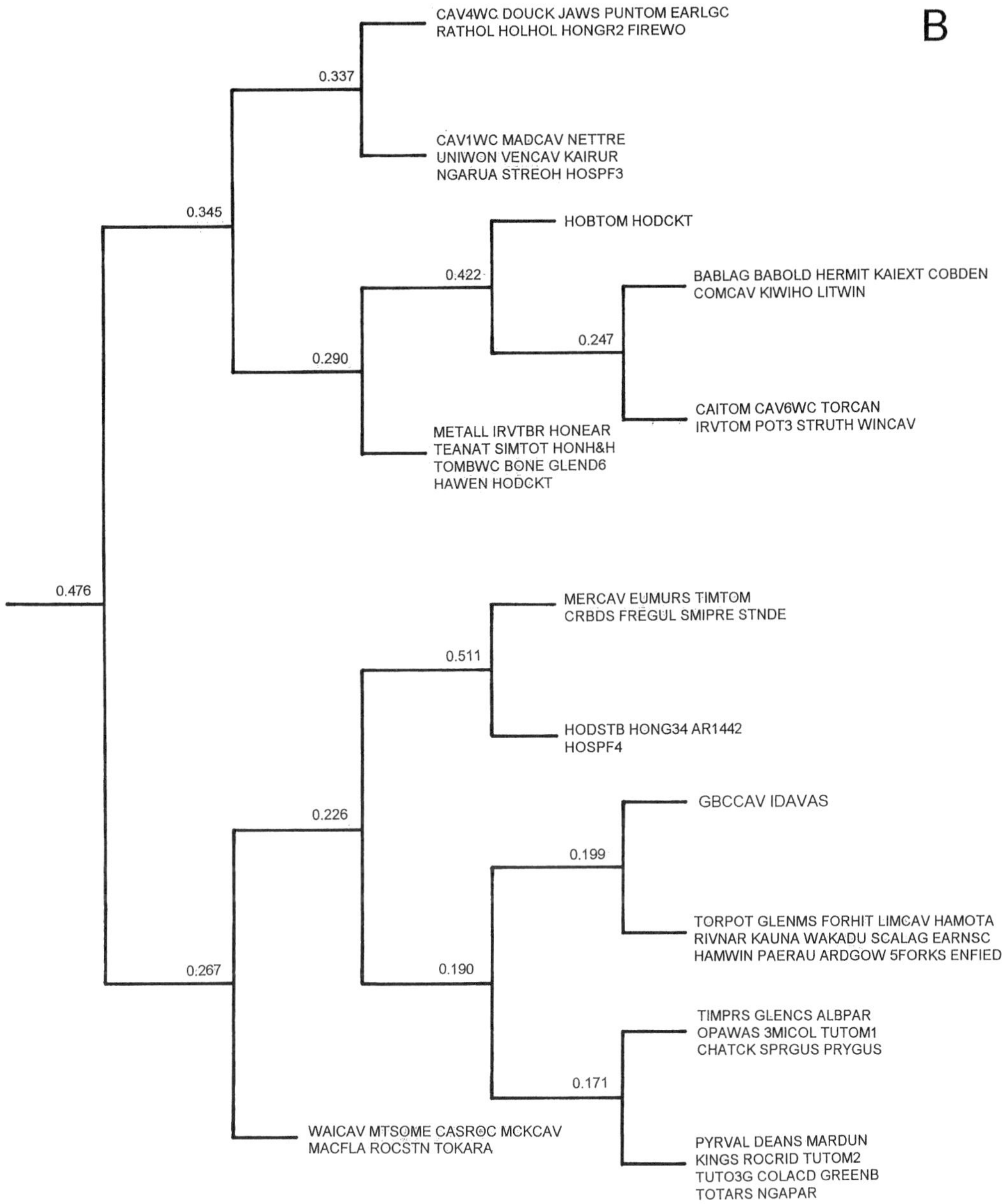

Rockshelter (Wai); GLENCS = Glencrieff Swamp (Wai); ALBPAR = Albury Park Swamp (SC); OPAWAS = Opawa Swamp (SC); 3MICOL = Three Mile Road colluvial deposit (SC); TUTOM1 = Tuarangi Tomo 1 (SC); CHATCK = Chatto Creek Sites (Oc); SPRGUS = Spring Gully Swamp (Oco); PRYGUS = Pryde's Gully Road Swamp (Oco); PYRVAL = Pyramid Valley Swamp (Wai); DEANS = The Deans (Wai); MARDUN = Marfells Beach Dunes (M); KINGS = Kings Cave (SC); ROCRID = Rocky Ridges (SC); TUTOM2 = Tuarangi Tomo 2 (SC); TUTO3G = Tuarangi Tomo 3 gravel (SC); COLACD = Colac Bay Dunes (So); GREENB = Greenhills Back Beach Dunes (So); TOTARS = Totara Swamp (Oco); NGAPAR = Ngapara (Oco); WAICAV = Waikari Cave (Wai); MTSOME = Mount Somer Quarry (Central Canterbury); CASROC = Castle Rocks (So); MCKCAV = McKerchar's Cave (So); MACFLA = Macrae's Flat deposit (Deep Dell Creek) (Oc); ROCSTN = Rockdale Station (SC); TOKARA = Tokarahi (Oco). Areas: M = Marlborough; MtA = Mount Arthur, northwest Nelson; MtC = Mount Cookson, North Canterbury; Oc = central Otago; Oco = coastal Otago; Op = Oparara, northwest Nelson; SC = South Canterbury; So = Southland; TH = Takaka Hill, northwest Nelson; TV = Takaka Valley, northwest Nelson; W = West Coast; Wai = Waikari area, North Canterbury.

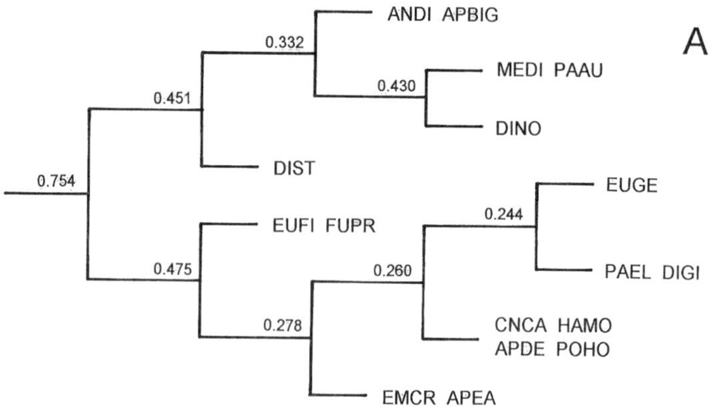

*Figure 14.5. (A) (above) TWINSPAN dendrogram showing classification of species and hierarchic relationships between the species groupings for presence or absence of the same species as in Figure 14.4, plus Finsch's duck (*Euryanas finschi*). Taxa are as follows: ANDI* = Anomalopteryx didiformis; *APBIG* = Apteryx australis *or* A. haastii; *APDE* = Aptornis defossor; *APEA* = Apteryx *eastern; APOW* = Apteryx owenii; *CNCA* = Cnemiornis calcitrans; *DIGI* = Dinornis giganteus; *DINO* = Dinornis novaezealandiae; *DIST* = Dinornis struthoides; *EMCR* = Emeus crassus; *EUFI* = Euryanas finschi; *EUGE* = Euryapteryx geranoides; *FUPR* = Fulica prisca; *HAMO* = Harpagornis moorei; *MEDI* = Megalapteryx didinus; *PAAU* = Pachyornis australis; *PAEL* = Pachyornis elephantopus; *POHO* = Porphyrio hochstetteri.

(B) TWINSPAN (opposite page) dendrogram showing classification of sites containing the same species as in A and the hierarchic relationships between the site groupings. Site abbreviations in order from top to bottom of diagram: DOUCRK = Doubtful Creek (W); BONE = Bone Cave (TH); HOLHOL = Holocene Hole (MtC); Honeycomb Hill Cave Graveyard Site layer 2 (Op); VENCAV = Venturer Cave (W); NGARUA = Ngarua Cave (TH); STEROH = Sterndale Road Overhang (SC); HOPF3 = Hospital Flat Site 3 (Oc); SIMTOT = Sim's Cave (TV); HONEAR = Honeycomb Hill Cave, Eagle's Roost Site (Op); HODTOM = Hodge Creek Tomo (MtA); HOTATO = Hodge Creek Takahe Tomo (MtA); GLEND6 = Glendhu Station Site 6 (Oc); BABLAG = Babylon Cave, Passage of Thoth, lag deposit (W); BABOLD, Babylon Cave, Passage of Thoth, old deposit (Pleistocene) (W); HERMIT = Hermit's Cave (W); COBDEN = Cobden Fissure Site (W); COMCAV = Commentary Cave (TV); IRVTOM = Irvine's Tomo (TV); HOBTOM = Hobson's Tomo (TH); KIWIHO = Kiwi Hole (TH); POT3 = Pothole 3 (W); CAITOM = Cairn's Tomo (W); KAIEXT = Kairuru Cave Extension (TH); LITWIN = Little Winter Cave (TH); WINCAV = Winter Cave (TH); AR1441 = AR144 Cave layer 1 (Op); Cave 1 (W); CAV6WC = Cave 6 (W); PUNTOM = Punakaiki Tomo (W); TORCAN = Tory Canyon Cave (W); UNIWON = Unique Wonder Cave (W); EARLGC = Earl Grey Cave (TH); Ratite Hole (TH); STRUTH = Struthoides Cave (TH); CAV4WC = Cave 4 (W); JAWS = Jaws Cave (W); NETTRE = Nettletrench Cave (W); MADCAV = Madonna-Equinox Cave System (W); METALL = Metro Cave (all sites) (W); TEANAT = Te Ana Titi [cave] (W); IRVTBR = Irvine's Tomo beyond rockfall (TV); HAWENT = Hawke's Cave Entrance (TH); GNCCAV = Golden Bay Cement Cave (TV); KAIRUR = Kairuru Cave (TH); TORPOT = Pothole by Torture Track (MtC); GLENMS = Glenmark Swamp (Wai); LIMCAV = Limehills Cave (So); HAMOTA = Hamilton Swamp (Oc); RIVNAR = Riverton Narrows Site (So); KAUNA = Kauana Swamp (So); WAKADU = Wakapatu Dunes (So); SCALAG = Scaife's Lagoon (Oc); IDAVAS = Ida Valley Swamp (Oc); HAMWIN = Hamilton's Swamp = Winton (So); PAERAU = Paerau Swamp (Oc); ARDGOW = Ardgowan Swamp (Oco); 5FORKS = Five Forks Swamp (Oco); ENFIED = Enfield Swamp (Oco); TIMPRS = Timpendean Rockshelter (Wai); GLENCS = Glencrieff Swamp (Wai); PYRVAL = Pyramid Valley Swamp (Wai); DEANS = The Deans (Wai); MARDUN = Marfells Beach Dunes (M); ALBPAR = Albury Park Swamp (SC); KINGS = Kings Cave (SC); OPAWAS = Opawa Swamp (SC); 3MICOL = Three Mile Road Colluvial Site (SC); TUTOM1 = Tuarangi Tomo 1 (SC); TUTO3C = Tuarangi Tomo 3 clay (SC); FORHIT = Forest Hills Tomos (So); GREENB = Greenhills Back Beach Dunes (So); CHATCK = Chatto Creek sites (Oc); MOACKS = Moa Creek Swamp (Oc); EARNSC = Earnscleugh Cave (Oc); TOTARS = Totara Swamp (Oco); SPRGUS = Spring Gully Swamp (Oco); NGAPAR = Ngapara (Oco); PRGUS = Pryde's Gully Road Swamp (Oco); HODSTB = Hodge Creek Cave Streambed (MtA); MERCAV = Merino Cave (MtC); FIREWO = Firewood Creek (Oc); HONGR3 = Honeycomb Hill Cave Graveyard Site layer 3 (Op); CRBDS = Craigmore Bird Rockshelter (SC); HONH&H = Honeycomb Hill Cave, His & Her Cave (Op); AR1442 = AR144 Cave layer 2; HOSPF4 = Hospital Flat Site 4 (Oc); TIMTOM = Timaru Tomo (SC); FREGUL = Frenchman's Gully Rockshelter (SC); ROCRID = Rocky Ridges (SC); 3MIPRE = Three Mile Road Predator Site (SC); TUTOM2 = Tuarangi Tomo 2 (SC); TUTO3G = Tuarangi Tomo 3 gravel (SC); MCKCAV = McKerchar's Cave (So); COLACD = Colac

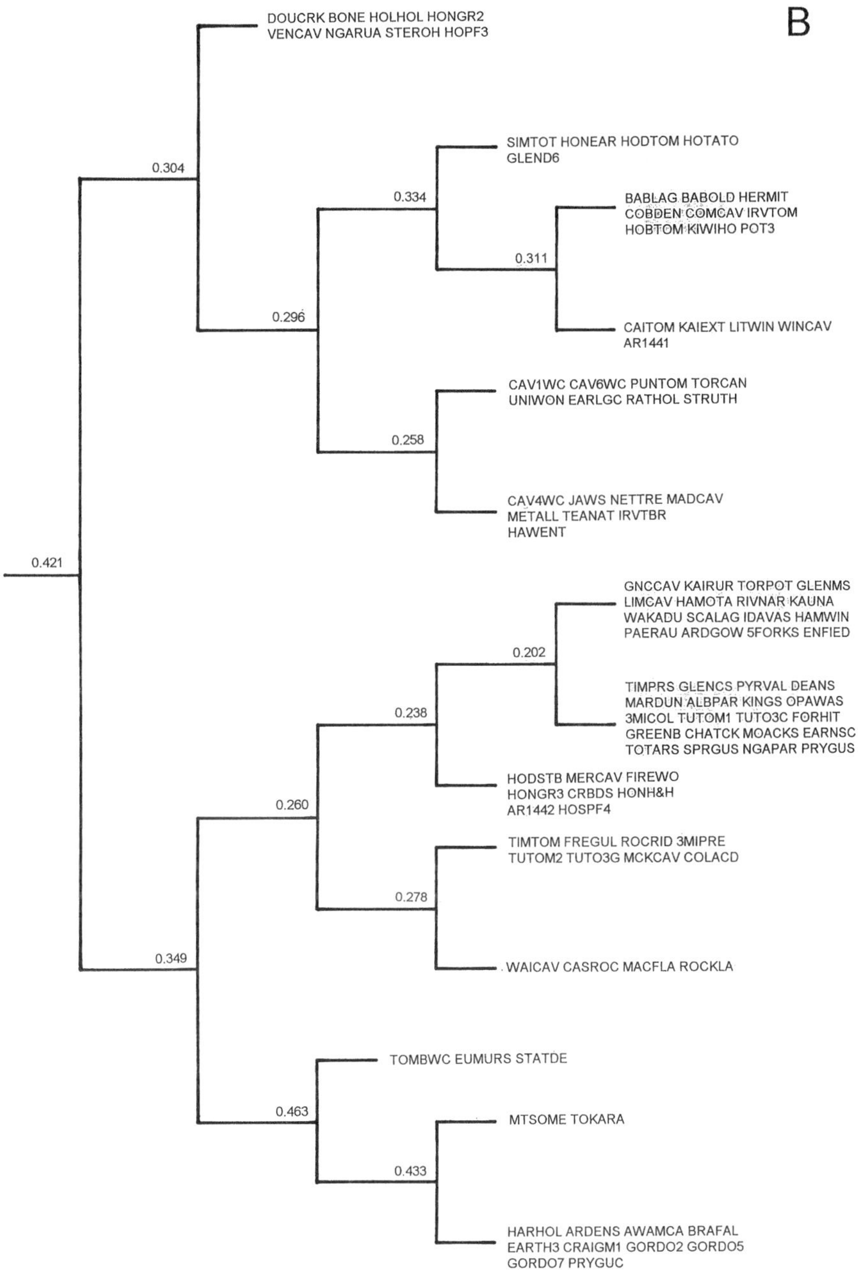

Bay Dunes (So); WAICAV = Waikari Cave (Wai); CASROC = Castle Rocks (So); MACFLA = Macrae's Flat deposit (Deep Dell Creek) (Oc); ROCKLA = Rocklands (Oco); TOMBWC = Tomo B (W); EUMURS = Euan Murchison's Rockshelter (Wai); STADE = Station Deposit (Oc); MTSOME = Mount Somer Quarry = Central Canterbury; TOKARA = Tokarahi Site (Oco); HARHOL = Harwood's Hole (TH); ARDENS = Ardenest (Wai); AWAMCA = Awamoko River Cave (Oco); BRAFAL = Braeburn Falcon Site (SC); EARTH3 = Earthquakes 3 (Oco); CRAIGM1 = Craigmore Station 1 (SC); GORDO2 = Gordon's Valley 2 (SC); GORDO5 = Gordon's Valley 5 (SC); GORDO7 = Gordon's Valley 7 (SC); PRYGUC = Pryde's Gully Road Cave (Oco). Areas: M = Marlborough; MtA = Mount Arthur, northwest Nelson; MtC = Mount Cookson, North Canterbury; Oc = central Otago; Oco = coastal Otago; Op = Oparara, northwest Nelson; SC = South Canterbury; So = Southland; TH = Takaka Hill, northwest Nelson; TV = Takaka Valley, northwest Nelson; W = West Coast; Wai = Waikari area, North Canterbury.

common before humans arrived, has been used as an indicator species for the eastern fauna (Worthy and Holdaway 1996b). When the analysis of moa and other large terrestrial species was repeated with *Euryanas finschi* included, the patterns of co-occurrence were changed slightly (Fig. 14.5A), partly because sites that contained the duck were added. In the western fauna, the pattern was the same except that *Megalapteryx didinus* was associated more with *Pachyornis australis* than with *Dinornis novaezealandiae.* Instead of being a companion of the eagle–*Euryapteryx geranoides* group, *Fulica prisca* was usually associated with *Euryanas finschi,* but both were at the base of the "eastern" branch, which indicates that they were in some places also associated with the species that made up the western fauna.

The whole eastern branch was reorganized by the inclusion of *Euryanas. Euryapteryx geranoides* was grouped with *Pachyornis elephantopus* and *Dinornis giganteus,* as indeed they were in the Pyramid Valley deposit (Holdaway and Worthy 1997), but the fourth species at that site, *Emeus crassus,* was more often associated with the eastern kiwi and not the other moa. The associations of *Harpagornis* were also altered: it grouped with *Cnemiornis calcitrans, Aptornis defossor,* and *Porphyrio hochstetteri* rather than with *Euryapteryx.* The retention of the two main branches, with only *Fulica prisca* swapping between them, provided, despite the rearrangements within the eastern branch, good support for the intuitive groupings of species on which much of the interpretation of the South Islands faunas had been based since 1993.

So it is no surprise that when the degrees of co-occurrence of species of moa and other large birds are considered by site (Figs. 14.4B, 14.5B), the sites containing the eastern (or "eagle") and "western" faunas are clearly separated by the Southern Alps. However, some western and eastern Pleistocene faunas were more similar to each other than they were to either of the western and eastern Holocene faunas. For example, the Hodges Creek, Merino Cave, Graveyard Layer 3, AR144, and His & Hers Cave sites were all grouped, as were Babylon, Hermit's Cave, Irvine's Tomo, and Hobson's Tomo (Fig. 14.5B). There were some eastern sites (Firewood Creek, Craigmore Bird Rockshelter, and Hospital Flat 4) among the first group, but the common factor there was the presence of *Megalapteryx* in the hill country of South Canterbury and Otago.

Although the fauna on the western side of the island during the Pleistocene was substantially the same as that in the east during the Holocene (Worthy and Holdaway 1995, 1996b), the results of the cluster analysis suggest that there were differences. For example, *Dinornis novaezealandiae* was part of the West Coast fauna during the Otiran glaciation, whereas it was not part of the Holocene eastern fauna, from which it was probably excluded by *Dinornis giganteus. D. novaezealandiae* was able to live in shrubland when that was the dominant vegetation, as it did in montane shrubland on both sides of the Southern Alps during the Holocene. Conversely, *Emeus crassus* has never been recorded west of the Alps. At the height of the Otiran glaciation, when greater areas of shrubland and forest–shrubland mosaic would have been available, shrubland and forest mosaic species would have been able to penetrate areas that were clothed by heavy forest during the previous interglacials. Other species in the western fauna may have survived in pockets of tall forest that survived where there was a favorable microclimate.

Unfortunately, there are few dates for deposits older than 25,000 years in New Zealand (the suite of dates to 38,000 years B.P. from Merino Cave is exceptional), so the detail of faunal distribution before that time is sketchy. Knowledge of the faunas of the last interglacial period relies much on inference of age from the depth of deposition and the presence of plant macrofossils, as in the buried forest at Motunau, but new technology—for example, that of optically stimulated luminescence, which dates the last time quartz grains saw light—may help researchers with the older colluvial sequences at places such as Cape Wanbrow.

Whole South Island Fauna

The intuitively defined faunas had some empiric basis for the larger species, but did those associations include definable groups among the smaller birds? When the association patterns of all 95 South Island species in deposits not accumulated by predators were analyzed by the same clustering technique, the eastern and western faunas appeared as the two major branches of the species dendrogram (Fig. 14.6A). As was noted for the limited analysis, larger species are more likely to be found together (and smaller species with each other) if site type is not factored out. That the species associations in the complete data set were not entirely driven by site properties was shown by the separation of the typically eastern and western moa in different major branches (A and B) in Figure 14.6A. The one apparent discrepancy in the groupings was the presence of *Dinornis novaezealandiae* near the base of the eastern fauna main branch (A). As we have noted already, that moa is usually considered to be characteristic of western areas (or at least those clothed with wet forest), but 1000–5000 years ago, it was present to the east of the Southern Alps faunas in hilly areas where soil moisture permitted closed canopy podocarp forest.

"Eagle" or Eastern Fauna

The 20 species in branch A (Fig. 14.6A) include the 10 that constitute the eastern ("eagle") fauna in Figures 14.4 and 14.5A. All the species in branch A presumably had similar habitats. Only two of the species are extant, but another one survived into historic times, and so three provide some information on the habitat preferences of the rest. The pipit prefers rough grassland, with low shrubland, but penetrates forest along roads and riverbeds (Heather and Robertson 1996). That taller shrubs and trees were present is indicated by the pigeon, which was most common in areas dominated by matai (*Prumnopitys taxifolia*) forest (Holdaway and Worthy 1997). But a seasonal use of kowhai (*Sophora microphylla*), lacebarks (*Hoheria* spp.), and wineberry (*Aristotelia serrata*), which are all shrubs of disturbed sites or forest edge, would also explain the species association. Such plants are common in

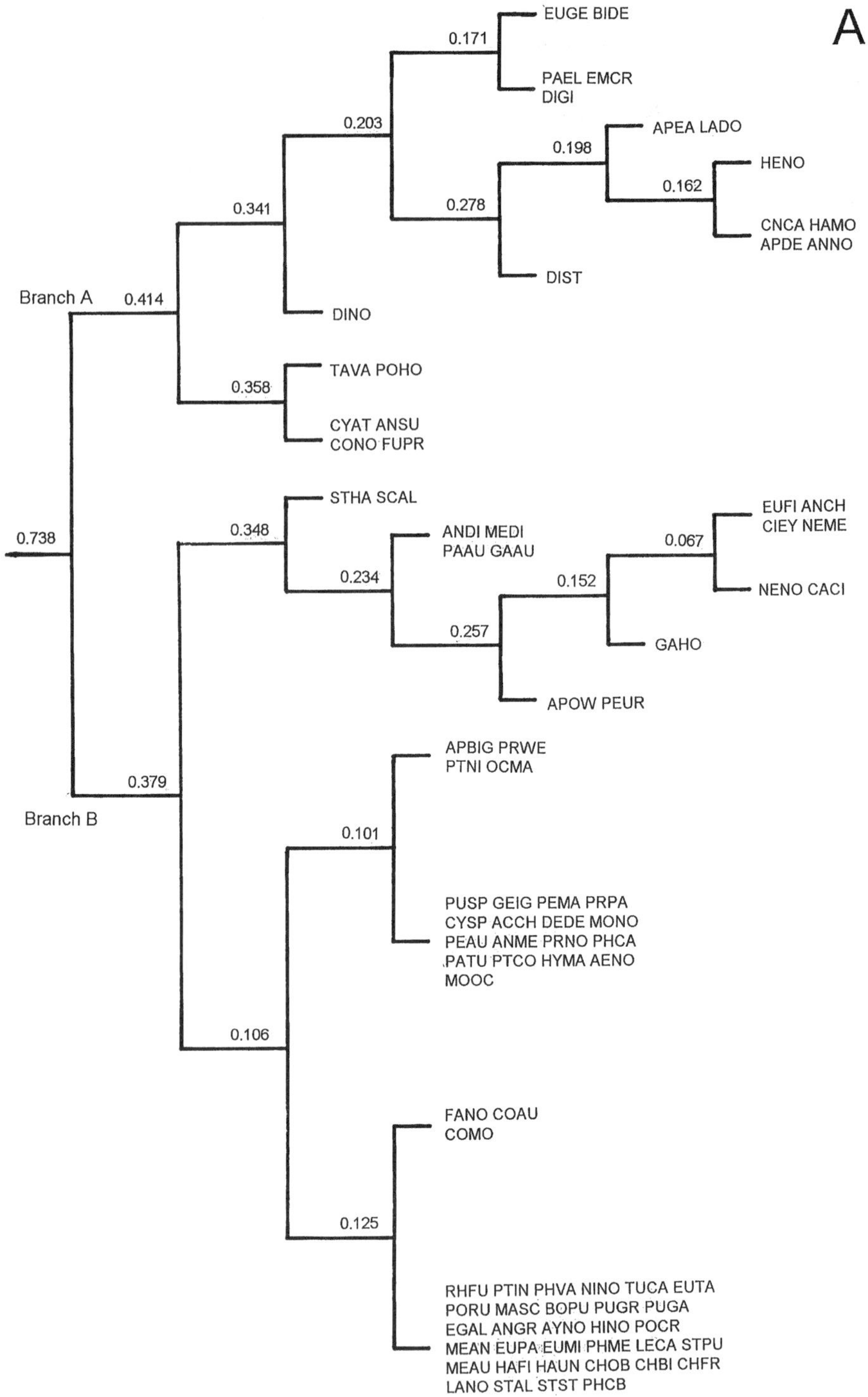
A
Branch A
Branch B
0.738
0.414
0.379
0.341
0.203
0.171
EUGE BIDE
PAEL EMCR
DIGI
0.278
0.198
APEA LADO
0.162
HENO
CNCA HAMO
APDE ANNO
DIST
DINO
0.358
TAVA POHO
CYAT ANSU
CONO FUPR
0.348
STHA SCAL
0.234
ANDI MEDI
PAAU GAAU
0.257
0.152
0.067
EUFI ANCH
CIEY NEME
NENO CACI
GAHO
APOW PEUR
0.106
0.101
APBIG PRWE
PTNI OCMA
PUSP GEIG PEMA PRPA
CYSP ACCH DEDE MONO
PEAU ANME PRNO PHCA
PATU PTCO HYMA AENO
MOOC
0.125
FANO COAU
COMO
RHFU PTIN PHVA NINO TUCA EUTA
PORU MASC BOPU PUGR PUGA
EGAL ANGR AYNO HINO POCR
MEAN EUPA EUMI PHME LECA STPU
MEAU HAFI HAUN CHOB CHBI CHFR
LANO STAL STST PHCB

Figure 14.6. (A) (opposite page) TWINSPAN dendrogram showing classification of species and hierarchic relationships between the species groupings for 95 species of bird represented in fossil deposits not accumulated by predators in the South Island. Taxa abbreviations: *ACCH* = Acanthisitta chloris; *AENO* = Aegotheles novaezealandiae; *ANCH* = Anas chlorotis; *ANDI* = Anomalopteryx didiformis; *ANGR* = Anas gracilis; *ANME* = Anthornis melanura; *ANNO* = Anthus novaeseelandiae; *ANSU* = Anas superciliosa; *APBIG* = Apteryx australis *or* A. haastii; *APDE* = Aptornis defossor; *APEA* = Apteryx *Eastern*; *APOW* = Apteryx owenii; *AYNO* = Aythya novaeseelandiae; *BIDE* = Biziura delautouri; *BOPU* = Bowdleria punctata; *CACI* = Callaeas cinerea; *CHBI* = Charadrius bicinctus; *CHFR* = Charadrius frontalis; *CHOB* = Charadrius obscurus; *CIEY* = Circus eylesi; *CNCA* = Cnemiornis calcitrans; *COAU* = Coenocorypha *sp.*; *COMO* = Corvus *Mainland*; *CONO* = Coturnix novaezelandiae; *CYAT* = Cygnus atratus; *CYSP* = Cyanoramphus *spp.*; *DEDE* = Dendroscansor decurvirostris; *DIGI* = Dinornis giganteus; *DINO* = Dinornis novaezealandiae; *DIST* = Dinornis struthoides; *EGAL* = Casmerodius *(=* Egretta*)* alba; *EMCR* = Emeus crassus; *EUFI* = Euryanas finschi; *EUGE* = Euryapteryx geranoides; *EUMI* = Eudyptula minor; *EUPA* = Eudyptes pachyrhynchus; *EUTA* = Eudynamys taitensis; *FANO* = Falco novaeseelandiae; *FUPR* = Fulica prisca; *GAAU* = Gallirallus australis; *GAHO* = Gallinula hodgenorum; *GAPH* = Gallirallus philippensis; *GEIG* = Gerygone igata; *HAFI* = Haematopus finschi; *HAMO* = Harpagornis moorei; *HAUN* = Haematopus unicolor; *HENO* = Hemiphaga novaeseelandiae; *HINO* = Himantopus novaezelandiae; *HYMA* = Hymenolaimus malacorhynchos; *LADO* = Larus dominicanus; *LANO* = Larus novaehollandiae; *LECA* = Leucocarbo carunculatus; *MASC* = Malacorhynchus scarletti; *MEAN* = Megadyptes antipodes; *MEAU* = Mergus australis; *MEDI* = Megalapteryx didinus; *MONO* = Mohoua novaeseelandiae; *MOOC* = Mohoua ochrocephala; *NEME* = Nestor meridionalis; *NENO* = Nestor notabilis; *NINO* = Ninox novaeseelandiae; *OCMA* = Oceanites maorianus; *PAAU* = Pachyornis australis; *PAEL* = Pachyornis elephantopus; *PATU* = Pachyptila turtur; *PAYA* = Pachyplichas yaldwyni; *PEAU* = Petroica australis; *PEMA* = Petroica macrocephala; *PEUR* = Pelecanoides urinatrix; *PHCA* = Philesturnus carunculatus; *PHCB* = Phalacrocorax carbo; *PHME* = Phalacrocorax melanoleucos; *PHVA* = Phalacrocorax varius; *POCR* = Podiceps cristatus; *POHO* = Porphyrio hochstetteri; *PORU* = Poliocephalus rufopectus; *PRNO* = Prosthemadera novaeseelandiae; *PRPA* = Procellaria parkinsoni; *PRWE* = Procellaria westlandica; *PTCO* = Pterodroma cookii; *PTIN* = Pterodroma inexpectata; *PTNI* = Pterodroma nigripennis; *PUGA* = Puffinus gavia *or* P. huttoni; *PUGR* = Puffinus griseus; *PUSP* = Puffinus spelaeus; *RHFU* = Rhipidura fuliginosa; *SCAL* = Sceloglaux albifacies; *STAL* = Sterna albostriata; *STHA* = Strigops habroptilus; *STPU* = Stictocarbo punctatus; *STST* = Sterna striata; *TAVA* = Tadorna variegata; *TRLY* = Traversia lyalli; *TUCA* = Turnagra capensis; *XESP* = Xenicus longipes *or* X. gilviventris.

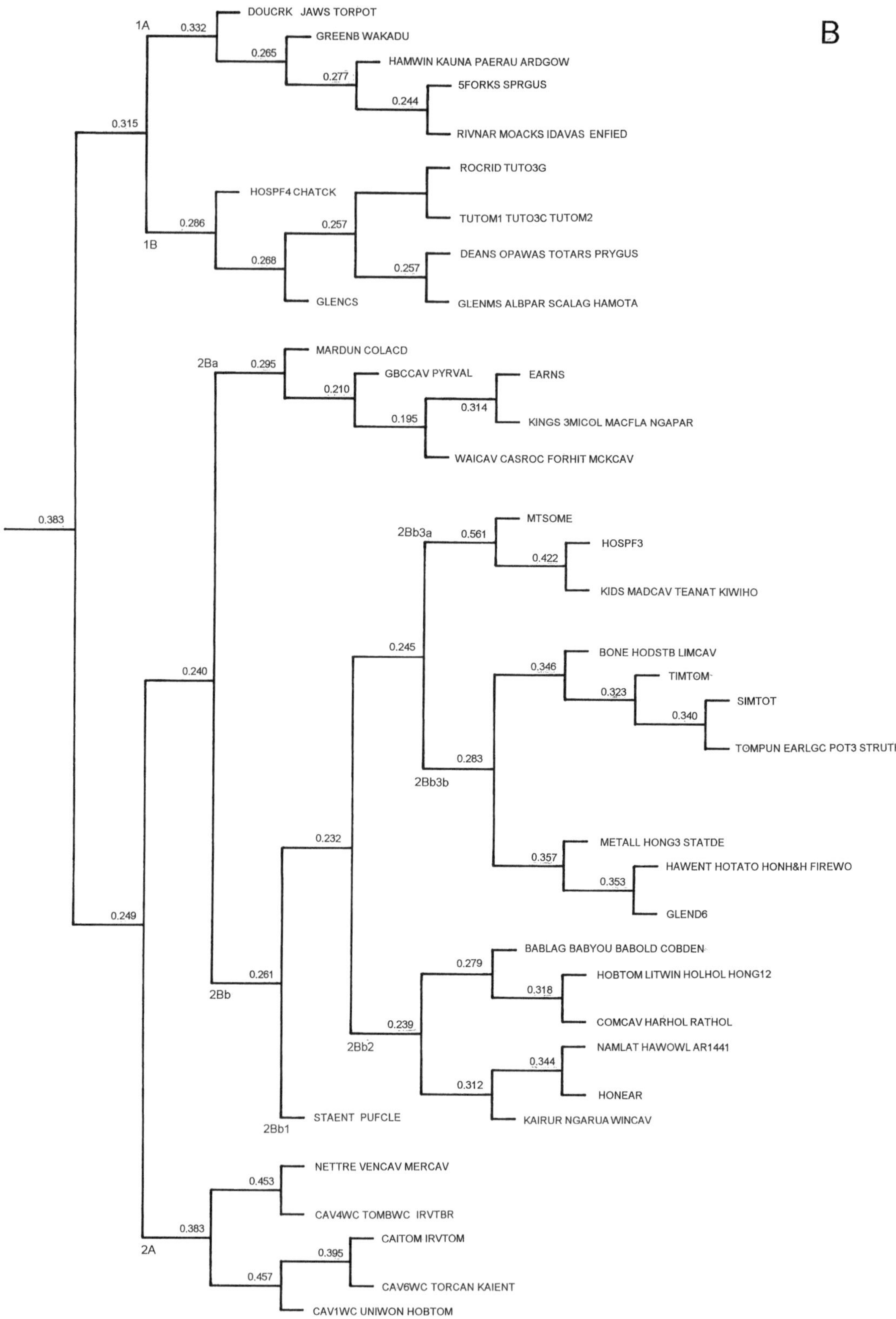
B
1A
0.332
DOUCRK JAWS TORPOT
0.265
GREENB WAKADU
0.277
HAMWIN KAUNA PAERAU ARDGOW
0.244
5FORKS SPRGUS
RIVNAR MOACKS IDAVAS ENFIED
0.315
ROCRID TUTO3G
HOSPF4 CHATCK
TUTOM1 TUTO3C TUTOM2
0.286
0.257
1B
DEANS OPAWAS TOTARS PRYGUS
0.268
0.257
GLENCS
GLENMS ALBPAR SCALAG HAMOTA
MARDUN COLACD
2Ba
0.295
GBCCAV PYRVAL
EARNS
0.210
0.314
0.195
KINGS 3MICOL MACFLA NGAPAR
WAICAV CASROC FORHIT MCKCAV
0.383
MTSOME
2Bb3a
0.561
HOSPF3
0.422
KIDS MADCAV TEANAT KIWIHO
0.245
BONE HODSTB LIMCAV
0.346
TIMTOM
0.240
0.323
SIMTOT
0.340
TOMPUN EARLGC POT3 STRUTH
0.283
2Bb3b
0.232
METALL HONG3 STATDE
0.357
HAWENT HOTATO HONH&H FIREWO
0.353
0.249
GLEND6
BABLAG BABYOU BABOLD COBDEN
0.279
HOBTOM LITWIN HOLHOL HONG12
0.261
0.318
2Bb
COMCAV HARHOL RATHOL
0.239
2Bb2
NAMLAT HAWOWL AR1441
0.344
0.312
HONEAR
KAIRUR NGARUA WINCAV
STAENT PUFCLE
2Bb1
NETTRE VENCAV MERCAV
0.453
CAV4WC TOMBWC IRVTBR
0.383
2A
CAITOM IRVTOM
0.395
0.457
CAV6WC TORCAN KAIENT
CAV1WC UNIWON HOBTOM

Figure 14.6. (B) (opposite page) TWINSPAN dendrogram showing classification of sites containing the same species as in A and the hierarchic relationships between the site groupings. Site abbreviations in order from top to bottom of diagram: DOUCRK = Doubtful Creek (W); JAWS = Jaws Cave (W); TORPOT = Tomo by Torture Track, Annandale (MtC); GREENB = Greenhills Back Beach Dunes (So); WAKADU = Wakapatu Dunes (So); HAMWIN = Hamilton's Swamp, Winton (So); KAUNA = Kauana Swamp (So); PAERAU = Paerau Moa Swamp (Oc); ARDGOW = Ardgowan Swamp (Oco); 5FORKS = Five Forks Swamp (Oco); SPRGUS = Spring Gully (Oco); RIVNAR = Riverton Narrows Site (So); MOACKS = Moa Creek Swamp (Oc); IDAVAS = Ida Valley Swamp (Oc); ENFIE = Enfield (Oco); ROCRID = Rocky Ridges (SC); TUTO3G = Tuarangi Tomo 3 (gravels) (SC); TUTOM1 = Tuarangi Tomo 1 (SC); TUTO3C = Tuarangi Tomo 3 (clay) (SC); DEANS = The Deans (NC); OPAWAS = Opawa Swamp (SC); TOTARS = Totara Swamp (Oco); PRYGUS = Pryde's Gully Road Swamp (Oco); GLENMS = Glenmark Swamp (Wai); ALBPAR = Albury Park Swamp (SC); SCALAG = Scaife's Lagoon (Oc); HAMOTA = Hamilton Swamp (Oc); MARDUN = Marfells Beach Dunes (M); COLACD = Colac Bay Dunes (So); GBCCAV = Golden Bay Cement Cave (TV); PYRVAL = Pyramid Valley Swamp (Wai); EARNS = Earnscleugh Cave (Oc); KINGS = Kings Cave (SC); 3MILCOL = Three Mile Road Colluvial deposit (SC); MACFLA = Macrae's Flat deposit (Deep Dell Creek) (Oc); NGAPAR = Ngapara (Oco); WAICAV = Waikari Cave (Wai); CASROC = Castle Rocks (So); FORHIT = Forest Hills Tomos (So); MCKCAV = McKerchar's Cave (So); MTSOME = Mount Somer's Quarry (Central Canterbury); HOSPF3 = Hospital Flat Site 3 (Oc); KIDS = Kids Cave (W); TEANAT = Te Ana Titi [cave] (W); KIWIHO = Kiwi Hole (TH); BONE = Bone Cave (TH); HODSTB = Hodge's Stream Cave Streambed (MtA); LIMCAV = Limehills Cave (So); TIMTOM = Timaru Tomo (SC); SIMTOM = Sim's Cave = Tomo Entrance (TV); TOMPUN = Punakaiki Tomo (W); EARLGC = Earl Grey Cave (TH); POT3 = Pothole 3 (W); STRUTH = Struthoides Cave (TH); METALL = Metro Cave (all sites) (W); HONG3 = Honeycomb Hill Cave, Graveyard Site, layer 3 (Op); STATDE = Station Deposit (Oc); HAWENT = Hawke's Cave Entrance (TH); HOTATO = Hodge Creek, Takahe Tomo (MtA); HONH&H = Honeycomb Hill Caves, His & Her Cave (Op); FIREWO = Firewood Creek (Oc); GLEND6 = Glendhu Station Site 6 (Oc); BABLAG = Babylon Cave, Passage of Thoth, lag deposit (W); BABYOU, Babylon Cave, Passage of Thoth, young (Holocene) sediments (W); BABOLD = Babylon Cave, Passage of Thoth, old (Pleistocene) sediments (W); COBDEN = Cobden Fissure Site (W); HOBTOM = Hobson's Tomo (TH); LITWIN = Little Winter Cave (TH); HOLHOL = Holocene Hole (MtC); HONG12 = Honeycomb Hill Cave, Graveyard Site, layers 1 and 2 (Op); COMCAV = Commentary Cave (TV); HARHOL = Harwood's Hole (TH); RATHOL = Ratite Hole (TH); NAMLAT = Name Later Cave (W); HAWOWL = Hawke's Cave Owl Site (TH); AR1441; AR144 Cave layer 1 (Op); HONEAR = Honeycomb Hill Cave = Eagle's Roost Site (Op); KAIRUR = Kairuru Cave (TH); NGARUA = Ngarua Cave (TH); WINCAV = Winter Cave (TH); NETTRE = Nettletrench Cave (W); VENCAV = Venturer Cave (W); MERCAV = Merino Cave (MtC); CAV4WC = Cave 4 (W); TOMBWC = Tomo B (W); IRVTBR = Irvine's Tomo, beyond rockfall (TV); CAITOM = Cairn's Tomo (W); IRVTOM = Irvine's Tomo (TV); CAV6WC = Cave 6 (W); TORCAN = Tory Canyon Cave (W); KAIENT = Kairuru Cave Extension (TH); CAV1WC = Cave 1 (W); UNIWON = Unique Wonder Cave (W); HOBTOM = Hobson's Tomo (TH). Areas: M = Marlborough; MtA = Mount Arthur, northwest Nelson; MtC = Mount Cookson, North Canterbury; Oc = Central Otago; Oco = coastal Otago; Op = Oparara, northwest Nelson; SC = South Canterbury; So = Southland; TH = Takaka Hill, northwest Nelson; TV = Takaka Valley, northwest Nelson; W = West Coast; Wai = Waikari area, North Canterbury.

eastern Holocene floras. The quail is another grassland and shrubland species (Worthy 1997b). The extinct eastern kiwi was also part of this association, but because brown kiwi (the group to which this species apparently belonged) can live in either grassland or forest, its presence provides no additional information on the dominant habitats used by the assemblage.

A likely reason for the separation of the smaller birds on a branch away from the four moa is that the different-size birds were likely to be trapped in different sites, with moa being heavily biased toward swamps and pitfall traps such as potholes in limestone terrain. In this instance, separation on the diagram does not mean separation in the environment. Within the groups, however, placement on different branches may indicate differences in microhabitat. An apparent preference by *Euryapteryx geranoides* for hilly terrain has already been mentioned, and this moa is separated from the other three in the terminal branches. The apparent sympatry of *Dinornis novaezealandiae* with *Harpagornis moorei* at Glenmark, the type locality for the eagle, may be an artifact of the use of presence–absence data rather than representation within sites. Glenmark is in an area where the fossil megafauna is otherwise represented only by the "eastern four": *Pachyornis elephantopus, Emeus crassus, Dinornis giganteus,* and *Euryapteryx geranoides. D. novaezealandiae* was rare at Glenmark and perhaps present only when the climate was somewhat wetter and warmer in the mid-Holocene.

In sites where there was shrubland and access to water, such as around lagoons, three waterfowl (black swan *Cygnus atratus,* paradise shelduck *Tadorna variegata,* and grey duck *Anas superciliosa*) were part of a group that included the New Zealand quail (*Coturnix novaezelandiae*), the South Island takahe (*Porphyrio hochstetteri*), and the extinct coot (*Fulica prisca*). Quail are usually found in deposits where the vegetation was shrubland, and they were found in tussock grassland in the 19th century. Apart from the swan and duck, these species seem to have preferred places where there was a mix of dense cover and more open areas. Takahe certainly bred near Pyramid Valley, which was then a small lake surrounded by podocarp–broadleaf forest dominated by matai, but with abundant lowland ribbonwood (*Plagianthis regius*), a tree that reaches 17 m in height and that does not grow under a canopy. Unlike most New Zealand trees, *P. regius* is deciduous, and pigeons favor the new leaf buds. A floristically rich ribbon of shrubland surrounded the lake, and similar vegetation clothed the upper slopes of the valley; there were also open areas formed as the lake level fluctuated seasonally.

A key to the relative richness of the eagle fauna (which included in places most of the western birds) may have been the input of guano from petrels nesting on the ridges. With a low rainfall and a higher pH level, most of the nutrients would have been retained in the soil, where some can still be detected (Hawke et al. 1999). In the west, where many petrels also bred (Worthy and Holdaway 1993), soils were leached and podzolized by the high rainfall. Absence of leaching and more available minerals probably allowed higher plant productivity that was in turn echoed in both a higher species diversity and higher numbers of individual birds. Today, with the petrel colonies long gone, New Zealand has had to turn to the residues left by seabirds elsewhere (formerly, mainly Nauru in the tropical Pacific) as sources of phosphate for the agricultural base for the New Zealand economy. Most of the New Zealand soils are deficient in trace elements (Campbell 1973) and need applications of fertilizer and elemental supplements to sustain intensive agriculture and extensive pastoralism. In the eastern South Island, the problem is exacerbated by the low solubility of the basement rock of Jurassic greywacke sandstone and older schists.

A Western Fauna?

In contrast to the eagle fauna, there are few species that are completely confined to west of the axial mountains in the Holocene. Although some species, such as kakapo (*Strigops habroptilus*) and little bush moa (*Anomalopteryx didiformis*), were definitely more abundant on the West Coast, only certain petrels had no known eastern populations.

Branch B in Figure 14.6A includes species found on both sides of the South Island during the Holocene. Hence, although it is possible to say that there was a western moa fauna, it is more accurate to describe the remaining species as forest species that lived, in varying abundance, wherever forest was available. For some tall forest specialists, that meant in the higher altitude, higher rainfall areas in the western ranges and in the south. The inclusion of the Mount Cookson and some western Otago and Southland sites with the western bloc in Figure 14.6B supports this conclusion. When conditions were drier, as in the Otiran glaciation, components of the eagle fauna spread west and overlaid the cosmopolitan fauna. Later, as the climate became damper and warmer, the distributions of the large species of the eagle fauna shrank back to the east, leaving those species able to cope with the forests on low-productivity soils to themselves again.

Beech Forests or Ecotonal Areas

Southern beech (*Nothofagus*) forests included lowland associations but most were, as the remnants are now, river terrace forests in the ranges, montane slope forests, and high-altitude moss forests. The guilds present in these regions are far less well known, principally because of the paucity of fossil deposits. Some indication of the beech forest fauna may be obtained from careful extrapolation from the Takaka Hill and Mount Cookson faunas, both of which were in mixed beech and podocarp forest and not pure *Nothofagus*. Few, if any, species seem to have been restricted to beech, although the long-billed wren (*Dendroscansor decurvirostris*) may have extended into the forest from upland shrublands, which seem to have been its main habitat. Bush wrens (*Xenicus longipes*) and South Island kokako (*Callaeas cinerea*) may also have been more abundant in beech forest. So far as we know, no moa or other extinct species was associated exclusively with *Nothofagus* forest.

Alpine Shrubland and Herb Fields

In the alpine shrubland, the species that are presently characteristic of that habitat—rock wren (*Xenicus gilviventris*) and kea (*Nestor notabilis*)—were more widespread in the past, and *Dendroscansor decurvirostris* was probably the only species confined more or less to highland areas. Other wrens, including Lyall's wren (*Traversia lyalli*), were also found at altitude, but they were probably separated by their feeding sites and techniques because the bills were decidedly different (Millener and Worthy 1991). However, eastern taxa such as Haast's eagle, the New Zealand coot, Finsch's duck, and Eyles's harrier all at least seasonally used the subalpine zone in karst areas of the northern South Island. They shared the fellfields and shrublands with *Megalapteryx didinus, Pachyornis australis,* and the occasional *Dinornis novaezealandiae* and *Dinornis struthoides.*

The situation from Southland is less clear because of the complex mix of habitats of lowland wet forest and shrublands, and drier hill forests with much beech. Species elsewhere typical of high altitudes such as *Pachyornis australis* reached the coastline in Southland. Castle Rocks, in western Southland, is the only site known where the moa *Anomalopteryx didiformis* was sympatric with Haast's eagle. It appears to be a region where several habitat faunas interdigitate along with the habitats. Across Foveaux Strait, the Stewart Island fauna differs from that of the adjacent mainland, because the forest on the island was largely composed of podocarps and other hardwoods, and *Nothofagus* is absent. Only the common generalist moa (*Dinornis struthoides*) seems to have been present on the island. Other flightless species seem to have been rare or absent on the island (Worthy 1998e). Consequently, the fauna there was a depauperate subset of the South Island wet forest fauna.

North Island

In the North Island, the assemblage of species (or North Island representatives of species pairs such as the adzebill *Aptornis otidiformis*) that in the South Island constituted the eagle fauna did not include the eagle itself. The common species of moa in the drier eastern areas had low body mass (*Pachyornis mappini* and *Euryapteryx curtus*). Larger species such as *Anomalopteryx didiformis* and *Dinornis struthoides* lived within dense forest, where the carrying capacity was probably low and the total populations were likely to have been small. Of these species, only *D. struthoides* approached the mass of the four common moa of the eastern South Island. *Pachyornis mappini* was apparently confined to swamp shrubland and damp shrubland associations. Areas of shrubland supporting *Euryapteryx curtus* tended to be smaller in the North Island, and larger areas were governed by the extent and frequency of volcanic tephra showers.

Dinornis giganteus was, as in the eastern South Island, present but uncommon and so would not have been a viable resource for Haast's eagle. There were few areas of the mosaic forest and shrubland habitat that seem to have been favored by the eagle and its usual prey of

Finsch's duck, other flightless carinates, and larger moa in the South Island. Taxa such as the adzebill and Finsch's duck may have been restricted more closely to rare microhabitats and hence confined to very specific geographic areas. Adzebills seem to have been more common at higher altitudes or on ridges than on valley floors. They were absent from the swamp forest habitat around Lake Poukawa during the Holocene. Farther south, they were relatively common at 500–1000 m, near ridge tops—for example, at Coonoor (Figs. 14.7–14.9).

Figure 14.7. Entrance to Coonoor Cave, Puketoi Range, southeastern North Island, in 1914, when the fossil deposit was excavated. Photograph courtesy of Canterbury Museum.

Eagles are absent from North Island dune deposits, whereas almost all dune deposits along the eastern and southern coasts of the South Island have produced eagle bones. Even the coastal shrublands of the southeastern North Island, which were within sight of the South Island and which had populations of two species of *Euryapteryx*, as well as of *Dinornis struthoides* and *D. giganteus*, lacked eagles. Why? The key may have been that although the same or analogous prey species might

Figure 14.8. Bones excavated from Coonoor Cave arrayed on farmhouse porch. Photograph courtesy of Canterbury Museum.

have been present, the available habitat was long and thin, and a sufficient territory size of 10–15 km^2 would be perhaps 200 m wide and 50 km long. Not enough of the habitat was available to maintain a viable population, which would need to consist of several pairs. Inland of the 200-m coastal strip was dense forest that supported the typical, sparsely distributed, wet forest fauna. The moa and other species could maintain contact with their own kind within that coastal strip, but an eagle would need to forage over too great a distance to be able to get enough to eat, much less feed and raise young. The more extensive areas of coastal shrubland in the far north of the North Island might have been able to support eagles if they had been able to reach there. That area was probably under heavy forest during the Pleistocene, and eagles were separated from there by large areas of wet forest with little favorable habitat and low prey numbers.

Relating Faunas to Habitats

Unlike most continental faunas, few members of the New Zealand avifauna appear to have been confined to a single habitat. Such an impression may result from blurring of distinctions by time-averaging and spatial compression in fossil sites, but it more likely has resulted from the low species diversity of most genera. However, some general

Figure 14.9. Skulls of birds found at Coonoor Cave site. Photograph courtesy of Canterbury Museum.

patterns in the distribution of species and habitats can be seen. In Figure 14.3, the broad distribution of vegetation types is plotted against altitudinal and rainfall gradients. Each of the major habitats seems to have had a distinct fauna, some of which (for example, the beech forest assemblage) are still poorly known because of the shortage of fossil sites in those areas. The relative areas of the different habitat types as constrained by climatic and geographic conditions were reflected in the overall abundance of the species that lived there. Areas of grassland were limited, and the geese were among the least common of the larger birds in the fauna.

One way of investigating the habitat requirements of extinct species is to compare their distribution to those of extant species whose habitats are known. One ordination technique that is applicable to ecologic data is known as detrended correspondence analysis (DECORANA; Hill and Gauch 1980). DECORANA transforms species presence and absence data for a range of sites with differing characteristics into scores on new, noncorrelated axes that represent environmental gradients specific to the system being studied.

For the 95 species of the South Island fossil fauna, in sites formed by all processes of both Pleistocene and Holocene age, the plots of ordination scores for the first three axes are shown in Figure 14.10. Scores on axis I are not very useful for discerning habitat because they are related to the taphonomy of the site; species from sites that best preserved large species such as moa have the highest scores, and those that were found only in predators' sites have the lowest. However, axis II reflects the geographic spread of the sites and hence both rainfall and

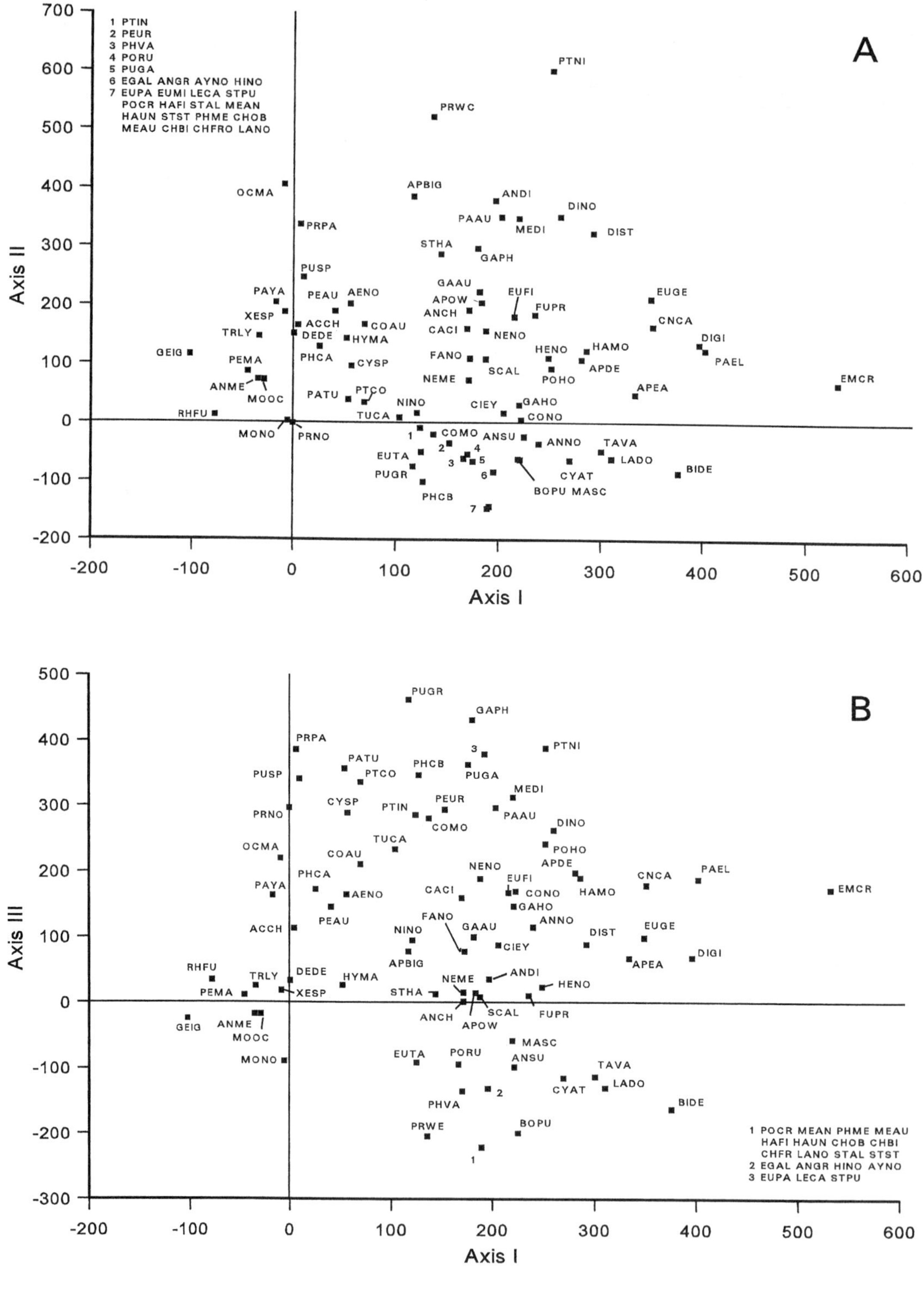
A
1 PTIN
2 PEUR
3 PHVA
4 PORU
5 PUGA
6 EGAL ANGR AYNO HINO
7 EUPA EUMI LECA STPU
POCR HAFI STAL MEAN
HAUN STST PHME CHOB
MEAU CHBI CHFRO LANO
Axis II
Axis I
PTNI
PRWC
OCMA
APBIG
ANDI
DINO
PAAU
MEDI
DIST
PRPA
STHA
GAPH
PUSP
GAAU
EUFI
EUGE
PAYA
PEAU
AENO
APOW
FUPR
ANCH
CNCA
XESP
ACCH
COAU
CACI
NENO
DIGI
TRLY
DEDE
HYMA
HENO
HAMO
GEIG
PHCA
FANO
PAEL
PEMA
CYSP
SCAL
APDE
POHO
NEME
EMCR
ANME
APEA
MOOC
PATU
PTCO
NINO
CIEY
GAHO
RHFU
TUCA
CONO
MONO
PRNO
COMO
ANSU
ANNO
TAVA
EUTA
LADO
BIDE
PUGR
CYAT
PHCB
BOPU MASC
B
Axis III
Axis I
PUGR
GAPH
PRPA
PTNI
PATU
PHCB
PUSP
PTCO
PUGA
MEDI
CYSP
PEUR
PRNO
PTIN
PAAU
COMO
DINO
TUCA
OCMA
POHO
COAU
NENO
APDE
PAEL
PHCA
EUFI
CNCA
PAYA
AENO
CACI
CONO
HAMO
EMCR
GAHO
PEAU
FANO
ANNO
ACCH
NINO
GAAU
DIST
EUGE
CIEY
APBIG
DIGI
RHFU
APEA
DEDE
ANDI
TRLY
HYMA
NEME
HENO
PEMA
XESP
STHA
SCAL
FUPR
ANCH
GEIG
ANME
APOW
MOOC
MASC
MONO
EUTA
PORU
ANSU
TAVA
CYAT
LADO
PHVA
BIDE
BOPU
PRWE
1 POCR MEAN PHME MEAU
HAFI HAUN CHOB CHBI
CHFR LANO STAL STST
2 EGAL ANGR HINO AYNO
3 EUPA LECA STPU

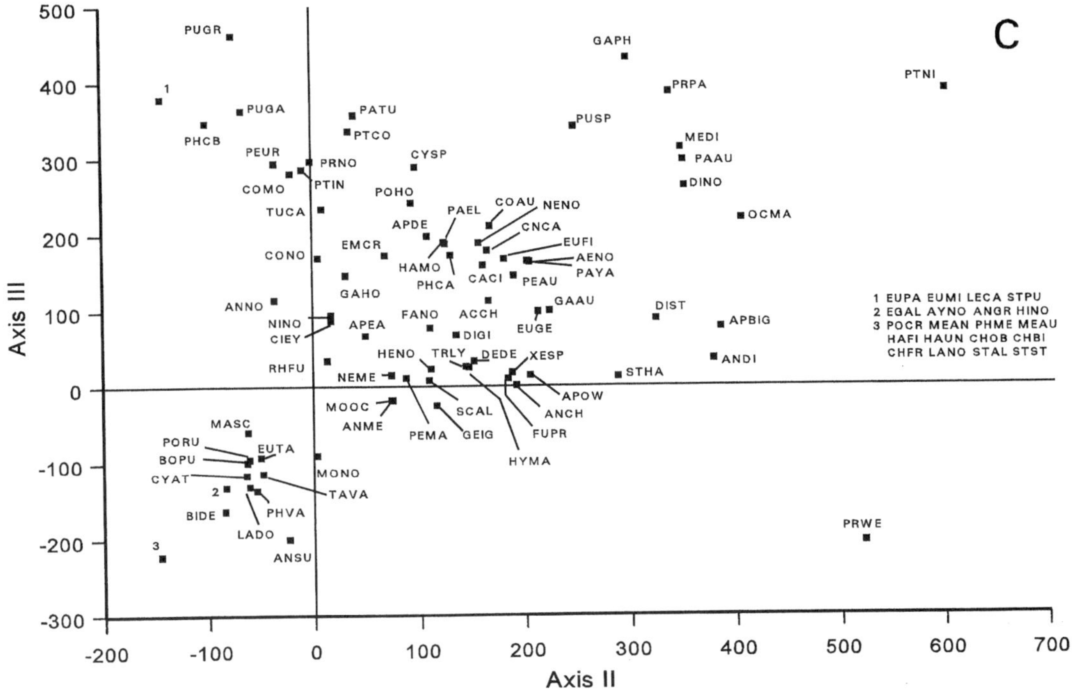

Figure 14.10. Position of species on the first three axes of a DECORANA ordination of presence–absence data for 95 species of birds in the South Island in sites of both Pleistocene and Holocene age. (A) (opposite page top) Axis II versus axis I. (B) (opposite page bottom) Axis III versus axis I. (C) (above) Axis III versus axis II. Taxa abbreviations: ACCH = Acanthisitta chloris; *AENO* = Aegotheles novaezealandiae; *ANCH* = Anas chlorotis; *ANDI* = Anomalopteryx didiformis; *ANGR* = Anas gracilis; *ANME* = Anthornis melanura; *ANNO* = Anthus novaeseelandiae; *ANSU* = Anas superciliosa; *APBIG* = Apteryx australis *or* A. haastii; *APDE* = Aptornis defossor; *APEA* = Apteryx *eastern*; *APOW* = Apteryx owenii; *AYNO* = Aythya novaeseelandiae; *BIDE* = Biziura delautouri; *BOPU* = Bowdleria punctata; *CACI* = Callaeas cinerea; *CHBI* = Charadrius bicinctus; *CHFR* = Charadrius frontalis; *CHOB* = Charadrius obscurus; *CIEY* = Circus eylesi; *CNCA* = Cnemiornis calcitrans; *COAU* = Coenocorypha *sp.*; *COMO* = Corvus *Mainland*; *CONO* = Coturnix novaezelandiae; *CYAT* = Cygnus atratus; *CYSP* = Cyanoramphus *spp.*; *DEDE* = Dendroscansor decurvirostris; *DIGI* = Dinornis giganteus; *DINO* = Dinornis novaezealandiae; *DIST* = Dinornis struthoides; *EGAL* = Casmerodius (= Egretta) albus; *EMCR* = Emeus crassus; *EUFI* = Euryanas finschi; *EUGE* = Euryapteryx geranoides; *EUMI* = Eudyptula minor; *EUPA* = Eudyptes pachyrhynchus; *EUTA* = Eudynamys taitensis; *FANO* = Falco novaeseelandiae; *FUPR* = Fulica prisca; *GAAU* = Gallirallus australis; *GAHO* = Gallinula hodgenorum; *GAPH* = Gallirallus philippensis; *GEIG* = Gerygone igata; *HAFI* = Haematopus finschi; *HAMO* = Harpagornis moorei; *HAUN* = Haematopus unicolor; *HENO* = Hemiphaga novaeseelandiae; *HINO* = Himantopus novaezelandiae; *HYMA* = Hymenolaimus malacorhynchos; *LADO* = Larus dominicanus; *LANO* = Larus novaehollandiae; *LECA* = Leucocarbo carunculatus; *MASC* = Malacorhynchus scarletti; *MEAN* = Megadyptes antipodes; *MEAU* = Mergus australis; *MEDI* = Megalapteryx didinus; *MONO* = Mohoua novaeseelandiae; *MOOC* = Mohoua ochrocephala; *NEME* = Nestor meridionalis; *NENO* = Nestor notabilis; *NINO* = Ninox novaeseelandiae; *OCMA* = Oceanites maorianus; *PAAU* = Pachyornis australis; *PAEL* = Pachyornis elephantopus; *PATU* = Pachyptila turtur; *PAYA* = Pachyplichas yaldwyni; *PEAU* = Petroica australis; *PEMA* = Petroica macrocephala; *PEUR* = Pelecanoides urinatrix; *PHCA* = Philesturnus carunculatus; *PHCB* = Phalacrocorax carbo; *PHME* = Phalacrocorax melanoleucos; *PHVA* = Phalacrocorax varius; *POCR* = Podiceps cristatus; *POHO* = Porphyrio hochstetteri; *PORU* = Poliocephalus rufopectus; *PRNO* = Prosthemadera novaeseelandiae; *PRPA* = Procellaria parkinsoni; *PRWE* = Procellaria westlandica; *PTCO* = Pterodroma cookii; *PTIN* = Pterodroma inexpectata; *PTNI* = Pterodroma nigripennis; *PUGA* = Puffinus gavia *or* P. huttoni; *PUGR* = Puffinus griseus; *PUSP* = Puffinus spelaeus; *RHFU* = Rhipidura fuliginosa; *SCAL* = Sceloglaux albifacies; *STAL* = Sterna albostriata; *STHA* = Strigops habroptilus; *STPU* = Stictocarbo punctatus; *STST* = Sterna striata; *TAVA* = Tadorna variegata; *TRLY* = Traversia lyalli; *TUCA* = Turnagra capensis; *XESP* = Xenicus longipes *or* X. gilviventris.

altitudinal effects. The highest scores on axis II are for those taxa that, like *Procellaria westlandica,* are or were confined to the wet forests of the West Coast, whereas the lowest scores are associated with taxa found solely in eastern coastal sites such as Marfells Beach in Marlborough.

Axis III (Fig. 14.10B, C) appears to represent the gradient between heavy forest and shrubland–forest mosaic, with the higher scores associated with more varied habitats with an admixture of grassland. When axes II and III are plotted against each other (Fig. 14.10C), the position of species with known habitat in the geographic–vegetation space requirements allow the possible habitat requirements of other species to be predicted. The petrels form a peripheral row with the highest average scores on axis III, which probably reflects their requirement for access to open sites, usually ridges. Others with high scores on axis III (but at the eastern end of axis II) include the penguins and marine shags found in coastal sites from Marlborough to Southland. The only terrestrial bird with a high score on axis III is *Gallirallus philippensis,* from the Pleistocene of Honeycomb Hill Cave, where the environment was rather open (Worthy and Mildenhall 1989). Conversely, the only petrel with a high "forest" score is the Westland petrel, which nests in dense West Coast rainforest.

The apparent anomaly of wetland species, including *Cygnus atratus* and *Himantopus novaezelandiae,* being grouped at the forest end of axis III is a result of their occurring in deposits with forest birds at, for example, Pyramid Valley, where the wetland habitat was surrounded by tall vegetation. Areas of short wetland vegetation were limited before Polynesian fires altered the landscape.

An interesting grouping on Figure 14.10C is the cluster of eastern species in shrubland–grassland–forest mosaic that included the raven (*Corvus* Mainland), pipit (*Anthus novaeseelandiae*), takahe (*Porphyrio hochstetteri*), quail (*Coturnix novaezelandiae*), eastern moa (*Emeus crassus*), and heavy-footed moa (*Pachyornis elephantopus*). The association of the prehistorically extinct species with the pipit and quail indicated a preference for more open vegetation. Close to that group were placed the South Island adzebill (*Aptornis defossor*), Haast's eagle (*Harpagornis moorei*), Hodgens' waterhen (*Gallinula hodgenorum*), pigeon (*Hemiphaga novaeseelandiae*), tui (*Prosthemadera novaeseelandiae*), and South Island goose (*Cnemiornis calcitrans*), all of which were slightly further toward the forest end of the range. With distributions that included the higher country to the west, species such as the South Island kokako (*Callaeas cinerea*) and kea (*Nestor notabilis*) also were within the range of scores on axis III, typical of mosaic habitats rather than dense forest. At higher altitudes still, and therefore well to the "west" on axis II, are the three upland moa, *Megalapteryx didinus, Pachyornis australis,* and *Dinornis novaezealandiae.* At the other end of the vegetation spectrum, but in the same geographic area (and hence close in altitude preference), is *Anomalopteryx didiformis,* almost alone at the bottom right-hand corner of the figure.

Most typical of the closed-canopy forest species between the scores of the kaka (*Nestor meridionalis*) and kakapo (*Strigops habroptilus*) years 1768–1771 was Joseph Banks, a man of independent wealth and wide experience. Later to become Sir Joseph Banks, the longest serving president of the most prestigious scientific society in England, the Royal Society of London, Banks slept during the voyage in a hammock over the chart table in the Great Cabin of His Majesty's Bark *Endeavour* because he was too tall to sleep comfortably in his own cabin. In the early morning of January 17, 1770, as *Endeavour* lay at anchor in Queen Charlotte Sound, Banks heard something that moved him to record in his diary:

> This morn I was awakd by the singing of the birds ashore from whence we are distant not a quarter of a mile, the numbers of them were certainly very great who seemd to strain their throats with emulation perhaps; their voices were certainly the most melodious wild musick I have ever heard, almost imitating small bells but with the most tuneable silver sound imaginable to which maybe the distance was no small addition. (Beaglehole 1962: 455)

He would be shocked at the silence that greets the dawn there now. He would be lucky to hear just a few of the bellbirds (*Anthornis melanura*) and tui (*Prosthemadera novaeseelandiae*) that enchanted him that morning, among the introduced blackbirds (*Turdus merula*) and song thrushes (*T. philomelos*). But usually he would have to strain to hear any birds at all—and certainly he would hear none of the other species that made up that chorus of birdsong: the piopio, kokako, and saddleback. And one of the reasons for the present silence was likely scampering beneath his hammock that morning. The arrival of Europeans ushered in a new phase of extinctions, but the catastrophe that robbed the forest of songsters after Cook sailed onward into history had begun nearly 1800 years before he sighted the southern land.

On his three voyages, Cook deliberately attempted to introduce mammals he deemed useful to the Māori population or to future European visitors, such as pigs and goats, and accidentally released another, the Norway rat. Pigs and Norway rats were the first new predatory mammals to arrive since people; they would not be the last.

It was obvious from the earliest days of European settlement that some birds had become extinct. As the most obvious example, there were moa bones but no moa. In the 1840s, extinction itself was a relatively new idea, given respectability by Baron Cuvier in France. Until the late 18th century, it was unthinkable that any species could cease to exist, because they had all been created at the same time and had been around since that creation. Extinction was a denial of Scripture. Mounting evidence for whole faunas of animals that were very different from those alive at present led scholars to first propose a series of creations, each destroyed by a flood and replaced by a new creation. That still did not allow for extinctions of species within a fauna. Examples of losses from the European fauna, and observations of the presence of the bones of extinct animals with the bones of humans, eventually showed that species can indeed become extinct. That was the worldview by the time moa were discovered, so the reason for their extinction was an issue for debate.

And from the second half of the 19th century, the discussion had to be extended when it became apparent that there were many other extinct species in New Zealand, from geese to crows. What could have caused them to become extinct along with the giant ground birds? Climate change, genetic senescence, and disease were all hypotheses favored by researchers at one time or other. The main difficulty with climate change in the New Zealand extinctions is that most of them happened at a time when the local climate was relatively stable. The average temperature in New Zealand has fluctuated by less than 1°C over the past 1000 years (Burrows and Greenland 1979). A fauna that was supposed to have been decimated by changes in vegetation and other aspects of climate change had survived changes of 4–5°C during the most severe ice age of the Pleistocene without losing a single species barely 10,000 years earlier.

Genetic senescence, or orthogenesis, its older cousin, has been an especially durable theme in debates on the New Zealand extinctions. The idea has some pedigree. Darwin had been one of the first to express this Northern Hemisphere view of Southern Hemisphere life when he commented on the inevitable decrease in the plants and animals in the face of competition from superior introduced species when commenting on his visit to the Bay of Islands in late 1835 (Darwin 1884). Much more recently, a senior New Zealand government biologist seriously proposed that all the species were in decline when humans arrived and that any hunting or habitat alteration merely served to finish off the remnants (Williams 1962). But in response, the eminent ornithologist and paleontologist Charles A. Fleming pointed out that genetic senility would have had to simultaneously affect groups as different as moa and wrens at the exact point in history when people arrived on the scene. The coincidence was too great (Fleming 1962b).

Although Oliver (1949) was a strong proponent of the genetic senescence hypothesis in the technical literature, Duff (1949b, 1951, 1952) was responsible for reinforcing it in public and scientific perception by promoting one of the most persistent misconceptions about the extinction of New Zealand's birds. By then, there was abundant evidence that humans had eaten the more spectacular species. For example, more than 100 years earlier, in June 1847, W. Mantell observed sites where people had eaten moa and *Aptornis* at Waingongoro (Mantell 1848a, 1848b, 1873) and he knew that *Pachyornis elephantopus* was the main fare at Awamoa (Mantell 1873). Haast had established that moa species, including the largest, *Dinornis giganteus,* were undoubtedly associated with people at the mouth of the Rakaia River (Haast 1872b, 1872d) and farther south at the mouth of the Shag River (Haast 1875a). Yet Duff stated in two highly influential and popular publications—one a bulletin for high school classes, the other reprinted after just three years—that *Aptornis, Cnemiornis, Emeus crassus, Pachyornis elephantopus,* and *Dinornis giganteus* had all become extinct before people arrived (Duff 1949b: 27; 1951: 142; 1952: 29). In each, he asked: "What caused most of the moa and other flightless birds to die out before the arrival of man?" and answered "[by] . . . marked alternations of climate. These might alternately cause the forest to spread too rapidly on the east coast and restrict grassland, on which the birds depended."

Other authors have repeated the same theme. Among them, Falla (1955) thought that many of the moa and other large flightless birds were extinct before human times, contending that they were naturally becoming extinct and would have all vanished even if people had never reached New Zealand. Some years later, when the issue was being raised again, he wrote that "Probably the nearest we would get to a correct answer [as to why moas died out] is to say that the moa family as a group were on their way out and that man hastened the process. It is quite unlikely that if moas had been adaptable and vigorous biologically that they would ever have been exterminated" (Falla 1962: 189). So Falla, although acknowledging that people did eat all kinds of moa, refused to believe that this alone could have caused their extinction. His views and those of his contemporaries influenced subsequent workers. For example, Simmons (1968: 119) could believe that "The fact that quite a large section of the bird fauna was heading for extinction in the post-Pleistocene period is demonstrated by 40 species of bird becoming extinct in the prehistoric period."

Gordon Williams, one of the figures to become pivotal in wildlife management in New Zealand, entered the extinction debate in the mid-1950s. His major contribution to the controversy came later (Williams 1962), when he too expressed serious doubts that many of the species extinct in the Holocene had survived until people arrived. He paid lip service to that fact by saying that some had "probably" survived. He was clearly reluctant to believe it, noting that since European contact and its apparently vastly greater environmental impact, there had been relatively few extinctions. He downplayed the obviously ongoing extinctions of the late 19th and early 20th centuries by not accepting extinctions on the main islands when small remnant populations of the same species had fortuitously survived on nearshore islands. In fact, some of the species that he listed as rare had already been effectively extinct for years.

Of his seven "endangered" species, three are now extinct (*Xenicus longipes, Turnagra capensis,* and *T. tanagra*), two are extinct apart from managed island populations (*Philesturnus carunculatus* and *P. rufusater*), and the orange-fronted parakeet (*Cyanoramphus malherbi*) survives as a single small mainland population. Only one, the stitchbird (*Notiomystis cincta*), has recovered well from low numbers in its sole remaining natural population on Little Barrier Island after the removal of the cats that were supposed to not have been harming it. Of the three he listed as "rare," one had not had a reputable sighting since the 1920s and is now extinct (the laughing owl *Sceloglaux albifacies*), and another (the kakapo parrot *Strigops habroptilus*) is extinct on the mainland and survives only as intensively managed populations on offshore islands. The third species, the takahe (*Porphyrio hochstetteri*), a large flightless rail, barely survives on the mainland as the subject of intensive management. Of the endangered subspecies he listed, a southern snipe

(*Coenocorypha aucklandica iredalei*) is now extinct. Obviously, the situation was much worse than Williams had portrayed. It is now known to be *very* much worse, with another 30 species now considered to be endangered (Molloy et al. 1994). Although noting that speculation about causes could not be useful at that stage, he tried to invalidate some of the causes that might, and indeed should, have been obvious.

Williams (1962) claimed to be looking for some pattern in the distribution of the birds that might explain why they were declining or why they had become extinct. The pattern was there: where there were rats, cats, and mustelids, most indigenous New Zealand birds were extinct or rare. On smaller islands, in the absence of the predators, bird populations had remained healthy. Even if an island had, of the suite of introduced predators, only *Rattus exulans*, many bird species were "missing." As with other workers between 1950 and 1990, Williams (1962) could not see causation in the circumstantial evidence. He noted that some declines could not be related to a particular cause (e.g., the piopio had declined before stoats were introduced), but made only one tentative suggestion: that it was possible that rats had something to do with the decline of some birds. He seems to have dismissed stoats as being irrelevant to the issue.

Williams (1962) maintained that the Polynesian rat (*Rattus exulans*) was harmless and ignored the observations of Oliver (1955), who stated that there could be no doubt that rats and stoats were among the worst enemies of the smaller native birds. Further, Oliver (1955) specifically described how on Raoul Island in the Kermadec group, he had seen many nests of tui (*Prosthemadera novaeseelandiae*) destroyed. Among other species Oliver knew to be vulnerable to rats were robins (*Petroica australis*), bellbirds (*Anthornis melanura*), and other passerines. He pointed out that their populations had been reduced by the destruction of their eggs and young. The near extinction of the piopios (*Turnagra capensis* and *T. tanagra*), stitchbird (*Notiomystis cincta*), and saddlebacks (*Philesturnus carunculatus* and *P. rufusater*) was also to be ascribed to rats and stoats (Oliver 1955).

The view that predators could not exterminate species was particularly strongly held by a whole generation of biologists. Even when just beginning his career, his experiences in the field led Dr. Don Merton, one of the foremost workers in the management and conservation of some of New Zealand's most endangered bird species, to the conclusion that predators exterminated New Zealand's birds. But that view was unpopular with his seniors. He remembers

> many years ago as a junior wildlife officer expounding this same theme in front of Gordon Williams and R. A. Falla, only to be told in no uncertain terms that predators and predation were a natural part of any system and did not cause extinctions! Subtle vegetational changes induced by climatic change, and not so subtle changes brought about by land clearance were the major causes of extinction in New Zealand I was told. . . . My own view then as now was why then did kakapo and a raft of others disappear from

> the 1.5 million ha Fiordland National Park and other apparently unmodified areas? Besides, they had obviously survived numerous glacial and interglacial periods and associated vegetational changes. (D. V. Merton, personal communication)

Falla doubted the effects of predation on the New Zealand biota to the end. Late in life, but when he was still chairman of the New Zealand Nature Conservation Council and hence a powerful voice in decisions on conservation priorities and practices, he wrote an article on rare birds and conservation in New Zealand (Falla 1974). Obviously moved by deep sentiments, he commented on the removal to predator-free islands of the last South Island saddlebacks (*Philesturnus carunculatus*) after *Rattus rattus* became established on the Big South Cape islands. Although the islands were the birds' last refuge and the rats were clearly destroying both the plant and animal life on the group (Bell 1978), Falla (1974: 132) wrote that as "An operation planned and carried out under what were regarded as emergency conditions it had several features which could be open to points of criticism, but it would be inappropriate to develop them here. For better or worse the job has been done, and the consensus of opinion that it was both necessary and successful has been fully documented in numerous articles published all over the world. If it is granted that the South Cape saddlebacks really were threatened, the species has certainly been saved."

There was one dissenting voice in those times. In 1951, Charles Fleming suggested that most Holocene extinctions must be attributed to people, because only the presence of people accounted for the changes in the environment necessary to cause extinctions (Fleming 1953). About a decade later, with the benefit of the first radiocarbon results, and apparently in reaction to Williams's contribution, Fleming (1962b) attacked the prevailing opinions. He demonstrated that all the moa listed by Oliver (1949), except the five species known only from unique specimens, had been found in Polynesian middens. The same was also true for other extinct species from Holocene deposits. He concluded with a stinging rebuttal of the views of Oliver, Falla, and Williams: "It seems we are reluctant to blame our fellow men for a prehistoric offence against modern conservation ideals and would rather blame climate or the animals themselves. The simplest explanation is to attribute all late Holocene extinction to the profound ecological changes brought about by man with fire, rats and dogs" (Fleming 1962b: 117). In contrast, only one marine invertebrate is known to have naturally become extinct during the Holocene, a bivalve mollusc (the cockle *Anadara*), which briefly established populations that failed to survive.

Despite Fleming's best efforts, arguments over what caused moa extinction and the extinction of other species have continued. One of the new suite of "reasons" is disease (MacPhee and Marx 1999). Much the same argument can be put against disease as against orthogenesis as a cause of the extinctions. Too many coincidences are required and too many inconsistencies in the evidence for disease to be a factor. Consider the example of the shearwaters in the genus *Puffinus*. At least five species (*P. assimilis, P. spelaeus, P. gavia, P. huttoni,* and *P. griseus*) bred on the mainland in the late Holocene. Only two (*P. huttoni* and *P. griseus*) do so today, *P. huttoni* as a relict population nesting above the tree line in the Seaward Kaikoura mountains and *P. griseus* in small and declining colonies at a few points around the coast. *P. assimilis* and *P. gavia* still breed on offshore islands that lack mammalian predators, but *P. spelaeus* is extinct. Any disease responsible for the elimination of *P. spelaeus* and the mainland populations of the other species must not have been transmissible to the populations that survived on the islands. Nor could it have affected the high-altitude population of *P. huttoni*, although the lowland populations have become extinct. And the disease must have spared some populations and species in the genus *Puffinus* while affecting those in other genera. Several species of storm petrel, a diving petrel, and a prion all became extinct on the mainland at the same time as the shearwaters. And populations of these also survived on islands near the former mainland colonies. No known disease has such a complex epidemiology.

If not genetic senescence or disease, then what? Falla (1974) held that subtle, imperceptible changes in the environment caused most extinctions. Falla postulated that changes that are difficult, if not impossible, to detect caused the extinction of several endemic species of bird and insect on Big South Cape Island at the time of an invasion by black rats in the early 1960s. According to Falla (1974), the rat could not have directly killed off the robins, snipe, and bush wrens on Big South Cape, but simultaneous fluctuations in the food supply for these very different species caused their rapid decline and extinction. Even apart from the blatant special pleading in Falla's "explanation," there are reasonable biological arguments why "subtle" effects were not responsible for either the decline in birds on Big South Cape or the vastly bigger main islands. Among these is the fact that even gross habitat destruction has not affected some species, and others have vanished when their habitat was apparently unaltered.

Again the example of the small shearwaters is instructive. *Puffinus huttoni* survived in alpine grasslands, whereas the similar but smaller *P. spelaeus* became extinct in the wet forests of the West Coast. Neither habitat had been altered, but one species became extinct and the other has survived. Almost all of the habitat destruction in the Polynesian period occurred in the drier eastern parts of both main islands, so the disappearance of shearwaters there might seem explicable, but *P. spelaeus* disappeared from the wet western lowlands in seemingly unaltered forest, whereas *P. huttoni* survived high in mountain grasslands that were also unaltered. Why? Petrels spend most of their lives at sea, where they feed. No changes are known in the marine environment around New Zealand in the past several thousand years. It is likely that any changes that had occurred would also have affected other species as well. The reasons for the extinction of *Puffinus spelaeus* and of the fairy prions, diving petrels, and storm petrels that also bred in those forests lie elsewhere, in the first of the predatory mammals introduced to New Zealand: the Pacific rat (Holdaway 1999b).

Perhaps the preference for nebulous factors and the reluctance to

Islands, where the tall *Poa litorosa* tussock grass and the fern *Polystichum vestitum* reach 1.8 m high (Warham and Bell 1979), and the Bounty Islands, where there are no terrestrial plants at all (Robertson and van Tets 1982). Soil depth, slope, and aspect are usually as important as the appropriate vegetation cover, as are places such as cliffs, tall rocks, or sloping trees from which to launch themselves into flight. Ultimately, the distribution of each species of petrel was, and is, governed by the oceanographic conditions of the adjacent waters. New Zealand straddles the subantarctic convergence, where two great water masses with different average water temperatures meet. At the convergence, there is upwelling of deep, nutrient-rich waters; north of it, the water is warmer, and to the south, it is markedly cooler. Different assemblages of petrel and other seabirds exist on each side of the convergences, and others exploit the upwellings at the front; these petrels may be separated at sea, but because most can cover large distances between the feeding grounds and the nest site, species of both groups can nest on the same island.

The petrel fauna of the main islands included storm petrels, diving petrels, shearwaters, and *Pterodroma* petrels. Breeding sites ranged from coastal cliffs and dunes to subalpine grasslands, and the colonies contributed significantly to the nutrient flux in the terrestrial ecosystems (Hawke et al. 1999). It is not clear whether any terrestrial bird species was associated with the petrel colonies, but kea parrots (*Nestor notabilis*) are known to dig out, kill, and eat chicks of Hutton's shearwater (*Puffinus huttoni*). Invertebrate and lizard populations are higher in the vicinity of extant colonies, so there might well have been some coincidence of distributions of some species and breeding petrels. These associations will probably be unrecoverable without much more data on the former distribution of the petrel colonies. The apparent association between tuatara (*Sphenodon punctatus*) and petrel burrows often reported from offshore islands does not hold with fossil faunas, so interpretations of associations that are entirely extinct must be made with extra care.

Penguins, too, nested beneath whatever coastal vegetation was thick enough to provide cover from predators and shelter from the weather.

Changes through Time

Although the Holocene faunas are best known, especially for the smaller taxa, the record for the Pleistocene faunas is good enough to allow changes through time in response to climatic and vegetation patterns to be followed. During the Otiran glaciation until about 10,000 years ago, the northwestern and western areas of the South Island had an avifauna that closely resembled that of the eastern South Island in the Holocene. The change parallels that in the vegetation, which responded rapidly to the warming climate and increased precipitation of the Late Glacial and early Holocene. *Dinornis novaezealandiae* survived on the West Coast during the Otiran glaciation, despite being typical of deposits formed in areas covered by tall, multistoried forest in the Holocene. Evidence from stable isotopes (R.N.H., unpublished data) suggests that it did so by obtaining food in more open vegetation. Because the species was also found in wet subalpine shrublands in the Holocene, it was probably a versatility of diet and habitat that allowed this middle-size *Dinornis* to compete with its larger congener *D. giganteus* and also allowed it to survive in the west during the Pleistocene.

The species in branch A of Figure 14.6A are never found in deposits of Holocene age on the West Coast of the South Island or in northwest Nelson, which were covered in dense, wet forests in the Holocene. Analysis of pollen data indicates that during the Otiran glaciation, the vegetation in those areas was rather like that of the east during the Holocene. Apart from *Emeus crassus,* all the eastern moa have been found in Otiran-aged deposits in the west and northwest, as have Haast's eagle, the goose, takahe, and the adzebill. However, the eastern kiwi has not been identified in any deposits west of the Main Divide. Its place was taken there by two species of large kiwi, which were present throughout the late Pleistocene and Holocene. Conversely, apart from the minor intrusions of *Megalapteryx didinus* and *Dinornis novaezealandiae* into wetter areas of the east, there appears to be no deposit on the eastern side of the South Island that contains a truly western fauna. It is likely that the rain-shadow effect of the Southern Alps has been sufficient to maintain a drier environment there throughout the period for which we have fossil evidence.

In the North Island, the available information suggests that areas such as that around the Waitomo Caves (at 38°S) supported more open vegetation during the Otiran glaciation. The fauna then included species typical of shrublands in the Holocene, including *Euryapteryx curtus* and *Pachyornis mappini,* with large cold-climate morphs of both present (Worthy 1987a). Moa species typical of later Holocene forests were absent or minor components of the fauna. To the east, and east of the axial ranges, in southern Hawke's Bay, similar but less marked changes in distribution occurred.

The distribution of *Euryapteryx curtus* in the central North Island is most interesting and results from the unique set of environmental factors governing the distribution of both vegetation and fauna there. The position and extent of the shrublands in the central North Island have fluctuated throughout the Quaternary. Repeated volcanism has driven, or at least modulated, the vegetation patterns in the area, particularly around the volcanoes and lakes and to the east on the Kaimanawa and other mountain ranges. Prevailing westerly winds carried most ash clouds to the east, where they fell as far as the coast and beyond. Occasional paroxysmal events, such as the Kawakawa eruption at about 24,000 calendar years ago (Naish et al. 1998), spread deep layers of ejecta as far as the southern North Island and lower Northland and obviously had major effects on the environment. That event happened near the coldest part of the Otiran glaciation, and its effects on the fauna are unknown. However, since then, major eruptions have occurred on average every 1000 years, although the spacing and magnitude of successive events have not been correlated (Wilson 1993). The most recent was the eruption from Lake Taupo nearly 2000 years ago, which was about 100 times the magnitude of the 1981 Mount Saint Helens eruption.

In Hawaii, where the frequent volcanism is basaltic, the eruptions cause damage mainly by the lava flows, which, although locally extensive, cover only small parts of each island at one time. In contrast, the most important of the New Zealand volcanoes erupt rhyolite, the most viscous and explosive of the volcanic materials. Consequently, large areas are affected by the airfall tephras and ignimbrites from major explosive eruptions (Wilson and Walker 1985). The random spacing of the events and the different distributions of ash mean that although vegetation in some areas was able to complete a transition from barren landscape to mature forest, in others, it was arrested and returned to sterility. The result was a patchwork of grassland, shrubland, and forest in space and time. During longer periods of quiescence, the area of shrubland may have shrunk to the point where the coastal shrublands were the major areas of that habitat in the North Island. At other times, extensive areas of shrubland probably extended from the interior plateau to the coast, providing direct corridors for reinvasion by shrubland birds and other animals.

The ashfalls created a succession of mobile habitats as revegetation led to new forests that were then overwhelmed in a different pattern by the next volcanic event. The fluidity of the environment can be inferred from the record of at least 26 eruptions from the Taupo Volcanic Centre alone over the past 26,000 years (Wilson 1993).

Overall vegetation changes in the central North Island caused by volcanism were of similar magnitude to and superimposed on changes induced by the astronomically forced climatic shifts of the Quaternary, yet were nearly an order of magnitude more frequent, at least over the past 30,000 years. The ebb and flow of flora and fauna was likely to have been so much more rapid than that in the South Island, which was unaffected by most of the North Island volcanism, that fine-scale stratigraphic analysis will be required to sort out the changes in each area through time. Much remains to be learned about the interactions of climate, volcanoes, vegetation, and faunas. The first sites have already shown a complexity that was not apparent in 10 years of research in the South Island.

15 • Extinction: The Final Word

Historical Perspectives on the New Zealand Holocene Extinctions: Or, Who, or What, Killed the Moa and Nearly Everything Else?

As soon as moa were discovered, people began to wonder why they had vanished. As the remains of more and more extinct birds appeared in swamps and caves, the mystery deepened. And it was only too apparent from the middle of the 19th century that many living species were rapidly following the moa into oblivion. But only recently has it dawned on us that the key to the present crisis of endangered species might lie in the extinctions of the past.

This book would not be necessary if extinction was not the dominating feature of the recent history of the New Zealand fauna. Instead of the tentative conclusions on biology and ecology of species and communities outlined in previous chapters, we could have drawn on a multitude of studies and publications to write a much different book on the amazing species and assemblages. Unfortunately, given the fashions in science and science funding, although there would have been many more techniques available for systematic research, there would not have been much room (or finance) for studies of the systematics of many endemic groups, so most effort would have been on the ecology of the major species. That conclusion is the surer because even if Europeans had been the first humans to have reached New Zealand, their impact on the environment would have threatened most of the communities, and conservation science would have been as important as it is now.

Today, New Zealand forests are quiet. But among the earliest accounts to be written of the islands there is one that shows how recently they were otherwise. The naturalist who accompanied Lieutenant James Cook on his first voyage of discovery to the South Seas during the accept the role of predators in the extinctions were partly a result of the body of theory on predator–prey relationships developed from theoretical models dating back to Lotka and Volterra in the 1920s and being taught in universities in New Zealand as elsewhere up to the 1980s and beyond. These models, and more refined versions that filled the literature for awhile, were applied to try and explain observations of species interactions in, mainly, boreal environments, including classic studies of lynxes and snowshoe hares in Canada, and tawny owls and voles in England (Krebs 1978).

Although the models often predict cycles in the abundance of predators and prey, Krebs (1978: 261) could still maintain that oscillations were "not common in the real world." Instead, predator–prey relationships were supposed to be stable. Prey were not driven to extinction. One goal of ecology was to understand the causes of their stability. One view of predators was that they were prudent (Slobodkin 1961) and took preferentially members of the prey population that were young, old, or sick (Slobodkin 1974). The relationship between predators and prey was a coevolutionary race in which the prey was always under pressure to improve its antipredator behaviors and reproductive success. Krebs (1978) could cite successes in using the prudent predator hypothesis in long-standing continental systems such as wolves and mountain sheep. However, none of the theorists seems to have considered the case of a completely novel predator in an environment that has always lacked them. Even the introduction of mink (*Mustela vison*) to Britain was not a valid analog of Pacific island systems because the herbivores preyed on by mink had been the prey of several other mustelids for much, if not most, of their evolutionary existence.

To give the punch line first, the simple, inescapable fact is that the New Zealand avifauna has been decimated by introduced mammalian predators, including people. Only predation explains how coevolved ecosystems that had survived many glacial–interglacial cycles vanished in a geologic instant, after humans arrived. Only predation could produce the patterns of extinction and survival evident in the New Zealand avifauna that depended on body size and habits of the birds rather than on their phylogenetic lineage (Holdaway 1999b).

Before Europeans brought their own collection of troublemakers to the archipelago, the two predators foreign to the New Zealand ecosystem were the Pacific rat *Rattus exulans* and the people who brought them here. The two predators concentrated on different ends of the size spectrum. Rats could tackle species of their own body mass (100–150 g) or less and those with eggs less than 60 mm long, whereas humans took the largest species, including all species of moa, down to takahe and large ducks in size. Polynesian dogs arrived with the first settlers, but there is no evidence that the breed ever became feral (Anderson 1981b). They were kept in the camps; on the evidence of their feces, they were fed fish and birds and were in turn used for food and their skins for capes. Their remains have never been found in pitfall trap caves—or indeed away from direct association with Polynesian sites. Much later, when European dogs did live in packs independently of people, most of the ground birds were extinct.

There is a clear pattern to the extinctions through time. The differential vulnerability of species of different sizes and habitats to predation by rats and humans meant that the order of introduction of the predators was critical. Although humans must have brought the Pacific rat to New Zealand, those first people ashore did not stay to influence the environment (Holdaway 1999b). So the first wave of extinctions in the terrestrial fauna included the smallest vertebrates, the largest invertebrates, and probably some plants. Flightless and nocturnal species were especially vulnerable.

The differences in vulnerability stemming from the different body sizes of the predators meant that extinctions happened in two phases. Their beginnings were separated by the delay between first contact (during which the Pacific rat was offloaded) and the settlement by Polynesians over a thousand years later (Holdaway 1999b). The first phase was the elimination or diminution of the smallest vertebrates and the large invertebrates as a result of predation by the Pacific rat (Holdaway 1999a), introduced by transient visitors nearly 2000 years ago (Holdaway 1996a, 1999a; Holdaway and Beavan 1999). The rat may have also damaged the flora, but information on changes in the prehuman flora is not yet as detailed as that of the vertebrate fauna. The second phase started when people arrived to settle, just before A.D. 1300. Both rats and people encountered a fauna that was naïve to mammalian predators. It was not that they had no predators—the eagle and harrier were formidable killers—but there were no predators that hunted by smell and in numbers at night. The predatory birds also hunted from perches or in flight. Any defense behaviors, including hiding colorful beaks, were directed toward avoiding danger that came from above. The new predators were on the ground.

The New Zealand Quaternary extinctions happened much later than the (in)famous extinctions of mammoths and saber cats in North America (Martin and Steadman 1999). There is no evidence that any New Zealand vertebrate vanished during the environmental convulsion that accompanied the end of the most recent ice age. Indeed, all the species of vertebrate known from New Zealand in the past 100,000 years survived to within the past 2000 years.

Which Predators and When Did They Arrive? Chronology of Introductions

Twelve species of mammalian predator (apart from humans) have been introduced to New Zealand over the past 2000 years (Table 15.1). As is obvious from the dates, the predators arrived in three stages. The first was the Pacific rat, introduced by transient human visitors. Then human settlers arrived, bringing more Pacific rats and their dog, and finally came Europeans with a far larger suite of attendant carnivores.

All the predators introduced to New Zealand, including people, have been implicated in the extinctions. Rats, cats, and mustelids are insidious killers. Rats especially are so subtle in their predation that until recently, their contribution to the disaster had been underestimated or rejected. They hunt by night, robbing nests and invading roosting

TABLE 15.1.
Mammalian predators successfully established on the main islands of New Zealand and the Chatham Islands since first human contact, with approximate calendar dates of arrival or establishment.[a]

Species	Common name	Island			
		North	South	Stewart	Chatham
Trichosurus vulpecula	Brush-tailed possum	1869–1915	1858–1895	1890	1911
Erinaceus europaeus	European hedgehog	1907–1912	1870	1930	?
Rattus exulans	Pacific rat	ca. 150 + ca. 1280	ca. 150 + ca. 1280	ca. 1350	ca. 1300
Rattus norvegicus	Norway rat	(1772) 1792–	(1772) 1792–	1800?	1830?
Rattus rattus	Black rat[b]	1855	1890	Before 1911	1890?
Mus musculus	House mouse	ca. 1830	1850s	1850s?	1850s?
Felis catus	Domestic cat	1830s	By 1850	By 1850?	Before 1840
Canis familiaris	Dog	ca. 1280	ca. 1280	ca. 1300	—
Mustela furo	Ferret	1882	1879	—	—
Mustela erminea	Stoat	1886	1885	—	—
Mustela nivalis	European weasel	1885	1885	—	—
Sus scrofa	Feral pig	(1773) 1792–	(1773) 1792–	1790s?[c]	Early 1800s?

[a] Dates in parentheses = potential earliest date; ? = unknown; — = never established. Data based on King (1990) and Atkinson (1973), with adjustments for more recent information.
[b] There is evidence that the black rat was established in the northern North Island by the early 1840s (Colenso 1843), but Atkinson (1973) suggests that it was not widespread before the mid-1850s.
[c] Extinct or exterminated.

sites, killing the sleeping adults (Brown 1997; Lovegrove 1996a), as is now shown repeatedly by night-vision, motion-triggered video photography of the nests of several different species, including petrels (Booth et al. 1996), kokako, and kakapo. Unlike raptorial birds, rats hunt by scent as well as by sight. On continents, they are merely one of many small predators, and the local birds, small mammals, lizards, and amphibians have evolved for many millions of years in company with an array of small nocturnal mammalian killers. On islands, particularly oceanic islands, introduced rodents are often the only mammals. The faunas they came into contact with evolved without them; the island faunas are naïve. For the smallest vertebrates, those with a body mass under 200 g, the new predators were a major problem.

The sequential introductions of predators are important for anyone wanting to find out what happened to the New Zealand fauna. Arriving at different times and in different places, their effects might be visible in differences in the geographic and temporal patterns of extinction. The predators also differed in the size of prey they could handle and in their habitat, so the species that they might affect should also differ in these factors in predictable ways. For example, if a predator can handle animals up to its own body mass and if it lives in dry habitats, then it would be expected to be a threat to species up to the same size that lived in dry habitats. A potential prey species in the correct size range but that lived in wetlands would be less vulnerable, as would a large species that lived in the predator's habitat. A range of attributes, including body size and probable habitat, can be determined with some confidence for extinct species as well as for living ones, so their vulnerability to different predators can be predicted and compared with the patterns seen in the record of extinctions (Holdaway 1999b). Congruence of pattern would be a strong argument for causation: if a new predator was the only new influence in the environment and it was capable of killing a species, then predation is the simplest explanation for the extinction of that species. Species vulnerable to predators that arrived earlier would have been either absent or present in much reduced numbers when larger or more capable predators arrived.

An important simplification must be made if the comparisons are to be useful. A simple list of extinctions for the whole fauna obscures much of the important pattern because there are so many exceptions. Species often survived as relict populations on small islands, even when they were eliminated over the majority of their original range. Data on the original faunas of most of the smaller islands are just not available. To gain a true measure of the extent of the extinctions, each ecologic and geographic unit, such as the North or South Islands, must be considered separately when counting extinctions. Only then will the survival of a species elsewhere on another island not bias the data. For example, Lyall's wren (*Traversia lyalli*) was exterminated in both the North and South Islands, but it survived on tiny Stephens Island in Cook Strait. It was not extinct when Europeans arrived, but it was extinct everywhere that had populations of Pacific rats. By considering only the fauna of a geographic unit that had a singular history of extinctions and introductions, the sequence of arrivals can be related to the time of extinctions and the attributes of the predator can be compared to the vulnerability of the species that survived or became extinct.

Extinctions in Time and Space

Recognition of the possibility that new factors appeared to influence the fauna in different places at different times has led to new insights into the causes of the New Zealand extinction event. Even differences in intensity of the interactions between indigenous and introduced species in different places at different times could produce different outcomes, thereby leaving patterns in the fossil record and in present distributions.

Patterns are indeed apparent in both the time and space dimensions of the extinction event. Those patterns are important not only for the evidence they hold of the reasons for the extinctions but also because they have set the scene for the present conservation and management environment in New Zealand. Briefly, the time of the elimination of species from the main islands, when correlated with the timing of new factors, can show what the cause or causes of the extinctions were. The patterns of survival of many species on small offshore islands provide additional information on causation but also present conservators with different problems. Without the islands, ecologists in New Zealand would have been studying a historical event, such as the extinction of sure from a small human population would have been more like persistent withdrawals from a finite balance. Moa were mined, not cropped.

The mining was done by snares and by simply walking up to the birds and clubbing them. In the absence of any previous experience with humans, moa and all New Zealand birds would have been as naïve as those of the Galápagos and other islands where human contact has been too brief for the fauna to have learned to recognize people as predators. The only predator on adult moa was Haast's eagle, which attacked from above. Polynesians used many methods for catching birds (Steadman 1997), but large birds with necks outstretched forward as they moved through dense vegetation must have seemed preadapted to snaring. The adults would have been killed too rapidly to have learned fear of humans—and certainly too rapidly for avoidance behaviors to have evolved. Moa hunters did not even have the time or need to develop a specialized hunting technology before the moa were extinct. So far as is known, they did not use specialized spears, and they lacked bows. The only big game processing tools seem to have been simple skinning or flensing blades struck from obsidian, silcrete, or porcellanite cores.

The model's prediction of a rapid decline of moa numbers was supported by archaeologic evidence that moa were rare or nonexistent in the diet of former moa hunters less than a century after human settlement began (Holdaway and Jacomb 2000). For most North Island areas, whose original vegetation was wet, multistory forest whose carrying capacity seems to have been rather less than eastern, drier areas (see Chapter 14), the period of moa hunting is likely to have been breathtakingly short. On the Coromandel Peninsula, the moa population may have survived for less than a decade after human settlement and perhaps for as few as 5 years. Throughout the North Island, the biological and archaeologic evidence together suggest that the period was briefer overall than in the South Island.

That difference was mainly a result of the larger areas of richer habitats in the South Island. The total population on the larger and more diverse landmass was also higher, but that was balanced by the greater ease with which the human population could reduce the amount of habitat by fire. It is accepted that anthropogenic fires destroyed at least 50% of the natural vegetation of the South Island (M. S. McGlone, personal communication), much of it probably within the first century of settlement. Radiocarbon dates for deforestation in coastal and inland sites all point to a brief, early period of firing (McGlone 1989; McGlone and Wilmshurst 1999; McGlone et al. 1994). Habitat loss was a major additional cause of extinction of the large moa populations in the east and south of the South Island (Holdaway and Jacomb 2000).

Although small groups of moa must have survived in more remote locations, the main populations were extinct by A.D. 1400. That would have increased the pressure on the relict populations, as more hunters sought fewer birds. Increased rarity may well have been a challenge that hunters could not ignore. There would have been great honor (*mana*) for the hunter who could catch something so rare and valuable. Nineteenth-century commentators gained the impression that people traveled many kilometers into the western mountains of Southland to hunt takahe (*Porphyrio hochstetteri*), which were already rare (Reid 1974), when they lived in the midst of pukeko (*P. melanotus*), which were just as good eating and were much more abundant.

Other Large Birds

Most of the argument for rapid extinction of moa applies also to the other large birds such as geese, swan, and adzebill. The geese and adzebill were probably not abundant for reasons of restricted habitat (geese) or diet (the adzebill was a predator). They and most birds that did not survive into European times in New Zealand have been found in few archaeologic sites. All those sites contain artifacts characteristic of the earliest stages of human settlement (C. Jacomb, personal communication). It is likely that the demographic Achilles' heel that spelled the end for the moa was also the main reason for the extinction of the larger flightless birds other than moa.

Marine birds other than petrels were not spared entirely. On the main islands, the king shag (*Leucocarbo carunculatus*) was eliminated from the North Island and from most of the South Island, except for a remnant in the extreme north on nearshore islands and a larger population, formerly recognized as a distinct species, in the southeast (Worthy 1996). Two penguins also underwent substantial range reductions after Polynesian settlement. Both the yellow-eyed penguin (*Megadyptes antipodes*) and the New Zealand crested penguin (*Eudyptes pachyrhynchus*) formerly bred as far north as Cook Strait, and probably in at least the lower North Island (Worthy 1997d). On the Chathams, another penguin, the undescribed Chathams *Eudyptes*, became extinct, and a population of northern royal albatrosses (*Diomedea sanfordi*) was apparently extinguished on Pitt Island (Millener 1999).

The effects on the marine environment were not confined to the removal of larger penguins and shags and the continuing diminution of the formerly immense populations of petrels. Seals and sea lions were eliminated as breeding species from large areas; the sea lion *Phocarctos hookeri* was completely extirpated on the main islands (Crawley, in King 1990; Gill 1998; Smith 1978; Worthy 1994c). The New Zealand fur seal (*Arctocephalus forsteri*) was reduced to relict populations in the south and west of the South Island. Both species bred north to North Cape before Polynesian settlement.

People, Rats, and the Environment

So far as we can tell, substantial, measurable damage to the environment started as soon as people settled in New Zealand (McGlone 1989; McGlone and Wilmshurst 1999). Evidence from pollen cores and macrofossil remains shows that firing of the vegetation, accidental and deliberate, altered the vegetation pattern, resulting in the establishment of large tracts of grassland and bracken fern in the eastern and southern South Island and the east and north of the North Island. In the South

Island alone, about 50% of the original forest and shrubland was converted permanently to tussock grassland, with scattered pockets of forest on moister sites.

Trickle Effect or Cascade of Death? Further Effects of Extinctions

Although not enough vegetation was lost in most areas to affect many species of bird, the loss of key species in the food chain would have seriously affected the predators. Other species may have depended on the habitat created by the activities of, for example, burrowing petrels. Then there were the effects of removing members of feeding flocks. Mixed-species feeding flocks were probably important for the smaller insectivorous species, as they are elsewhere. Such flock relationships are almost unknown in the present fauna, but vestiges of the relationships still exist on some offshore islands with more or less intact species groups. McLean (1984) observed the relationship between saddlebacks and fantails on Little Barrier Island, but the situation he described seems to have been simple in comparison to the common multispecies associations stressed by earlier observers such as McLean (1911). Because the species assemblages that have been observed are only fragments of the original system, it is reasonable to assume that other such linkages were broken with the extinction of particular species. For example, it appears that Finsch's duck (*Euryanas finschi*) was a major food of both large raptors and perhaps also of those strange gruiforms, the adzebills (*Aptornis*).

It has recently been postulated that some bird extinctions may have had far greater effects than just subtle shufflings of the food web (Hawke et al. 1999). When Pacific rats eliminated the small petrels, they cut off a major nutrient flow from the oceanic food chain to the relatively poor, young soils of New Zealand. Part of the local nutrient flow would have originated in the North Pacific, brought back by the millions of migratory shearwaters and petrels.

Plants

Petrel guano and petrel bodies contributed nutrients to the terrestrial ecosystem via the vegetation. Even today, the snow tussock grasslands around the *Puffinus huttoni* colonies at above 1500 m in the Seaward Kaikoura mountains are more luxuriant than in areas nearby where the petrels do not nest (R. Morris, personal communication). The invertebrate fauna and the lizards that feed on them are also richer than in sites where the ocean nutrients are not available. Removal of petrel populations is known to affect soils on small islands, and it is likely that the extinction of the huge mainland populations of shearwaters and prions resulted in profound changes in the vegetation. In the drier eastern areas, the original vegetation was almost totally destroyed by Polynesian fires about 700 years ago, along with any evidence for vegetation changes.

It seems likely that the Pacific rat also affected the distribution of some species of plants, both directly and indirectly, apart from any effects via removal of petrels. Campbell and Atkinson (1999) reported that there was lower recruitment of some tree species in coastal forests on northern offshore islands. They attributed the reduction to Pacific rats eating significant amounts of the seed of those trees. On the mainland, Campbell and Atkinson suggest that the reduction in the numbers of some trees, in addition to predation on insects and other invertebrates, could have directly affected birds such as kiwi, whose diet of invertebrates overlaps with that of the Pacific rat. The present composition of the coastal forests differs significantly from that of the forest before rats, and the rats will prevent the re-creation of that original forest (Campbell and Atkinson 1999). Other rats could affect the reproduction of other plants, which, after more than 200 years of occupation by Norway rats, suggests that the vegetation of New Zealand has been subjected to the kind of serial predation that has been demonstrated for the birds (Holdaway 1999b).

It is possible that the removal of pollinators and seed dispersers among the large invertebrates and small vertebrates (birds and lizards) has affected the distribution of other plants. It would be difficult to differentiate between the effects of rats and those of natural succession or slight climatic shifts with the pollen data presently available. The wholesale removal of vegetation in drier eastern districts by fire early in the Polynesian period undoubtedly biased present knowledge of the prehuman flora, and that effect too makes it difficult to determine whether the rat affected the vegetation. Whitaker (1987) called attention to the role played by geckos in the reproductive strategies of some common forest species. The abundance of geckos on small islands was once mirrored on the main islands, and there is little reason to doubt that the elimination or great reduction in numbers of arboreal geckos has altered the population dynamics of some trees and other plants in New Zealand.

Invertebrates

Evidence for considerable losses among the larger invertebrates rested for many years on the observation that many species presently have highly disjunct distributions on small islands and are absent or rare on the main islands. Another clue to former distributional changes are in the inverse relationships between presence of large and vulnerable invertebrates such as the giant flightless crickets called wetas, and the presence of Pacific rats and other rodents on offshore islands. In places where the owl deposits were preserved in dry sediments, they included abundant remains of large beetles. Many of the specimens could be identified to species. It became apparent that many of the present gaps in distributions were artificial, the results of removal of habitat, especially in eastern districts, and probably by predation by rats (Kuschel and Worthy 1996; Worthy 1997b; Worthy and Holdaway 1995). Filling in the distribution gaps has changed some views of insect biogeography in New Zealand; it is now apparent that most populations of large insects on small islands are relicts. Some species whose distributions did not include island refuges seem to have become extinct. Others have relict populations in the mountains, where the rats were less abundant.

the North Island but is still rare or absent north of Auckland, despite having survived in dense populations on islands a few kilometers offshore (Bartle and Sagar 1987). The decline in the South Island was never as severe as in the North Island, and the bellbird is one of the most common indigenous species there.

Other birds were also found to have declined earlier in the North Island, especially in lowland areas, with remnants surviving longest in the axial ranges of the southeast of the North Island. Still others had apparently vanished from the northern and western North Island before Europeans arrived. Conspicuous among these was the huia (*Heteralocha acutirostris*), which was confined to the southeastern section of the island in the 19th century.

In the South Island, declines were also noticed first in the lowlands and in the east, with populations of species such as the South Island kokako (*Callaeas cinerea*) and saddleback (*Philesturnus carunculatus*) and piopio (*Turnagra capensis*) surviving into the 20th century in the mountains of the northwest and southwest. When these patterns were being discussed in the 1950s and 1960s, predation was, of course, not seen to have been a factor until late in the 19th century, after mustelids were introduced, and the earlier declines were puzzling and thought to be related to changes in habitat and food availability.

It is now clear that the differences in extinction rates and the pattern between regions is a function of the interplay of the main factors in the extinctions, and mainly in the differences in the presence, numbers, and demography of the main predators. Before the arrival of Europeans, the main predators were, as we have seen, humans and the Pacific rat. In the northern peninsula, average temperatures are the highest of any part of the main islands, and the forest is floristically diverse, with mangrove swamps on the great harbors. It is the region of mainland New Zealand that is closest to being subtropical. The terrain has a low relief in comparison with most other regions. In both respects, the area is the part of New Zealand that most resembles the source islands for both people and rats.

The northern North Island had the densest Polynesian populations when European contacts began. Large areas of forest had been cleared for cultivation of the *kumara,* the sweet potato, and the local bracken fern. The benign climate and diverse flora would have allowed the rats to rear large litters. There would have been sufficient food at all seasons to maintain rat populations at levels that could affect the indigenous fauna. Coupled with the human disturbance, the steady predation pressure would have affected the species that were only marginally susceptible to predation by the rat. Hole-nesting species, such as the huia and rifleman, would have been especially vulnerable.

Farther south, and at higher altitudes there, less food and the colder climate, particularly in winter, would have limited peak rat populations, and with lower winter survival, the numbers in the spring breeding season would have been lower. At intervals, the southern beech forests could support large numbers of rats, as they do now of mice in the mast years, but at other times, the populations would have been lower and would have fluctuated more in the highly seasonal environment.

The geographic pattern of avifaunal decline and extinction within New Zealand is one of survival at the periphery of former ranges, which matches the pattern of geographic range collapse in endangered mammals in North America and Australia (Lomolino and Channell 1995). The extinction pattern appears to be counterintuitive. Populations are usually highest near the center of a distribution. It may be that the peripheral populations are more adaptable because they are continually stressed by suboptimal habitats, that the marginal habitats are even more marginal for predators causing the decline, or that the prey population is so sparse that any disease cannot sustain an epidemic. Because disease can be discounted in the New Zealand extinctions, it is most likely that the relict populations at the edges of the former ranges survive thanks to the low populations of introduced predators that can survive in the harsh environments of Fiordland.

Island Refugia

Islands were the first places where the effects of rodents on the indigenous fauna were seen in patterns of presence and absence of invertebrates, lizards, tuatara, birds, and rats (Crook 1973; Ramsay 1978; Whitaker 1978). Vulnerable species of invertebrates, lizards, tuatara, birds, and the indigenous bats survived on islands that the new predators never reached. The possible effects of rodents on the fauna were debated heatedly for many years. Some scientists still dispute that the Pacific rat affected the fauna at all (e.g., Craig 1986).

The small islands have for many years been used as natural and enhanced refuges for rare or endangered species, which either had existed naturally on the islands or had been relocated from other islands or the mainland. The technique of translocation has been used with impressive success in saving or securing from extinction species such as the saddlebacks and little spotted kiwi. Both the island faunas and the results of different translocation attempts are also being used to study the effects of habitat disruption and predation on a range of species (e.g., Lovegrove 1996a). Removing Pacific rats from most islands has also provided a series of "removal" experiments whereby both the ability of relict populations to rebound with the cessation of predation and the effects of predation itself may be quantified. The enhanced management practices developed in the past 30 years have added greatly to the traditional conservation values of the islands.

Reasons for Vulnerability: Why Was New Zealand's Fauna So Easy to Destroy?

When the predators first arrived, the amount of food available to a new predator in the environment was unimaginable in terms of the fauna now present. Predation was the principal factor in the extinction and extirpation of New Zealand's late Quaternary fauna; both invertebrates and vertebrates were naïve and were not adapted to predation by are species including the coot (*Fulica prisca*), little spotted kiwi (*Apteryx owenii*), and Lyall's wren (*Traversia lyalli*). Also among this group is the rare and still enigmatic long-billed wren (*Dendroscansor decurvirostris*).

It is obvious that the insights obtained from analyses such as these must be weighed against interpretations of other information such as the paleoecology of the areas in which the fossils of each species have been found, the morphology of the species, and biochemical and isotopic data, but the results are of use in framing questions that other techniques can be called on to answer.

Although the general impression from some of the more dramatic accumulations of moa bones such as Kapua and Enfield has been that the forests and shrublands of New Zealand harbored great numbers of moa and other birds, this was probably not so. The populations were limited by primary production in the different habitats. For example, little light penetrates to the understory of multitier forests, and most of the primary production is tied up in the trees themselves, with little available as browse within the reach of ground birds. Moa species living in the wet forests had to subsist on limited quantities of low-quality browse, with the best food sources probably located in gaps and along the edges of streams and land slips; large areas of habitat would have been occupied by relatively few individuals of each species.

If the natural pit-trapping experiments are anything to go by, population densities of forest moa were low. Strangely, few individuals of each species are usually found in potholes (pitfall traps) that have been open for at least the past 10,000 years. The same holes have often already captured several deer in the century or so since deer were introduced, when it took about 10,000 years for the same hole to trap a similar or lesser number of moa. It is likely that deer are at least as alert to the dangers of falling in one of these traps, so the numbers were not biased in that way. One conclusion would be that moa never reached anything like the densities that deer (with ruminant stomachs and in the expansion stage of a biologic invasion) have reached in New Zealand.

In cool-temperate rainforests in New Zealand, there were unlikely to have been more than one or two pairs per square kilometer and in many areas perhaps one pair to 5–10 km^2. If this seems to be counterintuitive, it should be remembered that cassowary (*Casuarius casuarius*) in the rainforests of North Queensland, Australia, have territories of several square kilometers, and okapi (*Okapia johnstoni*) in the Ituri Forest of central Africa are not herd animals like their relatives, the giraffe (*Giraffa camelopardalis*), on the open plains.

Marine Birds

The distribution of one group of birds—the petrels—largely transcended forest or other vegetation types in New Zealand. Other things being equal, the same species of petrel might nest beneath shrubland, forest, or grassland. Although about 86% of petrels nest in open habitats with low vegetation or none at all (Warham 1990), in New Zealand, the remaining colonies of many species are on islands with vegetation of at least tall shrubland. Major exceptions are the treeless Antipodes nocturnal, scent-hunting, fast-breeding, omnivorous mammals. The "omnivorous" is important, because it meant that the predator could maintain its numbers even if the animal prey was migratory and was absent for part of the year. The indigenous fauna was predominantly K selected—that is, the populations were always near the carrying capacity of the habitat, adults survived for a long time, and there was low recruitment to the populations. A few species, and those mainly recent additions to the fauna, such as small songbirds, tended toward *r* selection as a way of lowering the effects of predation pressures. Those species laid more eggs per clutch and often laid several clutches in a season, so they were able to produce replacements. Examples of these resilient species are the fantail (*Rhipidura fuliginosa*) and grey warbler (*Gerygone igata*), which are closely related to Australian species that survive in the presence of mammalian and reptilian predators.

In contrast, the K-selected species were poorly placed to resist the removal of adults from their populations. These species live for several, or many, years, must rear only two young to adulthood over that period, and can withstand the loss of their young in most of the breeding years, so long as they succeed occasionally. But if the adults themselves are killed, they cannot replace themselves, and the other adults in the population are fully committed to replacing only themselves, so the population declines. To a human observer on a human time scale, there is a characteristic pattern to this decline. For years, there does not seem to be a problem. There are apparently the same numbers of birds in the environment. Then, without detailed knowledge of the breeding success, there might seem to be a slight decline but no reason for concern. But then the surviving, unreplaced, adults begin to die of old age without issue, and the population suddenly collapses.

In recent years, this effect has been seen in the relict mainland populations of several New Zealand birds but perhaps most dramatically in the North Island kokako (*Callaeas wilsoni*). Its population, although small and composed of small groups scattered over large areas of the island, seemed to be relatively secure. As knowledge increased, it became apparent that breeding success was low; there were few recruits and many pairs were unsuccessful at breeding. Most were never seen to raise chicks. There were fears that the isolated populations were now so small that social factors that had formerly enhanced breeding success were no longer operating.

The answer was simpler. Energetic efforts to reduce the decline included observations at the few successful nests to see if predation was a factor. When it was shown to occur, measures were taken to protect individual nests, with the result that more chicks were raised. But most nests still failed. Then one breeding pair produced a clutch of young females, and the effect on the population was dramatic. As the young females left the family group, long-established pairs broke up, and the former partners mated with the new females. Many of the unsuccessful pairs had in fact been male–male bonds formed because of an unrecognized (by humans) shortage of females. Proportionately more females were being killed by predators at the nest, so there were too few to go around. A high proportion of males is a feature of the surviving poputhe mammoth faunas of Eurasia and North America, rather than even a small selection of living Pleistocene species. The same might be said for Hawaii, but there, the extinctions appear to be continuing at an even greater rate than has afflicted New Zealand since A.D. 1800.

Extinctions Chronology

Extinction dates are not well known even for extinctions occurring in historical times (Holdaway 1999b). A major marker for the timing of prehistoric extinctions is the presence of species in Polynesian sites, indicating that they survived until the settlement of New Zealand in the second half of the 13th century (Higham et al. 1999). But the association between fossils and the cultural sites has to be made carefully, because some sites are known to have mixed archaeologic and fossil faunas (Worthy 1999a, 1998e; Worthy and Holdaway 1996b).

Three stages or phases can be recognized in the New Zealand extinction event on the basis of the arrival times of the most important predators—the Pacific rat, people, and the suite of mammals imported by Europeans. It is probable that the Pacific rat arrived about 2000 years before the present (B.P.) (Holdaway 1996a, 1999a; Holdaway and Beavan 1999). Although the date is still contentious (Anderson 1996; Holdaway and Beavan 1999), that does not alter the fact that the Pacific rat was the only rodent—indeed, it was the only introduced mammal of consequence other than people—in New Zealand for a considerable period in the presence of an undisturbed fauna and flora. Most large taxa are found in early Polynesian sites, and their extinction comprises stage 2. The extinction of further species after Europeans arrived is the third stage and was precipitated by the suite of mammals they imported. Because they arrived simultaneously on the total time scale of the event and coincided with renewal of major environmental change, all the 19th-century introductions may be grouped together, although it is apparent that the species involved would have had different effects on the remaining fauna.

The Human Dimension

At the other end of the scale, any species unlucky enough to represent to a hungry Polynesian a large package of meat for a small amount of effort (mainly those species over 2–3 kg and flightless) was fair game for the human colonists. And it was probably more like supermarket shopping—or gathering—than hunting. Quite apart from their role in bringing predators to New Zealand, since the end of the 13th century, people have been removing the indigenous vegetation and squeezing out a large proportion of the fauna. Steadman (1997) pointed out that Polynesians throughout the South Pacific used many methods of catching birds for many uses, only one of which was for food. The people who settled in New Zealand would have been well used to birds as food and were practiced in many methods of obtaining them. From simply lifting individuals from a branch or off a rock to setting snares and creating elaborate petrel-gathering traditions, Polynesians were well equipped to slip easily into a hunting-gathering economy, in addition to their fishing and horticulture practices. That harvesting of the natural resources was carried beyond the capacity of the species involved to recoup their numbers (Steadman 1997). There is now no doubt that human predation caused the loss of most of the large flightless birds in New Zealand and the diminution of others. The characteristics of the avifauna that made certain species vulnerable to hunting pressure and habitat loss were explored by, among others, Cassels (1984) and Anderson (1984, 1989a, 1989c).

The evidence of often vast amounts of moa bone in early Polynesian sites shows that moa provided a substantial proportion of the diet of the people in the eastern South Island and a lesser proportion elsewhere (Anderson 1989a, 1989c). Some sites were so rich in bone that railroad wagons of bone were removed for fertilizer. Other large species, such as the geese, adzebills, takahe, and swan, are also present in early sites. Worthy (1999a) has surveyed the representation of extinct and living species in archaeologic sites. Of 177 assemblages, 103 contained moa bone from one or more species. Almost all assemblages were biased toward larger species. In many excavations, small bones were not collected because the sediment was passed through only a coarse sieve that failed to retain bones of songbirds and other small species. Larger species, such as moa, the swan, geese, adzebills, eagle, and albatrosses, were overrepresented among the extinct taxa that were found in more than six sites on both the North and South Islands (Worthy 1997a).

One handicap to studying the species representation and importance of birds in the Polynesian diet is that faunal material was not collected in many excavations, which concentrated on retrieving artifacts instead. Even where bird bone was collected, often little effort was made to collect adequate samples, so samples are small and not likely to be representative of the hunted fauna (Worthy 1999a). Despite the biases, it is obvious that extinct birds were important in the diet, because of 34 extinct taxa, only 5 were not represented in archaeologic sites. These were either very small (three kinds of wren), which the Pacific rat may have already exterminated, or only recently recognized as distinct and found in areas with few archaeologic sites (*Pachyornis australis* and *Puffinus spelaeus*). Many accounts state that the giant moa (*Dinornis giganteus*) was rare, but it has been reported in 22% (23 of 103) of the sites with adequate faunal lists (Worthy 1999a).

Apart from the early sites, most Polynesian sites contain few extinct species. For example, apart from the moa, the avifauna from the well-studied village at Shag River Mouth in northern Otago (which dates from the mid- to late 14th century) could have been collected today in some parts of the South Island, with few exceptions (*Charadrius, Coturnix,* and *Mergus*). The period of extinction was, as we will see, breathtakingly short.

Although most attention has been focused on the terrestrial birds, overharvesting also drastically affected the populations of some of the larger seabirds. The effects have been rather cryptic because problems of identification and separation of the remains of similar species had to be overcome before the pattern could be discerned. For example, most of the material of marine shags and freshwater cormorants had to be checked before it became apparent that the present distribution of the king shag *Leucocarbo carunculatus* is relict and not the result of limitation of a subantarctic species by lower food supplies at the present northern limit of its range (Falla 1974). Before Polynesian settlement, the king shag bred almost to the subtropics, in the far north of the North Island (Worthy 1996). In this, it matched two supposedly subantarctic pinnipeds, Hooker's sea lion (Gill 1998) and the New Zealand fur seal. Interpretations of present distributions that are based on inferred physiologic or ecologic limits derived (circularly) from the present breeding distributions must be reassessed in light of the new information on the former breeding range.

Whether the two large penguins in the South Island fauna also bred north of Cook Strait and have been similarly restricted must await reanalysis of the coastal faunas. Previous records of subantarctic species such as the rockhopper and erect-crested penguins (*Eudyptes filholi* and *E. sclateri*) in middens from the South Island coast have proved to be incorrect identifications of both the yellow-eyed penguin *Megadyptes antipodes* and the New Zealand crested penguin *E. pachyrhynchus* (Worthy 1997d), which are extant but threatened on the mainland of the South Island. Breeding is now inferred at least as far north as Cook Strait, and there is no obvious barrier to either species having bred farther north in the past. On the Chatham Islands, an undescribed species of *Eudyptes* became extinct in the Polynesian period (Holdaway et al. 2001; Tennyson and Millener 1994).

Even less is known about the former status of the Australasian gannet (*Morus serrator*), the spotted shag (*Stictocarbo punctatus*), or the other phalacrocoracids. Further research may well show the former existence of colonies of these species outside their range at European contact. There is strong circumstantial evidence that Polynesian hunters eliminated the northern royal albatross from Pitt Island in the Chathams. Exploitation of larger petrels in southern New Zealand began when moa were still available as food. People took joints of moa meat to Stewart Island to eat while they harvested sooty shearwaters (*Puffinus griseus*) on the comparatively sheltered northeastern parts (Worthy 1998e). Anderson (1997a) believes, however, that exploitation of petrel populations on the outlying southern islands did not begin in earnest until early European times, when whaleboats made voyages in the stormy seas south and west of Stewart Island comparatively safe. Although there is evidence that people actually reached these islands soon after settlement, there is no indication of settlement—or indeed regular use of the resources—in prehistoric times. Movement of rats to these islands might well have been delayed until harvesting the chicks of petrels—known as muttonbirding—became well established.

European Influences: Late to the Party

Ten of the 12 predators were introduced deliberately or accidentally by Europeans in the 19th century. But the general view that these latecomers caused the most damage ignores the fact that most of the extinctions occurred before Lieutenant James Cook arrived in late 1769. Almost all the extinctions of birds and all the frog extinctions happened when just two foreign predators—people and the Pacific rat—were the only new influences in the New Zealand environment (Holdaway 1999b). The Polynesian dog can be exonerated; it was kept so close to the camps that it is not a factor (Anderson 1981b). The period from the introduction of the Pacific rat to the arrival of human settlers coincides with the first phase of the New Zealand Quaternary extinctions.

Geographic Patterns

In addition to the temporal pattern to the extinctions, there were two obvious geographic patterns. First, there was the extinction of populations on the main islands (with or without survival on offshore islands). The mutually exclusive distribution of predators and potential prey species was one of the first indications of the overwhelming importance of predation in the extinctions (Bell 1978; Crook 1973; Ramsay 1978; Whitaker 1978).

Differential reductions within the main islands in particular areas and at different times have been historically more difficult to account for. The realization that predators as well as birds had preferred habitats and varied in abundance accordingly has thrown light on some survivals (including Hutton's shearwater). The importance of timing of predator abundance and prey breeding has been shown to be of the utmost importance in the kaka (*Nestor meridionalis*) (Wilson et al. 1998). Predator–rodent abundance cycles driven by mast seeding of southern beech *Nothofagus* have been shown to be critical for predation on nests and young of parakeets (*Cyanoramphus* spp.) and yellowheads (*Mohoua ochrocephala*) (O'Donnell 1996).

Biological Impact of the New Order

The Pacific rat reached New Zealand after a long odyssey from its homeland in Southeast Asia (Matisoo-Smith et al. 1998; Roberts 1991). This small (60–180 g) rat is a common commensal rodent from Burma to Indonesia. In prehistory, it accompanied the ancestors of the Polynesians to the edge of the vast ocean and then, over the past 4000 years, was spread by Polynesian voyagers to the extremities of the Polynesian triangle, from Hawaii in the north to Easter Island in the east, reaching its southernmost point on islands off New Zealand's Stewart Island. Despite its transoceanic range, it is a poor and reluctant swimmer and cannot cross water barriers more than 200 m wide (Atkinson and Moller 1990).

Although it prefers grasslands, the Pacific rat has successfully colonized habitats as different as coral atolls and rainforests (Atkinson and Moller 1990). In New Zealand, it was found throughout the main islands, on Chatham Island, and on many, but not all, of the smaller offshore islands. The distribution pattern is consistent with random transport by Polynesians in prehistory (Holdaway 1999a). In New Zealand, the Pacific rat was found in most habitats, from tussock grasslands to *Nothofagus* forests, at altitudes up to 1300 m (Atkinson and Moller 1990), but it was rare in wetter areas or wetland habitats.

Although the Pacific rat can climb, it is less arboreal than the black rat. It vanished from the North Island and from most of the South Island in the late 19th century, apparently as a result of competition from the three other introduced rodents (Table 15.1).

Long considered to be almost totally vegetarian (Atkinson and Moller 1990), the Pacific rat is now known to be a significant predator on invertebrates and small vertebrates as well as some plants (Campbell and Atkinson 1999). Whether it was effective enough as a predator to eliminate a substantial section of the New Zealand biota has been hotly debated. That it was the first mammalian predator to reach New Zealand cannot be disputed; being the only one until Europeans finally landed in 1769 means that it must be a strong candidate for that dubious honor.

Phase 1 Extinction

The New Zealand extinction event began about 2000 years B.P., with the arrival of the Pacific rat on the two main islands during a transient visit by early Pacific voyagers (Holdaway 1996a, 1999a). Predation by Pacific rats would be expected to have affected mainly the smallest vertebrates (<150 g), especially those in terrestrial habitats and those that lived on or near the ground. Because of the habitat preferences of the rat, species with geographic ranges confined to northern or lowland habitats would be more likely to have been affected than those that lived in wet forests or at high altitudes. Further effects of the presence of Pacific rats probably included damage to vegetation and to the populations of large invertebrates, which were probably important in the diet of some birds.

Details of the number and timing of extinctions in phase 1 of the event are still little known, partly because the possibility of a presettlement predation threat has been appreciated only recently (Holdaway 1996a, 1999a). Few terminal dates are known for taxa under 200 g, but former large populations of several species of small petrel are not represented in Polynesian sites (Holdaway 1999b), which is a criterion used for determining the time of extinction of larger taxa. We argue that the terminal date is not as important as the time of the initial population crash. Although a species may not have been completely extinct before the next phase began, the initial crash would have been irreversible, and extinction of that species was inevitable.

Phase 2 Extinction

After Polynesian people settled in New Zealand late in the 13th century, their living sites—at least those dating from the first decades of their occupation—contain the remains of almost all of the larger flightless species (Worthy 1999a). Later sites lack remains of these species, so it must be assumed that the larger birds became extinct soon after the human settlement, probably well within a century after settlement began. In the early years of settlement, large areas of woody vegetation were removed permanently from the eastern and southern areas of both islands. The forests and shrublands were replaced by extensive grasslands and fern land, which must have had dramatic effects on the distribution of all the species that were closely tied to the drier eastern areas.

A Sudden Collapse of the Moa Fauna: Attrition versus Cropping

There is no question that people ate moa: Anderson (1989c) mapped the distribution of more than 300 sites. In several sites in the eastern South Island, moa bones and earth ovens covered many hectares near river mouths. Only certain parts of the animals' bodies were used for food, but in some sites, there is evidence that many thousands of individuals were so used (Anderson 1989c). Standard models for the early human settlement of New Zealand (Anderson and McGlone 1992; Davidson 1984) envisage a protracted period when moa were at least part of the diet. Even the short chronology for settlement (Anderson 1991) and the accompanying model for moa extinction (Anderson 1991) assume a slow buildup in human activity and in hunting of moa. Even the latest analysis of a major moa-hunting site (Anderson et al. 1996) is predicated on a protracted period of moa exploitation, even if it is recognized that the returns were starting to decline in the late 14th century. The reality was probably very different.

Models of moa hunting have been based on the assumption that there was a surplus of birds available for cropping (e.g., Anderson 1989a, 1989c). The evidence for low clutch size and delayed maturation suggests that moa were like most other New Zealand birds in being K strategists, with the breeding population being at or near saturation most of the time and with high adult survivorship. Unfortunately, species with high adult survivorship are most vulnerable to losses from the breeding adults. They can withstand high levels of cropping of juveniles but are slow to replace breeding adults.

The initial human population of New Zealand has been estimated at about 200 individuals (Murray-McIntosh et al. 1998) on the basis of the number of women required to explain the present mitochondrial DNA complement of New Zealand Māori people. Estimates of catch rates based on minimal levels of consumption of moa by that number of people were used by Holdaway and Jacomb (2000) to model the time to economic extinction of the whole moa fauna by use of some basic assumptions on the population structure, survival rates, and fecundity of moa in general. Their model predicted a much faster decline to extinction than expected under existing attrition models. Such a rapid decline under predation of adults is consistent with modern studies of low-level human predation on adults of long-lived birds such as wandering albatrosses. Albatross populations subjected to apparently minimal losses (1% of adults per year) from long-line fishing at their feeding grounds have declined rapidly (Croxall 1979; Croxall et al. 1990; Weimerskirch and Jouventin 1987; Weimerskirch et al. 1987). Far from being a cropping of an annual surplus, even the minimal hunting pres-

Phase 3 Extinction: Europeans

At the time of European arrival, the pattern of vegetation developed after the Polynesian burning of the 14th century had probably stabilized and been relatively static for 200–300 years. But some of the birds, particularly the remaining larger flightless species (kakapo and takahe), were still in decline. There is also some evidence that others, including the saddlebacks (*Philesturnus carunculatus* and *Philesturnus rufusater*), were still declining as a result of predation by Pacific rats (Lovegrove 1996a). Hence, the new influences were superimposed on the continuing declines started by rats and Polynesians centuries before.

Just as the Polynesians seem to have done, Europeans arrived first as transient visitors, skirting the shores and observing, but not founding permanent settlements. After the visit of the Dutchman Abel Tasman in 1642, there was a long gap during which the nations of Europe busied themselves elsewhere. Then exploration really became the vogue, and a series of expeditions left Britain and France to chart and document the lands of the South Seas, where humans were supposed to live in sylvan peace and money was to be made by trade.

When Europeans finally reached New Zealand in the late 18th century, the main attractions for them were the marine mammals (whales and seals) that were still comparatively abundant in the south of the region. Whaling was conducted off the North and South Islands, and most sealing and whaling vessels also visited the northern North Island for refitting and provisioning. Both these activities led to further impacts on the environment, which began the third phase of the extinction event. The introductions began as soon as the Europeans returned in the late 1760s. In 1769, Jean François Marie de Surville in the *Saint Jean Baptiste* gave a Māori chief at Doubtless Bay in Northland two pigs, but they probably did not survive to populate the country. In his first voyage, in the *Endeavour,* Cook probably left only rats (as no doubt de Surville did too), but on his second and third voyages, he released pigs or gave them to the Māori, as did the sealers and whalers who followed him.

Settlers brought with them horses, cattle, sheep, and dogs. With little fencing, feral populations of the livestock were established in many areas of the main islands within a few decades of European settlement. Some populations of sheep and cattle survived the end of farming attempts on the windswept and inhospitable Campbell and Auckland Islands. Some of these populations have been finally removed only recently, and before that, they had permanently altered the vegetation of most of the islands in the New Zealand subantarctic.

Cats reached some of the most important of what are now reserve islands. On Little Barrier Island, they undoubtedly hastened the extinction of the snipe (Miskelly 1988) and the saddleback, and they contributed to the near extinction of the stitchbird (*Notiomystis cincta*) and Cook's petrel (*Pterodroma cookii*). The former species has recovered its numbers since the removal of the cats in the early 1980s (Veitch 1982), but Cook's petrel is still declining in the presence of Pacific rats. On

Stewart Island, feral cats nearly extinguished the last population of kakapo soon after it was rediscovered in the 1970s. More than 50% of the kakapo carrying radio tracking transmitters were killed within one year, despite intensive trapping of the cats. The kakapo were saved only by removal to islands free of predators. Paradoxically, the best publicized "example" of the extinction of a New Zealand bird by an introduced predator, that of the last population of the flightless Lyall's wren (*Traversia lyalli*) by the lighthouse keepers' cat on Stephens Island in Cook Strait, is probably an injustice to the cat. The cat certainly killed some of the birds, but after the new bird was discovered, it was a quarry for collectors and their masters (including the naturalist and professed conservationist Walter L. Buller), and the keepers cut down most of the forest on the island at the same time. Habitat loss and collecting probably contributed as much to the loss of the wren as it did to those of the island's kokako, piopio, and saddleback populations.

A striking example of the damage that cats *can* do, apart from the near catastrophe with the last kakapo population on Stewart Island, is the recently documented extinctions on Mangere Island in the Chatham group (Tennyson and Millener 1994). Sheep, goats, rabbits, and cats (to control the rabbits) were introduced to the 112-ha island in the late 19th century, and the island's vegetation of low coastal forest and shrubland was rapidly removed by burning and grazing. Tennyson and Millener (1994) found from an analysis of several fossil deposits that up to 22 species of bird had become extinct on Mangere Island in the past 450 years. At least seven of those are globally extinct, and at least 75% of the species found in the earliest site are now extinct either locally or globally. Of the 22 extinct species, 7 or 8 were extinct by 1871–1872, when Henry Travers collected there and observed the habits of some of the birds (Travers and Travers 1872).

Records from later collectors and reports from settlers indicate that a further 9–11 species became extinct between 1872 and 1923, 1–3 before 1961, and another between 1957 and 1988. Cats were responsible for the extinction of more than 12 species before they died out about 1950. Species differed in their vulnerability to the cats: the Chatham Island tit (*Petroica macrocephala*) survived, but the black robin (*Petroica traversi*) did not. The diving petrel (*Pelecanoides exsul*) was destroyed by the cats, but the fairy prion (*Pachyptila turtur*) is still abundant. The record from other islands shows that both diving petrels and the prions are vulnerable to rats (Holdaway 1999b), and Tennyson and Millener (1994) suggest that the difference in their responses to cats is related to factors such as patterns of nest attendance.

Sheep were removed from Mangere by 1968 as part of the island's transition to being a reserve, but the rabbits may have actually been eliminated by the cats. On the mainland, there were simply too may rabbits, and the feral cats could not stem the tide. The desire to make a new England in the South Seas led the settlers to introduce many of the sights and sounds of home—and some of the sport. Although foxes were forbidden, hares (*Lepus europaeus*) were introduced for the chase. Rabbits (*Oryctolagus cuniculus*) were introduced both as a reminder of home and as a source of meat for the poor people, but within 20 years of their arrival, they were a raging pest that threatened to turn the grazing lands into a desert. So in the early 1880s, farmers and businessmen introduced stoats (*Mustela erminea*), ferrets (*M. furo*), and European weasels (*M. nivalis*) in a vain and misguided attempt to control the rabbit population. All these species became established quickly. Although the weasel is not abundant anywhere, stoats are common in all habitats on both the North and South Islands, and ferrets are common in drier districts. They did not control the rabbits. Stoats reached the farthest, wettest part of the South Island within 20 years of their liberations on the East Coast (Henry 1903). Reaching the limits of the main islands, they then swam to and colonized most of the islands within 2–3 km of the coast. Stoats climb and swim extremely well, wander long distances, can kill birds several times their own weight, and can breed rapidly to take advantage of high prey populations. They climb so well that they can move through the forest from tree to tree (R. Morris, personal communication). A stoat will explore all the branches and trunk of a tree for nest holes, then move on to the next. They are also patient and persistent; in short, they are extremely efficient predators. Their staple diet is rodents, but they take birds, eggs, lizards, and large invertebrates as well (King 1990).

Ferrets do not climb well, but they are known to be efficient predators of ground birds, and they even take carrion, including road-killed ferrets (R.N.H., personal observation). They do not penetrate in any numbers into rainforest, but they have been implicated in the continuing decline of weka (*Gallirallus australis*) in shrublands in both the North and South Islands. Despite the presence of stoats, ferrets, and cats, constant poisoning and shooting, as well as the introduction of the diseases myxomatosis and rabbit calicivirus, rabbits are still a problem today, so the massive and ongoing damage to the New Zealand biota has been in vain. Only one of the introduced predators has given ground on the main islands. The Pacific rat was soon supplanted by the three new rodents and survives on the main islands only as a relict population in the southwestern corner of the South Island and on Stewart Island, where there are no mice. The only place that all four rodents are known to coexist is on main Chatham Island; otherwise, one or more of the mix is missing. The other rats, mice, cats, and mustelids could tolerate extreme habitats generally shunned by Pacific rats. The species that had been saved before by their preference for a place to live were no longer physically separated from the new predators. And safety in size vanished for most because most of the new predators were bigger and more powerful than any Pacific rat.

The smallest rodent, the house mouse, is known to eat eggs and small birds such as storm petrels, but its major contribution to the sorry tale of introduced predators in New Zealand is as the staple food of the stoat. In *Nothofagus* forests, mice have a cycle of abundance that is based on the occurrence of mast seedings of the *Nothofagus* trees. Stoat numbers increase dramatically during a peak year for the mouse populations. When the mouse population crashes, the extra stoats concentrate on birds to fill the gap. It is after mouse plagues that breeding

TABLE 15.2.

Mammalian herbivores successfully established (as feral populations) on the main islands of New Zealand and the Chatham Islands since first human contact, with approximate dates of arrival or establishment.[a]

Species	Common name	Island: North	South	Stewart	Chatham
Macropus eugenii	Dama wallaby	1870,[b] 1912	—	—	—
M. rufogriseus	Bennett's wallaby	—	1874	—	—
M. parma	Parma wallaby	1870[b]	—	—	—
Petrogale penicillata	Brush-tailed rock wallaby	1873[c]	—	—	—
Wallabia bicolor	Swamp wallaby	1870[b]	—	—	—
Trichosurus vulpecula	Brush-tailed possum	1869–1915	1858–1895	1890	1911
Oryctolagus cuniculus	European rabbit	1777–	1777–	1860s–1946	—
Lepus europaeus	European hare	1874	1867	—	—
Equus caballus	Horse	1814	1836[d]	—	—
Sus scrofa	Feral pig	(1773) 1792–	(1773) 1792–	1790s?[d]	Early 1800s?
Bos taurus	Cattle	1814	1820s	—	1841
Rupicapra rupicapra	Chamois	—	1907, 1914	—	—
Hemitragus jemlahicus	Himalayan tahr	—	1904	—	—
Capra hircus	Goat	(1777)–	(1777)–	—	—
Ovis aries	Sheep	Early 1800s	Early 1800s	—	Early 1800s
Cervus elaphus	Red deer/wapiti	1883–	1861–/1905	1901	—
C. nippon	Sika deer	1905	—	—	—
C. unicolor	Sambar	1875	—	—	—
C. timorensis	Rusa deer	1907	—	—	—
Dama dama	Fallow deer	1877	1864	—	—
Odocoileus virginianus	White-tailed deer	—	1905	1905	—
Alces alces	Moose	—	1900, 1909	—	—

[a] Dates in parentheses = potential earliest date; ? = unknown; — = never established. Data based on King (1990).
[b] On Kawau Island in the Hauraki Gulf.
[c] Islands in the Hauraki Gulf.
[d] Extinct or exterminated.

success of hole-nesting species such as kaka, yellowhead, and the parakeets is lowest and predation on adults is at a peak (O'Donnell 1996).

In addition to hares and rabbits, English settlers hungered for the larger game and sport fishing that had been denied to most by the hierarchic hunting laws and ownership of land and waterways in England. Acclimatization societies were formed, not only to bring in small birds to enliven the landscape, but to introduce deer (Table 15.2), trout, and salmon. They even tried wallabies from Australia. Various species of trout, salmon, and char from Britain and North America now provide one of the great sport fisheries in the world, but at the expense of the native fish, especially the grayling (*Prototroctes oxyrhynchus*), which is almost certainly extinct. The deer went from providing some of the finest trophy heads anywhere, to plague numbers that damaged forests and mountain grasslands, to farm animals in little more than a century. Their continuing effects on the vegetation include massive changes to the composition of understory plant flora, altering successions in forests, and competing with endangered bird species such as the takahe. In the mountains of the South Island, Himalayan tahr (*Hemitragus jemlahicus*) and chamois (*Rupicapra rupicapra*) cause damage in the mountain herb fields and shrublands. Control of all these herbivores has been not only difficult and dangerous but also controversial; there is a strong hunting fraternity that lobbies for maintaining and even extending the herds.

Most recently, the top pest has been shown to attack the New Zealand biota as both predator and herbivore. The Australian brush-tailed possum (*Trichosurus vulpecula*) was introduced to provide the basis for a fur industry. It is still in the expansion phase, aided by illegal transfers to new areas, and has a population reputed to total 70 million or more. Besides targeting and destroying canopy trees such as southern rata (*Metrosideros umbellata*) and being responsible for the probable extinction of a mistletoe (*Trilepidea adamsii*), brush-tailed possums also take eggs, nestlings, and adult birds, eat other vertebrates, and occupy kiwi nest and roost holes. They are also vectors of bovine tuberculosis. Programs of control use aerial drops of poisoned bait.

But the people who brought in the herbivores also added to the fragmentation and destruction of the vegetation, both directly and indirectly. They used the northern harbors for refitting and provisioning their ships, which encouraged the local Māori population to enter the trading economy. Apart from the trade in kauri (*Agathis australis*) for ship's spars, the initial European effects on the New Zealand vegetation were indirect. They imported potatoes, which were immediately adopted by the Māoris as a productive crop that they understood (their other crops were root crops as well, and it took time to adapt to the growing and harvesting of grain). Growing potatoes for local and international trade led to the Māori population clearing forest for fresh soil. Altogether, about 30% of the forest remaining in the North Island in 1800 is thought to have been sacrificed to the potato trade (Cameron 1964). Potatoes also allowed the expansion of the Māori population in southern areas where previously they had been limited because kumara (*Ipomoea batatas*) is frost tender and cannot be grown much south of Cook Strait, and the bracken fern (*Pteridium esculentum*) also produced an inferior crop in the South Island.

The direct effects started as soon as land was acquired and farming began. Timber was needed for buildings, and the trees stood on what the new settlers hoped was good farm soil. Clearance for farming continued into the 20th century. Most of the forest was burned deliberately or accidentally, and relatively little was cut for timber. New Zealand's forests, which formerly covered nearly 90% of the land area at the time of Polynesian settlement, are now restricted to about 20%, almost all of which is montane forest. The lowlands have mostly been cleared and turned to farming or exotic forestry, largely involving the Monterey pine (*Pinus radiata*), which, in contrast to its habit in its homeland in California, grows tall and straight and is a premium wood and pulp producer. Large-scale drainage of swamps began early in the European era, and wetlands are among the most endangered on New Zealand's remaining natural habitats.

Altogether, people and the introduced herbivores have wrought massive changes to the patterns and composition of the New Zealand vegetation. Those changes and the reduction in area have naturally led to losses in the fauna. But the losses would have been much smaller, and the lists of endangered species much shorter, if the predators had been left out.

Evidence for Predation

It is one thing to know that a predator has been introduced to an island and that it kills the indigenous fauna, but it is quite another to establish cause and effect between predation and extinction. Evidence for human predation on the large flightless birds was found almost as soon as the birds themselves were discovered and described. Mantell's observations at Waingongoro and Waikouaiti should have put the matter beyond doubt before 1850, but it was not until Haast's investigations at the Rakaia River mouth and Moa Bone Point Cave that the association was made to everyone's satisfaction.

But as important as the contents of the middens were in terms of assessing the species present—and they included almost all the extinct species as well as most of the extant ones (Worthy 1999a)—it was what none of them contained that ultimately might be most important. The lack of middens containing bones of the small petrels that formerly were abundant on the main islands is prima facie evidence that rats were present before human settlement and that rats were efficient predators on small birds. Regardless of the timing, the elimination of the smallest species of bird and reptiles such as the tuatara from the main islands before Europeans brought other predators shows that the Pacific rat, as the only fresh factor in the environment, must have caused those extinctions. The details of the pattern of loss then and now confirm it.

Differences within the Main Islands

Until recently, it has been accepted that New Zealand birds were once common throughout the main islands. Some species, such as the whitehead (*Mohoua albicilla*), the yellowhead (*Mohoua ochrocephala*), kea (*Nestor notabilis*), and brown creeper (*Mohoua novaeseelandiae*) were confined to one or other of the main islands, and the kea and rock wren (*Xenicus gilviventris*) were confined to the alpine areas, but most species were thought to have been found throughout.

Imposed on this general pattern were puzzling gaps. For example, some small passerines such as the bellbird (*Anthornis melanura*) (totally) and rifleman (*Acanthisitta chloris*) (almost) vanished from Northland early in the European period; indeed, the numbers of bellbirds apparently declined so dramatically in the 19th century that observers foresaw its early extinction. The bellbird recovered in the lower half of lations of endangered or rare species such as kakapo, kokako, kaka, and all the kiwi.

Breeding success for the new pairs with protected nests has been high, and the lesson must be that although low numbers might have some effect for some species with low populations, causes of reduction, extirpation, and extinction are often much simpler. Having determined that predation is the main factor in declines, the possibility of actually halting and reversing declines is then governed by the technology for removing or lowering the rates of predation—and the money to do it. The New Zealand Department of Conservation has now embarked on a program of developing "mainland islands" to provide secure areas, with lower predation rates from lower populations of introduced rodents, cats, and mustelids, so that populations of rare native species can be rehabilitated. The supply of real islands suitable for many species has become limiting. Although research into removal of the predators is continuing and is even being augmented as the problem is better appreciated, that aim is far from being achieved, and providing safe havens is seen as an interim—and indeed probably a long-term—solution. But no solution requiring continued human input is long term. Humans have already doomed these taxa to extinction.

The ability of a predator to affect a prey species depends on its "predatory scope," which may be measured by a combination of its size, normal diet, habitat, and propensity to climb or swim. The vulnerability of a species to a new predator depends on a complex series of interactions involving the ecology and behavior of the species, its body size in comparison to that of the predator, and the size and vulnerability (nesting site) of its eggs and chicks. The resident avifauna at the time of first human contact (Holdaway 1999b; Holdaway et al. 2001) was therefore partitioned by size, ecology, and preferred living space, and six groups with different vulnerabilities can be discerned (Holdaway 1999b). The avian vulnerability groups identified in the fossil fauna as well as among living species were as follows: (1) flightless species; (2) petrels (which, although marine, nest usually in burrows, often many kilometers inland); (3) coastal and freshwater species; (4) arboreal species; (5) volant species that live on or near the ground; and (6) predators.

Within each of these groups, size classes were identified that reflect the major size groups of the introduced predators. The size classes were not strictly allocated by adult body mass, but also took into account the typical length of each species' egg (as a measure of the ability of a rat to open it). Rats and other predators may affect a species by killing the adult if it is of a suitable body mass (usually the upper limit is about the body mass of the predator), or the eggs if within an accessible size range (as conveniently measured by the length of the egg). Some birds in higher body mass classes lay comparatively small eggs. The best example of such a species in New Zealand is the kakapo (*Strigops habroptilus*).

The time dimension is important because the predators arrived in a known sequence (even if the actual times of arrival of some are disputed). The clearest distinction is the sole presence of the Pacific rat

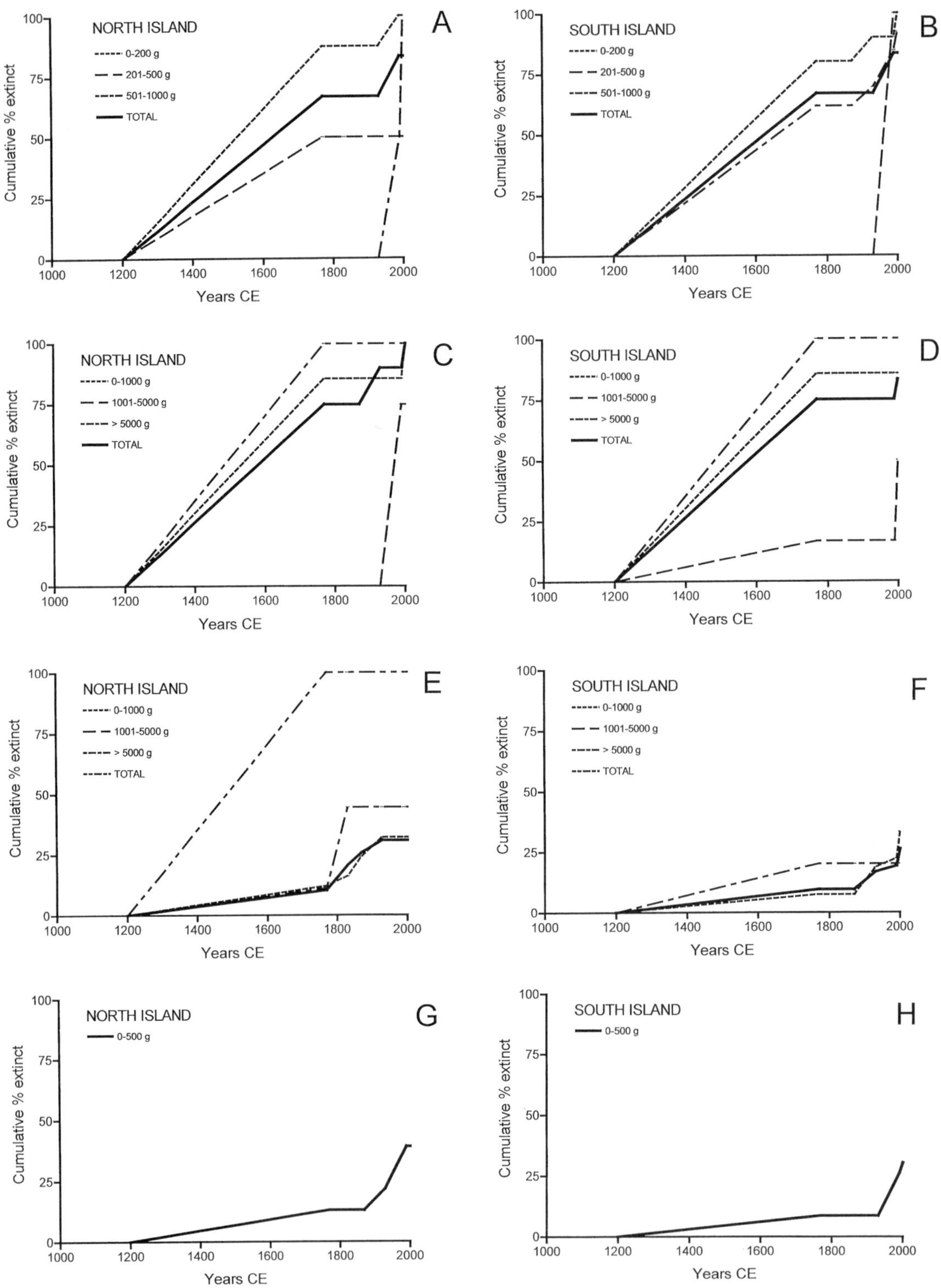
NORTH ISLAND
A
0-200 g
201-500 g
501-1000 g
TOTAL
Cumulative % extinct
Years CE
SOUTH ISLAND
B
0-200 g
201-500 g
501-1000 g
TOTAL
Cumulative % extinct
Years CE
NORTH ISLAND
C
0-1000 g
1001-5000 g
> 5000 g
TOTAL
Cumulative % extinct
Years CE
SOUTH ISLAND
D
0-1000 g
1001-5000 g
> 5000 g
TOTAL
Cumulative % extinct
Years CE
NORTH ISLAND
E
0-1000 g
1001-5000 g
> 5000 g
TOTAL
Cumulative % extinct
Years CE
SOUTH ISLAND
F
0-1000 g
1001-5000 g
> 5000 g
TOTAL
Cumulative % extinct
Years CE
NORTH ISLAND
G
0-500 g
Cumulative % extinct
Years CE
SOUTH ISLAND
H
0-500 g
Cumulative % extinct
Years CE
0
25
50
75
100
1000
1200
1400
1600
1800
2000

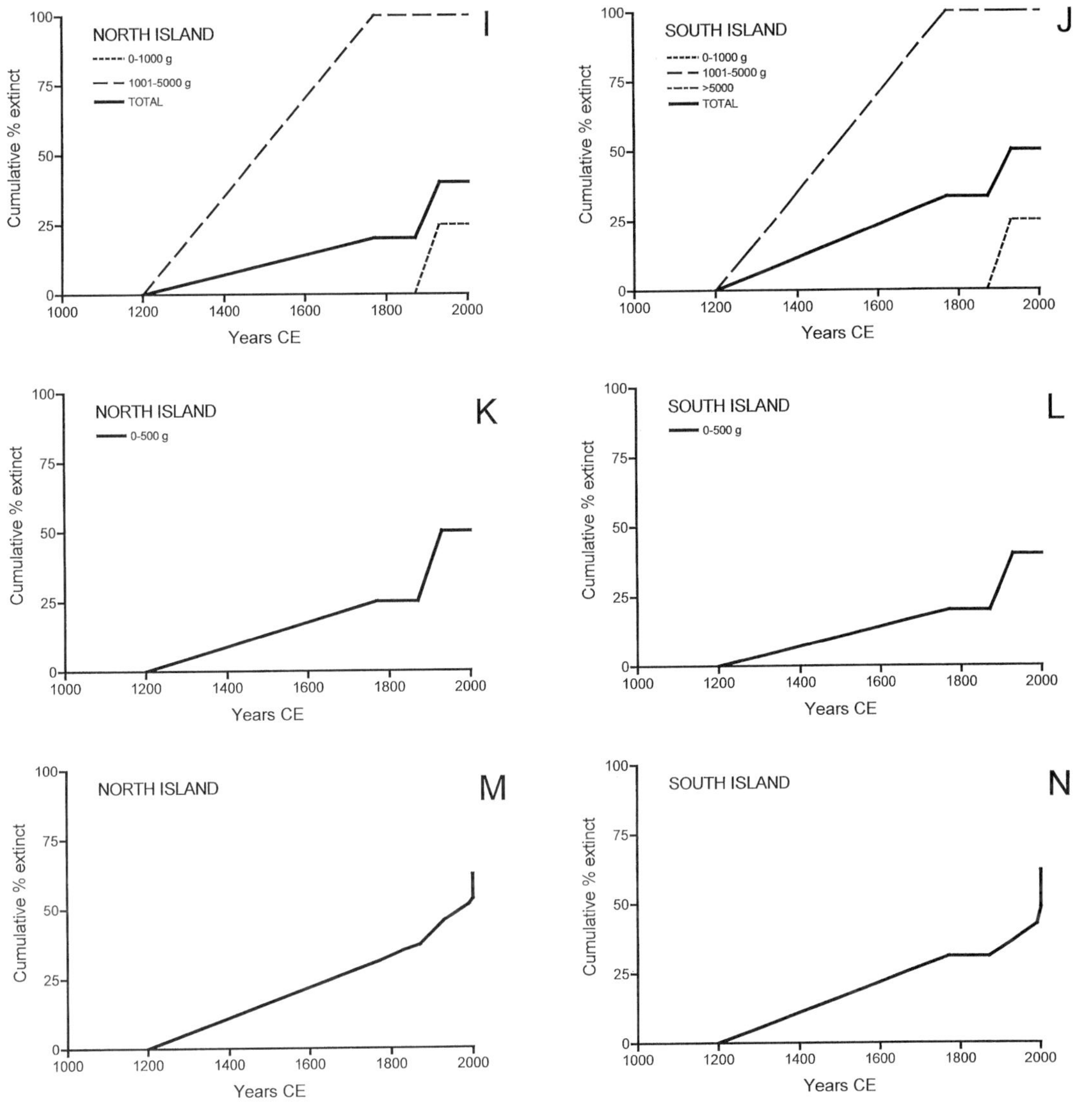

Figure 15.1. Cumulative percentage of breeding bird species in New Zealand at time of human contact extinct at or before critical dates in the sequence of colonization and environmental change, grouped by vulnerability group, adult body mass, and egg length (all taxa for predators), for both major islands. (A, B) Petrels. (C, D) Flightless birds. (E, F) Coastal and freshwater birds. (G, H) Arboreal, volant birds. (I, J) Ground-living, volant birds. (K, L) Predators. (M, N) Total extinctions; vertical line at 2000 indicates species surviving only with human assistance. After Holdaway (1999b: figs. 10, 11).

before 1770. When the times of extinction are grouped into pre-European (A.D. 1770) and post-European arrival, with further subdivisions made on the basis of the time of arrival of the later predators, it is apparent that the smaller petrels and small flightless species such as wrens suffered their greatest losses before 1770 (Fig. 15.1). The loss of large species cannot be attributed to rats, but the relative immunity imbued by habitat preferences may be the reason that few arboreal or aquatic species were lost before 1770.

The susceptibility of small birds to rat predation, shown by their elimination when only one species of small rat was present, is confirmed by the extinction of snipe (*Coenocorypha iredalei*), saddleback (*Philesturnus carunculatus*), bush wren (*Xenicus longipes*), and robin (*Petroica australis*) shortly after black rats reached Big South Cape Island, southwest of Stewart Island, in the early 1960s. In a more recent study of the results of the success of translocations of birds between islands with and without rats, Lovegrove (1996a) concluded that South Island saddlebacks may not be able to survive on small islands in the presence of Pacific rats, and that even on the largest islands, they would have been vulnerable. The preferential loss in phase 1 of the very species that were on size and behavioral and ecologic grounds vulnerable to the Pacific rat is strong circumstantial evidence that the rat was responsible for the extinctions of those species.

Aside from recent examples, the strongest case both for Pacific rats having caused the extinctions in phase 1, and indeed for its early arrival in the archipelago, involves the small petrels, of which several species bred in enormous numbers on the main islands before human contact (Worthy and Holdaway 1996b). A remnant and declining population of the one species, Hutton's shearwater *Puffinus huttoni,* that survives in the mountains of the northern South Island numbers about 150,000 pairs. The Pacific rat would have preyed on the small petrels from the moment of its arrival. But because of the sheer number of birds, the rat would have taken several centuries to exterminate them. Even the smallest petrels (apart from the diving petrels, *Pelecanoides* spp.) do not become sexually mature until they are 5–7 years old, and with their high adult life expectancy of 30–40 years, they would be resilient to predation of the eggs and young that were most vulnerable to rats. At the same time, there is no evidence that people helped the process in any way. No mainland Polynesian midden contains a significant number of small petrel bones (Holdaway 1999b; Worthy 1998e, 1999a), in contrast to middens on islands. Where they were available, as on Norfolk Island (Holdaway and Anderson 2001) and Chatham Island (Sutton 1979), small petrels were a significant part of the Polynesian diet.

The New Zealand extinction is a sad, macabre natural experiment that has allowed the effects of introduced rodents to be separated from other potential factors such as climate and habitat change, disease, and genetic problems. The experiment continues, with the enhanced range of predators still posing probably insuperable (in the long term) threats to the survival of many indigenous vertebrates on the main islands.

Naïveté

One of the major imponderables in the effects of predators on environments that formerly lacked them is the behavior of prey species in the presence of the new threats. In this respect, island species are characteristically naïve and often have no antipredator behaviors (Milberg and Tyberg 1993). When the species also has a low fecundity, being naïve has been a major factor in its reduction or extinction. Naïveté played a major role in assuring the extinction of many species before they could evolve either protective fear to avoid predators or reproductive strategies that would replace the inevitable losses. For example, kakapo are still easy to catch and handle, and most of the smaller petrels are absurdly tame after centuries of exposure to rats. Another large parrot, the kaka (*Nestor meridionalis*), has declined to the brink of extinction on the main islands. The main cause of the decline is predation by stoats (*Mustela erminea*) of the females in their nest holes (Wilson et al. 1997). Even though the female has a bill that could cut a stoat in two, she seldom fights back; she just sits on her eggs and waits for the stoat to bite her. Five females on seven nests monitored in northwest Nelson in 2000 were killed by stoats. Yet in the nearby Rotoiti nature recovery area, a "mainland island" (with 98% of predators removed), all kaka nesting attempts were successful.

Naïveté to mammalian predators is a difficult concept for biologists raised on continental faunas to grasp. But it was visible for those who chose to see what was around them. Charles Darwin, on his visit to the Galápagos, found that some of the birds were so "tame" that he could not get far enough away to shoot them. That "tameness" or naïveté is still a quality of Galápagos birds. A classic photograph in Grant (1998: fig. 1.3) illustrates the point, showing a descendant of the Galápagos hawks that Darwin found so difficult to shoot standing on Grant's head. The lack of understanding by many continent-oriented (and, it must be added, Northern Hemisphere–trained) biologists of both the importance of naïveté and the effects of rat and human predation in island faunas has held back the understanding of the cause of extinction on oceanic islands.

The inescapable conclusion is that when the first Polynesian settlers arrived in New Zealand, they met a naïve avifauna that was highly vulnerable to any new predator. None of the introduced mammals had any natural analogs on Pacific islands, where the only predators were raptors, along with reptiles on a few island groups. Even though most of the New Zealand avifauna can be shown to have originated in Australia, in most instances, they had had millions of years of evolution in situ, exposed solely to avian predators. It is likely that moa and other birds would have simply stood and watched the new arrivals. And there was to be no time for them to develop a fear of the newcomers.

The medium-size birds had survived the attentions of rats and Polynesians in better shape than very large and very small species. They were too large for the former to handle and too small to be worth it for the other (Holdaway 1999b). They began to suffer when the third wave of predators arrived. Takahe had survived (barely) into the 19th century on the North Island and into the 20th on the South Island. They presumably had large enough initial populations and were too large to be highly vulnerable to the Pacific rat to be eliminated rapidly. Steady attrition by hunting and habitat destruction had removed most of the populations, leaving the last living on a biological knife-edge in the Fiordland mountains. The new predators and competitive browsing mammals have almost proved too much.

The last surviving flightless herbivore, the kakapo, provides an excellent case study of the importance of life history strategy in survival and extinction of birds in New Zealand. The adult kakapo is large enough to be safe from rats, and its nocturnal behavior (an adaptation to avian predation) lessened direct predation by humans. An herbivore, its breeding success in the absence of predators depends largely on the food supply. In years when no trees (rimu *Dacrydium cupressinum* or beech *Nothofagus*) mast seeded, no young might be raised (Merton 1999; Merton et al. 1999; Powlesland et al. 1992), but the clutch size of two to four eggs might have allowed the population to have responded to human cropping of adults by earlier maturation and greater survival of the young, as has been described for other species. However, kakapo eggs (at 51 by 38 mm) and chicks are both small enough to be vulnerable to Pacific rats (Merton et al. 1999).

The long-lived adults were able to slow the rate of decline, especially in wetter areas and at higher altitudes where Pacific rats were less abundant, but they were not able to counter it entirely. The last populations in the North Island became extinct in the early 20th century, probably hastened by predation from packs of feral dogs and stoats. The last known wild bird in the South Island, a male, vanished about 1985. The females were more vulnerable because they alone incubate and feed the chicks.

The kakapo is one of several species in New Zealand that could be described as the "living dead." Their presence on the planet depends on the actions of humans, whose actions brought them to that parlous state. Kakapo, the Chatham black robin, South Island saddleback, and the little spotted kiwi all exist solely as translocated populations. Without that active management, all of them would be as extinct as the bush wrens and Iredale's snipe, which did not survive a last-ditch removal to islands without rats (Bell 1978).

Often it was only a small aspect of a species' ecology that determined its fate. Such an ecologic Achilles' heel may be the width of the entrance to the winter roost of the greater short-tailed bat *Mystacina robusta*. This species weighed about twice as much as the lesser short-tailed bat (*M. tuberculata*) and was bulkier in proportion. Although the greater short-tailed bat is common in some laughing owl sites, the abundance of its remains is in almost inverse proportion to the number of Pacific rat bones (Fig. 15.2). Whether the changes in representation merely mark the period when the bat was declining or mark cause and effect is not yet clear.

Subtleties often existed in the ecology of both the predator and prey, and the interactions between these have also been responsible for

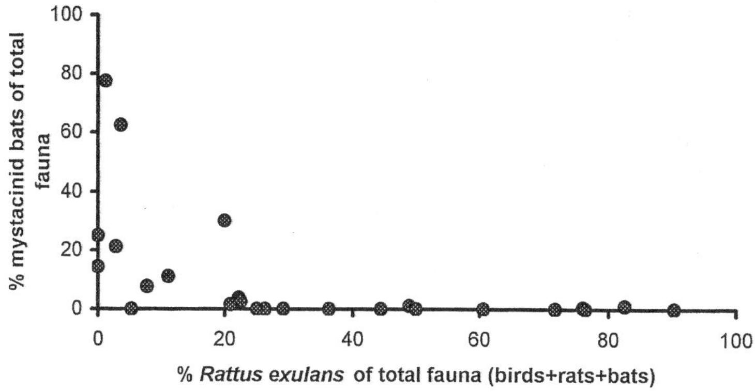

Figure 15.2. Percentage of minimum number of individuals (MNI) of mystacinid bats in the total MNI for birds, bats, and Pacific rats against percentage of MNI of Pacific rats in the total MNI for samples from predator sites in South Canterbury, South Island. After Worthy (1997b: fig. 14A).

declines and extinctions. Yellowheads (*Mohoua ochrocephala*), also known as mohua, declined precipitately in the mid-20th century. The birds nest in tree holes, usually high in *Nothofagus* trees, and the incubating adults and the young are highly susceptible to predation by stoats. The young birds beg loudly for food, which helps the predator find the nest. Stoat numbers are usually relatively low in beech forest, and in most years, some pairs of yellowheads (and breeding groups) can raise sufficient young to maintain the population. Periodically, the beech trees seed heavily, and during such a mast year, the rodent population rises sharply, followed with a lag by the stoat population. The high stoat numbers correlate well with population crashes in the yellowheads and in other hole-nesting species such as yellow-crowned parakeets (*Cyanoramphus auriceps*) (Elliott 1996a, 1996b; Elliott et al. 1996; O'Donnell 1996; O'Donnell and Phillipson 1996; O'Donnell et al. 1996).

Summary

In all, at least 76 species (64 land and freshwater species), or 31% of the 245 species formerly breeding in the New Zealand region (Appendix 1), have become locally or globally extinct since human arrival. At least 41% of the endemic species are extinct. The situation was very much worse for the mainland of New Zealand, where 49 (51%) of North Island and 53 (47%) of South Island birds have become extinct (Holdaway et al. 2001). Another 30 are presently listed as endangered, and many survive only with the presence of active (and expensive) predator control and habitat management.

The New Zealand extinction event was one of a series that affected islands in all oceans and in the Mediterranean in the past 10,000 years. They differed from the "classical" extinctions that affected continental faunas somewhat earlier, at the end of the Pleistocene, in that they affected small species as well as large and in that the extinctions were not accompanied by major climatic shifts. Therefore, the climatic–vegetation change hypothesis for continental extinctions cannot apply to the island groups extinctions. Indeed, the fossil record in New Zealand is so complete that there is no doubt that there were no extinctions during or immediately after the climate warming between 14,000 years and 10,000 years B.P. Nor is there any indication that species became extinct during the preceding cooling episode.

With no apparent effects of climatic change, apart from regional shifts in faunas, and with the evidence against other possible causes such as disease, the New Zealand faunal extinctions can have had no cause other than predation by people and introduced mammals, and possibly in a few instances by the removal of habitat by persistent firing of the vegetation. James and Olson (1991) reached almost the same conclusion for the fauna of Hawaii in that they attributed the extinctions to human interference. They stopped short of implicating rodents, doubting that the Pacific rat was responsible for the extinction of songbirds. But most recently, video evidence has been obtained that conclusively shows that rats are at work there, as they are in New Zealand and elsewhere. Steadman (1997), too, although seeing human influences as being the critical factor in the extinction of hundreds of species elsewhere in the Pacific, has downplayed the direct effects of predation on a naïve fauna. Lovegrove's analyses (1996a, 1996b) of the susceptibility of saddlebacks, endemic New Zealand passerines, are pivotal to understanding the impact of the Pacific rat on passerines. His work strongly supports the contention, made on the basis of the timing of extinctions in relation to body size and ecologic factors, that the Pacific rat was the only factor in the extinction of a large percentage of New Zealand's small vertebrate fauna over the past 2000 years (Holdaway 1999b). Such clear evidence of the importance of predation in the New Zealand extinctions may provide a model for further examination of extinctions on other islands and, indeed, on continents.

TABLE 15.3.
Summary of extinctions in the New Zealand geographic region (data derived from Appendix 1) by avian Order.

Order	Number globally extinct	Number locally extinct	Number of species formerly breeding
Dinornithiformes	11	—	11
Apterygiformes	1	—	5
Podicipediformes	—	—	2
Procellariiformes	3	1	57
Sphenisciformes	1	—	10
Pelecaniformes	—	—	14
Ciconiiformes	1	—	3
Anseriformes	9	1	18
Accipitriformes	2	1	3
Falconiformes	—	—	1
Galliformes	1	1	2
Gruiformes	12	1	19
Charadriiformes	4	—	32
Columbiformes	2	—	4
Psittaciformes	2	—	12
Cuculiformes	—	—	2
Strigiformes	1	—	2
Caprimulgiformes	1	—	1
Coraciformes	—	—	1
Passeriformes	15	—	46

Appendix 1

Systematic List of Late Holocene Avifauna

This systematic list of the late Holocene avifauna incorporates changes (n. comb. and new status) for which there is presently no nomenclatural precedent but that in the authors' opinion are justified (Holdaway et al. 2001; Worthy et al., in press). The list is presented as an attempt to identify the breeding species diversity of the avifauna of the New Zealand region at the time of first human contact. This list is the basis for the analysis of regional groupings in the avifauna. Taxa tentatively recorded breeding from a region, including taxa that have probably recently colonized that region, are excluded from the analysis, but those that are likely to have become extinct have been included. Migrants and other regular visitors are excluded. *Cygnus atratus* is recorded as extinct because it was exterminated before European arrival. Common names recommended by the authors are also listed.

We have used "island" in common names consistently when there are several species from different island groups in the fauna. Globally extinct species are shown (†), as are species locally extirpated from the region (†e). Species that still exist but that have no natural populations are shown (*) and are logged as extinct in the biogeographic analysis. Distribution indicates the geographical regions species are found breeding in: K = Kermadec Island; Nf = Norfolk Island; N = North Island; No = northern offshore islands; S = South Island; So = southern offshore islands (e.g., in the Marlborough Sounds and Fiordland); St = Stewart Island and islets; Ch = Chatham Group; and Sub = the subantarctic islands of New Zealand (Snares, Antipodes, Bounty, Campbell, Macquarie, and Auckland groups). Under "Author," the author of the species is given, or if it is an undescribed taxon, a reference providing its justification is given in all-capital letters.

Taxa	Author	Common name	Distribution
Dinornithiformes			
Family Emeidae			
Anomalopteryx didiformis†	(Owen, 1844)	Little bush moa	N S
Emeus crassus†	(Owen, 1846)	Eastern moa	S
Euryapteryx curtus†	(Owen, 1846)	Coastal moa	No N
Euryapteryx geranoides†	(Owen, 1848)	Stout-legged moa	?No N S
Megalapteryx didinus†	(Owen, 1843)	Upland moa	S
Pachyornis australis†	Oliver, 1949	Crested moa	S
Pachyornis elephantopus†	(Owen, 1856)	Heavy-footed moa	S
Pachyornis mappini†	Archey, 1941	Mappin's moa	N
Family Dinornithidae			
Dinornis struthoides†	Owen, 1844	Slender bush moa	No N S So St
Dinornis novaezealandiae†	Owen, 1843	Large bush moa	N S So
Dinornis giganteus†	Owen, 1844	Giant moa	N S So
Order Apterygiformes			
Family Apterygidae			
Apteryx "East South Island"†	WORTHY 1997	Eastern kiwi	S
Apteryx australis	Shaw & Nodder, 1813	Southern brown kiwi	S So St
Apteryx haastii	Potts, 1872	Great spotted kiwi	S
Apteryx mantelli	Bartlett, 1852	Northern brown kiwi	N No
*Apteryx owenii**	Gould, 1847	Little spotted kiwi	N S So
Order Podicipediformes			
Family Podicipedidae			
Podiceps cristatus	(Linnaeus, 1758)	Southern crested grebe	N S
Poliocephalus rufopectus	(G. R. Gray, 1843)	New Zealand dabchick	N S
Order Procellariiformes			
Family Diomedeidae			
Diomedea epomophora	Lesson, 1825	Southern royal albatross	Sub
Diomedea chionoptera	Salvin, 1896	Snowy albatross	Sub
Diomedea exulans	Linnaeus, 1758	Wandering albatross	Sub
Diomedea sanfordi	Murphy, 1917	Northern royal albatross	Ch
Phoebetria palpebrata	(Forster, 1785)	Light-mantled sooty albatross	Sub
Thalassarche bulleri	(Rothschild, 1893)	Southern Buller's albatross	St Sub
Thalassarche cauta	(Gould, 1841)	White-capped albatross	Sub
Thalassarche chrysostoma	(Forster, 1785)	Grey-headed albatross	Sub
Thalassarche eremita	Murphy, 1930	Chatham Island albatross	Ch
Thalassarche impavida	Mathews, 1912	Campbell Island albatross	Sub
Thalassarche salvini	(Rothschild, 1893)	Salvin's albatross	Sub
Thalassarche sp. 1 (Chatham form of *T. bulleri*)	ROBERTSON & NUNN 1997	Northern Buller's albatross	Ch
Family Procellariidae			
Daption capense	(Linnaeus, 1758)	Cape petrel	Sub
Halobaena caerulea	(Gmelin, 1789)	Blue petrel	Sub
Macronectes giganteus	(Gmelin, 1789)	Southern giant petrel	Sub
Macronectes halli	Mathews, 1912	Northern giant petrel	St Ch Sub
Pachyptila crassirostris	(Mathews, 1912)	Fulmar prion	Ch Sub
Pachyptila desolata	(Gmelin, 1789)	Antarctic prion	Sub
Pachyptila turtur	(Kuhl, 1820)	Fairy prion	No N S So St Ch Sub

Taxa	Author	Common name	Distribution
Pachyptila vittata	(G. Forster, 1777)	Broad-billed prion	S So St ?Ch Sub
Pelecanoides exsul	Salvin, 1896	Richdale's diving petrel	So St Ch Sub
Pelecanoides georgicus	Murphy & Harper, 1916	South Georgian diving petrel	St Ch Sub
Pelecanoides urinatrix	Gmelin, 1789	Northern diving petrel	No N S So
Procellaria aequinoctialis	Linnaeus, 1758	White-chinned petrel	Sub
Procellaria cinerea	Gmelin, 1789	Grey petrel	Sub
Procellaria parkinsoni	G. R. Gray, 1862	Parkinson's petrel	No N S
Procellaria westlandica	Falla, 1946	Westland petrel	S
Pterodroma alba†e	(Gmelin, 1789)	Phoenix petrel	K?
Pterodroma axillaris	(Salvin, 1893)	Chatham petrel	Ch
Pterodroma cervicalis	(Salvin, 1891)	White-naped petrel	K
Pterodroma solandri	(Gould, 1844)	Providence petrel	Nf
Pterodroma cookii	(G. R. Gray, 1843)	Cook's petrel	No N S St
Pterodroma inexpectata	(J. R. Forster, 1844)	Mottled petrel	N S So St Sub
Pterodroma lessonii	(Garnot, 1826)	White-headed petrel	Sub
Pterodroma macroptera	(A. Smith, 1840)	Grey-faced petrel	No N
Pterodroma magentae	(Giglioli & Salvadori, 1869)	Chatham Island taiko	Ch
Pterodroma neglecta	(Schlegel, 1863)	Kermadec petrel	Nf K
Pterodroma nigripennis	(Rothschild, 1893)	Black-winged petrel	?Nf K ?N
Pterodroma pycrofti	Falla, 1933	Pycroft's petrel	Nf No
Pterodroma sp. 1†	MILLENER 1999	Unnamed Chatham petrel	Ch
Puffinus assimilis	Gould, 1838	Little shearwater	Nf
Puffinus haurakiensis new status	Fleming & Serventy, 1943	North Island little shearwater	No N
Puffinus kermadecensis new status	Murphy, 1927	Kermadec Island little shearwater	K
Puffinus bulleri	Salvin, 1888	Buller's shearwater	No
Puffinus carneipes	Gould, 1844	Flesh-footed shearwater	No So
Puffinus elegans	Giglioli & Salvadori, 1869	Subantarctic little shearwater	Ch Sub
Puffinus gavia	(Forster, 1844)	Fluttering shearwater	No N S So
Puffinus griseus	(Gmelin, 1789)	Sooty shearwater	No N S So St Ch Sub
Puffinus huttoni	Mathews, 1912	Hutton's shearwater	S
Puffinus pacificus	Gmelin, 1789	Wedge-tailed shearwater	K Nf
Puffinus spelaeus†	Holdaway & Worthy, 1994	Scarlett's shearwater	S
Family Hydrobatidae			
Fregetta grallaria	(Vieillot, 1817)	White-bellied storm petrel	?Nf K
Fregetta tropica	(Gould, 1844)	Black-bellied storm petrel	Sub
Garrodia nereis	(Gould, 1841)	Grey-backed storm petrel	S Ch Sub
Oceanites maorianus†	(Mathews, 1932)	New Zealand storm petrel	N S
Pelagodroma marina	(Latham, 1790)	White-faced storm petrel	No N S So St Ch Sub
Pelagodroma albiclunis new status	Murphy & Irving, 1951	Kermadec storm petrel	?Nf K
Order Sphenisciformes			
Family Spheniscidae			
Aptenodytes patagonicus	J. F. Miller, 1778	King penguin	Sub
Pygoscelis papua	(J. R. Forster, 1781)	Gentoo penguin	Sub
Eudyptes "Chatham Island"†	MILLENER, 1999	Chatham crested penguin	Ch

Taxa	Author	Common name	Distribution
Eudyptes filholi	Hutton, 1879	Eastern rockhopper penguin	Sub
Eudyptes schlegeli	Finsch, 1873	Royal penguin	Sub
Eudyptes pachyrhynchus	G. R. Gray, 1845	New Zealand crested penguin	S So St
Eudyptes robustus	Oliver, 1953	Snares crested penguin	Sub
Eudyptes sclateri	Buller, 1888	Erected-crested penguin	Sub
Eudyptula minor	(J. R. Forster, 1781)	Little penguin	No N S So St Ch
Megadyptes antipodes	(Hombron & Jacquinot, 1841)	Yellow-eyed penguin	S So St Sub
Order Pelecaniformes			
Family Phaethontidae			
Phaethon rubricauda	Boddaert, 1783	Red-tailed tropicbird	Nf K
Family Sulidae			
Sula dactylatra	Lesson, 1831	Masked booby	Nf K
Morus serrator	(G. R. Gray, 1843)	Australasian gannet	No N So St
Family Phalacrocoracidae			
Leucocarbo campbelli	(Filhol, 1878)	Campbell Island shag	Sub
Leucocarbo carunculatus	(Gmelin, 1789)	King shag	N S So St
Leucocarbo colensoi	(Buller, 1888)	Auckland Island shag	Sub
Leucocarbo onslowi	(Forbes, 1893)	Chatham Island shag	Ch
Leucocarbo purpurascens	(Brandt, 1837)	Macquarie Island shag	Sub
Leucocarbo ranfurlyi	(Ogilvie-Grant, 1901)	Bounty Island shag	Sub
Phalacrocorax carbo	(Linnaeus, 1758)	Great cormorant	No N S So St Ch
Phalacrocorax melanoleucos	Gould, 1837	Little cormorant	No N S So St
Phalacrocorax varius	(Gmelin, 1789)	Pied cormorant	No N S So St
Stictocarbo featherstoni	(Buller, 1873)	Pitt Island shag	Ch
Stictocarbo punctatus	(Sparrman, 1786)	Spotted shag	No N S So St
Order Ciconiiformes			
Family Ardeidae			
Casmerodius albus	(Linnaeus, 1758)	Great white egret	N S
Egretta sacra	(Gmelin, 1789)	Reef heron	No N S So St Ch?
Ixobrychus novaezelandiae†	(Potts, 1871)	New Zealand little bittern	N S Ch
Order Anseriformes			
Family Anatidae			
Anas aucklandica	(G. R. Gray, 1844)	Auckland Island teal	Sub
Anas chlorotis	G. R. Gray, 1845	Brown teal	No N S So St Ch
Anas gracilis	Buller, 1869	Grey teal	N S
Anas nesiotis	J. H. Fleming, 1935	Campbell Island teal	Sub
Anas "Macquarie Island"†		Macquarie Island teal	Sub
Anas superciliosa	Gmelin, 1789	Grey duck	?Nf K No N S So St Ch Sub
Aythya novaeseelandiae	(Gmelin, 1789)	New Zealand scaup	No N S Ch
Biziura delautouri†	Forbes, 1892	New Zealand musk duck	N S
Cnemiornis calcitrans†	Owen, 1865	South Island goose	S
Cnemiornis gracilis†	Forbes, 1892	North Island goose	N
Cygnus atratus†e	(Latham, 1790)	Black swan	No N S So St Ch

Taxa	Author	Common name	Distribution
Euryanas finschi†	(Van Beneden, 1875)	Finsch's duck	N S
Hymenolaimus malacorhynchos	(Gmelin, 1789)	Blue duck	N S
Malacorhynchus scarletti†	Olson, 1977	Scarlett's duck	N S Ch
Mergus australis†	Hombron & Jacquinot, 1841	New Zealand merganser	No N S St Ch Sub
Pachyanas chathamica†	Oliver, 1955	Chatham Island duck	Ch
Tadorna "Chathams Island"†	MILLENER 1999	Chatham Island shelduck	Ch
Tadorna variegata	(Gmelin, 1789)	Paradise shelduck	N S So St
Order Accipitriformes			
Family Accipitridae			
Circus eylesi†	Scarlett, 1953	Eyles's harrier	N S
Accipiter cf. *fasciatus*	(Vigors & Horsfield, 1827)	Brown goshawk	Nf
Harpagornis moorei†	Haast, 1872	Haast's eagle	S
Order Falconiformes			
Family Falconidae			
Falco novaeseelandiae	Gmelin, 1788	New Zealand falcon	No N S So St Ch Sub
Order Galliformes			
Family Phasianidae			
Coturnix novaezelandiae†	Quoy & Gaimard, 1830	New Zealand quail	No N S St
Family Megapodiidae			
Megapodius sp.†e	CHEESEMAN 1891	Megapode	K
Order Gruiformes			
Family Aptornithidae			
Aptornis defossor†	Owen, 1871	South Island adzebill	S
Aptornis otidiformis†	(Owen, 1844)	North Island adzebill	N
Family Rallidae			
Cabalus modestus†	(Hutton, 1872)	Chatham Island rail	Ch
Capellirallus karamu†	Falla, 1954	Snipe rail	N
Diaphorapteryx hawkinsi†	(Forbes, 1892)	Hawkins' rail	Ch
Dryolimnas muelleri	(Rothschild, 1893)	Auckland Island rail	Sub
Fulica chathamensis†	Forbes, 1892	Chatham Island coot	Ch
Fulica prisca†	Hamilton, 1893	New Zealand coot	N S
Gallinula hodgenorum†	(Scarlett, 1955)	Hodgens' waterhen	N S
Gallirallus australis	Buller, 1888	Weka	No N S So St
Gallirallus dieffenbachii†	(G. R. Gray, 843)	Dieffenbach's rail	Ch
Gallirallus philippensis	(Linnaeus, 1766)	Banded rail	Nf K? No N S So St
Gallirallus "Norfolk Island"†		Norfolk Island rail	Nf
Gallirallus macquariensis†	(Hutton, 1879)	Macquarie Island rail	Sub
Porphyrio hochstetteri	(A. B. Meyer, 1883)	South Island takahe	S
Porphyrio mantelli†	(Owen, 1848)	North Island takahe	N
Porphyrio porphyrio†e	(Linnaeus, 1758)	Purple swamphen	Nf
Porzana pusilla	(Pallas, 1776)	Baillon's crake	N S Ch
Porzana tabuensis	(Gmelin, 1789)	Spotless crake	Nf K No N S Ch

Taxa	Author	Common name	Distribution
Order Charadriiformes			
Family Haematopodidae			
Haematopus chathamensis	Hartert, 1927	Chatham Island oystercatcher	Ch
Haematopus finschi	Martens, 1897	Finsch's oystercatcher	S
Haematopus unicolor	J. R. Forster, 1844	Variable oystercatcher	No N S So St
Family Recurvirostridae			
Himantopus novaezelandiae	Gould, 1841	Black stilt	N S
Family Charadriidae			
Charadrius bicinctus	Jardine & Selby, 1827	Banded dotterel	No N S St Ch
Charadrius exilis new status	Falla, 1978	Auckland Island banded dotterel	Sub
Charadrius frontalis	Quoy & Gaimard, 1830	Wrybill	S
Charadrius novaeseelandiae n. comb.	Gmelin, 1789	Shore plover	No N S Ch Sub?
Charadrius obscurus	Gmelin, 1789	New Zealand dotterel	No N S St
Family Scolopacidae			
Coenocorypha "Norfolk Island"†		Norfolk Island snipe	Nf
Coenocorypha "Campbell Island"		Campbell Island snipe	Sub
Coenocorypha aucklandica	G. R. Gray, 1845	Auckland Island snipe	Sub
Coenocorypha meinertzhageni	Rothschild, 1927	Antipodes Island snipe	Sub
Coenocorypha chathamica†	(Forbes, 1893)	Forbes' snipe	Ch
Coenocorypha huegeli	(Tristram, 1893)	Snares Island snipe	Sub
Coenocorypha iredalei†	Rothschild, 1921	South Island snipe	S St
Coenocorypha barrierensis† new status	Oliver, 1955	North Island snipe	No N
Coenocorypha pusilla	(Buller, 1869)	Chatham Island snipe	Ch
Family Stercorariidae			
Catharacta skua	Brünnich, 1764	Subantarctic skua	S St Ch Sub
Family Laridae			
Anous minutus	Boie, 1844	Black noddy	Nf K
Anous stolidus	(Linnaeus, 1758)	Brown noddy	Nf ?K
Gygis alba	(Sparrman, 1786)	White tern	Nf K
Larus bulleri	Hutton, 1871	Black-billed gull	N S
Larus dominicanus	Lichtenstein, 1823	Kelp gull	No N S So St Ch Sub
Larus novaehollandiae	Stephens, 1826	Red-billed gull	No N S So St Ch Sub
Procelsterna cerulea	(Bennett, 1840)	Grey noddy	Nf K
Sterna albostriata	(G. R. Gray, 1845)	Black-fronted tern	?N S
Sterna caspia	Pallas, 1790	Caspian tern	N S
Sterna fuscata	Linnaeus, 1766	Sooty tern	Nf K
Sterna nereis	(Gould, 1843)	Fairy tern	N S
Sterna striata	Gmelin, 1789	White-fronted tern	No N S So St Ch Sub
Sterna vittata	Gmelin, 1789	Antarctic tern	St Ch? Sub

Taxa	Author	Common name	Distribution
Order Columbiformes			
Family Columbidae			
Hemiphaga chathamensis	(Rothschild, 1891)	Chatham Island pigeon	Ch
Hemiphaga novaeseelandiae	(Gmelin, 1789)	New Zealand pigeon	K No N S So St
Hemiphaga spadicea†	(Latham, 1801)	Norfolk Island pigeon	Nf
Gallicolumba ?norfolciensis†	(Latham, 1801)	Norfolk Island ground dove	Nf
Order Psittaciformes			
Family Psittacidae			
Cyanoramphus cookii	(G. R. Gray, 1859)	Norfolk Island green parrot	Nf
Cyanoramphus auriceps	(Kuhl, 1820)	Yellow-crowned parakeet	No N S So St Sub
Cyanoramphus erythrotis	(Wagler, 1832)	Hochstetter's parakeet	Sub
Cyanoramphus forbesi	Rothschild, 1893	Forbes' parakeet	Ch
Cyanoramphus malherbi	Souance, 1857	Orange-fronted parakeet	No N S
Cyanoramphus novaezelandiae	(Sparrman, 1787)	Red-crowned parakeet	K No N S So St Ch Sub
Cyanoramphus unicolor	(Lear, 1831)	Antipodes Island parakeet	Sub
Nestor "Chatham Island"†	MILLENER, 1999	Chatham Island kaka	Ch
Nestor productus†	(Gould, 1836)	Norfolk Island kaka	Nf
Nestor meridionalis	(Gmelin, 1788)	Kaka	No N S So St
Nestor notabilis	Gould, 1856	Kea	S
*Strigops habroptilus**	G. R. Gray, 1845	Kakapo	N S So St
Order Cuculiformes			
Family Cuculidae			
Chrysococcyx lucidus	(Gmelin, 1788)	Shining cuckoo	Nf No N S So St Ch
Eudynamys taitensis	(Sparrman, 1787)	Long-tailed cuckoo	?Nf No N S So St
Order Strigiformes			
Family Strigidae			
Ninox novaeseelandiae	(Gmelin, 1788)	Morepork	Nf No N S So St
Sceloglaux albifacies†	(G. R. Gray, 1844)	Laughing owl	?No N S St
Order Caprimulgiformes			
Family Aegothelidae			
Aegotheles novaezealandiae†	(Scarlett, 1968)	New Zealand owlet-nightjar	N S So
Order Coraciiformes			
Family Alcedinidae			
Todiramphus sanctus	(Vigors & Horsfield, 1827)	Sacred kingfisher	Nf K No N S So St
Order Passeriformes			
Family Acanthisittidae			
Acanthisitta chloris	(Sparrman, 1787)	Rifleman	No N S So St
Dendroscansor decurvirostris†	Millener & Worthy, 1991	Long-billed wren	S
Pachyplichas yaldwyni†	Millener, 1988	Stout-legged wren	N S

Taxa	Author	Common name	Distribution
Traversia lyalli†	Rothschild, 1894	Lyall's wren	N S So
Xenicus gilviventris	Pelzeln, 1867	Rock wren	N S
Xenicus longipes†	(Gmelin, 1789)	Bush wren	N S St
Family Motacillidae			
Anthus aucklandicus	G. R. Gray, 1862	Subantarctic pipit	Sub
Anthus novaeseelandiae	(Gmelin, 1789)	New Zealand pipit	No N S So St
Anthus chathamensis new status	Lorenz-Liburnau, 1902	Chatham Island pipit	Ch
Family Campephagidae			
Lalage leucopyga†	(Gould, 1838)	Long-tailed triller	Nf
Family Muscicapidae			
Turdus poliocephalus†	Latham, 1801	Grey-headed blackbird	Nf
Family Sylviidae			
Bowdleria caudata	(Buller, 1894)	Snares Island fernbird	Sub
Bowdleria punctata	Quoy & Gaimard, 1830	New Zealand Fernbird	No N S So St
Bowdleria rufescens†	(Buller, 1869)	Chatham Island fernbird	Ch
Family Pachycephalidae			
Pachycephala pectoralis	(Latham, 1801)	Golden whistler	Nf
Mohoua albicilla	(Lesson, 1830)	Whitehead	No N
Mohoua ochrocephala	(Gmelin, 1789)	Yellowhead	S So St
Mohoua novaeseelandiae	(Gmelin, 1789)	Brown creeper	S So St
Family Acanthizidae			
Gerygone albofrontata	G. R. Gray, 1844	Chatham Island warbler	Ch
Gerygone igata	(Quoy & Gaimard, 1830)	Grey warbler	No N S So St
Gerygone modesta	Pelzeln, 1860	Norfolk Island warbler	Nf
Family Monarchidae			
Rhipidura fuliginosa	(Sparrman, 1787)	Grey fantail	Nf No N S So St Ch Sub
Family Petroicidae			
Petroica multicolor	(Gmelin, 1789)	Scarlet robin	Nf
Petroica longipes	(Garnot, 1827)	North Island robin	No S
Petroica australis	(Sparrman, 1788)	South Island robin	S So St
Petroica dannefaerdi	(Rothschild, 1894)	Snares Island tomtit	Sub
Petroica macrocephala	(Gmelin, 1789)	Yellow-breasted tomtit	S So St Ch
Petroica marrineri new status	(Mathews & Iredale, 1913)	Auckland Island tomtit	Sub
Petroica toitoi	(Lesson, 1828)	White-breasted tomtit	No N
*Petroica traversi**	(Buller, 1872)	Chatham Island robin	Ch
Family Zosteropidae			
Zosterops tenuirostris	Gould, 1837	Long-billed white-eye	Nf
Zosterops albogularis	Gould, 1837	White-breasted white-eye	Nf
Family Meliphagidae			
Anthornis melanocephala†	G. R. Gray, 1843	Chatham Island bellbird	Ch

Taxa	Author	Common name	Distribution
Anthornis melanura	(Sparrman, 1786)	New Zealand bellbird	No N S So St Sub
Notiomystis cincta	(Du Bus, 1839)	Stitchbird	No N
Prosthemadera novaeseelandiae	(Gmelin, 1788)	Tui	K No N S So St Ch Sub
Family Sturnidae			
Aplonis fusca†	Gould, 1836	Norfolk Island starling	Nf
Family Callaeatidae			
Callaeas cinerea?†	(Gmelin, 1788)	South Island kokako	S So St
Callaeas wilsoni	(Bonaparte, 1851)	North Island kokako	No N
Heteralocha acutirostris†	(Gould, 1837)	Huia	N
*Philesturnus carunculatus**	(Gmelin, 1789)	South Island saddleback	S So St
Philesturnus rufusater	(Lesson, 1828)	North Island saddleback	No N
Family Turnagridae			
Turnagra capensis†	(Sparrman, 1787)	South Island piopio	S So
Turnagra tanagra†	(Schlegel, 1865)	North Island piopio	N
Family Corvidae			
Corvus "mainland"†		New Zealand raven	No N S St Sub?
Corvus moriorum†	Forbes, 1892	Chatham Island raven	Ch

Appendix 2

Keys to the Major Limb Bones of Moa

Femur

1. Only one, occasionally complex, tuberosity on ventral shaft; distinct neck between femoral ball and trochanter, the articular surfaces not continuous between ball and trochanter; lateral surface of trochanter converges with ventral surface of proximal shaft region in continuous curve..ANOMALOPTERYGIDAE....................................2

Two distinct tuberosities on ventral shaft region separated by a smooth surface along the axis of the shaft; neck short, indistinct, not narrower than diameter of ball, articular surface of ball and trochanter continuous; lateral surface of trochanter converges with ventral surface of proximal shaft region in a distinct right angle..DINORNITHIDAE..9

2. Shaft relatively long (shaft width: total length 0.109–0.144; mean, 0.129; $n = 25$), curved dorsoventrally; axis of lateral condyle subparallel to shaft in dorsal view..3

Shaft short (shaft width: total length, 0.145–0.220; mean, 0.172; $n = 41$), robust, not curved dorsoventrally; axis of lateral condyle diverges widely from axis of shaft in dorsal view...4

3. Popliteal fossa equidistant with ectocondylar fossa from proximal end; broad rotular groove; ligament scar on dorsal pretrochanteric surface not deeply marked...*Anomalopteryx didiformis*..........Figures A.1B, A.2B

Popliteal fossa distal to ectocondylar fossa; narrow rotular groove...*Megalapteryx didinus*................Figures A.1C, A.2C

4. Large fossa at base of trochanter either ventrally or dorsally...5

No large fossa...6

5. Large fossa (>10 mm diameter) on dorsal surface at base of trochanter...*Pachyornis australis*...................Figures A.1H, A.2H

Large fossa (>10 mm diameter) on ventral surface at base of trochanter by neck...*Pachyornis mappini*...................Figures A.1G, A.2G

6. Crest of trochanter ends on ventral surface so that neck of femoral ball has an even curve on which nutrient foramen is sited; shaft cross section round or dorsoventrally compressed......................................7

Crest of trochanter extends as a ridge across the ventral surface of the neck to the femoral ball, passing above the nutrient foramen; shaft cross section usually laterally compressed so dorsoventral width is greater than lateral width...8

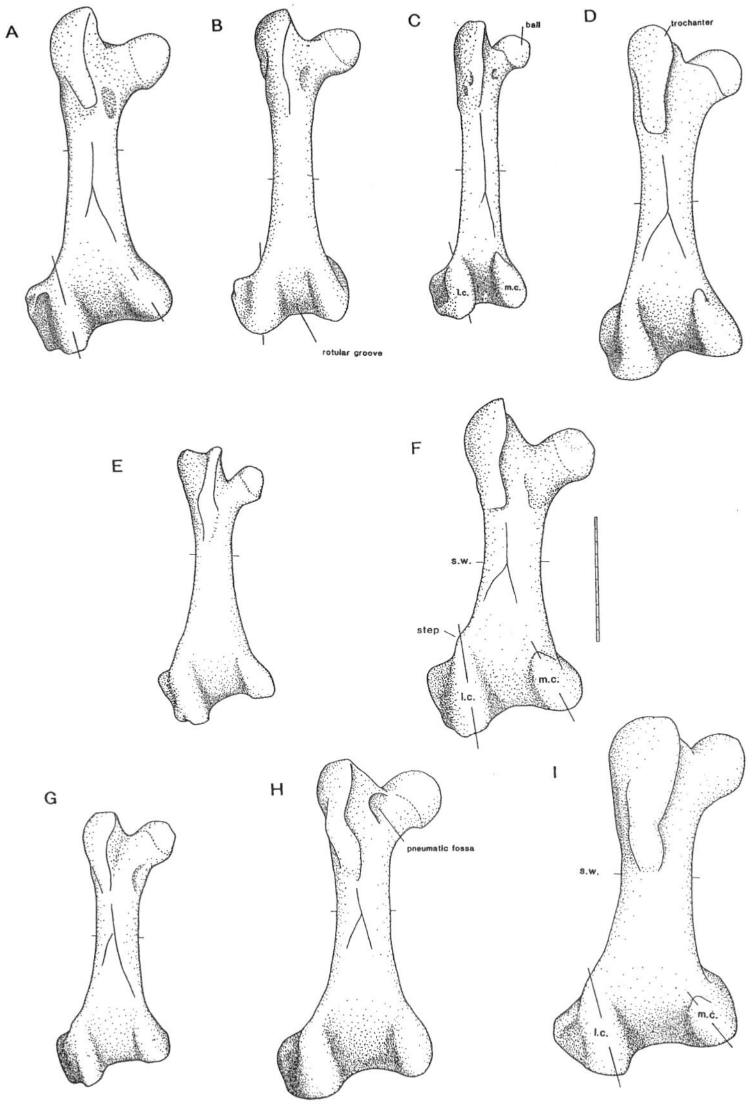

Figure A.1. Dorsal view of moa femora. (A) Emeus crassus. *(B)* Anomalopteryx didiformis. *(C)* Megalapteryx didinus. *(D)* Dinornis. *(E)* Euryapteryx curtus. *(F)* Euryapteryx geranoides. *(G)* Pachyornis mappini. *(H)* Pachyornis australis. *(I)* Pachyornis elephantopus. *Scale bar in centimeters.*

7. Length 150–225 mm, North Island only; shaft narrowest in proximal third.. ....*Euryapteryx curtus*......................Figures A.1E, A.2E

Length 215–300 mm. Larger in south; shaft narrowest at midlength.. ....*Euryapteryx geranoides*................Figures A.1F, A.2F

8. Length 265–335 mm, proximal and especially distal ends greatly expanded (distal width 46% total length; $n = 31$; range, 43%–50%), shaft length between trochanter and lateral condyle (dorsally) only 39% of total length..*Pachyornis elephantopus*..............Figures A.1I, A2.I

Length 215–300 mm; proximal and especially distal ends not greatly expanded (distal width 42% total length; $n = 17$; range, 38.6%–44.1%); shaft length between trochanter and lateral condyle about 47% of total length; ligament scar on pretrochanteric surface strongly marked, often a fossa........................... ....*Emeus crassus*............................Figures A.1A, A.2A

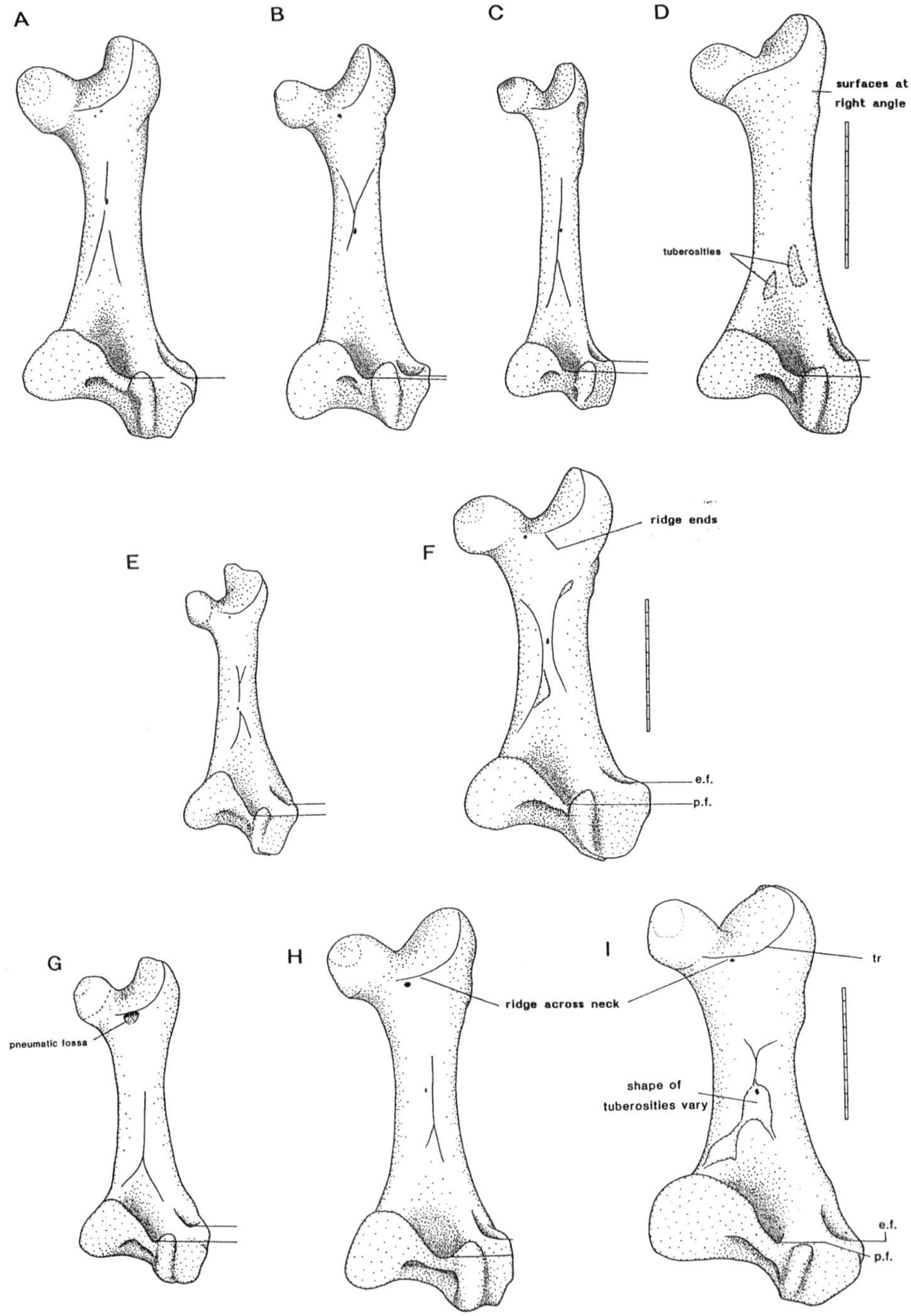

Figure A.2. Ventral view of moa femora. (A) Emeus crassus. *(B)* Anomalopteryx didiformis. *(C)* Megalapteryx didinus. *(D)* Dinornis. *(E)* Euryapteryx curtus. *(F)* Euryapteryx geranoides. *(G)* Pachyornis mappini. *(H)* Pachyornis australis. *(I)* Pachyornis elephantopus. *Scale bar in centimeters. e.f. = epicondylar fossa; p.f. = popliteal fossa.*

9. Size range: 230–315 mm....................................*Dinornis struthoides*..................Figures A.1D, A.2D
290–370 mm....................................*Dinornis novaezealandiae*
328–447 mm....................................*Dinornis giganteus*

Note. Size ranges of *Dinornis* species vary with geologic age and geographic location, but within single sites, there is usually a clear demarcation of species unless the site is time transgressive.

Tibiotarsus

1. Less than 600 mm long; shaft not very elongate; angle between ectocnemial ridge and ventral support strut less than or equal to 90°..................................ANOMALOPTERYGIDAE....................................2
Shaft very elongate; relatively slender; angle between ectocnemial ridge and its ventral support strut distinctly greater than 90°..DINORNITHIDAE..9

2. Procnemial ridge straight...................................*Megalapteryx didinus*............................Figure A.3C
Procnemial ridge deflected laterally where proximal end expands...3

3. Distal end flared medially, height of medial condyle usually > shaft width; proximal shaft swollen by extension of ectocnemial ridge, so shaft markedly convex between fibular crest and crest from procnemial ridge..4
Distal end not greatly flared medially, height of medial condyle ≤ shaft width; shaft flat between the fibular crest and crest from the procnemial ridge..6

4. Fibular crest long (23%–26%) total length; nutrient foramen level with distal end of fibular crest...........
....*Pachyornis australis*...............................Figure A.3H
Fibular crest short (18%–23%) total length; nutrient foramen always distal to end of fibular crest........5

5. North Island only, length <500 mm...................*Pachyornis mappini*...............................Figure A.3G
South Island only, length usually >500 mm...........*Pachyornis elephantopus*..........................Figure A.3I

6. Nutrient foramen usually just distal to fibular crest; proximodistal height of lateral condyle usually >60% distal width; profile of shaft is stepped above the medial condyle
....*Anomalopteryx didiformis*.....................Figure A.3B
Nutrient foramen usually proximal to end of fibular crest; proximodistal height of lateral condyle usually <55% distal width; profile of the shaft in anterior view concave or straight where it meets medial condyle, not stepped..7

7. Shaft relatively slender, may be laterally bowed, <380 mm length, North Island only...*Euryapteryx curtus*
....Figure A.3D
Shaft straight, robust; usually >380 mm..8

8. Proximodistal height of medial condyle < shaft width; in anterior view top of procnemial ridge flattened; behind and below the femur articular surface is a perforated hollow not bound by a distinct ridge medially; length 405–586 mm; North and South Islands.................................*Euryapteryx geranoides*
....Figure A.3E
Proximodistal height of medial condyle equals shaft width; crest rising from articular surface to top of procnemial ridge not squared off; behind and below the femur articular surface is a perforated hollow usually bound by a distinct ridge medially; length 340–495 mm; South Island only.............*Emeus crassus*
....Figure A.3A

9. Length, 450–620 mm..*Dinornis struthoides*
570–740 mm...*Dinornis novaezealandiae*
681–992 mm...*Dinornis giganteus*
....Figure A.3F

Note. Size ranges of *Dinornis* species vary with geologic age and geographic location, but within single sites, there is usually clear demarcation of species.

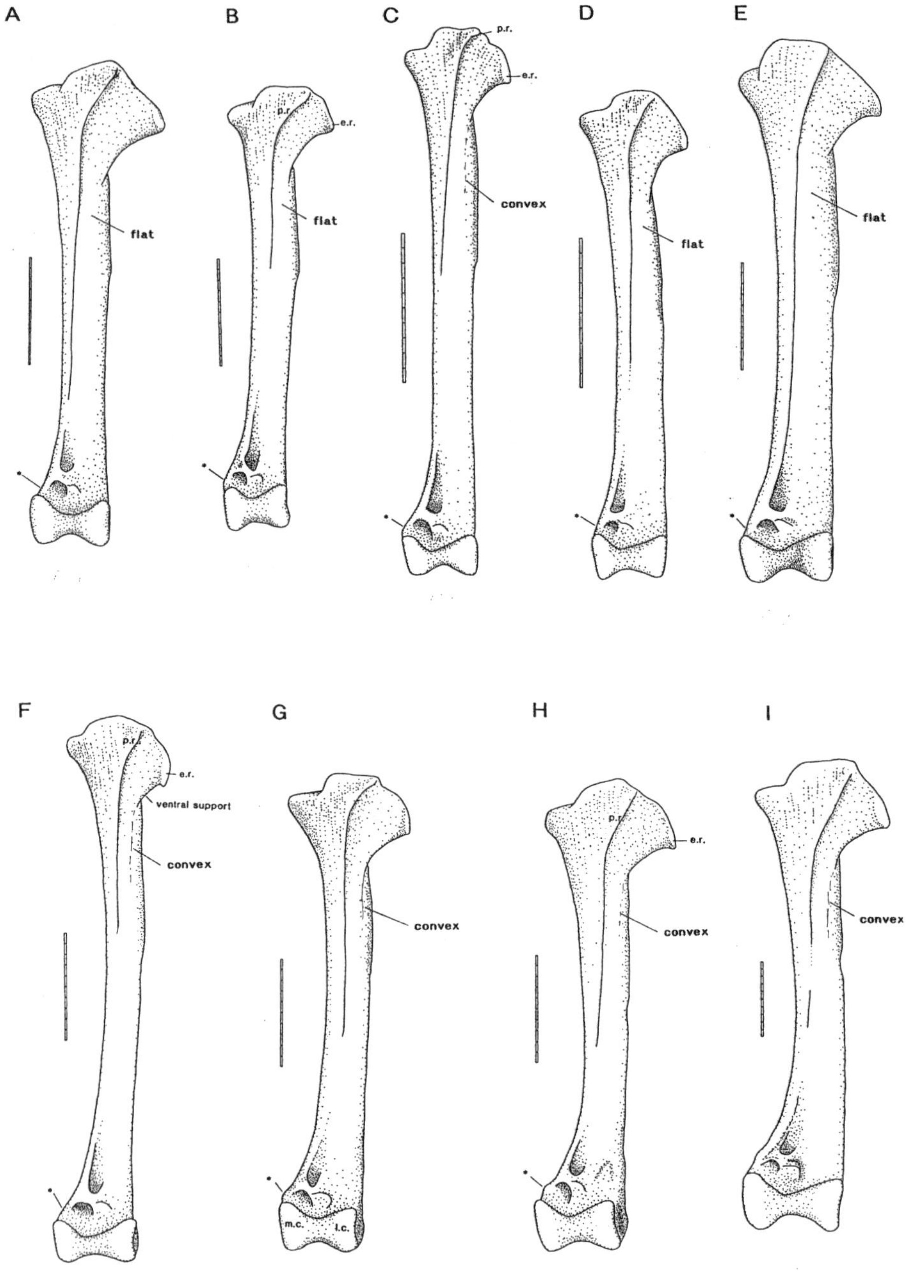

Figure A.3. Anterior view of moa tibiotarsi. (A) Emeus crassus. *(B)* Anomalopteryx didiformis. *(C)* Megalapteryx didinus. *(D)* Euryapteryx curtus. *(E)* Euryapteryx geranoides. *(F)* Dinornis. *(G)* Pachyornis mappini. *(H)* Pachyornis australis. *(I)* Pachyornis elephantopus. *Scale bars in centimeters. e.r. = ectocnemial ridge; p.r. = procnemial ridge.*

Tarsometatarsus

1. Length >2.5 times maximum width......................ANOMALOPTERYGIDAE....................................2
Length, 2.5–3.0 times maximum width..................DINORNITHIDAE..9

2. Length 2.0–2.5 times distal width, shaft between proximal and distal expansions relatively long......3
Length <2.1 times distal width; shaft relatively short..6

3. In medial aspect, high point of endotrochlea is above or over the low point; in anterior view, medial side of the endotrochlea pointed; middle trochlea arising at shallow angle from shaft...........................4
In medial aspect, high point of endotrochlea markedly distal to low point; in anterior view, medial side of endotrochlea not pointed, but flattened and subparallel to outer edge of ectotrochlea; middle trochlea arising abruptly from shaft..5

4. Lateral hypotarsal ridge higher than medial ridge; intercondylar ridge prominent; anterior margin of proximal end not flattened, so midwidth depth markedly more than 50% of width, length 160–227 mm. South Island only...*Emeus crassus*..............................Figures A.4A, A.5A
Lateral and medial hypotarsal ridges of equal height; intercondylar ridge not prominent; proximal surface is flattened so that the midwidth depth about 50% of width. North Island only, length 110–165 mm ....*Euryapteryx curtus*........................Figures A.4E, A.5E

5. Lateral hypotarsal ridge longer than medial ridge; the medial nutrient foramen not in distinct hollow bound by ridge proximally; length typically 2.2 times distal width............*Anomalopteryx didiformis* ....Figures A.4B, A.5B
Lateral hypotarsal ridge equal in length to medial ridge; medial nutrient foramen in distinct hollow bound by ridge proximally; length typically 2.4 times distal width...............................*Megalapteryx didinus* ....Figures A.4C, A.5C

6. Middle trochlea not arising abruptly from shaft; endotrochlea often elongated so change in point of curvature markedly distal to proximal extreme of adjacent intertrochlea notch; length 158–267 mm, North and South Islands, more common in eastern districts of both islands; length 1.8–2.0 times distal width..*Euryapteryx geranoides*................Figures A.4F, A.5F
Middle trochlea arising abruptly from shaft; change in curvature of endotrochlea adjacent to proximal extreme of intertrochlea notch; length usually <2.1 times distal width..7

7. North Island only...*Pachyornis mappini*.....................Figures A.4G, A.5G
South Island only...8

8. Very massive, average shaft width 26% length, average distal width 59% length or total length ≤1.8 times distal width; usually in lowlands.....................*Pachyornis elephantopus*................Figures A.4I, A.5I
Less robust, shaft width 23% length, distal width 52% length; usually in uplands..........*Pachyornis australis* ....Figures A.4H, A.5H

9. 240–330 mm length..*Dinornis struthoides*.....................Figures A.4D, A.5D
300–420 mm length...*Dinornis novaezealandiae*
410–550 mm length...*Dinornis giganteus*

Note. Size ranges of *Dinornis* species vary with geologic age and geographic location, but within single sites, there is usually clear demarcation of species.

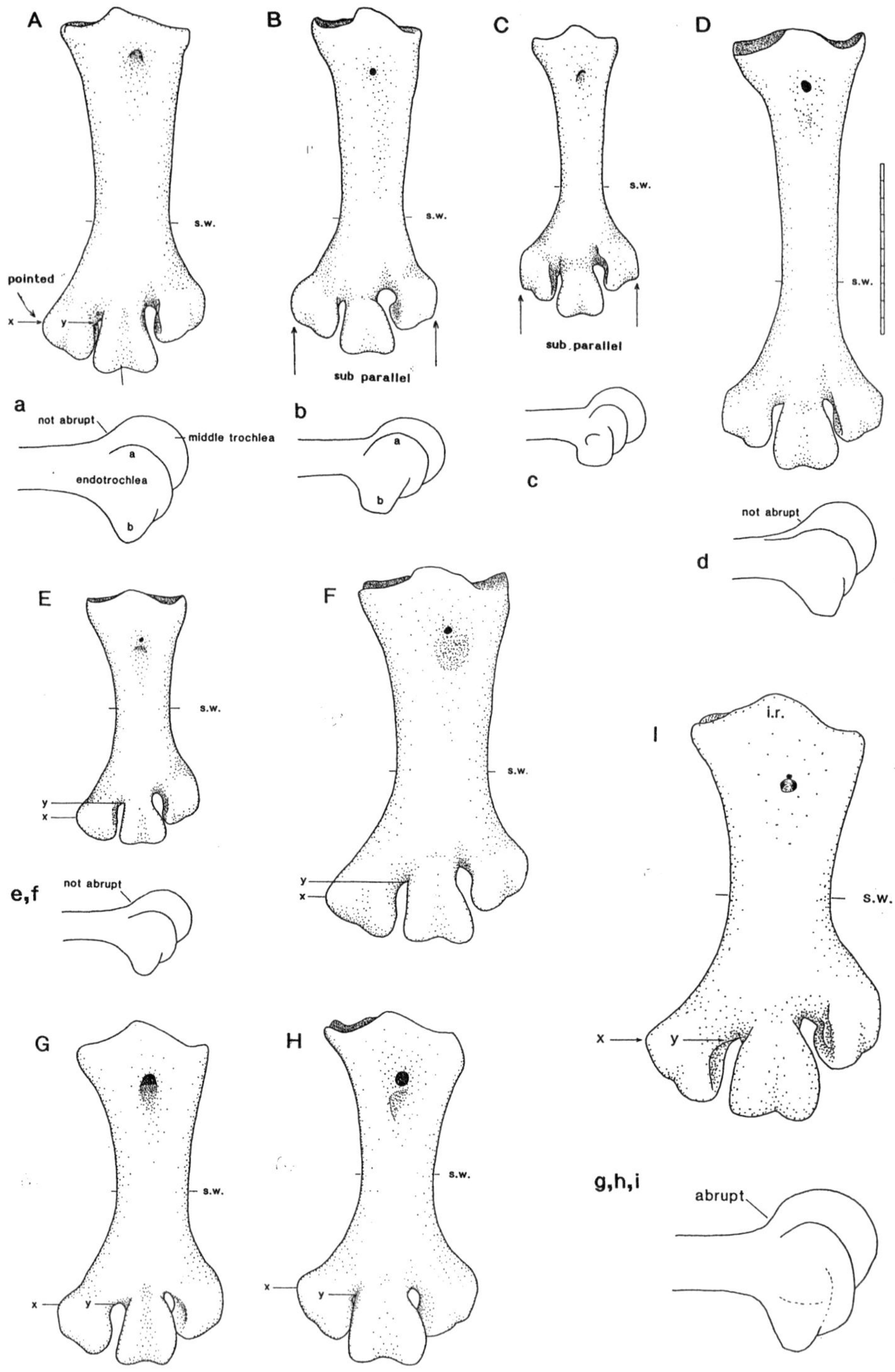

Figure A.4. Anterior view of moa tarsometatarsi. (A) Emeus crassus. *(B)* Anomalopteryx didiformis. *(C)* Megalapteryx didinus. *(D)* Dinornis. *(E)* Euryapteryx curtus. *(F)* Euryapteryx geranoides. *(G)* Pachyornis mappini. *(H)* Pachyornis australis. *(I)* Pachyornis elephantopus. *Scale bar in centimeters. All medial views are of the distal end.*

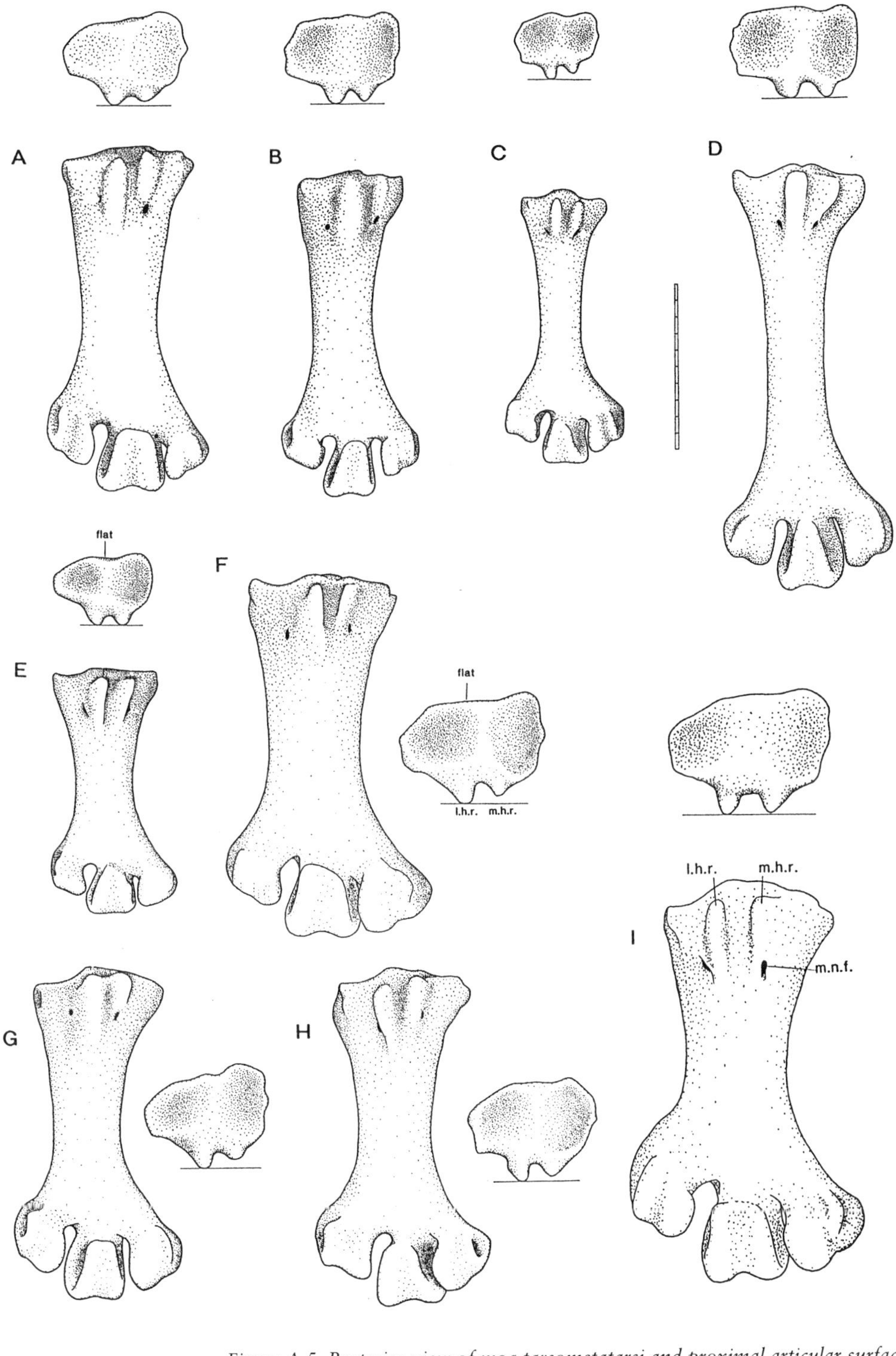

Figure A.5. Posterior view of moa tarsometatarsi and proximal articular surface. (A) Emeus crassus. *(B)* Anomalopteryx didiformis. *(C)* Megalapteryx didinus. *(D)* Dinornis. *(E)* Euryapteryx curtus. *(F)* Euryapteryx geranoides. *(G)* Pachyornis mappini. *(H)* Pachyornis australis. *(I)* Pachyornis elephantopus. *Scale bar in centimeters.*

Appendix 3

Morphologic Characters Used in Cladistic Analysis of Moa

Skull

1.	Cranium less than twice as broad as high	0
	Cranium more than twice as broad as high	1
2.	Postorbital width approximately equal to or slightly greater than squamosal width	0
	Width across postorbitals markedly greater than across squamosals	1
3.	Orbital margins parallel or semiparallel sided	0
	Orbital margins converge proximally	1
4.	Descending maxillary process of nasal lies parallel to lachrymal, large ventral expansion	0
	Descending maxillary process of nasal overlapped by lachrymal, no ventral expansion	1
	No maxillary process	2
5.	The maxillary antrum is absent or minute	0
	There is a small maxillary antrum	1
	There is a large maxillary antrum	2
6.	The prefrontal diverges from the rostrum at 90° and end is further laterad of origin than anterior to origin	0
	The prefrontal diverges from the rostrum at about 60–70°	1
	The prefrontal diverges from the rostrum at markedly less than 90°, so end is farther anterior of origin than laterad of it	2
7.	There is usually pitting on the dorsal surface of the cranium	0
	There is no pitting on the dorsal surface of the cranium	1
8.	Mamillar tuberosities with prominent ventral development in caudal view	0
	Mamillar tuberosities have weak ventral development	1
9.	The rostrum is rounded ventrally throughout its length	0
	The rostrum is compressed laterally so is V-shaped at the prefrontals	1
	The rostrum is compressed laterally in anterior part between the lateral processes	1

10.	Temporal ridges are in contact with lambdoidal ridge	0
	Temporal ridge is relatively close to lambdoidal ridges	1
	Temporal ridge is widely separated from lambdoidal ridge	2
11.	Marked tuberosities are laterad of occipital crest	0
	No marked tuberosities are laterad of occipital crest	1
12.	No marked hollows are above paroccipital process	0
	Marked hollows are above paroccipital process	1
13.	Basipterygoid processes are narrow, constricted at base	0
	Basipterygoid processes are broader, not constricted	1
14.	Basipterygoid processes are directed widely and so have shallow, concave anterior margin	0
	Basipterygoid processes are directed anteriorly and so have deeply concave anterior margin	1
15.	The occipital condyle is subspherical in ventrolateral view	0
	The occipital condyle is markedly wider than high in this view	1
16.	The occipital condyle has a pronounced neck when seen in ventral view	0
	The occipital condyle has no neck	1
17.	Quadrate with fossa found posteriorly on dorsal process under otic head	0
	Quadrate with no fossa found posteriorly on dorsal process	1
18.	Mandible has no prominent posterior process	0
	Mandible has prominent posterior process	1
19.	Turbinal is without spiral process	0
	Turbinal has spiral process with one twist	1
	Turbinal has spiral process with 2–2.5 twists	2
20.	Width of the premaxilla is much more than half zygomatic width	0
	Width of the premaxilla is less than or equal to half zygomatic width	1
21.	Premaxilla in dorsal view	
	—Straight convergent sides with near pointed tip, culmen straight	0
	—Convex convergent sides, culmen not steep, rounded tip	1
	—Convex convergent sides, culmen steep, broad rounded tip	2
	—Straight, slightly concave sides, narrow, robust pointed tip	3
	—Straight, elongate with rounded tip, culmen shallow	4
22.	Premaxilla—laterally compressed	0
	—Height at maxilla slightly less than width	1
	—Markedly dorsoventrally compressed	2
23.	Temporal fossa are small, not extending far dorsally over cranium	0
	Temporal fossa are large, not extending over dorsal surface of cranium	1
	Temporal fossa are large, extend over dorsal surface of crania	2
24.	Basitemporal platform forms angle about 90–100° with plane of occipital foramen	0
	Basitemporal platform forms angle about 120° with plane of occipital foramen	1

25. Olfactory capsules are small 0
Olfactory capsules are large 1

26. Frontals above orbit are smooth with no or only a weakly defined fossa 0
Frontals above orbit have pronounced fossa 1

27. In dorsal view, exoccipitals protrude caudally past occipital condyle 0
In dorsal view, exoccipitals do not protrude caudally past occipital condyle 1

28. Occipital crest is narrow 0
Occipital crest is swollen to form a prominent supraforaminal ridge 1

29. Cranium lacks pronounced interorbital elevation 0
Cranium has pronounced interorbital elevation[1] 1

30. The ridge dorsally bordering the tympanic cavity crosses the zygomatic process so that in caudal view, the zygomatic process is stepped 0
The ridge dorsally bordering the tympanic cavity does not cross the zygomatic process so that in caudal view, the zygomatic process is evenly curved 1

31. Mandible has slight down-curving to tip 0
Mandible has markedly down-curved anterior third 1

32. Zygomatic processes are small, shorter than width at base 0
Zygomatic processes are well developed, longer than width at base 1

Sternum

33. Lateral processes are straight, narrowly diverging 0
Lateral processes are straight, widely diverging 1
Lateral processes are curved, widely diverging 2
No lateral processes exist 3

34. Sterna are longer than wide 0
Sterna are as long as wide 1
Sterna are wider than long 2

35. Sterna have coracoid groove 0
Sterna lack coracoid groove 1

36. Sterna have two large, well-defined, multiperforated hollows on dorsal surface behind coracoid sulci 0
Sterna have a single concavity behind anterior margin with distinct small areas of multiperforated fossa behind coracoidal sulci 1
Sterna have single hollow on dorsal surface behind anterior margin, foramina scattered 2

37. Sterna have bilateral symmetry 0
Sterna have markedly deeper depression on left side 1

38. Lateral notch is <half length midline 0
Lateral notch is ≥half length midline 1

39. Xiphoid process is not notched 0
Xiphoid process is weakly notched 1
Xiphoid process is deeply notched 2

40.	Lateral process is same length as xiphoid	0
	Lateral process is slightly longer than xiphoid	1
	Lateral process is markedly longer than xiphoid	2
41.	Scapulocoracoid has well-defined components and glenoid cavity	0
	Scapulocoracoid have less well defined components, no glenoid cavity	1
	Scapulocoracoid is not present	2

Femur

42.	Single complex tuberosity present on ventral shaft	0
	Two distinct tuberosities present on ventral shaft	1
	Shaft lacks tuberosities, as in moa	2
43.	Lateral surface of trochanter curves evenly to central surface of proximal shaft	0
	Lateral surface of trochanter forms a plane that meets proximoventral shaft region in right angle	1
44.	Ball is separated by distinct constricted neck	0
	Ball is not separated by constricted neck	1
45.	Shaft width is 0.109–0.144 length	0
	Shaft width is 0.145–0.220 length	1
46.	Shaft is dorsoventrally curved	0
	Shaft is straight	1
47.	Lateral condyle diverges widely from axis of shaft	0
	Lateral condyle is subparallel to shaft	1
48.	Popliteal fossa is equidistant with ectocondylar fossa from proximal end	0
	Popliteal fossa is farther distad than ectocondylar fossa from proximal end	1
49.	Large multiperforated fossa exist dorsally between trochanter and neck	0
	No large fossa exist dorsally between trochanter and neck	1
50.	Large fossa exist ventrally between trochanter and neck	0
	No large fossa exist ventrally between trochanter and neck	1
51.	Rotular groove is wide	0
	Rotular groove is narrow	1
52.	Ventrally, the trochanteric ridge is continuous across neck	0
	Ventrally, the trochanteric ridge is not continuous across neck	1
53.	The linea obliqua longitudinalis superior has one crest in central section	0
	The linea obliqua longitudinalis superior has two crests in central section	1
54.	Shaft is narrowest at midlength	0
	Shaft is narrowest in proximal third	1

Tibiotarsus

55.	Tibiotarsus is twice femur length	0
	Tibiotarsus is 1.6–1.8 times femur length	1

56. Procnemial ridge is straight with little or no lateral deflection 0
Procnemial ridge is markedly deflected laterally 1

57. Ectocnemial ridge and ventral support strut form angle about 90° 0
Ectocnemial ridge and ventral support strut form angle markedly greater than 90° 1

58. Distal end is not markedly flared medially 0
Distal end is markedly flared medially 1

59. Fibular crest is 23%–26% total length 0
Fibular crest is <23% total length 1

60. Medulloarterial foramen is at or distal to end of fibular crest 0
Medulloarterial foramen is proximal to end of fibular crest 1

61. Ridge above medial condyle is usually stepped in anterior view 0
In anterior view, there is no step above the medial condyle 1

62. Anterior shaft between fibular crest and procnemial ridge is markedly convex 0
Anterior shaft between fibular crest and procnemial ridge is flat 1

63. Posterior surface between lateral and medial condyles is deeply concave 0
Posterior surface between lateral and medial condyles is more or less flat 1

Tarsometatarsus

64. Tarsometatarsus is longer than femur 0
Tarsometatarsus is shorter than femur 1

65. Length is 2.5–3.0 maximum width 0
Length is 2.1–2.5 maximum width 1
Length is <2.1 maximum width 2
Length is >4 times maximum width 3

66. Intertrochlear grooves extend proximally unequal length 0
Intertrochlear grooves extend proximally equal length 1

67. Medial margin of trochlea 1 is angular with maximum width distal to maximum depth of intertrochlear space 0
Medial margin of trochlea 1 is angular with maximum width level with maximum depth of intertrochlear space 1
Medial margin of trochlea 1 is straight and parallel with the lateral margin of trochlea 3 2

68. Middle trochlea arises gradually from shaft 0
Middle trochlea rises abruptly from shaft 1

69. Medial nutrient foramen is on flat distal surface 0
Medial nutrient foramen is in distinct fossa on distal surface 1

70. Lateral and medial hypotarsal ridges are always of unequal height in proximal view 0
Lateral and medial hypotarsal ridges are variably of equal or unequal height 1
Lateral and medial hypotarsal ridges are always of equal height in proximal view 2

71.	Ungual phalanges are more than twice as long as broad, marked ventral curvature	0
	Ungual phalanges are less than twice as long as broad, fairly flat ventrally	1
	Ungual phalanges are more than twice as long as broad, flat ventrally	2
72.	Phalangeal formula is 2:3:4:5	0
	Phalangeal formula is 2:3:4:4	1

Pelvis

73.	Dorsal profile ilia are straight between anterior and caudal extremes	0
	Dorsal profile ilia are curved	1
74.	Lateral processes in escutcheon diverge at about 45° in straight line	0
	Lateral processes in escutcheon diverge at <45°, then turn posteriorly	1
75.	Acetabuli meets synsacrum ventrally in curve	0
	Acetabuli meets synsacrum ventrally in right angle	1
76.	Escutcheon, lateral iliac portions retain width posteriorly or are slightly convergent	0
	Escutcheon, lateral iliac portions narrow markedly posteriorly	1
	Escutcheon is very compressed	2
77.	Femoral socket is anterior of midlength	0
	Femoral socket is at midlength	1
	Femoral socket is in posterior half	2

Vertebrae, ribs

78.	Vertebrae, 27 presacral	0
	Vertebrae, 29 or 30 presacral	1
	Vertebrae, <27 presacral	2
79.	Cervical vertebrae 3–5 with bifid neural spine	0
	Cervical vertebrae 3–5 with single neural spine	1
80.	Ribs 5 pairs presacral ribs	0
	Ribs 6 pairs presacral	1
	Ribs 7–8 pairs presacral	2
81.	Pneumatic fossae between dorsal and ventral zygapophyses of ribs is large	0
	Pneumatic fossae on ribs is small, obsolete, only on biggest ribs	1
82.	Trachea has no body loop	0
	Trachea has body loop of large ossified rings	1

1. This is not to be confused with intertemporal elevation seen in *E. curtus*, for example.

Coded State of Characters Used in Cladistic Analysis of Moa Characters

Moa taxa are shown by acronyms derived from the first two letters of their current binomials.

Char	1	1111111112	2222222223	3333333334	4444444445	5555555556	6666666667	777777777888
No.	1234567890	1234567890	1234567890	1234567890	1234567890	1234567890	1234567890	123456789012
Andi	000022n000	1010111111	3020101n01	0101020000	1000001011	0000110000	0101102n00	0010011002n0
Pama	011022n01n	1n1n111111	0120101101	0112020000	1000110110	0000110110	0001201000	000001000?10
Paau	011022n012	1n10100111	01101000n1	0112n20000	1000110101	0000110100	0101201000	000001100200
Medi	0111111001	1101001101	0121101001	0101100000	100000n111	1000101001	0001102010	001011200n10
Pael	000022100n	1n10101111	01201010n1	0112n20000	1000110011	0000110110	n001201001	000001100?10
Eucu	010?021111	10111111?1	2101001010	0100110?n1	2000110111	0101110001	1111200102	110101000111
Euge	011202101n	1011011111	2100000001	0100110010	2000110111	0110110001	11n120n101	111101100211
Emcr	0112121102	1110111111	1101010110	01001001n1	2000110011	0011110001	1101101100	111001n00?01
Dist	110020002n	0010001020	1220100001	1022001022	1111110101	0010001000	1000012102	000000111100
Dino	1100200102	1010111020	1220100001	1022001022	1111110101	0010001000	1100012102	000000111110
Digi	11002000n2	0010101020	1220100001	1022001022	1111110101	0010001000	1100012102	000000111100
Rhea	00120?0100	1000110001	4221000101	01300?0?0?	0210000010	101011?010	1?00312110	20??12021000
Emu1	0112111112	1000010100	0201000001	00301?0?0?	0210000010	1100110010	1000302100	001?12021200

Appendix 4

The following maps are Holocene distributions (solid circles) for a selection of species in the South Island. The distributions are extracted from data for 301 fossil sites (hollow circles) and plotted on the metric map grid for the NZMA 260, 1:50,000 series. The Main Divide of the Southern Alps is shown by the line in the central part of the map; some of the major lakes are shown in black.

Megalapteryx didinus

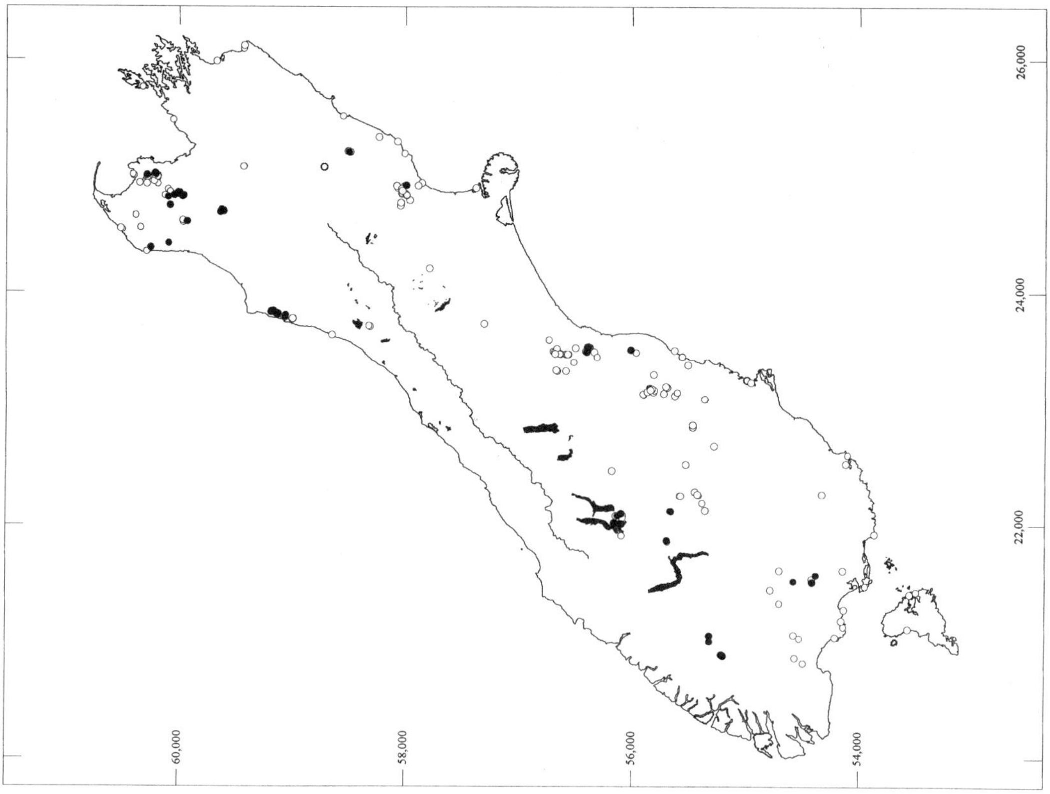

Apteryx owenii

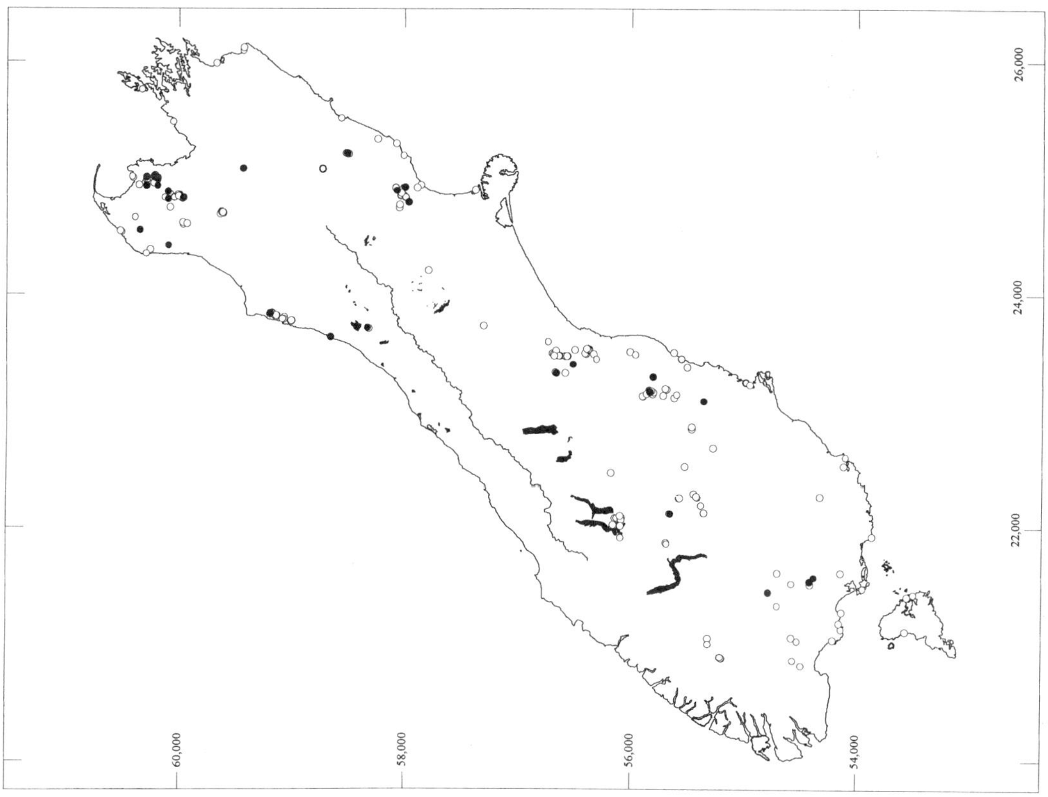

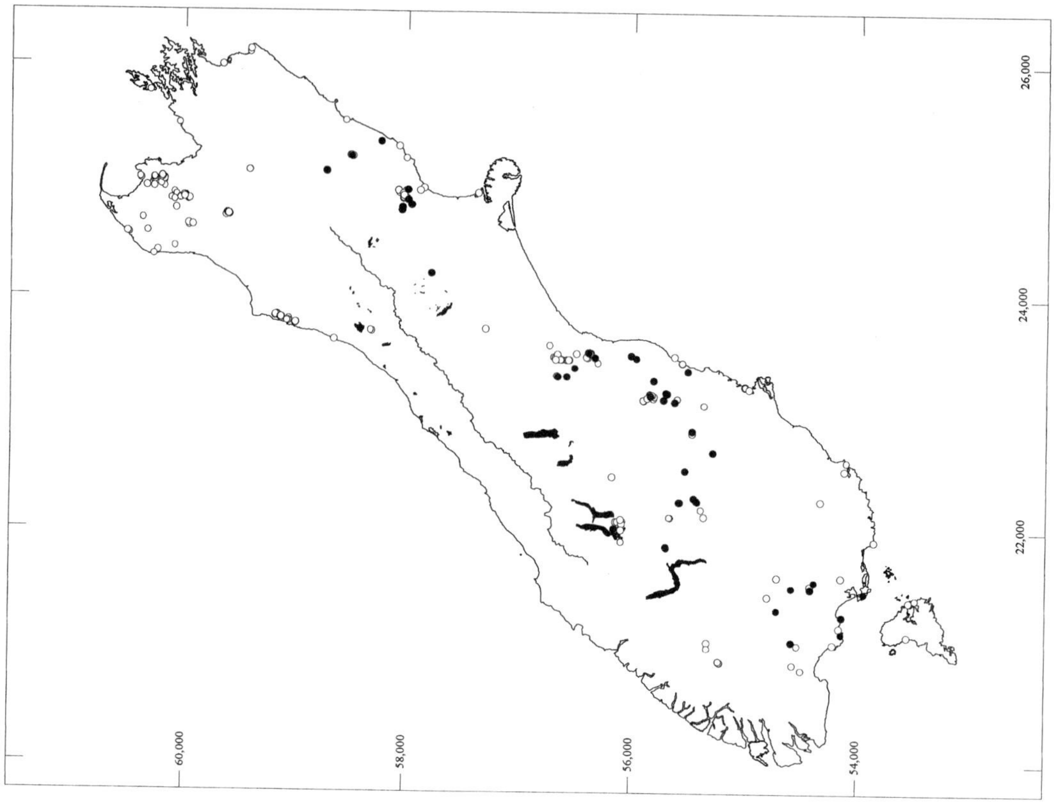

Pachyornis elephantopus

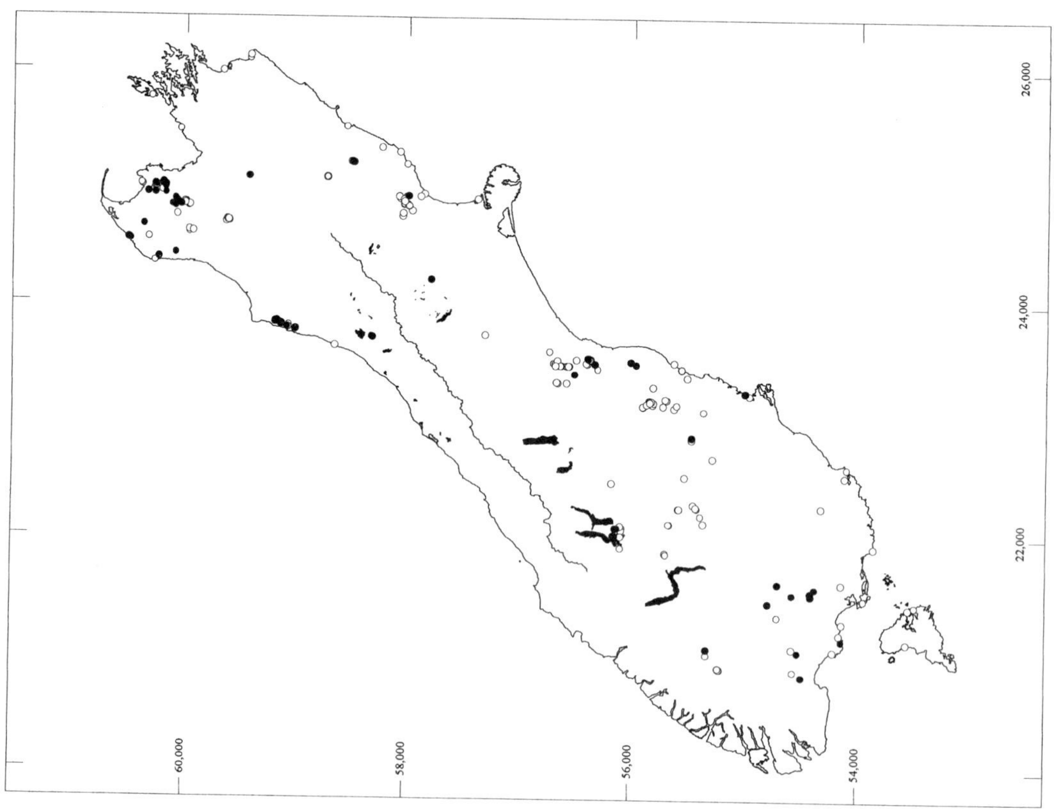

Anomalopteryx didiformis

Euryapteryx geranoides

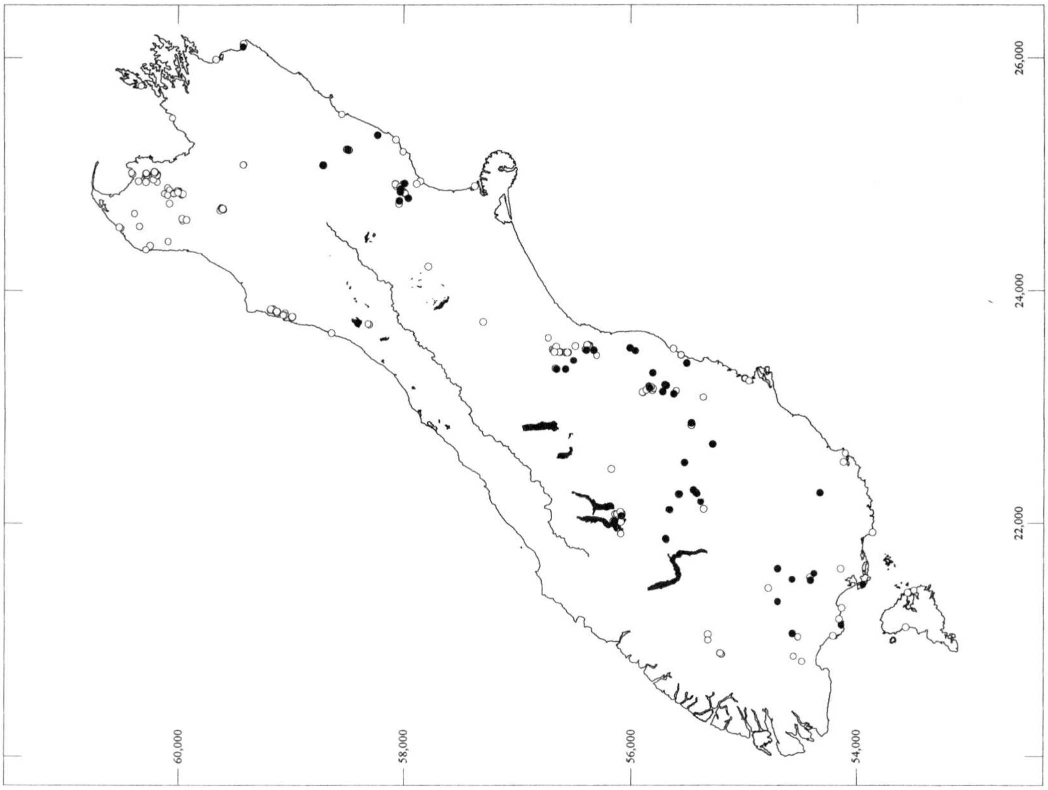

Pachyornis australis

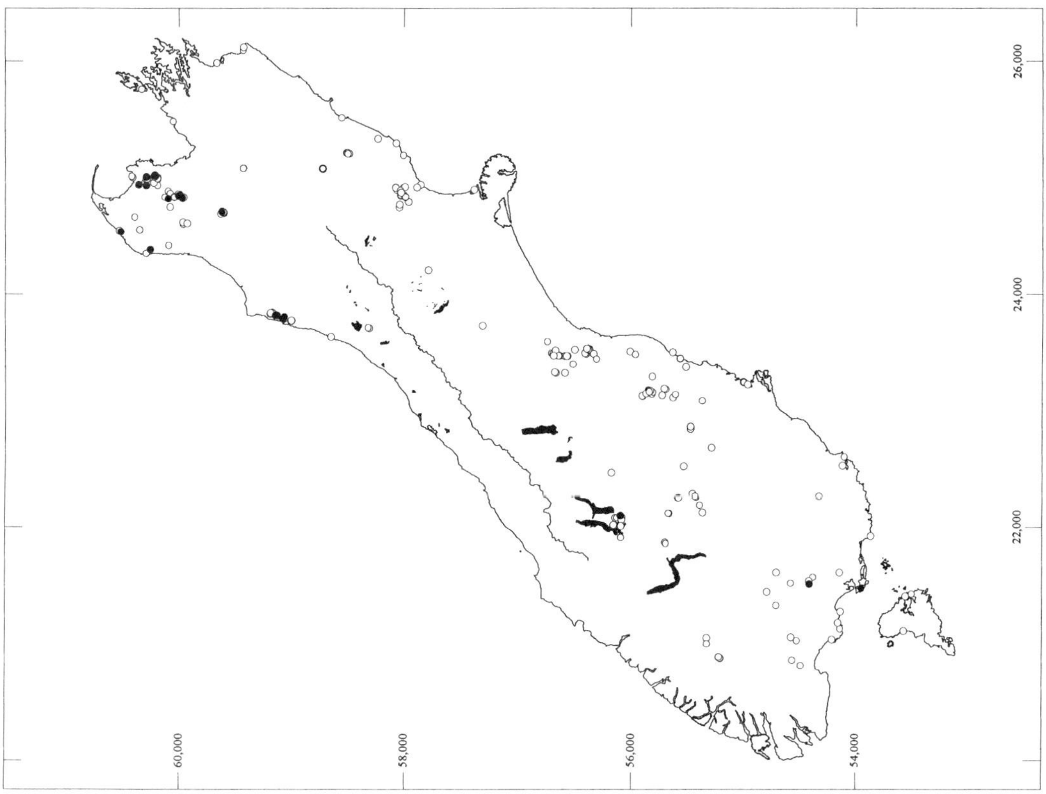

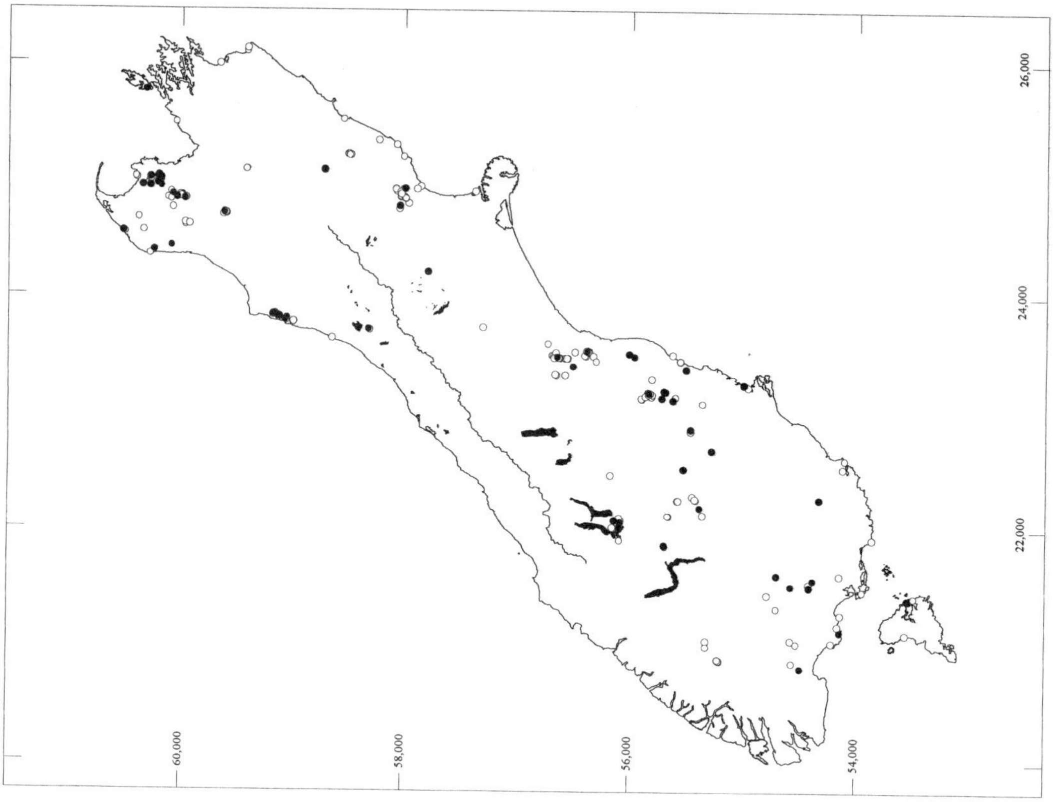

Dinornis struthoides

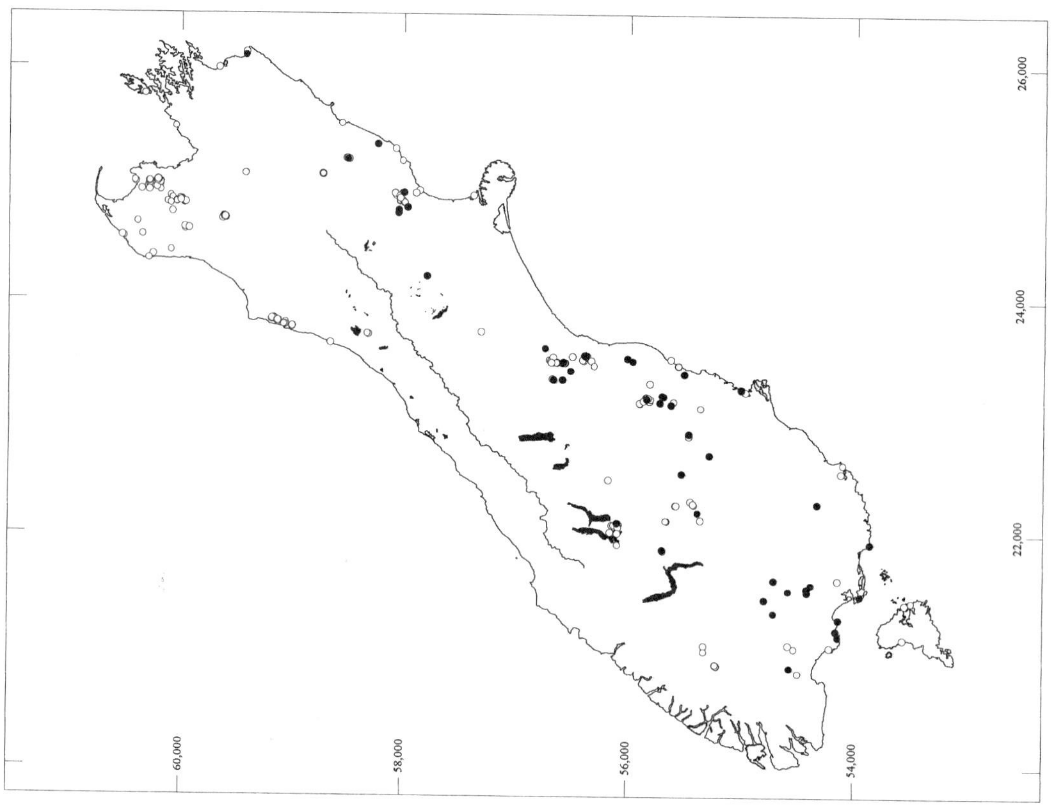

Emeus crassus

Dinornis giganteus

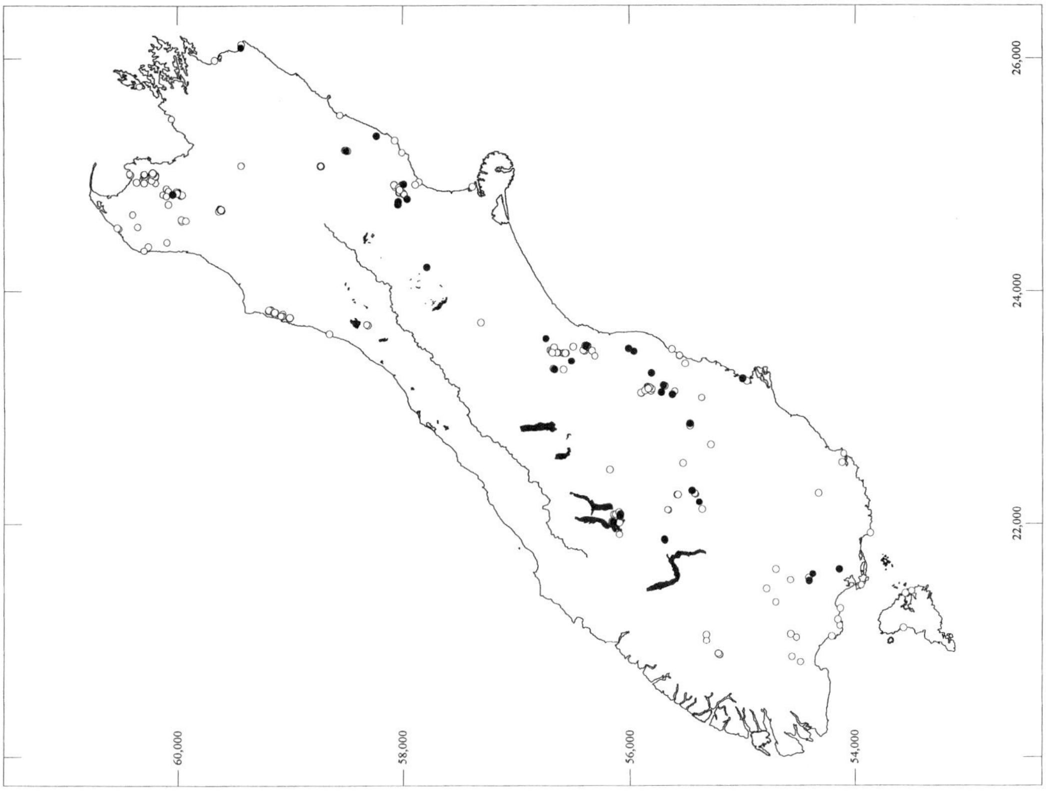

Dinornis novaezealandiae

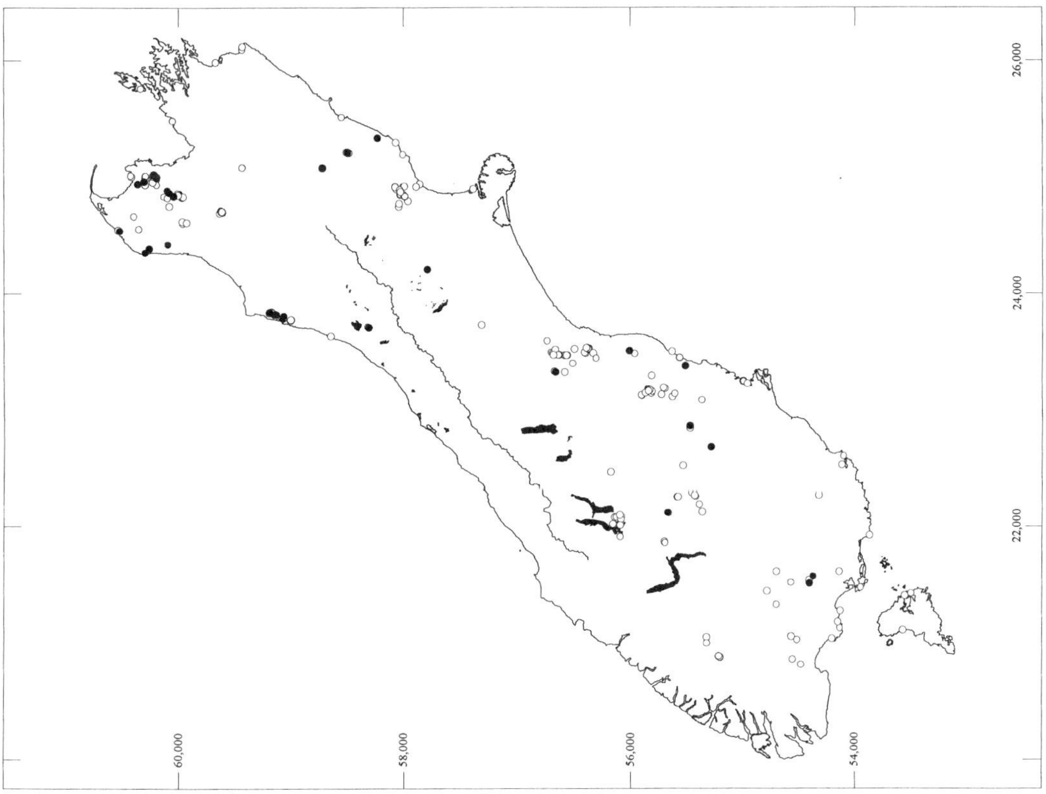

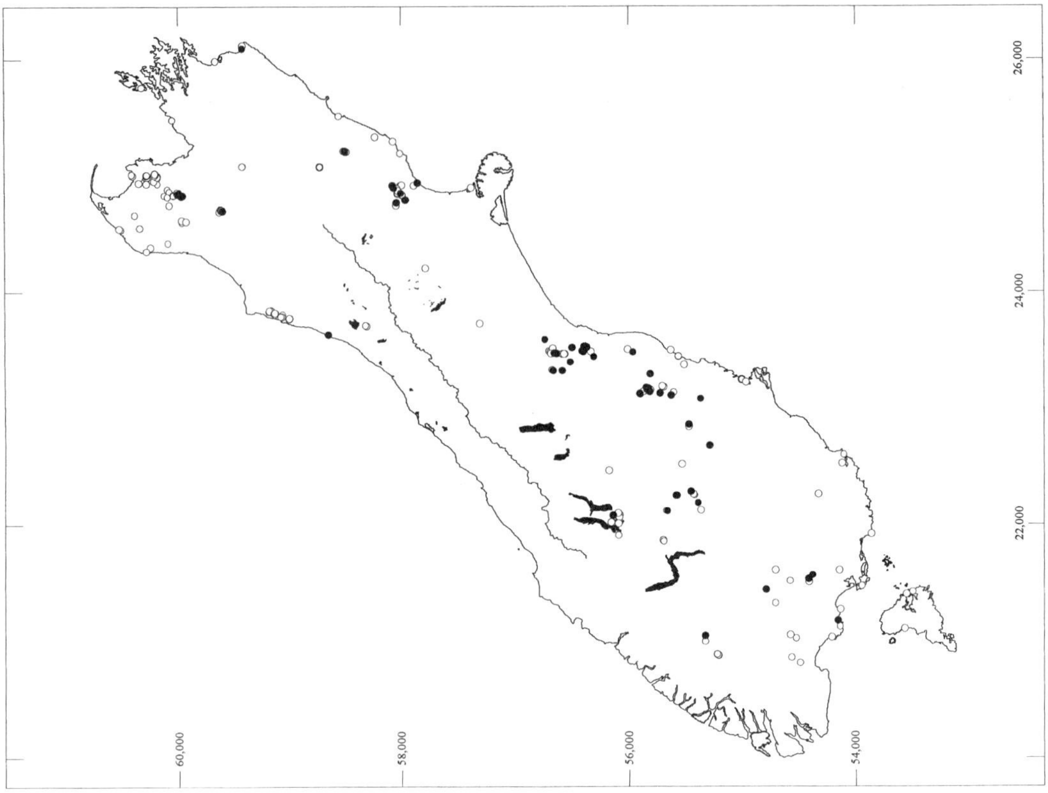

Euryanas finschi

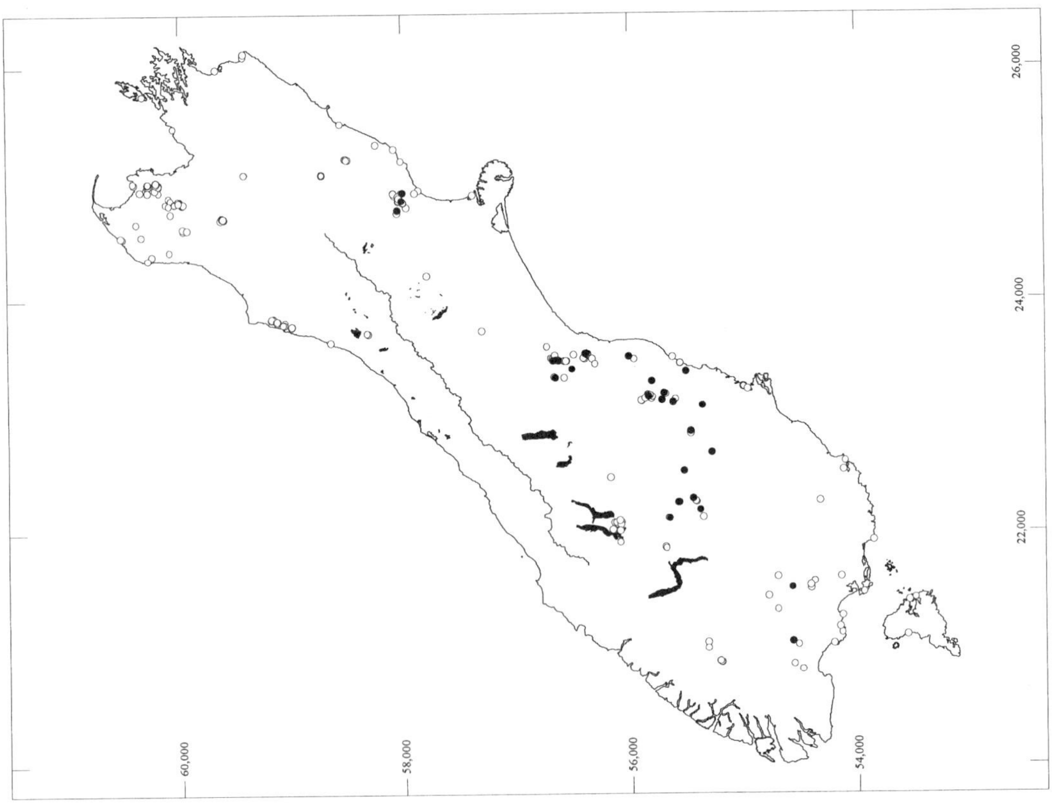

Cnemiornis calcitrans

Coturnix novaezelandiae

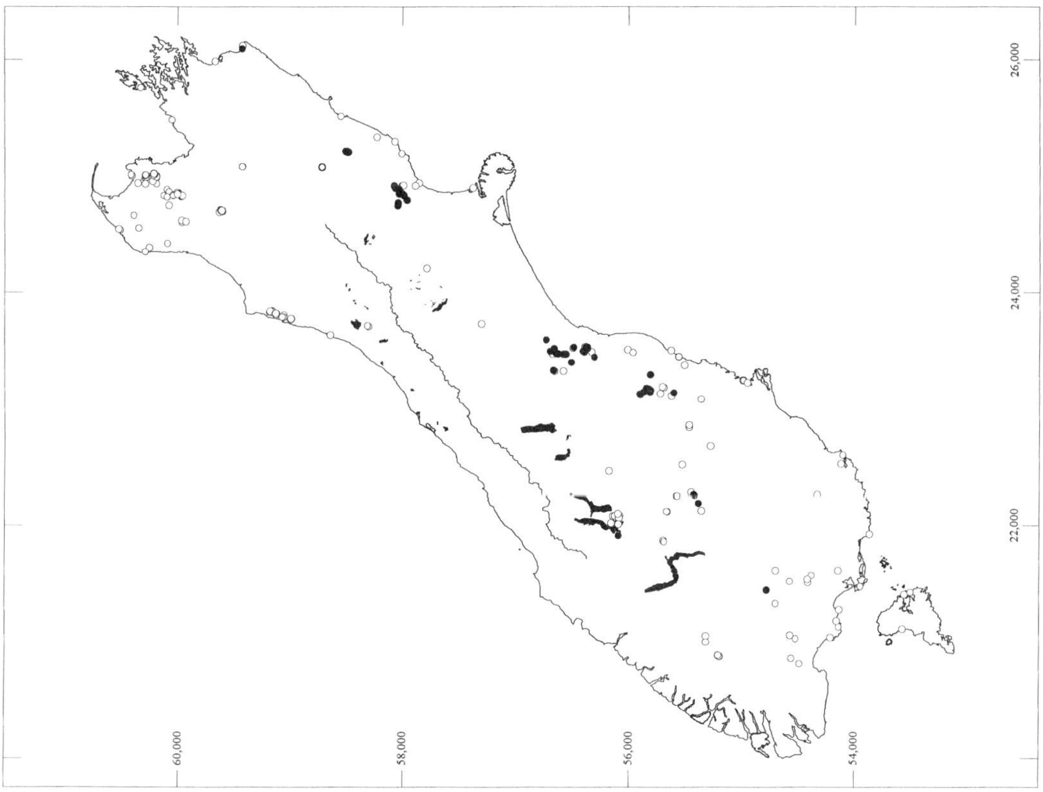

Coenocorypha cf *C. aucklandica*

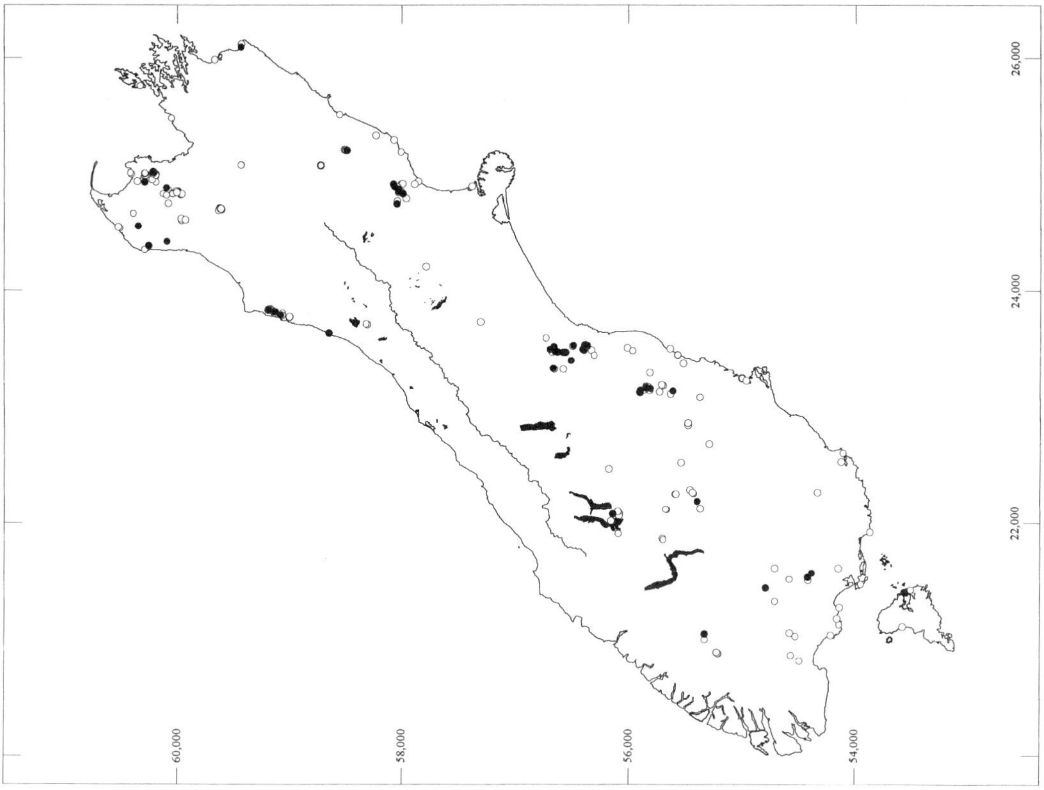

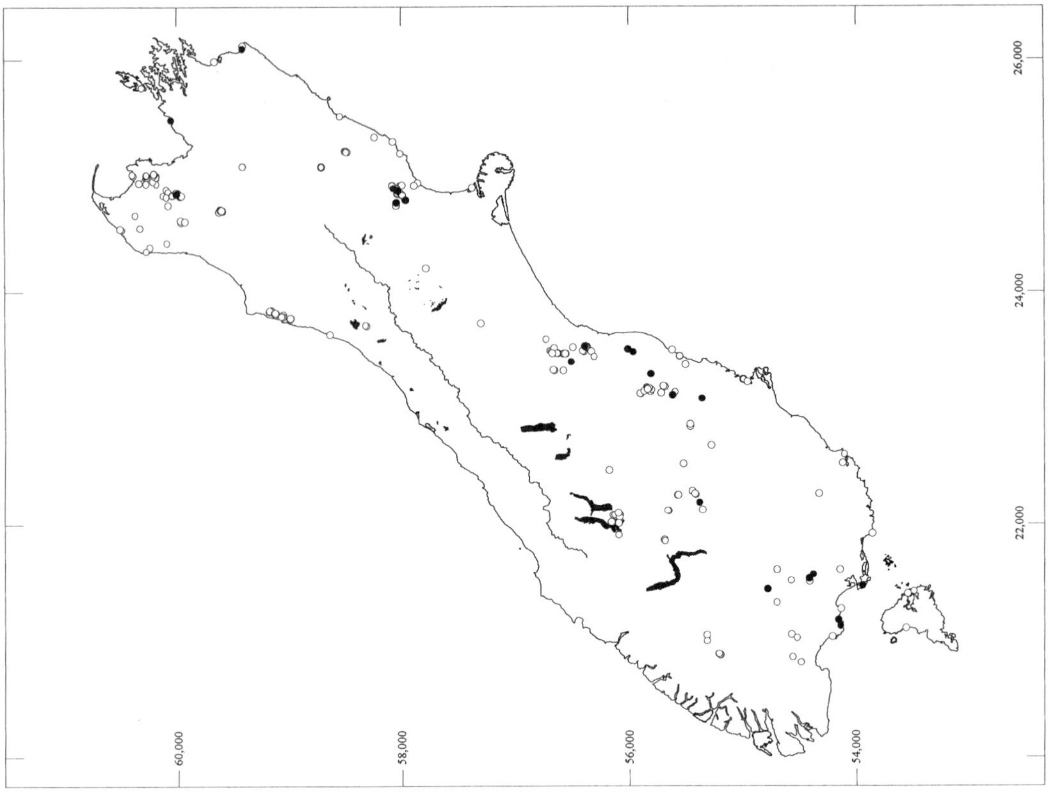

Porphyrio hochstetteri

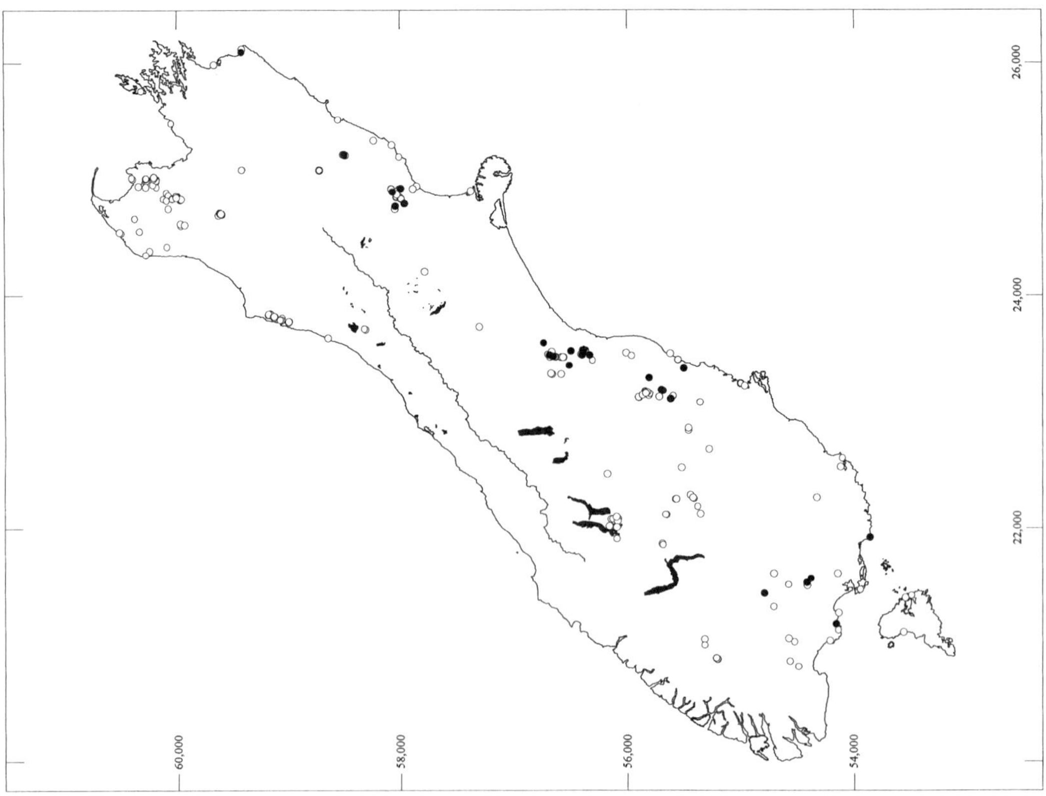

Aptornis defossor

Fulica prisca

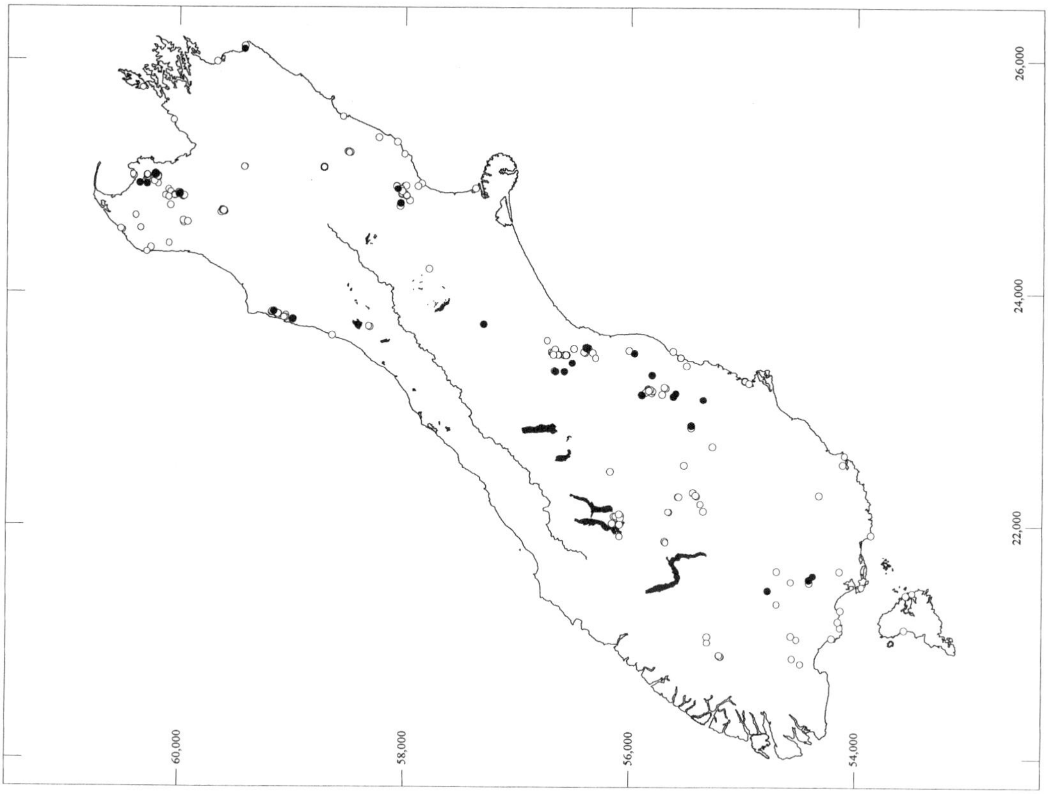

Gallinula hodgenorum

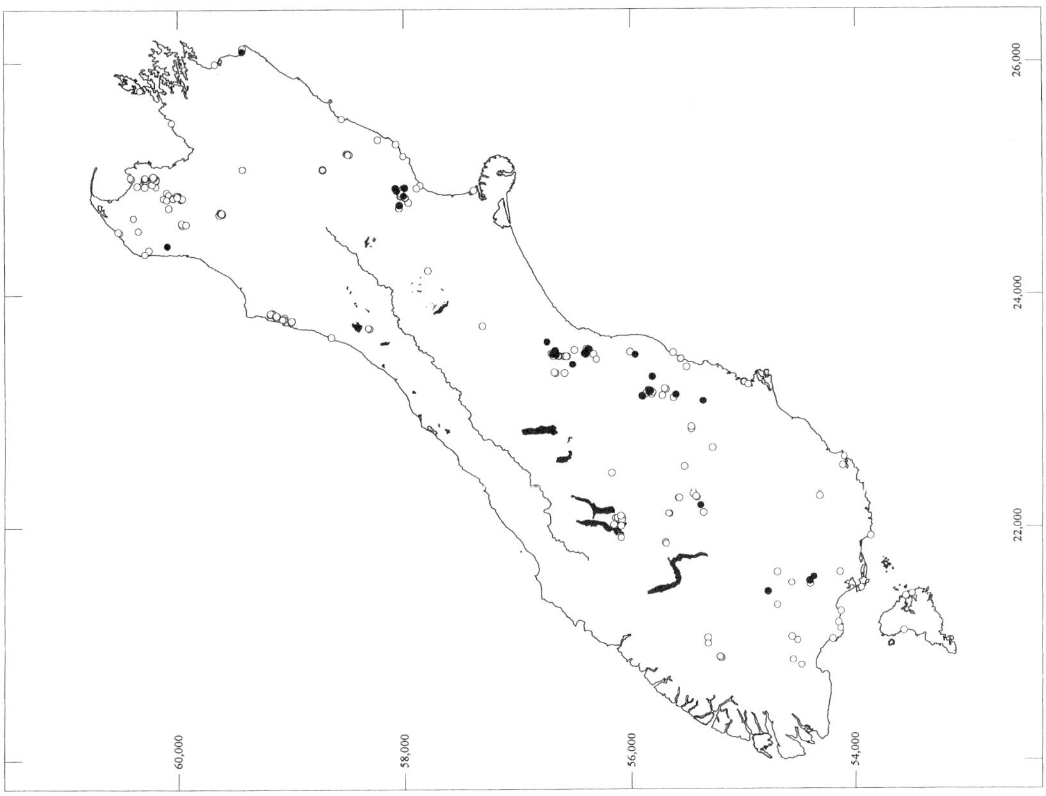

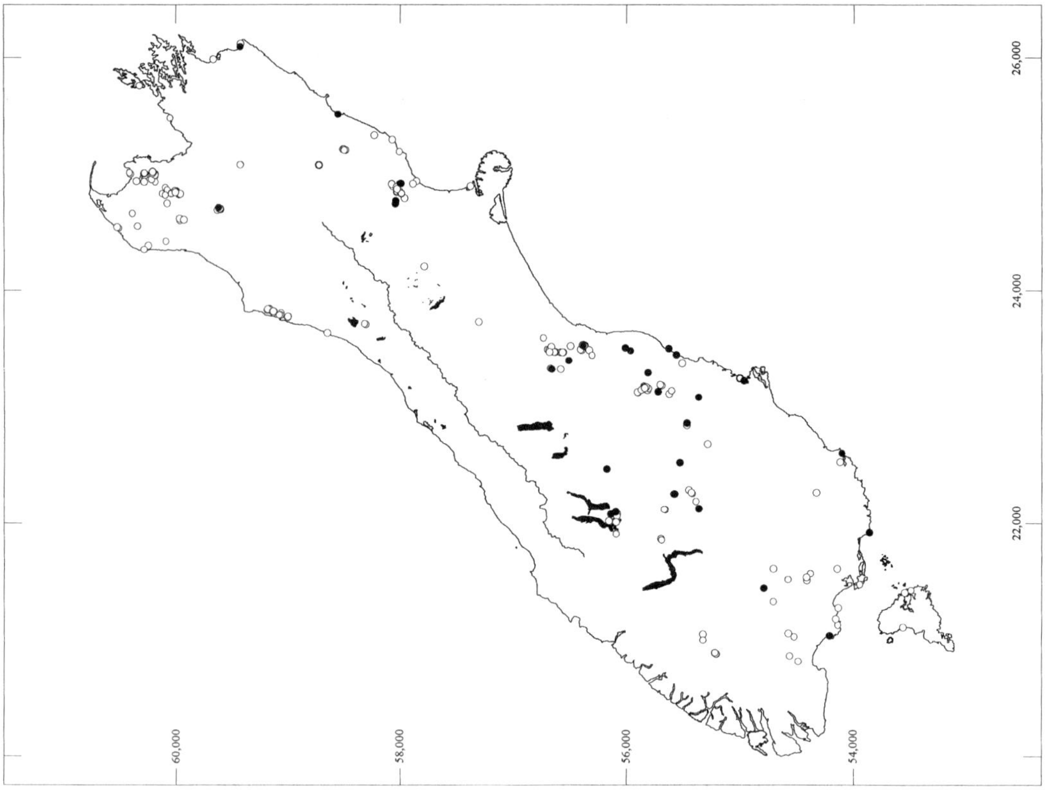

Harpagornis moorei

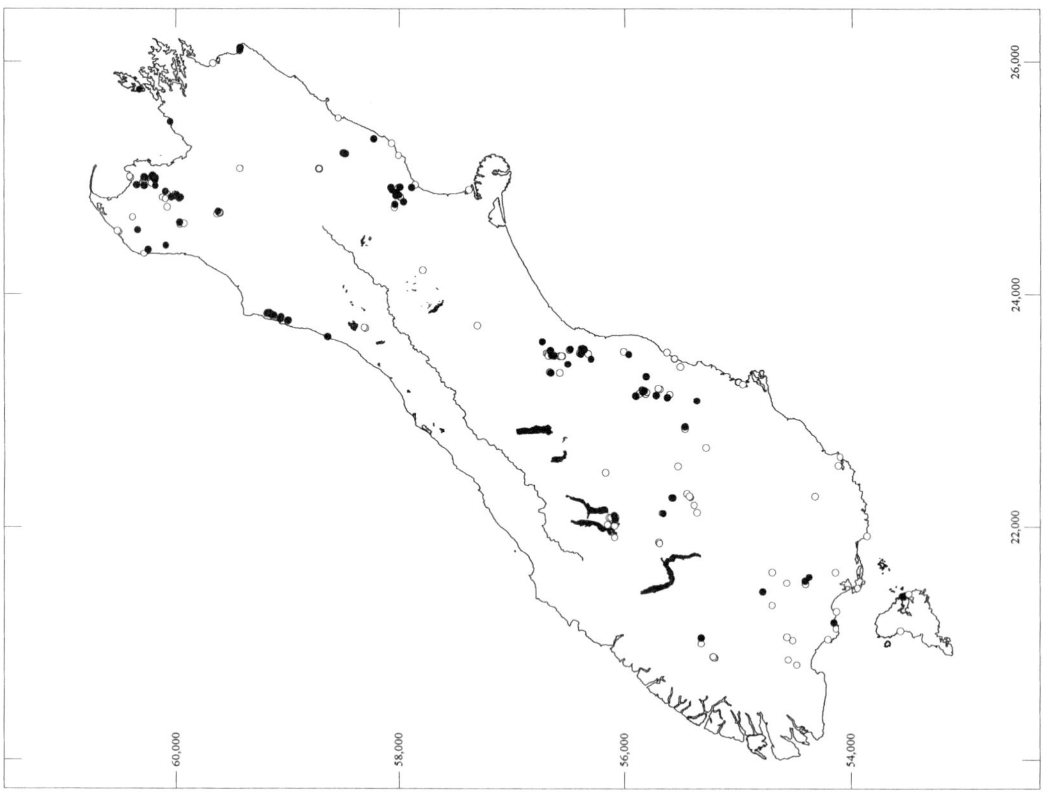

Gallirallus australis

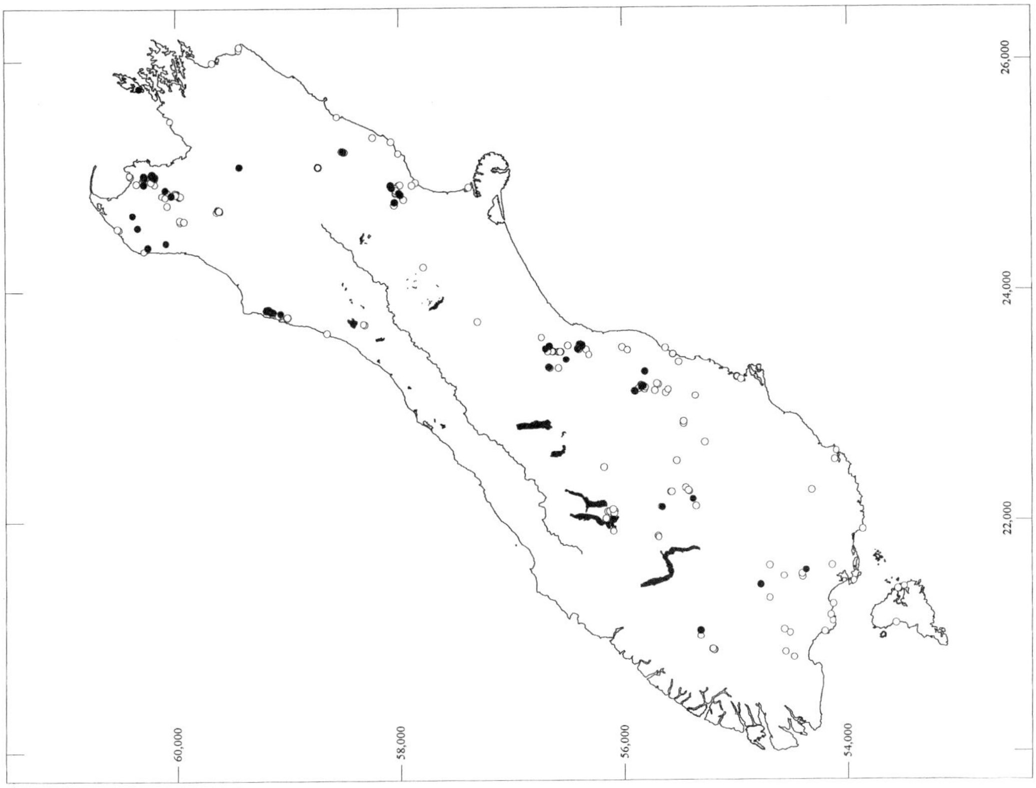

Aegotheles novaezealandiae

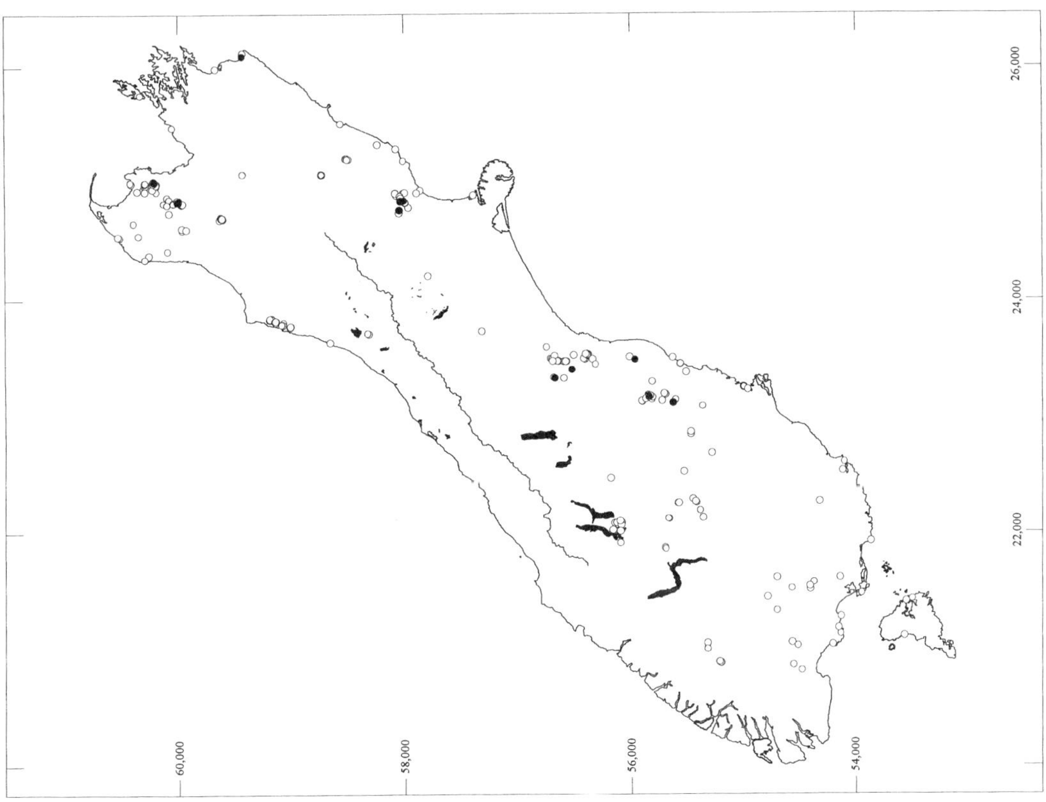

Circus eylesi

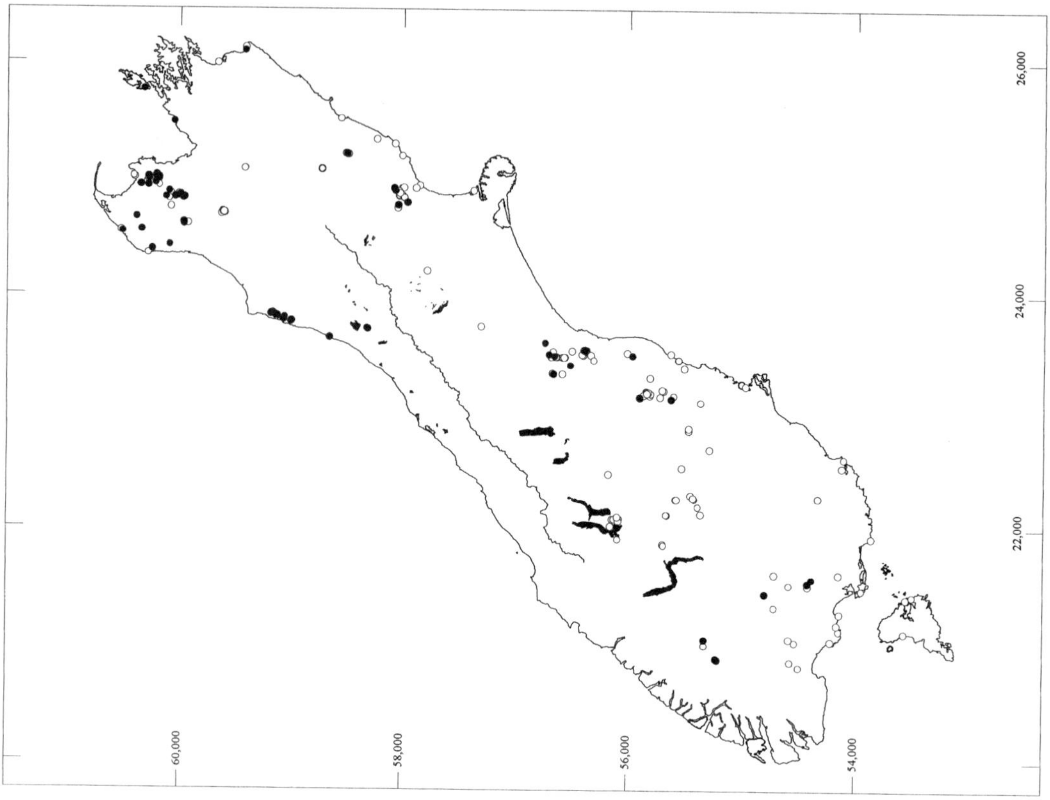

Strigops habroptilus

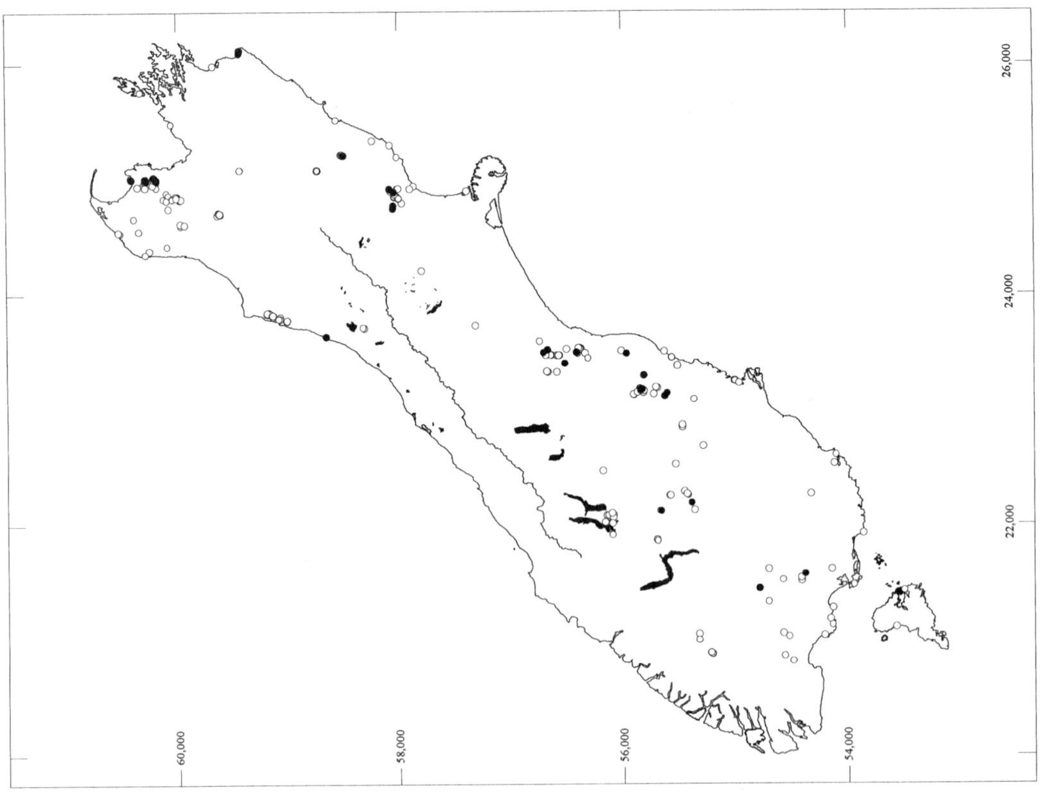

Sceloglaux albifacies

Pachyplichas yaldwyni

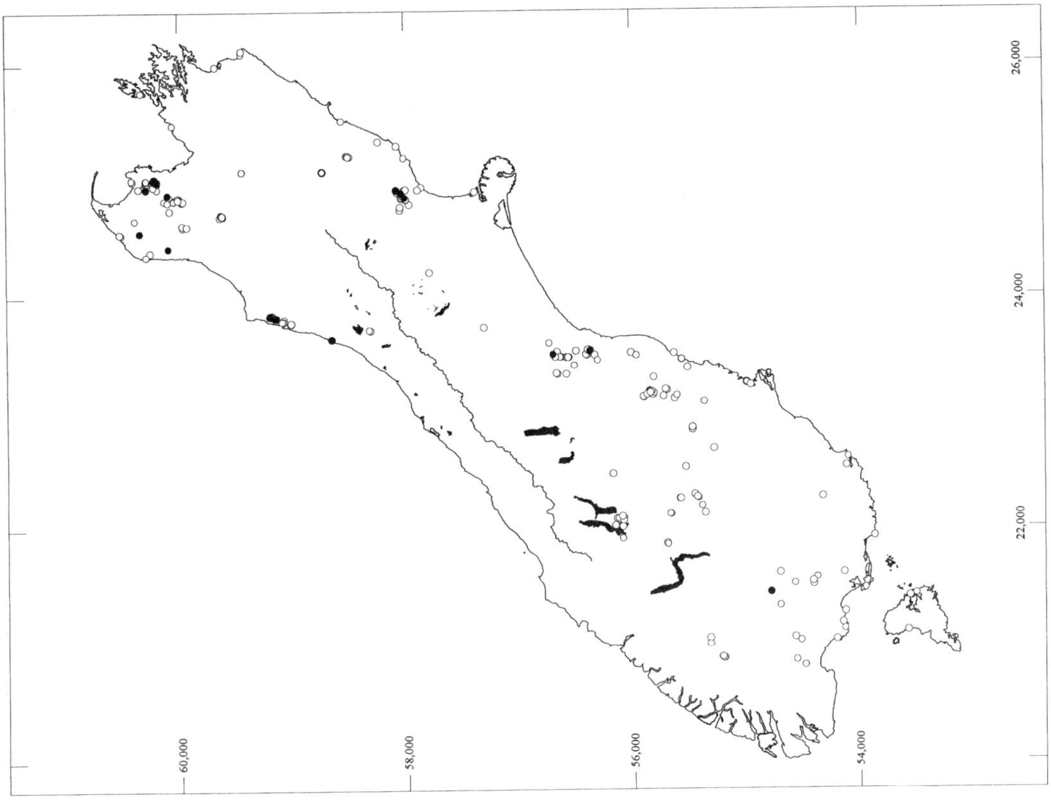

Cyanoramphus spp.

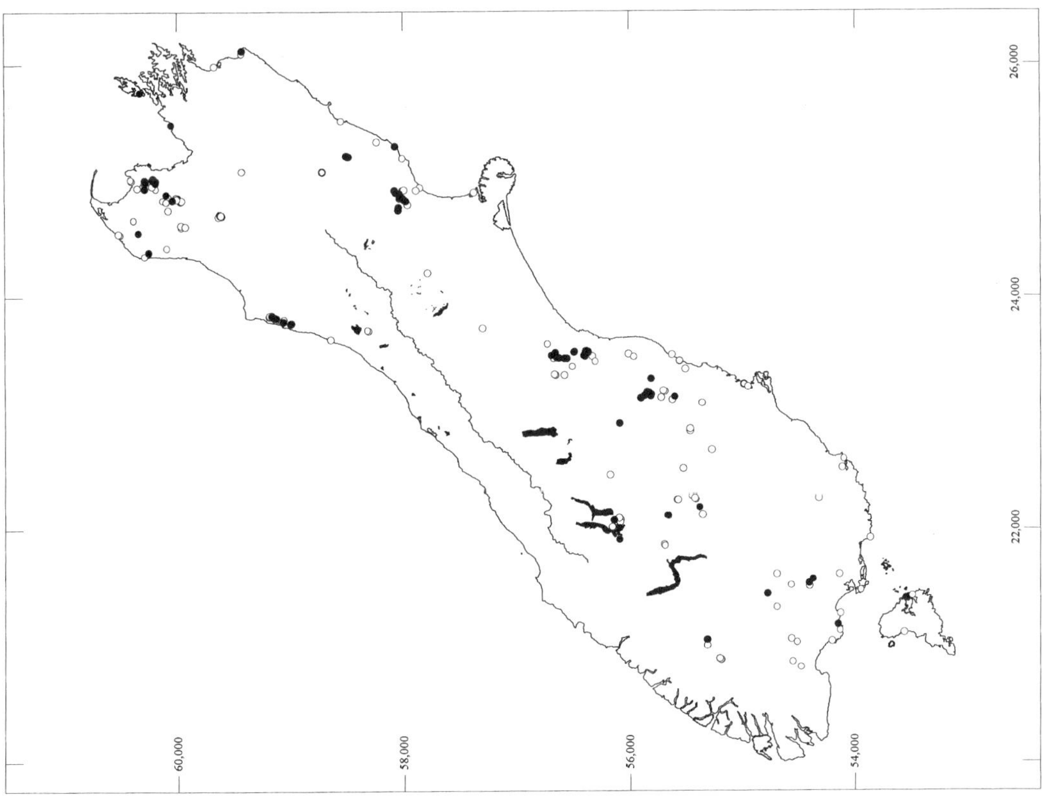

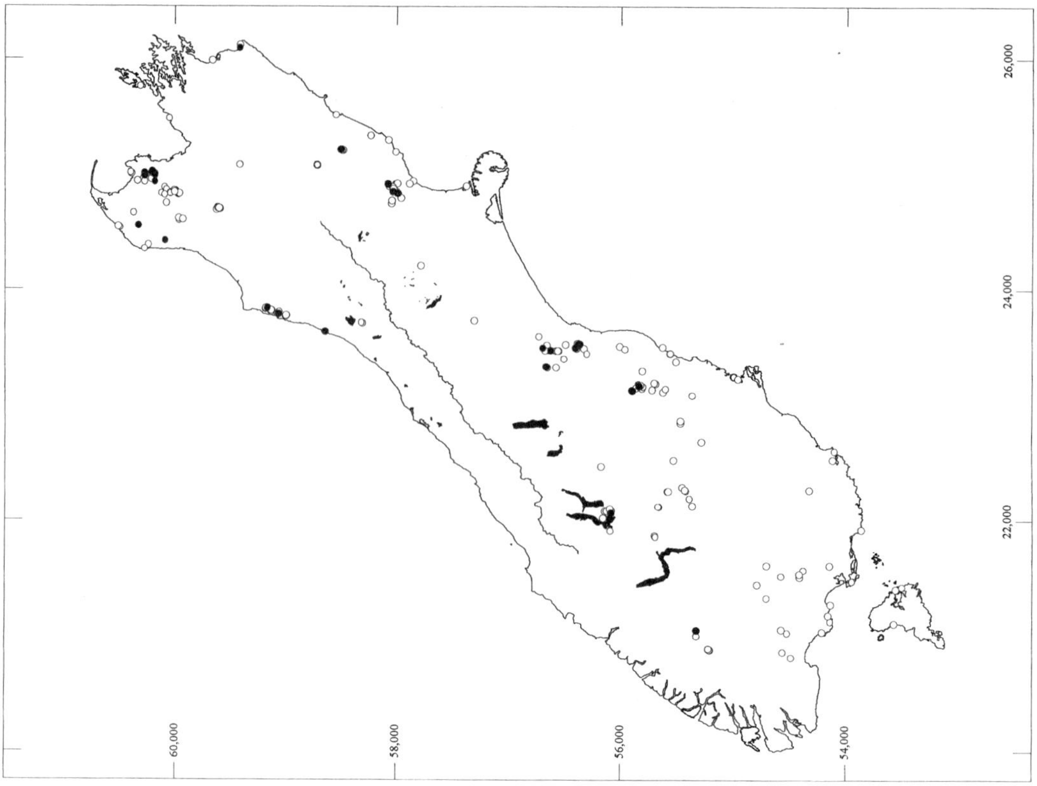

Mohoua ochrocephala

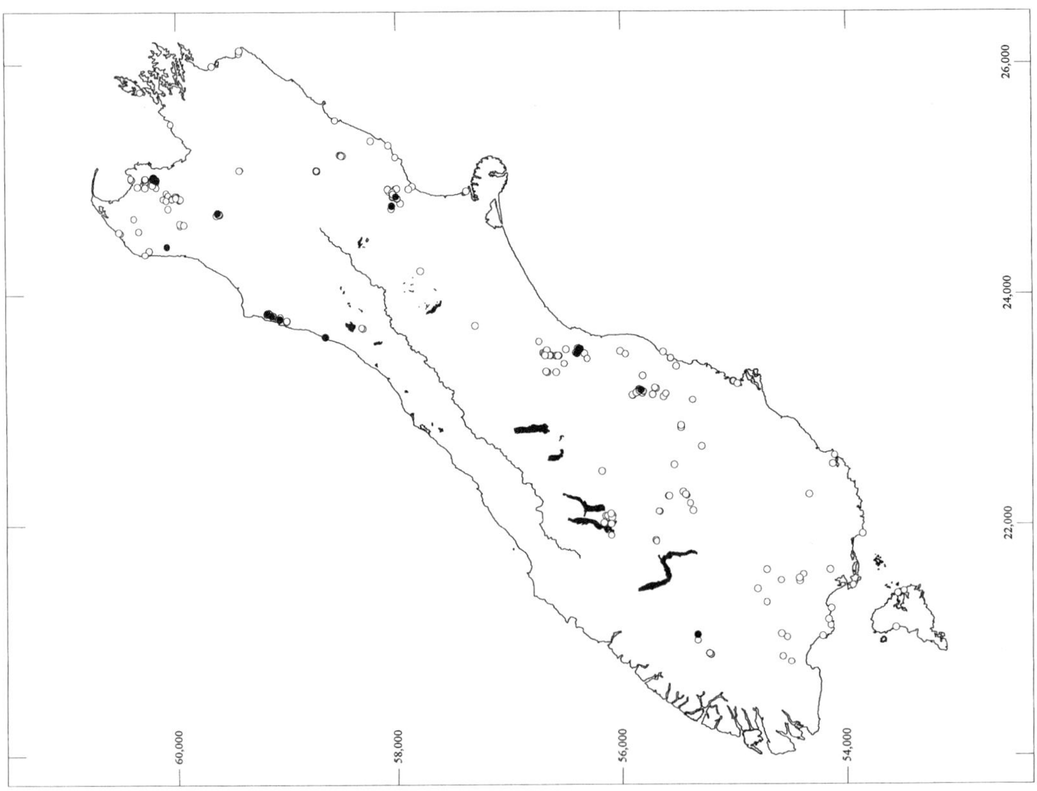

Traversia lyalli

Turnagra capensis

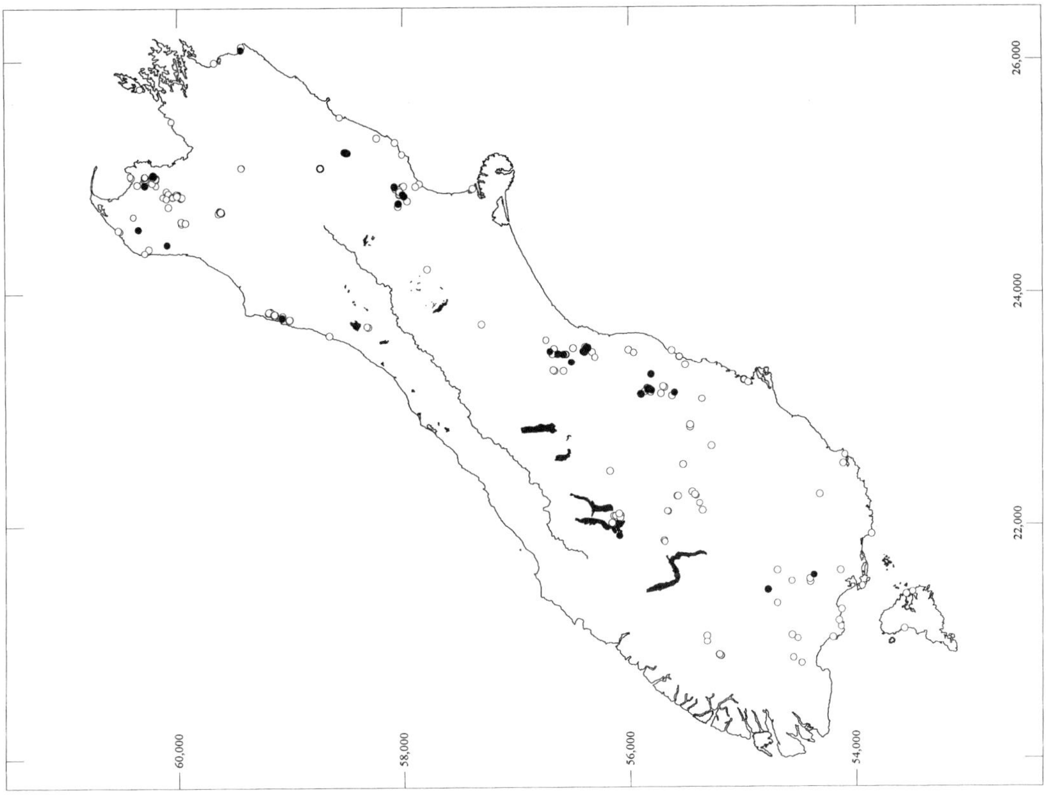

Callaeas cinerea

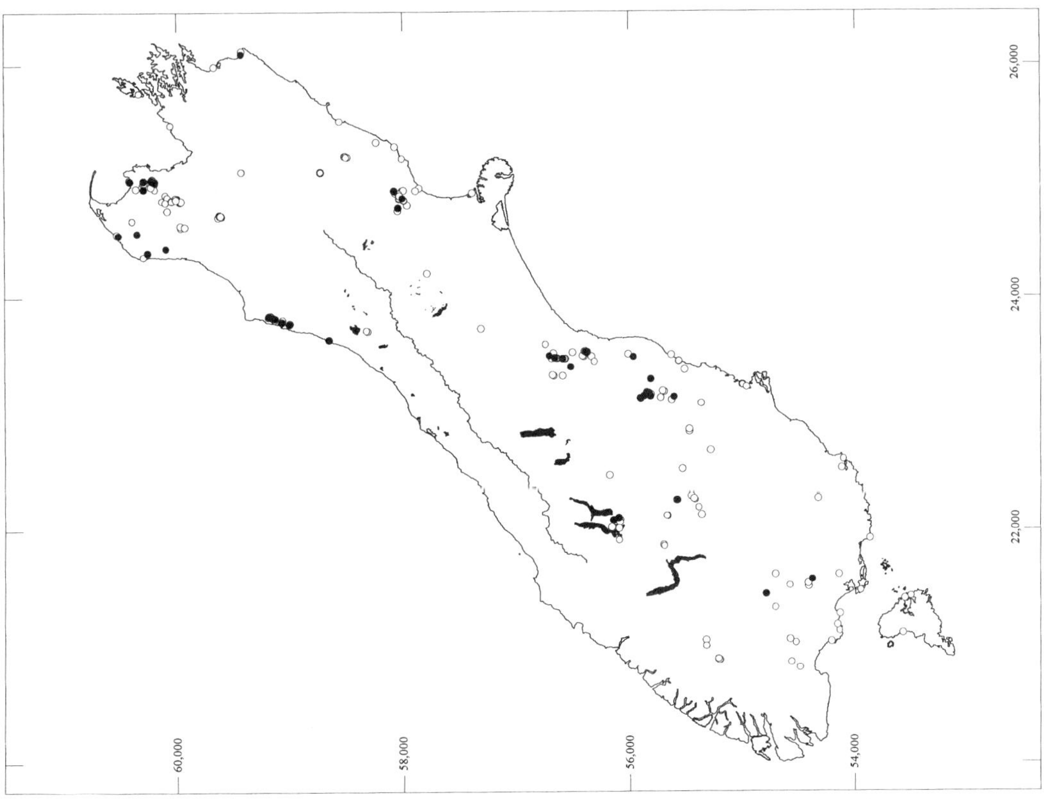

Puffinus spelaeus • and *P. gavia* or *P. huttoni* ■

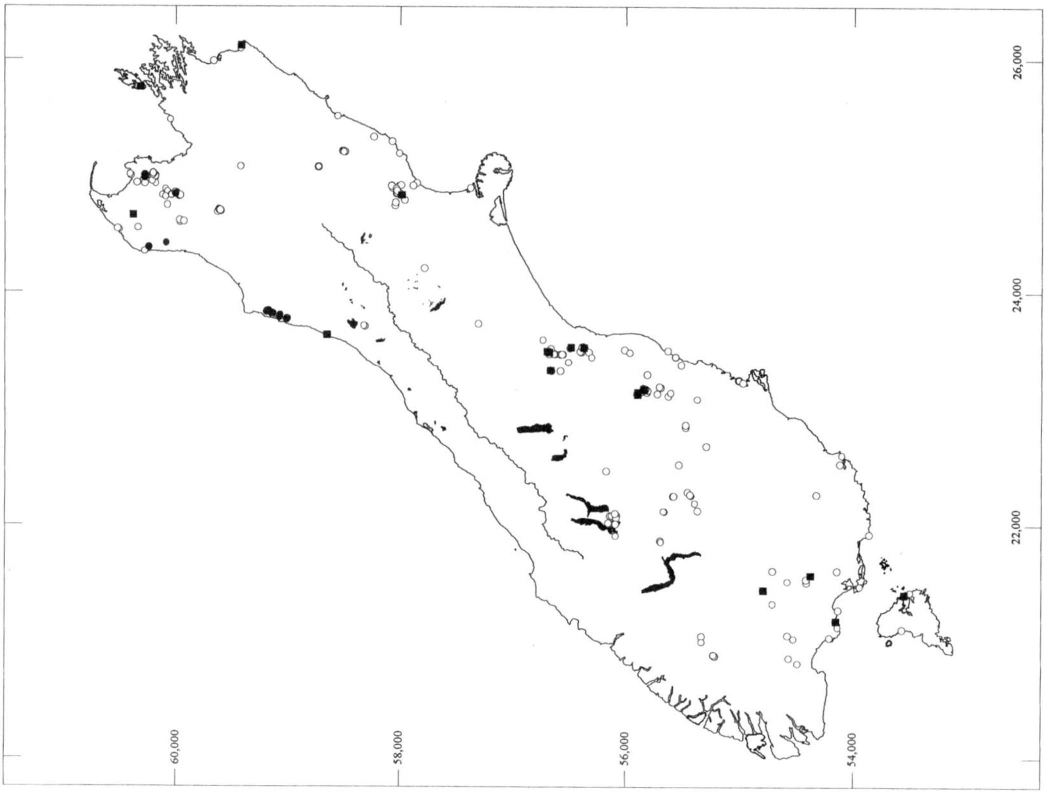

Philesturnus carunculatus

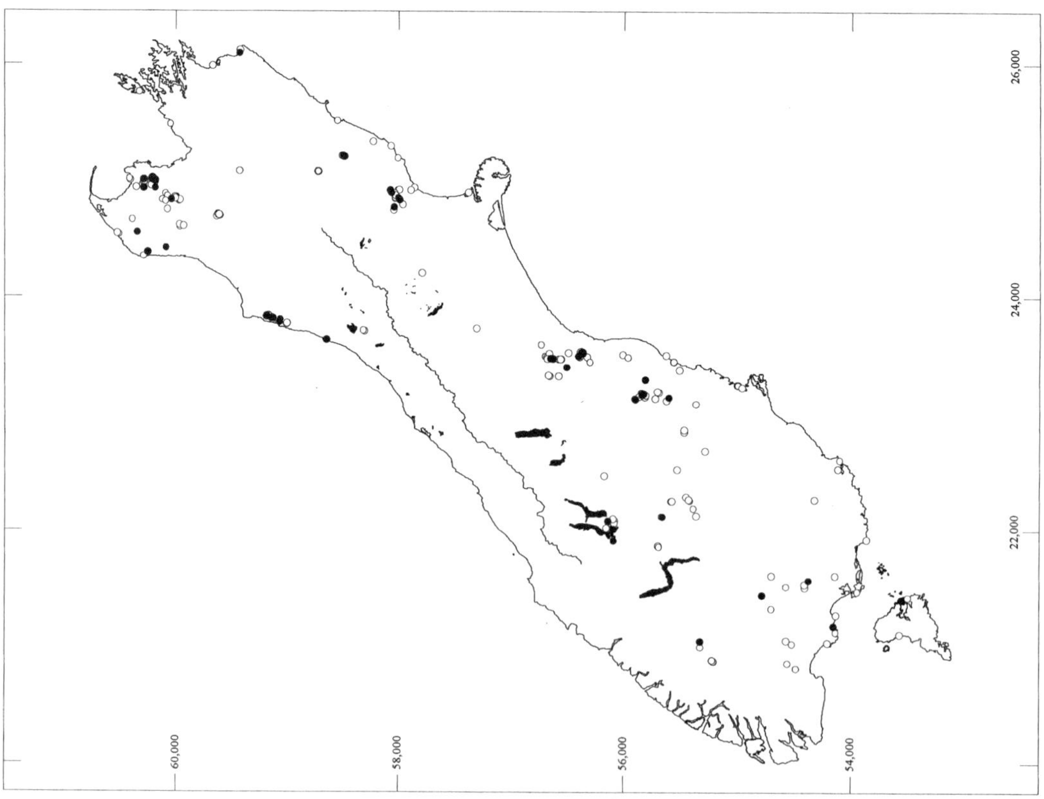

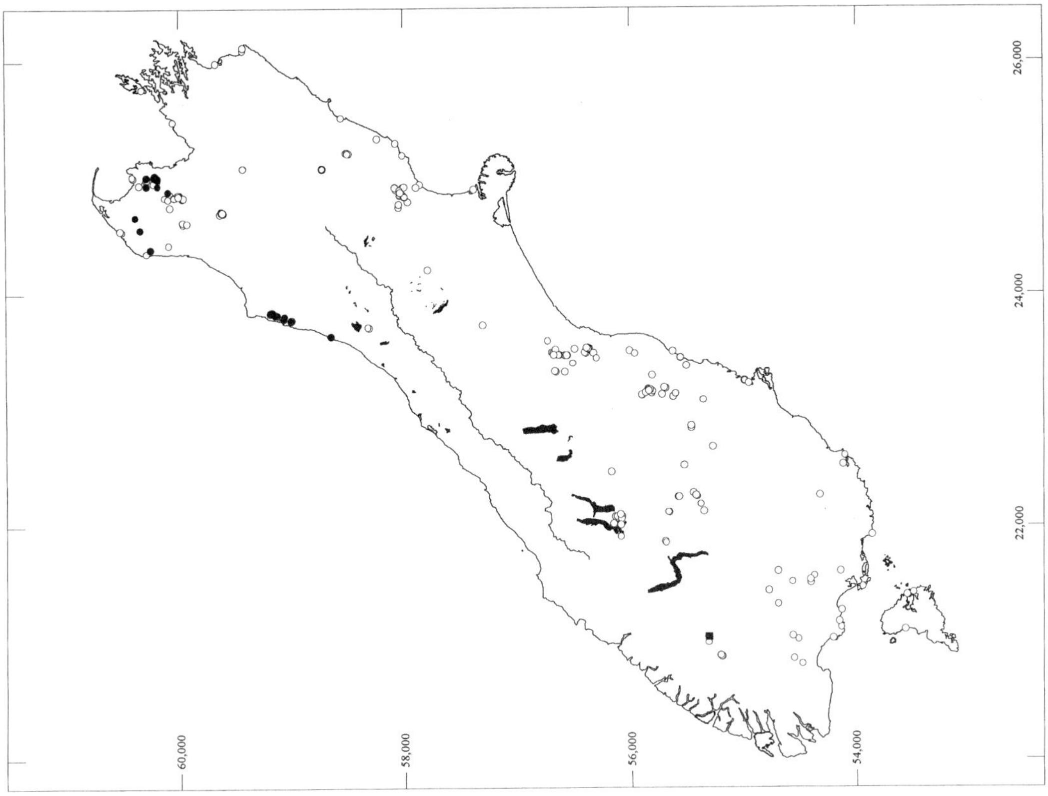

Leiopelma markhami • and *L. auraraensis* ■

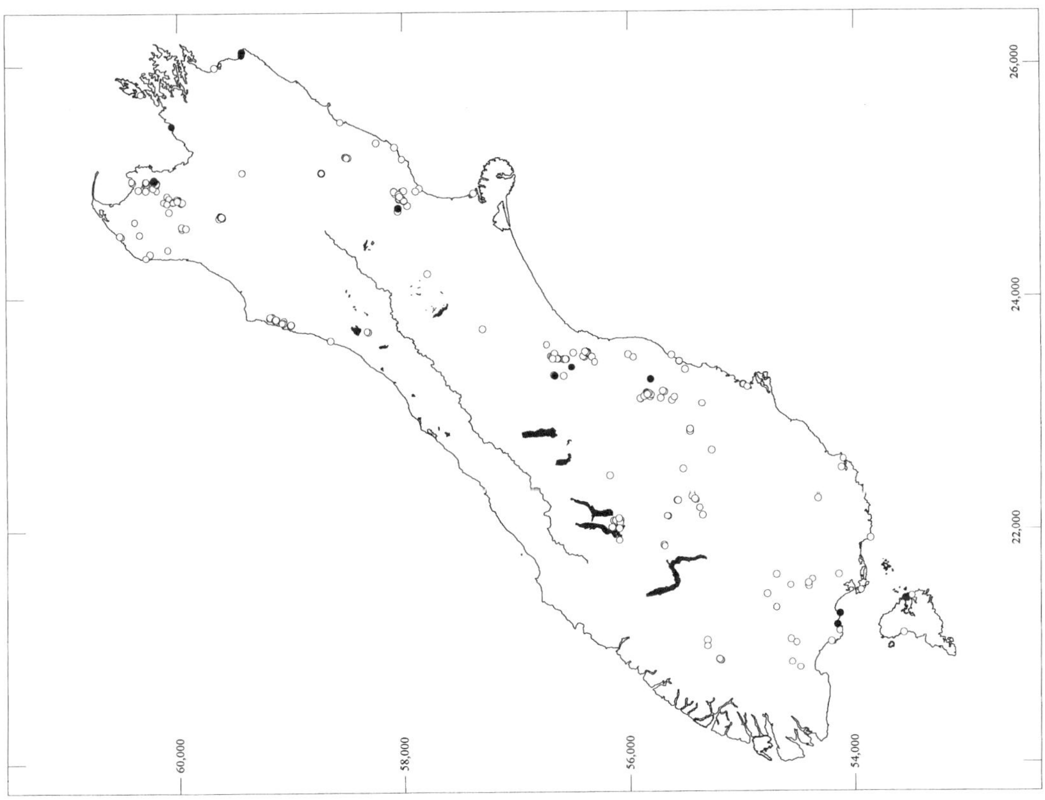

Corvus moriorum

Mystacina robusta • and *M. tuberculata* ▲

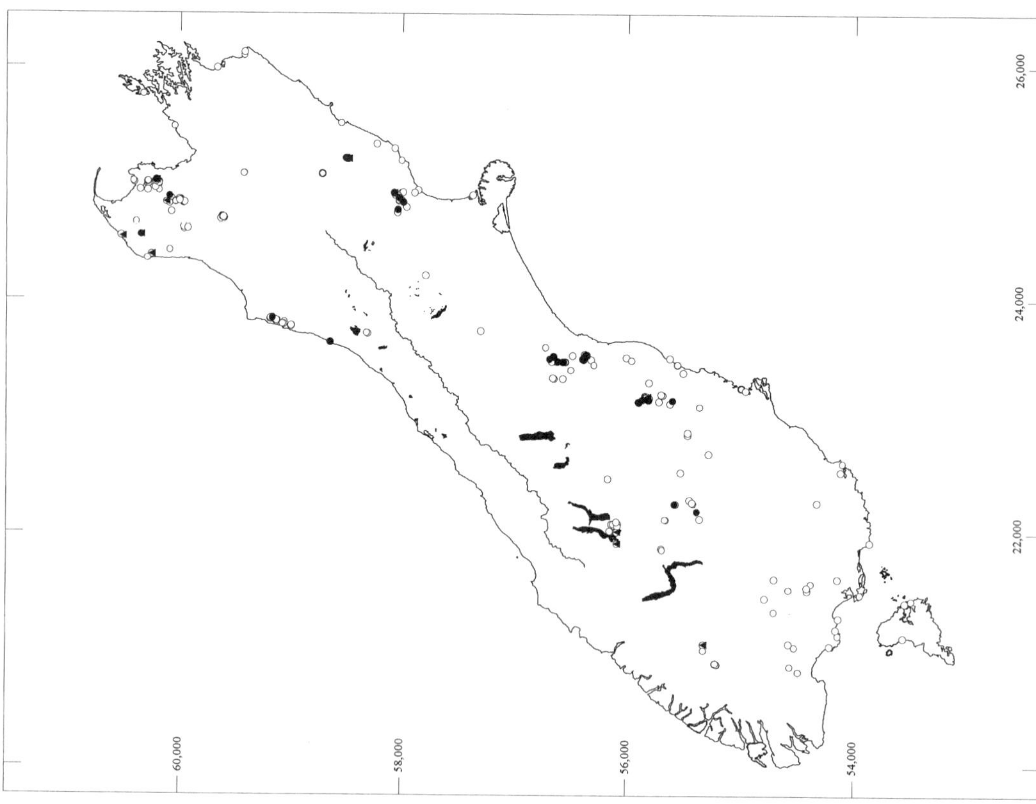

Sphenodon punctatus

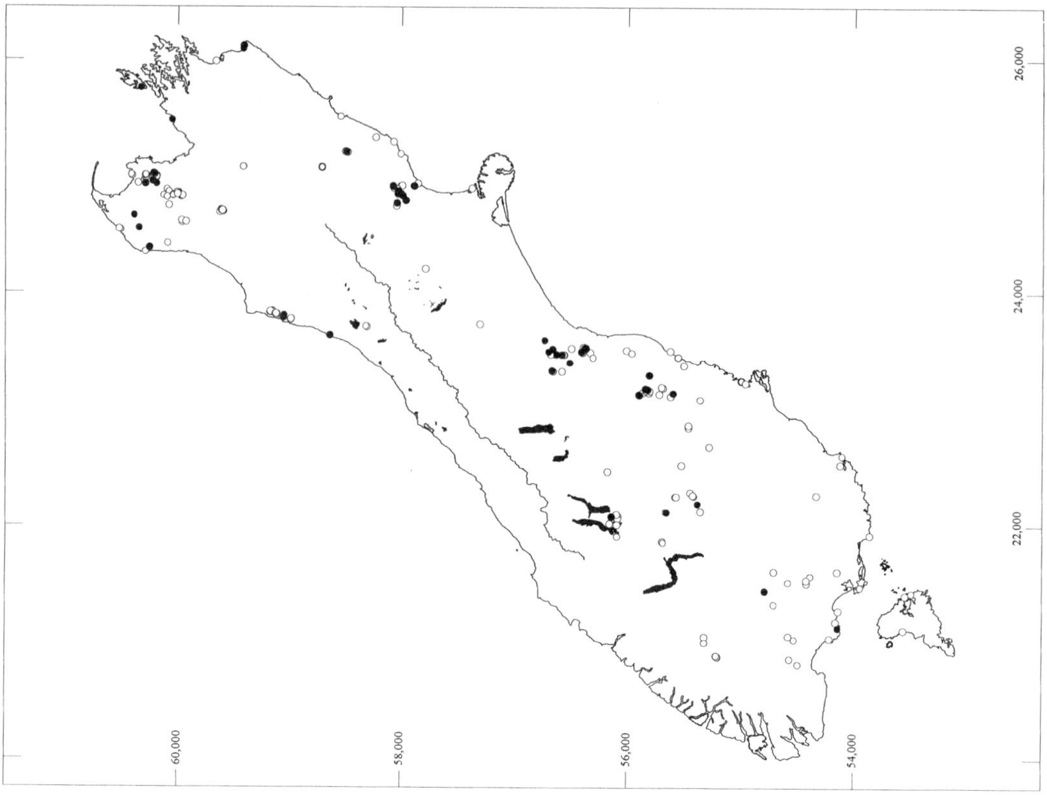

Reference List and Selected Bibliography of New Zealand (Mainly Quaternary) Paleobiology

The literature of the New Zealand Quaternary is large and diverse, and much of it is in, to workers outside New Zealand, relatively obscure, often old, publications. Some reviews (e.g., Baker 1991) have omitted large portions of the recent literature, which deals with many of the long-standing problems of the New Zealand Quaternary biota, thereby giving a biased account of the current state of knowledge. This bibliography attempts to bring together a large, but by no means complete, summary of the available material so that points raised in the other chapters but not explicitly referenced may be followed up.

Anonymous. 1911. The Maori and the moa. *Journal of the Polynesian Society* 20: 54–59.

Anonymous. 1940. More about the mysterious moa. *Ward's Natural Science Bulletin* 14(3–4): 1–3.

Alexander, R. McN. 1976. Estimates of speed for dinosaurs. *Nature* 261: 129–130.

———. 1983a. Allometry of the leg bones of moas (Dinornithes) and other birds. *Journal of Zoology, London* 200: 215–231.

———. 1983b. On the massive legs of a moa (*Pachyornis elephantopus* Dinornithes). *Journal of Zoology, London* 201: 363–376.

———. 1989. Mechanics of fossil vertebrates. *Journal of the Geological Society, London* 46: 41–52.

Allan, R. S., E. Percival, R. S. Duff, and R. A. Falla. 1941. Preliminary report on excavations at Pyramid Valley Swamp, Waikari, North Canterbury. *Records of the Canterbury Museum* 4: 325–353.

Allis, T. 1864a. The recent moa. *Zoologist,* Series 1, 22: 9195–9197.

———. 1864b. Skeleton of a moa. *Zoologist,* Series 1, 22: 9114.

———. 1865a. Notice of a nearly complete skeleton of a *Dinornis*; presented by Mr Gibson to the Museum of the Yorkshire Philosophical Society. *Journal of the Proceedings of the Linnaean Society* 8: 50–52.

———. 1865b. Further note on a skeleton of *Dinornis robustus,* Owen, in the York Museum. *Journal of the Proceedings of the Linnaean Society* 8: 140–1.

Alroy, J. R. 1999. Putting North America's end-Pleistocene megafaunal extinction in context: Large-scale analyses of spatial patterns, extinction rates, and size distributions. In R. D. E. MacPhee (ed.), *Extinc-*

tions in Near Time: Causes, Contexts, and Consequences, pp. 105–143. New York: Kluwer Academic/Plenum Publishers.

Amadon, D. 1947. An estimated weight of the largest known bird. *Condor* 49: 159–163.

Ames, A. L., E. S. Van Fleet, and W. M. Sackett. 1996. The use of stable carbon isotope analysis for determining the dietary habits of the Florida manatee, *Trichechus manatus latirostris*. *Marine Mammal Science* 12: 555–563.

Andersen, J. C. 1926. *Bird-Song and New Zealand Song Birds*. Auckland: Whitcomb & Tombs.

Anderson, A. J. 1981a. A fourteenth-century fishing camp at Purakanui Inlet, Otago. *Journal of the Royal Society of New Zealand* 11: 201–221.

———. 1981b. Pre-European hunting dogs in the South Island, New Zealand. *New Zealand Journal of Archaeology* 3: 15–20.

———. 1982a. Habitat preferences of moa in central Otago, A.D. 1000–1500, according to palaeobotanical and archaeological evidence. *Journal of the Royal Society of New Zealand* 12: 321–336.

———. 1982b. Maori settlement in the interior of southern New Zealand from the early 18th to late 19th centuries A.D. *Journal of the Polynesian Society* 91: 53–80.

———. 1982c. A review of economic patterns during the archaic phase in southern New Zealand. *New Zealand Journal of Archaeology* 4: 45–75.

———. 1983a. Faunal depletion and subsistence change in the early prehistory of southern New Zealand. *Archaeology in Oceania* 18: 1–10.

———. 1983b. The prehistoric hunting of moas (Aves: Dinornithidae) in the high country of southern New Zealand. In C. Grigson and J. Clutton-Brock (eds.), *Animals and Archaeology*, pp. 33–52. Vol. 2, *Shells, Middens and Birds*. Oxford: B.A.R.

———. 1984. The extinction of moa in southern New Zealand. In P. S. Martin and R. G. Klein (eds.), *Quaternary Extinctions: A Prehistoric Revolution*, pp. 728–740. Tucson: Arizona University Press.

———. 1987. The first recorded name for moa. *Journal of the Royal Society of New Zealand* 17: 421–422.

———. 1988. Moa extinctions in southern New Zealand: A reply to Sutton. *Archaeology in Oceania* 23: 78–79.

———. 1989a. Mechanics of overkill in the extinction of New Zealand moas. *Journal of Archaeological Science* 16: 137–151.

———. 1989b. On evidence for the survival of moa in European Fiordland. *New Zealand Journal of Ecology* 12(Suppl.): 39–44.

———. 1989c. *Prodigious Birds: Moas and Moa-Hunting in Prehistoric New Zealand*. Cambridge: Cambridge University Press.

———. 1990. The last archipelago: 1000 years of Maori settlement in New Zealand. In A. J. Anderson et al. (eds.), *Towards 1990: Seven Leading Historians Examine Significant Aspects of New Zealand History*, pp. 1–19. Wellington: Government Printer.

———. 1991. The chronology of colonization in New Zealand. *Antiquity* 65: 767–795.

———. 1996. Origins of Procellariidae hunting in the Southwest Pacific. *International Journal of Osteoarchaeology* 6: 403–410.

———. 1997a. Historical and archaeological aspects of muttonbirding in New Zealand. *New Zealand Journal of Archaeology* 17(1995): 35–55.

———. 1997b. Prehistoric Polynesian impact on the New Zealand environment: Te Whenua Hou. In P. V. Kirch and T. L. Hunt (eds.), *Histori-*

cal Ecology in the Pacific Islands: Prehistoric Environmental and Landscape Change, pp. 271–283. New Haven, Conn.: Yale University Press.

Anderson, A. J., and M. S. McGlone. 1992. Living on the edge—Prehistoric land and people in New Zealand. In J. Dodson (ed.), *The Naive Lands: Prehistory and Environmental Change in Australia and the Southwest Pacific,* pp. 199–241. London: Longman Cheshire.

Anderson, A. J., and R. McGovern-Wilson. 1990. The pattern of prehistoric Polynesian colonisation in New Zealand. *Journal of the Royal Society of New Zealand* 20: 41–63.

Anderson, A. J., and I. G. Smith. 1992. The Papatowai site: New evidence and interpretations. *Journal of the Polynesian Society* 101: 129–157.

Anderson, A. J., T. H. Worthy, and R. McGovern-Wilson. 1996. Moas remains and taphonomy. In A. Anderson, B. Allingham, and I. G. Smith (eds.), *Shag River Mouth: The Archaeology of an Early Maori Village,* pp. 200–213. Research Papers in Archaeology and Natural History, No. 27. Canberra: ANH Publications, RSPAS, Australian National University.

Anderson, C. C., and J. H. Sorenson. 1944. Chronic osteomyelitis occurring in *Emeus crassus. Transactions of the Royal Society of New Zealand* 74: 182–184.

Anderson, J. F., H. Rahn, and H. D. Prange. 1979. Scaling of supportive tissue mass. *Quarterly Review of Biology* 54: 139–148.

Anderson, J. F., A. Hall-Martin, and D. A. Russell. 1985. Long-bone circumference and weight in mammals, birds and dinosaurs. *Journal of Zoology, London* 207: 53–61.

Anderson, S. H. 1992. Shearwaters and saddlebacks as prey at a morepork ground nest. *Notornis* 39: 69–70.

Andors, A. V. 1992. Reappraisal of the Eocene groundbird *Diatryma* (Aves: Anserimorphae). In K. E. Campbell (ed.), *Papers in Avian Paleontology Honoring Pierce Brodkorb,* pp. 109–125. Natural History Museum of Los Angeles County Science Series, No. 36. Los Angeles, Calif.: Natural History Museum of Los Angeles County.

Andrews, C. W. 1896a. On the extinct birds of the Chatham Islands. Part 1.—The osteology of *Diaphorapteryx hawkinsi. Novitates Zoologicae* 3: 73–84.

———. 1896b. On the extinct birds of the Chatham Islands. Part 2.—The osteology of *Palaeolimnas chathamensis* and *Nesolimnas* (gen. nov.) *dieffenbachii. Novitates Zoologicae* 3: 260–271.

———. 1896c. Note on a nearly complete skeleton of *Aptornis defossor* (Owen). *Geological Magazine,* n.s., 3: 241–242.

Andrews, J. R. H. 1986. *The Southern Ark: Zoological Discoveries in New Zealand 1769–1900.* Auckland: Century Hutchinson.

Andrews, P. 1990. *Owls, Caves and Fossils. Predation, Preservation, and Accumulation of Small Mammal Bones in Caves, with an Analysis of the Pleistocene Cave Faunas from Westbury-Sub-Mendip, Somerset, UK.* Chicago: University of Chicago Press.

Archey, G. 1927. On a moa skeleton from Amodeo Bay and some moa bones from Karamu. *Transactions of the New Zealand Institute* 58: 151–156.

———. 1931. Skeleton of an Australian pelican found with moa bones. *Records of the Auckland Institute and Museum* 1: 116–121.

———. 1941. *The Moa: A Study of the Dinornithiformes.* Bulletin of the Auckland Institute and Museum, No. 1. Auckland: Unity Press.

Arredondo, O. 1976. The great predatory birds of the Pleistocene of Cuba. *Smithsonian Contributions to Paleobiology* 27: 169–187.

Atkinson, I. A. E. 1973. Spread of the ship rat (*Rattus r. rattus* L.) in New Zealand. *Journal of the Royal Society of New Zealand* 3: 457–472.

———. 1977. A reassessment of factors, particularly *Rattus rattus* L., that influenced the decline of endemic forest birds in the Hawaiian Islands. *Pacific Science* 31: 109–133.

———. 1983. The spread of commensal species of *Rattus* to oceanic islands, and their effects on island avifaunas. In P. J. Moors (ed.), *Conservation of Island Birds,* pp. 35–81. ICBP Technical Publication 3.

———. 1989. Introduced mammals and extinctions. In D. Western and M. C. Pearl (eds.), *Conservation for the Twenty-First Century.* New York: Oxford University Press.

Atkinson, I. A. E., and R. M. Greenwood. 1980. Divaricating plants and moa browsing: A reply. *New Zealand Journal of Ecology* 3: 165–167.

———. 1989. Relationships between moas and plants. *New Zealand Journal of Ecology* 12(Suppl.): 67–96.

Atkinson, I. A. E., and P. R. Millener. 1991. An ornithological glimpse into New Zealand's pre-human past. *Acta XX Congressus Internationalis Ornithologici* 1: 129–192.

Atkinson, I. A. E., and H. Moller. 1990. Kiore. In C. M. King (ed.), *The Handbook of New Zealand Mammals,* pp. 175–192. Auckland: Oxford University Press.

Baker, A. J. 1991. A review of New Zealand ornithology. *Current Ornithology* 8: 1–67.

Baker, A. J., C. H. Daugherty, R. Colbourne, and J. L. McLennan. 1995. Flightless brown kiwis of New Zealand possess extremely subdivided population structure and cryptic species like small mammals. *Proceedings of the National Academy of Sciences USA* 92: 8254–8258.

Ballmann, P. 1969. Die Vogel aus der altburdigalen Spaltenullung von Wintershof (West) bei Eichstatt in Bayern. *Zitteliana* 1: 5–60.

———. 1973. Fossile Vogel aus dem Neogen der Halbinsel Gargano (Italien). *Scripta Geologica* 17: 1–75.

Balouet, J. C., and S. L. Olson. 1989. Fossil birds from late Quaternary deposits in New Caledonia. *Smithsonian Contributions to Zoology* 469: 1–38.

Bartle, J. A. 1968. Observations on the breeding habits of Pycroft's petrel. *Notornis* 15: 70–99.

Bartle, J. A., and P. M. Sagar. 1987. Intraspecific variation in the New Zealand bellbird *Anthornis melanura. Notornis* 34: 253–306.

Bartlett, A. D. 1852. On the genus *Apteryx. Proceedings of the Zoological Society of London for 1850,* Part XVIII, no. ccxviii, 274–276.

Batcheler, C. L. 1989. Moa browsing and vegetation formations, with particular reference to deciduous and poisonous plants. *New Zealand Journal of Ecology* 12(Suppl.): 57–65.

Bauer, A., and P. Russell. 1986. *Hoplodactylus delcourti* n. sp. (Reptilia: Gekkonidae), the largest known gecko. *New Zealand Journal of Zoology* 13: 141–148.

Baumel, J. J., and L. M. Witmer. 1993. Osteologia. In J. J. Baumel, A. S. King. J. E. Breazile, H. E. Evans, and J. C. Van den Berge (eds.). *Handbook of Avian Anatomy: Nomina Anatomica Avium,* pp. 45–132. 2nd ed. Publications of the Nuttall Ornithological Club 23. Cambridge, Mass.: Nuttall Ornithological Club.

Baumel, J. J., A. S. King, J. E. Breazile, H. E. Evans, and J. C. Van Den Berge (eds.). 1993. *Handbook of Avian Anatomy: Nomina Anatomica Avium.* 2nd ed. Publications of the Nuttall Ornithological Club 23. Cambridge, Mass.: Nu ttall Ornithological Club.

Beaglehole, J. C. (ed.) 1962. *The* Endeavour *Journal of Joseph Banks 1768–1771*. Vol. 1. Sydney: Trustees of the Public Library of New South Wales and Angus and Robertson.

Beale, G. 1985. A radiological study of the kiwi (*Apteryx australis mantelli*). *Journal of the Royal Society of New Zealand* 15: 187–200.

———. 1991. The maturation of the skeleton of a kiwi (*Apteryx australis mantelli*)—A ten year radiological study. *Journal of the Royal Society of New Zealand* 21: 219–220.

Beattie, H. 1958. *The Moa. When Did It Become Extinct? Maori Traditions and Pakeha Conjectures.* Dunedin: Otago Daily Times and Witness Newspapers.

Beauchamp, A., and T. H. Worthy. 1988. The takahe's decline in distribution—A re-examination. *Journal of the Royal Society of New Zealand* 18: 103–112.

Beddard, F. E. 1889 (June). On certain points in the anatomy of the Accipitres, with reference to the affinities of *Polyboroides. Proceedings of the Zoological Society of London for 1889,* Part I: 77–82.

———. 1898. *The Structure and Classification of Birds.* London: Longmans, Green.

Beggs, J. R., and P. R. Wilson. 1987a. Energetics of South Island kaka *Nestor meridionalis meridionalis* feeding on the larvae of kanuka longhorn beetles *Ochrocydus huttoni. New Zealand Journal of Ecology* 10: 143–147.

———. 1987b. The kaka *Nestor meridionalis,* a New Zealand parrot endangered by introduced wasps and mammals. *Biological Conservation* 56: 23–38.

Bell, B. (Brian) D. 1978. The Big South Cape Islands rat irruption. In P. R. Dingwall, I. A. E. Atkinson, and C. Hay (eds.), *The Ecology and Control of Rodents in New Zealand Nature Reserves,* pp. 33–40. New Zealand Department of Lands and Survey Information Series, No. 4.

Bell, B. (Ben) D. 1991. Recent avifaunal changes and the history of ornithology in New Zealand. *Acta XX Congressus Ornithologici* 1: 195–230.

Bell, G. B., Jr., M. W. Caldwell, R. Holmes, J. Wiffen, and J. McKee. 1999. Sea monsters of the South Pacific: On the Late Cretaceous mosasaurs of New Zealand [abstract]. In Abstracts of Papers, Fifty-Ninth Annual Meeting, Society of Vertebrate Paleontology. *Journal of Vertebrate Paleontology* 19(3, Suppl.): 32A.

Benham, W. B. 1902. Note on an entire egg of a moa, now in the museum of the University of Otago. *Transactions and Proceedings of the New Zealand Institute* 34: 149–151.

———. 1910. The discovery of moa remains on Stewart Island. *Transactions and Proceedings of the New Zealand Institute* 42: 354–356.

———. 1934. The skeleton of a small moa, *Emeus huttoni* Owen. *Transactions and Proceedings of the New Zealand Institute* 64: 87–103.

Berger, M., and J. S. Hart. 1974. Physiology and energetics of flight. In D. S. Farner and J. R. King (eds.), *Avian Biology 4,* pp. 360–415. New York: Academic Press.

Best, E. 1896. The Maori and the moa. *Journal of the Polynesian Society* 5: 121–122.

Best, E. 1977. Forest lore of the Maori. *Dominion Museum Bulletin* No. 14. Government Printer, Wellington, 421 pp.

Blackburn, A. 1968. The birdlife of Codfish Island. *Notornis* 15: 51–65.

Bledsoe, A. H. 1988. A phylogenetic analysis of postcranial skeletal characters of the ratite birds. *Annals of the Carnegie Museum* 57: 73–90.

Bochenski, Z., and T. Tomek. 1994. Fossil and subfossil bird remains from five Austrian caves. *Acta Zoologica Cracov* 37: 347–358.

Bocherens, H., D. Billiou, M. Patou-Mathis, D. Bonjean, M. Otte, and A. Mariotti. 1997. Paleobiological implications of the isotopic signatures (^{13}C, ^{15}N) of fossil mammal collagen in Scladina Cave (Sclayn, Belgium). *Quaternary Research* 48: 370–380.

Bock, W. J. 1963. The cranial evidence for ratite affinities. In C. G. Sibley (ed.), *Proceedings of the 13th International Ornithological Congress,* pp. 39–54.

Bock, W., and A. McEvey. 1969. The radius and relationship of owls. *Wilson Bulletin* 81: 55–68.

Bonaparte, C. S. A. 1853. Classification ornithologique par series. *Compte Rendu des Séances de L'Académie des Sciences* 37: 641–646.

———. 1854. Conspectus systematus ornithologie. *Annales des Sciences Naturelles Zoology et Biologie Animale* 1: 105–152.

———. 1856. Classification ornithologique par series. *Compte Rendu des Séances de L'Académie des Sciences* 43: 841.

Bonser, R. H. C., and C. A. Walker. 1993. Some notes on type material of moas (Aves: Dinornithidae). *Bulletin of the British Ornithologists' Club* 113: 257–259.

Boon, W. M., C. H. Daugherty, and G. K. Chambers. 2001. The Norfolk Island green parrot and New Caledonian red-crowned parakeet are distinct species. *Emu* 101: 113–121.

Boon, W. M., J. C. Kearvell, C. H. Daugherty, and G. K. Chambers. 2000. Molecular systematics of New Zealand *Cyanoramphus* parakeets: Conservation of orange-fronted and Forbes's parakeets. *Bird Conservation International* 10: 211–239.

Booth, B. S. 1875. Description of the moa swamp at Hamilton. *Transactions and Proceedings of the New Zealand Institute* 7: 123–138.

———. 1877. On a second discovery of moa bones at Hamilton. *Transactions and Proceedings of the New Zealand Institute* 9: 365–366.

Booth, A. M., E. O. Minot, R. A. Fordham, and J. G. Innes. 1996. Kiore (*Rattus exulans*) predation on the eggs of the little shearwater (*Puffinus assimilis haurakiensis*). *Notornis* 43: 147–153.

Bourne, W. R. P. 1967. Subfossil petrel bones from the Chatham Islands. *Ibis* 109: 1–7.

Brathwaite, D. H. 1992. Notes on the weight, flying ability, habitat, and prey of Haast's eagle (*Harpagornis moorei*). *Notornis* 39: 239–247.

Brathwaite, D. H., and R. N. Holdaway. 1987. A preliminary reconstruction of *Harpagornis moorei*: Top predator in the moa's world [abstract]. *New Zealand Journal of Ecology* 10: 162.

Brewster, B. 1987. The moa: The life and death of New Zealand's unique bird. Nelson, New Zealand: Nikau Press.

Brodie, J. W. 1950. Moa remains at Castlepoint. *New Zealand Science Review* 8(9–10): 87–89.

Brodkorb, P. 1963. Catalogue of fossil birds, part 1 (Archaeopterygiformes through Ardeiformes). *Bulletin of the Florida State Museum (Biological Sciences)* 7: 180–293.

———. 1964. Catalogue of fossil birds, part 2 (Anseriformes through Galliformes). *Bulletin of the Florida State Museum (Biological Sciences)* 8: 195–335.

———. 1978. Catalogue of fossil birds, part 5 (Passeriformes). *Bulletin of the Florida State Museum (Biological Sciences)* 23: 139–228.

Brooke, M. de L. 1995. The breeding biology of the gadfly petrels *Pterodroma* spp. of the Pitcairn Islands: Characteristics, population sizes and controls. In T. G. Benton and T. Spencer (eds.), *The Pitcairn Islands: Biogeography, Ecology and Prehistory,* pp. 213–231. London: Academic Press. Reprinted from *Biological Journal of the Linnean Society* 56 (parts 1, 2).

Brown, L. H. 1982. Genus *Stephanoaetus.* In L. H. Brown, E. K. Urban, and K. Newman (eds.), *The Birds of Africa,* pp. 429–433. London: Academic Press.

Brown, K. P. 1997. Predation at nests of two New Zealand endemic passerines; implications for bird community restoration. *Pacific Conservation Biology* 3: 91–98.

Brown, L. H., and D. Amadon. 1968. *Eagles, Hawks and Falcons of the World.* 2 vols. London: Country Life.

Brown, L. H., E. K. Urban, and K. Newman (eds.). 1982. *The Birds of Africa.* London: Academic Press.

Buick, T. L. 1931. *The Mystery of the Moa, New Zealand's Avian Giant.* New Plymouth, New Zealand: Thomas Avery and Sons.

———. 1936. *The Discovery of* Dinornis. New Plymouth, New Zealand: Thomas Avery and Sons.

———. 1937. *The Moa-Hunters of New Zealand.* New Plymouth, New Zealand: Thomas Avery and Sons.

Buist, A. G., and J. C. Yaldwyn. 1960. An "articulated" moa leg from an oven excavated at Waingongoro, South Taranaki. *Journal of the Polynesian Society* 69: 76–88.

Bull, P. C., P. D. Gaze, and C. J. R. Robertson. 1985. *The Atlas of Bird Distribution in New Zealand.* Wellington: Ornithological Society of New Zealand.

Buller, W. L. 1869. Essay on the ornithology of New Zealand. *Transactions and Proceedings of the New Zealand Institute* 2nd ed. (1875), 213–231.

———. 1874. On certain species of New Zealand birds, *Rallus modestus. Transactions and Proceedings of the New Zealand Institute* 6: 124–125.

———. 1875. [Proceedings of Eighth Annual General Meeting]. *Transactions and Proceedings of the New Zealand Institute* 7: 511.

———. 1878. Notes on the ornithology of New Zealand. *Transactions and Proceedings of the New Zealand Institute* 10: 191–192.

———. 1879. Additions to list of species, and notices of rare occurrences, since the publication of "The Birds of New Zealand." *Transactions and Proceedings of the New Zealand Institute* 11: 361–366.

———. 1888. *A History of the Birds of New Zealand.* 2nd ed. London: The author.

———. 1893. Note on the flightless rail of the Chatham Islands (*Cabalus modestus*). *Transactions and Proceedings of the New Zealand Institute* 25: 52–53.

Burrows, C. J. 1960. Recent changes in vegetation of the Cass area, Canterbury. *New Zealand Geographer* 16: 57–70.

———. 1980a. Diet of New Zealand Dinornithiformes. *Naturwissenschaften* 67: 151–153.

———. 1980b. Some empirical information concerning the diet of moas. *New Zealand Journal of Ecology* 3: 125–130.

———. 1982. On New Zealand climate within the last 1000 years. *New Zealand Journal of Archaeology* 4: 157–167.

———. 1983. Radiocarbon dates from late Quaternary deposits in the Cass district, Canterbury, New Zealand. *New Zealand Journal of Botany* 21: 443–454.

———. 1984. Problems of dating, correlation and environmental interpretation of the New Zealand Quaternary. In W. C. Mahaney (ed.), *Correlation of Quaternary Chronologies,* pp. 317–337. Norwich: Geo Books.

———. 1988. Late Otiran and early Aranuian radiocarbon dates from South Island localities. *New Zealand Natural Sciences* 15: 25–36.

———. 1989a. Aranuian radiocarbon dates from moraines in the Mount Cook region, New Zealand. *New Zealand Journal of Geology and Geophysics* 32: 205–216.

———. 1989b. Moa browsing: Evidence from the Pyramid Valley mire. *New Zealand Journal of Ecology* 12(Suppl.): 51–56.

———. 1995. A macrofossil flora from sediments in a lagoon marginal to Lake Coleridge, Canterbury, New Zealand. *New Zealand Journal of Botany* 33: 519–522.

———. 1996. Radiocarbon dates for Holocene fires and associated events, Canterbury, New Zealand. *New Zealand Journal of Botany* 34: 111–121.

Burrows, C. J., and D. E. Greenland. 1979. An analysis of the evidence for climatic change in the last thousand years: Evidence from diverse natural phenomena and from instrumental records. *Journal of the Royal Society of New Zealand* 9: 321–373.

Burrows, C. J., and N. T. Moar. 1996. A mid Otira Glaciation palaeosol and flora from the Castle Hill Basin, Canterbury, New Zealand. *New Zealand Journal of Botany* 34: 539–545.

Burrows, C. J., and J. B. Russell. 1975. Moraines of the upper Rakaia Valley. *Journal of the Royal Society of New Zealand* 5: 463–477.

———. 1990. Aranuian vegetation history of the Arrowsmith Range, Canterbury. I. Pollen diagrams, plant macrofossils, and buried soils from Prospect Hill. *New Zealand Journal of Botany* 28: 323–345.

Burrows, C. J., B. McCulloch, and M. M. Trotter. 1981. The diet of moas based on gizzard contents samples from Pyramid Valley, North Canterbury and Scaifes Lagoon, Lake Wanaka, Otago. *Records of the Canterbury Museum* 9: 309–336.

Burrows, C. J., M. J. McSavenay, R. J. Scarlett, and B. Turnbull. 1984. Late Holocene forest horizons and a *Dinornis* moa from an earthflow on North Dean, North Canterbury. *Records of the Canterbury Museum* 10: 1–8.

Burrows, C. J., P. Randall, N. T. Moar, and B. G. Butterfield. 1993. Aranuian vegetation history of the Arrowsmith Range, Canterbury, New Zealand. III. Vegetation changes in the Cameron, upper South Ashburton, and Public Hill Creek catchments. *New Zealand Journal of Botany* 31: 147–174.

Burton, P. J. K. 1972. Some anatomical notes on the wrybill. *Notornis* 19: 26–32.

———. 1974. Anatomy of head and neck in the huia (*Heteralocha acutirostris*) with comparative notes on other Callaeidae. *Bulletin of the British Museum (Natural History), Zoology* 27: 3–48.

Caccamise, D. F., and R. S. Hedin. 1985. An aerodynamic basis for selecting transmitter loads in birds. *Wilson Bulletin* 97: 306–318.

Calder, W. A. 1974. Consequences of body size for avian energetics. In R. A. Paynter (ed.), *Avian Energetics,* pp. 86–151. Publications of the Nuttall Ornithological Club, No. 15.

Cameron, R. J. 1961. Maori impact upon the forests of New Zealand. *Whakatane and District Historical Society Review* 9: 131–141.

———. 1964. Destruction of the indigenous forests of New Zealand for Maori agriculture during the nineteenth century. *New Zealand Journal of Forestry* 9: 98–109.

Campbell, J. D. 1965. New Zealand Triassic saurians. *New Zealand Journal of Geology and Geophysics* 8: 505–509.

Campbell, A. G. 1973. Agricultural ecology. In G. R. Williams (ed.), *The Natural History of New Zealand: An Ecological Survey*, pp. 204–228. Wellington: A. H. and A. W. Reed.

Campbell, H. 1996. Geology. In *The Chatham Islands: Heritage and Conservation*, pp. 34–48. Canterbury: Canterbury University Press and the New Zealand Department of Conservation, Christchurch.

Campbell, D. J., and I. A. E. Atkinson. 1999. Effects of kiore (*Rattus exulans* Peale) on recruitment of indigenous coastal trees on northern offshore islands of New Zealand. *Journal of the Royal Society of New Zealand* 29: 265–290.

Campbell, K. E., and L. Marcus. 1992. The relationship of hindlimb bone dimensions to body weight in birds. In K. E. Campbell (ed.), *Papers in Avian Paleontology Honoring Pierce Brodkorb*, pp. 395–412. Natural History Museum of Los Angeles County Science Series, No. 36.

Campbell, K. E., and E. P. Tonni. 1983. Size and locomotion in teratorns (Aves: Teratornithidae). *Auk* 100: 390–403.

Carroll, A. L. K. 1963. Food habits of the North Island weka. *Notornis* 10: 289–300.

Cassels, R. 1979. Whatever happened to New Zealand's moas? *Wildlife* 21: 31–33.

———. 1984. The role of prehistoric man in the faunal extinctions of New Zealand and other Pacific islands. In P. S. Martin and R. G. Klein (eds.), *Quaternary Extinctions: A Prehistoric Revolution*, pp. 741–767. Tucson: Arizona University Press.

Cassels, R. J. S., K. L. Jones, A. Walton, and T. H. Worthy. 1988. Late prehistoric subsistence practices at Parewanui, lower Rangitikei River, New Zealand. *New Zealand Journal of Archaeology* 10: 109–128.

Caughley, G. 1977. The taxonomy of moas. *Tuatara* 23: 20–25.

———. 1988. The colonisation of New Zealand by the Polynesians. *Journal of the Royal Society of New Zealand* 18: 245–270.

———. 1989. New Zealand plant–herbivore systems: Past and present. *New Zealand Journal of Ecology* 12(Suppl.): 3–10.

Chapman, F. R. 1885. Notes on moa remains in the Mackenzie Country and other localities. *Transactions and Proceedings of the New Zealand Institute* 17: 172–178.

Cheeseman, T. F. 1876. Moa remains discovered at Ellerslie, near Auckland. *Transactions and Proceedings of the New Zealand Institute* 8: 427–429.

———. 1891. On the birds of the Kermadec Islands. *Transactions and Proceedings of the New Zealand Institute* 23: 217–226.

Clark, G. R., P. Petchey, M. S. McGlone, and P. Bristow. 1996. Faunal and floral remains from Earnscleugh Cave, Central Otago, New Zealand. *Journal of the Royal Society of New Zealand* 26: 363–380.

Clarke, R. 1790. [Diary in manuscript.] Sydney: Mitchell Library.

Clarkson, B. R., R. N. Patel, and B. D. Clarkson. 1988. Composition and structure of forest overwhelmed at Pureora, central North Island, New Zealand, during the Taupo eruption (c. 130 AD). *Journal of the Royal Society of New Zealand* 18: 417–436.

Clarkson, B. R., B. D. Clarkson, and R. N. Patel. 1992. The pre-Taupo (c.

AD 130) forest of the Benneydale–Pureora district, central North Island, New Zealand. *Journal of the Royal Society of New Zealand* 22: 61–76.

Clarkson, B. R., M. S. McGlone, D. J. Lowe, and B. D. Clarkson. 1995. Macrofossils and pollen representing forests of the pre-Taupo volcanic eruption era (c. 1850 yr BP) at Pureora and Benneydale, central North Island, New Zealand. *Journal of the Royal Society of New Zealand* 25: 263–281.

Climo, F. M. 1975. The land snail fauna. In G. Kuschel (ed.), *Biogeography and Ecology in New Zealand,* pp. 459–492. The Hague: W. Junk.

Clout, M. N. 1986. Habitat use and spring movements of New Zealand pigeons at Lake Rotoroa, Nelson Lakes National Park. *Notornis* 33: 37–44.

Clout, M. N., and J. R. Hay. 1989. The importance of birds as browsers, pollinators and seed disperses in New Zealand forests. *New Zealand Journal of Ecology* 12(Suppl.): 27–33.

Clout, M. N., P. D. Gaze, J. R. Hay, and B. J. Karl. 1986. Habitat use and spring movements of New Zealand pigeons at Lake Rotoroa, Nelson Lakes National Park. *Notornis* 33: 37–44.

Clout, M. N., B. J. Karl, and P. D. Gaze. 1991. Seasonal movements of New Zealand pigeons from a lowland forest reserve. *Notornis* 38: 37–47.

Clunie, F. 1981. Nesting season of the Fiji goshawk. *Notornis* 28: 136–137.

Cockayne, L. 1912. Observation concerning evolution derived from ecologic studies in New Zealand. *Transactions and Proceedings of the New Zealand Institute* 44: 1–50.

———. 1928. *Vegetation of New Zealand.* 2nd ed. Leipzig: Engelmann.

Cockburn-Hood, T. H. 1874. Remarks upon the footprints of the *Dinornis* in the sand rock at Poverty Bay, New Zealand, and upon its recent extinction. *Proceedings of the Royal Society of Edinburgh* 8: 236–240.

———. 1875. Notes upon the probable changes that have taken place in the physical geography of New Zealand since the arrival of the Maori. *Transactions and Proceedings of the New Zealand Institute* 7: 112–120.

Colenso, W. 1843 [October 6]. An account of some enormous fossil bones, of an unknown species of the class Aves, lately discovered in New Zealand. *Tasmanian Journal of Natural Science* 2(7): 81–107.

———. 1844. An account of some enormous fossil bones of an unknown species of the class Aves, lately discovered in New Zealand. *Annals and Magazine of Natural History* 14: 81–96.

———. 1880. On the moa. *Transactions and Proceedings of the New Zealand Institute* 12: 63–108.

———. 1894. Notes and observations on M. A. de Quartrefages' paper "On moas and moa-hunters" republished in Vol. XXV TNZI. *Transactions and Proceedings of the New Zealand Institute* 26: 498–513.

Cooper, A. 1994a. Ancient DNA sequences reveal unsuspected phylogenetic relationships within New Zealand wrens (Acanthisittidae). *Experientia* 50: 558–563.

———. 1994b. Moas, gizzard stones, and New Zealand plants: Further comments. *New Zealand Science Review* 51: 59, 62.

———. 1997. Ancient DNA and avian systematics: From Jurassic Park to modern island extinctions. In D. Mindell (ed.), *Avian Molecular Evolution and Molecular Systematics,* pp. 345–373. New York: Academic Press.

Cooper, A., and R. A. Cooper. 1995. The Oligocene bottleneck and New Zealand biota: Genetic record of a past environmental crisis. *Proceedings of the Royal Society of London, B* 261: 293–302.
Cooper, R. A., and P. R. Millener. 1993. The New Zealand biota: Historical background and new research. *Trends in Ecology and Evolution* 8: 429–433.
Cooper, A., C. Mourer-Chauviré, G. K. Chambers, A. Haeseler, A. C. Wilson, and S. Pääbo. 1992. Independent origins of New Zealand moas and kiwis. *Proceedings of the National Academy of Sciences USA* 89: 8741–8744.
Cooper, A., C. Lalueza-Fox, S. Anderson, A. Rambaut, J. Austin, and R. Ward. 2001. Complete mitochondrial genome sequences of two extinct moas clarify ratite evolution. *Nature* 409: 704–707.
Cox, J. E., and C. B. Mead. 1963. Soil evidence relating to post glacial climate on the Canterbury Plains. *Proceedings of the New Zealand Ecological Society* 10: 28–38.
Cracraft, J. 1974a. Continental drift and vertebrate distribution. *Annual Review of Ecology and Systematics* 5: 215–261.
———. 1974b. Phylogeny and evolution of the ratite birds. *Ibis* 116: 494–521.
———. 1976a. Covariation patterns in the post-cranial skeleton of the moas (Aves, Dinornithidae): A factor analytic study. *Paleobiology* 2: 166–173.
———. 1976b. The hindlimb elements of the moas (Aves, Dinornithidae): A multivariate assessment of size and shape. *Journal of Morphology* 150: 495–526.
———. 1976c. The species of moas (Aves: Dinornithidae). *Smithsonian Contributions to Paleobiology* 27: 189–205.
———. 1980. Moas and the Maoris. *Natural History* 10/80: 28–36.
———. 1982. Phylogenetic relationships and trans-Atlantic biogeography of some gruiform birds. *Geobios Memoires Speciale* 6: 393–402.
Craig, J. L. 1986. The effects of kiore on other fauna. In A. E. Wright and R. E. Beever (eds.), *The Offshore Islands of Northern New Zealand*, pp. 75–83. New Zealand Department of Lands and Survey Information Series, No. 16.
Cranwell, L. M., and L. von Post. 1936. Post-Pleistocene pollen diagrams from the Southern Hemisphere. 1. New Zealand. *Geografiska Annaler* 3–4: 308–347.
Christidis, L., P. R. Leeton, and M. Westerman. 1996. Were bowerbirds part of the New Zealand fauna? *Proceedings of the National Academy of Sciences USA* 93: 3898–3901.
Clunie, F. 1981. Nesting season of the Fiji goshawk. *Notornis* 28: 136–137.
Crockett, D. E. 1994. Rediscovery of the Chatham Island taiko *Pterodroma magentae*. *Notornis* 41(Suppl.): 49–60.
Crook, I. G. 1973. The tuatara, *Sphenodon punctatus* Gray, on islands with and without populations of the Polynesian rat, *Rattus exulans* (Peale). *Proceedings of the New Zealand Ecological Society* 20: 115–120.
Croxall, J. P. 1979. Distribution and population changes in the wandering albatross *Diomedea exulans* at South Georgia. *Ardea* 67: 15–21.
Croxall, J. P., P. Rothery, S. P. C. Pickering, and P. A. Prince. 1990. Reproductive performance, recruitment and survival of wandering albatrosses *Diomedea exulans* at Bird Island, South Georgia. *Journal of Animal Ecology* 59: 775–796.

Cumberland, K. B. 1962. Moas and men, New Zealand about A.D. 1250. *Geographical Review* 52: 151–173.

Dallas, W. S. 1865. 2. On the feathers of *Dinornis robustus* O., a skeleton in the Yorkshire Museum. *Proceedings of the Zoological Society of London for 1865,* Part I: 265–268. Reprint, *Annals and Magazine of Natural History* 3(16): 66–69. Reprint, *Annals of Science and Nature* 4: 292.

Daniel, M. J. 1979. The New Zealand short-tailed bat *Mystacina tuberculata,* a review of current knowledge. *New Zealand Journal of Zoology* 6: 357–370.

———. 1990. Order Chiroptera. In C. M. King (ed.), *The Handbook of New Zealand Mammals,* pp. 114–137. Auckland: Oxford University Press.

Darwin, C. 1884. *A Naturalist's Voyage: Journal of Researches into the Natural History and Geology of the Countries Visited during the Voyage of H.M.S.* "Beagle" *Round the World under the Command of Capt. FitzRoy, R.N.* London: John Murray.

Daugherty, C. H., L. R. Maxon, and B. D. Bell. 1982. Phylogenetic relationships within the New Zealand frog genus *Leiopelma*—Immunological evidence. *New Zealand Journal of Zoology* 9: 239–242.

Daugherty, C. H., A. Cree, J. M. Hay, and M. B. Thompson. 1990a. Neglected taxonomy and continuing extinctions of tuatara (*Sphenodon*). *Nature* 347: 177–179.

Daugherty, C. H., G. B. Patterson, C. J. Thorn, and D. C. French. 1990b. Differentiation of the members of the New Zealand *Leiolopisma nigriplantare* species complex (Lacertilia: Scincidae). *Herpetological Monographs* 4: 61–76.

Daugherty, C. H., M. Williams, and J. M. Hay. 1990c. Genetic differentiation, taxonomy and conservation of Australasian teals *Anas* spp. *Bird Conservation International* 9: 29–42.

Daugherty, C. H., G. B. Patterson, and R. A. Hitchmough. 1994. Taxonomic and conservation review of the New Zealand herpetofauna. *New Zealand Journal of Zoology* 21: 317–323.

Davidson, J. 1978. Archaeological salvage excavations at Paremata, Wellington, New Zealand. *National Museum of New Zealand Records* 1: 203–236.

———. 1979. Archaic middens of the Coromandel region: A review. In A. J. Anderson (ed.), *Birds of a Feather,* pp. 183–202. New Zealand Archaeological Association Monograph II; B.A.R. International Series 62.

———. 1984. *The Prehistory of New Zealand.* Auckland: Longman Paul.

Davies, G. H. 1907. The name moa. *Journal Polynesian Society* 16: 106.

———. 1910. The moa. *Journal Polynesian Society* 19: 223.

Davis, A. 1994. Status, distribution, and population trends of the New Zealand shore plover *Thinornis novaeseelandiae. Notornis* 41(Suppl.): 179–194.

Davison, W. M. 1870. The *Dinornis. Nature* 1: 604.

Dawson, E. W. 1949. Subfossil bird remains from Lake Grassmere. *New Zealand Bird Notes* 21: 219–220.

———. 1952. Subfossil *Nestor* (Psittacidae) from New Zealand. *Emu* 42: 259–272.

———. 1958a. Advances in the knowledge of the small subfossil birds of New Zealand, and their biogeographical significance [paper 22]. In *Programs and Abstracts of the 15th International Congress on Zoology,* London, section 5.

———. 1958b. Re-discoveries of the New Zealand subfossil birds named by H. O. Forbes. *Ibis* 100: 232–237.

———. 1961. An extinct eagle in the Chatham Islands. *Notornis* 9: 171–172.

Dawson, L., and M. L. Augee. 1997. Late Quaternary sediments and fossil vertebrate fauna from Cathedral Cave, Wellington Caves, New South Wales. *Proceedings of the Linnean Society of New South Wales* 117: 51–78.

De Boer, L. E. M. 1980. Do the chromosomes of the kiwi provide evidence for a monophyletic origin of the ratites? *Nature* 287: 84–85.

De Niro, M. J., and S. Epstein. 1978a. Carbon isotopic evidence for different feeding patterns in two hyrax species occupying the same habitat. *Science* 201: 906–908.

———. 1978b. Influence of diet on the distribution of carbon isotopes in animals. *Geochimica et Cosmochimica Acta* 42: 495–506.

De Vis, C. W. 1884. The moa (*Dinornis*) in Australia. *Proceedings of the Royal Society of Queensland* 1: 23–28.

———. 1891. The moa in Australia. *New Zealand Journal of Science* 1: 97–101.

De Quatrefages, A. 1893. The moas and moa-hunters. *Transactions and Proceedings of the New Zealand Institute* 25: 17–49.

Dell, R. J., and R. A. Falla. 1972. The Kaikoura moa egg. *Dominion Museum Records in Ethnology* 2: 97–104.

Diamond, J. M. 1984. Distributions of New Zealand birds on real and virtual islands. *New Zealand Journal of Ecology* 7: 37–55.

———. 1989a. The present, past and future of human-caused extinctions. *Philosophical Transactions of the Royal Society, London, B.*

———. 1989b. Quaternary megafaunal extinctions: Variations on a theme by Paganini. *Journal of Archaeological Science* 16: 167–175.

———. 1990a. Bob Dylan and moa's ghosts. *Natural History* 10/90: 26–31.

———. 1990b. Biological effects of ghosts. *Nature* 345: 769–770.

Diamond, J. M., and C. R. Veitch. 1981. Extinctions and introductions in the New Zealand avifauna: Cause and effect? *Science* 211: 499–501.

Dodson, J. R., N. J. Enright, and R. F. McLean. 1988. A late Quaternary vegetation history for far northern New Zealand. *Journal of Biogeography* 15: 647–656.

Dowding, J. E., and E. C. Murphy. 1993. Decline of the Stewart Island population of the New Zealand dotterel. *Notornis* 40: 1–13.

Downes, T. W. 1916. New light on the period of the extinction of the moa (according to Maori record). *Transactions and Proceedings of the New Zealand Institute* 48: 426–434.

———. 1926. Notes on the moa, as contributed by natives of the Wairarapa district. *Journal of the Polynesian Society* 35: 36–37.

Duff, R. S. 1941. Notes on the excavations at Pyramid Valley, Waikari. *Records of the Canterbury Museum* 4: 330–338.

———. 1949a. *The Moa-Hunter Period of Maori Culture.* Canterbury Museum Bulletin, No. 1.

———. 1949b. *Pyramid Valley.* Reprint, 1952: Christchurch: Pegasus Press for the Association of Friends of the Canterbury Museum.

———. 1949c. Moas and man (part 1). *Antiquity* 23: 172–179.

———. 1950. Moas and man (part 2). *Antiquity* 24: 72–84.

———. 1951. Moas and moa-hunters. *New Zealand Education Department Post-Primary School Bulletin* 5(7): 129–164.

———. 1955. Further report on the excavations at Pyramid Valley Swamp,

Waikari, North Canterbury. *Records of the Canterbury Museum* 6: 253–255.
———. 1956. *The Moa-Hunter Period of Maori Culture*. 2nd ed. Wellington: Government Printer.
———. 1963. *The Problem of Moa Extinction*. Thomas Cawthron Memorial Lecture, No. 38. Nelson, New Zealand: Cawthron Institute.
———. 1977. *The Moa Hunter Period of Maori Culture*. 3rd ed. Wellington: Government Printer.
Dugdale, J., and R. Emberson. 1996. Insects. In *Chatham Islands: Heritage and Conservation*. Christchurch: Canterbury University Press.
Duncan, F. M. 1937. On the dates of publication of the Society's "Proceedings," 1859–1926. With an Appendix containing the dates of publication of "Proceedings," 1830–1858, compiled by the late F. H. Waterhouse, and of the "Transactions," 1833–1869, by the late Henry Peavot, originally published in P. 2. S. 1893, 1913. *Proceedings of the Zoological Society of London* 107: 71–84.
Duncan, K. W. 1994. *Terrestrial Talitridae (Crustacea: Amphipoda)*. Fauna of New Zealand, No. 31. Auckland: Manaaki Whenua Press.
Duncan, K., and R. Holdaway. 1989. Footprint pressures and locomotion of moas and ungulates and their effects on the New Zealand indigenous biota through trampling. *New Zealand Journal of Ecology* 12(Suppl.): 97–101.
Dunning, J. B. (ed.). 1993. *CRC Handbook of Avian Body Masses*. Boca Raton, Fla.: CRC Press.
Dunwiddie, P. W. 1979. Dendrochronological studies of indigenous New Zealand trees. *New Zealand Journal of Botany* 17: 251–266.
Dwyer, P. D. 1962. Studies of the two New Zealand bats. *Zoology Publications from Victoria University of Wellington* 28: 1–28.
East, R., and G. R. Williams. 1984. Island biogeography and the conservation of New Zealand's indigenous forest-dwelling birds. *New Zealand Journal of Ecology* 7: 27–35.
Elliott, G. P. 1996a. Productivity and mortality of mohua (*Mohoua ochrocephala*). *New Zealand Journal of Zoology* 23: 229–237.
———. 1996b. Mohua and stoats: A population viability analysis. *New Zealand Journal of Zoology* 23: 239–247.
Elliot, M. B., B. Striewski, J. R. Flenley, and D. G. Sutton. 1995. Palynological and sedimentological evidence for a radiocarbon chronology of environmental change and Polynesian deforestation from Lake Taumatawhana, Northland, New Zealand. *Radiocarbon* 37: 899–916.
Elliott, G. P., P. J. Dilks, and C. F. J. O'Donnell. 1996. Nest site selection by mohua and yellow-crowned parakeets in beech forest in Fiordland, New Zealand. *New Zealand Journal of Zoology* 23: 267–278.
Emslie, S. D. 1988. The fossil history and phylogenetic relationships of condors (Ciconiiformes: Vulturidae) in the New World. *Journal of Vertebrate Paleontology* 8: 212–228.
Ewen, C. A. 1896. On the discovery of moa remains on Riverton Beach. *Transactions and Proceedings of the New Zealand Institute* 28: 651–654.
Falla, R. A. 1941. The avian remains. Preliminary report on excavations at Pyramid Valley Swamp, Waikari, North Canterbury. *Records of the Canterbury Museum* 4: 325–353.
———. 1953. The Australian element in the avifauna of New Zealand. *Emu* 53: 36–46.
———. 1954. A new rail from cave deposits in the North Island of New Zealand. *Records of the Auckland Institute and Museum* 4: 241–244.

———. 1955. *New Zealand Bird Life, Past and Present.* Thomas Cawthron Memorial Lecture, No. 29. Nelson, New Zealand: Cawthron Institute.
———. 1962. The moa, zoological and archaeological. *New Zealand Archaeological Association Newsletter* 5(3): 189–191.
———. 1973. Obituary—R. C. Murphy. *Notornis* 20: 174–176.
———. 1974. Rare birds and conservation. In C. J. R. Robertson (ed.), *Birds in New Zealand,* pp. 121–134. Wellington: A. H. and A. W. Reed.
Farlow, J. O. In press. Toebone proportions of the "prodigious birds": Implications for dinosaur ichnology? In D. Wolberg (ed.), *Dinofest III.* Dinofest International.
Feduccia, A. 1980. *The Age of Birds.* Cambridge: Harvard University Press.
Field, H. C. 1882. On the extinction of the moa. *Transactions and Proceedings of the New Zealand Institute* 14: 540.
———. 1892. Discoveries of moa bones. *Transactions and Proceedings of the New Zealand Institute* 24: 558–561.
———. 1894. The date of extinction of the moa. *Transactions and Proceedings of the New Zealand Institute* 26: 560–568.
Fisher, H. I. 1946. Adaptations and comparative anatomy of the locomotor apparatus of New World vultures. *American Midland Naturalist* 35: 545–727.
Fitch, W. T. 1999. Acoustic exaggeration of size in birds via tracheal elongation: Comparative and theoretical analyses. *Journal of Zoology, London* 248: 31–48.
Flannery, T. F. 1995. *Mammals of the South-West Pacific and Moluccan Islands.* Sydney: Reed Books.
Fleming, C. A. 1950. New Zealand flycatchers of the genus *Petroica* Swainson. *Transactions of the Royal Society of New Zealand* 78: 14–47, 127–160.
———. 1953. Materials for a recent geochronology of New Zealand. In M. A. Black (ed.), *Seventh Science Congress, Christchurch, May 15–21, 1951,* pp. 114–123.
——— (ed.). 1959. *Ferdinand von Hochstetter, Geology of New Zealand. Contributions to the Geology of the Provinces of Auckland and Nelson.* Wellington: Government Printer.
———. 1962a. History of the New Zealand land bird fauna. *Notornis* 9: 270–274.
———. 1962b. The extinction of moas and other animals during the Holocene period. *Notornis* 10: 113–117.
———. 1963. A moa-bone from the sea-floor in Cook Strait. *Records of the Dominion Museum* 4: 231–233.
———. 1966. Sir James Hector. In A. H. McLintock (ed.), *An Encyclopaedia of New Zealand,* vol. 2, pp. 11–13. Wellington: New Zealand Government Printer.
———. 1968. New Zealand fossil seals. *New Zealand Journal of Geology and Geophysics* 11: 1184–1187.
———. 1969. Rats and moa extinction. *Notornis* 16: 210–211.
———. 1975. The geological history of New Zealand and its biota. In G. Kuschel (ed.), *Biogeography and Ecology in New Zealand,* pp. 1–86. The Hague: Junk.
———. 1977. The history of life in New Zealand forests. *New Zealand Journal of Forestry* 22: 249–262.
———. 1979. *The Geological History of New Zealand and Its Life.* Auckland: Auckland University Press and Oxford University Press.

Fleming, T. H. 1995. The use of stable isotopes to study the diets of plant-visiting bats. *Symposia of the Zoological Society of London* 67: 99–110.

Fleming, C. A., D. R. Gregg, and S. P. Welles. 1971. New Zealand ichthyosaurs—A summary, including new records from the Cretaceous. *New Zealand Journal of Geology and Geophysics* 14: 734–741.

Fleming, C. A., D. C. Mildenhall, and N. T. Moar. 1976. Quaternary sediments and plant microfossils from Enderby Island, Auckland Islands. *Journal of the Royal Society of New Zealand* 6: 433–458.

Fleming, T. H., R. A. Nuñez, and L. da Silva Lobo Sternberg. 1993. Seasonal changes in the diets of migrant and non-migrant nectarivorous bats as revealed by carbon stable isotope analysis. *Oecologia* 94: 72–75.

Flux, J. E. C. 1989. Biogeographic theory and the number and habitat of moas. *New Zealand Journal of Ecology* 12(Suppl.): 35–37.

Forbes, H. O. 1890a (January 2). [Extinct swan from New Zealand.] *Nature* 41(1053): 209.

———. 1890b. New extinct swan from New Zealand. *Ibis* 2, Series 6, 32: 264–265.

———. 1891a. On avian remains found under a lava-flow near Timaru, in Canterbury. *Transactions and Proceedings of the New Zealand Institute* 23: 366–373.

———. 1891b. On the disappearance of the moa. *Transactions and Proceedings of the New Zealand Institute* 23: 373–375.

———. 1891c (October 2). Some extinct birds of New Zealand. *The Christchurch [New Zealand] Press* 48(8000): 2.

———. 1892a. New extinct rail [telegram]. *Nature* 45: 252.

———. 1892b (March). On a recent discovery of the remains of extinct birds in New Zealand. *Nature* 45(1166): 416–418.

———. 1892c. On a recent discovery of the remains of extinct birds in New Zealand. *Science* 19(476): 163–165.

———. 1892d. Preliminary notice of additions to the extinct avifauna of New Zealand [abstract]. *Transactions and Proceedings of the New Zealand Institute* 24: 185–189.

———. 1892e. *Aphanapteryx* in the New Zealand region. *Nature* 45: 580.

———. 1892f. *Aphanapteryx* and other remains in the Chatham Islands. *Nature* 46: 252–253.

———. 1892g. [Exhibition and remarks on the osteological remains of birds from the Chatham Islands, and a description of the new genera *Diaphorapteryx* and *Palaeocorax*.] *Bulletin of the British Ornithologists' Club* 1: 21–22.

———. 1892h. Remarks on a series of extinct birds of New Zealand, recently discovered. *Report of the British Association* 1892: 760.

———. 1892i. Evidence of a wing in *Dinornis*. *Nature* 45: 257.

———. 1893a. [Return of *Diaphorapteryx* to *Aphanapteryx*.] *Bulletin of the British Ornithologists' Club* 1: 50–51.

———. 1893b. The moas of New Zealand. *Natural Science* 2: 374–380.

———. 1893c. On *Anomalopteryx antiqua* Hutton, and other species of moa from Enfield, New Zealand. *Natural Science* 3: 318–319.

———. 1893d. Antarctica: A supposed former southern continent. *Natural Science* 3: 54–57.

———. 1893e. [Repeats 1892g.] *Ibis* (6)5: 253–254.

———. 1893f. A list of the birds inhabiting the Chatham Islands. *Ibis* (6)5: 521–546.

———. 1900. Catalogue of the lizard-tailed (Saururae), the toothed (Odontornithes) and the ostrich-like (Struthiones) birds, and of the tinamous (Tinami) and the divers (Colymbi) in the Derby Museum. *Bulletin of the Liverpool Museums* 3: 25–34.

Fordyce, R. E. 1982. The fossil vertebrate record of New Zealand. In P. Vickers-Rich and E. M. Thompson (eds.), *The Fossil Vertebrate Record of Australasia,* pp. 629–698. Clayton, Victoria: Monash University Offset Printing Unit.

———. 1985. Late Eocene archaeocete whale (Archaeoceti: Dorudontinae) from Waihao, South Canterbury, New Zealand. *New Zealand Journal of Geology and Geophysics* 28: 351–357.

———. 1991. A new look at the fossil vertebrate record of New Zealand. In P. Vickers-Rich, J. M. Monaghan, R. F. Baird, and T. H. Rich (eds.), *Vertebrate Palaeontology of Australasia,* pp. 1191–1316. Melbourne: Pioneer Design Studio and Monash University Publications Committee.

———. 1994. *Waipatia maerewhenua,* new genus and new species (Waipatiidae, new family), an archaic Late Oligocene dolphin (Cetacea: Odontoceti: Platanistoidea) from New Zealand. *Proceedings of the San Diego Museum of Natural History* 29: 147–176.

Fordyce, R. E., and L. G. Barnes. 1994. The evolutionary history of whales and dolphins. *Annual Reviews of Earth and Planetary Sciences* 22: 419–455.

Fordyce, R. E., and C. M. Jones. 1990. Penguin history and new fossil material from New Zealand. In J. Darby (ed.), *Penguin Biology,* pp. 419–446. San Diego: Academic Press.

Forrest, R. M. 1987. A partially mummified skeleton of *Anomalopteryx didiformis* from Southland. *Journal of the Royal Society of New Zealand* 17: 399–408.

Fox, N. C. 1978a. The biology of the New Zealand falcon (*Falco novaeseelandiae* Gmelin 1788). Ph.D. thesis. University of Canterbury, Christchurch.

———. 1978b. The distribution and numbers of New Zealand falcons (*Falco novaeseelandiae*). *Notornis* 25: 317–331.

———. 1988. A taxonomic redescription of the New Zealand falcon, *Falco novaeseelandiae* Gmelin, 1788. *Notornis* 35: 270–272.

Fraser, T. 1873. A description of the Earnscleugh moa cave. *Transactions and Proceedings of the New Zealand Institute* 5: 102–105.

Friedmann, H. 1950. The birds of North and Middle America. Part XI. Order Falconiformes: Diurnal birds of prey. *Bulletin of the United States National Museum* 50: 1–527.

Frith, H. J. (ed.). 1969. *Birds in the Australian High Country.* Sydney: A. H. and A. W. Reed.

Froggatt, P. C., and D. J. Lowe. 1990. A review of late Quaternary silicic and some other tephra formations from New Zealand: Their stratigraphy, nomenclature, distribution, volume, and age. *New Zealand Journal of Geology and Geophysics* 33: 89–109.

Fuller, E. 1986. *Extinct Birds.* London: Rainbird/Viking.

Fürbringer, M. 1888. *Untersuchungen zur Morphologie und Systematik der Vogel II.* Amsterdam: Allgemeiner Theil Van Holkema.

Gage, M. 1958. Late Pleistocene glaciations of the Waimakariri Valley, Canterbury, New Zealand. *New Zealand Journal of Geology and Geophysics* 1: 123–155.

———. 1977. Glacial geology. In C. J. Burrows (ed.), *Cass: History and*

Science in the Cass District, Canterbury, New Zealand. Christchurch: Department of Botany, University of Canterbury.

Galbreath, R. 1989. *Walter Buller: The Reluctant Conservationist.* Wellington: GP Books.

Gaze, P. D., and B. M. Fitzgerald. 1982. Food of honeyeaters on Little Barrier Island. *Notornis* 29: 209–213.

George, J. C., and A. J. Berger. 1966. *Avian Myology.* New York: Academic Press.

Gibb, J. G. 1986. A New Zealand regional holocene eustatic sea level curve and its application to determination of vertical tectonic movements. A contribution to IGCP project 200. *Royal Society of New Zealand Bulletin* 24: 377–395.

Giffin, E. B. 1990. Gross spinal anatomy and limb use in living and fossil reptiles. *Paleobiology* 16: 448–458.

Gill, B. J. 1985. Subfossil bones of a large skink (Reptilia: Lacertilia) from Motutapu Island, New Zealand. *Records of the Auckland Institute and Museum* 22: 69–76.

———. 1989. Bones of contention—What do we know about the posture of moas? *AGMANZ Journal* 20(1): 30–32.

———. 1996a. A fossil bone of the rifleman (*Acanthisitta chloris*) from Cape Reinga. *Notornis* 43: 113–114.

———. 1996b. Geographical variation in the bone length of laughing owls (*Sceloglaux albifacies*). *Notornis* 43: 85–90.

———. 1998. Prehistoric breeding sites of New Zealand sea lions (*Phocarctos hookeri,* Carnivora: Otariidae) at North Cape. *Records of the Auckland Museum* 35: 55–64.

———. 2000. Morphometrics of moa eggshell fragments (Aves: Dinornithiformes) from Late Holocene dune-sands of the Karikari Peninsula, New Zealand. *Journal of the Royal Society of New Zealand* 30: 131–145.

Gill, B. J., and B. D. Heather. 1990. *A Flying Start.* Auckland: Random Century, in association with the Ornithological Society of New Zealand.

Gill, B., and P. J. Martinson. 1991. *New Zealand's Extinct Birds.* Auckland: Random Century.

Gillies, T. B. 1872. On the occurrence of footprints of the moa at Poverty Bay. *Transactions and Proceedings of the New Zealand Institute* 4: 127–128.

Given, D. 1996. Flora. In *The Chatham Islands: Heritage and Conservation,* pp. 80–92. Christchurch: Canterbury University Press.

Goh, K. M., and B. P. J. Molloy. 1979. Contaminants in charcoals used for radiocarbon dating. *New Zealand Journal of Science* 22: 39–47.

Goh, K. M., B. P. J. Molloy, and T. A. Rafter. 1977. Radiocarbon dating of Quaternary loess deposits, Banks Peninsula, Canterbury, New Zealand. *Quaternary Research* 7: 177–196.

Golson, J. 1957. New Zealand archaeology, 1957. *Journal of the Polynesian Society* 66: 271–290.

Goodman, S. M. 1999. Holocene bird subfossils from the sites of Ampasambazimba, Antsirabe and Ampoza, Madagasacar: Changes in the avifauna of south central Madagascar over the past few millennia. In N. J. Adams and R. H. Slowtow (eds.), *Proceedings of the 22nd International Ornithological Congress* (Durban), pp. 3071–3083. Johannesburg: BirdLife South Africa.

Goslow, G. E. 1971. The attack and strike of some North American raptors. *Auk* 88: 815–827.

———. 1972. Adaptive mechanisms of the raptor pelvic limb. *Auk* 89: 47–64.

Gould, S. J. 1998. An awful, terrible dinosaurian irony. *Natural History* 2/98: 24–26, 61–68.

Graham, R. W. 1993. Processes of time-averaging in the terrestrial vertebrate record. In S. M. Kidwell and A. K. Behrensmeyer (eds.), *Taphonomic Approaches to Time Resolution in Fossil Assemblages,* pp. 102–117. Short Courses in Paleontology, No. 6. Knoxville: Department of Geological Sciences, University of Tennessee.

Graham, R. W., and E. C. Grimm. 1990. Effects of global climate change on the patterns of terrestrial biological communities. *Trends in Ecology and Evolution* 5(9): 289–293.

Graham, R. W., and E. L. Lundelius. 1984. Coevolutionary disequilibrium and Pleistocene extinctions. In P. S. Martin and R. G. Klein (eds.), *Quaternary Extinctions: A Prehistoric Revolution,* pp. 223–249. Tucson: Arizona University Press.

Grant, P. J. 1965. Major regime changes of the Tukituki River, Hawke's Bay, since about 1650 A.D. *Journal of Hydrology (New Zealand)* 4: 17–30.

———. 1985. Major periods of erosion and alluvial sedimentation in New Zealand during the late Holocene. *Journal of the Royal Society of New Zealand* 15: 67–121.

Grant, P. R. (ed.). 1998. *Evolution on Islands.* Oxford: Oxford University Press.

Grant-Mackie, E. J., and J. A. Grant-Mackie. 1994. A New Zealand evolutionary story: The wrens. *New Zealand Science Teacher* 76: 17–22.

Grant-Mackie, J. A., and G. G. Simpson. 1973. Tertiary penguins from the North Island of New Zealand. *Journal of the Royal Society of New Zealand* 3: 441–452.

Gray, G. R. 1845. Birds. In J. Richardson and J. E. Gray (eds.), *The Zoology of the Voyage of H.M.S.* Erebus *and* Terror, *under the Command of Captain Sir James Clark Ross, R.N., F.R.S., during the years 1839 to 1843.* 2 vols. London: E. W. Janson.

———. 1855. *Catalogue of the Genera and Subgenera of Birds.* London: Longman.

Greene, T. C. 1999. Aspects of the ecology of Antipodes Island parakeet (*Cyanoramphus unicolor*) and Reischek's parakeet (*C. novaezelandiae hochstetteri*) on Antipodes Island, October–November 1995. *Notornis* 46: 301–310.

Greenewalt, C. H. 1962. Dimensional relationships for flying animals. *Smithsonian Miscellaneous Collections* 144(2): 1–45.

———. 1975. The flight of birds: The significant dimensions, their departure from the requirements for dimensional similarity, and the effects on flight aerodynamics of that departure. *Transactions of the American Philosophical Society* 65: 5–67.

Greenway, J. C. 1958. *Extinct and Vanishing Birds of the World.* New York: American Committee for International Wildlife Protection.

Greenwood, R. M., and I. A. E. Atkinson. 1977. Evolution of divaricating plants in New Zealand in relation to moa browsing. *New Zealand Ecological Society Proceedings* 24: 21–23.

Gregg, D. R. 1966. New dates for Pyramid Valley moas. *New Zealand Archaeological Association Newsletter* 9: 155–159.

———. 1972. Holocene stratigraphy and moas at Pyramid Valley, North Canterbury, New Zealand. *Records of the Canterbury Museum* 9: 151–158.

Grey, G. 1873. Description of the extinct gigantic bird of prey Hokioi, by a Maori. *Transactions and Proceedings of the New Zealand Institute* 5: 435.

Griffiths, G. A. 1979. High sediment yields from major rivers of the western Southern Alps, New Zealand. *Nature* 282: 61–63.

Gurney, J. H., Sr. 1854. Note on *Nestor productus,* the extinct parrot of Philip Island. *Zoologist* 1854: 4298.

Gurr, L. 1952. A skeleton of *Notornis hochstetteri* Meyer from Waitati, Otago, and notes on distribution of the bird in the South Island, New Zealand. *Transactions of the Royal Society of New Zealand* 80: 19–21.

Guthrie, R. D. 1984. Mosaics, allelochemics and nutrients: An ecological theory for late Pleistocene extinctions. In P. S. Martin and R. G. Klein (eds.), *Quaternary Extinctions: A Prehistoric Revolution,* pp. 259–298. Tucson: Arizona University Press.

Guthrie-Smith, H. 1925. *Bird Life on Islands and Shore.* Edinburgh: Wm. Blackwood and Sons.

Haast, J. 1869. On the measurements of *Dinornis* bones, obtained from excavations in a swamp situated at Glenmark, on the property of Messrs. Kermode and Co., up to 15th February 1868. *Transactions and Proceedings of the New Zealand Institute* 2nd ed. (1875), 80–89.

———. 1870 (June). On cooking-pits and kitchen-middens in Canterbury [letter]. *Proceedings of the Zoological Society of London for 1870, Part 1,* 4: 53–56.

———. 1872a. Additional notes. *Transactions and Proceedings of the New Zealand Institute* 4: 90–94.

———. 1872b. Moas and moa-hunters. Address to the Philosophical Institute of Canterbury. *Transactions and Proceedings of the New Zealand Institute* 4: 66–90.

———. 1872c. Notes on *Harpagornis moorei,* an extinct gigantic bird of prey, containing description of femur, ungual phalanges, and rib. *Transactions and Proceedings of the New Zealand Institute* 4: 192–196.

———. 1872d. Third paper on moas and moa-hunters. *Transactions and Proceedings of the New Zealand Institute* 4: 94–110.

———. 1874a. On *Harpagornis,* an extinct genus of gigantic raptorial birds of New Zealand. *Transactions and Proceedings of the New Zealand Institute* 6: 62–75.

———. 1874b. Presidential address to the Philosophical Institute of Canterbury. *Transactions and Proceedings of the New Zealand Institute* 6: 419–432.

———. 1874c. Remarks on the extinct birds of New Zealand. *Ibis* 1874: 209–220.

———. 1875a. Notes on a moa-hunter encampment at Shag Point, Otago. *Transactions and Proceedings of the New Zealand Institute* 7: 91–98.

———. 1875b. Researches and excavations carried on, in, and near the Moa Bone Point Cave, Sumner Road, in the year 1872. *Transactions and Proceedings of the New Zealand Institute* 7: 54–85.

———. 1877. Observations on Captain Hutton's paper "Notes on the Maori cooking-places at the mouth of the Shag River" (Trans. N.Z. Inst., VIII., p. 103). *Transactions and Proceedings of the New Zealand Institute* 9: 670–672.

———. 1879. *Geology of the Provinces of Canterbury and Westland, New Zealand. A Report Comprising the Results of Official Explorations.* Christchurch: Times.

———. 1881. On *Harpagornis* [third paper]. *Transactions and Proceedings of the New Zealand Institute* 13: 232–234.
———. 1884. On the leg-bones of a new species of kiwi (*Megalapteryx hectori*). *Transactions and Proceedings of the New Zealand Institute* 16: 576–577.
———. 1885a (October). Preliminary notice of paper on *Dinornis oweni. Proceedings of the Zoological Society of London for 1885, Part 3*, 31: 482.
———. 1885b. Preliminary notice of paper on *Megalapteryx hectori. Proceedings of the Zoological Society of London for 1885*, Part III: 541.
———. 1886a. On *Dinornis oweni*, a new species of the Dinornithidae, with some remarks on *D. curtus. Transactions of the Zoological Society of London* 12: 171–182.
———. 1886b. On *Megalapteryx hectori*, a new gigantic species of apterygian bird. *Transactions of the Zoological Society of London* 12: 161–169.
Haast, H. F. 1948. *The Life and Times of Sir Julius von Haast.* Wellington: The author.
Hamel, G. E. 1979. The breeding ecology of moas. In A. J. Anderson (ed.), *Birds of a Feather,* pp. 61–66. New Zealand Archaeological Association Monograph II; B.A.R. International Series 62.
Hamilton, A. 1889. Notes on a deposit of moa-bones in the Te Aute Swamp, Hawke's Bay. *Transactions and Proceedings of the New Zealand Institute* 21: 311–318.
———. 1892a. Notes on moa gizzard stones. *Transactions and Proceedings of the New Zealand Institute* 24: 172–175.
———. 1892b. On the genus *Aptornis,* with more especial reference to *Aptornis defossor,* Owen. *Transactions and Proceedings of the New Zealand Institute* 24: 175–182.
———. 1893. On the fissures and caves at the Castle Rocks, Southland; with a description of the remains of the existing and extinct birds found in them. *Transactions and Proceedings of the New Zealand Institute* 25: 88–106.
———. 1894a. Materials for a bibliography of the Dinornithidae, the great extinct birds of New Zealand. *Transactions and Proceedings of the New Zealand Institute* 26: 229–257.
———. 1894b. Results of a further exploration of the bone fissure at the Castle Rocks, Southland. *Transactions and Proceedings of the New Zealand Institute* 26: 226–228.
———. 1895a. Further contributions towards a bibliography of the Dinornithidae, the great extinct birds of New Zealand, usually called moas. *Transactions and Proceedings of the New Zealand Institute* 27: 228–231.
———. 1895b. On the feathers of a small species of moa (*Megalapteryx*) found in a cave at the head of the Waikaia River; with a notice of a moa-hunter's camping place on the Old Man Range. *Transactions and Proceedings of the New Zealand Institute* 27: 232–238.
———. 1904a. Notes on the remains of some of the extinct birds of New Zealand found near Ngapara. *Transactions and Proceedings of the New Zealand Institute* 36: 474–477.
———. 1904b. Second supplement to the "Materials for a bibliography of the Dinornithidae." *Transactions and Proceedings of the New Zealand Institute* 26: 471–473.
Hancock, R. G. V., M. D. Grynpas, and K. P. H. Pritzker. 1989. The abuse

of bone analyses for archaeological dietary studies. *Archaeometry* 31: 169–179.

Hand, S. J., P. Murray, D. Megirian, M. Archer, and H. Godthelp. 1998. Mystacinid bats (Microchiroptera) from the Australian Tertiary. *Journal of Paleontology* 72: 538–545.

Harper, P. C. 1983. Biology of the Buller's shearwater (*Puffinus bulleri*) at the Poor Knights Islands, New Zealand. *Notornis* 30: 299–318.

Harrison, C. J. O. (ed.). 1978. *Bird Families of the World.* New York: Harry M. Abrams.

Harrison, C. J. O., and C. A. Walker. 1970. The extinct musk duck (*Biziura*) of New Zealand: A re-appraisal of *B. lautouri*. *Bulletin of the British Ornithologists' Club* 90: 6–10.

———. 1973. An undescribed extinct fish eagle from the Chatham Islands. *Ibis* 115(2): 274–277.

———. 1976. A review of the bony-toothed birds (Odontopterygiformes) with descriptions of some new species. *Tertiary Research Special Papers* 2: 1–62.

Hartman, F. A. 1961. Locomotor mechanisms of birds. *Smithsonian Institution Miscellaneous Collections* 143: 1–91.

Hartree, W. H. 1999. A preliminary report on the nesting habits of moas on the East Coast of the North Island. *Notornis* 46: 457–460.

Hawke, D. J., R. N. Holdaway, J. E. Causer, and S. Ogden. 1999. Soil indicators of pre-European seabird breeding in New Zealand at sites identified by predator deposits. *Australian Journal of Soil Research* 37: 103–113.

Heather, B. D., and H. A. Robertson. 1996. *The Field Guide to the Birds of New Zealand.* Auckland: Viking.

Hector, J. 1868 (April). Notice of an egg of the great moa (*Dinornis gigantea*), containing remains of an embryo, found in the province of Otago, New Zealand. *Proceedings of the Zoological Society of London for 1867, Part 3,* 6: 991–992.

———. 1870. Preliminary notes on the bones of a fossil penguin [abstract]. *Transactions and Proceedings of the New Zealand Institute* 2: 403.

———. 1871. On recent moa remains in New Zealand. *Nature* 4: 184–6.

———. 1872a. On recent moa remains in New Zealand. *Transactions and Proceedings of the New Zealand Institute* 4: 110–120.

———. 1872b. On the remains of a gigantic penguin (*Palaeeudyptes antarcticus,* Huxley) from the Tertiary rocks on the west coast of Nelson. *Transactions and Proceedings of the New Zealand Institute* 4: 341–346.

———. 1873. Further notice of bones of a fossil penguin (*Palaeeudyptes antarcticus,* Huxley). *Transactions and Proceedings of the New Zealand Institute* 5: 438–439.

———. 1874. On *Cnemiornis calcitrans,* Owen, showing its affinity to the Lamellirostrate Natatores. *Transactions and Proceedings of the New Zealand Institute* 6: 76–84.

———. 1883. Remarks on some bones lately discovered by Mr H. T. Wharton in caves at Highfield, Canterbury. *Transactions and Proceedings of the New Zealand Institute* 15: 526.

———. 1893. Summary report on moa, 1st meeting of Otago Institute. *Transactions and Proceedings of the New Zealand Institute* 25: 555–558.

———. 1895. On the discovery of bones of *Dinornis giganteus* on the surface. *Transactions and Proceedings of the New Zealand Institute* 27: 655–656.

Hemming, F. 1954. Opinion 229. Acceptance of the lectotype selected by Lydekker (1891) for the nominal species *Dinornis novaezealandiae* Owen, 1843 (class Aves). *Opinions and Declarations Rendered by the International Commission on Zoological Nomenclature* 4(19): 221–230.

Hendy, C. 1969. Isotopic studies of speleothems. *New Zealand Speleological Bulletin* 4(71): 306–319.

Henry, R. 1903. *The Habits of the Flightless Birds of New Zealand; with Notes on Other New Zealand Birds.* Wellington: John McKay, New Zealand Government Printer.

Hertel, F. 1992. Morphological diversity of past and present New World vultures. In K. E. Campbell (ed.), *Papers in Avian Paleontology Honoring Pierce Brodkorb*, pp. 413–418. Natural History Museum of Los Angeles County Science Series, No. 36.

———. 1995. Ecomorphological indicators of feeding behavior in recent and fossil raptors. *Auk* 112: 890–903.

Higgins, P. J. (ed.). 1999. *Handbook of Australian, New Zealand and Antarctic Birds.* Vol. 4, *Parrots to Dollarbirds.* Melbourne: Oxford University Press.

Higgins, P. J., and S. J. J. F. Davies. (eds.). 1996. *Handbook of Australian, New Zealand and Antarctic Birds.* Vol. 3, *Snipe to Pigeons.* Melbourne: Oxford University Press.

Higham, T. 1994. Radiocarbon dating New Zealand prehistory with moa eggshell: Some preliminary results. *Quaternary Geochronology (Quaternary Science Reviews)* 13: 163–169.

Higham, T. F. G., and A. G. Hogg. 1997. Evidence for late Polynesian colonisation of New Zealand: University of Waikato radiocarbon measurements. *Radiocarbon* 39: 149–192.

Higham, T., A. Anderson, and C. Jacomb. 1999. Dating the first New Zealanders: The chronology of Wairau Bar. *Antiquity* 73: 420–427.

Hill, H. 1889. Discovery of fossil moa-feathers in rocks of Pliocene age. *Transactions and Proceedings of the New Zealand Institute* 21: 318–320.

———. 1895. On the occurrence of moa footprints in the bed of the Manawatu River near Palmerston North. *Transactions and Proceedings of the New Zealand Institute* 27: 476–477.

———. 1914. The moa—Legendary, historical and geological: Why and when the moa disappeared. *Transactions and Proceedings of the New Zealand Institute* 46: 330–351.

Hill, M. O. 1979. TWINSPAN—A FORTRAN program for arranging multivariate data in an ordered two-way table by classification of the individuals and attributes. Ithaca, N.Y.: Cornell University Press.

Hill, J. E., and M. J. Daniel. 1985. Systematics of the New Zealand short-tailed bat *Mystacina* Gray, 1843 (Chiroptera: Mystacinidae). *Bulletin of the British Museum of Natural History (Zoology)* 48: 279–300.

Hill, M. O., and H. G. Gauch. 1980. Detrended correspondence analysis: An improved ordination technique. *Vegetatio* 42: 47–58.

Hillis, D. M., and J. P. Huelsenbeck. 1992. Signal, noise, and reliability in molecular phylogenetic analyses. *Journal of Heredity* 83: 189–195.

Hobson, K. A., and S. G. Sealy. 1991. Marine protein contributions to the diet of northern saw-whet owls on the Queen Charlotte Sound Islands: A stable-isotope approach. *Auk* 108: 437–440.

Hobson, K. A., J. L. Sease, R. L. Merrick, and J. F. Piatt. 1997. Investigating trophic relationships of pinnipeds in Alaska and Washington using stable isotope ratios of nitrogen and carbon. *Marine Mammal Science* 13: 114–132.

Hochstetter, F. von 1867. *New Zealand: Its Physical Geography, Geology and Natural History, with Special Reference to Results of Government Expeditions in the Provinces of Auckland and Nelson.* Stuttgart: J. G. Cotta.

Holdaway, R. N. 1989. New Zealand's pre-human avifauna and its vulnerability. *New Zealand Journal of Ecology* 12(Suppl.): 11–25.

———. 1990. *Harpagornis assimilis* Haast, 1874, a synonym of *Harpagornis moorei* Haast, 1872 (Aves: Accipitridae). *New Zealand Natural Sciences* 17: 39–47.

———. 1991. Systematics and palaeobiology of Haast's eagle (*Harpagornis moorei* Haast, 1872) (Aves: Accipitridae). Ph.D. thesis. University of Canterbury, Christchurch.

———. 1994. An exploratory phylogenetic analysis of the genera of the Accipitridae, with notes on the biogeography of the family. In B.-U. Meyburg and R. D. Chancellor (eds.), *Raptor Conservation Today,* pp. 601–649. Berlin: World Working Group on Birds of Prey and Owls and The Pica Press.

———. 1995. A fossil record of the black stilt *Himantopus novaezelandiae* Gould, 1841. *New Zealand Natural Sciences* 22: 69–74.

———. 1996a. Arrival of rats in New Zealand. *Nature* 384: 225–226.

———. 1996b. Stephens Island—A chance lost. *New Zealand Geographic* 32: 81–82.

———. 1999a. A spatio-temporal model for the invasion of the New Zealand archipelago by the Pacific rat (*Rattus exulans*). *Journal of the Royal Society of New Zealand* 29: 91–105.

———. 1999b. Introduced predators and avifaunal extinction in New Zealand. In R. D. E. MacPhee (ed.), *Extinctions in Near Time: Causes, Contexts, and Consequences,* pp. 189–238. New York: Kluwer Academic/Plenum Publishers.

Holdaway, R. N., and A. J. Anderson. 1998. ^{14}C AMS dates on *Rattus exulans* bones from natural and archaeological contexts on Norfolk Island south-west Pacific. *Archaeology in New Zealand* 41: 195–198.

———. 2001. The avifauna of the Emily Bay archaeological site, Norfolk Island, south-west Pacific: A preliminary account. *Records of the Australian Museum,* supplement 27.

Holdaway, R. N., and N. R. Beavan. 1999. Reliable ^{14}C AMS dates on bird and Pacific rat *Rattus exulans* bone gelatin, from a $CaCO_3$-rich deposit. *Journal of the Royal Society of New Zealand* 29: 185–211.

Holdaway, R. N., and D. H. Brathwaite. 1987. New Zealand's prehuman avian ecology: A scenario (without moas) [abstract]. *New Zealand Journal of Ecology* 10: 161.

Holdaway, R. N., and C. Jacomb. 2000. Rapid extinction of the moas (Aves: Dinornithiformes): Model, test, and implications. *Science* 287(5461): 2250–2254.

Holdaway, R. N., and R. V. J. Varman. In preparation. A probable predator-derived deposit of vertebrate fossils from the dunes at Cemetery Bay, Norfolk Island, Southwest Pacific.

Holdaway, R. N., and T. H. Worthy. 1993. First North Island fossil record of Kea, and morphological and morphometric comparison of Kea and Kaka. *Notornis* 40: 95–108.

———. 1994. A new fossil species of shearwater *Puffinus* from the late Quaternary of the South Island, New Zealand, and notes on the biogeography and evolution of the *Puffinus gavia* superspecies. *Emu* 94: 201–215.

———. 1996. Diet and biology of the laughing owl *Sceloglaux albifacies* (Aves: Strigidae) on Takaka Hill, Nelson, New Zealand. *Journal of Zoology, London* 239: 545–573.

———. 1997. A reappraisal of the late Quaternary fossil vertebrates of Pyramid Valley Swamp, North Canterbury, New Zealand. *New Zealand Journal of Zoology* 24: 69–121.

Holdaway, R. N., T. H. Worthy, and A. J. T. Tennyson. 2001. A working list of breeding bird species of the New Zealand region at first human contact. *New Zealand Journal of Zoology* 28 (in press).

Hood, T. H. C. 1874. *Dinornis* footprints and recent extinction. *Proceedings of the Royal Society of Edinburgh* 8: 236–240.

Horn, P. L. 1980. Probable occurrence of the black bittern *Dupetor flavicollis* (Linnaeus) in New Zealand. *Notornis* 35: 153–157.

———. 1983. Subfossil avian remains from Poukawa, Hawke's Bay, and the first record of *Oxyura australis* (blue-billed duck) from New Zealand. *Journal of the Royal Society of New Zealand* 13: 67–78.

Horner, J. R., A. de Ricqles, and K. Padian. 2000. Long bone histology of the hadrosaurid dinosaur *Maiasaura peeblesorum*: Growth dynamics and physiology based on an ontogenetic series of skeletal elements. *Journal of Vertebrate Paleontology* 20: 115–129.

Houde, P. 1986. Ostrich ancestors found in the Northern Hemisphere suggest new hypothesis of ratite origins. *Nature* 324: 563–565.

———. 1988. Palaeognathous birds from the early Tertiary of the Northern Hemisphere. *Publications of the Nuttall Ornithological Society* 22: 1–148.

Houde, P., and H. Haubold. 1987. *Palaeotis weigelti* restudied: A small Eocene ostrich (Aves: Struthioniformes). *Palaeovertebrata* 17: 27–42.

Houde, P., A. Cooper, E. Leslie, A. E. Shand, and G. A. Montano. 1997. Phylogeny and evolution of 12S rDNA in gruiformes (Aves). In D. P. Mindell (ed.), *Avian Molecular Evolution and Systematics,* pp. 117–154. San Diego: Academic Press.

Howard, H. 1932. *Eagles and Eagle-Like Vultures of the Pleistocene of Rancho La Brea.* Contributions to Palaeontology. Carnegie Institute of Washington Publication, No. 429, 1–111.

———. 1964. Fossil Anseriformes. In J. Delacour (ed.), *Waterfowl of the World,* pp. 233–326. Vol. 4. London: Country Life.

Howard, H., and S. L. Warter. 1969. A new species of bony-toothed bird (family Pseudodontornithidae) from the Tertiary of New Zealand. *Records of the Canterbury Museum* 8: 345–357.

Huq, L. N., S. M. Rambaud, L. The, A. D. Davies, B. McCulloch, M. M. Trotter, and G. E. Chapman. 1985. Immunochemical detection and characterisation of osteocalcin from Moa bone. *Biochemical and Biophysical Research Communications* 129: 714–720.

Hutton, F. W. 1872a. Notes on some birds from the Chatham Islands collected by H. H. Travers, Esq.; with descriptions of two new species. *Ibis,* Series 3, 2: 243–250.

———. 1872b. On some moa feathers. *Transactions and Proceedings of the New Zealand Institute* 4: 172–173.

———. 1872c. On the microscopical structure of the egg-shell of the moa. *Transactions and Proceedings of the New Zealand Institute* 4: 166–167.

———. 1873a. Notes on some of the birds brought by Mr Henry Travers

from the Chatham Islands, with descriptions of new species. *Transactions and Proceedings of the New Zealand Institute* 5: 223–224.
———. 1873b. On the geographical relations of the New Zealand fauna. *Transactions and Proceedings of the New Zealand Institute* 5: 227–256.
———. 1874. On a new genus of Rallidae. *Transactions and Proceedings of the New Zealand Institute* 6: 108–110.
———. 1875. On the dimensions of *Dinornis* bones (in Hamilton Swamp). *Transactions and Proceedings of the New Zealand Institute* 7: 274–278.
———. 1877. Remarks on Dr von Haast's classification of the moas. *Transactions and Proceedings of the New Zealand Institute* 9: 363–365.
———. 1891a. On the classification of the moas [abstract]. *New Zealand Journal of Science*, n.s., 1: 247–249.
———. 1891b. History of the moas. *New Zealand Journal of Science*, n.s., 1: 261–263.
———. 1892a. On the origin of the struthious birds of Australasia. *Report of the Fourth Meeting of the Australasian Association for the Advancement of Science* (Hobart) 1892, pp. 365–369.
———. 1892b. The history of moas, or extinct flightless birds of New Zealand, 1. *Natural Science* 1: 588–592.
———. 1892c. The moas of New Zealand. *Transactions and Proceedings of the New Zealand Institute* 24: 93–172.
———. 1893a. The moas of New Zealand. *Natural Science* 3: 317–318.
———. 1893b. On new species of moa. *Transactions and Proceedings of the New Zealand Institute* 25: 6–13.
———. 1893c. On *Anomalopteryx antiqua*. *Transactions and Proceedings of the New Zealand Institute* 25: 14–16.
———. 1894. On *Dinornis* (?)*queenslandiae*. *Proceedings of the Linnaean Society of New South Wales* 8: 7–10.
———. 1895a. On the axial skeleton in the Dinornithidae. *Transactions and Proceedings of the New Zealand Institute* 27: 157–173.
———. 1895b. On the occurrence of a pneumatic foramen in the femur of a moa. *Transactions and Proceedings of the New Zealand Institute* 27: 173–174.
———. 1896a. On a deposit of moa-bones at Kapua. *Transactions and Proceedings of the New Zealand Institute* 28: 627–644.
———. 1896b. On the moa bones from Enfield. *Transactions and Proceedings of the New Zealand Institute* 28: 645–650.
———. 1896c. On the discovery of moa remains on Riverton Beach. *Transactions and Proceedings of the New Zealand Institute* 28: 651–652.
———. 1897a. On the leg bones of *Meionornis* from Glenmark. *Transactions and Proceedings of the New Zealand Institute* 29: 558–560.
———. 1897b. On two moa skulls in the Canterbury Museum. *Transactions and Proceedings of the New Zealand Institute* 29: 561–564.
———. 1897c. The moas of the North Island of New Zealand. *Transactions and Proceedings of the New Zealand Institute* 29: 541–557.
———. 1899. On the footprint of a kiwi-like bird from Manaroa (Pelorus Sound). *Transactions and Proceedings of the New Zealand Institute* 31: 486.
———. 1906. On a skeleton of *Emeus crassus* from the North Island. *Transactions and Proceedings of the New Zealand Institute* 38: 66–67.
Hutton, F. W., and M. Coughtrey. 1875a. Descriptions of some moa remains from the Knobby Ranges. *Transactions and Proceedings of the New Zealand Institute* 7: 266–273.

———. 1875b. Notes on the Earnscleugh Cave. With remarks on some of the more valuable moa remains found in it by Professor Millen Coughtrey. *Transactions and Proceedings of the New Zealand Institute* 7: 138–144.

Huxley, T. H. 1867 (October). On the classification of birds; and on the taxonomic value of the modifications of certain of the cranial bones observable in that class. *Proceedings of the Zoological Society of London for 1867,* Part II: 415–472.

Imber, M. J. 1978. The effects of rats on the breeding success of petrels. In P. R. Dingwall, I. A. E. Atkinson, and C. Hay (eds.), *The Ecology and Control of Rodents in New Zealand Nature Reserves,* pp. 67–71. New Zealand Department of Lands and Survey Information Series, No. 4.

Innes, J., and R. Hay. 1995. The nesting of the North Island kokako (*Callaeas cinerea wilsoni*)—Review of accounts from 1880 to 1989. *Notornis* 42: 79–93.

International Commission on Zoological Nomenclature. 1985. *International Code of Zoological Nomenclature.* London: International Trust for Zoological Nomenclature.

Irwin, G. 1992. *The Prehistoric Exploration and Colonisation of the Pacific.* Cambridge: Cambridge University Press.

Jäger, G. 1863. Bericht über ein fast vollständiges skelet von *Palapteryx ingens* über dessen restauration und die davon angefortigten Gypsabgusse, mit einigen Bemerkungen über die Aufstellung der Vogelskelete überhaupf. Vienna: Wilhelm Braunmüller.

———. 1864. Bericht über einen fast vollständingen Schädel von *Palapteryx. Reis der Novara, Geologie II,* Part 2: 305–318.

Jaksić, F. M., and H. E. Braker. 1983. Food-niche relationships and guild structures of diurnal birds of prey: Competition versus opportunism. *Canadian Journal of Zoology* 61: 2230–2241.

James, H. F., and S. L. Olson. 1991. Descriptions of thirty-two new species of birds from the Hawaiian islands: Part II. Passeriformes. *Ornithological Monographs* 46: 1–88.

Jamieson, I. G., and H. G. Spencer. 1996. The bill and foraging behaviour of the huia (*Heteralocha acutirostris*): Were they unique? *Notornis* 43: 14–18.

Johnson, K. P., and M. D. Sorenson. 1999. Phylogeny and biogeography of dabbling ducks (genus *Anas*): A comparison of molecular and morphological evidence. *Auk* 116: 792–805.

Jollie, M. 1976. A contribution to the morphology and phylogeny of the Falconiformes, part 1. *Evolutionary Theory* 1: 285–298.

———. 1977a. A contribution to the morphology and phylogeny of the Falconiformes, part 2. *Evolutionary Theory* 2: 115–208.

———. 1977b. A contribution to the morphology and phylogeny of the Falconiformes, part 3. *Evolutionary Theory* 2: 209–300.

———. 1977c. A contribution to the morphology and phylogeny of the Falconiformes, part 4. *Evolutionary Theory* 3: 1–141.

Jones, C. M., and A. Mannering. 1997. New Paleocene fossil bird material from the Waipara Greensand, North Canterbury, New Zealand. *Geological Society of New Zealand Miscellaneous Publication* 95a: 88.

Jones, D. N., R. W. R. J. Dekker, and C. S. Roselaar. 1995. *The Megapodes, Megapodiidae.* Oxford: Oxford University Press.

Kelly, D., J. J. Ladley, A. W. Robertson, J. Edwards, and D. C. Smith. 1996. The birds and the bees. *Nature* 384: 615.

Kemp, A. 1989. Estimation of biological indices for little-known African owls. In B.-U. Meyburg and R. D. Chancellor (eds.), *Raptors in the*

Modern World, pp. 441–450. Berlin: World Working Group on Birds of Prey and Owls.

Kennedy, M., and H. G. Spencer. 2000. Phylogeny, biogeography and taxonomy of Australasian teals. *Auk* 117: 154–163.

Kennedy, M., A. M. Paterson, J. C. Morales, S. Parsons, A. Winnington, and H. G. Spencer. 1999. The long and the short of it: Branch lengths and the problem of placing the New Zealand short-tailed bat, *Mystacina. Molecular Phylogenetics and Evolution* 13: 405–416.

Keyes, I. W. 1977. How many species of moa? *New Zealand Archaeological Association Newsletter* 20(3): 195–199.

Keyes, I. W., and P. Froggatt. 1978. A second record of a moa bone from the sea-floor of Cook Strait, New Zealand. *New Zealand Journal of Geology and Geophysics* 21: 663–664.

King, P. G. 1786–1790. Manuscript journal. In Mitchell Library, Sydney. Microfilm copy CY reel 13 of C115 (not seen; cited in Schodde et al. 1983).

King, J. E. 1973. Pleistocene Ross seal (*Ommatophoca rossi*) from New Zealand. *New Zealand Journal of Marine and Freshwater Research* 7: 391–397.

———. 1983. The Ohope skull—A new species of Pleistocene sea lion from New Zealand. *New Zealand Journal of Marine and Freshwater Research* 17: 105–120.

King, C. M. (ed.). 1990 *The Handbook of New Zealand Mammals.* Auckland: Oxford University Press.

Kinsky, F. C. (ed.). 1970. *Annotated Checklist of the Birds of New Zealand Including the Birds of the Ross Dependency.* Wellington: A. H. and A. W. Reed, for the Ornithological Society of New Zealand.

Kitchener, A. C. 1993. On the external appearance of the dodo, *Raphus cucullatus* (L. 1758). *Archives of Natural History* 20: 279–301.

Kline, T. C., J. J. Goering, O. A. Mathisen, P. H. Poe, P. L. Parker, and R. S. Scalan. 1993. Recycling of elements transported upstream by runs of Pacific salmon. II. δ^{15}N and δ^{13}C evidence in the Kvichak River watershed, Bristol Bay, southwestern Alaska. *Canadian Journal of Fisheries and Aquatic Sciences* 50: 2350–2365.

Koch, P. L., M. L. Fogel, and N. Tuross. 1994. Tracing the diet of fossil animals using stable isotopes. In K. Lajtha and R. H. Michener (eds.), *Stable Isotopes in Ecology and Environmental Science,* pp. 63–92. Oxford: Blackwell Scientific.

Köhler, R. 1995. A new species of the fossil turtle *Psephophorus* (order Testudines) from the Eocene of the South Island, New Zealand. *Journal of the Royal Society of New Zealand* 25: 371–384.

Köhler, R., and R. E. Fordyce. 1997. An archaeocete whale (Cetacea: Archaeoceti) from the Eocene Waihao Greensand, New Zealand. *Journal of Vertebrate Paleontology* 17: 574–583.

König, C. 1982. Zur systematischen Stellung der Neuweltgeier (Cathartidae). *Journal für Ornithologie* 123: 259–267.

Krebs, C. J. 1978. *Ecology: The Experimental Analysis of Distribution and Abundance.* 2nd ed. New York: Harper.

Krösche, O. 1963. Die Moa-Strausse, Neuseelands ausgestorbene Riesenvögel. Wittenberg Lutherstadt: A. Ziemsen Verlag.

Kurtén, B. 1955. Sex dimorphism and size trends in the cave bear, *Ursus spelaeus* Rosenmuller and Heinroth. *Acta Zoologica Fennica* 90: 1–48.

Kuschel, G., and T. H. Worthy. 1996. Past distribution of large weevils

(Coleoptera: Curculionidae) in the South Island, New Zealand, based on Holocene fossil remains. *New Zealand Entomologist* 19: 15–22.

Ladley, J. J., and D. Kelly. 1996. Dispersal, germination and survival of New Zealand mistletoes (Loranthaceae): Dependence on birds. *New Zealand Journal of Ecology* 20: 69–79.

Lajtha, K., and J. D. Marshall. 1994. Sources of variation in the stable isotopic composition of plants. In K. Lajtha and R. H. Michener (eds.), *Stable Isotopes in Ecology and Environmental Science*, pp. 1–21. Oxford: Blackwell Scientific.

Lamberton, C. 1934. Ratites subfossiles de Madagascar, Les Mullerornithidae. In Lémuriens et Ratites: Contribution à la connaissance de la Faune subfossile de Madagascar. *Mémoires de L'Académie Malgache* 17: 125–168.

Lambrecht, K. 1933. *Handbuch der Palaeornithologie*. Berlin: Gebrüder Borntraeger.

Latter, J. H., E. F. Lloyd, I. E. M. Smith, and S. Nathan. 1992. *Volcanic Hazards in the Kermadec Islands, and at Submarine Volcanoes between Southern Tonga and New Zealand*. New Zealand Ministry of Civil Defence Volcanic Hazards Information Series, No. 4. Wellington: Ministry of Civil Defence.

Law, G. 1972. Sources of moas and moa-hunters. *New Zealand Archaeological Association Newsletter* 15: 4–15.

———. 1973. Tokoroa moa-hunter site N75/1. *New Zealand Archaeological Association Newsletter* 16: 150–164.

Lawver, L. A., L. M. Gahagan, and M. F. Coffin. 1992. The development of paleoseaways around Antarctica. *American Geophysical Union, Antarctic Research Series* 56: 7–30.

Leaghy, A. 1976. Whakamoenga Cave, Taupo, N94/7. A report on the ecology, economy and stratigraphy. *Records of the Auckland Institute and Museum* 13: 29–75.

Lee, K., J. Feinstein, and J. Cracraft. 1997. The phylogeny of ratite birds: Resolving conflicts between molecular data and morphological data sets. In D. Mindell (ed.), *Avian Molecular Evolution and Systematics*, pp. 173–211. New York: Academic Press.

Lewis, K. B., L. Carter, and F. J. Davey. 1994. The opening of Cook Strait: Interglacial tidal scour and aligning basins at a subduction to transform plate edge. *Marine Geology* 116: 293–312.

Ligon, J. D. 1967. Relationships of the cathartid vultures. *Occasional Papers of the Museum of Zoology, University of Michigan* 651: 1–26.

Lintott, W. H., and C. J. Burrows. 1973. A pollen diagram and macrofossils from Kettlehole Bog, Cass, South Island, New Zealand. *New Zealand Journal of Botany* 11: 269–282.

Liversidge, F. C. S. 1881. An analysis of moa eggshell. *Transactions and Proceedings of the New Zealand Institute* 13: 225–227.

Livezey, B. C. 1986. A phylogenetyic analysis of recent anseriform genera using morphological characters. *Auk* 105: 681–698.

———. 1989a. Phylogenetic relationships and incipient flightlessness in the extinct Auckland Islands merganser. *Wilson Bulletin* 101: 410–435.

———. 1989b. Phylogenetic relationships of several subfossil anseriformes of New Zealand. *Occasional Papers of the Museum of Natural History of the University of Kansas* 128: 1–25.

———. 1990. Evolutionary morphology of flightlessness in the Auckland Islands teal. *Condor* 92: 639–673.

———. 1991. A phylogenetic analysis and classification of Recent dabbling ducks (Tribe Anatini) based on comparative morphology. *Auk* 108: 471–507.

———. 1992. Morphological corollaries and ecological implications of flightlessness in the kakapo (Psittaciformes: *Strigops habroptilus*). *Journal of Morphology* 213: 105–145.

———. 1994. The carpometacarpus of *Apterornis*. *Notornis* 41: 51–60.

———. 1996a. A phylogenetic analysis of geese and swans (Anseriformes: Anserinae), including selected fossil species. *Systematic Biology* 45: 415–450.

———. 1996b. A phylogenetic analysis of modern pochards (Anatidae: Aythyini). *Auk* 113: 74–93.

———. 1996c. A phylogenetic reassessment of the tadornine–anatine divergence (Aves: Anseriformes: Anatidae). *Annals of the Carnegie Museum* 65: 27–88.

———. 1997a. A phylogenetic analysis of basal Anseriformes, the fossil *Presbyornis*, and the interordinal relationship of waterfowl. *Zoological Journal of the Linnaean Society* 121: 361–428.

———. 1997b. A phylogenetic analysis of modern shelducks and sheldgeese (Anatidae: Tadornini). *Ibis* 139: 51–66.

———. 1997c. A phylogenetic classification of waterfowl (Aves: Anseriformes), including selected fossil species. *Annals of the Carnegie Museum* 67: 457–496.

———. 1998. A phylogenetic analysis of the Gruiformes (Aves) based on morphological characters, with an emphasis on the rails (Rallidae). *Philosophical Transactions of the Royal Society of London: Biological Sciences B* 353(1378): 2077–2151.

Lomolino, M. V., and R. Channell. 1995. Splendid isolation: Patterns of geographic range collapse in endangered mammals. *Journal of Mammalogy* 76: 335–347.

Long, J. A. 1998. *Dinosaurs of Australia and New Zealand and Other Animals of the Mesozoic*. Cambridge, Mass.: Harvard University Press.

Loomis, L. M. 1918. A review of the albatrosses, petrels, and diving petrels. *Proceedings of the California Academy of Sciences* 2(2), 12: 1–187.

Lovegrove, T. G. 1996a. Island releases of saddlebacks *Philesturnus carunculatus* in New Zealand. *Biological Conservation* 77: 151–157.

———. 1996b. A comparison of the effects of predation by Norway (*Rattus norvegicus*) and Polynesian rats (*R. exulans*) on the saddleback (*Philesturnus carunculatus*). *Notornis* 43: 91–112.

Lowe, P. R. 1915. Studies on the Charadriiformes. II. On the osteology of the Chatham Island snipe (*Coenocorypha pusilla* Buller). *Ibis* 1915: 690–716.

———. 1926. More notes on the quadrate as a factor in avian classification. *Ibis* 68: 152–188.

———. 1928. Studies and observations bearing on the phylogeny of the ostrich and its allies. *Proceedings of the Zoological Society of London for 1928*, Part I, xi: 185–247.

Lowry, J. B. 1980. Evolution of divaricating plants in New Zealand in relation to moa browsing. *New Zealand Journal of Ecology* 3: 165.

Lydekker, R. 1891. *Catalogue of fossil birds in the British Museum (Natural History)*. London: British Museum (Natural History).

———. 1892a. On a new species of moa. *Proceedings of the Zoological Society of London for 1891*, Part I, xxxiii: 479–482.

———. 1892b. The history of the moas or extinct flightless birds of New Zealand. *Natural Science* 1: 588–595.

MacPhee, R. D. E., and P. A. Marx. 1999. The 40,000-year plague: Humans, hyperdisease, and first first-contact extinctions. In S. Goodman and G. B. Patterson (eds.), *Natural Change and Human Impact in Madagascar,* pp. 169–217. Washington, D.C.: Smithsonian Institution Press.

Mair, W. G. 1890. On the disappearance of the moa. *Transactions and Proceedings of the New Zealand Institute* 22: 70–75.

———. 1893. On the antiquity of the moa. *Transactions and Proceedings of the New Zealand Institute* 25: 534–535.

Maloney, R., and D. Murray. 1999. Present and future research and management of black stilts [abstract]. *Notornis* 46: 410–411.

Mantell, G. A. 1848a (February 2). On the fossil remains of birds collected in various parts of New Zealand by Mr Walter Mantell, of Wellington. *The Quarterly Journal of the Geological Society of London* 4: 225–238.

———. 1848b (February 23). Additional remarks on the geological position of the deposits in New Zealand which contain bones of birds. *Quarterly Journal of the Geological Society of London* 4: 238–241.

———. 1850. Notice of the remains of the *Dinornis* and other birds, and of fossils and rock-specimens, recently collected by Mr. Walter Mantell in the Middle Island of New Zealand: With additional notes in the Northern Island. With note on fossiliferous deposits in the Middle Island of New Zealand, by E. Forbes. *Quarterly Journal of the Geological Society of London* 6: 319–343.

———. 1851. *Petrifactions and Their Teachings: Or, a Handbook to the Gallery of Organic Remains of the British Museum.* London: Henry G. Bohn.

Mantell, W. B. D. 1873. On moa beds. *Transactions and Proceedings of the New Zealand Institute* 5: 94–97.

Marchant, S., and P. J. Higgins (eds.). 1990. *Handbook of Australian, New Zealand and Antarctic Birds.* Vol. 1, *Ratites to Ducks.* Melbourne: Oxford University Press.

———. 1993. *Handbook of Australian, New Zealand and Antarctic Birds.* Vol. 2, *Raptors to Lapwings.* Melbourne: Oxford University Press.

Marples, B. J. 1946. Notes on some neognathous bird bones from the Early Tertiary of New Zealand. *Transactions of the Royal Society of New Zealand* 76: 132–134.

———. 1949. Vertebrate palaeontology in New Zealand. *Tuatara* 2: 103–108.

———. 1952. Early Tertiary penguins of New Zealand. *New Zealand Geological Survey Palaeontological Bulletin* 20.

. 1962a. A fossil penguin from the Late Tertiary of North Canterbury. *Records of the Canterbury Museum* 7: 185–195.

———. 1962b. Observations on the history of penguins. In G. W. Leeper (ed.), *The Evolution of Living Organisms,* pp. 408–416. Melbourne: Melbourne University Press.

Marples, B. J., and C. A. Fleming. 1963. A fossil penguin bone from Kawhia, New Zealand. *New Zealand Journal of Geology and Geophysics* 6: 189–192.

Marshall, P. 1919. Occurrence of fossil moa-bones in the Lower Wanganui strata. *Transactions and Proceedings of the New Zealand Institute* 51: 250–253.

Martin, G. 1990. *Birds by Night*. London: T. and A. D. Poyser.

Martin, H. 1886. The protection of native birds. *Transactions and Proceedings of the New Zealand Institute* 18:112–117.

Martin, P. S. 1984. Prehistoric overkill: The global model. In P. S. Martin and R. G. Klein (eds.), *Quaternary Extinctions: A Prehistoric Revolution*, pp 354–403. Tucson: Arizona University Press.

Martin, P. S., and R. G. Klein. 1984. *Quaternary Extinctions: A Prehistoric Revolution*. Tucson: Arizona University Press.

Martin, P. S., and D. W. Steadman. 1999. Prehistoric extinctions on islands and continents. In R. D. E. MacPhee (ed.), *Extinctions in Near Time: Causes, Contexts, and Consequences*, pp. 17–55. New York: Kluwer Academic/Plenum Publishers.

Marwick, J. 1937. Mud-holes containing moa-bones at Upokongaro. *New Zealand Journal of Science and Technology* 19: 329–336.

Matheus, P. E. 1995. Diet and co-ecology of Pleistocene short-faced bears and brown bears in eastern Beringia. *Quaternary Research* 44: 447–453.

Mathews, W. H., and G. H. Curtis. 1966. Date of the Pliocene–Pleistocene boundary in New Zealand. *Nature* 212(5066): 979–980.

Matisoo-Smith, E. 1994. The human colonisation of Polynesia. A novel approach: Genetic analyses of the Polynesian rat (*Rattus exulans*). *Journal of the Polynesian Society* 103: 75–87.

———. In press. The kiore story: The origin of New Zealand *Rattus exulans*. In D. G. Sutton (ed.), *The Origin of the First New Zealanders*. 2nd ed. Auckland: Auckland University Press.

Matisoo-Smith, E., R. M. Roberts, G. J. Irwin, J. S. Allen, D. Penny, and D. M. Lambert. 1998. Patterns of prehistoric human mobility in Polynesia indicated by mtDNA from the Pacific rat. *Proceedings of the National Academy of Sciences USA* 95: 15145–15150.

Mayr, E. 1967. In J. A. Paynter, Jr. (ed.), *Checklist of Birds of the World*, vol. 12, p. 52. Cambridge, Mass.: Museum of Comparative Zoology.

———. 1986. [Letter to editor.] *Notornis* 33: 268.

McAllan, I. A. W., and M. B. Bruce. 1988. *The Birds of New South Wales*. Turramurra, Australia: Biocon Research Group.

McCallum, J. 1982. The penetration of exotic passerines into modified forests on Little Barrier Island, northern New Zealand. *Tane* 28: 37–51.

McCulloch, B. 1982. *No moa*. Christchurch: Canterbury Museum.

———. 1991 (January 3). New Zealand eagle: Killer or carrion eater? *[Christchurch] Press*, p. 13.

———. 1992a. *Moas, Lost Giants of New Zealand*. Auckland: Harper Collins.

———. 1992b. Unique, dark olive-green moa eggshell from Redcliffe Hill, Rakaia Gorge, Canterbury. *Notornis* 39: 63–65.

———. 1994. A likely association of the New Zealand owlet-nightjar (*Megaegotheles novaezealandiae* Scarlett) with early humans in New Zealand. *Notornis* 41: 209–210.

McCulloch, B., and M. M. Trotter. 1975a. The first twenty years. Radiocarbon dates for the South Island moa-hunter sites, 1955–74. *New Zealand Archaeological Association Newsletter* 18: 2–17.

———. 1975b. South Island radiocarbon dates. *New Zealand Archaeological Association Newsletter* 19: 110–111.

———. 1979a. How big is a moa egg? *New Zealand Archaeological Association Newsletter* 26: 271–273.

———. 1979b. Some radiocarbon dates for moa remains from natural deposits. *New Zealand Journal of Geology and Geophysics* 22: 277–279.
McDonnell [Lt. Col.]. 1889. The ancient moa-hunters at Waingongoro. *Transactions and Proceedings of the New Zealand Institute* 21: 438–441.
McDowall, R. M. 1969. Extinction and endemism in New Zealand land birds. *Tuatara* 17: 1–12.
———. 1978. *New Zealand Freshwater Fishes. A Guide and Natural History.* Auckland: Heinemann.
———. 1990. *New Zealand Freshwater Fishes: A Natural History and Guide.* 2nd ed. Auckland: Heinemann Reed and MAF Publishing Group.
McDowell, S. 1948. The bony palate of birds. Part 1. The Palaeognathae. *Auk* 65: 520–549.
McFadgen, B. G. 1979. The antiquity of moa at Lake Poukawa, New Zealand. *Journal of the Royal Society of New Zealand* 19: 375–382.
———. 1982. Dating New Zealand archaeology by radiocarbon. *New Zealand Journal of Science* 25: 379–392.
———. 1985. Late Holocene stratigraphy of coastal deposits between Auckland and Dunedin, New Zealand. *Journal of the Royal Society of New Zealand* 15: 27–65.
———. 1987. Beach ridges, breakers and bones: Late Holocene geology and archaeology of the Fyffe site, S49/46, Kaikoura Peninsula, New Zealand. *Journal of the Royal Society of New Zealand* 17: 381–394.
———. 1994. Archaeology and holocene sand dune stratigraphy on Chatham Island. *Journal of the Royal Society of New Zealand* 24: 17–44.
McGahan, J. 1973. Flapping flight of the Andean condor in nature. *Journal of Experimental Biology* 58: 239–253.
McGlone, M. S. 1973. Pollen analysis. In D. M. Leslie (ed.), *Relic Periglacial Landforms at Clarke's Junction, Otago, New Zealand. New Zealand Journal of Geology and Geophysics* 16: 575–578.
———. 1978. Forest destruction by early Polynesians, Lake Poukawa, Hawke's Bay, New Zealand. *Journal of the Royal Society of New Zealand* 8: 275–281.
———. 1981. Forest fire following Holocene tephra fall. In R. Howorth, P. Froggatt, C. G. Vucetich, and J. D. Collen (eds.), *Proceedings of Tephra Workshop, June 30–July 1, 1980,* pp. 80–86. Wellington: Geology Department of Victoria University.
———. 1983a. History of New Zealand lowland forest since the last glaciation. In K. Thompson, A. P. H. Hodder, and A. S. Edmonds (eds.), *Lowland Forests in New Zealand,* pp. 1–17. Hamilton, New Zealand: Centre for Continuing Education and Environmental Studies Unit, University of Waikato.
———. 1983b. Holocene pollen diagrams, Lake Rotorua, North Island, New Zealand. *Journal of the Royal Society of New Zealand* 13: 53–65.
———. 1983c. Polynesian deforestation of New Zealand: A preliminary synthesis. *Archaeology in Oceania* 18: 11–25.
———. 1985. Plant biogeography and the late Cenozoic history of New Zealand. *New Zealand Journal of Botany* 23: 723–749.
———. 1988. New Zealand. In B. Huntley and T. Webb (eds.), *Vegetation History,* pp. 557–599. New York: Kluwer Scientific Publishers.
———. 1989. The Polynesian settlement of New Zealand in relation to

environmental and biotic changes in New Zealand. *New Zealand Journal of Ecology* 12(Suppl.): 115–129.
———. 1995. Late Pleistocene and Holocene vegetation history, Central Otago, South Island, New Zealand. *Journal of the Royal Society of New Zealand* 25: 1–22.
McGlone, M. S., and N. T. Moar. 1977. The *Ascarina* decline and post-glacial climatic change in New Zealand. *New Zealand Journal of Botany* 15: 485–489.
———. 1997. Pollen–vegetation relationships on the subantarctic Auckland Islands, New Zealand. *Review of Palaeobotany and Palynology* 96: 317–338.
McGlone, M. S., and N. W. Topping. 1977. Aranuian (post-glacial) pollen diagrams from the Tongariro region, North Island, New Zealand. *New Zealand Journal of Botany* 15: 749–760.
McGlone, M. S., and C. J. Webb. 1981. Selective forces influencing the evolution of divaricating plants. *New Zealand Journal of Ecology* 4: 20–28.
McGlone, M. S., and J. M. Wilmshurst. 1999. Dating initial Maori environmental impact in New Zealand. *Quaternary International* 59: 5–16.
McGlone, M. S., A. J. Anderson, and R. N. Holdaway. 1994. An ecological approach to the Polynesian settlement of New Zealand. In D. G. Sutton (ed.), *The Origins of the First New Zealanders,* pp. 136–163. Auckland: Auckland University Press.
McGlone, M. S., R. N. Holdaway, and J. M. Wilmshurst. In press. Ecological dimensions to Polynesian settlement of New Zealand: Cataclysm and accommodation. In D. G. Sutton (ed.), *Origins of the First New Zealanders.* 2nd ed. Auckland: Auckland University Press.
McGlone, M. S., R. Howorth, and W. A. Pullar. 1984. Late Pleistocene stratigraphy, vegetation and climate of the Bay of Plenty and Gisborne regions, New Zealand. *New Zealand Journal of Geology and Geophysics* 27: 327–350.
McGlone, M. S., A. F. Mark, and D. Bell. 1995. Late Pleistocene and Holocene vegetation history, Central Otago, South Island, New Zealand. *Journal of the Royal Society of New Zealand* 25: 1–22.
McGlone, M. S., D. C. Mildenhall, and M. S. Pole. 1996. History and paleoecology of New Zealand *Nothofagus* forests. In T. T. Vebeln, R. S. Hill, and J. Read (eds.), *The Ecology and Biogeography of* Nothofagus *Forest,* pp. 83–130. New Haven, Conn.: Yale University Press.
McGlone, M. S., P. Wardle, and T. H. Worthy. In press. Environmental change since the last glaciation. In J. Darby and A. Mark (eds.), *The Natural History of Southern New Zealand.*
McGowan, C. 1979a. The hind limb musculature of the brown kiwi *Apteryx australis mantelli. Journal of Morphology* 160: 33–74.
———. 1979b. The wing musculature of the brown kiwi *Apteryx australis mantelli* and its bearing on ratite affinities. *Journal of Zoology* 197: 173–219.
———. 1984. Evolutionary relationships of ratites and carinates: Evidence from the ontogeny of the tarsus. *Nature* 307: 733–735.
———. 1986. The wing musculature of the weka (*Gallirallus australis*), a flightless rail endemic to New Zealand. *Journal of the Zoological Society of London A* 210: 305–346.
McGraw, J. D. 1975. Quaternary airfall deposits in New Zealand. In R. P. Suggate and M. M. Cresswell (eds.), *Quaternary Studies: Selected*

Papers from IX INQUA Congress, Christchurch, New Zealand, pp. 35–44. Royal Society of New Zealand Bulletin 13.

McKay, A. 1882. On a deposit of moa bones near Motunau, North Canterbury. *New Zealand Geological Survey Reports of Geological Explorations* 15: 76–79.

McKee, J. W. A. 1985. A pseudodontorn (Pelecaniformes: Pelagornithidae) from the middle Pliocene of Hawera, Taranaki, New Zealand. *New Zealand Journal of Zoology* 12: 181–184.

———. 1987. The occurrence of the Pliocene penguin *Tereingaornis moisleyi* (Sphenisciformes: Spheniscidae) at Hawera, Taranaki, New Zealand. *New Zealand Journal of Zoology* 14: 557–561.

McLean, J. C. 1911. Bush-birds of New Zealand. *Emu* 11: 1–17, 65–77, 171–187, 223–236.

McLean, I. G. 1984. Feeding association between fantails and saddlebacks; who benefits. *New Zealand Journal of Ecology* 7: 165–168.

McLean, I. G., C. Hölzer, and P. M. Sagar. 1994. Niche overlap and foraging ecology of island *Petroica* species. *Notornis* 41(Suppl.): 39–48.

McLennan, J. A. 1988. Breeding of North Island brown kiwi, *Apteryx australis mantelli*, in Hawke's Bay, New Zealand. *New Zealand Journal of Ecology* 10: 97–107.

———. 1997. *Ecology of Brown Kiwi and Cause of Population Decline in Lake Waikaremoana Catchment.* Conservation Advisory Science Notes 169. Wellington: New Zealand Department of Conservation.

McLennan, J. A., and A. J. McCann. 1991. Ecology of great spotted kiwi, *Apteryx haastii*. New Zealand Department of Scientific and Industrial Research Land Resources Contract Report 91/48. Department of Conservation Investigation, No. 509.

McLennan, J. A., M. A. Potter, H. A. Robertson, G. C. Wake, R. Colbourne, L. Dew, L. Joyce, A. J. McCann, J. Miles, P. J. Miller, and J. Reid. 1996. Role of predation in the decline of kiwi, *Apteryx* spp., in New Zealand. *New Zealand Journal of Ecology* 20: 27–35.

McLeod, H. N. 1902. Caves in the Martinborough District, and moa bones found therein. *Transactions and Proceedings of the New Zealand Institute* 34: 562–563.

McLeod, D., and C. J. Burrows. 1977. History of the Cass District. In C. J. Burrows (ed.), *Cass: History and Science in the Cass District, Canterbury*. Christchurch: Department of Botany, University of Canterbury.

McSaveney, M. J., and I. E. Whitehouse. 1989. Anthropic erosion of mountain land in Canterbury. *New Zealand Journal of Ecology* 12(Suppl.): 151–163.

Medway, D. G. 1967. Avian remains from new caves in the Taumatamaire District. *Notornis* 14: 158–160.

———. 1968. Records of the huia, North Island thrush and North Island kokako from the diaries of Joseph Robert Annabell (1857–1924). *Notornis* 15: 177–192.

———. 1971. Subfossil avian remains from the Awakino-Mahoenui area. *Notornis* 18: 218–219.

Melvin, J. 1871. Extinction of moa. *Nature* 4: 306.

Meredith, C. W., 1985. The fossil fauna of Norfolk Island, south-west Pacific, and a review of the phylogeny of the genus *Pterodroma*. Ph.D. thesis. Monash University, Victoria, Australia.

———. 1991. Vertebrate fossil faunas from islands in Australasia and the southwest Pacific. In P. Vickers-Rich, J. M. Monaghan, R. F. Baird, and T. H. Rich (eds.), *Vertebrate Palaeontology of Australasia,* pp. 1345–1381. Melbourne: Pioneer Design Studio and Monash University Publications Committee.

Meredith, C. W., J. R. Specht, and P. Vickers-Rich. 1985. A minimum date for Polynesian visitation to Norfolk Island, southwest Pacific, from faunal evidence. *Search* 16(9–12): 304–305.

Merton, D. V. 1999. Kakapo. In P. J. Higgins (ed.), *Handbook of Australian, New Zealand and Antarctic Birds,* vol. 4, pp. 633–646. Melbourne: Oxford University Press.

Merton, D., R. B. Morris, and I. Atkinson. 1984. Lek behaviour in a parrot: The kakapo *Strigops habroptilus* of New Zealand. *Ibis* 126: 277–283.

Merton, D. V., C. Reed, and D. Crouchley D. 1999. Recovery strategies and techniques for three free-living, critically-endangered New Zealand birds: Kakapo (*Strigops habroptilus*), black stilt (*Himantopus novaezelandiae*) and takahe (*Porphyrio mantelli*). *Proceedings of the 7th World Conference on Breeding Endangered Species, Cincinnati, Ohio,* 151–162.

Meyer, A. B. 1883. *Abbildungen von Vogel-Skeletten.* 2 vols. Berlin.

Milberg, P., and T. Tyrberg. 1993. Naïve birds and noble savages—A review of man-caused prehistoric extinctions of island birds. *Ecography* 16: 229–250.

Mildenhall, D. C. 1974. The record of the rocks (Miocene kiwi footprint). *New Zealand Nature Heritage* 1: 43–46.

Mildenhall, D. C., and P. R. Moore. 1983. A late Holocene pollen sequence at Turakirae Head, and climatic and vegetational changes in the Wellington area in the last 10 000 years. *New Zealand Journal of Science* 26: 447–459.

Millener, P. R. 1980. The taxonomic status of extinct New Zealand coots *Fulica chathamensis* subspp. (Aves: Rallidae). *Notornis* 27: 363–376.

———. 1981a. The Quaternary avifauna of the North Island, New Zealand. Ph.D. thesis. University of Auckland.

———. 1981b. The subfossil distribution of New Zealand coots *Fulica chathamensis* subspp. *Notornis* 28: 1–9.

———. 1982. And then there were twelve: The taxonomic status of *Anomalopteryx oweni* (Aves: Dinornithidae). *Notornis* 29: 165–170.

———. 1983. *Tyto alba* (Aves: Strigidae): A deletion from the New Zealand subfossil record. *Notornis* 30: 15–21.

———. 1984 (January 24). National Museum notebook: *Harpagornis* was big in pre-pakeha skies. *Evening Post [Wellington].*

———. 1987. Fossil birds in the British Museum: Corrections to Lydekker's (1891) catalogue. *Bulletin of the British Ornithologists Club* 107: 168–172.

———. 1988. Contributions to New Zealand's late Quaternary avifauna. I: *Pachyplichas,* a new genus of wren (Aves: Acanthisittidae), with two new species. *Journal of the Royal Society of New Zealand* 18: 383–406.

———. 1989. The only flightless passerine: The Stephens Island wren (*Traversia lyalli*: Acanthisittidae). *Notornis* 36: 280–284.

———. 1990. New light on old bones—Recent research on New Zealand's subfossil birds [abstract]. In H. J. Campbell, B. W. Hayward, and D. C. Mildenhall (eds.), *Geological Society of New Zealand, Annual Conference Programme and Abstracts,* p. 97. Lower Hutt: DSIR, Geology and Geophysics.

———. 1991. The Quaternary avifauna on New Zealand. In P. Vickers-Rich, J. M. Monaghan, R. F. Baird, and T. H. Rich (eds.), *Vertebrate Palaeontology of Australasia,* pp. 1317–1344. Melbourne: Pioneer Design Studio and Monash University Publications Committee.

———. 1996. Extinct birds. In *The Chatham Islands: Heritage and Conservation,* pp. 113–120. Canterbury: Canterbury University Press and the New Zealand Department of Conservation, Christchurch.

———. 1999. The history of the Chatham Islands' bird fauna of the last 7000 years—A chronicle of change and extinction. Proceedings of the 4th International Meeting of the Society of Avian Paleontology and Evolution (Washington, D.C., June 1996). *Smithsonian Contributions to Paleobiology* 89: 85–109.

Millener, P. R., and R. Cassells. 1985. Moa. In B. Campbell and E. Lack (eds.), *Dictionary of Birds,* pp. 355–357. London: British Ornithologists' Union.

Millener, P. R., and R. G. Powlesland. 2001. The Chatham Islands pigeon (Parea) deserves full species status; *Hemiphaga chathamensis* (Rothschild, 1891); Aves; Columbidae. *Journal of the Royal Society of New Zealand* 31 (in press).

Millener, P. R., and C. J. Templer. 1982. The subfossil deposits of Paryphanta (Mac's Quarry) Cave, Waitomo, New Zealand. *Journal of the Royal Society of New Zealand* 11: 157–166.

Millener, P. R., and T. H. Worthy. 1991. Contributions to New Zealand's late Quaternary avifauna. II: *Dendroscansor decurvirostris,* a new genus and species of wren (Aves: Acanthisittidae). *Journal of the Royal Society of New Zealand* 21: 179–200.

Miller, L. H. 1937. Skeletal studies on the tropical hawk genus *Harpagus. Condor* 39: 219–221.

Miller, A. H. 1963. Fossil ratite birds of the late Tertiary of South Australia. *Records of the South Australian Museum* 14: 413–420.

Mills, J. A. 1969. The distribution of breeding red-billed gull colonies in New Zealand in relation to areas of plankton enrichment. *Notornis* 16: 180–185.

———. 1974. Preliminary results of research into the present status of takahe (*Notornis mantelli*) in the Murchison Mountains. *Notornis* 21: 312–317.

Mills, J. A., R. B. Lavers, and W. G. Lee. 1984. The takahe—A relict of the Pleistocene grassland avifauna of New Zealand. *New Zealand Journal of Ecology* 7: 57–70.

———. 1988. The post-Pleistocene decline of takahe (*Notornis mantelli*): A reply. *Journal of the Royal Society of New Zealand* 18: 112–118.

Milne-Edwards, A. 1896. Sur les ressemblances qui existent entre la faune des Iles Mascareignes et celle de certaines Iles de l'Océan Pacifique Australe. *Annales des Sciences Naturelles (Zoology),* Series 8, 2: 117–136.

Minagawa, M., and E. Wada. 1984. Stepwise enrichment of ^{15}N along food chains: Further evidence and the relation between $\delta^{15}N$ and animal age. *Geochimica et Cosmochimica Acta* 48: 1135–1140.

Miskelly, C. M. 1988. The Little Barrier Island snipe. *Notornis* 35: 273–281.

Mivart, G. 1874. On the axial skeleton of the ostrich (*Struthio camelus*). *Transactions of the Zoological Society, London* 8: 385–451.

———. 1877. On the axial skeleton of the Struthionidae. *Transactions of the Zoological Society, London* 10: 1–52.

Mizutani, H., M. Fukuda, Y. Kabaya, and E. Wada. 1990. Carbon isotope ration of feathers reveals feeding behavior of cormorants. *Auk* 107: 400–403.

Moar, N. T. 1958. Contributions to the Quaternary history of the New Zealand flora 1. Auckland Island peat studies. *New Zealand Journal of Science* 1: 449–465.

———. 1970. Recent pollen spectra from three locations in the South Island, New Zealand. *New Zealand Journal of Botany* 8: 210–221.

———. 1971. Contributions to the Quaternary history of the New Zealand flora. 6. Aranuian pollen diagrams from Canterbury, Nelson and North Westland, South Island. *New Zealand Journal of Botany* 9: 80–145.

———. 1973. Contributions to the Quaternary history of the New Zealand flora. 7. Two Aranuian pollen diagrams from Central South Island. *New Zealand Journal of Botany* 11: 291–304.

———. 1980. Late Otiran and early Aranuian grasslands in central South Island. *New Zealand Journal of Ecology* 3: 4–12.

Moar, N. T., and D. C. Mildenhall. 1988. Pollen assemblages from late Quaternary deposits in Canterbury. *New Zealand Journal of Geology and Geophysics* 31: 305–335.

Molloy, B. P. J. 1977. The fire history. In C. J. Burrows (ed.), *Cass: History and Science in the Cass District, Canterbury, New Zealand.* Christchurch: Department of Botany, University of Canterbury.

Molloy, L. 1994. *Wild New Zealand.* London: New Holland.

Molloy, B. P. J., and J. E. Cox. 1972. Subfossil forest remains and their bearing on forest history in the Rakaia catchment, Canterbury, New Zealand. *New Zealand Journal of Botany* 10: 267–276.

Molloy, B. P. J., C. J. Burrows, J. E. Cox, J. A. Johnston, and P. Wardle. 1963. Distribution of subfossil forest remains, eastern South Island, New Zealand. *New Zealand Journal of Botany* 1: 68–77.

Molloy, J., A. Davis, and C. Tisdall. 1994. *Setting Priorities for the Conservation of New Zealand's Threatened Plants and Animals.* Wellington: New Zealand Department of Conservation.

Molnar, R. E., and M. Pole. 1997. A Miocene crocodilian from New Zealand. *Alcheringa* 21: 65–70.

Molnar, R. E., and J. Wiffen. 1994. Polar dinosaur faunas from New Zealand. *Cretaceous Research* 15: 689–706.

Monnier, L. 1913. Paléontologie de Madagascar VII, les *Aepyornis. Annales de Paléontologie* 8: 125–172.

Moorhouse, R. 1996. The extraordinary bill dimorphism of the huia (*Heteraclocha* [*sic*] *acutirostris*: Sexual selection or intersexual competition? *Notornis* 43: 19–34.

Murison, W. D. 1872. Notes on moa remains. *Transactions and Proceedings of the New Zealand Institute* 4: 120–124.

Murray, P. 1991. The Pleistocene megafauna of Australia. In P. Vickers-Rich, J. M. Monaghan, R. F. Baird, and T. H. Rich (eds.), *Vertebrate Palaeontology of Australasia,* pp. 1071–1164. Melbourne: Pioneer Design Studio and Monash University Publications Committee.

Murray, P., and D. Megirian. 1998. The skull of the dromornithid birds: Anatomical evidence for their relationship to Anseriformes. *Records of the South Australian Museum* 31: 51–97.

Murray-McIntosh, R. P., B. J. Scrimshaw, P. J. Hatfield, and D. Penny. 1998. Testing migration patterns and estimating founding population size in Polynesia by using human mtDNA sequences. *Proceedings of the National Academy of Sciences USA* 95: 9047–9052.

Naish, T. R., S. T. Abbott, B. V. Alloway, A. G. Beu, R. M. Carter, A. R. Edwards, T. D. Journeaux, P. J. J. Kamp, B. J. Pillans, G. Saul, and K. J. Woolfe. 1998. Astronomical calibration of a southern hemisphere Plio-Pleistocene reference section, Wanganui Basin, New Zealand. *Quaternary Science Reviews* 17: 695–710.

Nelson, C. S., and J. A. Grant-Mackie. 1980. Significance of carbonate concretionary growths in a moa bone in Holocene tidal flat deposits beneath Auckland City, New Zealand. *Records of the Auckland Institute and Museum* 17: 105–112.

Nelson, C. S., P. J. Cooke, C. H. Hendy, and A. M. Cuthbertson. 1993. Oceanographic and climate changes over the past 160,000 years at deep sea drilling project site 594 off southeastern New Zealand, southwest Pacific Ocean. *Paleoceanography* 8: 435–458.

Nevinson, J. 1991. Wedge-tailed eagle kills emu chick. *Australian Raptor Association News* 12(1): 10.

Newnham, R. M., D. J. Lowe, and J. D. Green. 1989. Palynology, vegetation and climate of the Waikato lowlands, North Island, New Zealand, since c. 18,000 years ago. *Journal of the Royal Society of New Zealand* 19: 127–150.

Newnham, R. M., P. J. de Lange, and D. J. Lowe. 1995. Holocene vegetation, climate and history of a raised bog complex, northern New Zealand, based on palynology, plant macrofossils and tephrachronology. *Holocene* 5: 267–282.

Newnham, R. M., D. J. Lowe, M. S. McGlone, J. M. Wilmshurst, and T. F. G. Higham. 1998. The Kaharoa Tephra as a critical datum for earliest human impact in New Zealand. *Journal of Archaeological Science* 25: 533–544.

Newton, A. 1885. Ornithology. *Encyclopaedia Brittanica,* vol. 18, p. 2. 9th ed. Cambridge: Cambridge University Press.

Nixon, A. J. 1981. The external morphology and taxonomic status of the orange-fronted parakeet. *Notornis* 28: 292–300.

———. 1994. Feeding biology of hybridizing parakeets on Mangere Island, Chatham Islands. *Notornis* 41(Suppl.): 5–18.

O'Donnell, C. F. J. 1996. Predators and the decline of New Zealand forest birds: An introduction to the hole-nesting bird and predator programme. *New Zealand Journal of Zoology* 23: 213–219.

O'Donnell, C. F. J., and P. J. Dilks. 1986. Forest birds in South Westland: Status, distribution and habitat use. *New Zealand Wildlife Service Occasional Publication* 10: 1–179.

———. 1989. Sap-feeding by the kaka (*Nestor meridionalis*) in South Westland, New Zealand. *Notornis* 36: 65–71.

———. 1994. Foods and foraging of forest birds in temperate rainforest, South Westland, New Zealand. *New Zealand Journal of Ecology* 18: 87–107.

O'Donnell, C. F. J., and S. M. Phillipson. 1996. Predicting the incidence of mohua predation from the seedfall, mouse, and predator fluctuations in beech forests. *New Zealand Journal of Zoology* 23: 287–293.

O'Donnell, C. F. J., P. J. Dilks, and G. P. Elliott. 1996. Control of a stoat (*Mustela erminea*) population irruption to enhance mohua (yellowhead) (*Mohoua ochrocephala*) breeding success in New Zealand. *New Zealand Journal of Zoology* 23: 279–286.

O'Donnell, C. F. J., J. Christie, C. Corben, J. A. Sedgeley, and W. Simpson. 1999. Rediscovery of short-tailed bats (*Mystacina* sp.) in Fiordland, New Zealand: Preliminary observations of taxonomy, echolocation calls, population size, home range, and habitat use. *New Zealand Journal of Ecology* 23: 21–30.

Ogden, J., R. M. Newnham, J. G. Palmer, R. G. Serra, and N. D. Mitchell. 1993. Climatic implications of macro–microfossil assemblages from Late Pleistocene deposits in northern New Zealand. *Quaternary Research* 39: 107–119.

Okazaki, M., Y. Oshida, R. Malony, and J. Warham. 1993. Effects of sooty shearwaters *Puffinus griseus* on surface soils on Motuara Island, New Zealand. *Journal of the Yamashina Institute for Ornithology* 25: 137–143.

Oliver, W. R. B. 1928. The flora of the Waipaoa Series (Later Pliocene) of New Zealand. *Transactions and Proceedings of the New Zealand Institute* 59: 287–303.

———. 1930. *New Zealand Birds.* Wellington: Fine Arts.

———. 1945a. Avian evolution in New Zealand and Australia (part 1). *Emu* 45: 55–77.

———. 1945b. Avian evolution in New Zealand and Australia (part 2). *Emu* 45: 119–152.

———. 1949. *The Moas of New Zealand and Australia.* Dominion Museum Bulletin 15. Wellington, New Zealand.

———. 1955. *New Zealand Birds.* Wellington: A. H. and A. W. Reed.

Olson, S. L. 1975. A review of the extinct rails of the New Zealand region (Aves, Rallidae). *National Museum of New Zealand Records* 1(3): 63–79.

———. 1976. Extinct rails [letter to editor]. *Notornis* 23: 79.

———. 1977a. Notes on the subfossil Anatidae from New Zealand, including a new species of pink-eared duck *Malacorhynchus. Emu* 77: 123–135.

———. 1977b. A synopsis of the fossil Rallidae. In S. D. Ripley (ed.), *Rails of the World: A Monograph of the Family Rallidae,* pp. 339–373. Boston: David R. Godine.

———. 1982. The distribution of fused phalanges of the inner toe in the Accipitridae. *Bulletin of the British Ornithologists' Club* 102: 8–12.

———. 1984. The relationships of the extinct Chatham Island eagle. *Notornis* 31: 273–277.

———. 1985. The fossil record of birds. In D. S. Farner, J. R. King, and K. C. Parkes (eds.), *Avian Biology,* vol. 8, pp. 79–256. New York: Academic Press.

———. 1986. Emendation of the name of the fossil rail *Rallus hodgeni* Scarlett. *Notornis* 33: 32.

———. 1987a. More on the name *Rallus hodgenorum. Notornis* 34: 167–168.

———. 1987b. Variation in the procoracoid foramen in the Accipitridae. *Rivista Italiana Ornitologia, Milano* 57(3–4): 1–4.

———. 1990a. Osteology and systematics of the fernbirds (*Bowdleria*: Sylviidae). *Notornis* 37: 161–171.

———. 1990b. The prehistoric impact of man on biogeographical patterns of insular birds. In *International Symposium on Biogeographical Aspects of Insularity,* pp. 45–51. Atti dei convegni lincei 85. Rome: Accademia Nazionale dei Lincei.

———. 1991. Patterns of avian diversity and radiation in the Pacific as seen through the fossil record. In E. C. Dudley (ed.), *The Unity of Evolutionary Biology. Proceedings of the Fourth International Congress of Systematic and Evolutionary Biology,* 1: 314–318. Portland, Ore.: Dioscorides Press.

Olson, S. L., and A. Feduccia. 1979. Flight capability and the pectoral girdle of *Archaeopteryx. Nature* 278(5701): 247–248.

Olson, S. L., and W. B. Hilgartner. 1982. Fossil and subfossil birds from the Bahamas. *Smithsonian Contributions to Paleobiology* 48: 22–60.

Olson, S. L., and H. F. James. 1982. Prodromus of the fossil avifauna of the Hawaiian Islands. *Smithsonian Contributions to Zoology* 365: 1–59.

———. 1991. Descriptions of thirty-two new species of birds from the Hawaiian Islands: Part I. Non-passerines. *Ornithological Monographs* 45: 1–88.

Olson, S. L., and P. Jouventin. 1996. A new species of small flightless duck from Amsterdam Island, southern Indian Ocean (Anatidae: *Anas*). *Condor* 98: 1–9.

Olson, S. L., J. C. Balouet, and C. J. Fisher. 1987. The owlet-nightjar of New Caledonia, *Aegotheles savesi,* with comments on the systematics of the Aegothelidae. *Gerfaut* 37: 341–352.

Olson, S. L., K. C. Parkes, M. H. Clench, and S. R. Borecky. 1983. The affinities of the New Zealand passerine genus *Turnagra. Notornis 30*: 319–346.

Orchiston, D. W., 1977. Discovery of the bones of the extinct giant rail, *Aptornis otidiformis,* at an archaic site near Needles Point, Marlborough. *New Zealand Archaeological Association Newsletter* 20: 256–262.

Owen, R. 1840 (March).[1] [On the bone of an unknown struthious bird from New Zealand, meeting of November 12, 1839.] *Proceedings of the Zoological Society London for 1839,* Part VII, no. lxxxiii: 169–171.

———. 1842. Notice of a fragment of femur of a gigantic bird of New Zealand. *Transactions of the Zoological Society of London* 3: 29–33.

———. 1843a (July). [Proceedings of a meeting of January 24, 1843: Part 1 of Owen's first memoir on *Dinoris*; contains original proposal of *Dinornis novaezealandiae*]. *Proceedings of the Zoological Society of London for 1843,* Part XI, no. cxx: 8–10.

———. 1843b (July). [Letter proposing to substitute the name of *Dinornis* for that of *Megalornis*]. *Proceedings of the Zoological Society of London for 1843,* Part XI, no. cxxi: 19.

———. 1843c (July). [Letter from Rev. W. C. Cotton to R. Owen announcing Rev. Williams's moa collection, read to the meeting January 10, 1843.] *Proceedings of the Zoological Society of London for 1843,* Part XI, no. cxx: 1–2.

———. 1844a. On *Dinornis,* an extinct genus of tridactyle struthious birds, with descriptions of portions of the skeleton of five species which formerly existed in New Zealand (part I). *Transactions of the Zoological Society of London* 3: 235–275.

———. 1844b (March). [Part 2 of Owen's memoir on *Dinornis,* read November 28, 1843]. *Proceedings of the Zoological Society of London for 1843,* Part XI, no. cxxix: 144–146.

———. 1846a. On *Dinornis* (part II), containing descriptions of portions of the skull, the sternum and other parts of the skeleton of the species previously determined, with osteological evidences of three additional

1. The publication date of this and successive Owen publications is taken from Duncan (1937). The Proceedings of the Zoological Society of London were published in sections, often over more than one year.

species, and a new genus, *Palapteryx*. *Transactions of the Zoological Society of London* 3: 307–329.
———. 1846b (July). [Proceedings of a meeting: Owen read his second memoir on various *Dinornis* remains]. *Proceedings of the Zoological Society of London for 1846,* Part XIV, no. clx: 46–49.
———. 1848a. On *Dinornis* (part III): Containing a description of the skull and beak of that genus, and of the same characteristic parts of *Palapteryx,* and of two other genera of birds, *Notornis* and *Nestor*; forming part of an extensive collection of ornithic remains discovered by Mr. Walter Mantell at Waingongoro, North Island of New Zealand. *Transactions of the Zoological Society of London* 3: 345–378.
———. 1848b (April 13). On the remains of the gigantic and presumed extinct wingless or terrestrial birds of New Zealand (*Dinornis* and *Palapteryx*), with indications of two other genera (*Notornis* and *Nestor*) [read January 11, 1848]. *Proceedings of the Zoological Society of London for 1848,* Part XVI, no. clxxx: 1–11.
———. 1850 (November 12). [Proceedings of a meeting: Memoir on gigantic wingless birds of New Zealand communicated; read February 26, 1850]. *Proceedings of the Zoological Society of London for 1850,* Part XVIII, no. cciii: 45–46.
———. 1851a. On *Dinornis* (part IV); containing the restoration of the feet of that genus and of *Palapteryx,* with a description of the sternum in *Palapteryx* and *Aptornis*. *Transactions of the Zoological Society of London* 4: 1–20.
———. 1851b (February 28). [Proceedings of a meeting: Paper read on the cranium of *D. giganteus* and *D. ingens*; read November 12, 1850]. *Proceedings of the Zoological Society of London for 1850,* Part XVIII, no. ccxiii: 208–209.
———. 1852. On *Dinornis* (part V): Containing a description of the skull and beak of a large species of *Dinornis,* of the cranium of an immature specimen of *Dinornis giganteus* (?) and of crania of species of *Palapteryx*. *Transactions of the Zoological Society of London* 4: 59–68.
———. 1855 (April 11). On the bones of the leg of *Dinornis (Palapteryx) struthioides* and the *Palapteryx gracilis* [read November 14, 1854]. *Proceedings of the Zoological Society of London for 1854,* Part XXII, no. cclxxv: 244–248.
———. 1856a. On *Dinornis* (part VI): Containing a description of the bones of the leg of *Dinornis (Palapteryx) struthioides* and of *Dinornis gracilis,* Owen. *Transactions of the Zoological Society of London* 4: 141–147.
———. 1856b (July 30). On *Dinornis* (part VII): Containing a description of the bones of the leg and foot of the *Dinornis elephantopus,* Owen [read April 8, 1856]. *Proceedings of the Zoological Society of London for 1856,* Part XXIV, no. ccciv: 54–61.
———. 1858a. On *Dinornis* (part VII): Containing a description of the bones of the leg and foot of the *Dinornis elephantopus,* Owen. *Transactions of the Zoological Society of London* 4: 149–158.
———. 1858b. On *Dinornis* (part VIII): Containing a description of the skeleton of *Dinornis elephantopus,* Owen. *Transactions of the Zoological Society of London* 4: 159–164.
———. 1865 (May). [Proceedings of a meeting: Paper read on skull, atlas, and scapulo-coracoid of *Dinornis robustus*]. *Proceedings of the Zoological Society of London for 1864,* Part III, no. xli: 648–649.

———. 1866a. On *Dinornis* (part IX): Containing a description of the skull, atlas, and scapulo-coracoid bone of *Dinornis robustus. Transactions of the Zoological Society of London* 5: 337–358.
———. 1866b. On *Dinornis* (part X): Containing a description of part of the skeleton of a flightless bird indicative of a new genus and species (*Cnemiornis calcitrans,* Ow.). *Transactions of the Zoological Society of London* 5: 395–404, pls. 63–67.
———. 1868 (April). [Proceedings of a meeting: Two papers presented (i) *D. robustus*; (ii) *D. maximus*]. *Proceedings of the Zoological Society of London for 1867,* Part III, no. lvii: 891.
———. 1868 (September). [Proceedings of a meeting: Papers on *D. elephantopus, D. rheides, crassus, casuarinus*]. *Proceedings of the Zoological Society of London for 1868,* Part II, no. xxvi: 404.
———. 1869a. On *Dinornis* (part XI): Containing a description of the integument of the sole, and tendons of a toe, of the foot of *Dinornis robustus,* Owen. *Transactions of the Zoological Society of London* 6: 495–496.
———. 1869b. On *Dinornis* (part XII): Containing a description of the femur, tibia, and metatarsus of *Dinornis maximus,* Owen. *Transactions of the Zoological Society of London* 6: 497–500.
———. 1869c (June). [Proceedings of a meeting: Paper read on craniology of *Dinornis*]. *Proceedings of the Zoological Society of London for 1869,* Part I, no. iv: 59.
———. 1870a. On *Dinornis* (part XIII): Containing a description of the sternum in *Dinornis elephantopus* and *D. rheides,* with notes on that bone in *D. crassus* and *D. casuarinus. Transactions of the Zoological Society of London* 7: 115–122.
———. 1870b. On *Dinornis* (part XIV): Containing contributions to the craniology of the genus, with a description of part of the fossil cranium of *Dasiornis londinensis,* Ow., from the London Clay of Sheppey. *Transactions of the Zoological Society, London* 7: 123–150.
———. 1870c. On *Dinornis* (part XV): Containing a description of the skull, femur, tibia, fibula, and metatarsus of *Aptornis defossor,* Owen, from near Oamaru, Middle Island, New Zealand; with additional observations on *Aptornis otidiformis,* on *Notornis mantelli,* and on *Dinornis curtus. Transactions of the Zoological Society, London* 7: 353–380.
———. 1870d (June). [Notes on a letter from Dr. J. Haast (Haast, 1870)]. *Proceedings of the Zoological Society of London for 1870,* Part I, no. iv: 56–57.
———. 1870e (June). [Preliminary notice of a paper, to contain (inter alia) description of *D. curtus*]. *Proceedings of the Zoological Society of London for 1870,* Part I, no. viii: 128.
———. 1871a. On *Dinornis* (part XVI): Containing notices of the internal organs of some species, with a description of the brain and some nerves and muscles of the head of *Apteryx australis. Transactions of the Zoological Society of London* 7: 381–396.
———. 1871b (August). [Preliminary notice of 17th memoir on *Dinornis*]. *Proceedings of the Zoological Society of London for 1871,* Part II, no. xxxii: 506.
———. 1872a. On *Dinornis* (part XVII): Containing a description of the sternum and pelvis, with an attempted restoration, of *Aptornis defossor,* Ow. *Transactions of the Zoological Society, London* 8: 119–126.

———. 1872b (August). [Preliminary notice of 18th memoir to include proposal of *D. gravis*]. *Proceedings of the Zoological Society of London for 1871,* Part II, no. xxxviii: 605.
———. 1873a. On *Dinornis* (part XVIII): Containing a description of the pelvis and bones of the leg of *Dinornis gravis. Transactions of the Zoological Society of London* 8: 361–380.
———. 1873b. On *Dinornis* (part XIX): containing a description of a femur indicative of a new genus of large wingless bird (*Dromornis australis* Owen) from a post-Tertiary deposit in Queensland, Australia. *Transactions of the Zoological Society of London* 8: 381–384.
———. 1875a (June). [Note on a new locality of Dinornithidae (Hamilton Swamp)]. *Proceedings of the Zoological Society of London for 1875,* Part I, no. vi: 88.
———. 1875b. On *Dinornis* (part XX): Containing a restoration of the skeleton of *Cnemiornis calcitrans,* Ow., with remarks on its affinities in the Lamellirostral group. *Transactions of the Zoological Society, London* 9: 253–272, pls. 35–39.
———. 1875c (October). [Notice of Owen's own account of the Glenmark material.] *Proceedings of the Zoological Society London for 1875,* Part III: 470.
———. 1876 (April). [Preliminary notice of 22nd memoir (on *D. maximus*)]. *Proceedings of the Zoological Society of London for 1875,* Part IV, no. xl: 634.
———. 1877. On *Dinornis* (part XXI): Containing a restoration of the skeleton of *Dinornis maximus,* Owen. *Transactions of the Zoological Society of London* 10: 147–186.
———. 1879. *Memoirs on the Extinct Wingless Birds of New Zealand, with an Appendix on Those of England, Australia, Newfoundland, Mauritius and Rodriguez.* 2 vols. London: John van Voorst.
———. 1882a (June). [Proceedings of a meeting: Notice of 23rd memoir: *D. parvus, nom. nud.*]. *Proceedings of the Zoological Society of London for 1882,* Part I: 1–2.
———. 1882b (October). [Proceedings of a meeting: Notice of 24th memoir: *D. didinus, nom. nud.*]. *Proceedings of the Zoological Society of London for 1882,* Part III, no. xxxvi: 549.
———. 1883a. On *Dinornis* (part XXIII): Containing a description of the skeleton of *Dinornis parvus,* Owen. *Transactions of the Zoological Society, London* 11: 233–256.
———. 1883b. On *Dinornis* (part XXIV): Containing a description of the head and feet, with their dried integuments, of an individual of the species *Dinornis didinus,* Owen. *Transactions of the Zoological Society, London* 11: 257–261.
———. 1884 (August). [Proceedings of a meeting: Notice of 25th memoir: *D. didinus, nom. nud.*]. *Proceedings of the Zoological Society of London for 1884,* Part II, no. xii: 176.
———. 1886. On *Dinornis* (part XXV): Containing a description of the sternum of *Dinornis elephantopus. Transactions of the Zoological Society, London* 12: 1–3.
Owen, H. G. 1984. *Atlas of Continental Displacement, 200 Million Years to the Present.* Cambridge: Cambridge University Press.
Palmer, J. G., J. Ogden, and R. N. Patel. 1988. A 426-year floating tree-ring chronology from *Phyllocladus trichomanoides* buried by the Taupo eruption at Pureora, central North Island, New Zealand. *Journal of the Royal Society of New Zealand* 18: 407–415.
Park, J. 1888. On the exploration of moa-bone deposit at Patangata

Swamp, Te Aute. *Colonial Museum and Geological Survey of New Zealand Reports of Geological Explorations* 19: 88–90.
Parker, W. K. 1866. On the structure and development of the skull in the ostrich tribe. *Philosophical Transactions of the Royal Society, London* 156: 113–183.
Parker, T. J. 1893a. On the classification and mutual relations of the Dinornithidae. *Transactions and Proceedings of the New Zealand Institute* 25: 1–3.
———. 1893b. On the classification and mutual relations of the Dinornithidae. *Nature* 2(March): 431.
———. 1893c. On the presence of a crest of feathers in certain species of moa. *Transactions and Proceedings of the New Zealand Institute* 25: 3–6.
———. 1895. On the cranial osteology, classification, and phylogeny of the Dinornithidae. *Transactions of the Zoological Society of London* 8: 373–428.
———. 1897. [Report on specimen of *Harpagornis* from McCrae's to Otago Institute, fifth meeting, September 8, 1896.] *Transactions and Proceedings of the New Zealand Institute* 30 (1896): 574.
Parkes, K. C., and G. A. Clark. 1966. An additional character linking ratites and tinamous, and an interpretation of their monophyly. *Condor* 66: 459–471.
Pascoe, J. (ed.). 1957. *Mr. Explorer Douglas.* Wellington: A. H. and A. W. Reed.
Pennycuick, C. J. 1975. Mechanics of flight. In D. S. Farner and J. R. King (eds.), *Avian Biology 5,* pp. 1–73. New York: Academic Press.
Petchey, F. J. 1999. New Zealand bone dating revisited: A radiocarbon discard protocol for bone. *New Zealand Journal of Archaeology* 19(1997): 81–124.
Peters, R. H. 1983. *The Ecological Implications of Body Size.* Cambridge: Cambridge University Press.
Phillipps, W. J. 1959. The last (?) occurrence of *Notornis* in the North Island. *Notornis* 8: 93–94.
Pickrill, R. A. 1976. The evolution of coastal landforms of the Wairau Valley. *New Zealand Geographer* 32: 17–29.
Pierce, R. J. 1979. Foods and feeding of the wrybill (*Anarhynchus frontalis*) on its riverbed breeding grounds. *Notornis* 26: 1–21.
Pierson, E. D., V. M. Sarich, J. M. Lowenstein, M. J. Daniel, and W. E. Rainey. 1986. A molecular link between the bats of New Zealand and South America [errata, *Nature* 324: 282]. *Nature* 323: 60–63.
Pillans, B. 1991. New Zealand Quaternary stratigraphy: An overview. *Quaternary Science Reviews* 10: 405–418.
Pillans, B., M. McGlone, A. Palmer, D. Mildenhall, B. Alloway, and G. Berger. 1993. The last glacial maximum in central and southern North Island, New Zealand: A palaeoenvironmental reconstruction using Kawakawa Tephra Formation as the chronostratigraphic marker. *Palaeogeography, Palaeoclimatology, Palaeoecology* 101: 283–304.
Pizzey, G. 1980. *A Field Guide to the Birds of Australia.* Sydney: Collins.
Plant, J. 1877. Footprints of "*Dinornis*," bones of "*Dinornis*" and "*Dodo.*" *Proceedings, Literary and Philosophical Society of Manchester* 16: 181–182.
Plomley, N. J. B. 1969. The Tasmanian Journal of Natural Science. *Papers and Proceedings of the Royal Society of Tasmania* 103: 13–15.
Plótnick, R 1956. Posicion sistematica del genero *Heterospizias. Hornero* 10: 136–139.

Polack, J. S. 1838. *New Zealand, Being a Narrative of Travels and Adventures.* 2 vols. London: Richard Bentley.

Poole, E. 1938. Weights and wing areas of 143 species of North American birds. *Auk* 55: 511–517.

Porter, S. C. 1975. Equilibrium-line altitudes of late Quaternary glaciers in the Southern Alps, New Zealand. *Quaternary Research* 5: 27–47.

Potts, T. H. 1871. On the birds of New Zealand, part II. *Transactions and Proceedings of the New Zealand Institute* 3: 59–109.

———. 1882. *Out in the Open: A Budget of Scraps of Natural History, Gathered in New Zealand.* Christchurch: Lyttelton Times.

Powlesland, R. G. 1987. The foods, foraging behaviour and habitat use of North Island kokako in Puketi State Forest, Northland. *New Zealand Journal of Ecology* 10: 117–128.

Powlesland, R. G., B. D. Lloyd, H. A. Best, and D. V. Merton. 1992. Breeding biology of the kakapo *Strigops habroptilus* on Stewart Island, New Zealand. *Ibis* 134: 361–373.

Prange, H. D., J. F. Anderson, and H. Rahn. 1979. Scaling of skeletal mass to body mass in birds and mammals. *American Naturalist* 113: 103–122.

Pycraft, W. P. 1900. On the morphology and phylogeny of the Palaeognathae (Ratitae and Crypturi) and Neognathae (Carinatae). *Transactions of the Zoological Society of London* 15: 149–290.

———. 1902. Contributions to the osteology of birds. Part V, Falconiformes. *Proceedings of the Zoological Society of London for 1902,* 1: 277–320.

Ramsay, G. W. 1978. A review of the effects of rodents on the New Zealand invertebrate fauna. In P. R. Dingwall, I. A. E. Atkinson, and C. Hay (eds.), *The Ecology and Control of Rodents in New Zealand Nature Reserves,* pp. 89–95. New Zealand Department of Lands and Survey Information Series, No. 4.

Ramsay, M. A., and K. A. Hobson. 1991. Polar bears make little use of terrestrial food webs: Evidence from stable-carbon isotope analysis. *Oecologia* 86: 598–600.

Rando, J. C., M. Lopez, and B. Segui. 1999. A new species of flightless passerine (Emberizidae: *Emberiza*) from the Canary Islands. *Condor* 101: 1–13.

Rattenbury, M. S., R. A. Cooper, and M. R. Johnston (eds.). 1998. *Geology of the Nelson Area* [map]. Institute of Geological and Nuclear Sciences 1:250,000 geological map 9. Lower Hutt, New Zealand: Institute of Geological and Nuclear Sciences.

Rau, G. H., A. J. Mearns, D. R. Young, R. J. Olson, H. A. Schafer, and I. R. Kaplan. 1983. Animal $^{13}C/^{12}C$ correlates with trophic levels in pelagic food webs. *Ecology* 64: 1314–1318.

Rayner, J. M. V. 1979a. A new approach to animal flight mechanics. *Journal of Experimental Biology* 80: 17–54.

———. 1979b. A vortex theory of animal flight. Part 1. The vortex wake of a hovering animal. *Journal of Fluid Mechanics* 91: 697–730.

———. 1979c. A vortex theory of animal flight. Part 2. The forward flight of birds. *Journal of Fluid Mechanics* 91: 731–763.

Rea, H. A. 1983. Cathartid affinities: A brief overview. In S. R. Wilbur and J. A. Jackson (eds.), *Vulture Biology and Management,* pp. 26–54. Berkeley: University of California Press.

Reichenbach, H. G. L. (1849–)1850(–1852). *Avium Systema Naturale. Das Natürliche System der Vögel, &c.—Ornithologie Méthodique, &c.* Berlin.

Reid, B. 1974. Sightings and records of the takahe (*Notornis mantelli*) prior to its "official rediscovery" by Dr. G. B Orbell in 1948. *Notornis* 21: 277–295.
Reif, W., and H. Silyn-Roberts. 1985. On the robustness of moa's leg bones. An exercise in functional morphology of extinct organisms. *Konstruktionsmorphologie* 193: 155–159.
Rich, T. H. 1975. Potential pre-Pleistocene fossil tetrapod sites in New Zealand. *Mauri Ora* 3: 45–54.
Rich, T. H., P. Vickers-Rich, R. E. Fordyce, P. Gatehouse, and R. J. Scarlett. 1979. A deceptive terrestrial vertebrate fossil site on the Waipara River, North Canterbury, New Zealand. In A. J. Anderson (ed.), *Birds of a Feather*, pp. 25–52. New Zealand Archaeological Association Monograph II; B.A.R. International Series 62.
Richards, R. 1986. *Which Pakeha Ate the Last Moa?* Wellington: Paremata Press.
Ricker, K. E., T. J. Chinn, and M. J. McSaveney. 1993. A late Quaternary moraine sequence dated by rock weathering rinds, Craigieburn Range, New Zealand. *Canadian Journal of Earth Science* 30: 1861–1869.
Ridpath, M.G. 1969. Dusky moorhen. In H. J. Frith (ed.), *Birds in the Australian High Country*, pp. 164–165. Sydney: A. H. and A. W. Reed.
Ripley, S. D. 1977. *Rails of the World: A Monograph of the Family Rallidae*. Boston: David R. Godine.
Ritchie, N. 1982. Two sub-fossil faunal deposits uncovered near Cromwell, Central Otago. *New Zealand Archaeological Association Newsletter* 25(2): 97–105.
Roberts, W. H. S. 1875. Letter on moa. *Transactions and Proceedings of the New Zealand Institute* 7: 548–549.
Roberts, M. 1991. Origin, dispersal routes, and geographic distribution of *Rattus exulans*, with special reference to New Zealand. *Pacific Science* 45: 123–130.
Robertson, C. J. R., and G. B. Nunn. 1997. Towards a new taxonomy for albatrosses. In G. Robertson and R. Gales (eds.), *Albatross Biology and Conservation*, pp. 13–19. Chipping Norton: Surrey Beatty and Sons.
Robertson, C. J. R., and G. F. van Tets. 1982. The status of birds at the Bounty Islands. *Notornis* 29: 311–336.
Robertson, A. W., D. Kelly, J. J. Ladley, and A. D. Sparrow. 1999. Effects of pollinator loss on endemic New Zealand mistletoes (Loranthaceae). *Conservation Biology* 13: 499–508.
Robson, C. H. 1876. On moa remains at Cape Campbell. *Transactions and Proceedings of the New Zealand Institute* 8: 95–97.
———. 1877. Further notes on moa remains. *Transactions and Proceedings of the New Zealand Institute* 9: 279–280.
Rothschild, W. 1907. *Extinct Birds*. London: Hutchison.
Rowley, G. D. 1878. Remarks on the extinct birds of Madagascar and New Zealand. *Ornithological Miscellany* 3: 237–247.
Rupke, N. A. 1994. *Richard Owen, Victorian Naturalist*. New Haven, Conn.: Yale University Press.
Russell, I. C. 1877. The giant birds. *American Naturalist* 11: 11–21.
Sagar, P. M., U. Shankar, and S. Brown. 1999. Distribution and numbers of waders in New Zealand 1983–1994. *Notornis* 46: 1–43.
Scarlett, R. J. 1953. A sub-fossil hawk from New Zealand. *Records of the Canterbury Museum* 6: 245–252.
———. 1955a. Further report on bird remains from Pyramid Valley. *Records of the Canterbury Museum* 6: 261–264.

———. 1955b. A new rail from South Island swamps in New Zealand. *Records of the Canterbury Museum* 6: 265–266.

———. 1957a. Former faunal areas: Some sub-fossil evidence. *Proceedings of the New Zealand Ecological Society* 4: 17–18.

———. 1957b. Sub-fossil bones of the Australian pelican from the South Island. *Notornis* 7: 114.

———. 1962. Sub-fossil records of the little grey kiwi in the North Island. *Notornis* 10: 84–85.

———. 1966. A pelican in New Zealand. *Notornis* 13: 204–217.

———. 1967a. Further North Island records of *Apteryx oweni. Notornis* 14: 225.

———. 1967b. A sub-fossil record of a barn owl in New Zealand. *Notornis* 14: 218–219.

———. 1968a. An owlet-nightjar from New Zealand. *Notornis* 15: 254–266.

———. 1968b. A second North Island locality for *Pachyornis elephantopus* (Owen). *Notornis* 15: 36.

———. 1969a. Moas and other extinct birds. In G. A. Knox (ed.), *The Natural History of Canterbury,* pp. 565–568. Wellington: A. H. and A. W. Reed.

———. 1969b. The occurrence of the musk duck, *Biziura lobata* (Shaw), in New Zealand. *Notornis* 16: 64–65.

———. 1969c. On the alleged Queensland moa, *Dinornis queenslandiae* De Vis. *Memoirs of the Queensland Museum* 15: 207–212.

———. 1970a. The genus *Capellirallus. Notornis* 17: 303–319.

———. 1970b. A small woodhen from New Zealand. *Notornis* 17: 68–74.

———. 1972a. Bone of a presumed odontopterygian bird from the Miocene of New Zealand. *New Zealand Journal of Geology and Geophysics* 15: 269–274.

———. 1972b. *Bones for the New Zealand Archaeologist.* Canterbury Museum Bulletin, No. 4.

———. 1974. Moas and man in New Zealand. *Notornis* 21: 1–12.

———. 1975a. Classification of the Rallidae. *Notornis* 22: 184.

———. 1975b. Thyroid bones of moas. *Records of the Canterbury Museum* 9: 247–258.

———. 1976a. Extinct rails [letter to editor]. *Notornis* 23: 78.

———. 1976b. Short-tailed shearwater: A new subfossil record from the Chatham Islands [letter to editor]. *Notornis* 23: 266.

———. 1976c. Further records of the grey petrel on Chatham Island [letter to editor]. *Notornis* 23: 178.

———. 1976d. King penguins at the Chatham Islands. *Notornis* 23: 355.

———. 1979. Avifauna and man. In A. J. Anderson (ed.), *Birds of a Feather,* pp. 75–90. New Zealand Archaeological Association Monograph II; B.A.R. International Series 62.

———. 1983. *Tereingaornis moisleyi*—A new Pliocene penguin. *New Zealand Journal of Geology and Geophysics* 26: 419–428.

Schodde, R., P. Fullagar, and N. Hermes. 1983. *A Review of Norfolk Island Birds: Past and Present.* Canberra: Australian Parks and Wildlife Service Special Publication, No. 8.

Schoeninger, M. J., and M. J. De Niro. 1984. Nitrogen and carbon isotopic composition of bone collagen from marine and terrestrial animals. *Geochimica et Cosmochimica Acta* 48: 625–639.

Schönwetter, M. 1960–92. *Handbuch der Oologie.* Vols. 1–4. Berlin: Akademie Verlag.

Sedgeley, J. A., and C. F. J. O'Donnell. 1999. Roost selection by the long-tailed bat, *Chalinolobus tuberculatus,* in temperate rainforest and its implications for the conservation of bats in managed forests. *Biological Conservation* 88: 261–276.

Sherley, G. 1992. Monitoring Hutton's shearwater. *Notornis* 39: 249–261.

Shufeldt, R. W. 1896. On the affinities of *Harpagornis*: A letter dated 25th January, 1895 to Professor T. Jeffery Parker by R. W. Shufeldt, M.D., Smithsonian Institution, Washington. *Transactions and Proceedings of the New Zealand Institute* 28 (1895): 665.

Sibley, C. G., and J. E. Ahlquist. 1981. The phylogeny and relationships of the ratite birds as indicated by DNA-DNA hybridization. In G. G. E. Scudder and J. L. Reveal (eds.), *Evolution Today,* Proceedings of the 2nd International Congress on Systematics and Evolutionary Biology, 301–335. Pittsburgh: Hunt Institute Botanic Document, Carnegie-Mellon University.

———. 1990. *Phylogeny and Classification of Birds.* New Haven, Conn.: Yale University Press.

Sibley, C. G., and B. L. Monroe, Jr. 1990. *Distribution and Taxonomy of Birds of the World.* New Haven, Conn.: Yale University Press.

Simmons, D. R. 1968. Man, moa and the forest. *Transactions of the Royal Society of New Zealand, General* 2: 115–127.

———. Economic change in New Zealand prehistory. *Journal of the Polynesian Society* 78(1): 3–34.

Simpson, L. O. 1955. A note on a green moa egg from Chatto Creek, Central Otago. *Transactions of the Royal Society of New Zealand* 83: 223–226.

Simpson, G. G. 1970. Ages of fossil penguins in New Zealand. *Science* 168(3929): 361–362.

———. 1971. A review of the pre-Pliocene penguins of New Zealand. *Bulletin of the American Museum of Natural History* 144: 321–378.

———. 1972. Pliocene penguins from North Canterbury, New Zealand. *Records of the Canterbury Museum* 9: 159–182.

———. 1975. Fossil penguins. In B. Stonehouse (ed.), *The Biology of Penguins,* pp. 19–41. London: Macmillan.

Slobodkin, L. B. 1961 *Growth and Regulation of Animal Populations.* New York: Holt, Rinehart and Winston.

———. 1974. Prudent predation does not require group selection. *American Naturalist* 108: 665–678.

Smith, W. W. 1884a. On moa and other remains from the Tengawai River, Canterbury. *New Zealand Journal of Science* 2: 293–295.

———. 1884b. *Sceloglaux albifacies* (laughing owl). *New Zealand Journal of Science* 2: 86–88.

———. 1891. On the occurrence of moa and other remains at Albury. *New Zealand Journal of Science* (new series) 1: 193–198.

Smith, I. W. G. 1978. Seasonal sea mammal exploitation and butchering patterns in an archaic site (Tairua, N44/2) on the Coromandel Peninsula. *Records of the Auckland Institute and Museum* 15: 17–26.

———. 1999. Settlement permanence and function at Pleasant River Mouth, East Otago, New Zealand. *New Zealand Journal of Archaeology* 19(1997): 27–79.

Sondaar, P. Y. 1977. Insularity and its effect on mammal evolution. In M. K. Hecht, P. C. Goody, and B. M. Hecht (eds.), *Major Patterns in Vertebrate Evolution,* pp. 671–707. New York, London: NATO Advanced Studies Institute Series: Series A, Life Sciences, Vol. 14.

Soons, J. M., and C. J. Burrows. 1978. Dates for Otiran deposits, including plant microfossils and macrofossils, from Rakaia Valley. *New Zealand Journal of Geology and Geophysics* 21: 607–615.

Soper, M. F. 1972. *New Zealand Birds.* Wellington: Whitcombe and Tombs.

———. 1976. *New Zealand Birds.* 2nd ed. Christchurch: Whitcoulls.

Sparks, R. J., W. H. Melhuish, J. W. A. McKee, J. Ogden, J. G. Palmer, and B. P. J. Molloy. 1995. ^{14}C calibration in the southern hemisphere and the date of the last Taupo eruption: Evidence from tree-ring sequences. *Radiocarbon* 37: 155–163.

Srami, M., L. Christidis, S. Easteal, P. Horn, and C. Collet. 1996. Molecular relationships within Australasian waterfowl (Anseriformes). *Australian Journal of Zoology* 44: 47–58.

Stapel, S. O., A. M. Leunissen, M. Versteeg, J. Wattel, and W. W. de Jong. 1984. Ratites as oldest offshoots of avian stem—Evidence from α-crystaline A sequences. *Nature* 311: 257–259.

Steadman, D. W. 1989a. Extinction of birds in eastern Polynesia: A review of the record, and comparisons with other Pacific island groups. *Journal of Archaeological Science* 16: 177–205.

———. 1989b. New species and records of birds (Aves: Megapodiidae, Columbidae) from an archaeological site on Lifuka, Tonga. *Proceedings of the Biological Society of Washington, DC* 102: 537–552.

———. 1991. Fossil birds and biogeography in Polynesia. *Acta XIX Congressus Internationalis Ornithologici* 2: 1526–1534.

———. 1993a. Biogeography of Tongan birds before and after human impact. *Proceedings of the National Academy of Sciences USA* 90: 818–822.

———. 1993b. Bird bones from the To'aga site: Prehistoric loss of seabirds and megapodes. In P. V. Kirch and T. L. Hunt (eds.), *The To'aga Site,* pp. 217–228. Contributions of the University of California Archaeological Research Facility, No. 51.

———. 1995. Prehistoric extinctions of Pacific Island birds: Biodiversity meets zooarchaeology. *Science* 267: 1123–1131.

———. 1997. Extinctions of Polynesian birds: Reciprocal impacts of birds and people. In P. V. Kirch and T. L. Hunt (eds.), *Historical Ecology in the Pacific Islands: Prehistoric Environmental and Landscape Change,* pp. 51–79. New Haven, Conn.: Yale University Press.

Steadman, D. W., and P. S. Martin. 1984. Extinctions of birds in the late Pleistocene of North America. In P. S. Martin and R. G. Klein (eds.), *Quaternary Extinctions: A Prehistoric Revolution,* pp. 466–477. Tucson: University of Arizona Press.

Stevens, G. R. 1980a. *New Zealand Adrift—The Theory of Continental Drift in a New Zealand Setting.* Wellington: A. H. and A. W. Reed.

———. 1980b. Southwest Pacific faunal palaeobiogeography in Mesozoic and Cenozoic times: A review. *Palaeogeography, Palaeoclimatology, Palaeoecology* 31: 153–196.

Stevens, G., M. McGlone, and B. McCulloch. 1988. *Prehistoric New Zealand.* Auckland: Reed.

Stidolph, R. H. D. 1921. Fossil birds in New Zealand. *Emu* 21: 313–314.

———. 1946. [Report of moa footprint.] *New Zealand Bird Notes* 2(1): 16.

Stout, J. D. 1973. Soil ecology. In G. R. Williams (ed.), *The Natural History of New Zealand: An Ecological Survey,* pp. 131–154. Wellington: A. H. and A. W. Reed.

St. Paul, J. W. 1974. The kokako (*Callaeas cinerea wilsoni*) in the Hunua Ranges. *Notornis* 21: 205–218.

Stresemann, E., and D. Amadon. 1979. Falconiformes. In E. Mayr and G. W. Cottrell (eds.), *Checklist of Birds of the World*, Vol. 1, 2nd ed., pp. 271–425. Cambridge, Mass.: Museum of Comparative Zoology.

Suggate, R. P. 1985. *The Glacial/Interglacial Sequence of North Westland, New Zealand.* New Zealand Geological Survey Records, No. 7.

———. 1990. Late Pleistocene and Quaternary glaciations in New Zealand. *Quaternary Science Reviews* 9: 175–197.

Suggate, R. P., and N. T. Moar. 1970. Revision of the chronology of the Late Otira Glaciation. *New Zealand Journal of Geology and Geophysics* 13: 742–746.

Suschkin, P. 1900. Systematische Ergebnisse osteologischer Untersuchungen einiger Tagraubvögel. *Zoologischer Anzeiger* 23: 269–277.

———. 1905. Zur Morphologie des Vögelskeletets. Vergleichende Osteologie der Normalen Tagraubvögel (Accipitres) und die Fragen der Classification. Teil 1. Grundeinteilung der Accipitres. *Nouvelle Memoires Société Imperial Naturaliste, Moscou* 16: 1–65.

Sutton, D. G. 1979. Island and coastal fowling strategies of the prehistoric Moriori. In A. Anderson (ed.), *Birds of a Feather*, pp. 123–139. New Zealand Archaeological Association Monograph II. BAR International Series 62.

———. 1987. A paradigmatic shift in Polynesian prehistory: Implications for New Zealand. *New Zealand Journal of Archaeology* 9: 135–155.

Swofford, D. L. 1993. *PAUP: Phylogenetic Analysis Using Parsimony*, version 3.1.1 [software]. Champaign, Ill.: Illinois Natural Survey.

Sy, M. 1936. Funktionall-anatomische Untersuchungen am Vögelflugel. *Journal für Ornithologie* 84: 199–296.

Szabo, M. 1993. Geonews. *New Zealand Geographic* 18(April–June): 7–9, 126.

Taborsky, M. 1988. Kiwis and dog predation: Observations in Waitangi State Forest. *Notornis* 35: 197–202.

Tambussi, C. P. 1995. The fossil Rheiformes from Argentina. *Courier Forschung Institut Senckenberg* 181: 121–129.

Tambussi, C. P., J. Noriega, A. Gazdzicki, A. Tatur, M. A. Reguero, and S. F. Vizcaino. 1994. The first ocurrence of ratite birds in the Paleocene of Antarctica. In S. M. Zalewki (ed.), *XXI Polar Symposium*, pp. 45–58. Warsaw: Polish Academy of Sciences.

Taylor, R. 1873. An account of the discovery of moa remains. *Transactions and Proceedings of the New Zealand Institute* 5: 97–101.

Taylor, R. H. 1975. Some ideas on speciation in New Zealand parakeets. *Notornis* 22: 110–121.

Tennyson, A. J. D., and P. R. Millener. 1994. Bird extinctions and fossil bones from Mangere Island, Chatham Islands. *Notornis* 41(Suppl.): 165–178.

Thomson, A. S. 1854 (April). Description of two caves in the North Island of New Zealand, in which were found bones of the large extinct wingless bird, called by the natives, moa, and by naturalists *Dinornis*; with some general observations on this genus of birds. *Edinburgh New Philosophical Journal* 56: 1–30.

Thompson, M. B., C. H. Daugherty, A. Cree, D. C. French, J. C. Gillingham, and R. E. Barwick. 1992. Status and longevity of the tuatara, *Sphenodon guntheri*, and Duvaucel's gecko, *Hoplodactylus duvaucelii*, on North Brother Island, New Zealand. *Journal of the Royal Society of New Zealand* 22: 123–130.

Thoresen, A. C. 1967. Ecological observations on Stanley and Green Islands, Mercury Group. *Notornis* 14: 182–200.

Thorne, G. 1876. Notes on the discovery of moa and moa-hunters remains at Pataua River, near Whangarei. *Transactions and Proceedings of the New Zealand Institute* 8: 83–94.

Tieszen, L. L., T. W. Boutton, K. G. Tesdahl, and N. A. Slade. 1983. Fractionation and turnover of stable carbon isotopes in animal tissues: Implications for $\delta^{13}C$ analysis of diet. *Oecologia* 57: 32–37.

Tieszen, L. L., D. Hein, S. A. Qvortrup, J. H. Troughton, and S. K. Imbamba. 1979. Use of $\delta^{13}C$ values to determine vegetation selectivity in East African herbivores. *Oecologia* 37: 351–359.

Tonkin, P. J., and L. R. Basher. 1990. Soil-stratigraphic techniques in the study of soil and landform evolution across the Southern Alps, New Zealand. *Geomorphology* 3: 547–575.

Townsend, J. I. 1961. A record of tuatara remains from the Takaka District. *New Zealand Speleological Bulletin* 2(37): 154–157.

Travers, W. T. L. 1873. On the birds of the Chatham Islands, by H. H. Travers, with introductory remarks on the avifauna and flora of the islands in their relation to those of New Zealand. *Transactions and Proceedings of the New Zealand Institute* 5: 212–222.

———. 1876. Notes on the extinction of the moa etc. *Transactions and Proceedings of the New Zealand Institute* 8: 58–83.

Travers, H. H., and W. T. L. Travers. 1872. On the birds of the Chatham Islands, with introductory remarks on the avi-fauna and flora on the islands in their relation to those of New Zealand. *Transactions and Proceedings of the New Zealand Institute* 5: 212–222.

Tregear, E. 1888. The Maori and the moa. *Journal of the Anthropological Institute of Great Britain and Ireland* 17: 292–305.

———. 1892. The extinction of the moa. *Journal of the Polynesian Society* 25: 413–426.

Trewick, S. A. 1996a. The diet of kakapo (*Strigops habroptilus*), takahe (*Porphyrio mantelli*) and pukeko (*P. porphyrio melanotus*) studied by faecal analysis. *Notornis* 43: 79–84.

———. 1996b. Morphology and evolution of two takahe: Flightless rails of New Zealand. *Journal of Zoology, London* 238: 221–237.

———. 1997a. Sympatric flightless rails *Gallirallus dieffenbachii* and *G. modestus* on the Chatham Islands, New Zealand; morphometrics and alternative evolutionary scenarios. *Journal of the Royal Society of New Zealand* 27: 451–464.

———. 1997b. Flightlessness and phylogeny amongst endemic rails (Aves: Rallidae) of the New Zealand region. *Philosophical Transactions of the Royal Society, London, B* 352: 429–446.

———. 1997c. On the skewed sex ratio of the kakapo *Strigops habroptilus:* sexual and natural selection in opposition? *Ibis* 139: 652–663.

Trewick, S. A., and T. H. Worthy. 2001. Origins and prehistoric ecology of takahe based on morphometric, molecular, and fossil data. Chapter 4 in *The Takahe, Fifty Years of Conservation Management and Research*. Ed. W. G. Lee and I. G. Jamieson, 31–48.

Triggs, S., and C. H. Daugherty. 1996. Conservation and genetics of New Zealand parakeets. *Biological Conservation International* 6: 89–101.

Trotter, M. M. 1965. Avian remains from North Otago archaeological sites. *Notornis* 12: 176–178.

———. 1967a. Investigations of a moa-hunter site at Redcliffs, Sumner. *Records of the Canterbury Museum* 8: 251–254.

———. 1967b. Radiocarbon dates from North Otago. *New Zealand Archaeological Association Newsletter* 10: 137–142.

———. 1970. A Holocene section at Scaifes Lagoon, Wanaka. *Records of the Canterbury Museum* 8: 463–467.
———. 1972. A moa-hunter site near the mouth of the Rakaia River, South Island. *Records of the Canterbury Museum* 9: 129–150.
———. 1975. Archaeological investigations at Redcliffs, Canterbury, New Zealand. *Records of the Canterbury Museum* 9: 189–220.
———. 1979. Tai Rua: A moa-hunter site in North Otago. In A. J. Anderson (ed.), *Birds of a Feather,* pp. 203–230. New Zealand Archaeological Association Monograph II; B.A.R. International Series 62.
———. 1980. Archaeological investigations at Avoca Point, Kaikoura. *Records of the Canterbury Museum* 9: 277–288.
———. 1988. Julius von Haast's moas and moa hunters. *Historic Places in New Zealand* 21: 18–20.
Trotter, M. M., and B. McCulloch. 1984. Moa, men and middens. In P. S. Martin and R. G. Klein (eds.), *Quaternary Extinctions: A Prehistoric Revolution,* pp. 708–727. Tucson: Arizona University Press.
Tucker, V. A. 1973. Bird metabolism during flight: Evaluation of a theory. *Journal of Experimental Biology* 58: 689–709.
———. 1974. Energetics of natural avian flight. In R. A. Paynter (ed.), *Avian Energetics,* pp. 298–327. Publications of the Nuttall Ornithological Club, No. 15.
Turbott, E. G. 1967. *Buller's Birds of New Zealand.* Christchurch: Whitcoulls.
———, (ed.). 1990a. *Checklist of the Birds of New Zealand and Ross Dependency, Antarctica.* 3rd ed. Auckland: Ornithological Society of New Zealand and Random Century.
———. 1990b. Frederick Wollaston Hutton 1836–1905. In B. J. Gill and B. D. Heather (eds.), *A Flying Start.* Auckland: Random Century and the Ornithological Society of New Zealand.
Tyler, C. 1957. Some chemical, physical and structural properties of moa egg shells. *Journal of the Polynesian Society* 66: 110–130.
———. 1964. *Wilhelm von Nathusius 1821–1899 on Avian Eggshells. A Translated and Edited Version of His Work.* Reading, UK: University of Reading.
———. 1966. A study of the egg shells of the Falconiformes. *Journal of Zoology, London* 150: 413–425.
Tyler, C., and K. Simkiss. 1960. A study of the eggshells of ratite birds. *Proceedings of the Zoological Society of London* 133: 201–243 + pls.
Tyrberg, T. 1995. Palaeobiogeography of the genus *Lagopus* in the West Palearctic. *Courier Forschung Institui Senckenberg* 181: 275–291.
van Beneden, P. J. 1875. Un oiseau fossile nouveau des cavernes de la Nouvelle Zelande. *Annales de la Société géologique de Belgique* 2: 123–130.
———. 1877. A new fossil bird, *Anas finschi*, from the Earnscleugh Caves, Otago, New Zealand. *Transactions and Proceedings of the New Zealand Institute* 9: 599–602.
van Tets, G. F., C. W. Meredith, P. J. Fullagar, and P. M. Davidson. 1988. Osteological differences between *Sula* and *Morus* and a description of an extinct new species of *Sula* from Lord Howe and Norfolk Islands, Tasman Sea. *Notornis* 35: 35–57.
Veitch, C. R. 1982. Eradication of cats from Little Barrier Island. *Landscape* 11: 27–29.
———. 1983. A cat problem removed. *Wildlife—A Review* 12: 47–49.
Vickers-Rich, P. 1979. *The Dromornithidae, an Extinct Family of Large*

Ground Birds Endemic to Australia. Australian Bureau of Natural Resources, Geology and Geophysical Bulletin, No. 184.

———. 1980. The Australian Dromornithidae: A group of extinct large ratites. *Contributions in Science, Natural History Museum of Los Angeles County* 330: 93–103.

———. 1990. The Mesozoic and Tertiary history of birds on the Australian Plate. In P. Vickers-Rich, J. M. Monaghan, R. F. Baird, and T. H. Rich (eds.), *Vertebrate Palaeontology of Australasia,* pp. 721–808. Melbourne: Pioneer Design Studio and Monash University Publications Committee.

Vickers-Rich, P., and J. C. Balouet. 1984. The waifs and strays of the bird world or the ratite problem revisited, one more time. In M. Archer and G. Clayton (eds.), *Vertebrate Zoogeography and Evolution in Australasia: Animals in Time and Space,* pp. 447–455. Melbourne: Hesperian Press.

Vickers-Rich, P., and R. J. Scarlett. 1977. Another look at *Megaegotheles,* a large owlet nightjar from New Zealand. *Emu* 77: 1–8.

Vickers-Rich, P., and G. F. van Tets. 1981. The fossil pelicans of Australasia. *Records of the South Australian Museum* 18: 235–264.

Vickers-Rich, P., P. Trusler, M. J. Rowley, A. Cooper, G. K. Chambers, W. J. Bock, P. R. Millener, T. H. Worthy, and J. C. Yaldwyn. 1995. Morphology, myology, collagen and DNA of a mummified upland moa, *Megalapteryx didinus* (Aves: Dinornithiformes) from New Zealand. *Tuhinga: Records of the Museum of New Zealand Te Papa Tongarewa* 4: 1–26.

Voous, K. 1969. Predation potential in birds of prey from Surinam. *Ardea* 57: 117–147.

Voy, C. D. 1880. On the occurrence of footprints of *Dinornis* at Poverty Bay, New Zealand. *American Naturalist* 14: 682–684.

Wakelin, M. D. 1993. Contents of blue duck faeces from the Tongariro River. *Notornis* 40: 205–212.

Wardle, P. 1963. Evolution and distribution of the New Zealand flora, as affected by Quaternary climates. *New Zealand Journal of Botany* 1: 3–17.

Warham, J. 1977. Wing loadings, wing shapes, and flight capabilities of Procellariiformes. *New Zealand Journal of Zoology* 4: 73–83.

———. 1990. *The Petrels: Their Ecology and Breeding Systems.* London: Academic Press.

Warham, J., and B. D. Bell. 1979. The birds of Antipodes Island, New Zealand. *Notornis* 26: 121–169.

Warham, J., and G. J. Wilson. 1982. The size of the sooty shearwater population at The Snares Islands, New Zealand. *Notornis* 29: 23–30.

Webb, C. J., and D. Kelly. 1993. The reproductive biology of the New Zealand flora. *Trends in Ecology and Evolution* 8: 442–447.

Weber, E., and A. Hesse. 1995. The systematic position of *Aptornis,* a flightless bird from New Zealand. *Courier Forschung Institut Senckenberg* 181: 293–301.

Weber, E., and F.-T. Krell. 1995. Case 2879: *Aptornis* Owen, [1848] (Aves): Proposed conservation as the correct original spelling. *Bulletin of Zoological Nomenclature* 52: 170–174.

Weetman, S. 1886. Notes on some moa remains found at the Great Barrier Island during February 1886. *Transactions and Proceedings of the New Zealand Institute* 19: 193–194.

———. 1890. Notes on the Great Barrier Island. *Transactions and Proceedings of the New Zealand Institute* 22: 79–84.

Weick, F., and L. H. Brown. 1980. *Birds of Prey of the World.* London: Collins.
Weimerskirch, H., and P. Jouventin. 1987. Population dynamics of the wandering albatross, *Diomedea exulans,* of the Crozet Islands: Causes and consequences of the population decline. *Oikos* 49: 315–322.
Weimerskirch, H., J. Clobert, and P. Jouventin. 1987. Survival in five southern albatrosses and its relationship with their life history. *Journal of Animal Ecology* 56: 1043–1055.
Weis-Fogh, T., and R. McN. Alexander. 1977. The sustained power output from striated muscle. In T. J. Pedley (ed.), *Scaling of Terrestrial Locomotion,* pp. 511–525. London: Academic Press.
Weller, M. W. 1974. Habitat selection and feeding patterns of brown teal (*Anas castanea chlorotis*) on Great Barrier Island. *Notornis* 21: 25–35.
Welles, S. P., and D. R. Gregg. 1971. Late Cretaceous marine reptiles of New Zealand. *Records of the Canterbury Museum* 9: 1–111.
Wellman, H. 1994. Moas, gizzard stones and New Zealand plants: A geologist's viewpoint. *New Zealand Science Review* 51: 56–59.
West, J. A. 1994. Chatham petrel (*Pterodroma axillaris*)—An overview. *Notornis* 41(Suppl.): 19–26
Weston, R. J., C. A. Repenning, and C. A. Fleming. 1973. Modern age of supposed Pliocene seal, *Arctocephalus caninus* Berry (= *Phocarctos hookeri* Gray) from New Zealand. *New Zealand Journal of Science* 16: 591–598.
Whitaker, A. H. 1978. The effects of rodents on reptiles and amphibians. In P. R. Dingwall, I. A. E. Atkinson, and C. Hay (eds.), *The Ecology and Control of Rodents in New Zealand Nature Reserves,* pp. 75–86. New Zealand Department of Lands and Survey Information Series, No. 4.
———. 1987. The roles of lizards in New Zealand plant reproductive strategies. *New Zealand Journal of Botany* 25: 315–328.
White, T. 1876. Notes on moa caves, etc., in the Wakatipu district. *Transactions and Proceedings of the New Zealand Institute* 8: 97–102.
———. 1886. Remarks on the feathers of two species of moa. *Transactions and Proceedings of the New Zealand Institute* 18: 83–84.
———. 1895. On the bird moa and its aliases. *Transactions and Proceedings of the New Zealand Institute* 27: 262–273.
———. 1897. On the Poua and other extinct birds of the Chatham Islands. *Transactions and Proceedings of the New Zealand Institute* 29: 162–168.
White, J. 1925. The moa in Maori tradition. *Journal of the Polynesian Society* 34: 170–174.
Wiffen, J. 1980. *Moanasaurus,* a new genus of marine reptile (Family Mosasauridae) from the Upper Cretaceous of North Island. *New Zealand Journal of Geology and Geophysics* 23: 507–528.
———. 1990. New mosasaurs (Reptilia; Family Mosasauridae) from the Upper Cretaceous of North Island. *New Zealand Journal of Geology and Geophysics* 33: 67–85.
Wiffen, J., and W. L. Moisley. 1986. Late Cretaceous reptiles (families Elasmosauridae and Pliosauridae) from the Mangahouanga Stream, North Island. *New Zealand Journal of Geology and Geophysics* 29: 205–252.
Wiffen, J., and R. E. Molnar. 1988. First pterosaur from New Zealand. *Alcheringa* 12: 53–59.
Williams, W. L. 1872. On the occurrence of footprints of a large bird,

found at Turanganui, Poverty Bay. *Transactions and Proceedings of the New Zealand Institute* 4: 124–127.
Williams, G. R. 1956. The kakapo (*Strigops habroptilus,* Gray): A review and re-appraisal of a near-extinct species. *Notornis* 7: 29–56.
———. 1960. The takahe (*Notornis mantelli*) a general survey. *Transactions of the Royal Society of New Zealand* 88: 235–258.
———. 1962. Extinction and the land and freshwater inhabiting birds of New Zealand. *Notornis* 10: 15–32.
———. 1981. Aspects of avian island biogeography in New Zealand. *Journal of Biogeography* 8: 439–456.
Williams, M. J. 1971. The distribution and abundance of the paradise shelduck (*Tadorna variegata,* Gmelin) in New Zealand from pre-European times to the present day. *Notornis* 18: 71–86.
———. 1979. Detecting the presence of blue duck from faecal sign. *Notornis* 26: 306–307.
———. 1981. *The Duckshooter's Bag.* Wellington: Wetland Press.
———. 1995. Social structure, dispersion and breeding of the Auckland Island teal. *Notornis* 42: 219–262.
Williams, G. R., and M. Harrison. 1972. The laughing owl *Sceloglaux albifacies* (Gray, 1844): A general survey of a near-extinct species. *Notornis* 19: 4–19.
Williams, G. R., and M. W. Weller. 1974. Unsuccessful search for the Auckland Islands merganser (*Mergus australis*). *Notornis* 21: 247–249.
Wilmshurst, J. M. 1997. The impact of human settlement on vegetation and soil stability in Hawke's Bay, New Zealand. *New Zealand Journal of Botany* 35: 97–111.
Wilmshurst, J. M., M. S. McGlone, and T. R. Partridge. 1997. A late Holocene history of natural disturbance in lowland podocarp/hardwood forest, Hawke's Bay, New Zealand. *New Zealand Journal of Botany* 35: 79–96.
Wilson, K. 1913. Footprints of the moa. *Transactions and Proceedings of the New Zealand Institute* 45: 211–212.
Wilson, C. J. N. 1985. The Taupo eruption, New Zealand II. The Taupo ignimbrite. *Philosophical Transactions of the Royal Society, London, A* 314: 229–310.
———. 1993. Stratigraphy, chronology, styles and dynamics of late Quaternary eruptions from Taupo volcano, New Zealand. *Philosophical Transactions of the Royal Society, London, A* 343: 205–306.
Wilson, K.-J. 1997. Extinct and introduced vertebrate species in New Zealand: A loss of biodistinctiveness and gain in biodiversity. *Pacific Conservation Biology* 3: 301–305.
Wilson, C. J. N., and G. P. L. Walker. 1985. The Taupo eruption, New Zealand II. General aspects. *Philosophical Transactions of the Royal Society, London, A* 314: 199–228.
Wilson, P. R., B. J. Karl, R. J. Toft, J. R. Beggs, and R. H. Taylor. 1998. The role of introduced predators and competitors in the decline of kaka (*Nestor meridionalis*) populations in New Zealand. *Biological Conservation* 83: 175–185.
Wiman, C. 1937. On supernumary metapodials in *Aepyornis,* the moas and some other birds. *Proceedings of the Zoological Society of London, B,* Part 2: 245–256, pls. 1–5.
Wiman, C., and T. Edinger. 1941. Sur les crânes et les encéphales d'*Aepyornis* et de *Mullerornis. Bulletin de l'Academie Malgache* 24: 1–47, pls. 1–6.

Wodzicki, K. A., and R. H. Taylor. 1984. Distribution and status of the Polynesian rat *Rattus exulans*. *Acta Zoologica Fennica* 172: 99–101.

Woolfenden, G. E. 1961. Postcranial osteology of the waterfowl. *Bulletin of the Florida State Museum (Biological Sciences)* 6: 1–129.

Worthy, T. H. 1982. Subfossil remains from the Fred Catchment, Waitomo. *New Zealand Speleological Bulletin* 122: 46–48.

———. 1983a. *Moas and Other Subfossils.* Waitomo: Waitomo Museum.

———. 1983b. Subfossil insects: A key to past diversity. *Weta* 6: 42–43.

———. 1984a. Faunal and floral remains from F1, a cave near Waitomo. *Journal of the Royal Society of New Zealand* 14: 367–377.

———. 1984b. An extensive subfossil deposit in Gardner's Gut Cave, Waitomo. *New Zealand Speleological Bulletin* 130: 257–262.

———. 1984c. *Honeycomb Hill Cave System—A Description of the Cave, Its Structure and Development with Comments on Its Fauna Both Recent and Subfossil.* Nelson: New Zealand Forest Service.

———. 1985a. The subfossil fauna of Aurora—Te Ana-au Cave. In P. W. Williams (ed.), *Aurora—Te Ana-au Cave, A Survey and Interpretation of Scientific Resources*, pp. 14–21. Auckland: Geography Department, Auckland University.

———. 1985b. Subfossil deposits from caves on Mt. Luxmore, Fiordland. Report to the Fiordland National Park Board, New Zealand.

———. 1986. *Subfossil bones of the frog* Leiopelma *in New Zealand.* M.Sc. thesis. Victoria University.

———. 1987a. Sexual dimorphism and temporal variation in the North Island moa species *Euryapteryx curtus* (Owen) and *Pachyornis mappini* Archey. *National Museum of New Zealand Records* 3(6): 59–70.

———. 1987b. Osteology of *Leiopelma* (Amphibia: Leiopelmatidae) and descriptions of three new subfossil *Leiopelma* species. *Journal of the Royal Society of New Zealand* 17: 201–251.

———. 1987c. Palaeoecological information concerning members of the frog genus *Leiopelma*: Leiopelmatidae in New Zealand. *Journal of the Royal Society of New Zealand* 17: 409–420.

———. 1987d. Osteological observations on the larger species of the skink *Cyclodina* and the subfossil occurrence of these and the gecko *Hoplodactylus duvaucelii* in the North Island, New Zealand. *New Zealand Journal of Zoology* 14: 219–229.

———. 1987e. The subfossil fauna of Honeycomb Hill Cave: I. A study of Otiran deposits in the graveyard; II. A reassessment of *Megalapteryx benhami* Archey. Report to the National Museum of New Zealand.

———. 1987f. Subfossil snipe [letter]. *Notornis* 34: 343.

———. 1988a. Opening Day Cave, Stubb's Farm, Waitomo. *New Zealand Speleological Bulletin* 7(140): 588–595.

———. 1988b. Loss of flight ability in the extinct New Zealand duck *Euryanas finschi*. *Journal of Zoology, London* 215: 619–628.

———. 1988c. *An Illustrated Key to the Main Leg Bones of Moas (Aves: Dinornithiformes).* National Museum of New Zealand Miscellaneous Publications 17.

———. 1988d. A re-examination of the genus *Megalapteryx*. *Notornis* 35: 99–108.

———. 1989a. An analysis of moa bones (Aves: Dinornithiformes) from three lowland North Island swamp sites: Makirikiri, Riverlands and Takapau Road. *Journal of the Royal Society of New Zealand* 19: 419–432.

———. 1989b. Validation of *Pachyornis australis* Oliver (Aves; Dinornithiformes), a medium sized moa from the South Island, New Zealand. *New Zealand Journal of Geology and Geophysics* 32: 255–266.

———. 1989c. Aspects of the biology of two moa species (Aves: Dinornithiformes). *New Zealand Journal of Archaeology* 11: 77–86.

———. 1989d. Number of presacral vertebrae in *Dinornis. Notornis* 36: 170.

———. 1989e. Moas of the subalpine zone. *Notornis* 36: 191–196.

———. 1989f. The glossohyal and thyroid bone of *Aptornis otidiformis. Notornis* 36: 248.

———. 1989g. Mummified moa remains from Mt. Owen, northwest Nelson. *Notornis* 36: 36–38.

———. 1990. An analysis of the distribution and relative abundance of moa species (Aves; Dinornithiformes). *New Zealand Journal of Zoology* 17: 213–241.

———. 1991a. Fossil skink bones from Northland, New Zealand, and description of a new species of *Cyclodina,* Scincidae. *Journal of the Royal Society of New Zealand* 21: 329–348.

———. 1991b. An overview of the taxonomy, fossil history, biology and extinction of moas. *Acta XX Congressus Internationalis Ornithologici* 1: 555–562.

———. 1991c. Honeycomb Hill Cave. *New Zealand Speleological Bulletin* 8(157–158): 483–499.

———. 1992a. Fossil bones of Hooker's sea lions in New Zealand caves. *New Zealand Natural Sciences* 19: 31–39.

———. 1992b. A re-examination of the species *Euryapteryx geranoides* (Owen) including comparisons with other emeiin moas (Aves: Dinornithiformes). *Journal of the Royal Society of New Zealand* 22: 19–40.

———. 1993a. A review of fossil bird bones from loess deposits in eastern South Island, New Zealand. *Records of the Canterbury Museum* 10(8): 95–106.

———. 1993b. *Fossils of Honeycomb Hill.* Wellington: Museum of New Zealand/Te Papa Tongarewa.

———. 1993c. Submarine fossil moa bones from the northern South Island. *Journal of the Royal Society of New Zealand* 22: 255–256.

———. 1994a. Reappraisal of *Dinornis* (Aves: Dinornithiformes) species—A morphometric analysis. *New Zealand Journal of Zoology* 21: 113–134.

———. 1994b. Late Quaternary changes in the moa fauna on the West Coast of the South Island, New Zealand. *Records of the South Australian Museum* 27: 125–134.

———. 1994c. Holocene breeding of sea lions (*Phocarctos hookeri*) at Delaware Bay, Nelson, South Island, New Zealand. *New Zealand Natural Sciences* 21: 61–70.

———. 1995. Description of some post-cranial bones of *Malacorhynchus scarletti,* a large extinct pink-eared duck from New Zealand. *Emu* 95: 13–22.

———. 1996. Holocene populations of shags *Leucocarbo* spp. in the far north, New Zealand. *New Zealand Journal of Zoology* 23: 89–95.

———. 1997a. A mid-Pleistocene rail from New Zealand. *Alcheringa* 21: 71–78.

———. 1997b. Quaternary fossil fauna of South Canterbury, South Island, New Zealand. *Journal of the Royal Society of New Zealand* 27: 67–162.

———. 1997c. A survey of historical laughing owl (*Sceloglaux albifacies*) specimens in museum collections. *Notornis* 44: 241–252.

———. 1997d. The identification of fossil *Eudyptes* and *Megadyptes* bones at Marfell's Beach, Marlborough, South Island. *New Zealand Natural Sciences* 23: 71–85.
———. 1997e. Fossil deposits in the Hodges Creek Cave System, on the northern foothills of Mt. Arthur, Nelson, South Island, New Zealand. *Notornis* 44: 111–124.
———. 1998a. A remarkable fossil and archaeological fauna from Marfells Beach, Lake Grassmere, South Island, New Zealand. *Records of the Canterbury Museum* 12(1): 79–176.
———. 1998b. Quaternary fossil faunas of Otago, South Island, New Zealand. *Journal of the Royal Society of New Zealand* 28: 421–521.
———. 1998c. Fossils indicate *Pelecanoides georgicus* had large colonies at Mason Bay, Stewart Island, New Zealand. *Notornis* 45: 229–246.
———. 1998d. Quaternary fossil faunas of Southland, South Island, New Zealand. *Journal of the Royal Society of New Zealand* 28: 537–589.
———. 1998e. Fossil avifaunas from Old Neck and Native Island, Stewart Island—Polynesian middens or natural sites? *Records of the Canterbury Museum* 12: 49–82.
———. 1999a. What was on the menu—Avian extinctions in New Zealand. *New Zealand Journal of Archaeology* 19(1997): 125–160.
———. 1999b. Changes induced in the New Zealand avifauna by climate during the last glacial–interglacial period. Proceedings of the 4th International Meeting of the Society of Avian Paleontology and Evolution (Washington, D.C., June 1996). *Smithsonian Contributions to Paleobiology* 89: 111–123.
———. 2000. Two late-glacial avifaunas from eastern North Island, New Zealand—Te Aute Swamp and Wheturau Quarry. *Journal of the Royal Society of New Zealand* 30(1): 1–26.
Worthy, T. H., and R. Brassey. 2000. New Zealand pigeon (*Hemiphaga novaeseelandiae*) on Raoul Island, Kermadec Group. *Notornis* 47: 36–38.
Worthy, T. H., and R. N. Holdaway. 1991. Lost in time [moa]. *New Zealand Geographic* 12: 51–68.
———. 1993. Quaternary fossil faunas from caves in the Punakaiki area, West Coast, South Island, New Zealand. *Journal of the Royal Society of New Zealand* 23: 147–254.
———. 1994a. Scraps from an owl's table—Predator activity as a significant taphonomic process newly recognised from New Zealand Quaternary deposits. *Alcheringa* 18: 229–245.
———. 1994b. Quaternary fossil faunas from caves in Takaka valley and on Takaka Hill, northwest Nelson, South Island, New Zealand. *Journal of the Royal Society of New Zealand* 24: 297–391.
———. 1995. Quaternary fossil faunas from caves on Mt. Cookson, North Canterbury, South Island, New Zealand. *Journal of the Royal Society of New Zealand* 25: 333–370.
———. 1996a. Taphonomy of two Holocene microvertebrate deposits, Takaka Hill, Nelson, New Zealand, and identification of the avian predator responsible. *Historical Biology* 12: 1–24.
———. 1996b. Quaternary fossil faunas, overlapping taphonomies, and palaeofaunal reconstruction in North Canterbury, South Island, New Zealand. *Journal of the Royal Society of New Zealand* 26: 275–361.
———. 1999. Introduction to "A preliminary report on the nesting habits of moas on the East Coast of the North Island," by W. H. Hartree. *Notornis* 46: 457–460.
———. 2000. Terrestrial fossil vertebrate faunas from inland Hawke's Bay,

North Island, New Zealand. Part 1. *Records of the Canterbury Museum* 14: 89–154.

Worthy, T. H., and D. C. Mildenhall. 1989. A late Otiran–Holocene paleoenvironment reconstruction based on cave excavations in northwest Nelson, New Zealand. *New Zealand Journal of Geology and Geophysics* 32: 243–253.

Worthy, T. H., A. R. Edwards, and P. R. Millener. 1991. The fossil record of moas (Aves: Dinornithiformes) older than the Otira (last) glaciation. *Journal of the Royal Society of New Zealand* 21: 101–118.

Worthy, T. H., M. J. Daniel, and J. E. Hill. 1996. An analysis of skeletal size variation in *Mystacina robusta* Dwyer, 1962 (Chiroptera: Mystacinidae). *New Zealand Journal of Zoology* 23: 99–110.

Worthy, T. H., R. N. Holdaway, M. D. Sorenson, and A. C. Cooper. 1997. Description of the first complete skeleton of the extinct New Zealand goose *Cnemiornis calcitrans* (Aves: Anatidae) and a reassessment of the relationships of *Cnemiornis*. *Journal of Zoology, London* 243: 695–723.

Worthy, T. H., R. N. Holdaway, A. J. D. Tennyson, and J. A. Bartle. In press. First contact to the present, documenting changes in diversity in birds (class Aves). *Proceedings of the Species 2000 Conference.*

Yaldwyn, J. C. 1956. A preliminary account of the sub-fossil avifauna of the Martinborough caves. *Records of the Dominion Museum* 3: 1–7.

———. 1958. Notes on the environment and age of the sub-fossil deposits of the Martinborough Caves. *Records of Dominion Museum* 3: 129–133.

———. 1959. Moa remains from the Wellington District. *New Zealand Archaeological Association Newsletter* 2: 20–25.

———. 1966 (June). Moas and man in New Zealand. *Australian Natural History* 1966: 184–189.

———. 1979. The types of W. R. B. Oliver's moas and notes on Oliver's methods of measuring moa bones. In A. J. Anderson (ed.), *Birds of a Feather,* pp. 1–24. New Zealand Archaeological Association Monograph II; B.A.R. International Series 62.

Yaldwyn, J. C., G. J. Tee, and A. P. Mason. 1997. The status of Gideon Mantell's "first" iguanodon tooth in the Museum of New Zealand Te Papa Tongarewa. *Archives of Natural History* 24: 397–421.

Index

TREVOR H. WORTHY
is a research associate at the Museum of New Zealand Te Papa Tongarewa (Birds Department), Wellington, and is an independent researcher contracted to the Foundation for Research, Science and Technology (New Zealand government). He has published extensively, with emphases on taxonomy (variously addressing moa, other birds, frogs, lizards, and bats) and paleofaunal reconstruction in New Zealand and the Pacific.

RICHARD N. HOLDAWAY
is an extinction biologist specializing in the effects of human settlement and introduced mammals on pristine faunas. He is an independent researcher contracted to the Foundation for Research, Science and Technology (New Zealand government), and an Adjunct Senior Fellow in the Department of Zoology, University of Canterbury. He has published on the systematics of hawks and eagles, the paleobiology of New Zealand birds, and Quaternary extinctions.